1st International Conference on Energy and Sustainable Environment (ICESE 2019)

IOP Conference Series: Earth and Environmental Science
Volume 331

Ota, Nigeria
18 – 20 June 2019

ISBN: 978-1-5108-9666-6
ISSN: 1755-1307

Printed from e-media with permission by:

Curran Associates, Inc.
57 Morehouse Lane
Red Hook, NY 12571

proceedings.com

Some format issues inherent in the e-media version may also appear in this print version.

Institute of Physics
Dirac House, Temple Back
Bristol BS1 6BE UK

Phone: 44 1 17 929 7481
Fax: 44 1 17 920 0979

techtracking@iop.org

Additional copies of this publication are available from:

Curran Associates, Inc.
57 Morehouse Lane
Red Hook, NY 12571 USA
Phone: 845-758-0400
Fax: 845-758-2633
Email: curran@proceedings.com
Web: www.proceedings.com

Table of contents

Volume 331

International Conference on Energy and Sustainable Environment

18–20 June 2019, Covenant University, Nigeria

Accepted papers received: 21 August 2019
Published online: 16 October 2019

Preface

Papers

IOP Conf. Series: Earth and Environmental Science **331** (2019) 011001 doi:10.1088/1755-1315/331/1/011001

PREFACE

Moses Eterigho Emetere

This is the first proceeding of the 1st International Conference on Energy and Sustainable Environment (ICESE 2019). Four major research clusters in Covenant University were the organizers of the conference. A total of ninety-one (91) papers were presented in the four technical sessions – environment, energy, energy management and built environment. The percentage of presentation includes: 33.7% in environment session; 30.1% in energy session; 19.3% in energy management session; and 16.9% in built environment session. The conference sessions had a total of one hundred and fifty three (153) participants.

The lead speaker (Prof. Pierluigi Leone) gave the first speech on "The Energy Center Initiative as a multi- stakeholder hub for sustainable energy solutions". Prof Pierluigi Leone emphasize on the need for scientist from all part of the globe to synergize in solving environmental problems. The discussants were Prof. U. Uwuigbe and Prof. O. Ohunakin. On the second day of the conference, there was side event that took place alongside the technical sessions. The title of the event is "The next production revolution for Nigeria". The side event was anchored by the EAIL Team Nigeria. The panelists for the event were Mr. Tony Okpanachi, Ms. Damilola Ogunbiyi and Mr. Segun Ajayi.

The organizer of ICESE 2019 wishes to appreciate the Institute of Physics and their officials at the IOP conference series. The organizer also wishes to acknowledge the team from EAIL led by Prof Pierluigi Leone. The contributions and support of the Covenant University management is much appreciated. All the conference speakers and technical session presenters are much appreciated for sharing their research discoveries. The conference had salient recommendations on sustainable development in developing countries.

Peer review statement

All papers published in this volume of *IOP Conference Series: Earth and Environmental Science* have been peer reviewed through processes administered by the proceedings Editors. Reviews were conducted by expert referees to the professional and scientific standards expected of a proceedings journal published by IOP Publishing.

Evaluation of Sustainable Urban Renewal Strategies in an evolving Residential District of Lagos Island, Nigeria.

F O Jegede [1], foluke.jegede@covenantuniversity.edu.ng
B A Adewale [2], bukola.adewale@covenantuniversity.edu.ng
O D Olaniyan [3], olaniyanseun92@yahoo.com
1,2,3, Covenant University, Ota, Nigeria.

Abstract. Population growth and Urbanization are part of the attributes of a developing city. These attributes impact on the environment and the living condition of its inhabitants. The need to have a mechanism for sustainable development of cities is of necessity. One of such mechanism is the creation of a smart city. This development plan is an agenda for cities such as Lagos State, Nigeria. A study was done on some selected existing buildings found in the residential district areas of Lagos Island with the aim to identifying the housing and planning characteristics found in these residential areas to aid smart city implementation. The history of the buildings and factors that have influenced the change in housing types was studied; how these can influence and affect any agenda of smart city implementation or reform was explored. Survey methodology was used in carrying out the research with the use of questionnaire administered to residents; observation and photographic materials were also used as research instrument. Statistical methods using frequency and percentages were used to analyze the data collected. The result showed that a large percentage of the buildings investigated, had not been renovated in the last ten (10) years and are mixed used buildings. It was observed that the present state of housing in Lagos Island has been greatly influenced by increasing commercial activities in the area. Also, the result showed that the purpose and types of housing found in this area had evolved over time, a once resting place for a man and his family is now a place of several functions. Sustainable urban renewal and planning were suggested to enable the implementation of a smart city.

1. Introduction

Sustainable urban development and growth in the face of developing the economy has been an issue of concern to government and policy makers in different countries, especially in developing countries such as Nigeria. This is due to the continuous influx of people into urban centers, which is usually an outcome of the push of the rural areas to the pull of the urban area, seeking for a better living condition. This creates a rapid population growth, which depends on housing developments and facilities available in these urban areas. Now the population of people living in cities has increased, making the global urban population to be put at 54% in 2014 and projected to rise to 70% by 2050 (UN-Habitat III, 2015). This is coupled with the emergence of new cities, as the number of urban population is expected to increase. The increasing population in the urban areas will need shelter, which would increase the housing deficits and would require a proper organization and planning of the cities for human ideal habitation as stated by Jegede, Ibem and Oluwatayo (2018) that home environment needs to be planned, designed, constructed, and managed in such a way that it protects the residents from dangerous insects, animals, and human intruders. However, housing is not only about a place of abode, but housing requires the totality of both the house and the environment as explained by Ajanlekoko (2001). He opined that housing is not only the individual dwelling units or group of dwelling units, but also the entire neighbourhood system. Its components include both the physical infrastructures, services and the people. These are transportation networks for both pedestrian and vehicular movement, power supply, water supply, waste disposal, drainage, commercial outlets and health facilities. In line with these, the issue of

IOP Conf. Series: Earth and Environmental Science **331** (2019) 012001 doi:10.1088/1755-1315/331/1/012001

planning and re-planning of cities, requires not only solving the fundamental needs of shelter by housing provision, but also the need for comfort, serene environment and security within the housing environments.

The growth of towns and cities in Nigeria urban area and metropolitan areas has risen, with diverse planning problems such as waste management, pollution, slum, shanty towns, housing availability, poverty and several others as described by Ibem, Anosike and Azuh (2011) on the challenges in public housing provision in Nigeria. The need for proper management of these problems and creating sustainable solutions to solving them is of necessity.

Lagos Island is the heartbeat of the Lagos metropolis. In the 17th, 18th and 19th centuries, it was the living place for different people from different continents of the world, precisely during the slave trade activities of the 17th century. Though, the activities of the Europeans in the slave trade business is no more, the indelible marks left by these activities of the slave trade is still visible. These marks have become part and parcel of the history of Lagos State, one of such is in the housing types and designs, street layouts found in Lagos Island and some of these marks have become an unkempt remains. Various ideas and strategies had been used in the past by the authorities concerned, especially in Lagos State metropolis, such as Lagos Island. Lagos Executive Development Board (LEDB), which was initially given the responsibility of clearing slums and ghettos in Lagos and transforming them into a planned and habitable environment in the late 1920s. Other government agencies such as Ikeja Area Planning Authority (IAPA) and Epe Town Planning Authority (ETPA) were involved with the planning issues at the time (Lagos State Bureau of Statistics, 2013).

Also, different governments in Lagos State in the last five decades had used measures such as slum clearance, urban renewal, urban development amongst others, in bringing a serene urban development to the city, as explained in studies such as Agbola and Jinadu (1997); Gandy (2006); Ilesanmi (2010). In recent time, the issue of smart city agenda is on the increase. This study undertook a documentation of the characteristics of the existing buildings in the residential district areas of Lagos Island. Considering the changes in the building types found in these areas from their initial construction. Studying how these buildings characteristics fit into the smart city agenda. Also, the study investigated the factors that have influenced the changes in the building types, how it can affect smart city agenda, it implementations and reforms. This is viewed in correspondence with the strategy of urban renewal on the existing building types found in the selected Lagos Island residential district areas.

2. Literature review

2.1. The History and Architecture of Lagos Island.

Lagos Island is located within Lagos State. The state was created on May 27, 1967, although Lagos Island had existed since the 17th century, an island discovered by the Portuguese. Lagos Island is a traditional Central Business District (CBD) of Lagos State and is the birthplace of purpose-built office buildings in Lagos during the colonial era. This was due to its unique position of being sited off the coast of the Atlantic Ocean and proximity to the Lagos Apapa port. Lagos Island is popularly known as "*Isale- Eko*" in the Yoruba dialect (Akinlose, 2016).

Lagos Island architecture is a mix of different types, styles and periods, which features traditional/vernacular architecture, old European-styled buildings, Portuguese architecture and Brazilian styles. Some of these buildings dated back to the 17th century (Lagos State Bureau of Statistics, 2013).The modernization of Lagos, especially in the built environment was greatly influenced by the ex-slaves known as *Creoles*, they are the returnees from Sierra Leone, Freetown, the West Indies and Brazil. This had contributed to the diffusion of these styles in architecture from these returnees, which are seen in Lagos Island up till today (Lagos State Bureau of Statistics, 2013). Some of these notable building are spread out across the streets of Lagos Island, such as in Broad street, Marina, Ikoyi, Tinubu square, Lewis , Igbosere, Tom Jones, Bamgbose and others. Figure 1a and 1b shows the map of Lagos Island and some of the streets.

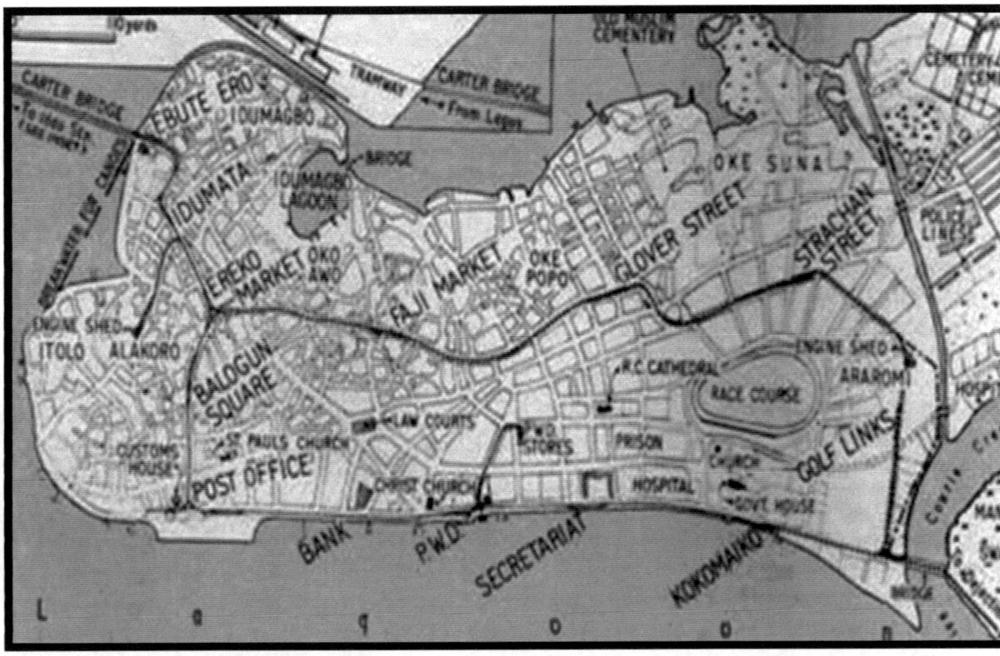

Figure 1a: Map of Lagos Island showing some streets layout.
Source: nairaland.com (Accessed 2019).

Figure 1b: Map of Lagos Island with some of the streets.
Source: www.google.com. (Accessed 2018).

2.2. *Evolution of Housing types in Lagos Island.*

Houses types in Lagos Island have gradually evolved, from different epoch. The evolution of housing characteristics in Nigeria especially Lagos Island is one that has not been critically documented over the years. However, Nigerian houses can be traced to specific geographical areas, stylistic trends or periods.

The prevailing influence in the houses in *Isale-Eko* area of Lagos, being the oldest inhabited part of the Lagos Island, is the Portuguese. This is noticeable in the arched doorways and windows as shown in figure 2 and figure 3. These houses have over time evolved to accommodate some forms of modernism with the availability of modern building materials and the increasing commercial activities in the area, shown in figure 4 and 5 where the walls are curtain walls, although, some of the buildings still retained its building styles. However, the changes in the building types in the area and the factors that have influenced these changes in the housing types and its surroundings would impress on how it can affect smart city agenda implementations and reforms.

Figure 2.

Figure 3.

Figure 2 and Figure 3: Brazilian styles of arched doorways and window of building on Lagos Island.
Source: tour2nigeria.com. (Accessed 2018).

Figure 4.

Figure 5.

Figure 4 and 5: Use of modern building material of Glass facing (curtain wall) for building on Lagos Island.
Source: Figure 4: primelocation.com, Figure 5: agoda.com (accessed 2019).

2.3. The Concept of Smart City.

There are different definitions of smart city as postulated by different authors. Basiri, Azim and Farrokhi (2017) observed that alternative adjectives such as "intelligent" or "digital" are substituted for the word smart by some authors and also noted that multi-variant templates of framing a smart city do exist. Likewise, Cocchia (2014) had earlier used words such as knowledge city, ubiquitous city, and sustainable city for smart city. Bakıcı, Almirall and Wareham (2013) also had agreed that there is no definite description about what smart cities are, but it can be said to be those cities that make use of information and communication technologies (ICT) to increase quality of life of their inhabitants while providing sustainable development. Also, the idea can be referred to as the safe, secure, environmental and efficient urban centre of the future (Bakıcı *et al*., 2013). Dameri, (2013) also stated that a very clear definition of smart city is still lacking, not only in the academic studies, but also in empirical applications of smart concepts and projects. It is a contested concept without a unified definition.

However, generally smart city as it were, can be summarised as an urban area that make use of different electronic data collection sensors to disseminate and manage information by collecting, processing and analysing data, used in traffic managements, transportation system, power plants, waste management, water supply, the law enforcement services, schools, hospitals, information system, libraries and many other facilities within the city. The concept includes both the social, Information and Communication Technology (ICT), design and planning of an area. The concept could be defined from the ICT view or from urban reform and renewal perspectives. However, in actualization of these processes, the involvement of the government is apparent in the implementation of smart city as noted by Bakıcı *et al.* (2013).The popular definition of the concept involves implementation and deployment of information and communication technology infrastructures to support social and urban growth through improving and upgrading the economy, citizens' involvement and governmental efficiency. Therefore, smart city can be summarised and described as a form of sustainable city or urban development method. In addition, Giffinger *et al.* (2007); Anthopoulos and Fitsilis (2014) presented a model for smart city that contained the following six characteristics, which interrelate and comprise the entire urban intelligence, which are smart people, smart economy, smart environment, smart living, smart governance and smart mobility. This is illustrated in Figure 6. Examples of cities where smart city are being implemented are Amsterdam, Singapore, Dubai, China, Milton Keynes, Barcelona, Madrid, Stockholm and New York cities.

2.4. Requirements and operations for smart city implementation.

The geographical nature and the physical environments of the area where smart city projects are to be executed and operated is very important as stated by Dameri (2013). This is because smart city technology allows for city officials and managers to interact directly with both the community where it is effected, and the city's infrastructures. This is to monitor the activities and happening in the city, how the city is changing and evolving. Based on this premise, the principle requires that some basic aspects of the community be in the desired shape for proper running of a smart city project. Example of such aspects are good planning system, the state of the buildings and the environment, proper census of community occupants using facilities in the community, the use of buildings, regular supply of electricity to accommodate ICT characteristics of smart city concept.

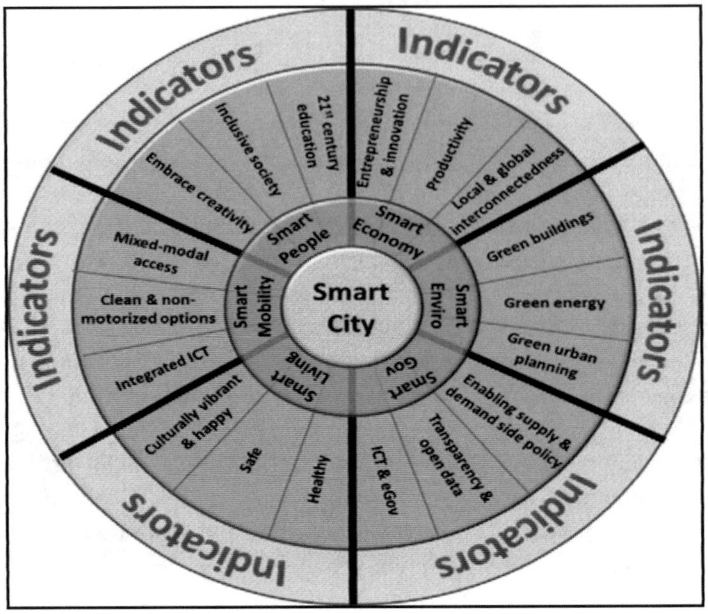

Figure 6: Illustration of the subjected components for smart city actualization. Source: enterpriseresillienceblog.typedad.com. (Accessed 2018).

Furthermore, Basiri *et al.* (2017) highlighted some of the enabling characteristics of making a smart city, which are: (i) the ability of the city and its systems to be instrumental and give allowance for collection of data about the city life form different sources according to the need of the city. Having a proper knowledge about the city, knowing which technical system or person would be right to fulfil their role or achieve their goals, which would be within the context of the overall effective functioning and ability of the city, (ii) The ability to make the city automated to enable appropriate city functions to be delivered reliably, and effectively, not relying on the need of direct human intervention, (iii) A network of collaborative space, which is to enable a dynamic communities that will spur innovation, growth and enhance the well-being of the citizen and (iv) there should be a continues interaction between the digital world and the physical to enable the process of decision open and inclusiveness, by having this, the residents, businesses and the government will be able to work together to make the smart city idea work. However, in the pursuit of these, there is the need to input urban development and renewal that would foster the proper managements, collaborative infusion of all the necessary entities that would make a city smart, which is explained in Figure 7. Here, the infrastructure, human factor (capacities), planning and management, work together to give a smarter building and urban planning, good environment, government and agency administration that works, public safety, a social, healthcare and education programme, that incorporate the whole community, and provision of good transportation, energy and water supply.

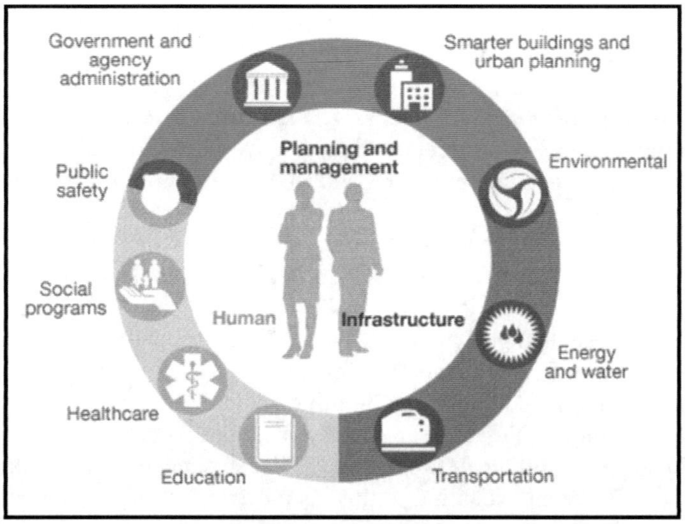

Figure 7: Compositions of smart city to enable functioning
Sources: www.enterrasolution.com. (Accessed 2018).

2.5. Housing and Urban Planning Characteristics that Depicts Smart City.

The concept of smart city varies from cities to cities. This is determined by the level of development in each city and the likelihood of changes, reforms and developments that the city and it residence embraced. To actualize smart city, some studies such as Giffinger *et al.* (2007) are of the opinion that there is need for the development of the entire urban system, which are the physical nature, the social infrastructure and the economic infrastructure. Such features would enable application of a smart city, which are the city:

1. Having flexibility in land use and the ability of the building regulations and bye-law to adapt to change.
2. Having an inclusive housing schemes, in this case, making housing available to everyone, whereby the circumstances of slums and shanties are eliminated.
3. Planning that develops open space into parks, recreational spaces and play ground. Planting of trees within these areas to reduce urban heat and noise effects.
4. Community having a walkable environment, where there is reduced congestion of persons and vehicles on the roads, good road network, motorable tarred roads in every area for vehicles and pedestrians and a good, well manage public transport system in operation.
5. Creation of a reliable online system and internet facility that brings governance closer to the people through the use of phone and mobile gargets. This reduces people going always to the municipal offices.
6. Promotion of the city identity, the distinct features that describes that city, its arts and crafts, culture, sport, health services, local cuisine, textile and furniture.

Another study, Angelidou (2017) had dwelled on this last feature as being very important. It advocated for traditional or old cities where smart city is to be incorporated, to incorporate smart city into the existing forms of such city. In the 15 cities studied by Angelidou, Amsterdam, seems to be different from others because the local needs of this area were not properly captured in the implementation of smart city, thereby leading to other problems such as security challenges. Also, the study by Dhingra and Chattopadhyay (2016) investigated the application of smart city into traditional planned and grown cities in India and Arab cities. It concluded that there is need for inclusiveness and interactive ICT and urban engineering solutions to make smart city work in this area, due to the nature of the city which included the presence of old buildings and streets, close community dwelling pattern. This case is similar to Lagos Island that is an old, traditional area.

2.6 Housing Developments and Planning in Lagos Island and the Concept of Smart city
Development and upgrading of a city is an agenda for any government in order to move forward in planning or developmental strategies. The case of Lagos Island requires that the various problems with its housing and environmental planning condition be considered. This is to give enabling platform for implementing policies, agenda, and programme for its developments, even with the continuous rise in the population of Lagos Island, which is also a global phenomenon. The world's population living in urban areas is forecast to rise to 75% by 2050 as against 50% in 2010 (UN-Habitat III, 2015). Due to this escalating population increase, governments are required to come up with how to create future spaces and its management for the citizens. This is the reason governments base their economic development policies on building advanced infrastructures management strategies such as smart city projects. Not only are policy makers involved in this initiative but also citizens are engaged in the project, as their future quality of life depends on this (Bakıcı, Almirall and Wareham, 2013). Lagos state is not an exception to this, as more and more people come into the state every day.
Planning for housing and setting standards for the regulation of buildings and its construction is a task that seems to overwhelming to the relevant government authorities in Nigeria and to many parts of the developing world. Rapid urbanization create an urgency for cities to find smarter ways to manage the accompanying challenges like high crime rate, traffic congestion, difficulty in waste management and many others of such problems (Nam and Pardo, 2011). The concept of smart city is emerging as a new idea and approach to alleviate and cure current urban problems as stated and to make urban development more sustainable (Alawadhi, Armando and Hafedh, 2012). The process of physical development in Lagos State is one that had over time witnessed various adoptions of concepts and strategies and ideas with various authors and publications propounding great ideas to foster these developments. Smart city development is another area in which most cities delve into. However, in the light of this developmental agenda, it is good to examine how cities fit into the concept of smart city programme; how the housing characteristics and the planning nature of the area can be adapted to a smart city.

3. Research methodology
This study used survey method to derive its data, which were primary data. Questionnaires were distributed to residents of 100 buildings within the residential district areas of Lagos Island, with one representative for each building. The study took place between February and March 2017, it was administered and retrieved by hand. The buildings were randomly selected from 11 streets, which are Tokunbo, Bamgbose, Mobolaji Bank Anthony, Campbell, Kakawa, Campos, Odunlami, Glover, Igbosere, Joseph and St Catholic Mission Streets of Lagos Island. The questionnaire was sectioned into two parts. First to collect data on the respondents profile and their relationship with the building and the other part of the questionnaire collected data on the building characteristics. Out of the 100 questionnaires distributed in these buildings, 77 were returned valid for further analyses. In addition to the research instrument of the questionnaire, photographic material and observation schedule were used to carry out the survey. The data were processed and analysed using descriptive statistical analytical methods with the use of statistical package for social science (SPSS) software.

4. Findings and discussion
The result showed that about 44 (57%) of the respondents were males, while 33 (43%) were females. This shows that the household heads are mostly male than female, either as house owners or tenants represented the house owners. Many of the respondents had stayed for more than ten years in the buildings, which validated the claim that most of them or their late parents are the owners of the buildings, or there were certain activity or reasons which had made them stayed in these buildings for

that length of time. Also, many of them had chosen to remain in those buildings for that number of years because of the good locations, a suggestion that this could be for commercial activities as presented from the response to the question on if shops are attached to the buildings or not, having 56 (72.7%) positive response, shown in Table 2. A considerable number of 10 (13.2%) also preferred to live in those buildings due to security reason. Other reasons are closeness to commercial activities, closeness to social amenities, closeness to friends and religious houses, which are other reasons given. (Table 1). Figure 8 shows the display of respondents stay in the building.

Table 1: Respondents' characteristics

Respondents' Characteristics	Variables	Frequency (N=77)	Percentage
Sex	Male	44	57.1
	Female	33	42.9
Length of stay	1-5 years	9	11.7
	6-10	13	16.9
	11-15	25	32.5
	Over 15 years	30	39.0
Reasons for living in the house	Good location	39	50
	Security	10	13.2
	Nearness to friends	5	6.6
	Others	23	30.3

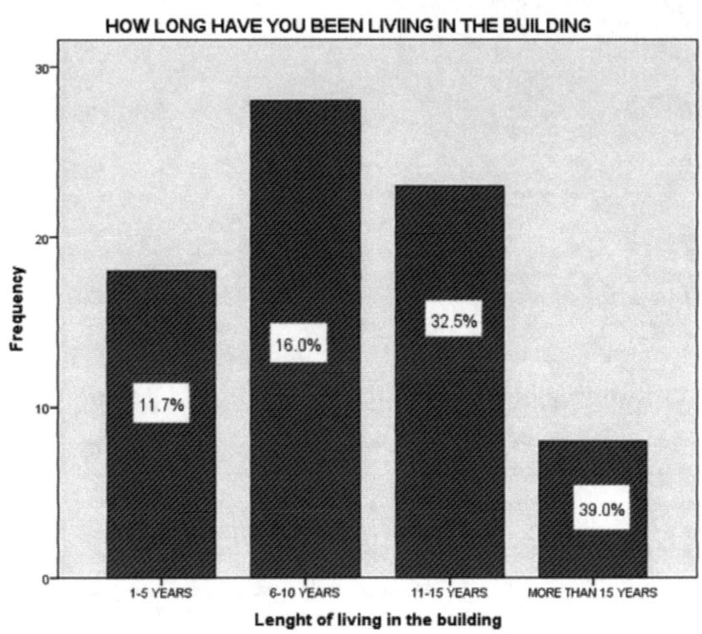

Figure 8: Respondents length of living in the building.
Source: Authors field study.

A survey of the buildings showed that 55 (71.5%) of the buildings had existed for over 10 years; 13 (16.9%) for 6-10 years, and 9 (11.7%) for 1-5 years. Only about 8 (10.4%) of the buildings have been renovated in the last 10 years, which are very few number as shown in figure 9. This is evident in the appearance of the buildings components (Table 2). It should be noted that most of these components (especially roofs, doors and windows) have not been changed in the last 10 years. Also, only 14 (18.2%) of the buildings have been painted in the last 5 years. It could be inferred that the buildings lack proper maintenance, they are left to the point where corrective maintenance is expedient. This is in variance with one of the agenda of smart city development as propounded by Hall (2000); Nam and Pardo (2011), according to these studies, a smart city should plan its preventive maintenance activities. Another characteristic of the surveyed buildings that is worth noting is the mixed use property. Majority 56 (72.7%) of the buildings had shops attached to them. This, of course, is in tandem with the opinion of Giffinger *et al.* (2007) that a smart city should perform well in a forward looking way, in economy, people, and environment, amongst others. Asides the 56 (72.7%) that are mixed-use buildings, 11 (14.3%) are exclusively for commercial activities like shops, banks, offices, church, mosque, clinic and schools. This is shown in figure 10 and figure 11. These activities are part of what make for life satisfaction and happiness. Also, as Guan (2012) had posited that a smart city must be set to provide conditions for a healthy and happy community.

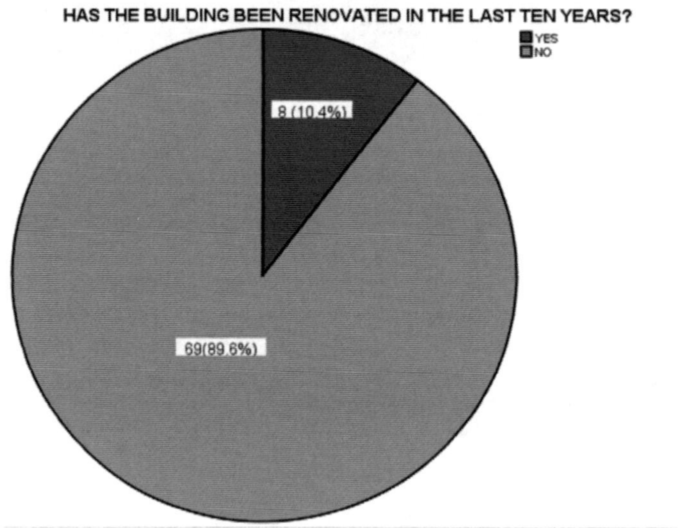

Figure 9: Proportion of building renovated in the last ten years.
Source: Authors field study.

Table 2: Building characteristics

Building Characteristics	Variables	Frequency (N=77)	Percentage
No of years the building has been existing	1-5 years	9	11.7
	6-10	13	16.9
	11-15	25	32.5
	Over 15 years	30	39.0
Renovation in the last ten years	Yes	8	10.4
	No	69	89.6
Shops attached to the building	Yes	56	72.7
	No	21	27.3
Number of shops in the building	None	21	27.3
	1-2	31	40.3
	3-5	18	23.4
	More than 5 shops	7	9.1

Other use of building	Yes	11	14.3
	No	66	85.7
Other uses of the building aside residential	None	66	85.7
	Business centre	1	1.3
	Clinic	1	1.3
	Commercial	1	1.3
	Legal practitioner	1	1.3
	Microfinance bank	1	1.3
	Office	1	1.3
	Religious (mosque)	1	1.3
	Religious us (church)	1	1.3
	School	2	2.6
	Shop	1	1.3
Change of building use in the last ten years	Yes	13	16.9
	No	64	83.1
Reason for change of use	**None**	64	83.1
	Commercial	9	11.7
	Storage	2	2.6
	Others (Clinic and school)	2	2.6
Change of entrance door	Yes	5	6.5
	No	72	93.5
Change of windows	Yes	11	14.3
	No	66	85.7
Change of roof	Yes	9	11.7
	No	68	88.3
The last time the building was painted	Never	3	3.9
	1-5 years ago	14	18.2
	6-10	34	44.2
	11-15	24	31.2
	Over 15 years	2	2.6

Figure 10: Figure: 11

Figure 10 and Figure 11: Buildings at Igbosere, used for both residential and commercial activities.

Source: Authors field study.

5. Conclusion, and recommendations

This study was on identification of the housing and planning characteristics found in Lagos Island residential district. The history of the buildings and factors that have influenced the changes in housing types and how they can influenced the agenda of smart city implementation or reform were examined. It could be concluded that the study area does not exhibit much of smart city characteristics elements. However, some characteristic of the areas that exhibits activities that make for life satisfaction and happiness could aid smart city program, but others characteristics, which are negative such as low maintenance of building could hinder the implementation of smart city agenda. It is therefore recommended that there is need for the transformation of the city for the operation of a smart city. Comprehensive urban renewal should first be implemented, like in the case of Barcelona, which has undertaken significant reforms and has a special significance, which is as a result of apparent tendency in its urban policies and reforms. Making it to become a leading smart city among the European cities. Hence, an assessment of the smart city initiative should cast light on current urban renewal policies and reforms which would make the implementation of smart city possible. However, developmental efforts should add value to any existing area, especially residential areas of the city. The study results showed a mixed use dwellings for both commercial and residential activities, which has become the life style of the dwellers. In view of this, the application of smart city in Lagos Island should be done without disturbing the economic, social and the daily activities of its dwellers. This is due to the peculiarity of the area, considering some of the identified problems of Lagos Island as stated by Aluko (2011) which ranges from uncontrolled use of property, non-adherence to approved designs and standards, and bouycourting of building regulations, this also, needs to be addressed for proper implementation of smart city agenda. A large percentage 69 (89.6%) of the building investigated in this study had not been renovated in the last ten (10) years and are mixed used buildings, greatly influenced by increasing commercial activities in the area. The buildings are in need of renovations and the environment needs reforms that would enable the mixed-use building nature of the houses work to implement smart city norms. Basic infrastructure requirements necessary for implementing of smart city as described in Figure 6 and figure 7 is found to be lacking in the studied area. Also, the commercial activities was found to influence the changes in the building found in the area, which in turn can affect the smart city agenda if the appropriate urban renewal and reforms process are not embarked on. Sustainable urban renewal and planning should be implemented by the government. Studies on the human capacity and the literacy level of the residents of this study area, which can make them to be active participatory partners to the implementations of smart city agenda is recommended for further study.

6. Acknowledgements

The authors would like to acknowledge the Management of Covenant University for encouraging and facilitating research through the Covenant University Center for Research, Innovation, and Discovery (CUCRID). We also acknowledge anonymous reviewers whose comments and suggestions shaped the final version of this article.

Reference

[1] Agbola, T. and Jinadu, A. (1997). Forced eviction and forced relocation in Nigeria: The Experience of those evicted from Maroko in 1990. *Environment and Urbanization, 9*(2), pp 271-288.

[2] Ajanlekoko, J. S. (2001). Sustainable Housing Development in Nigeria–the Financial and Infrastructural Implication. *International Conference on Spatial Information for Sustainable Development in Nairobi, Kenya*. pp. 2-5.

[3] Akinlose, R. (2016). Focus on the Lagos Island Office Construction Market Report. https://www.slideshare.net/Rotimi**Akinlose**/focus-on-the-**lagos**. Accessed: June 21, 2018.

[4] Alawadhi, S., Aldama-Nalda, A., Chourabi, H., Gil-Garcia, J. R., Leung, S., Mellouli, S. and Walker, S. (2012). Building Understanding of Smart City Initiatives. *International Conference on Electronic Government*. pp. 40-53. Springer, Berlin, Heidelberg.

[5] Aluko, O. (2011). Development Control in Lagos State: An Assessment of Public Compliance to Space Standards for Urban Development. *African Research Review*, *5*(5), pp169-184.

[6] Angelidou, M., and Psaltoglou, A. (2017). An Empirical Investigation of Social Innovation Initiatives for Sustainable Urban Development. *Sustainable Cities and Society*, *33*, pp113-125.

[7] Anthopoulos, L., and Fitsilis, P. (2014). Exploring Architectural and Organizational Features in Smart Cities. In *Advanced Communication Technology (ICACT), 16th International Conference on*. pp. 190-195. IEEE.

[8] Bakıcı, T., Almirall, E. and Wareham, J. (2013). A Smart City Initiative: The Case of Barcelona. *Journal of the Knowledge Economy*, *4*(2), 135-148.

[9] Basiri, M., Azim, A. and Farrokhi, M. (2017). Smart City Solution for Sustainable Urban Development. *European Journal of Sustainable Development*, 6, 1, 71-84, ISSN: 2239-5938

[10] Brazilian styles of Arched doorways and window of building on Lagos Island: tour2nigeria.com. Accessed: June 20, 2018.

[11] Cocchia, A. (2014). Smart and Digital City: A Systematic Literature Review. In *Smart City*. pp. 13-43. Springer, Cham.

[12] Compositions of smart city to enable functioning. www.enterrasolution.com. Accessed: June 20, 2018.

[13] Components for smart city actualization. enterpriseresillienceblog.typedad.com. Accessed: June 20, 2018)

[14] Dameri, R. P. (2013). Searching for Smart City Definition: A Comprehensive Proposal. *International Journal of Computers and Technology*, *11*(5), 2544-2551.

[15] Dhingra, M. and Chattopadhyay, S. (2016). Advancing Smartness of Traditional Settlements- Case Analysis of India and Arab Old Cities. *International Journal of Sustainable Built Environment*. (5)2, pp 549-563.

[16] Giffinger, R., Fertner, C., Kramar, H., Kalasek, R., Pichler-Milanovic, N. and Meijers, E. (2007). Smart Cities: Ranking of European Medium-sized Cities. Vienna. *Austria: Centre of Regional Science (SRF), Vienna University of Technology*.

[17] Gandy, M. (2006). Planning, Anti-planning, and the Infrastructure Crisis Facing Metropolitan Lagos. In *Cities in contemporary Africa*. pp. 247-264. Palgrave Macmillan, New York.

[18] Guan, L. (2012). Smart Steps to a Better City. *Government News*, *32*(2), 24.
Hall, R. (2000). The Vision of a Smart City." Proceeding of the 2nd International Life Extension Technology Workshop, Paris, France, 2000.

[19] Ibem, E.O; Anosike, M.N. and Azuh, D.E (2011). Challenges in public housing provision in the post- independence era in Nigeria. *International Journal of Human Sciences,* **8(2):** 421-443

[20] Ilesanmi, A. (2010). Urban Sustainability in the Context of Lagos Mega-city. *Journal of Geography and Regional Planning*, *3*(10), 240.

[21] Jegede, F.O, Ibem, E.O, Oluwatayo, A.A (2018), Manifestation of Defensible Space in Lagos State Development and Property Cooperation Housing Estates, Lagos, Nigeria. I*nternational Journal of Civil Engineering and Technology* (IJCIET) **9(12)**, pp. 491–505.

[22] Lagos State Bureau of Statistics (2013). Digest of Statistics. www.mepb.lagosstate.gov.ng. Accessed: June 20, 2018.

[23] Map of Lagos Island. www.google.com. Accessed: June 19, 2018.

[24] Nam, T. and Pardo, T. A. (2011). Conceptualizing Smart City with Dimensions of Technology, People, and Institutions. In *Proceedings of the 12th annual international digital government research conference: digital government innovation in challenging times*. pp. 282-291. ACM

[25] Smart City. enterpriseresillienceblog.typedad.com. Accessed: June 25, 2018.

[26] UN-Habitat III (2015), 21 – New York Smart Cities, Paper has been co-/led by UN-Habitat, UNDP and ITU with contributions from CBD.

[27] What is Smart City? smartcities.gov.in/content/innerpage/smart-city-features.php. Accessed: June 23, 2018.

IOP Conf. Series: Earth and Environmental Science **331** (2019) 012002 doi:10.1088/1755-1315/331/1/012002

Agricultural Exports and Economic Growth in Nigeria: An Econometric Analysis

R Osabohien D Akinpelumi2 O Matthew3 V Okafor4 E Iku5 T Olawande6 U Okorie7

1,2,3,4,5,7. Department of Economics and Development Studies, Covenant University, Ota, Nigeria
6. Department of Sociology, Covenant University, Ota, Nigeria

Abstract. It has been observed that export promotion drive boosts economic growth of developing countries (Nigeria inclusive) through the multiplier effects that the income earned from exports has on the economy. Prior to the discovery of oil, agriculture has been the main contributor to the Nigerian external sector. It is in the light of this that this study sets out to examine the impact of agricultural export on Nigeria's economic growth. This study used the Autoregressive Distribution Lag (ARDL) econometric technique to analyze the long run relationship and the impact of agricultural exports on Nigeria's economic growth. Economic growth is the dependent variable, and is proxied by the real gross domestic product, the explanatory variables include: agricultural export, foreign direct investment, inflation rate and the labour force. The results from the ARDL technique revealed that agricultural exports significantly affect Nigeria's economic growth; this suggests that, a 1percent increase in - agricultural export will boost economic growth in Nigeria by approximately 25percent.Therefore, from the results, the study recommended that agricultural export need to be promoted through the increase in agricultural production base.

1. Introduction

The performance of the sub-Saharan African (SSA) economies (Nigeria inclusive) has shown an improvement for the past two decades, this is reflected in upward trend of the Gross Domestic Product [1]. The Gross Domestic Product (GDP) of SSA countries grew by approximately 60%, with the middle class growing by 90% [2]. The SSA economies have been observed to be the fastest growing economy than any other region of the developing countries of the World, apart from South Asia and the growth of SSA economies have brought about the demand for agricultural commodities [2]
The SSA agricultural imports have also shown a rapid growth, this shows an average of 13 percent increase per year. In 2014, total imports of agricultural commodities was estimated at $485billion (United States Dollars), representing the second of the largest in the world. As a comparison, Asian countries, especially; India, imported only 17.6 billion USD in agricultural commodities, irrespective of the Indian population, which is about 300 million people more than that of the SSA countries [2]. The SSA total agricultural imports are about 11.7 billion USD from other countries within the region and 36.8 billion USD from other countries and regions of the world [2]

Economists, scholars and researchers across the world have drawn attention to the study of the connection between the horticultural yields and the financial advancement of an economy. It has been theoretically argued that the difference in the rate of export rates could lead to differences in economic growth [3]. The export growth is recognized as the major contributor to productivity and the growth of employment of an economy evidenced in GDP growth rate [3]. The favourable climatic conditions combined with good vegetation enable SSA countries to produce crops and livestock in particular [4, 3]. The increase in agricultural exports has greatly benefited Nigeria; therefore, the relevance of exports to the financial development and progress of a country cannot be overemphasised as it helps to accelerate a country's overall progress [5]. In addition, exports are a means of foreign exchange earnings, due to the fact that trade activities between countries are settled in foreign exchange [6] Job openings are created for their citizens consequently reducing the social costs of unemployment [7]

In the work of [8] they opined that the drive towards export promotion will boost an emerging nation's economic growth through the multiplier effects of export on the income level of the country; this is because income earned through exports help to enhance the aggregate demand in the country. Over the years, trade openness had attained a high degree in the Nigerian economy, thus, its efficiency should be promoted via the advancement of the efficiency of external trade. Primarily, agriculture has been the main dominant sector of the Nigerian external sector [9]

Agriculture has the distinguished features of low prices, unfavourable terms of trade earnings resulting from export instability [10]. The low earnings from export of agricultural commodities are due to decrease in production of agricultural commodities that made emerging economies rely on importation of food. Among the very important objectives of emerging economies in general and Nigeria to be specific, is the achievement of accelerated overall wellbeing of the nation. In this regard, exports are seen as a promoting factor for economic growth [10, 11, 12]

Two schools of thought (Adam Smith and Ricardian) are of the opinion that foreign trade remains as the major determinants of economic growth, through which higher benefits via specialization of a specific good. From the view of export led growth theory; exports as the main determinants of the growth of the economy, which have theoretical justifications like; first, in the Keynesian theory, higher exports generate higher earnings in the short term through the effect of foreign exchange multipliers. Second, higher exchange rises through higher exports, making purchases of machinery, electrical and transportation equipment. It is also used to buy fuel and food that can serve as a catalyst for any country's economic growth. Third, the economic growth of a country is promoted indirectly via competition improvement, scale of economies, development of technical know-how among others; fourth, externalities which are positive in nature such as the effective control of inefficiencies in organisations, production management and control, this can be done by learning from foreign firms and experts who are knowledgeable about product design to enhance productivity [11, 12]

Two schools of thought with different views about agricultural export are considered in this study: the first school of thought focused on the elements, which are exogenously determined to the individual country; such elements are minimal growth rate resulting from low production of primary products market and the terms of trade, which are unfavourable. On the other hand, the second school of thought focused on the elements, which are endogenously determined by a country, those factors are, the local institutions which have hampered agricultural export [4, 1, 13]. There is an obvious situation where no country is self-sufficient, nations inevitably trade with several others in order to enjoy goods and services for which it has a comparative weakness in its production while at the same time giving those goods a comparative advantage in its production, for example, Nigeria, where the masses are engaged in the agrarian section, while some are engaged in the manufacturing and tertiary sections.

However, recent information has it that Nigerian agricultural exports have experienced decreases in revenue because of price instability at the international market [14]. Compared to manufactured products imported from advanced countries around the world, these agricultural products are made less competitive, resulting in trade deficits. This has had a major impact on the share of agricultural products in Nigeria's economic growth. It is pertinent to study the previous and recent trend of such agricultural exports. With the above issue raised, coupled with agriculture-related issues have necessitated the need for a study in that regard. This study is therefore divided into five sections; following this introductory section is section two, which is the review of related literature, section three provides the data source and method used for the study, section four presents the results and discussions, and the fifth section presents the study's summary and conclusion.

2. Literature Review and Stylised Facts

2.1 Literature Review

As argued, agricultural exports have undoubtedly played a crucial role in boosting Nigerian economy growth as well as other sub-Saharan countries, contributing to foreign-exchange earnings in Nigeria and the SSA countries used to promote other capital projects. [15] Based on the Ordinary Least Square (OLS) regression model, the study analyzed the link between exports of agricultural commodities and economic growth in Nigeria. OLS results showed that agricultural exports in Nigeria have a positive impact on GDP. The agricultural sector was said to contribute about 34.4 percent to GDP growth between 1970 and 2010. Exports of agricultural products accounted for about 75 percent of total annual exports in the 1960s, according to studies of [16, 17].

In addition, Nigeria had been known to be the major producer and exporter of certain major produce in the world in the pre-independence period. Nigeria was the highest palm producing exporter (oil and kernel) while ranking second in cocoa to Ghana and third in groundnut production. [16] It is also stated that a long-term link exists between agricultural exports and Nigeria's economic growth. Akin to [16, 1], using the ARDL approach unveils the various factors, which determine the food production level in Nigeria. These included technical know-how, others entails savings, change in technology, infrastructure and so on.

In the empirical work of [9], they used the propensity score matching approach and they noted that social protection laws and regulation driven towards the agricultural sector will help boost the production capacity of the sector and in return, economic growth. This is because, the agricultural sector is the major driving force of the economy, through the provision job opportunities and means of livelihood for the teeming population, especially in rural areas in which about 85 percent of the population depends solely on agriculture. Similarly, [4] Using qualitative discourse, it was pointed out that the institutional quality of local communities affects the production, processing and distribution of agricultural commodities, in particular rice production, the main food consumed by more than 90% of Nigerian households.

These institutions and social protection programmes could be in form of controlling emissions greenhouse gases, which negatively affects farmers' health status and causes damages to controls and livestock alike [11] In line with this, it is good to point out that as at independence in 1960, the agricultural sector accounts for more than 50 percent of the Nigerian economy's GDP. Nonetheless, the sector's growth rate has been declining over the years due to the crude oil discovery.

According to [18], it is generally agreed that food production in Nigeria is adequate and that economic liberalization has had a positive impact on food production. There were substantial benefits in aggregate and per capita indices of agricultural food production, following the economic deregulation between 1980-1989 and 1990-1998 periods, all major food products in the economy recorded substantially positive output trends, with a significantly higher percentage change in production for the post-SAP era. The era of pre-deregulation has seen negative growth in almost all agricultural sub-sectors. A very high rate of growth in agricultural yields was experienced immediately following the economic deregulation, most of which occurred in the staple food sub-sector.

Evidence from previous studies has shown that the recent global warming and emissions of greenhouse gasses have affected health status of farmers and the production level of agricultural commodities which has led to the reduction in the level of food production [11, 6]. In an attempt to satisfy the ever growing food and non-food requirements as a result of population growth, fertile soils, fossil fuel are indiscriminatingly been depleted across the globe to meet their demands via the adoption of ICTs and good institutions [19, 20, 21, 22]

According to [16], they examined the contribution of agricultural export commodities to economic growth and found that agricultural export commodities accounted for about 75% of total annual merchandise exports in 1960. In the pre-independence period, Nigeria was also known to be the world's leading producer and exporter of major food crops. Nigeria was the largest exporter of palm oil and palm kernel products, only second in cocoa to Ghana and third in groundnut production [17]. They also found a long-term relationship between Nigeria's agricultural exports and economic growth. Evidence from [11] has shown a positive impact on economic growth in less developed countries, including Nigeria, from agricultural exports.

2.2 Stylised Facts

This study sub-section presents the facts of stylization as shown in Figure 1.

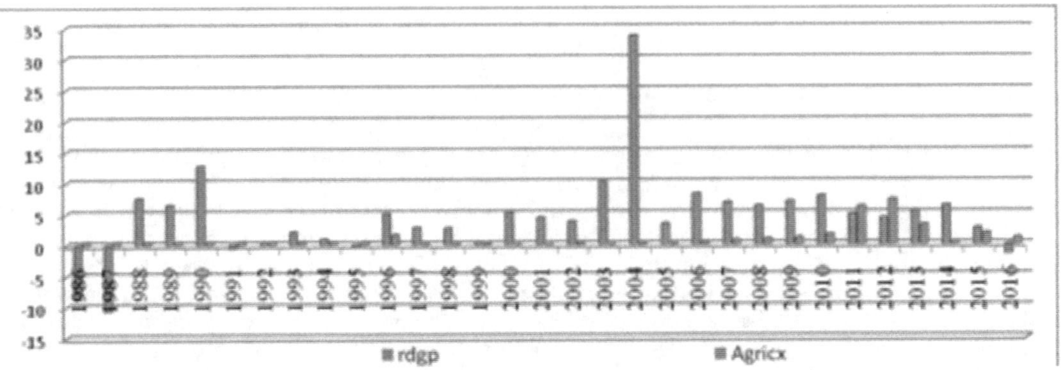

Figure 2.1. Trend of Agricultural Export and Economic Growth of Nigeria
Source: Authors' Computation

Figure 1 showed the trend of agricultural export between the periods of 1986 and 2016. The agricultural exports signify the percentage of agricultural raw materials as a fraction of total merchandise export. In 1986, the percentage contribution of agricultural raw materials to total merchandise export was approximately 41percent, which later increased to 51percent in 1987. However, from 1988 to 1990, the agricultural export average percentage of merchandise was 47percent. This further increased to an average of 52percent from the period of 1991 to 1995. The continuous rise in agricultural export can be traced to the effect of the Structural Adjustment Programme (SAP) in 1986. Agricultural export percentage of merchandise export increased to over a hundred percent in 1996. In the same period, the real gross domestic product (rgdp) grew at an increasing rate.

However, from 1996 to 2006 the agricultural export percentage of merchandise export experienced a continuous decline. By 2007, Agricultural exports accounted for only 7percent of export. According to [23], Nigeria became an importer of the basic food items formerly exported. This could also be explained by the shift of focus to oil exportation which led to the agrarian section been neglect. Nonetheless, the contribution of agriculture to export has been increasing although at a low rate since 2007 as seen in figure1. This can be as a result of the various intervention programmes such as social protection programmes and policies, [1] that have surfaced over the years for the purpose of reviving the agriculture sector. Notwithstanding, the agricultural sector is still growing at a low rate and is not utilising its full potential.

3. Methodology

3.1 Model Specification

Based on empirics, this study adopts the vent surplus theory adapted from the study of [17]. The vent surplus hypothesis featured the operation of remote exchange as promoted by Adam Smith [17]. The hypothesis recommended that in rising economies, the nearness of surplus assets and being worked on of human capital has prevented their productivity [11, 17]. Moreover, [24, 25] opined that the vent surplus theory concentrated on the effectiveness of creation strategies, with the goal that the resultant yield surpasses the underlying contributions because of surplus generation. Consequently, the model is specified as:

$$rgdp \quad = \beta_0 + \sum_{t=1}^{n} \beta X + e \qquad (1)$$

Where; rgdp represents gross domestic product economic growth as used in the studies of [25, 26]. β is consistent; X speaks to the factors exogenously determined in the model and e speaks to the error term capturing the other logical variables not in the model Equation (1) is implicitly and explicitly specified in equations (2) and (3) respectively:

RGDP = f(AGRICXPORT,FDI, REALXR, INFLR,LAB,) (2)

$$loggrdp = \alpha_0 + \alpha_1\, logagricxport + \alpha_2 logfdi + \alpha_3\, logrealxr + \alpha_4 loginflr + \alpha_4 logabf + e \quad (3)$$

Where: rgdp represents economic growth, agricxport represents rural fare as used in the studies of [17, 27, 28]. fdi represents foreign direct investment. Realxr represents real exchange rate, inflr represents inflation rate and labf represents labour force

3.2 Data Sources and Description

Table1: Variables Definition and Source

Variable Name	*Identifier*	*Source of Data*	*Definition and Measurement*
Real gross domestic product	Rgdp	WDI, 2017	In this study, real gross domestic product (GDP) measures the inflation-adjustment showing the total volume and value of goods and services produced in an economy.
Agricultural export	Agricxport	WTI, 2017	Agricultural raw materials exports (% of merchandise exports)
Foreign direct investment	Fdi	WDI, 2017	Foreign direct investment refers to foreigners making investments in a host country. It is measured as net (percentage of GDP) inflows.
Real exchange rate	Realxr	WDI, 2017	This is the official rate of exchange (LCU per US$, average period). The real exchange rate measures the ratio of foreign currency's price level and local currency's domestic price level if the foreign price level is converted via the current nominal exchange rate into domestic currency units.
Inflation rate	Inflr	WDI, 2017	The rate of inflation refers to the percentage increase in goods and services prices. It is measured as (yearly percent) consumer prices.
Labour force	Labforce	WDI, 2017	The rate of participation in the labor force refers to the number of people working in the agricultural sector. It is measured as a percentage of the population over 15 years of age. (Estimate model of the International Labor Organization)

Source: Authors' Compilation, 2019

4. Results and Discussion

4.1 Econometric Technique

The first step of the econometric analysis is to conduct the summary statistics of the variables selected as shown in Table 2. For analysis, descriptive statistics are used. Table 2 illustrates the results for Nigeria's descriptive statistics.

Table 2: Summary Statistic of the Variables

Variable	*Mean*	*Standard Deviation*	*Minimum*	*Maximum*
Rgdp	141399.7	185425.6	872.868	552097.4
Agricx	1.04169	1.659659	0.0059455	7.268343
Fdi	2.95e+09	2.69e+09	1.93e+08	8.84e+09
Realxcr	88.68143	70.20631	1.754523	253.4922
Inflrate	20.28164	18.82513	5.382224	72.8355
Labforce	53.09946	9.455727	15.69685	56.964

Source: Authors' Computation, 2019.

4.2 Unit Root Test

[29] It was suggested that the use of slacks in the Augmented Dickey-Fuller (ADF) was traded for the stationary unit root test, which in error terms was self-assertive of the sequential connection [30]. Another test for the unit root was carried out and the report confirmed that at their first difference all variables that were non-stationary at levels were stationary. That is, the application one, I (1)

incorporates all factors. Since a similar request co-ordinates all arrangements, a vital condition is met for the co-integration of the test investigation.

A co-integration test was conducted to determine the presence of a long-run relationship between variables to ensure that the variables relapse is significant and non-deceptive. It would also appear if the variables have a long-term connection. In the event that the follow measurement and the Max-Eigen measurement are more prominent than the 5% basic qualities, the invalid speculation of no co-coordination will be dismissed for the elective theory at that dimension [1]. The Max-Eigen measurement showed that a long-run relationship is co-integrated among the variables. This shows that agricultural exports have a long-term relationship with economic growth in Nigeria (see Table 2). This was also based in the unit root test (see Table 3).

Table 3: Unit Root Test Results

Series	*ADF Statistic*	*CV @ 1%*	*CV @ 5%*	*CV @ 10%*	*Order of Integration*	*Remark*
Rgdp	-4.456	-3.723	-2.989	-2.625	I(1)	Stationary
Agricxport	-4.460	-3.723	-2.989	-2.625	I(1)	Stationary
Fdi	-6.219	-3.723	-2.989	-2.625	I(1)	Stationary
Realxcr	-5.644	-3.723	-2.989	-2.625	I(1)	Stationary
Inflr	-4.781	-3.723	2.989	-2.625	I(1)	Stationary
Labforce	-8.912	-3.723	2.989	-2.625	I(1)	Stationary

Source: Authors' Computation using STATA 13, 2019. Note: CV is the critical value.

The co-integration test was evaluated with a slack length of 1 using the Johansen co-integration technique, and the unrestricted co-integration rank test with no intercept or pattern in co-integrating condition (CE) was conducted with follow-up insights showing 5 co-integrating vectors and most extreme self-value measurements showing 5 co-integrating vectors with a noteworthy dimension of 5 percent. The cointegration test by Johansen is the result as shown in Table 4.

Table 4: Co-integration Test Results

Maximum Rank	Eigen Value	Trace Statistics	Critical Value (5%)
0	-	128.9644	94.15
1	0.78959	83.7621	68.52
2	0.74938	43.6317	47.21
3	0.53291	21.5562	29.68
4	0.39266	7.0949	15.41
5	0.15268	2.2901	3.76

Source: Authors

Table 5: Cointegrating Equation Result

Rgdp	*Agricxport*	*Fdi*	*Realxr*	*inflrate*	*Labforce*
Coefficient	146234.3*	0.00025	1543.708**	241.13*	28817.66*
(P-value)	(0.000)	(0.213)	(0.025)	(0.000)	(0.000)

Source: Authors' Computation, 2019. Note: *, ** mean significant at 1 and 5% levels respectively. The Lag Selection was based on the Akaike Information Criterion (AIC), Schwarz Information Criterion (SIC), and Hannan-Quinn Information Criterion (HQIC).

The ECM was used to address disequilibrium in the co-incorporating relationship (see Table 6). This system filled in with its long-term conduct as a method for accommodating short-term disequilibrium behavior of an economic variable of enthusiasm. The parameter coefficient and the t-statistical or probability estimate are the two parameters used in the model for error correction (ECM). The coefficient is relied on to have negative sign, demonstrating that after each time of disequilibrium, a union of factors returns to balance. The p-value is used to check the essence of 5 percent (0.05) testing factors as shown in Table 6.

Table 6: Estimates from Vector Error Correction Model (VECM)

Regressand → Regressors	*D_rgdp*	*D_ agricxport*	*D_fdi*	*D_ realxcr*	*D_inflrate*	*D_labforce*
ECterm	0.0350***	-2.83e-06*	-10.82917	-.0000118	-9.44e-07	-0.0000*
	0.019840	4.36e-07	657.3528	.0000115	.0000113	1.71e-06
	(0.078)	(0.000)	(0.987)			
				(0.306)	(0.933)	(0.000)
rdp(LD)	0.042429	0.00003*	-16435.77**	0.0000384	-0.000022	0.0000185
	0.204662	4.50e-06	6780.959	.000119	0.000112	0.000018
	(0.836)	(0.000)	(0.015)	(0.746)	(0.850)	(0.293)
agricxport(LD)	-1019.199	.4539935*	4.51e+08**	0.6450548	-.2744291	5.272116*
	5479.1	.1203691	1.82e+08	3.177664	3.108375	0.4716604
	(0.852)	(0.000)	(0.013)	(0.839)	(0.930)	(0.000)
fdi(LD)	7.86e-06	-8.04e-11	-.2212347	-3.08e-09	-1.49e-09	2.05e-09*
	5.29e-06	1.16e-10	.175266	3.07e-09	3.00e-09	4.55e-10
	(0.137)	(0.489)	(0.207)	(0.315)	(0.619)	(0.000)
realxr (LD)	895.6451**	-.0017212	-1.94e+07	.1677323	.0532155	0.0219126
	441.419	0.0096974	1.46e+07	0.256006	0.2504237	0.0379989
	(0.042)	(0.859)	(0.184)	(0.512)	(0.832)	(0.564)
inflrate (LD)	45.78822	0.0102778	3718235	0.058061	.0768034	0.023178
	380.5723	0.008361	1.26e+07	0.22072	0.2159044	0.032761
	(0.904)	(0.219)	(0.768)	(0.793)	(0.722)	(0.479)
labforce (LD)	-808.9527	-.0032429	-2.19e+07	0.18572	0.08751	-0.7180526*
	909.3257	0.0199768	3.01e+07	0.527374	0.515874	0.078278
	(0.374)	(0.871)	(0.467)	(0.725)	(0.865)	(0.000)
Adj. R-sq	*0.4654*	*0.7796*	*0.4201*	*0.2870*	*0.0268*	*0.9182*
AIC	*93.90622*					
HQIC	*94.68883*					
SBIC	*96.40507*					

Source: Author's Computation, 2019. Note: *, ** means significant at 1% and 5%levels of significance respectively. LD signifies that they were lagged and differenced. The probability values are in parenthesis.

4.3 Discussion of Results

Statistics showed Nigeria earned $730m (N143bn) from its agricultural export in the fourth quarter of 2015. This is a pointer to the fact that Nigeria's agricultural exports are still underused; this is because the country is still heavenly dependent on crude oil. When most of the population of a country starts mainly in agriculture, it is generally necessary to increase agricultural productivity to generate transformative income growth. [1, 31]. Evidence from previous studies has shown that the prospect of agriculture in sub-Saharan African countries is a major determinant of economic growth. From 1990 to 2013, agricultural production increased by approximately 120%, crop production as a share of the total production of agricultural commodities, on the average, contributed more than 75%.

However, this contribution varies across region, ranging from Southern Africa (53%) to more than 90% in West Africa, as confirmed in [1]. In line with the study of [32, 33, 34, 35] who also used the approach of co-integration to examine the long-term contribution of agricultural exports to Nigeria's economic growth, he pointed out that exports of agricultural products were much more widespread among tropical African countries. Generally speaking, the increase in agricultural exports has been a significant success story and has brought many benefits to Africa. Agricultural exports are usually, in most developing countries, a rational policy for expanded foreign exchange earnings in the non-oil sector. This is confirmed in this study, as shown in Table 5; long-term increases in agricultural exports will contribute 0.25% to economic growth; while foreign direct investment, effective exchange rate and labour force will contribute 0.29%, 0.185% and 0.034% respectively to Nigerian economy growth.

The political implications of these findings are; first, this study reveals that agricultural exports have a long-term relationship with economic growth, meaning that the agricultural sector will contribute to economic growth in the long run. As agricultural exports boost economic growth, the implication of this result is that it is important that the government should encourage more people to practice agriculture which will in turn increase agricultural output, when output is increased the foreign exchange earned from exports will increase, and hence bring about the growth of the Nigerian economy. Second, the implication of the positive relationships between foreign direct investment (FDI), exchange rate and economic growth in Nigeria is that FDI boosts economic growth, when agricultural output increases, this will attract foreign investors into Nigeria to establish firms that will make use of the agricultural produce as raw materials, and this will bring about an improvement in the exchange rate as exports will become cheaper and imports will be more expensive. Finally, the implication of the positive relationship between labour-power and economic growth is that increasing farm labour-power would increase agricultural output, exports, and economic growth.

5. Summary and Conclusion

This study looked at the impact of agricultural exports on economic growth in Nigeria. It found that increasing agricultural exports will have a positive impact on Nigeria's economic growth, as well as foreign direct investment, effective exchange rates and labor will also have positive effects on economic growth. Against this backdrop, agricultural productivity growth in Africa and Asia is widely viewed as a major catalyst for the continent's structural transformation process. Agriculture is the backbone of socio-economic development in Nigeria as it serves as a catalyst for job creation and a major foreign exchange earner. To this end, the study supported the hypothesis of Nigeria's economic growth in agricultural exports. The domestic agro-processing sector should be encouraged in order for Nigeria to experience a favourable trade balance in agricultural trade. This would reduce the country from over-reliance on foreign food and increase the country's self-sufficiency, export, and economic growth rate of agricultural production.

References

[1] Osabohien R, Afolabi A and Godwin A 2018. Agricultural credit facility and food security in Nigeria: An analytic review. *The Open Agriculture J.*12 227-39.

[2]. Global Insight (2015). Productivity Report http://www.global-insight.net/

[3] Ramos RF 2001. Exports, Imports and Econmic growth in Portugal: evidence from causality and cointegration analysis. *Economic modelling.* 18(4): 613-23.

[4] Osabuohien E, Okorie U and Osabohien R2018. Rice production and processing in Ogun State, Nigeria: Qualitative insights from Farmers' Association. In Obayelu, E. (Eds).*Food Systems Sustainability and Environmental Policies in Modern Economics* (pp. 188-215), Hershey, PA: IGI Global.

[5] Abou-Stait F 2005. Are exports the engine of economic growth? An application of cointegration and causality analysis for Egypt, 1977-2003. *Economic Research Working Paper No. 76.*

[6] Matthew OA, Ede UC, Osabohien R, Ejemeyovwi J, Fasina FF and Akinpelumi D 2018. Electricity Consumption and Human Capital Development in Nigeria: Exploring the implications for economic growth, *Int'l J. of Energy Econs and Policy*, 8(6):1-8

[7] Osabohien R and Bamigbola A2017. Sustainable agricultural investment and employment generation in Nigeria: The role of institutional framework. *FUTA J. of Management and Tech*2(1):113-23

[8] Usman O and Salami A 2008. The contribution of Nigerian export-import (NEXIM) Bank towards export (non-oil) growth in Nigeria. *International Business Management.* 2(3): 85-90.

[9] Osabohien R and Osuagwu E2017. Social protection policies and agricultural output in Nigeria: Empirical investigation using household survey data. *Presented at the 4th Covenant University International Conference on E-Governance in Nigeria (CUCEN),* Covenant University, Ota, Nigeria, 7-9 May 2017.

[10] Iyoha Mand Oriakhi D 2002. *Explaining African Economic Growth Performance: The case of Nigeria.* African Economic Consortium Research.

[11] Matthew AO, Fasina FF, Olowe O and Adegboye FB 2010. Empirical modeling of the impact of financial innovation on the demand for money in Nigeria, *International Research J. of Finance and Econs* 58 73-90, December.

[12] Adeleye N,Osabuohien E, Bowale E, Matthew O andOduntan E 2017. Financial reforms and credit growth in Nigeria: Empirical insights from ARDL and ECM techniques, *Int'l Review of Applied Econs* 1-14.

[13] Amin S 1996. The challenge of globalization. *Review of Int'l Political Economy.* 3(2): 216-59.

[14] Osabuohien, E., Obiekwe, E., Urhie, E. and Osabohien, R. (2018). The inflation rate, exchange rate volatility and exchange rate pass-through nexus: The Nigerian experience, *Journal of Applied Economic Sciences, Volume XIII, Spring,* 2(56):574-585

[15] Olajide O, Ajayi BH and Tijani A2012. Agriculture resources and economic growth in Nigeria. *European Science J. 8(6):*12-25

[16] Ekpo A and Egwaikhide F 1994. Exports and Economic growth in Nigeria: A Reconsideration of the evidence. *J.of Economic Management.* 1(1): 100-15.

[17] Ijirshar VU 2015. The empirical analysis of agricultural exports and economic growth in Nigeria. *J. of Dev and Agric Econs* 2(3):24-36

[18] Idachaba F 2000. Food policy in Nigeria. *Agric. Research Bulletin,*1(162):16-24.

[19] Abrahamson, W 1989. Nutritional quality of animal dispersed fruits in Florida sandridge Habitats. *Bulletin of the Torrey Botanical club. Volume 116, No. 3*, 215-228.

[20] Ehrlich P R and Ehrlich AH 1990. How the rich can save the poor and themselves: lessons from the global warming In Global warming and climate change: perspectives from developing countries. *Tata Research Institute.*

[21] Ejemeyovwi JO, Osabuohien ES and Osabohien R 2018. ICT investments, human capital development and institutions in ECOWAS, *Int'l J. of Econs and Bus Research,* 15(4):463-74.

[22] Matthew O, Osabohien R, Fagbeminiyi F and Fasina A 2018. Greenhouse gas emission and health outcomes in Nigeria: Empirical insight from ARDL technique, *Int'l J. of Energy Econs and Policy,* 8(3):43-50

[23] Bakare A S2011. The crowding-out effects of corruption in Nigeria: An empirical study. *J. of Bus. Management and Econs*2(2): 059-068.

[24] Michael R 2001. Crude Oil Politics. *Atlantic* 3(5):19-30.

[25] Ayodele M 2004. *The elusive curse of oil foreign policy in focus.* Washington DC: Silver City University Press.

[26] Gilbert NA, Linyong SGand Divine GM2013. Impact of agricultural export on economic growth in cameroon: case of Banana, Coffee and Cocoa. *International J. of Bus. and Management Review* 44-71

[27]] Verter N and Becvarova V2016. The impact of agricultural exports on economic growth in Nigeria. *ACTA Universitatis Agriculturae Et Silviculturae Mendelianae Brunensis.*

[28] Gbaiye O G, Ogundipe A, Osabuohien E, Olugbire O O, AdeniranOA, Bolaji-Olutunji KA, Awodele OD and Aduradola O 2013. Agricultural exports and economic growth in Nigeria. *J. of Econs and Sustainable Dev.* 4(16):11-24.

[29] Nyong MO 2003. Predicability and Volatility of stock return in three emerging markets: Nigeria, South africa and Brazil. *Nigeria Journal of Economic and Development matters.* 2(1): 12-9.

[30] Gujarati D Nand Porter DC2009. *Basic Econometrics, (5th Edition).* Singapore: McGraw-Hill.

[31] Jayne T S, Chamberlin J and Benfica R 2018. Africa's Unfolding Economic Transformation

[32] Gralla F, Abson D J, Møller A P, Lang D J and von Wehrden H2017. Energy transitions and national development indicators: A global review of nuclear energy production. *Renewable and Sustainable Energy Reviews*, *70* 1251-65.

[33] Ojo E J, Awe I and Ogunjobi OJ 2014. Agricultural export and economic growth in Nigeria: A multivariateJjohansen cointegration analysis. *International Journal of Arts and Commerce.* 3(3):20-34.

[34] Osabohien R, Matthew O, Aderounmu B and Olawande T 2019. Greenhouse gas emissions and crop production in West Africa: Examining the mitigating potential of social protection, *Int'l J. of Energy Econa and Policy* 9(1):57-66.
[35] Osabohien R, Osabuohien E and Urhie E 2017. Food security, institutional framework and technology: Examining the nexus in Nigeria using ARDL approach. *Current Nutrition and Food Science.* 13(4):1-10.

NORM, a Health concern to Personnel exposed to Formation Drill Cuttings – Regulation issue in Nigeria

E E Okoro[1], S E Sanni[2], and E M Emetere[3,4]

[1] Petroleum Engineering Department, Covenant University Ota, Nigeria
[2] Chemical Engineering Department, Covenant University Ota, Nigeria
[3] Physic Department, Covenant University Ota, Nigeria
[4] Department of Mechanical Engineering and Science, University of Johannesburg, APK, South Africa.
E-mail: emeka.okoro@covenantuniversity.edu.ng

Abstract. During the drilling operations, formation rocks or shales drilled are removed from the wellbore as drill cuttings. These cuttings are real-time representative of the formation being drilled at that section of the wellbore, and it represents the formation of compositions and properties. Regulations for most oil well developments stipulate that the methods used for managing drill cuttings are dependent on the nature and type of drilling mud system used for the drilling operation. When the formation with these relatively high NORM contents is drilled using the drilling mud, these NORM traces are transported to the surface as drill cuttings; thus, exposing the personnel to health dangers and the immediate environment. The short and long-term health effects of NORM exposure include skin burns, harmful tissues reactions, increase the risk of lung cancer, mutation of somatic cells and leukemia. DPR has some standards and guidelines established to effectively police and regulate E&P industry activities in Nigeria. These guidelines and standards also take into consideration, the need for advanced pollution control strategies and technological treatments. However, the presence and harmful consequences of NORM in these drill cuttings are not considered in the current form of these regulations and guidelines. This study proposes the inclusion of NORM as a means of updating the already established guidelines and regulations for handling and treating these drill cuttings because they constitute the majority of the waste and toxic materials generated during drilling operations.

1. Introduction

Drilling fluid is a generic name for all compositions used to aid the production and removal of cuttings from a wellbore; this broad definition intentionally places no restriction on the objectives or type and source of the materials employed. The simplest drilling fluid is a dirty mixture of water, clay, and additives, often referred to as drilling mud. Drilling muds are fluids which are circulated through a well so as to remove formation drill-cuttings from the borehole. This definition is adopted in line with the objective of this research. These drilling muds are expected to perform some functions when in use during drilling operations, and for this purpose, some additives are added to optimize the drilling mud systems for effective suspension and transportation of formation drill-cuttings. B S Kinigoma [1] and E E Okoro et al., [2] conducted a study on the effects of these additives on field environments using Niger Delta as a case study. The soil, reserve pits in the field, plant growth and other biomasses were assessed. The study observed that the toxic effect of drilling fluids on the field environments contributed to poor plant growth. During the drilling operations, formation rocks or shales drilled are removed from the wellbore as drill cuttings. These cuttings are products of real-time fragmentations of the formation being drilled at that section of the wellbore and are considered parts of the formation compositions and properties, although, they also contaminate the drilling mud system. Regulations for most oil well developments stipulate that the methods used for managing drill cuttings are dependent on the nature and type of drilling mud system used for the drilling operation. Example of a sample drilling operation waste is given in Table 1.

Table 1: Types of wastes generated during drilling operations

S/N	Types of Waste	Main Components	Constituents
1	Spacers	Base fluids (oil/water), surfactants	Aromatic hydrocarbons, salts, chemicals
2	Spent drilling fluids	Whole mud, base fluids (water/oil), biodegradable matters	Heavy metals, organic materials, inorganic salts, chemicals
3	Drill Cuttings	Formation solids, base fluids (water/oil)	Solids/cuttings, heavy metals, organic materials, inorganic salts, chemicals
4	Drilling fluid additives	Bentonite, barites, salts, synthetic chemicals	Heavy metals, hydrocarbons, organic materials, solids, chemicals

On-site burial of cuttings is deemed acceptable when water-based mud is used, however, cuttings generated from the use of oil-based mud system must be contained in a closed-loop system, after which they can be treated and disposed. The treatment usually does not consider the possibility of Naturally Occurring Radioactive Materials (NORM). This is a gap that should be captured and used to modify existing regulations for wastes generated during drilling operations. Over a century ago, J Elster and H Geitel [3] and, H F Burton [4] observed the presence of high concentrations of Naturally Occurring Radioactive Materials (NORM) in wastes obtained from drilling operations and crude oil. Currently, research has been focused on investigating the presence of NORM in the exploitation, exploration and production sectors of the petroleum industry [5]. This study identifies and proposes measures to remedy the lapses in the current regulations for handling drill cuttings and their possible health impacts to the personnel in contact with them.

2. Empirical Review

According to A M D Sharif et al. [6], virtually all oil and gas exploration, production and storage related activities involve the generation of one form of waste or the other, and majority of these wastes are associated with drill cuttings from the wellbore. They have the potential to harm personnel and environment when contact is made through accidental leaks from storage tanks, spills from the roll-off container and other drilling operations. Y Shen et al. [7] recorded the presence of high NORM content in mudstones, coals and sandstone reservoirs. These traces are considered to be a potential source of radioactive contamination. When a formation with these relatively high NORM content is drilled using drilling mud, the particulates are transported to the surface as drill cuttings; thus, exposing the personnel to health and environmental dangers. Their study also identified the uranium as a major contributor to waste generation/NORM that contaminates tapped hydrocarbons from source rocks. K Al-Nabhari et al. [5] noted the importance and need for further studies on NORM and Technically Enhanced Naturally Occurring-Nuclear Radioactive Materials (TENO-NRM) that have not been addressed in details. Their study was focused on identifying and examining the occurrence of radioactivity in the Exploration and Production (E&P) industry. The investigation can also be used to establish the potential health hazards of these materials to man and the environment. Literature has shown that during oil and gas production, some of the technical processes applied at different stages can enhance the production of these radioactive materials; they concluded that there is need to establish regulatory controls and quantitative risk assessments for occupational exposure pathways in E&P industry operations and waste management (Table 2). Uranium and thorium are radioactive

IOP Conf. Series: Earth and Environmental Science **331** (2019) 012003 doi:10.1088/1755-1315/331/1/012003

materials which are most likely associated with shale, sandstones, bentonites, carbonate rocks [8, 9]. Edmonson et al. [10] also affirmed that these NORMs can be technologically enhanced by E&P industry operations. A S Paschoa and J M Godoy [11] in their study identified areas that have a significant amount of NORM in the wastes generated during drilling operations. A study in Nigeria by G O Avwiri and C P Ononugbo [12] to ascertain the presence of NORM during hydrocarbon exploration and production in a field in Niger Delta affirmed the presence of NORM at the production site. They used samples of soil from the host community and field sediments for their analysis. The experimental results showed that the concentration of NORM in the field was higher than the control samples from a non-oil producing community. Furthermore, the result shows some potential health threats to the workers, public health, and the environment. Occupational Safety and Health Administration (United States Department of Labor) has identified sludge, drilling fluids and cuttings as potential sources of harm to E&P industry workers especially when they are contaminated with NORM. Improper handling and disposal of NORM from site to site can cause short and long-term health effects such as skin burns, tissue reactions, increased risk of lung cancer, mutation of somatic cells and leukemia; hence, there is a pressing need to consider the careful handling of NORM when handling and treating drill cuttings. In addition, during the treatment operations, emphasis should not only be placed largely on the type of drilling fluid used to generate these cuttings but also on the likelihood of the radioactive wastes that may result from the operations. Exposure of E & P workers and inspectors, the public, environment to NORM will have a significant deterministic and stochastic health impact.

Table 2: Uranium (U) and Thorium (Th) mean and range concentration in parts-per-million (ppm) present in some Formations and Bentonite

Rock/ Mineral	U Mean	U range	Th Mean	Th range
Average/ Common Shale	3.5 – 3.7	1 – 13	12	2 – 47
Sandstone	2.4 – 2.7	0.2 – 0.6	8 – 8.8	0.7 – 2.0
Bentonites	5	1 - 21	24	6 - 44

3. Petroleum (Drilling and Production Amendment) Regulations in Nigeria

Environmental Guidelines and Standards for the Petroleum Industry in Nigeria (EGASPIN) is one of the statutory functions of the Department of Petroleum Resources (DPR). They ensure that the E&P industry operators do not degrade and pollute the environment in the course of their activities. DPR has some standards and guidelines established to effectively police and regulate E&P industry activities in Nigeria. These guidelines and standards cover is technically designed to incorporate recent advancements in pollution control technologies and treatments. The current list of qualified waste management companies provides information on the type of facility available in each company and their location within the country. For drilling fluids and associated-drill cuttings-waste treatments, the emphases have been on the drilling fluid constituents coating the cuttings. When coated with water-based drilling fluid, the cuttings can be treated on-site and disposed, but when these coatings are associated with oil-based drilling fluids, the cuttings are stored in a closed box and transported to treatment plants. DPR documentation requirements for the E&P industry operators have been limited to the presence and level of aromatic hydrocarbons, extractable acids, heavy metals and removable organics that may be toxic and hazardous to the environment after disposal. The possibility of NORM in these drill cuttings was not considered in the current form of the regulation and guidelines which is the reason behind this study. M I F Mkpaoro et al., [13] in their study reviewed the current DPR approved drill cuttings treatment and disposal techniques in Nigeria. The study outlined the inherent challenges encountered when utilizing these techniques, and they proposed the use of biochemical treatment as a solution because of its benefits. However, the analyses and discussion of the biochemical treatment technique mentioned in their study did not consider the possibility of NORM in the drill cuttings. The good news is that the DPR ECASPIN regulation and guidelines can be updated periodically with new knowledge. This study proposes that the knowledge of the harmful consequences of NORM be alarmed and used to review, update and establish an effective guideline and regulation for handling and treating these drill cuttings that constitutes the majority of the waste generated during exploitation operations.

4. Conclusion

NORM has a significant deterministic and stochastic health impact to the E&P industry workers, regulation inspectors and the public when not properly handled. Current ECASPIN regulations and guidelines for treating drill cuttings are focused on the drilling fluid constituents, aromatic hydrocarbons, and heavy metals. This study is designed to highlight the risks associated with NORM and its possible impact on the drilling crew and the environment. The knowledge of NORM can help to inform the need for improved strategies by both regulators (DPR) and E&P industry operators for effective handling and treatment techniques for drill cuttings. Furthermore, the advocacy here is that drilling wastes should no longer be handled, transported from site to site and discharged on the ground without conducting appropriate NORM tests.

Acknowledgment
The authors wish to appreciate the Management of Covenant University for their financial support.

References

[1] Kinigoma B S 2001 Effect of Drilling Fluid Additives on the Niger Delta Environment: - A Case Study of the Soku Oil Fields. J. Appl. Sci. Environ. Mgt., Vol. 5 (1) 57-61.
[2] Okoro E E, Dosunmu A and Oriji B 2016 Effect of Reversible Invert Emulsion Fluid Additives on The Environment – An Environmental Concern During Drilling Operation. International Journal of Advanced Research in Education & Technology (IJARET), Vol. 3, Issue 2, pp. 230-234.
[3] Elster J and Geitel H 1904 Über die radioactive Substanz, deren Emanation in der Bodenluft und der Atmosphäre enthalten ist. Phys. Z 5 (1), 11-20.
[4] Burton H F 1904 Radioactive emanation. Phys. Zs 55, 210-213.
[5] AL-Nabhani K, Khan F, and Yang M 2015 Technologically Enhanced Naturally Ocurring Radioactive Materials in Oil and gas production: A silent killer. Process Safety and Environmental Protection 99: 237-247.
[6] Sharif A M D, Nagalakshmi N V R, Srigowri R S, Vasanth G and Sankar U K 2017 Drilling Waste Management and Control the Effects. Journal of Advanced Chemical Engineering, 7:1. DOI: 10.4172/2090-4568.1000166
[7] Shen Y, Guo Y, Li Z, Wei X, Xue L and Liu J 2017 Distribution of radioactive elements (U, Th) in the upper Paleozoic coal-bearing strata of the eastern Ordos basin. Journal of Petroleum Science and Engineering 157: 1130–1142.
[8] Fisher R S 1995a Naturally Occurring Radioactive Materials (NORM) in Produced Water and Scale from Texas Oil, Gas and Geothermal Wells: Geographic
[9] Fisher R S 1995b Geologic and Geochemical Controls on Naturally Occurring Radioactive Materials (NORM) in Produced Water from Oil, Gas, and Geothermal Operations. The University of Texas at Austin, Texas.
[10] Edmonson R M, Jeliffe M R and Holwand K N 1998 Chapter 2. In: Naturally Occurring Radioactive Material (NORM). Jackson, Mississippi, USA.
[11] Paschoa A S and Godoy J M 2002 The areas of high natural radioactivity and TENORM wastes. Int. Congr. Ser. 1225, 3-8.
[12] Avwiri G O and Ononugbo C P 2011 Assessment of the Naturally Occurring Radioactive Material (NORM) Content of Hydrocarbon Exploration and Production Activities in Ogba/Egbema/Ndoni Oil/Gas Field, Rivers State, Nigeria, pp. 572-580.
[13] Mkpaoro M I F, Okpokwasili G C, and Joel O F 2015 A Review of Drill-Cuttings Treatment and Disposal Methods in Nigeria – The Gaps and way Forward. SPE-178325-MS, paper presented at the Nigeria Annual International Conference and Exhibition Lagos, Nigeria, 4-6 August.

IOP Conf. Series: Earth and Environmental Science **331** (2019) 012004 doi:10.1088/1755-1315/331/1/012004

Importance of Rural Biomass Post-Harvest Waste – An Overview

Okoro Emeka Emmanuel[1], Sanni E. Samuel[2], Emetere E. Moses[3,4]

[1]Petroleum Engineering, Covenant University Ota, Nigeria
[2]Chemical Engineering, Covenant University Ota, Nigeria
[3]Chemical Engineering, Covenant University Ota, Nigeria
[4]Department of Mechanical Engineering and Science, University of Johannesburg, APK, South Africa
E-mail: emeka.okoro@covenantuniversity.edu.ng

Abstract. This study rather than focus on food security examines the potential in utilizing Post-Harvest Waste (P-HW) materials and their benefits to the immediate rural communities where lignocellulose waste are found in abundance. It was observed that researchers have proposed a lot of unique approaches to managing solid waste generation; with emphasis on perfect waste collection and disposal, but these approaches are very expensive considering the tight budgeting expenses Nigeria is faced with. This study shows that some rural communities experience huge post-harvest organic wastes from their farms and the cost of these losses has been estimated, which then forms a good spot to start the implementation process for this initiative. The research highlights the overview benefits of adopting microbial and pyrolysis approach in converting the lignocellulosic aspect of these waste to economically viable products that will have a direct positive impact on the immediate community. This study proposed a modified fast pyrolysis approach in the conversion process to reduce the production of pyrolytic gas and optimize the production of biochar and pyroligneous acid. Adopting these approaches will reduce the amount of carbon emitted into the atmosphere when burning these waste and produce bio-products instead of synthesized chemical products that have health implications.

1. Introduction

The Glossary of the United Nations Statistics Division on Environment Statistics [1] describes waste as "non-essential materials" (that is, products intended for the market) for which the generator no longer finds useful for production, processing or consumption. There are several types of wastes identified by modern waste management systems, as shown in Table 1. According to A A Adedibu [2], wastes differ in chemical and physical and composition. This study will consider agro waste. Waste management in Nigeria has attracted private organization, due to the lack of lasting solutions despite several attempts by successive governments in this direction. It is common to see or meet piles of waste dumps across the cities in almost every corner.

2. Progress so Far

Nigeria Government has invested largely in agriculture mostly in the Northern parts of the country for food sufficiency [3]. Figure 1 illustrates government effort and investments in the agricultural sector and rural community development over the years [4]. Table 1 shows the percentage of the budget allocated to the agricultural sector for some selected African countries, and it is compared with Nigeria's eight (8) years average agriculture and rural development percentage allocation from national budgets from 2011 – 2018.

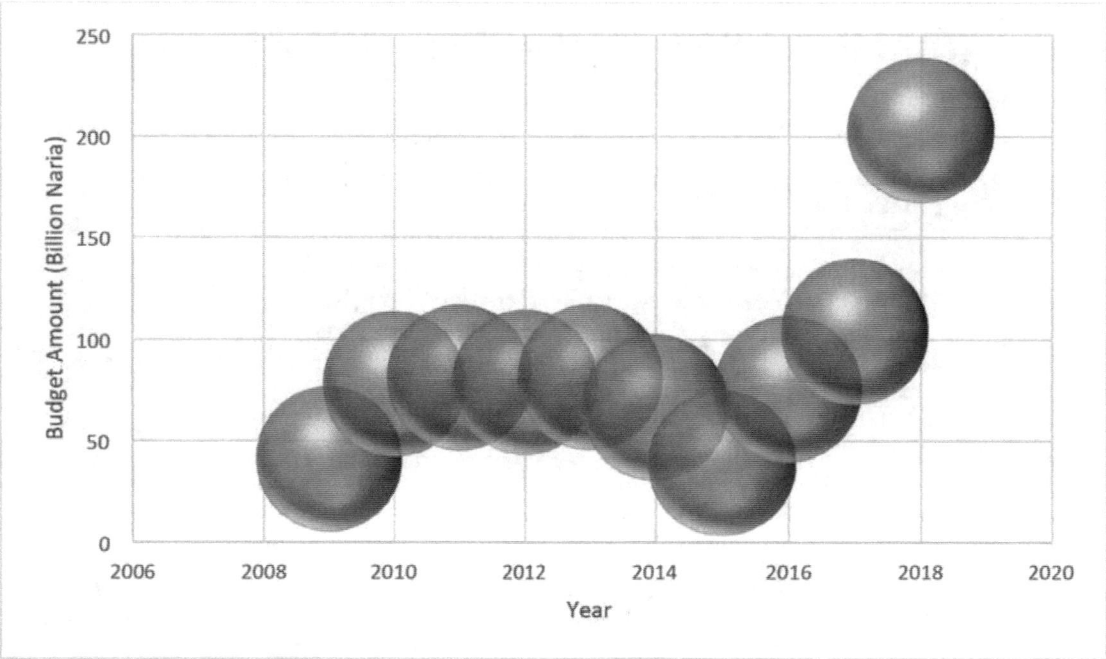

Figure 1: Nigeria Government allocation to Agriculture and Rural Development

Table 1: Percentage of Budget Allocated to the Agricultural Sector

S/N	COUNTRY(S)	PERCENTAGE ALLOCATION OF BUDGET
1	Malawi	27%
2	Zambia	10%
3	Burundi	10%
4	Mali	10%
5	Niger	13%
6	Sierra Leone	3%
7	Nigeria	1.6%

The Federal Institute of Industrial Research in 2017 puts the country's post-harvest losses at $ 9 billion. A survey on the six geo-political zones of the country as obtained from the Food and Agriculture Organization Fact Sheet also confirms the estimated value on post-harvest losses in Nigeria. The Institute observed that the quantity of crop loss is about 51.3 metric tonnes with huge amounts recorded in the rural communities. Post-harvest Engineering Director of NSPRI has acknowledged handling of these post-harvest waste as a major challenge in Nigeria's agricultural sector. The agency has researched and developed post-harvest technologies, but the generated wastes have not known much improvements in terms of their conversion for use or management within the communities (the source). According to the study by B T Abur et al., [6] it was observed that most local governments in Nigeria still record high post-harvest waste generations and they often have problems in discarding them from their farm and market areas [7, 8]. There is also a need for the government and all stakeholders to bridge the technological gap in the intensive commercialization of research products on wastes to bio-products. This research is a guild for in-country utilization of organic solid waste to stabilize the challenges of energy and clean environment which have a direct impact on the Nigerian economy and its citizens. The solution proposed in this study is an intensive waste to viable product initiative. The aim of this research is to highlight the benefits of microbial process in converting lignocellulosic biomass waste components into some by-products that have direct advantages to the agricultural sectors of rural communities.

3. Post-Harvest Waste

The focus of this study is not on food security, but on how to utilize approximately 1.4 billion tons per year of wastes generated globally from production and postharvest stages [9]. In Nigeria, one-quarter of our expected outputs from farming are lost to postharvest activities which pose a concern to the rural area were this waste is generated [10]. F M Bolarin and S O Bosa [10] outlined six major factors that contribute to this huge amount of wastes and proposed strategies to minimize these wastes. Most rural areas depend solely on the agriculture sector for their economic growth and development. The fact remains that these wastes have not been eliminated, and there is need to convert them into economically viable products that will not only eliminate the environmental pollution but also bring in useful products that can be used in their natural state to improve agricultural activities [11]. The generation of agricultural post-harvest wastes in the South-East region of Nigeria was considered in this study. An investigative approach was adopted and current published literature was also considered. Straw is the major lignocellulosic waste in this region because of the type of crops planted. Some of these wastes are; bamboos, corn Stover, oil palm trunk, cassava pulp residue, and other crop straws. Figure 2 shows the results of analysis from the samples collected during field observations in the region. These waste materials are biomass feed stocks and mostly organic matters. They are the major materials of focus in this study due to the inexpensive routine conversions of biomass to useful products. The production of valuable products and chemicals from these agricultural crop residues would partly solve waste disposal issues and provide some more environmentally friendly alternatives to synthesized chemical products.

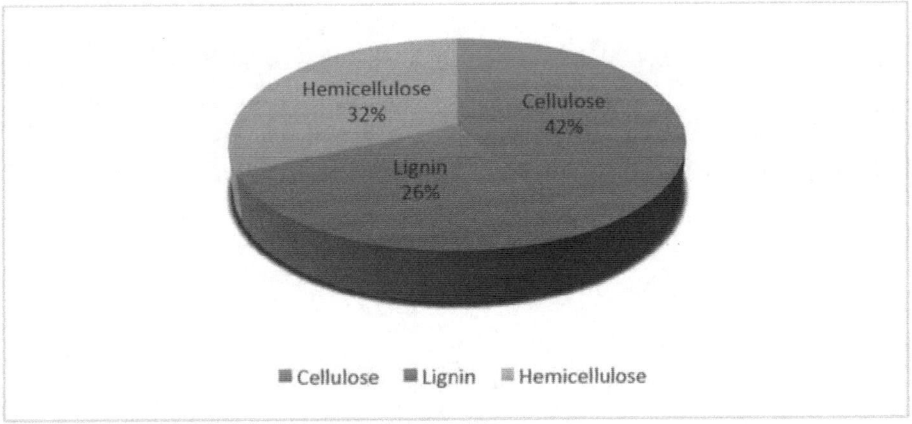

Figure 2: Post Harvest Waste Field Sample Composition

4. Utilization of Lignocellulosic Waste and its Economic benefits

Literature has shown the typical microbial technique adopted for the conversion of lignocellulosic agro-wastes. This technique has been applied to producing agro-food, beverages, textiles, and pharmaceutical products. Recent advancements in process engineering have shown that it is possible to optimize the production of micro-organisms and their metabolites when used to handle lignocellulosic agro-wastes. Bio-ethanol can be produced from lignocellulosic wastes with starch and simple sugar as by-products, and also a feedstock for the next stage of reaction (Fig. 3).

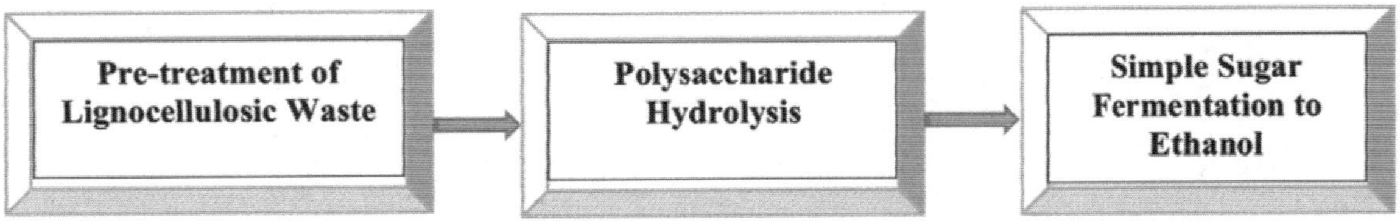

Figure 3: Steps in the production of Bio-Ethanol from Lignocellulosic Waste

One approach often adopted in the conversion of biomass feedstock into valuable chemicals is pyrolysis using a batch reactor process. The approach utilizes the principle of carbonization of biomass. This process produces charcoal and pyroligneous acid which can improve the growth, yield, and quality of crops at a lower cost. The aforementioned acid can be used to also produce acetic acid that serves as a preservative/additive for woods to discourage the growth of molds and fungi, thus reducing the health and environmental impacts of synthetic chemicals used as biocides and pesticides. A drive for the massive production of this preservative should be an action point for all stakeholders and government. This study proposes the use of modified fast pyrolysis in the production of pyroligneous acid from Lignocellulosic waste (Fig. 4 and 5). This process will significantly reduce the production of pyrolytic gases and optimize the production of pyroligneous acid and bio-char which has numerous application in farming, wastewater treatment, electronics, building, clothing, decontamination of soil, and other range of consumer products. This biochar when treated through a chemical process to improve its chemical and physical properties, can be applied as activated carbon for absorbing different types of pollutants [12]. Bio-char can be used in its natural form as acid soil neutralizer and micro-organisms activity promoter. Thus, acting as a nutrient to the soil [12].

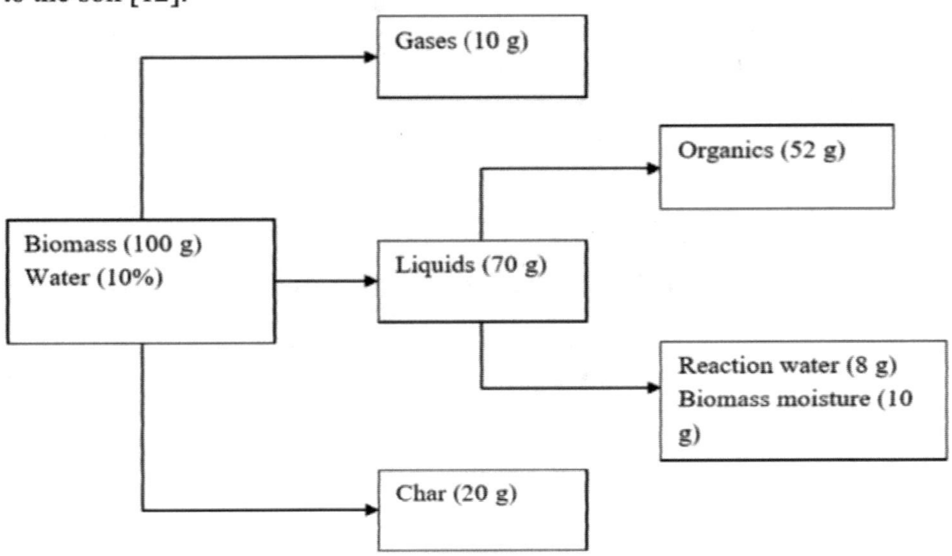

Figure 4: Fast Pyrolysis Mass Balance during Production (Fivga, [13])

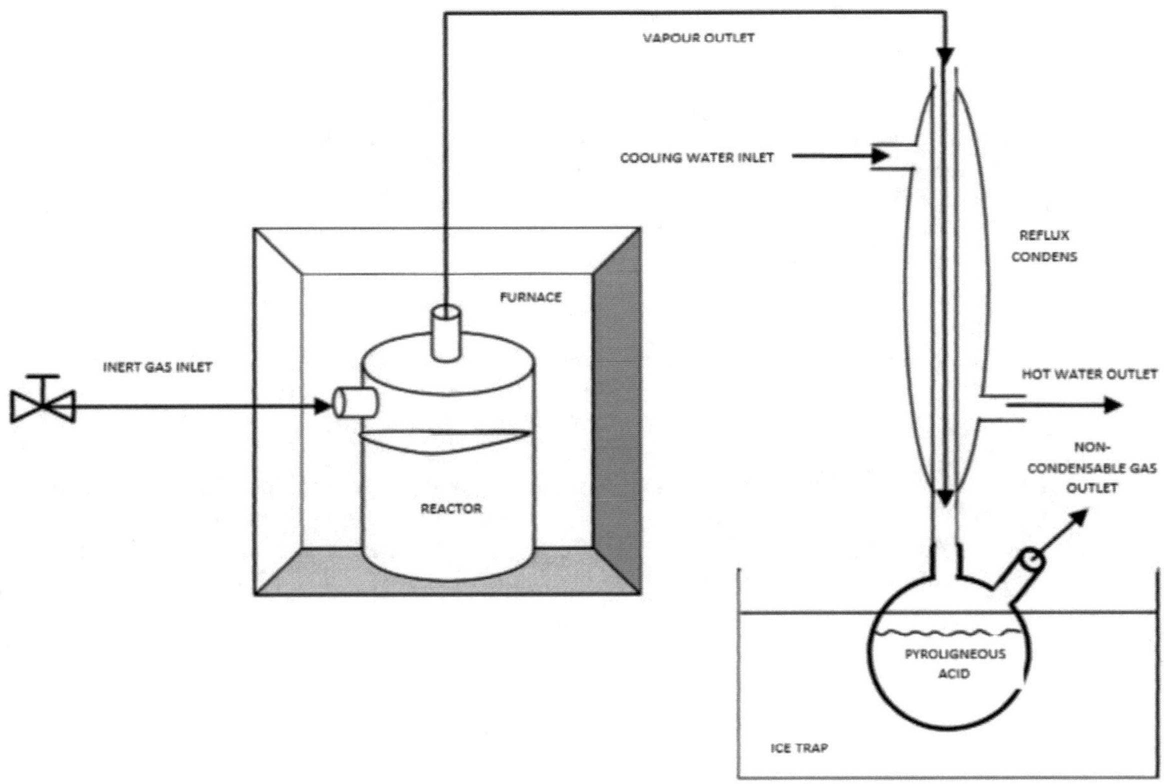

Figure 5: Modified Fast Pyrolysis Schematic Set-up

5. Conclusion

Wastes generated from the agricultural sector in Nigeria though huge and considered worthless, can serve as raw materials for viable products which are capable of sustaining and supporting wealth creation in the country. The study demonstrates the potential therein and benefits of utilizing huge post-harvest waste biomass materials in South-Eastern rural communities of Nigeria. Literature has shown that some agro waste that has biomass can produce 70-95 percent of lignocellulose. Thus, there is need to further research in this area.

Acknowledgment

The authors wish to appreciate the Management of Covenant University for their financial support.

References

[1] UNSD, “Glossary of Environment Statistics. Updated web version 2001 "Sources of waste," [Online]. Available: https://www.eschooltoday.com/waste-recycling/sources-of-waste.html. [Accessed 2018].
[2] Adedibu A A 1985 A comparative analysis of solid waste composition and generation in two cities of a developing nation. Environmentalist 5: 123. https://doi.org/10.1007/BF02235981.
[3] Bakare W 2016 Solid Waste Management in Nigeria. BioEnergy Consult [Online]. Available: https://www.bioenergyconsult.com/solid-waste-nigeria/
[4] Oghenefejiri B, Nwaogazie I L and Agunwamba J C 2016 Exploratory Factor Analysis & Assessment of Energy Potential of Generated Solid Waste in Nigeria. International Journal of Civil Engineering and Technology (IJCIET), Volume 7, Issue 1, pp. 274-289.
[5] Hoornweg D and Thomas L 1999 What A Waste: Solid Waste Management in Asia, Urban Development Sector Unit. East Asia and Pacific Region. Page 5.

[6] Abur B T, Oguche E E and Duvuna G A 2014 Characterization of Municipal Solid Waste in the Federal Capital Abuja, Nigeria. Global Journal of Science Frontier Research: H Environment & Earth Science Volume 14 Issue 2.

[7] Adeoye P A, Adebayo S E and Musa J J 2011 Agricultural Post-Harvest Waste Generation and Management for Selected Crops in Minna, Niger State, North Central Nigeria. Journal of Applied Sciences in Environmental Sanitation, 6 (4):427-435.

[8] Anyanwu C N, Ibeto C N, Eze I S and Ezeoha S L 2013 Present and prospective energy use potentials of selected agricultural wastes in Nigeria. J. Renewable Sustainable Energy 5, 032703 (2013); doi: 10.1063/1.4808046

[9] Chege C and Carson M (2016). Food Loss and Waste Africa, Intelligence Report. The ROCKEFELLER Foundation.

[10] Bolarin F M and Bosa S O 2015 Post-Harvest Losses: A Dilemma in ensuring Food Security in Nigeria. Journal of Natural Sciences Research, Vol. 5, No. 7.

[11] Mmojieje J 2016 Pyroligneous Acid: A farmer's flexible friend? Journal of Crop improvement, Vol. 30, Issue 3, pp. 341-351.

[12] Cha J S, Park S H, Jung S C, Ryu C, Jeon J K, Shin M C and Park Y K 2016 Production and utilization of biochar: A review. Journal of Industrial and Engineering Chemistry 40, 1–15.

[13] Fivga, A. (2011). Comparison of the effect of pre-treatment and catalysis on liquid quality from fast pyrolysis of biomass. PhD Thesis. Aston University.

INVESTIGATING THE ENDOTHERMIC NATURE OF THE REACTIONS OF HYBRID HYDROXIDE-PEROXIDE WITH LIGNIN IN CORN STOVER AND NAPIER GRASS FOR SIMPLE SUGAR PRODUCTION

S E Sanni[1*], J A Omoleye[1], E M Emetere[2,3] and E E Okoro[4]

[1] Chemical Engineering Department, Covenant University Ota, Nigeria
[2] Physic Department, Covenant University Ota, Nigeria
[3] Department of Mechanical Engineering and Science, University of Johannesburg, APK, South Africa.
[4] Petroleum Engineering Department, Covenant University Ota, Nigeria
E-mail: adexz3000@yahoo.com

Abstract. Napier grass and Corn Stover samples were collected as waste materials for sugar production. The samples were pretreated and hydrolyzed using hydrogen peroxide and *Trichoderma ressei* cellulase respectively. Chemical kinetics of the delignification process shows that the reduction reactions of lignin during pretreatment are dependent on the nature of raw material, temperature, concentration of the hydrogen peroxide and their activation energies. The calculated activation energies show that the reactions of lignin in both biomasses are though endothermic, but less energy intensive and more economically viable for Napier grass when compared to corn Stover. Based on the results, in order to obtain higher sugar yield from Napier grass and corn stover under the investigated conditions, both samples should be processed at 105 ^{0}C for 96 hours using 0.3 M and 0.1M hydrogen peroxide concentrations respectively.

Key words: Biomass; Delignification; Kinetics; hydrogen peroxide pretreatment; Alkali pretreatment

1. Introduction

Lignocellulosic biomasses are currently the most preferable sources of reducing sugars. Lignocellulose consists of cellulose, hemicellulose and lignin, [1-3]. The sugar chains in cellulose can be hydrolyzed to their monomeric forms [4]. Cellulose can be chemically split into, glucose units by treating it with concentrated acids [5, 6] or enzymes. Hemicellulose is a copolymer of different amounts of several saccharide molecules [7]. It is less chemically and thermally stable compared to cellulose because of its non-crystalline nature [8]. Lignin is a polymer linked by phenylpropane units [9]. It offers resistance to degradation and provides hydrolytic stability and structural robustness in plants [10-18] hence, its removal is pertinent for easy biomass digestion by enzymes/acids during hydrolysis. In this paper, chemical pretreatments and enzymatic hydrolysis of Napier grass and corn stover were investigated, in order to ascertain the endothermic nature of lignin reactions with hybrid-hydroxide peroxide mix. This is because, oxidizing agents, such as hydrogen peroxide and ozone have been reported to have high delignification efficiencies [19-20]. Also, alkali pretreatment gives delignification efficiencies of 60-80% [4]. Till date, besides the proposal by Sanni et al. [21] concerning the reaction between lignin in Napier grass and NaOH, no work has considered the nature of the reactions of hybrid NaOH-H_2O_2 mix with lignin in biomass.

2. Materials and Equipment

30% pure H_2O_2 manufactured by Sigma Aldrich, sodium hydroxide (500g NaOH) manufactured by J.T Baker Inc. Philipsburg USA, sulphuric acid (2.5L H_2SO_4 of 98% purity, manufactured by J N Chemicals), distilled water (99% pure) produced by all glassware distiller Stwart W400 and *Trichoderma ressei* (Fungi)

Digital magnetic hotplate stirrer: MS7-H550-S LED (manufactured by SCILOGEX), autoclave with capacity 75-200L, string IIB Bench top steam sterilizer (made by Celitron –Azteca), Shaking incubator: VS-8480sr-I with D.C motor and PID controller (manufactured by Vision Scientific Co. Ltd.), petri dishes (manufacturer; J Sill Borosilicate), Oven Vs-1202d3 (Vision Scientific Co. Ltd., material of construction is steel with powder coating), UV mass spectrophotometer: 7310 visible Scanning Spectrophotometer fitted with 10 x 10 mm cuvette holder, 320 - 1000nm wavelength (Manufactured by Jenway), pH meter: Orion^{TM} Dual Star^{TM} pH, ISE Mv, ORP with temperature dual channel Bench top meter (manufactured by Thermo Fisher Scientific), soxhlet extractor (Amfield UK), Furnace, with maximum temp. of 1100-1300^{0}C and 5 - 23L chamber volume (made by Carbolite Gero), vacuum filtration setup comprising of vacuum pump, filter flask and Buchner funnel, electronic weighing balance (Manufactured by Swastic Systems and Services, Shahdara, Delhi, India) and water bath: 12 L capacity, 120V/60Hz, ELED 2837 (made by Thermo Scientific).

3. Method

Three months old Corn stover and Napier grass were both harvested from Adewale farm field in Atan, Ota, Ogun state, Nigeria. After being washed and sun dried for 3 days, they were crushed, milled and stored in air tight containers for 2 days in readiness for biomass pretreatment and hydrolysis.

3.1. Preparation of Napier Grass and Corn Stover Biomasses

Napier grass and Corn stover-impurities such as leaves/grass, sand, stones and dirt) were sorted out. The raw biomasses were then weighed and washed; the waters were kept to determine the presence of lignin in water from the washed biomasses. The biomass samples were air-dried and subsequently oven dried to completely remove moisture. The weight of each biomass after drying was obtained using a digital weighing balance. Thereafter the samples were milled and sieved with a 150 μm sieve to obtain uniform particle size.

3.2. Characterization of Biomasses

5 g of each sample was weighed and put into 2 different crucibles of known weight. Their weights were measured with the aid of the weighing balance. The crucibles were put in the oven at 105 °C for 3 hours at an interval of 30 minutes. After oven drying, the samples were kept in the desiccator for 15mins to cool before the weight of the samples were taken and recorded. In order to be sure that the samples were void of moisture, they were further dried in the oven for another hour at 105 °C and at same time interval until a constant weight was obtained on the weighing balance. The samples were then cooled in the desiccator for 15mins. The moisture contents of the samples were calculated using (1).

$$\% \text{ (w/w) Moisture Content} = \frac{Wa-Wb}{Wa} \ x\ 100 \quad (1)$$

$W_a = W_s - W_c$, $W_d = W_{fc} - W_d$, W_a = Weight of raw sample, Ws= Weight of raw sample + Weight of crucible.

W_c = Weight of crucible, W_{fc} = Weight of dry sample + Weight of crucible, W_b= Weight of dry sample ($w_{fc} - w_c$).

3.2.1. Determination of extractive contents in the biomasses

The above dried samples of constant weights were collected on filter papers with their dry weights noted. The water bath was switched on and its temperature was adjusted to 70 °C with a volumetric flask. 150 ml of acetone was measured and transferred into a round bottom flask. The Soxhlet extractor was set up, 2.5g of sample was measured and placed in the cellulose thimble, after which it was immersed in the water bath set at 70 °C for 4 h. After extraction, the sample was air dried for 1 h and put into the oven set at 105 °C until a constant weight was achieved. The samples were weighed and the values were recorded. % Extractives was obtained from (2). Of primary concern in this analysis, is the removal of majority of the extractives without classifying them.

$$\% \text{ (w/w) Extractives} = \frac{WT-Wse}{Wa} \; x\; 100 \qquad (2)$$

$W_b = W_x - W_y$, $W_T = W_z - W_y$, W_{se} = Weight of sample after extraction, W_y = Weight of filter paper, W_x = Weight of filter paper + Weight of Sample after extraction, W_z = Weight of dry sample + filter before extraction, W_a = Weight of dry sample.

3.2.2. Procedure for Determination of Lignin and Ash Contents

This experiment was done in order to know the amount of lignin and ash present in the samples. There are 2 types of lignin to be accounted for i.e. soluble and insoluble lignin.

i. Insoluble lignin:

300 mg each of the extracted sample was weighed and kept inside a test tube. 3 ml of 98% H_2SO_4 was added. The mixtures were shaken gently, to ensure complete mixing. The samples were kept at room temperature for 2 h at an interval of 30 mins with continuous shaking so as to allow for complete hydrolysis. The samples were transferred to separate Erlenmeyer flasks and 84 ml of distilled water was added. The samples were autoclaved for 1 h at 121 °C and cooled to room temperature. The samples were filtered and then the residue was dried at 105 °C in an oven for about 4 h till a constant weight was achieved. The residue was ashed in a muffle furnace at 575 °C for 6 h. The mass of insoluble residue was obtained from calculation using (3).

$$\% \text{ (w/w) Insoluble lignin} = \frac{W_{IL}}{W_{DW}} \times 100\% \qquad (3)$$

W_{DW} = Oven dry weight sample = 0.3g, W_{IL}= Weight of insoluble residue = Wo – Wc, Wo = Weight of Sample + Weight of crucible, Wc = Weight of empty crucible

ii. Soluble lignin:

3 ml of the filtrate from (i) above was taken to the UV-Spectrometer to determine the absorbance of the sample. The Soluble lignin was also obtained from calculation. Equation 3 was used to obtain the ASL in the samples.

$$\% \text{ (W/W) Soluble Lignin} = \frac{uv\ absorbance\ x\ volume\ of\ filterate}{\varepsilon \times\ ODW\ sample} \times 100\% \qquad (4)$$

$\boldsymbol{\varepsilon}$ = 30 (absorptivity of biomass at specified wavelength) l/gcm, ODW Sample = Oven-dried sample weight= 0.3g

Volume of filtrate = 0.087 L

To determine the ash content, the residue (Klason lignin + insoluble ash) in " i " above, was kept in the muffle furnace for 6 h at 575 °C. The percent ash was and % insoluble lignin were obtained from (5) through to (7).

$$\% \text{ (w/w) Ash} = \frac{W_K}{W_{DA}} \times 100\% \qquad (5)$$

W_{DA} = Oven dry weight sample, Wk = Weight of Ash = Wf – We, Wc = Weight of empty crucible, Wf = Weight of crucible + ash (After burning insoluble residue in furnace)

% (w/w) Insoluble Lignin = % Insoluble Residue - % Ash (6)

% (w/w) Total Lignin = % Insoluble Lignin + % Soluble lignin (7)

3.3. Alkali-Hydrogen Peroxide Pretreatment of the Samples

100 ml of 1 – 3 % H_2O_2 and distilled water mixture was made from the 30% lab grade hydrogen peroxide. 1.8 g NaOH pellets was added to the mixture in order to raise the pH to 11.5 so as to prevent the formation of furfural, enhance hemicellulose degradation of hemicellulose. 6 g of each dried sample of Napier grass and corn Stover was added to separate conical flasks containing hydrogen peroxide and distilled water. The soaked samples were kept for 3 days to allow for good contact times. After 3 days, each mixture was poured into two other flasks and heated on hotplate with magnetic stirrer at temperatures of 85 and 105 °C for 1-2 h. The wet weights of the residues were then recorded. 2 g each of the wet-pair of treated biomass was dried at 105°C for 4 h so as to estimate the total dry solid left in the pretreated sample.

3.3.1 Determination H_2O_2 Volume for Pretreatment

To obtain 1 - 3 % H_2O_2 in 100 ml of hydrogen peroxide and distilled water mixture from 30% (v/w) H_2O_2

$$C_aV_a = C_bV_b \quad (8)$$

Ca = 1% H_2O_2, Va = 100 ml, Cb = 30% H_2O_2, Vb =?

Vb = 3.33 ml

The same procedure was repeated for 2% and 3% H_2O_2. After the pretreatment, the samples were separated by vacuum filtration. Hollocellulose was seen as pale yellow precipitate for Corn stover and as pale green precipitate for Napier grass. The supernatant solutions were adjusted to a pH of 5.0 by addition of a 72% w/w dilute H_2SO_4 in order to solubilize the acid soluble lignin. The amounts of acid soluble lignin removed, and that present in the filtrate were analyzed with UV spectrometer. The amount of Klason lignin or acid insoluble lignin removed was obtained as discussed in [22].

3.3.2 Determination of cellulose/hemicellulose contents of the biomass substrates

Borrowing a leaf from the extraction of hemicellulose by Aravantinos-Zafiris and Oreopoulou [1], for 1 g of each sample treated with 72% w/w H_2SO_4, the ASL, AIL and ash contents were estimated. 0.5 M was made. Two water baths were filled with distilled water and set at a temperature of 100 °C. 1 g of each sample was kept inside separate Erlenmeyer flasks. With a volumetric flask, 150 ml of NaOH was poured into the flasks containing the samples. The Erlenmeyer flasks were placed in the water baths and allowed to boil for 3.5 h in order to bring them to the desired temperature. Vacuum filtration of samples was done with the aid of the buchner filter and 64 buchner flask. The samples were washed with distilled water until the solution gave a pH value of 7 i.e. Each sample was then dried in the oven to a constant weight and the hemicellulose content was obtained from (9).

$$\%\ (w/w)\ \text{Hemicellulose} = \left(\frac{Wo-Wp}{Wo}\right) - AIL \times 100\% \quad (9)$$

Wo = 1g, Wp =Wa –Wc, Wa = Weight of sample + weight of crucible, Wc = Weight of crucible. However, the % cellulose = 100% - % (extractives + ASL + AIL + moisture + ash + hemicellulose) were calculated; see Figure 1.

3.4. Enzyme Hydrolysis of Treated Biomass

In order to further process, the alkali-peroxide pretreated biomasses, cellulase enzyme was prepared and added to the pretreated samples for hydrolysis. One gram each, of all four pretreated samples was added to 300 ml of distilled water in four different conical flasks, and the pH was adjusted to 4.8. The enzyme loadings were 5.0

FPU/g dry biomass of cellulase. The flasks were stirred at a temperature of 50 °C and autoclaved. The samples were inactivated by increasing the temperature to and maintaining it at 80°C for 15 minutes. Finally, the concentration of glucose and total reducing sugars (TRS) were determined using a kit based on the glucose oxidase reaction (reagent GOD PAD) and the dinitrosalicylic acid method (DNS), respectively. Cellulase activity (*Trichoderma reesei*, produced by Sigma Aldrich) was 64.1 FPU/ml and the experiment lasted for 96 h.

3.4.1 DNS reagent calibration

Standard glucose concentrations in the range of 0.2-1 mg/ml were prepared. 0.5 ml of each sample was put in separate testtubes. 1.5 ml of DNS reagent was added to the contents in the testtubes. The mixtures were shaken properly in a fume cupboard to prevent the formation/release of phenol vapour. The samples were boiled in a water bath for 15 mins. They were cooled for a few minutes and 8 ml of distilled water was added. The absorbance was taken with the spectrometer at 550 nm. The calibration curve was then obtained; see Figure 7.

3.4.2 Glucose analysis

Samples hydrolyzed for 72 and 96 h were taken for glucose analysis since the maximum yield is expected to be at both times. The RANDOX glucose kit contains a buffer with a pH of 7.0 and the enzyme glucose oxidase. The glucose concentration was determined after enzymatic hydrolysis and the hydrogen peroxide which was used in pretreating the samples reacted with phenol which is one of the constituents of the RANDOX buffer to form a red-violet quinoneimine dye as indicator. The enzymatic digestibilities of the pretreated Napier grass and corn stover samples were measured by removing 0.5 ml aliquot at 2, 24, 72 and 96 h. 0.5 ml of each sample was taken and placed in a water bath for 15minutes at 100 °C in order to denature the enzymes. The test tubes were withdrawn. 1.5 ml of the DNS reagent was added to each test-tube using the pipette and the mixture was shaken thoroughly. The test-tubes containing the mixtures were placed back into the water bath whose temperature was adjusted to 100 °C for another 15 minutes. 1ml of 40% Rochelle salt solution was added to the warm mixture in the test-tubes, after which they were cooled for 3 minutes and 6 ml of distilled water was added. The absorbance of each sample was obtained. Water was used as the blank reagent. The reducing sugar concentrations were calculated using the calibration graph of Figure 7.

$$\text{Glucose Concentration (C) (g/l)} = \frac{Absorbance\ of\ sample}{Absorbance\ of\ standard\ solution} \quad (10)$$

3.4.3 Reducing Sugar yield

3 test tubes were labelled appropriately. Test tube A was for the standard glucose solution and the other two were for the Napier and corn Stover samples. 1 ml of the reagent was poured into each testtube. 10 µl each of glucose solution and test samples were poured into their appropriate testtubes. The mixtures were shaken properly and incubated for 5 minutes at 37 °C. They were then filtered with their absorbance measured at a wavelength of 550 nm. The sugar yields were estimated using (11).

Reducing Sugar Yield:

$$Y = S \times D \times \frac{V}{W} \quad (11)$$

Y = Reducing sugar yield from enzymatic hydrolysis (mg/g of dry biomass), S = Sugar concentration in diluted sample (mg equivalent glucose/ml), D = Dilution factor, V = Working Volume (20 ml), W = Weight of dry biomass (g). Absorbance of Sample = 0.145, S = Concentration of the reducing sugar from the calibration graph or by interpolation using values in Table 10 = 0.119 mg/ml, V = Working Volume = 20 ml, D = Dilution Factor = 5ml enzyme to 20 ml distilled water. W = 0.928g (dry weight of the biomss)

IOP Conf. Series: Earth and Environmental Science **331** (2019) 012005 doi:10.1088/1755-1315/331/1/012005

3.5. Delignification Kinetics of Corn Stover and Napier Grass

According to [25], lignin degradation occurs in the initial (rapid) phase, bulk (dominant) phase, and the final residual (slow) phase. Parameters for these phases are presented in (12):

$$\frac{L_K}{L_{ko}} = a_i \exp(-k_i \times t) + a_b \exp(-k_b \times t) + a_r \exp(-k_r \times t) \qquad (12)$$

t is the reference time during which data was collected; L_K = mass of Klason lignin at time t; L_{ko} = initial mass of Klason lignin at t = 0; a_i = maximum fraction of lignin fragments released at the initial stage; a_b = maximum fraction of lignin fragments released in the bulk phase; a_r = maximum fraction of lignin fragments released at the residual stage; and k_i, k_b, k_r are the reaction rate constants for the initial, bulk, and residual delignification stages, respectively [24]. In the case of lignin removal during soda pulping of bagasse, the delignification lacks an initial phase. The alkaline delignification of bagasse proceeds with a fast and high rate of lignin removal which is not a characteristic of the bulk phase [10]. Also, under these conditions, this work assumes that delignification of the biomasses took place only in the bulk phase hence, the model presented in (12) reduces to (13).

$$lnk = -k_b \times t + a^b \, lnk_0 \qquad (13)$$

4. Results and Discussions

According to literature, lignocellulosic biomass is composed of major structural components of Cellulose, hemicellulose, lignin, extractives, ash and moisture. The results obtained for both samples are as contained in Table 11 and Figure 1 gives an illustration of the compositions of the constituents of the biomasses.

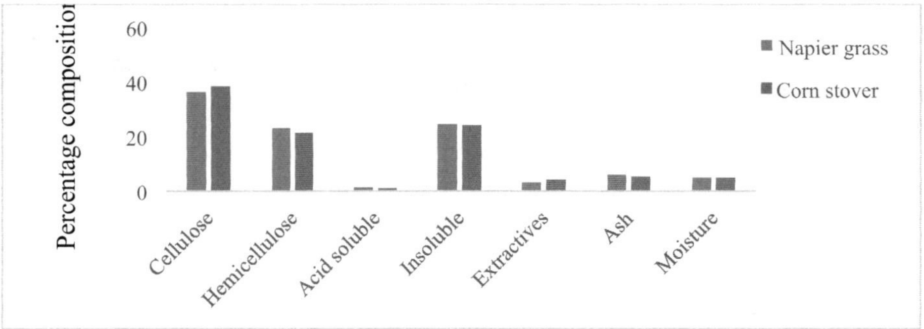

Figure 1: Percentage composition by weight (% w/w) of Napier Grass and Corn stover

From Figure 1, the ratio of cellulose to hemicellulose in the Napier grass sample (i.e. 36%: 21%) appears to be lower than that of corn stover whose cellulos: hemicellulose is 38%: 21%. Figures 2 and 3 give illustrations of the ASL removed between 0 and 60 minutes for both samples at 105 ^{0}C. Figures 4 and 5 show the amount of Klason (Acid insoluble lignin- AIL) lignin left in the biomasses after pretreatment at 85 ^{0}C and 105 ^{0}C. Since the amount of lignin removed after 60 and 120 mins at 85 and 105 ^{0}C were found to be lowest for 0.1 M pretreatment and highest for 0.3 M, the delignification kinetics of both pretreatment situations were then carried out in order to determine the kinetic parameters for the delignification processes. Kinetic data generated for the pretreated corn stover and Napier grass samples are as given in Table 1.

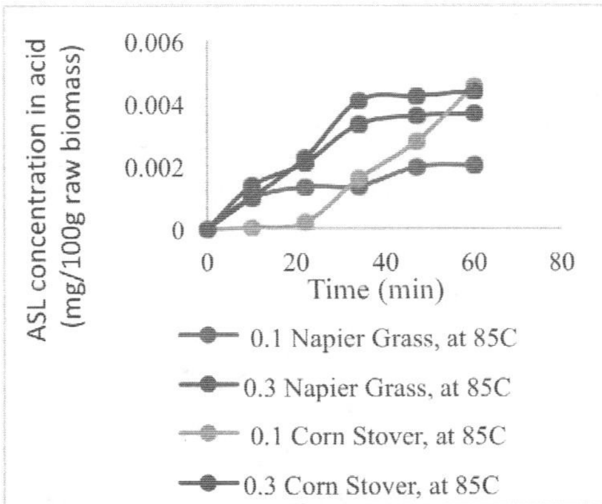

Figure 2: Variation of ASL with Time at 85 ^{0}C

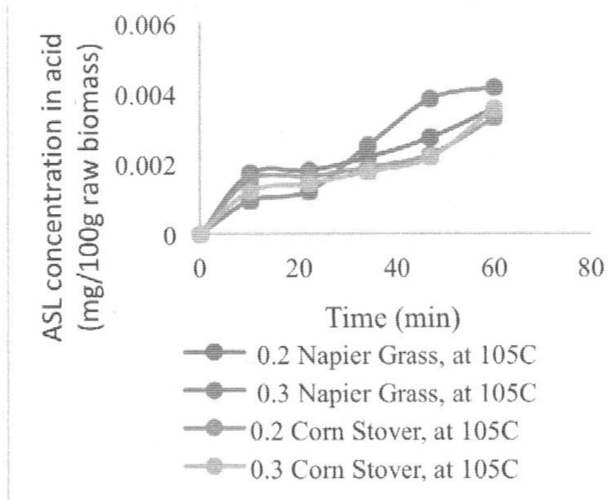

Figure 3: Variation of ASL with Time at 105 ^{0}

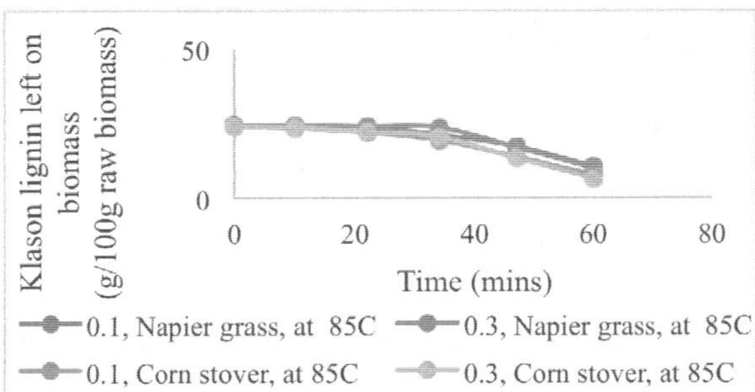

Figure 4: Lignin in Napier Grass and Corn Stover at 85 ^{0}C

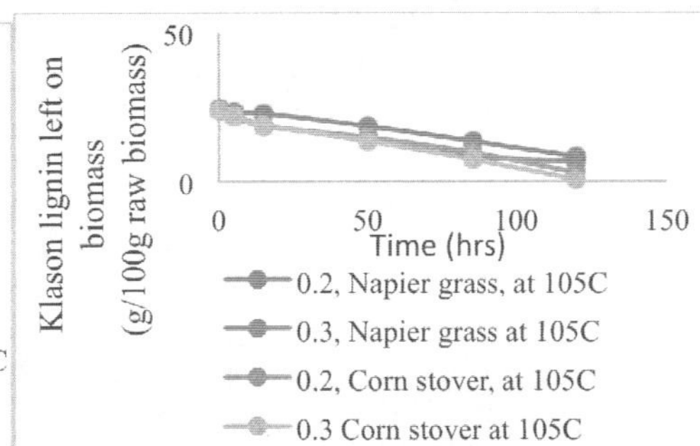

Figure 5: Lignin in Napier Grass and Corn Stover at 105 ^{0}C

Based on the data, graphs of lignin concentration versus time were obtained; Based on curve fitting tests conducted, the curves have good R^2 values. From the graphs, plots of ln R versus ln C_A were made; where R is the rate of lignin removal and C_A is the concentration of the lignin in solution. Using Differential Method of Analysis (DMA), the resulting polynomials were differentiated to obtain the rate. Equation 13 was then used to obtain the kinetic constants at 85 ^{0}C and 105 ^{0}C for both corn stover and Napier grass samples respectively (Figures 6a-h).

IOP Conf. Series: Earth and Environmental Science **331** (2019) 012005 doi:10.1088/1755-1315/331/1/012005

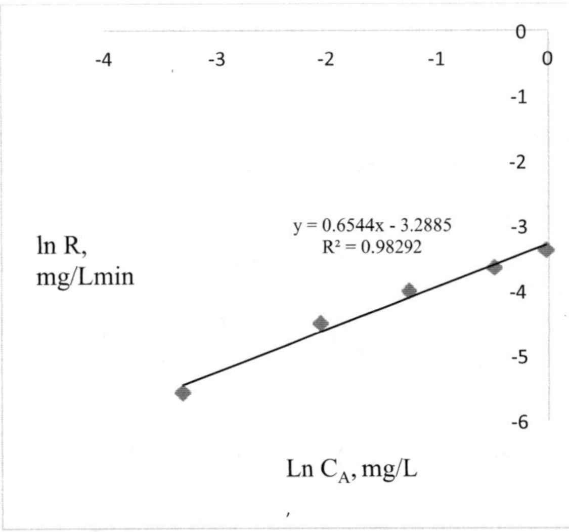

Figure 6a: ln R vs ln C_A for corn stover for 0.1M pretreatment at 85 ^{0}C

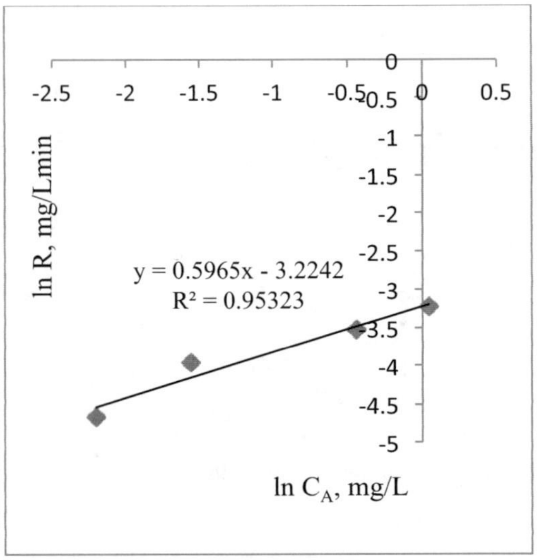

6b: ln R vs ln C_A for corn stover for 0.3M peroxide pretreatment at 85 ^{0}C

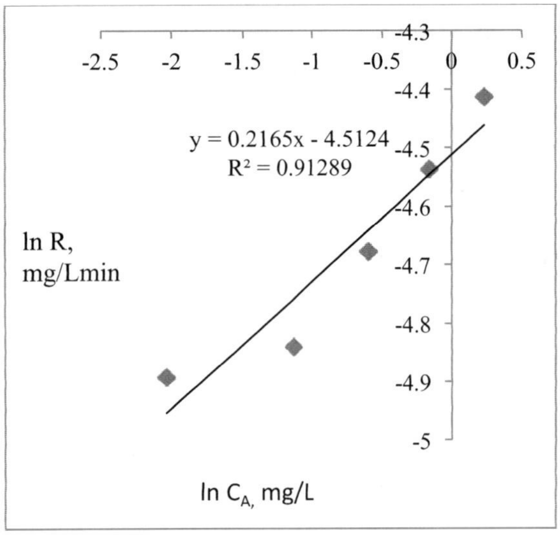

6c: ln R vs ln C_A for corn stover for 0.1 M peroxide pretreatment at 105 ^{0}C

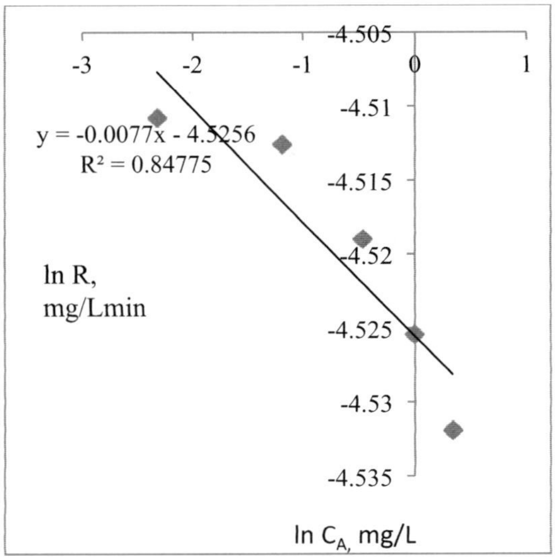

6d: ln R vs ln C_A for corn stover for 0.3M peroxide pretreatment at 105 ^{0}C

As shown in Table 1, k_1 is the rate constant for the reaction at 85^{0}C while k_2 is the rate constant for the reaction at 105^{0}C. The rate constants were seen to decrease at higher temperature which is justified by the Arhenius equation. The highest rate constant was obtained for 0.1 M hydrogen peroxide pretreatment of the grass. The rate constants are functions of the higher activation energies recorded at higher concentrations of the alkali. The concentration values obtained as well as the estimated rates of reaction were used to estimate the average rate constants in order to ensure accuracy. However, for 0.3 M peroxide pretreatment of Napier grass, the reaction order was zero, hence, the delignification of Napier under such conditions is independent of the lignin concentration in the biomass.

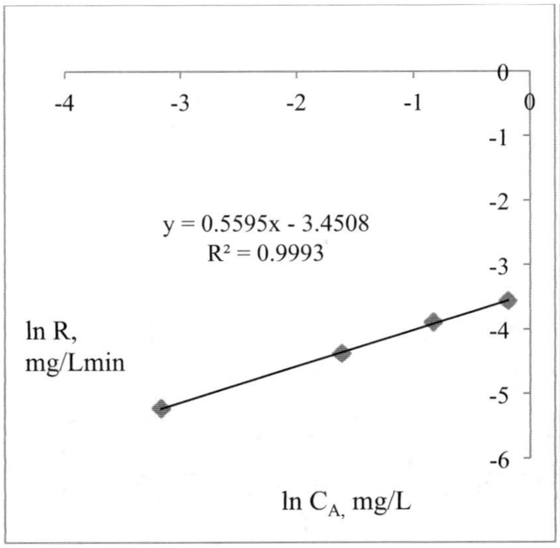

6e: ln R vs ln C_A for Napier grass for 0.1M pretreatment at 85 ^{0}C

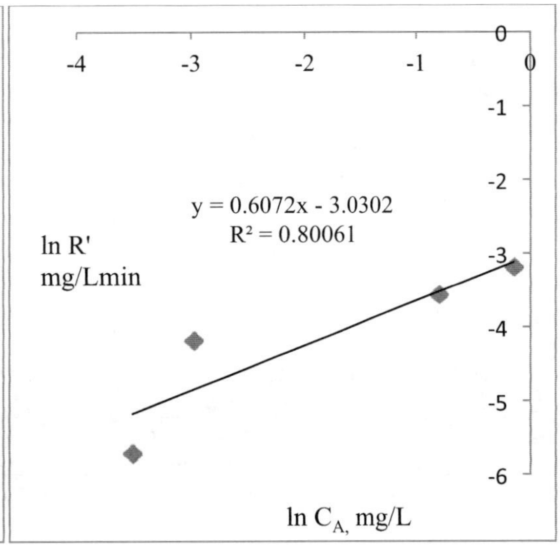

6f: ln R vs ln C_A for Napier grass for 0.3M peroxide peroxide pretreatment at 85 ^{0}C

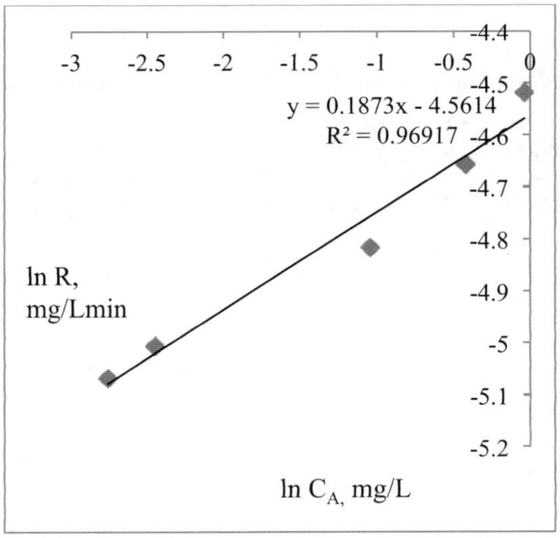

Figure 6g: ln R vs ln C_A for Napier grass for 0.1M peroxide pretreatment at 105 ^{0}C

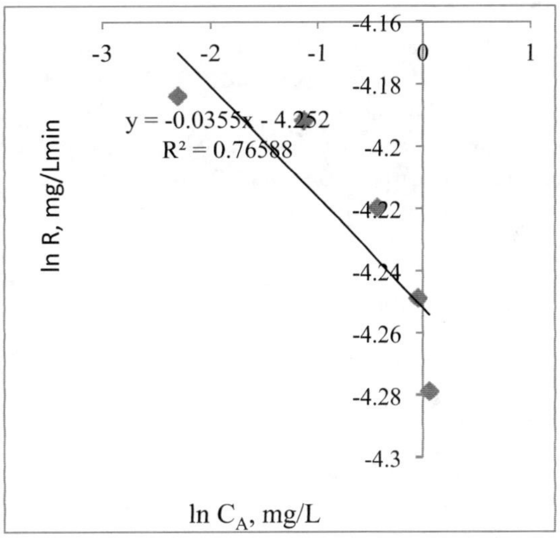

Figure 6h: ln R vs ln C_A for Napier grass for 0.3M peroxide pretreatment at 105 ^{0}C

From (13), the activation energies are 161.67 and 163.29 kJ/mol for 0.1M and 0.3M H_2O_2 pretreatments of corn stover respectively, and 64.76 and 71.66 kJ/mol for 0.1 M and 0.3M H_2O_2 pretreatments of Napier grass respectively. Based on the estimated activation energies, hydrogen peroxide pretreatment of Napier grass is preferred to corn stover for simple sugar production because, it is less energy intensive relative to corn stover; in this work, the results show that the effect of the peroxide in NaOH causes a change in the nature of the reactions of lignin in both substrates which implies that, based on the suggestions in [22] the reaction of lignin and hybrid

NaOH at higher concentrations i.e. > 0.1 M is altered by the presence of the peroxide thus making it endothermic.

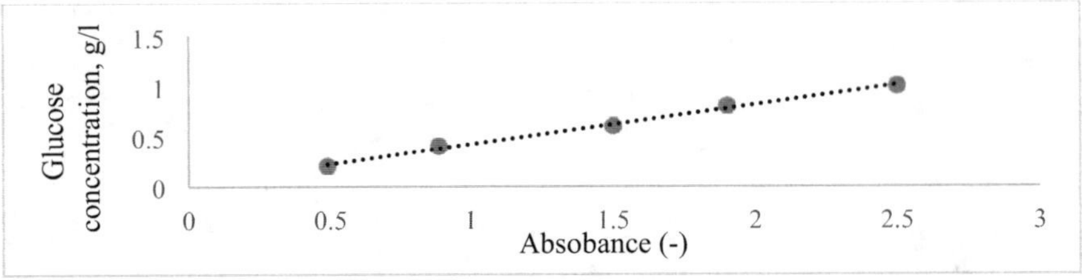

Figure 7: Glucose concentration vs absorbance

Table 1: Rate constant for the reactions at different Temperatures

Sample	**T ^{0}C**	**Rate constant**	**0.1M H_2O_2**	**0.3M H_2O_2**
Corn stover	85 ^{0}C	k_1	0.040mg$^{0.3}$/L$^{0.3}$min	0.040 mg$^{0.4}$/L$^{0.4}$min
	105 ^{0}C	k_2	0.0023 mg$^{0.8}$/L$^{0.8}$min	0.0022 mg/L min
Napier grass	85 ^{0}C	K_1	0.034 mg$^{0.4}$/L$^{0.4}$min	0.0520 mg$^{0.4}$/L$^{0.4}$min
	105 ^{0}C	K_2	0.011 mg$^{0.8}$/L$^{0.8}$min	0.015 mg/L. min

5. Conclusion

Peroxide pretreatment of Napier and corn stover increases access to their cellulose during enzyme hydrolysis. The activation energies obtained are 161.67 and 163.29 kJ/mol for 0.1M and 0.3 M H_2O_2 pretreatments respectively, for corn stover, while, for Napier grass, they are 64.76 and 71.66 kJ/mol for 0.1 M and 0.3 M H_2O_2 pretreatments respectively, hence, the reactions between lignin and the hybridized hydroxide-peroxide mix are endothermic for both substrates. Based results from the kinetic model, both delignification processes, raise an argument in favour of Napier grass for optimal production of simple sugars. The estimated reaction orders are in the range of 0 - 0.7.

Acknowledgment

The authors appreciate the management of Covenant University for their support all through this research.

References

[1] Aravin-Safaris, G and Oreopoulou, V. The Effect of Nitric Acid Extraction on Orange Pectin, *J. Sci. Food Agric*. 60:127-129 (1992).

[2] Granda, C.C.B. Sugarcane Juice Extraction and Preservation and Long-Term Lime Pretreatment of Bagasse, Ph.D Dissertation, Texas, A and M University (2004).

[3] Herbert, S. Handbook of Pulp and Paper. Wiley-VCH Verlag Gmb & Co, Weinheim, Germany (2006).

[4] Rabelo, S.C., Filho, R.M and Costa, A.C. A Comparison between Lime and Alkaline Hydrogen Peroxide Pretreatments of Sugarcane Bagasse for Ethanol Production, *Applied Biochemistry and Biotechnology*, 144(1): 87-100 (2008).

[5] Klemm, D., Heublein, B, Fink, H.P and Bohn, A. Cellulose: A Fascinating Biopolymer and Sustainable Raw Material. Angew Chem. Int. England, vol. 44, Series (Book 22), pp. 3358-3393 (2005).

[6] Nishiyama, Y., Langan, P and Chanzy, H. Crystal Structure and Hydrogen-Bonding System in Cellulose Iβ from Synchrotron X-Ray and Neutron Fiber Diffraction, *Journal of American Chemical Society,* 124(31): 9074-9082 (2002).

[7] Yang, T.C. Plant Fiber Chemistry, China Light Industry Press, Beijing, China, pp. 50-56 (2008).

[8] Mohammed, A.H. Conversion of Lignocellulosic Material into Fermentable Sugars. Ph.D Thesis, University of Berlin, pp. 1-49 (2012).

[9] Sjostron, E. (1993). Wood Chemistry Fundamentals and Applications, 2nd Edition, Academic Press, Orlando, USA, pp. 1-293 (1993).

[10] Katzen, R and Schell, D.J. Lignocellulosic Feed Stock Biorefinery: History and Plant Development for Biomass Hydrolysis. (*Eds*. Kamm, G, Grubber, P and Kamm, M) *In:* Biorefineries Industrial Processes and Products: Status Quo and Future Direction, Wiley-VCH, Weihein, 1: 129-138 (2006).

[11] Sarkanen, K.V and Hergert, H.L. Lignin: Occurrence, Formation, Structure and Reactions. (*Eds.* Sarkanen, K.V and Ludwig, C.H), Wiley Interscience, Chapter Four pp.95-99 (1971).

[12 Wood, T.M and Garcia-Campayo, V. Enzymology of Cellulose Degradation (*Eds.* Colin Ratledge) *In:* Physiology of Biodegradable Microorganisms. *Biodegradation,* 1: 147-161 (1990).

[13] Teymouri, F., Laureano-Perez, E., Alizadeh, H and Dale, B.E. Ammonia Fiber Explosion Treatment of Corn stover, *Applied Biochemistry and Biotechnology*, pp. 113-116: 951-963 (2004).

[14] Sun, Y and Cheng, J. Hydrolysis of Lignocellulosic Materials for Ethanol Production: A Review. *Bioresource Technology,* 83(1): 1-11 (2002).

[15] Bensah, E.C and Moses, M. Chemical Pretreatment Methods for the Production of Cellulosic Ethanol: Review Article. *International Journal of Chemical Engineering-Hindawi,* pp. 1-21 (2013).

[16] Gottschalk, G. Bacterial Metabolism. 2nd Edition, Springer-Verlag, New York, pp. 1-359 (1986).

[17] Carpeda, S.C. Introduction to Biomass and Energy Conversions, CRC Press, Taylor and Francis Group, pp. 1-645 (2013).

[18] Zhang, S. Revealing White Matter Fiber Structure with Diffusion Imaging. Ph.D Thesis, pp. 1-137 (2006).

[19] Chang, V. Burr, B and Holtzapple, M. Lime Pretreatment of Switch Grass. *Applied Biochemistry and Biotechnology,* 1: 63-65 (1997).

[20] Yang, T.C., Chou, C.C and Li, C.F. (2002). Preparation, Water Solubility and Rheological properties of the N-Alkylated Mono or Disaccharide Chitosan Derivatives, *Food Res. Int.* 35: 707-713.

[21] Sanni, S.E., Akinrinola, O., Yusuf, O.E., Fagbiele, O., Agboola, O. Chemical Kinetics of Alkaline Pretreatment of Napier Grass Prior Enzymatic Hydrolysis, Open Chemical Engineering Journal, 12: 36-56 (2018).

[21] Chang, V and Holtzapple, M. Fundamental Factors Affecting Biomass Enzymatic Reactivity, *Applied Biochemistry and Biotechnology*. 1: 84-86 (2000).

[22] Cardon, C.A., Sanchez, O.J and Gutierrez, L.F. Process Synthesis for Fuel Ethanol Production, Series (Book 32), pp. 1-415 (2009).

[23] Drapcho, C.M., Nhuan, P.N and Walker, T. Biofuels Engineering Process Technology, 1st Edition, Mc-Graw Hill, pp.1-371 (2008).

IOP Conf. Series: Earth and Environmental Science **331** (2019) 012005 doi:10.1088/1755-1315/331/1/012005

[24] Mosier et al. Features of Promising Technologies for Pretreatment, *Bioresource Technology,* 96(2005): 673-683 (2005).

[25] Drapcho, C.M., Nhuan, P.N and Walker, T. Biofuels Engineering Process Technology, 1st Edition, Mc-Graw Hill, pp.1-371 (2008).

IOP Conf. Series: Earth and Environmental Science **331** (2019) 012006 doi:10.1088/1755-1315/331/1/012006

Assessment of Energy Wastage and Saving Potentials for Higher Educational Institutional Buildings in South Western Nigeria

B N Obaju[1, 2 *], J S Adeleke[2], T O Yusuf[2], T A Buari[2] and C A Tokede[3]

[1] Department of Building Technology, Covenant University, Km 10, Idiroko Road, Ota, Ogun State.
[2] Department of Building Technology, Federal Polytechnic Ede, Osun State
[3] Department of Architectural Technology, Federal Polytechnic Ede, Osun State

*E-mail: Obaju82@gmail.com

Abstract. The study assessed the energy wastage areas in Higher Education Institutions (HEIs) Buildings in South Western Nigeria; factors influencing the energy wastage and energy saving opportunities available in the study area with a view to promote energy conservation and cost savings. The study adopts the use of stratified simple random sampling techniques and the administration of two hundred (200) open ended questionnaire on the respondents who are mainly students, teaching and non-teaching staff in the physical planning, works and maintenance service department. Data obtained from respondents was analyzed using descriptive statistics and mean score index (MSI). Findings revealed that energy wastage are common with lighting consumption (MSI 4.59), poor controls and regulators for both lighting and other appliances in HEIs (MSI 4.57). Lighting related operations also exert significant impacts on energy wastages as six out of the thirteen major energy wastage areas are connected to it. The major factor influencing energy wastage are due to poorly designed HEIs buildings (MSI 4.62) and all other factors aside this need to be adequately considered in future designs for effective energy management. Findings further revealed that HEIs buildings require sound energy policy, improvement on passive building designs principles are critical to the reduction of energy wastage and hence should be adhered to since it has potential to saves cost. The study recommends the usage of metering system to determine energy consumed in each facilities/building, using scheduling devices with intelligent building designs for power supply and controls during peak and off peak period, implementing standard Energy Policies and creating awareness for students and staffs on energy usage and cost savings benefits and provisions.

1. Introduction

Energy usage in buildings account for the highest energy consumption compared to energy usage in other sectors [1]. Energy is used in buildings basically as operational energy; for the maintenance and servicing of a building throughout its useful life, and the embodied energy; that is, the total energy required for the production of buildings using plants, equipment and various building materials [2]. Energy used during service of buildings is for various purposes such as lightings, heating, ventilation and air conditioning, hot water heater, security, cooking and for powering of appliances [3].

In recent years, Educational Institution Buildings in Nigeria is faced with dynamics of energy demand in the sub-sector as a result of rapid growth of the energy user. This remarkable growth of the sub-sector is due to population increase as well as the need for improved performance of key academic functions is responsible for high increase in energy consumption when compare with other sectors of the economy. This sub-sector depends on energy supply for virtually all activities and operations that includes teaching and learning aids, lightings, heating, air conditioning systems, laboratory equipment and powering of machinery used for researches, training, practical and demonstration as well as other support services [4, 5].

Recent reports indicated that institutional buildings, mostly universities and other institutions consumes copious amount of energy daily and that energy consumption by these institutions has grown to be comparable to commercial sector energy usage [6, 7, 8, 9]. According to [8], the amount of energy used in educational institutional buildings primarily depends on the building design, population of staff and students, the types of lamps and electrical appliances used. The diverse of ways in which energy is used in the institutional buildings is changing rapidly, but of particular interest is the pattern of energy consumption by the university community, especially staff and students who mostly use basic institutional fittings such as lightings, fans and air conditioning systems in addition to personal appliances which includes (hot plates, electric cookers, computers, micro waves, scanners, copiers, rechargeable appliances, refrigerators, television and pressing iron) for personal comfort and academic support activities. The characteristic increase in population of the academic institution community is having tremendous effect thus leading to increase in energy consumption over the past couple of decades.

However, [10] estimated that 25 percent of energy used in schools is wasted. [11] pointed out that unconcerned and unconscious behavior as well as bad attitude of both students and staff towards energy wastage is increasing directly in response to population increase. He further stated that un-implemented energy policies in schools, use of in-efficient design technologies, lack of awareness and information on energy usage and cost, poorly design buildings, poor controls and settings and most importantly non-payment of energy bills by building occupants in schools contribute to the problems of energy consumption in institutional buildings, resulting to increasing running and administrative cost in most institutions [12].

Higher Educational Institutions energy saving is a giant innovation, rich of huge potentials to reduce energy wastage; it is an effective option to reduce the energy bottleneck constraints and can significantly reduce energy shortages. Going by the current rate of development of Nigeria educational sector, effort towards energy savings possesses a huge potential through reduced energy bills thereby contributing to ameliorating the current economic challenges. In view of this scenario, this study assessed the existing spectrum of energy wastages in education institutional buildings. It also examined the factors influencing energy wastages and assessed the areas of energy saving opportunities for educational institution buildings in Nigeria with the benefits to achieving cost savings and environmental sustainability.

2. Methodology

The research used stratified simple random sampling, a method of probability sampling that allows for improved precision. The population was divided into non-overlapping groups and then simple random sampling was performed to select appropriate sample size along a relevant dimension, which included federal, state and privately owned tertiary institutions and this was based on the number of group that exists in the population. Ten (10) institutions were selected within the study area as sample population, a total of two hundred (200) numbers open ended questionnaires were used as sample size was administered mainly to students, teaching and non-teaching staff in the works and maintenance department and physical planning unit of each institution to provide information. Data obtained from respondents was analyzed using descriptive statistics and mean score which is also referred to as Users Agreement Score (UAS) using the following formulae

Formulae for Mean Score:

$$\text{Mean score} = \frac{\Sigma WV}{N}$$

Where, ΣWV = Summation of weight value

Σ (Numbers of Respondents X Rating Weight)

$\Sigma (1 \times F1) + (2 \times F2) + (3 \times F3) + (4 \times F4) + (5 \times F5)$

F1 to F5 = Frequency of Response (1 to 5) on the Likert scale

(5 =Very Common, 4 = Common, 3 = Indifference, 2 = Less common, 1 = Not common)

N = Total number of respondents.

IOP Conf. Series: Earth and Environmental Science **331** (2019) 012006 doi:10.1088/1755-1315/331/1/012006

3. Data Presentation and Analysis

The following data in Tables 1 to 5 are presented to give insight to effort made to obtain reliable data from respondents.

Table 1 below revealed the gender representation of the respondents to be 65% male and 35% female, thus showing that the percentage representation of male respondents is more than that of the female in the selected sample size.

Table 1. Sex Distribution of the Respondents

Sex of Respondents	Frequency	Percentage (%)
Male	130	65
Female	70	35
Total	200	100

It is shown in Table 2 that 32% of the respondents were HND holders, while 25% possess Bachelor's Degree (B.Sc/B.Tech), 19% hold Master Degree (M.Sc./M.Tech) and 5% of the respondents had PhD degrees and both undergraduate and postgraduate students make up 19% of the respondents

Table 2. Academic Qualifications of Respondents

Academic Qualifications	Frequency	Percentage (%)
HND	64	32
B.Sc/B.tech	50	25
M.sc/M.tech	38	19
Ph.D	10	5
Undergraduate	38	19
Total	200	100

It can be seen in Table 3 showing the Age Distribution of Respondents that none of the respondents age falls below twenty one years, 15% fall between twenty one years and twenty nine years of age, 40% were between thirty and thirty nine years, 36% were between age forty and forty nine, and 9% were above the age of fifty. The frequency and percentage distribution above therefore, clearly indicate that the available respondents have the prerequisite knowledge and are capable of providing detail information for this research.

Table 3. Age Distribution of Respondents

Age Range/Description	Frequency	Percentage (%)
Below 21 years	0	0
21 -29 years	30	15
30- 39 years	80	40
40-49 years	72	36
Above 50 years	18	9
Total	200	100

The placement of respondents in the organization /institution shown in Table 4 revealed that, 15% of the respondents are teaching staff, 65% are non- teaching staff, 10% are undergraduate and post graduate students and 10% are business and ventures owners within the institutions environment. The result shows that Non-teaching staff has the highest percentage representation with 130 target audience who were mainly staff from Works, Maintenance and Services Departments and Physical Planning Units who have direct contact with energy usage in the studied institutions on a routine basis.

Table 4. Placement of Respondents in the Organization / Institution

Status in Campus	Frequency	Percentage (%)
Academic staffs	30	15
Non - academic staffs	130	65
Students	20	10
Business Venture owners	20	10
Total	200	100

Data on residence of respondents in the institution in Table 5 reveals that 45% of the respondents resides in the institutions staff quarters, 10% stays in the institution hostel accommodation, 5% stays in the post graduate hostels, while 40% of the respondents lives off campuses probably due to lack of provisions of staff quarters. The 45% of the respondents living in staff quarters having the highest percentage representation makes the information supplied adequately reliable as a result of relative continuous stay on campus enabling the acquisition of relevant information required.

Table 5. Residence of Respondents in the Institution

Location of Stay in Campus	Frequency	Percentage (%)
Staffs quarters	90	45
Hostels	20	10
Post graduate hostels	10	5
Off campus	80	40
TOTAL	200	100

Duration of residence on Campus by respondents is contained in Table 6. It shows the number of years stayed in campuses by the respondents. A total of 19% of the respondents have stayed between 0 and 4 years. This group of respondents is mainly students. It is further observed that 30% have stayed in the campus between 5 to 10 years while 21% have stayed in the campus between 10 to 15years. Also, it is observed that 15% of the respondents have been on campus for 16 to 20 years, while 15% have also stayed in campus for 26 years and above. The relevant background of respondents therefore, strongly suggests the reliability of information gathered from the experienced respondents that have practical presence on the campuses.

Table 6. Duration of Residence on Campus

Numbers of Years	Frequency	Percentage (%)
0-4 years	38	19
5-10 years	60	30
10-15 years	42	21
16-20 years	30	15
26 years above	30	15
Total	200	100

3.1 Energy Wastage areas in Educational Institutions Buildings

The identified and evaluated thirteen (13) means of energy wastage in educational institution buildings in Nigeria are contained in Table 7. After due analysis of data, the results contained in Table 7 revealed that, the most critical area of energy wastage in the higher education institutions in South Western Nigeria is leaving lights turned on in an unoccupied spaces mostly in Lecture halls and classrooms and hostels accommodations with the highest mean score (MS = 4.59).

The line of wastage was observed to continue in decreasing order for cases of leaving fans running in an un-occupied space with (MS = 4.57), leaving Air Conditioners (AC) on in empty offices with (MS = 4.53), high level of usage of fans and ACs over natural ventilation (MS = 4.27), leaving security lights turned on during day time with (MS = 4.26), Faulty lighting controls and switch in buildings (MS= 4.23), running fans with no control regulators with (MS = 4.22), simultaneous running of multiple appliances in hostels and staff quarters with (MS= 4.08), leaving plug appliances in socket unused (MS = 4.07). The results further revealed that while energy wastages are mostly common with lighting consumption, poor controls and regulators for both lighting and other appliances in Higher Education Institutions, lighting related operations exert significant impacts on energy wastages as six out of the thirteen major energy wastage areas (1st, 5th, 6th, 10th, 11th and 13th) are connected with lighting, more so that the highest ranking wastage area has to do with leaving lights on in unoccupied spaces.

Table 7. Area of Energy Wastage in Educational Institutions Buildings in Nigeria

Variables	Mean Score	Rank
Leaving lights on in an empty space (lecture hall)	4.59	1st
Leaving fans running in empty space (lecture halls)	4.57	2nd
Leaving Air conditions on in empty offices	4.53	3rd
High-level usage of fans & ACs over natural ventilation	4.27	4th
Leaving security and street lights on in day time	4.26	5th
Faulty lighting controls and switches in buildings	4.23	6th
Running fans with bad regulators and controls	4.22	7th

Simultaneous running of multiple appliances at a time	4.08	8th
Leaving plug and unused appliances in socket (hostels)	4.07	9th
The use of artificial lighting over day lighting in day time	4.03	10th
Switching outdoor lighting on in broad day light	4.01	11th
Insufficient use of energy efficiency installations	3.70	12th
Dominant usage of incandescent bulb	3.33	13th

3.2 Factors Influencing Energy Wastages

The results of evaluation of factors influencing energy wastage in the educational institutions buildings in South Western Nigeria are contained Table 8. The factors are critical for energy consumption consideration in the adoption of smart building technologies. Ten (10) variables were identified and each indicator was rated by respondents using the five point Likert scale: The table revealed that, the major dominant factors influencing energy wastage in the study area was as a result of poorly designed HEIs buildings with average (MS = 4.62), followed by lack of improved sensors to monitor and regulate lightening consumption.

The other factors of energy wastage are in decreasing order with lack of information and awareness on energy usage and its cost implication with (MS = 4.54), Non metering of each facilities and buildings (MS = 4.52), In-effective energy policies on buildings design and usage (MS = 4.52), lack of controls and behavioral attitudes towards energy savings (MS = 4.42), lack of energy regulations and policies for students living in hostels accommodation (MS = 4.38), lack of improved software for institutional buildings design and operations and usage of inefficient appliances (MS = 4.26), lack of passive glazing materials for building designs was the least with (MS = 4.25).

All other factors aside poorly designed HEIs buildings need to be adequately considered during design phase for effective energy wastage management. This perhaps explains the highest ranking of poor designs of HEIs buildings. Furthermore, the major factors as agreed by the respondents from the study areas that are influencing energy wastage could reduce wastages when considered along with adoption of smart building technologies in institutional buildings especially with a view to minimize the energy consumption during peak period.

Table 8. Factors Influencing Energy Wastage in Educational Institutional Buildings in South West Nigeria

Variables	Mean Score	Rank
Poorly designed HEIs buildings	4.62	1st
Lack of improved sensor to monitor energy consumption	4.61	2nd
Lack of information and awareness on energy usage and cost	4.54	3rd
Non-metering of each facilities to measure energy usage	4.52	4th
Ineffective energy policies in HEIs buildings	4.52	5th
Poor controls and behavioral attitudes of occupants	4.42	6th
Lack of energy usage rules for students living in hostels	4.38	7th
Lack of improved software for HEIs design & operations	4.30	8th
High usage of inefficient appliances in HEIs	4.26	9th
Lack of passive glazing materials for windows and buildings	4.25	10th

3.3 Evaluation of Energy Saving Opportunities for Educational Institutional Buildings

In Table 9, existing energy saving opportunities that are deemed to be peculiar to higher education institutional buildings are examined. Fifteen (15) identified variables were ranked using five point Likert scale. The table shows that the most preferred potential savings technique to reduce energy wastage is the implementation of energy saving policies. The other energy savings opportunities that could be adopted decreased in ranking from adopting independent metering system for each buildings and facilities, hostels, staffs quarters and ventures (MS = 4.63), the use of improved switch, sensors and controls systems to regulate lightening consumption (MS = 4.59), encouraging energy savings behaviors among staffs and students (MS = 4.51), adopting natural ventilation and lightening over mechanical ventilation.

Table 9. Energy Saving Opportunities to be utilized in Higher Education Institutional Buildings

Variables	Mean Score	Rank

Implementation of energy savings policies in institutions	4.66	1st
Metering each department & facilities for energy usage	4.63	2nd
Improved sensors to regulate lighting consumption	4.59	3rd
Encouraging energy savings behaviour among students & staffs	4.51	4th
Adopting natural ventilation and lighting over mechanical	4.44	5th
Efficient lightening upgrades to energy saving lamps	4.40	6th
Using scheduling devices for power supply	4.38	7th
Passive building designs	4.36	8th
Passive glazing materials for building and windows	4.35	9th
Introducing energy savings education into teaching	4.35	10th
Good landscaping design and building envelope	4.34	11th
Switching from electricity streetlights to solar control sensors	4.34	12th
Replacing resistance regulators ceiling fans with electronic regulators	4.32	13th
Efficient heating and cooling technology with efficient appliances	4.30	14th
Selecting offices equipment with reduced heat emission	4.29	15th

(MS = 4.44), efficient upgrades from incandescent bulb to compact florescent lamp (MS = 4.40), uses of scheduling devices for power supply during vacations and off peak hours (MS = 4.38), passive building design for comfortable indoor environment (MS = 4.36), passive glazing materials for buildings and windows (MS = 4.35), introducing energy saving education into teaching (MS = 4.35), good landscaping design and building envelope (MS = 4.34), switching from electricity running street lights to solar with time control sensors (MS = 4.34), replacing resistance regulators ceiling fans with Electronic Regulators (MS = 4.32), efficient heating and cooling technologies with efficient appliances (MS = 4.30), scored least was selecting offices equipment with reduce heat equipment (MS = 4.29). From the respondents point of view it can be deduced that higher education institution energy policy, improvement on passive building designs principles are critical to the reduction of energy wastage and hence should be adhered to since it saves cost while other factors which involves the adoption of energy efficient technology for HSIs buildings are very costly to acquire.

4. Conclusion and Recommendation

The following conclusions are drawn from findings on the assessment of energy wastage and saving potentials for educational institutional buildings in South Western Nigeria;

- In appraising existing spectrum of energy wastes in institutional buildings of selected tertiary institutions in South western Nigeria, wastages associated with switching artificial light on in an unoccupied room spaces especially lecture halls and class rooms was the most critical means of energy wastages, while leaving appliances plugged in sockets unused was responsible for the least energy wastage. Furthermore, in the list of means through which energy is wasted in buildings, lighting related means collectively accounted substantially for energy wastages.
- Examination of factors influencing energy wastages revealed that poorly designed HEI building mostly affected energy wastages
- The most preferred potential savings technique to reduce energy wastage is the implementation of energy saving policies.

In the lights of the findings, the following recommendations are made:

- The usage of metering system to determine energy bills in each facilities/ department should be implemented to know which areas consumes energy most.
- Using scheduling devices with intelligent building designs for Power supply and controls during sessions and vacations
- Encouraging building management System (B.E.M.S) in institutions
- Provisions of standards controls to all lightening and ventilating equipments.
- Switching off all unused lightening during the days and utilization of natural lightening during peak period.

- Avoiding the usage of electric cookers and stoves in student's hostels accommodation and implementing standard Energy Policies.
- Awareness should be provided for students and staffs on energy usage and cost savings benefits attached in cutting energy bills and wastage to minimum.
- Provisions of passive glazing materials for building and windows and selecting offices equipment with reduced heat equipment.

References

[1] Energy Information Administration 2016 *International Energy Outlook 2016 report.* Report Number: DOE/EIA-0484(2016)
[2] Venkantarma B V and Jagadish K S 2003 *Energy and Buildings* **35**(**2**), 129-137
[3] Energy Information Administration (EIA) 2005 *World Energy Overview 1993-2003* Energy information Administration. Washington DC.
[4] Adelaja A O, Damisa O, Oke S A., Ayoola A B. *and* Ayeyemitan A O 2008 *Maejo. International Journal of Science and Technology* **2** (02), 331-344.
[5] Nwachukwu, G.O 2010 *Journal of Energy in Southern Africa* **21** (1) 2-10.
[6] Nigerian Electricity Regulatory Commission 2014 *File Repository*
[7] National Bureau of Statistics 2015 *Total Energy Generated and sent out*
[8] Oyedepo, S O, Adeleke T, Leramo R O., Kilanko O, Babalola O P, Balogun A O. *and* Akhibi M O 2015 *International Conference on African Development Issues* Pp. 203 – 211.
[9] Kim D W, Jung J W, Seok H T and Yang J H 2017 *International Conference on Sustainable Building Asia* CIB17386
[10] United State Environmental Protection Agency (USEPA) 2010 *Environmental technology Verification programme*
[11] Al-mumin A., Khittab, O and Sridhar 2003 *Energy and Buildings*, **35**(6), 549-559
[12] Noranai, Z and Azman, ADF 2017 IOP Conf. Series *Materials Science and Engineering* **243** 012023

Hybridization energy systems for a rural area in Nigeria

O G Olasunkanmi[1], O A Roleola[2], P O Alao[3], O Oyedeji[4] and F Onaifo[5]

[1,3,5]Department of Electrical and Electronic Engineering, Olabisi Onabanjo University, Ogun State, Nigeria.
[2]Advad Limited, Ikeja Energy Audit Unit, Lagos State, Nigeria.
[4]Department of Computer Engineering Olabisi Onabanjo University, Ogun State, Nigeria.

Email: grace.olasunkanmi@oouagoiwoye.edu.ng

Abstract. Mankind cannot live without one form of energy or another. Energy (Electrical) is needed to lighten our environment, power our homes, schools, hospitals, offices, businesses, and for industrialization. It is generally known that the economic growth of a nation depends on electricity. This paper is assessing the possibility of hybridizing the various renewable energy sources in Nigeria and evaluating its sustainability, challenges and benefits of hybridizing the combinations of the renewable energy sources (PVs and wind turbine) that are abundantly available in Nigeria. The evaluation of these combinations is to determine the best for the region (rural setting). We are also looking at efforts being put at utilizing these renewable sources for energy generation in Nigeria. The hybrid energy system of a mix of locally available abundant energy resources is a good solution to the persistent epileptic power supply in Nigeria. This will minimize atmospheric degradation as quality of life will be improved.

1. Introduction

In Power Engineering System, the term 'Hybrid' can be defined as a combination of power or energy backup system for the use of mankind. Electrical power producers used in hybrid energy systems are photovoltaic, wind turbines, and different kinds of generators such as diesel, gas and petrol. Hybrid Energy System entails a renewable energy device such as PV solar cells that is balanced through a backup form of generation such as a diesel/gas/petrol generator, wind turbine, fuel cell or battery storage system to ensure continuous supply of electricity.

In Nigeria today, efficient power supply to the populace has been a major challenge which has adversely affected the economy and increased the pressure on the government in terms of energy demand. For a nation to develop technologically, reliable energy is needed. This will attract investors which will in turn boost the economy. The convectional form of power generation (hydro, thermal) can be complemented using a hybrid system (PVs and wind turbines) to generate electrical power as a form of backup in the eventuality of power failure. Efforts made by various administrations/ government at addressing this problem (which have included the privatization of its power sector) are yet to yield desirable outcomes [1]. This paper is based on hybridized renewable energy supply system as a feasible alternative to the convectional mode of power generation in a rural area of low energy consumption. It is a fact that Nigeria is blessed with abundant sunshine all year round and there is also large vegetation that covers the country. Maren in (2013), stated that Nigeria was endowed with an annual average daily sunshine of 6.25 hours ranging between 3.5 hours at the coastal region and 9.0 hours in the northern region which receives about 5.08 x 1012 kWh of energy per day from the sun and if solar energy appliances with just 5% efficiency are used to cover only 1% of the country's surface area then 269.24 MWh of electrical energy can be obtained from solar energy. This amount of electrical energy from the sun is equivalent to 4.66 million barrels of oil per day. This gives rooms for Nigeria to have a greater accessibility and availability of solar energy to develop her solar energy technology [11]. There are a lot of potential for energy generation from the wind. These potential varies with wind

IOP Conf. Series: Earth and Environmental Science **331** (2019) 012007 doi:10.1088/1755-1315/331/1/012007

speed and is available in Nigeria at annual average speeds of about 2.0m at the coastal region and 4.0 m at the northern region of the country with an average annual wind speed of not less than 5 m/s at a height of 10 m above ground level [10]. We need to assess the possibility of hybridizing the various renewable energy sources in Nigeria and evaluating the sustainability of this hybridization in the rural area.

The price of oil has been unstable and for the grid to be expanded there have been the problem of right of way (ROW) non the less it is not a cost effective way out as well. Therefore, the integration of renewable energy source for a rural area becomes an important alternative solution [6].

Nigeria has been blessed with a wealth of natural resources that can be utilized to generate electricity in which the hybridization of solar and wind with proper market and regulatory mechanisms put in place could be utilized to provide a reliable supply of electricity to all residential in the rural area at cost-reflective and affordable tariffs [12]. There are several possible combinations that could be made but the combination of the solar and wind which is based on locally available abundant energy resource [2] was considered here as this often yield greater economic and environmental returns than wind, solar or geothermal stand-alone systems [13].

The condition of living in the rural area is an unhealthy one just because they lack some basic amenities like electricity, water supply, good roads, education etc. Bringing people in this area in mainstream and preventing large scale migration from the rural area to urban area, development of some sources that can fulfill the basic needs among vital classification to be is electricity. Therefore, the hybridization of energy system can prove to be a major cutting edge solution [13].

2. Hybridization of Wind and Solar Electric Systems for rural area

The investigation of Renewable Energy Experts, a hybridization of electric system then combines wind electric and solar PV technologies will offers several merits over a single system. In the United States of America, the speed of wind is low during the summer when the sun shines brightest and longest. Moreover, the wind is strong in the winter when less sunlight is available. As a result of this, the peak operating times for wind and solar energies occur at different times of the day per year [3]. Hybrid power systems produce power as at when needed, this made the integration of wind and solar energy more attractive and been used for the substitution of oil produce [4].

Hybrid electrical systems are stand-alone which operate "off-grid" i.e., it is not connected to an electricity distribution system. When either of the wind or solar system is producing electricity, power can be provided through batteries or engine generator powered by conventional fuel such as diesel, gas and petrol. When the batteries run low, the engine generator can provide power and recharge the batteries. It is essential that the storage capacity must be large enough to supply electrical load demand during non-charging periods. Battery banks should have enough capacity so that it can supply the electric load continuously to the community [7].

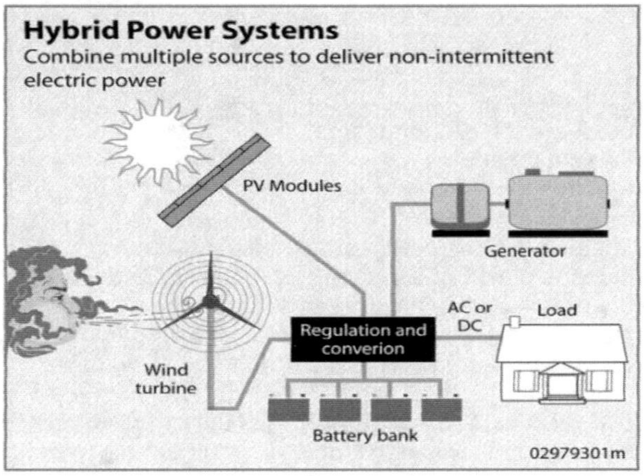

Fig. 1: Hybridization of Wind & Solar PV Systems

2.1. Off-Grid or Stand-Alone Renewable Hybrid Energy Systems for a Rural Area

According to the US Department of Energy, off-grid or stand-alone electrical systems can be more cost-effective than connecting to the grid in remote locations. A rural area of low energy consumption as shown in Figure 2 will benefit and enjoy hybridization of wind and solar. Domestic loads consumers or small businesses using a renewable energy system that is not connected to the electricity grid called a stand-alone system makes economic sense and appeals to their environmental values. Successful stand-alone systems take advantage of a combination of techniques and technologies to generate and distribute reliable power, reduce costs, and minimize inconvenience. Some of these strategies include using fossil fuel or renewable hybrid systems and reducing the amount of electricity required to meet the load demand. In addition to acquire photovoltaic panels or wind turbine, there is the need to invest in some other equipment called **balance-of-system** to safely transmit the electricity to the consumers. This equipment includes:

- Batteries
- Charge controller
- Power conditioning equipment
- Safety equipment
- Meters and instrumentation

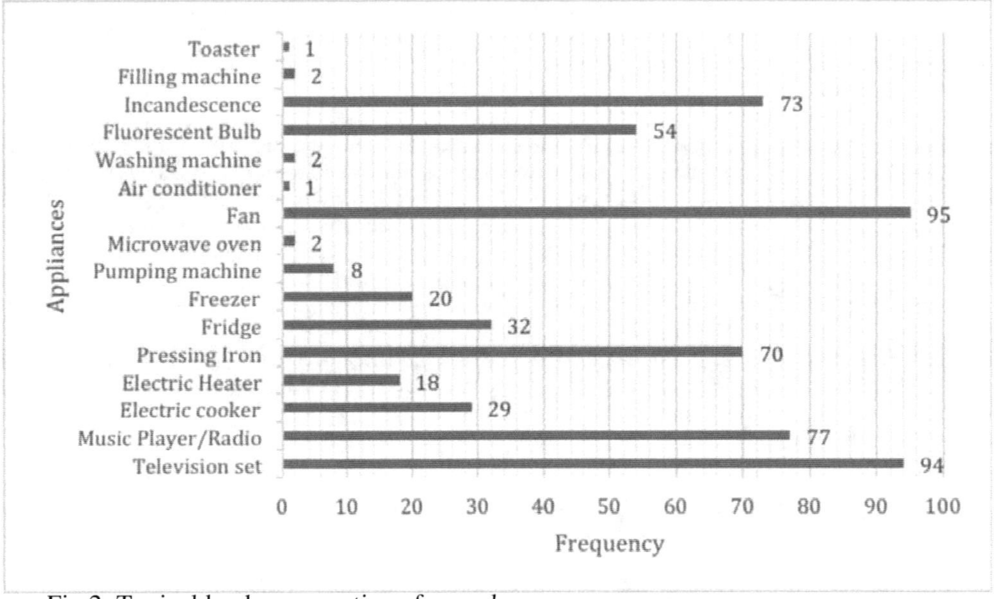

Fig 2: Typical load consumption of a rural area
Source: Authors' Field Survey, 2018

3. Sustainability of Hybrid Energy Systems in rural area

Sustainability is a phenomenon aimed at achieving a ''balance between human endeavor such as cost, health, comfort and environmental concerns (resource use and ecological degradation)'' [8]. So far so good, there is no formal sustainability analysis of the few existing hybrid energy systems in Nigeria. There have been some reports by the few hybrid energy developers in Nigeria which include that the system has helped them reduce their energy cost, enable them provide stable services and reduce their carbon footprints [5]. However, sustainability evaluation goes beyond the attributes of a system's perceived benefits. According to Barber (2014), there are some initiatives that have been aimed at resources conservation such as recycling of waste materials that end up having more negative effects on the environment which has contradicted its intended purpose. The overall effects would not be observable or noticed without a comprehensive analysis of the system. It is therefore expedient to have a critical and quantitative analysis of hybrid electrical system in a way that would enable stakeholders to have a good understanding of the extent to which the system is sustainable [1].

3.1. Challenges of Hybridizing Energy Systems in Nigeria

The few cases of hybrid energy supply projects have shown great potential as a solution pathway to addressing Nigerian energy supply systems [10]. It may be impossible to identify all the problems that could be encountered in the course of developing and operating all types of hybrid energy systems. However, there are some sustainability issues that would need to be considered regardless of the hybrid system. Few of the major challenges of hybridizing energy systems in Nigeria rural area are:

- Integrating the system in such a way that will ensure availability of sufficient amount of energy on a continuous basis with/without interruption.
- Ensuring that the system is simple enough in its design such that it can be maintained/even upgradable by the locally available technicians/artisans.
- Technically, the wind energy system may not be viable because of the low wind speed depending on the location.

3.2. Benefits of Hybrid Energy Systems

It provides assurance of steady supply of energy 24 hours. When either one is not available in bsufficient quantity, the other is available to compliment it.

- Since the hybrid system is premised on the abundant local resources, the problem of depending on external supply for energy generation which could be disrupted by vandalism is eliminated.
- It can be utilized at remote locations/terrains where grid connections are difficult or expensive.
- It requires members of the local communities to provide requisite services for its operation and thereby boost business opportunities.
- It increases employment in the locality which in turn would reduce rural urban migration.
- Steady supply of hybrid electric energy would eliminate the need for expensive and polluting fossil fuel powered generators by reducing business operating cost, idleness, and increase productivity.
- It reduces pollution that causes health problems and improve standard of living.
- It enables locally manufactured products to compete on the market.
- It makes local businesses to flourish.
- It increases employment opportunities, reduced social vices, promote peace and economic stability.

4. Conclusion

This discourse has showed that the development of hybrid energy systems in rural area is at an embryonic stage. Hybrid energy systems based on a mix of locally available abundant energy resources seems to be a good solution to the persistent epileptic power supply in Nigeria. For it will improve the quality of life and minimize atmospheric degradation.

It will also reduce the migration to the cities thereby reduce overpopulation. Getting it right for the first time would require a balance of human and environmental concerns. This study has provided a balanced criterion for evaluating the sustainability of a proposed/existing hybrid energy system. The next stage would be case studies that utilized the set of proposed criteria in evaluating the sustainability of a number of hybrid energy systems in Nigeria. Hybrid energy systems investors, regulators, researchers and developers would find it useful in their decision making regarding the choice of a hybrid energy systems option (wind, solar, biomass etc.) that are at their disposal.

IOP Conf. Series: Earth and Environmental Science **331** (2019) 012007 doi:10.1088/1755-1315/331/1/012007

References

[1] Barber D A 2014 Nigeria Moving into Solar Energy to Spur Business Growth, https://moguldom.com.
[2] Dunmade I 2016 Hybridizing Renewable Energy Systems in Nigeria: A Contextual Framework for Their Sustainability Assessment. *European J. Engineering and Technology.* Vol **4**. No.5. ISSN 2056-5860
[3] Lal D K, Dash B B and Akella A K 2011 Optimization of PV/Wind/Micro-Hydro/Diesel Hybrid Power System in HOMER for the Study Area. *Internation J. Electrical Engineering and informatics.* Vol **3**, No 3
[4] Nema P, Nema R K and Rangnekar S 2009 A current future state of art development of hybrid energy system using wind and PV-solar. *Renewable and sustainable Energy Review.*
[5] Okere R 2015 BPE clarifies status of N213bn capital injection to power firms
[6] Olatomiwa L, Mekhilef S, Huda A S N and Ohunakin 2015 Economic Evaluation of hybrid energy system for rural electrification in six geo-political zones in Nigeria.J *Renewable Energy,* Vol 83, pp 435-446.
[7] Simonyan K and Fasina O 2013 Biomass resources and bioenergy potentials in Nigeria. *African J. Agricultural Research,* vol **8**, No 40.
[8] Stanford University 2005 Guidelines for Lifecycle Cost Analysis – Land and buildings
[9] US Department of Energy at http://energy.gov/energysaver/
[10] Vincent, E. N. and Yusuf, S.D. (2014) Integrating Renewable Energy and Smart Grid Technology into the Nigerian Electricity Grid System. *Smart Grid and Renewable Energy* **5**, 220-238.
[11] Maren, I. Borok, Agontu, J. Agandu, Mangai. M. Morgan, (2013). Energy Security in Nigeria: Challenges and Way Forward. *International J. Engineering Science Invention, Vol 2 Issue 11, PP.01-06*
[12] Federal Ministry Of Power, Works And Housing 2015, Rural Electrification Agency (REA) Final *Draft Rural Electrification Strategy & Plan for Implementation.*
[13] Srivastava S D and Banerjee R 2015 Hybrid Renewable Energy Systems & their Suitability in Rural Regions. *Journal of Mechanical and Civil Engineering.* Vol **12**, PP 117- 120.

Air quality investigation over Moundou Chad

Emetere M.E.[1,3], **and Oladimeji T.E.**[2]
[1]Department of Physics, Covenant University Canaan land, P.M.B 1023, Ota, Nigeria.
[2]Department of Chemical Engineering, Covenant University Canaan land, Ota, Nigeria.
[3]Department of Mechanical Engineering and Science, University of Johannesburg, APK, South Africa

E-mail: emetere@yahoo.com

Abstract. The air quality over Moundou-Chad was examined using satellite dataset. Fifteen years aerosol optical depth (AOD) was obtained from the Multi-angle Imaging Spectro-Radiometer (MISR). The AOD dataset is statisticaly stable for further prediction. The highest AOD mean was observed in 2008. The standard deviation, average deviation and coefficient of variation all agrees that the AOD dataset is reliable. It is shown that the aerosol loading over Moundou is high and the size of the aerosol is very tiny to deposit in the human lungs.

Keywords: air quality, aerosol, pollution, deposition

1. Introduction

Pollution is one of the major environmental problems confronting Chad area, this pollution ranges from air pollution to water pollution as well as land pollution. Outdoor air pollution involves the reaction of particulate matter, biological materials as well as chemicals from different sources. Some of the industries that have the potential to affect the pollution level in Moundou include oil, sodium carbonate production, cotton textiles, soap, brewing, construction and cigarettes production industries among others (1). Some of the important Pollutants generated from these industries include particulate matter (PM), sulphur (IV) oxide (SO_2) and oxides of nitrogen (NO_x). Other generators of air pollution in this country are the transport sector, pollution result from increase in number of vehicles, use of old and dirty fuel as well as poorly managed public transport (2-3). Also, uncontrolled waste burning is one of the practice in this country which result in deterioration of the air quality in the country. In addition, air pollution from indoor sources is the largest contributor to the negative health effects of air pollution in Chad. Wood and charcoal provide 90% of the energy consumed in Chad which is the main contributor to indoor pollution, this result in estimated 9,600 premature deaths every year as reported by WHO, (4).

The research site is fast developing with the presence of industries (5). Hence, there are possibilities that among the anthropogenic emission, there may be industrial pollution. Industrial pollution is common in developing countries because there is inadequate enforcement of refuse disposal and particulates or gaseous emission laws. Another potential source of air pollution is from the agricultural sector via bush burning, emission of bio aerosols etc. Also, the desertification and dryness of the region between 1972–1973 and 1984–1985) are sources of pollution (6). Some of the air pollutants deposit over water ways are sources of heavy metal introduction. There is a typical case on the metallic pollution of Lake Chad and Chari River beside Sarh and N'Djamena cities (7).

In this paper, the air pollution over Mondou, chad was investigated to ascertain the level of pollution that may have arisen from previous air pollution sources. Air pollutants or aerosols over the research

IOP Conf. Series: Earth and Environmental Science **331** (2019) 012008 doi:10.1088/1755-1315/331/1/012008

area are commonly from anthropogenic pollution and Sahara dust pollution. However, there is the need to understand the accumulated aerosol deposits in the atmosphere. It is upon this background that aerosol retention was used in this study as a salient parameter for understanding the level of air pollution over a geographical area.

2. Methodology

Moundou is the second largest city in Chad and is the capital of the region of Logone Occidental. The city lies on the Mbéré River some 475 kilometres south of the capital N'Djamena. Moundou is located in Chad on latitude and longitude of 8.5915° N, 16.0758° E

Figure 1: Satellite map of Moundou Chad

Fifteen years aerosol optical depth dataset was obtained from Multi-angle Imaging Spectro-Radiometer (MISR) (8). The dataset was treated using the excel programme. The aerosol loading calculation was accomplished using the West African regional scale dispersion model (WASDM) expressed below (9-10).

$$\psi(\lambda) = a_1{}^2\cos\left(\frac{n_1\pi\tau(\lambda)}{2}x\right)\cos\left(\frac{n_2\pi\tau(\lambda)}{2}y\right) + \cdots \ldots a_n{}^2\cos\left(\frac{n_n\pi\tau(\lambda)}{2}x\right)\cos\left(\frac{n_n\pi\tau(\lambda)}{2}y\right) \quad (1)$$

a is atmospheric constant gotten from the fifteen years aerosol optical depth (AOD) dataset from MISR, n is the tunning constant, $\tau(\lambda)$ is the AOD of the area and $\psi(\lambda)$ is the aerosol loading.

The angstrom exponent is used to estimate the particle size of atmospheric aerosols or clouds. It was calculated using

$$\alpha = \frac{\log\frac{\tau_{\lambda 1}}{\tau_{\lambda}}}{\log\frac{\lambda 1}{\lambda 2}} \quad (2)$$

where τ_{λ} is the optical thickness at wavelength λ and $\tau_{\lambda 0}$ is the optical thickness at the reference wavelength λ_0

3. Results and Discussion

The statistical analysis of the AOD dataset after treatment is presented in Table 1. Unlike most cities in West Africa that have 44% data loss, Moundou has data loss of 17%. This means that there are more dataset to analyse without extrapolation. The highest AOD mean was observed in 2008. The standard deviation, average deviation and coefficient of variation all agrees that the dataset is reliable. Hence, the average monthly AOD dataset was used to calculate the aerosol loading over Moundou as presented in Figure 2. The aerosol loading at the moment is extremely high. Hence, the possibility of increased non cureable diseases and deaths is inevitable if not controlled. In order to have a better understanding of the aerosol size, we calculated the Angstrom exponent over Moundou as presented in Figure 3. It is observaed that the Angstrom exponent ranges between 0.05 to 0.3. However, there is still the need to know in more specific terms the radius of the atmospheric aerosols over Moundou. The radius of the aerosol is presented in Figure 4.

Table 1: statistical analysis of AOD over Moundou

	2000	2001	2002	2003	2004	2005	2006
Number of values	9	11	10	11	11	11	11
Number of missing values	3	1	2	1	1	1	1
Mean	0.296	0.284	0.358	0.343	0.344	0.330	0.327
First quartile	0.227	0.216	0.227	0.245	0.256	0.264	0.225
Third quartile	0.347	0.392	0.508	0.402	0.449	0.409	0.368
Standard error	0.045	0.030	0.052	0.039	0.035	0.027	0.051
95% confidence interval	0.103	0.068	0.119	0.087	0.078	0.061	0.114
99% confidence interval	0.150	0.096	0.170	0.124	0.112	0.087	0.162
Variance	0.018	0.010	0.027	0.017	0.014	0.008	0.029
Average deviation	0.100	0.089	0.135	0.100	0.100	0.076	0.112
Standard deviation	0.134	0.101	0.166	0.130	0.117	0.091	0.169
Coefficient of variation	0.451	0.354	0.463	0.377	0.339	0.276	0.518
Skew	1.337	0.187	0.848	1.225	0.536	-0.428	1.949
Kurtosis	0.913	-1.794	-0.378	1.779	-0.948	-0.961	5.049
Kolmogorov-Smirnov stat	0.300	0.234	0.204	0.187	0.219	0.163	0.285
Critical K-S stat, alpha=.10	0.387	0.352	0.369	0.352	0.352	0.352	0.352
Critical K-S stat, alpha=.05	0.430	0.391	0.409	0.391	0.391	0.391	0.391
Critical K-S stat, alpha=.01	0.513	0.468	0.489	0.468	0.468	0.468	0.468
	2007	2008	2009	2010	2011	2012	2013
Number of values	10	11	11	11	11	12	10
Number of missing values	2	1	1	1	1	0	2
Mean	0.333	0.387	0.369	0.352	0.353	0.296	0.376
First quartile	0.257	0.307	0.241	0.232	0.240	0.203	0.232
Third quartile	0.394	0.438	0.483	0.531	0.439	0.411	0.518
Standard error	0.039	0.049	0.054	0.052	0.050	0.037	0.056
95% confidence interval	0.089	0.108	0.121	0.116	0.111	0.081	0.127
99% confidence interval	0.128	0.154	0.172	0.165	0.158	0.115	0.182

IOP Conf. Series: Earth and Environmental Science **331** (2019) 012008 doi:10.1088/1755-1315/331/1/012008

Variance	0.015	0.026	0.032	0.030	0.027	0.016	0.031
Average deviation	0.095	0.115	0.146	0.137	0.129	0.106	0.145
Standard deviation	0.124	0.161	0.180	0.172	0.165	0.128	0.177
Coefficient of variation	0.372	0.417	0.488	0.489	0.468	0.432	0.471
Skew	1.044	0.733	0.692	0.699	1.386	0.633	0.885
Kurtosis	1.260	0.446	- 0.692	-0.871	1.421	- 0.946	-0.706
Kolmogorov-Smirnov stat	0.200	0.179	0.183	0.268	0.258	0.175	0.210
Critical K-S stat, alpha=.10	0.369	0.352	0.352	0.352	0.352	0.338	0.369
Critical K-S stat, alpha=.05	0.409	0.391	0.391	0.391	0.391	0.375	0.409
Critical K-S stat, alpha=.01	0.489	0.468	0.468	0.468	0.468	0.449	0.489

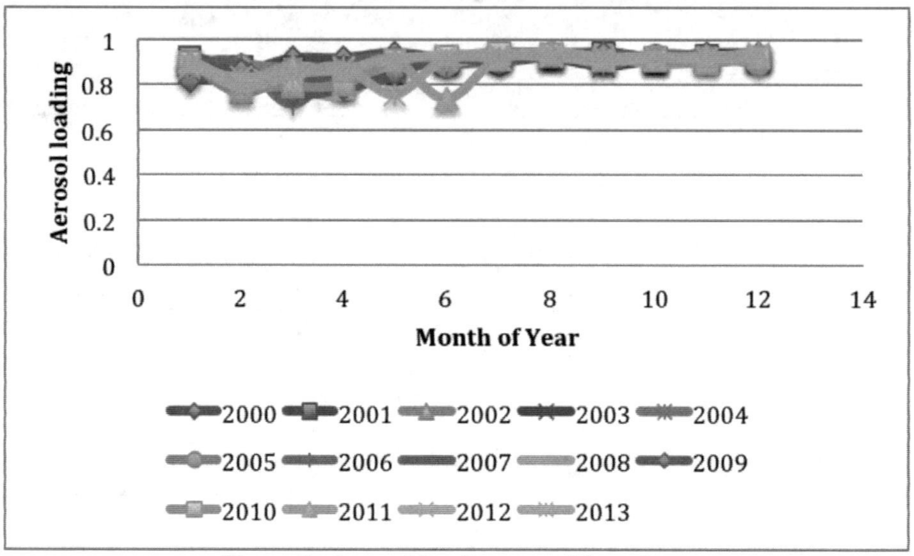

Figure 2: Aerosol loading over Moundou

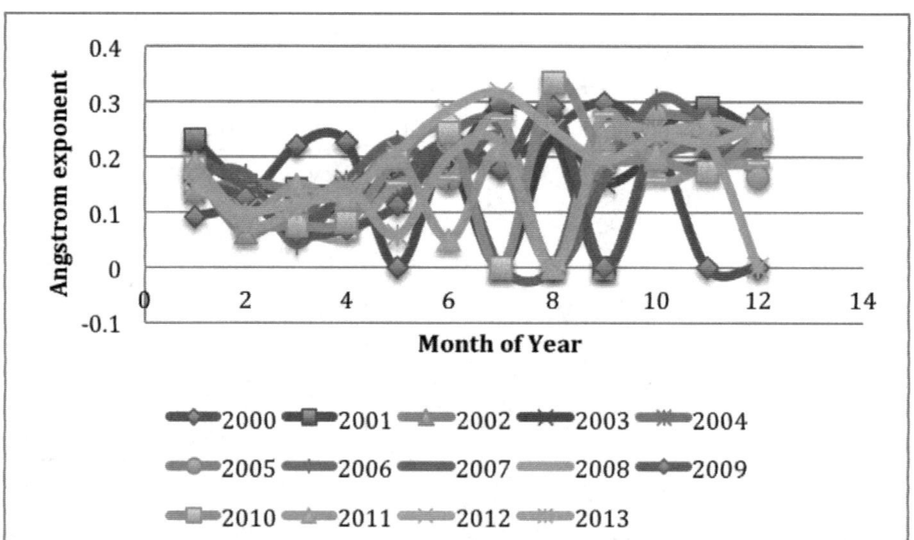

Figure 3: Angstrom exponent over Moundou

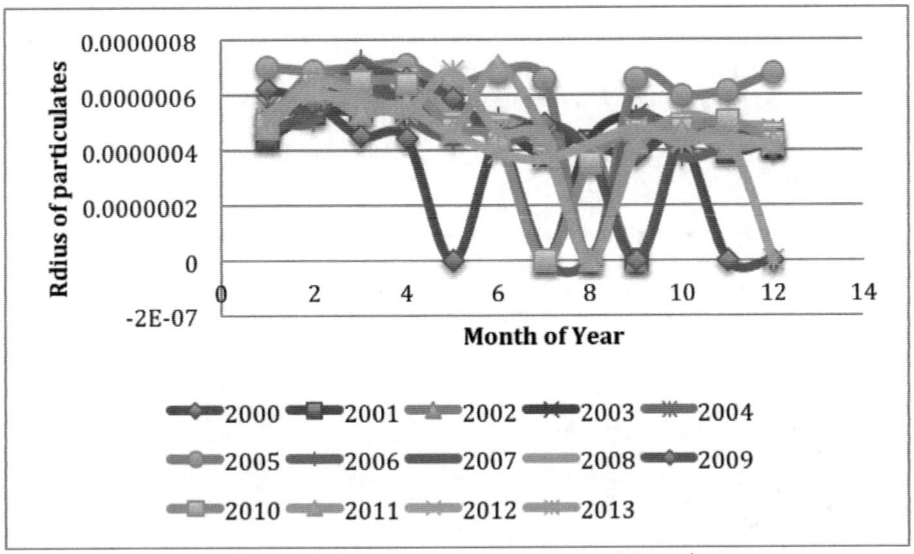

Figure 4: Angstrom exponent over Moundou

From Figure 4, it can be inferred that the radius of the aerosols in Moundou are fine minute particulates that are possible deposition chances in the human lungs.

4. Conclusion

The AOD dataset is statisticaly stable for further prediction. The standard deviation, average deviation and coefficient of variation all agrees that the AOD dataset is reliable. It is shown that the aerosol loading over Moundou is high and the size of the aerosol is very tiny to deposit in the human lungs.

Acknowledgement

The authors appreciate Covenant University for partial sponsorship. The authors acknowledge NASA for primary dataset.

5. References

[1]. Haywood, J.M., Pelon, J., Formenti, P., Bharmal, N.A., Brooks, M., Capes, G., Chazette, P., Chou, C., Christopher, S., Coe, H., Cuesta, J., Derimian, Y., Desboeufs, K., Greed, G., Harrison, M., Heese, B., Highwood, E.J., Johnson, B.T., Mallet, M., Marticorena, B., Marsham, J., Milton, S., Myhre, G., Osborne, S.R., Parker, D.J., Rajot, J.-L., Schulz, M., Slingo, A., Tanre, D. and Tulet, P., Overview of the dust and biomass burning experiment and African monsoon, multidisciplinary analysis special observing period-0. Journal of Geophysical Research 113 (2008). 1-12.

[2]. Emetere M.E., Numerical modelling of west africa regional scale aerosol dispersion. A doctoral thesis submitted to Covenant University, Nigeria. (2016) 65-289

[3]. Emetere, Moses Eterigho, Statistical Examination of the Aerosols Loading Over Mubi-Nigeria: The Satellite Observation Analysis, Geographica Panonica, 20(1) (2016) 42-50

[4]. WHO, (2018a). Air pollution, http://www.who.int/airpollution/en/ (accessed 5th July, 2018)

[5]. Zakaria, H.; Abderaman, M.A. *Profil National du Tchad sur la Gestion des Produits Chimiques*; MEE-UNITAR: Manhattan, NY, USA, 2002; p. 103.

[6]. N'garam Nambatingar, Yohann Clement, Alain Merle, Tchadanaye New Mahamat, and Pierre Lanteri (2017). Heavy Metal Pollution of Chari River Water during the Crossing of N'Djamena (Chad). Toxics,5(4):(26-33)

[7]. Massang Tchoroun, Guy Bertrand Noumi, New Mahamat Tchadanaye, and Jean Marie Dikdim (2015). Heavy Metals Pollution Level in Water, Fish and Sediments from

the Logone River Within Moundou City (Chad). International Journal of Environmental Monitoring and Analysis, 3(5): 275-281
[8]. McGill, M., CATS/ISS L2O D-M7.1 05kmPro HDF-5 File - Version 1.05 [Data set]. NASA Langley Research Center Atmospheric Science Data Center DAAC. (2016). https://doi.org/10.5067/iss/cats/l2o_d-m7.1-v1-05_05kmpro
[9]. Emetere M.E., Generation Of Atmospheric Constants Over Some Locations In West Africa: A Theoretical Aid for Measuring Instruments Design, International Journal of Engineering Research in Africa, 27 (2016) 119 – 146
[10]. Moses Eterigho Emetere, Investigations on aerosols transport over micro- and macro-scale settings of West Africa, Environ. Eng. Res., 22(1) (2017) 75-86

Analysis of aerosol optical depth and aerosol loading Over Voinjama: A projected estimation of atmospheric corrosion

Emetere M.E.[1,3], Sanni S.E.[2] and Okoro S.E.[4]
[1]Department of Physics, Covenant University Canaan land, P.M.B 1023, Ota, Nigeria.
[2]Department of Chemical Engineering, Covenant University Canaan land, Ota, Nigeria.
[3]Department of Mechanical Engineering and Science, University of Johannesburg, APK, South Africa
[4] Department of Petroleum Engineering, Covenant University Canaan land, Ota, Nigeria.

E-mail: emetere@yahoo.com

Abstract. The World Health Organization (WHO) warning on air pollution is a verified in this research by corroborating it with fifteen years primary (aerosol optical depth) dataset that was obtained from the Multi-angle Imaging Spectro-Radiometer (MISR) over Voinjama-Liberia. The detailed statistical and computational analysis was carried out using computational softwares. The spatial amd surface plot the pollution over the geographical location may constantly be dispersed from same sources of pollution. The implication of the aerosol optical depth (AOD) on the atmospheric corrosion was corroborated using mathematical estimations. The highest corrosion rate is given as 2.6 and the average corrosion rate over Voinjama is 0.5.

Keywords: aerosol loading, aerosol, Voinjama, Liberia, model

1. Introduction

The World Health Organization (WHO) released a report that "92% of people worldwide breathe unhealthy air, resulting in about 6.5 million deaths annually" (1). The west African region in this light had shown through satellite imagery that the pollution source is enormous at the moment (2). The properties of the atmospheric aerosols are found to be made-up of basically sand dust, sulfur dioxide, carbon ash, nitriogen dioxide and VOC. The health implication on the human and animal life-form is sickness and death. As a matter of fact, WHO claimed that one-third of deaths from the leading non communicable diseases (stroke, lung cancer, heart attacks and chronic obstructive pulmonary disease) are due to air pollution (1).

In this research, the area of interest is the analysis of the aerosol optical depth and aerosol loading over Vionjama to determine the atmospheric corrosion rate. The corrosivity of atmosphere refers to the extent of corrosion suffered by a metal caused by exposing it in the atmosphere (see Figure 1). Atmospheric corrosion in west Africa is of great concern because the sand and sahara dust that are predominant contains salt content which is one of the basic agent for metallic corrosion. Moreso, the sulfur dioxide is another corrosion stimulant present in the west Africa atmosphere. The corrosivity of atmospheres are of several stages depending on the type of metal. Table 1 shows the ISO classification of corrosivity of the atmospheres over zinc metal.

IOP Conf. Series: Earth and Environmental Science **331** (2019) 012009 doi:10.1088/1755-1315/331/1/012009

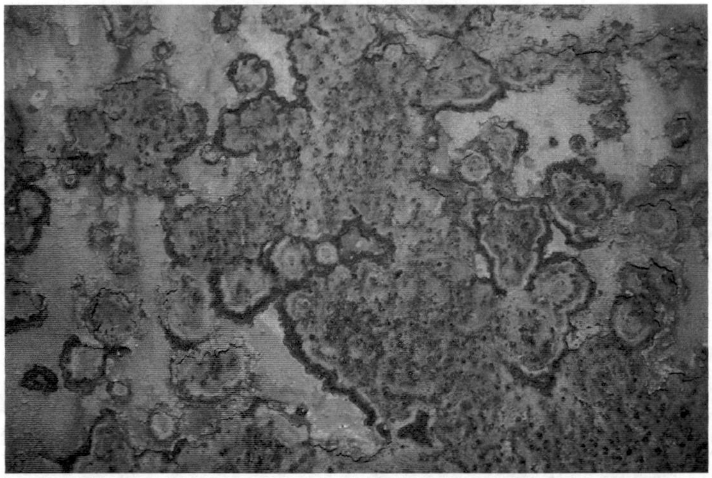

Figure 1: metallic corrosion caused by atmospheric aerosols

Table 1: ISO classification of corrosivity of the atmosphere (3).

C1	R≤0.1 µm/y	R≤0.0023σz/ft^2/y	R≤0.71 g/m^2/y
C2	0.1 <R≤0.7	0.0023 <R≤0.016	0.71 <R≤4.9
C3	0.7 <R≤2.1	0.016 <R≤0.049	4.9 <R≤15
C4	2.1 <R≤4.2	0.049 <R≤0.097	15 <R≤30
C5	4.2 <R≤8.4	0.097 <R≤0.19	30<R≤59

The corrosivity of the atmosphere depends on the content of the atmospheric aerosols. The significance of the research is the huge cost of maintenance of zinc roof in buildings such as schools, residential houses, churches and mosques. In developed countries, the cost due to corrosion is known. For example, Gerhardus et al. (4) revealed that corrosion of metals cost about $276 billion annually in the United State.

In this research, the analysis of the aerosol optical depth as well as the aerosol loading was carried out to estimate the atmospheric corrosion rate over the research area.

2. Experimental Design, Materials and Methods

Voinjama is located in Liberia on latitude and longitude of 8.4202° N and 9.7539° W respectively. The satellite map of the research location is presented in Figure 2. The research site is located in the hilly region, north of Liberia and near the Guinean border. The average human population of the area is about 60,000. The community is an agrarian community with the presence of some industries. At the moment, there are no ground measurement of air pollution in the research area. Hence, fifteen years aerosol optical depth (AOD) satellite dataset was obtained from the Multi-angle Imaging Spectro-Radiometer (MISR)(5).

The raw dataset was treated using excel programme. The statistical and computational analysis of the treated dataset was carried-out using the CERN Root C++ code. The aerosol loading was calculated using the West African regional scale dispersion model (WASDM). The WASDM is represented as (6-8):

$$\psi(\lambda) = a_1{}^2\cos\left(\frac{n_1\pi\tau(\lambda)}{2}x\right)\cos\left(\frac{n_1\pi\tau(\lambda)}{2}y\right) + \cdots \ldots a_n{}^2\cos\left(\frac{n_n\pi\tau(\lambda)}{2}x\right)\cos\left(\frac{n_n\pi\tau(\lambda)}{2}y\right) \quad (1)$$

a is atmospheric constant gotten from the fifteen years aerosol optical depth (AOD) dataset from MISR, n is the tuning constant, $\tau(\lambda)$ is the AOD of the area and $\psi(\lambda)$ is the aerosol loading.

The atmospheric corrossion rate of metals (9) over the Yekepa was calculated using the Faraday equation. It is given as :

$$CR\left(\frac{\mu m}{yr}\right) = k\,\frac{i_{corr}}{d}\,EW \qquad (2)$$

Where k is a conversion factor (3.27 x 106 $\mu m \cdot g \cdot A^{-1} \cdot cm^{-1} \cdot yr^{-1}$), icorr is the corrosion current density in $\mu A/cm^2$ (calculated from the measurements of Rp), EW is the equivalent weight, and d is the density of Alloy 22 (8.69 g/cm^3).

Figure 2: Satellite map of Voinjama

Based on equation (4), the modification in the work is the inclusion of aerosol loading.

$$CR\left(\frac{\mu m}{yr}\right) = k\,\frac{i_{corr}}{d}\,EW/\exp\left(\frac{\Delta W \tau \varphi(\lambda)}{2.32}\right) \qquad (3)$$

In this study, the corrosion current density of iron was considered and it is given as 3.2×10^{-3} $\mu A/cm^2$. The EW of iron is given as 27.9225.

3. Results and Discussion

The daily AOD dataset is presented in Figure 3. The peaks of the diagrams denoted the maximum AOD in the year. The maximum AOD occurs towards August. The diagram shows the four spectra bands of the MISR. The four spectra bands are used to estimates AOD over land and ocean. In this case, since marine aerosols pollution travels over long distances, the use of all spectra bands is justified for an indepth analysis of the atmospheric corrosion effect over the study area.

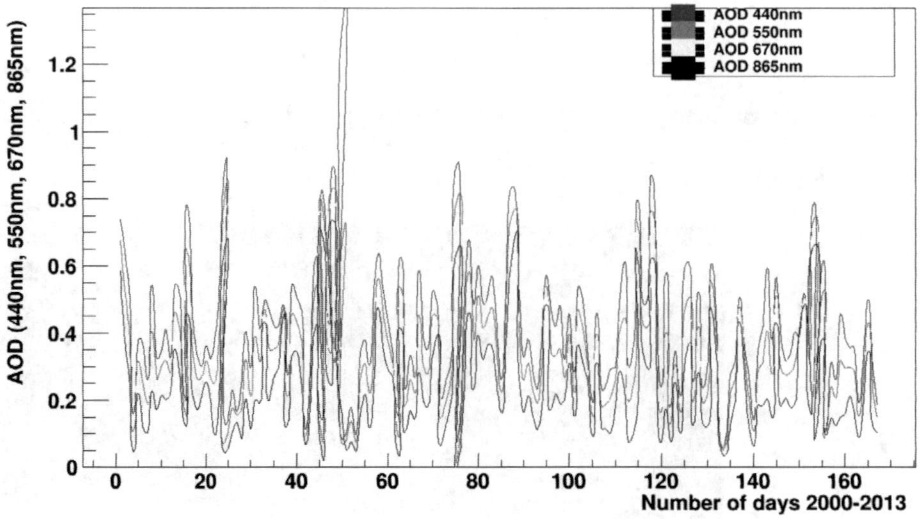

Figure 3: AOD daily dataset between 2000 and 2013

The spatial distribution of the AOD is presented using the contour map. It shows that aside the ussual anthropogenic pollution, there are influx of certain pollution suspected to be Sahara dust (Emetere, 2016b). The significance of the sahara dust influx is obvious in the surface plot shown in Figure 5,

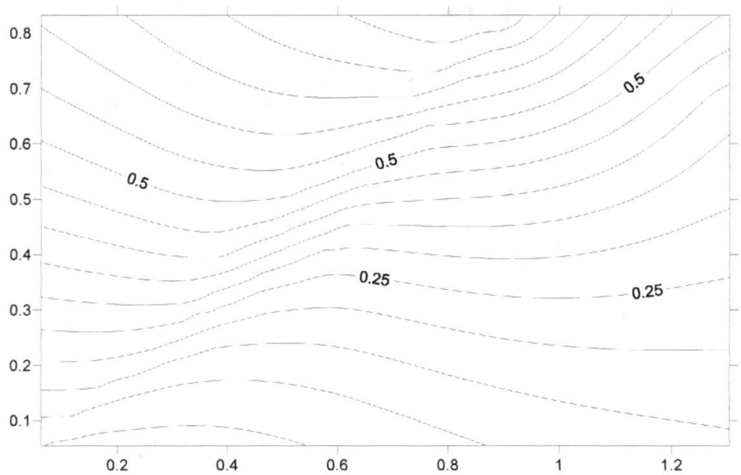

Figure 4: Contour plot of AOD distribution over Voinjama

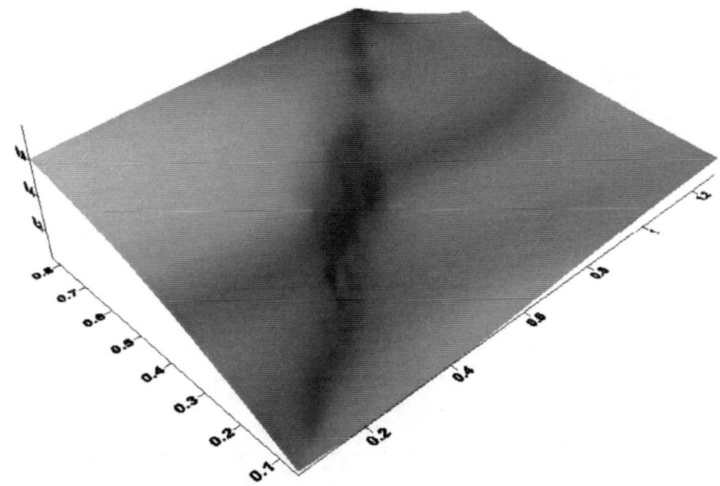

Figure 5: Surface plot of AOD distribution over Voinjama

The 3D plot of the interdependency of the spectra band is presented in Figure 6. This diagram is very important as it graphical shows that the satellite AOD dataset is quite reliable for derivative calculation such as aerosol loading. The reliability of the dataset was further tested using statistical treatment.

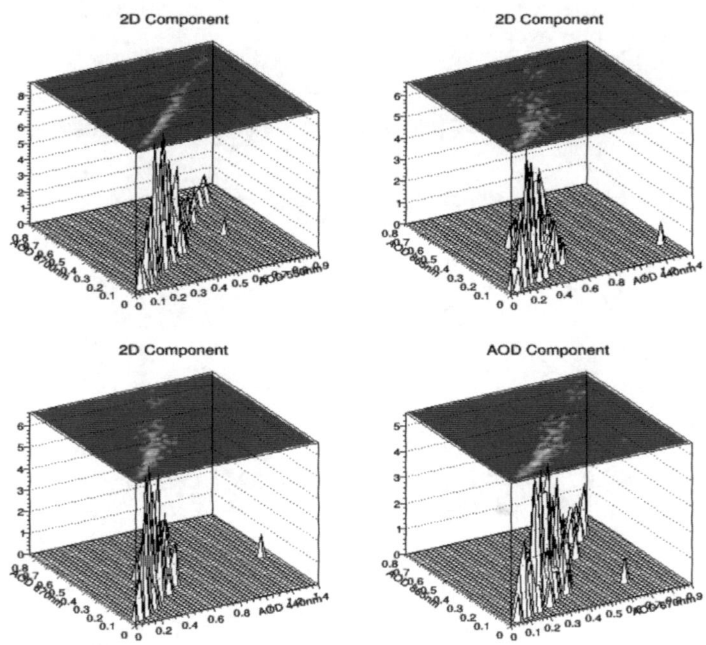

Figure 6: 3D dataset reliability test

The summation of the univariate statistics, inter-variable correlation, inter-variable covariance, planar regression, inter-parameter correlations and ANOVA show that the AOD of Yekepa has almost exceeded the limits described by World Health Organization (10).

IOP Conf. Series: Earth and Environmental Science **331** (2019) 012009 doi:10.1088/1755-1315/331/1/012009

Table 2: Univariate Statistics over Voinjama

	X	Y	Z
Minimum:	0.062	0.055	0.05
25%-tile:	0.297	0.223	0.173
Median:	0.382	0.315	0.262
75%-tile:	0.521	0.429	0.367
Maximum:	1.304	0.832	0.793
Midrange:	0.683	0.4435	0.4215
Range:	1.242	0.777	0.743
Interquartile Range:	0.224	0.206	0.194
Median Abs. Deviation:	0.117	0.1	0.093
Mean:	0.41468862275449	0.33985628742515	0.29180239520958
Trim Mean (10%):	0.40487417218543	0.33090728476821	0.28128476821192
Standard Deviation:	0.18694005905354	0.16098031550915	0.15353999418595
Variance:	0.034946585678942	0.025914661981426	0.023574529814622
Coef. of Variation:			0.52617797765394
Coef. of Skewness:			1.0165717081232

Table 3: Inter-Variable Correlation over Voinjama

	X	Y	Z
X:	1.000	0.961	0.897
Y:		1.000	0.984
Z:			1.000

Table 4: Inter-Variable Covariance over Voinjama

	X	Y	Z
X:	0.034946585678942	0.028909805550575	0.025745243859586
Y:		0.025914661981426	0.024323055434042
Z:			0.023574529814622

Table 5: Planar Regression: Z = AX+BY+C over Voinjama

Fitted Parameters

	A	B	C
Parameter Value:	-0.51531037174422	1.5134512696895	-0.0088601861562642
Standard Error:	0.0081776226650521	0.009496349035516	0.0010335305726505

Table 6: Inter-Parameter Correlations over Voinjama

	A	B	C
A:	1.000	0.961	-0.281
B:		1.000	-0.029
C:			1.000

Table 7: ANOVA Table over Voinjama

Source	df	Sum of Squares	Mean Square	F
Regression:	2	3.93200964689	1.966004823445	65310
Residual:	164	0.0049368321519374	3.0102635072789E-005	
Total:	166	3.9369464790419		

Coefficient of Multiple Determination (R^2): 0.99874602507852

Using equation (1) and (3), the corrosion rate was estimated from the aerosol loading over Voinjama as presented in Figure 7.

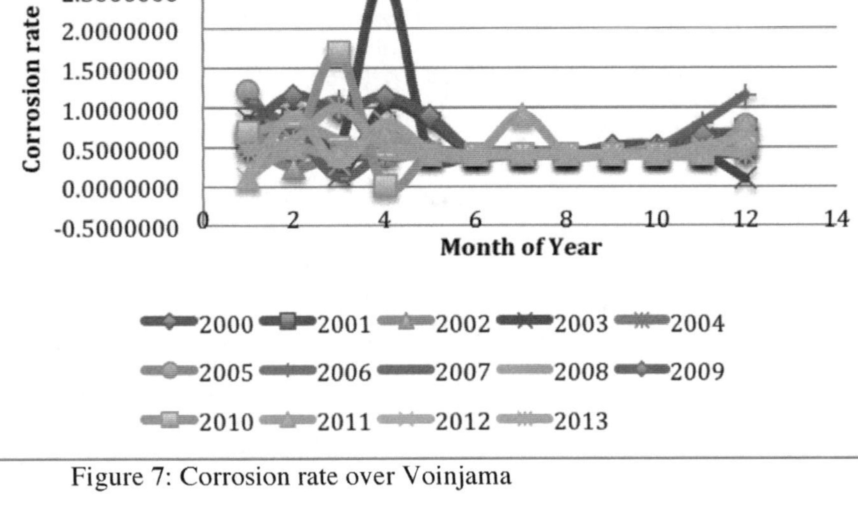

Figure 7: Corrosion rate over Voinjama

IOP Conf. Series: Earth and Environmental Science **331** (2019) 012009 doi:10.1088/1755-1315/331/1/012009

It is observed that the corrosion rate is highest between December and April. More so, the highest corrosion rate is given as 2.6. It is observed that it is only in two cases the corrosion rate in Voinjama exceeded unity i.e. 1. The average corrosion rate over Voinjama is 0.5. Hence, it can be inferred that the corrosion rate is directly proportional to the aerosol loading over Voinjama.

4. Conclusion

From the dataset reliability test, it was observed that the AOD satellite dataset has significant correlation and an R-squared calculate the aerosol loading and corrosion rate over Voinjama. The highest corrosion rate is given as 2.6 and the average corrosion rate over Voinjama is 0.5. This means that the atmospheric corrosion over Voinjama is significantly high. This portends higher cost of replacing corroded metallic surfaces in Voinjama.

Acknowledgement

The authors appreciate Covenant University for partial sponsorship. The authors acknowledge NASA for primary dataset.

5. References

[1]. WHO 2018a WHO's First Global Conference on Air Pollution and Health, http://www.who.int/airpollution/events/conference/WHO_Conference_on_Air_Pollution_and_Health-Concept_Note-final1.pdf (accessed 26 February, 2018).
[2]. Emetere, Moses Eterigho 2016 Statistical Examination of the Aerosols Loading Over Mubi-Nigeria: The Satellite Observation Analysis, Geographica Panonica, 20(1) 42-50
[3]. Galvinfo 2018 Atmospheric Corrosion, http://www.galvinfo.com:8080/zclp/Definition.htm (accessed on 25 February, 2018).
[4]. Gerhardus H. Koch, Michiel P.H. Brongers, and Neil G. Thompson 2018 Corrosion costs and preventive strategies in the United States. https://www.nace.org/uploadedFiles/Publications/ccsupp.pdf (accessed 27 February, 2019)
[5]. McGill, M., CATS/ISS L2O D-M7.1 05kmPro HDF-5 File - Version 1.05 [Data set]. NASA Langley Research Center Atmospheric Science Data Center DAAC. 2016 https://doi.org/10.5067/iss/cats/l2o_d-m7.1-v1-05_05kmpro
[6]. Emetere M.E. 2016 Generation Of Atmospheric Constants Over Some Locations In West Africa: A Theoretical Aid for Measuring Instruments Design, International Journal of Engineering Research in Africa 27 119 – 146
[7]. Moses Eterigho Emetere 2017 Investigations on aerosols transport over micro- and macro-scale settings of West Africa, Environ. Eng. Res., 22(1) 75-86
[8]. Emetere M.E. 2016 Numerical modelling of West Africa regional scale aerosol dispersion. A doctoral thesis submitted to Covenant University, Nigeria. 65-289
[9]. ASTM. 2003. Standard practice for preparing, cleaning, and evaluating corrosion test specimens. ASTM G-1-03. West Conshohocken, PA: American Society for Testing and Materials.
[10].WHO 2018b. Air pollution, http://www.who.int/airpollution/en/ (accessed 5th July, 2018)

IOP Conf. Series: Earth and Environmental Science **331** (2019) 012010 doi:10.1088/1755-1315/331/1/012010

Household waste burning techniques and correlated respiratory indicators in Northern Nigeria: Case study of Gwagwalada Area Council, Abuja Nigeria.

Igibah, Ehizemhen. C; Agashua, Lucia. O; Sadiq, Abubakar. A

Department of Civil Engineering, university of Abuja, F.C.T, Nigeria.
*corresponding author: igibahchrist1@gmail.com; mobile: +234 7052275257

Abstract. The household waste burning practice and impacts of the waste burning emission on respiratory health of the populace in the Gwagwalada Local Council Area happens frequently, though not well recorded. The goal of this study was to quantify the waste burning occurrence as well as scrutinize the correlation between waste burning practice and respiratory problem in Northern Nigeria with Gwagwalada as case study because is an example of rural settlement growing to urbanites. Questionnaire was utilized to accumulate information from houses in Gwagwalada Area Council Community, Abuja. Total of 150 participants were enrolled and their retorts on household waste burning besides respiratory symptoms were investigated. The participants indicated that their neighbors burn waste frequently, with 67 (44.7%) of them from Gwagwalada Local government (rural zone). The local waste experts collected refuse from 97.0% of these participants at least once per week in the enlightened zone and central but rural is once monthly or not at all. The prevalent lower respiratory sign connected with waste burning was dehydrated cough (31.4%). For Lower and Upper health signs, the participants who participated in the practice of waste burning had a faintly higher commonness of breath shortness (56%) and Water itchy eyes (49%) respectively. For each of lower and upper health signs, the trend is such that 6 month above is greater than immediately, which is greater than those of a month after. These results imply that waste or bush burning effect on health is on the long – term basis.

1. Introduction

The practice of both bush and waste burning serve as supportive part to the problem of open-air effluence in the environment whereas, smoke particles released from this process encompass tiny particles with aerodynamic distances which are mostly significantly less than 2.5 μm [1]. The chemical components of this particular matter are also varied which have been assessed to contain elemental carbon like water-soluble organic carbon (WSOC), polycyclic aromatic hydrocarbons (PAHs) and oxalate, potassium (K), Zinc (Zn), Chlorine (Cl) and so on [2]. Plant-based material has long been assumed through combustion or other processes serve as cause of fine particle organics in the air [3-4]. A host of gases such as CO2, CH4, NMHC (nonmethane hydrocarbon), CO and are associated with waste burning, as well as explosive and semivolatile organic compounds [5]. The fine particles have been revealed to infiltrate deep into the lower respiratory system (LRS) through the lungs besides threating health of the Family. Waste or bush burning method denotes household burning of garbage by inhabitants at their own property which is a very common throwing away technique utilized in various developing nations [6]. This rubbish can encompass plastics, paper, cardboard, yard trimmings, and several other resources [7-8]. In the combustion system emissions are spontaneously released straight into the air without filtration or treatment so as to confiscate particulate matter, besides the emissions are extremely variable, subject to the situations under which burning occurs [9] . U.S Environmental Protection Agency [6, 10} ascribe that waste burning is detrimental to both human wellbeing and the environment, emissions from this practice can intensify respiratory disease like asthma besides increase the risk of heart sickness. Lemieux et al., [11], reveal that the method could also generate a group of extremely toxic chemicals recognized as dioxins which might then rest on crops or deposit in seaways and disturb the well-being of those who would consume adulterated crops or water. This research will cover waste burning and health impacts study in the Gwagwalada Area council community, Abuja. The research restrictions will embrace the scheme elements of the unified waste burning system and health impact through using unified waste burning(UWB) as an assessment device.

2. Material and Method

Study Area

In this study Gwagwalada town in Abuja was chosen as an example of rural settlement growing to cosmopolitans. Gwagwalada was being ruled by the Kwali Province of the prior Abuja emirate presently Suleja emirate before the creation of Federal Capital Territory. Gwagwalada Area Council was established on 15th October 1984 with the official population number of 158618, inhabitants at the 2006 census. This Area council is one among the five (5) Local Government Area (LGAs) Councils of Federal Capital Territory (FCT) of Nigeria, others are Kuje, Abaji, kwali, Bwari and Abuja city itself. Due to Government seat relocation from Lagos to Abuja in 1992 besides latest demolition of unauthorized buildings in the Federal Metropolitan Center brought massive invasion of people into the Area Council. Being among the rapidest growing urbanized centers in the FCT, the populace to over three million people (3,000,000), the locations are shown in Fig. 1.

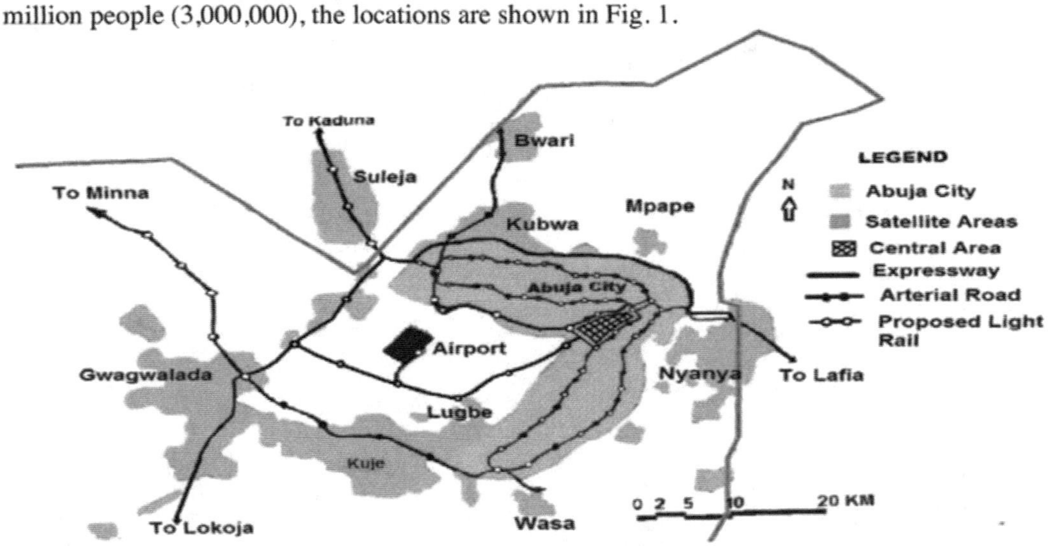

Figure 1: Gwagwalada maps

Research Methods and Sampling approaches

In this study, expert personnel collected information by means of a standardized questionnaire from houses in the Gwagwalada Local government community, Northern Nigeria. The sample size calculation comprised of an extra fifteen percent reserve population to address any likely sample population loss. WHO endorsed Kish style was utilized for the family participant selection (Kish, 1995). Health related questions respiration were created based on prevailing validated standard surveys by the American Thoracic Society, Respiratory Diseases Survey [12]. The questionnaire was pretested by the means of a pilot study in a diverse public regions in Northern Nigeria. The questionnaire comprised of enquiries on demographics, waste burning method, physician-diagnosed sicknesses and environmental fitness-related history, was given to participants at their home.

3. Results and Discussions

Demographic Analysis

Sex and Occupation

The demographic investigation (Gender and Profession) formed using the three zones Questionnaire that were dispensed to fifty (50) households selected randomly making the total of 150 families are presented in Figure 3.

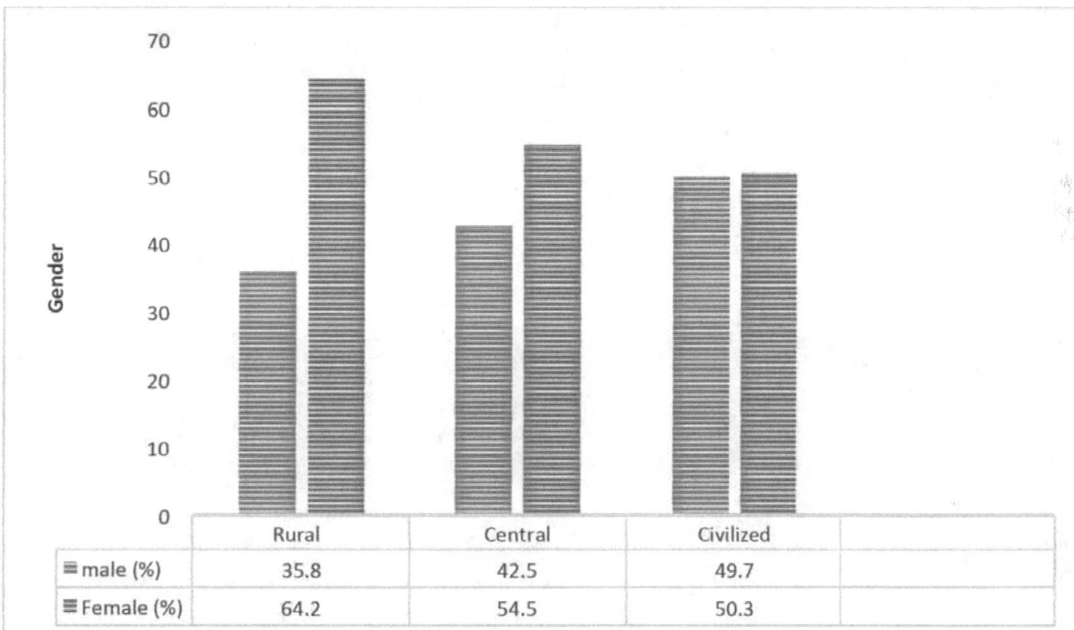

Figure 2: Gender respondent in Gwagwalada, Abuja.

B Age Group

The Age group formed using the three zones Questionnaire that were dispensed to fifty (50) households selected randomly making the total of 150 families are presented in Figure 4.

IOP Conf. Series: Earth and Environmental Science **331** (2019) 012010 doi:10.1088/1755-1315/331/1/012010

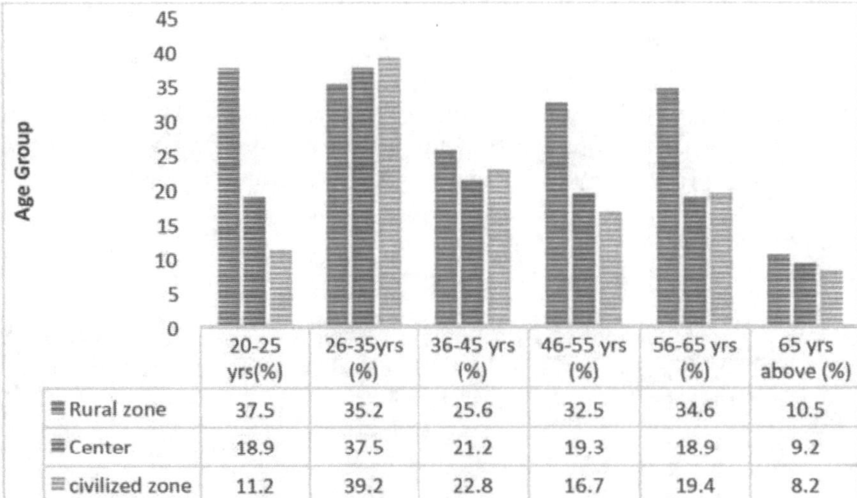

	20-25 yrs(%)	26-35yrs (%)	36-45 yrs (%)	46-55 yrs (%)	56-65 yrs (%)	65 yrs above (%)
Rural zone	37.5	35.2	25.6	32.5	34.6	10.5
Center	18.9	37.5	21.2	19.3	18.9	9.2
civilized zone	11.2	39.2	22.8	16.7	19.4	8.2

Figure 3: Age group for Gwagwalada, Abuja Resident

Figure 2 – 3 shows the demographic figure (gender and age) for Gwagwalada community, total of 150 participants responded to the questionnaire given with a partaking rate of 98.1%. Contributors were mostly female (50.3 – 64.2%) while men are (35.8 – 49.7%). Age group (26-35 years) has the highest values whereas the lowest fall within 65 years above.

Bush Burning

The Bush burning Frequency formed using the three zones Questionnaire that were dispensed to fifty (50) households selected randomly making the total of 150 families are presented in Figure 4.

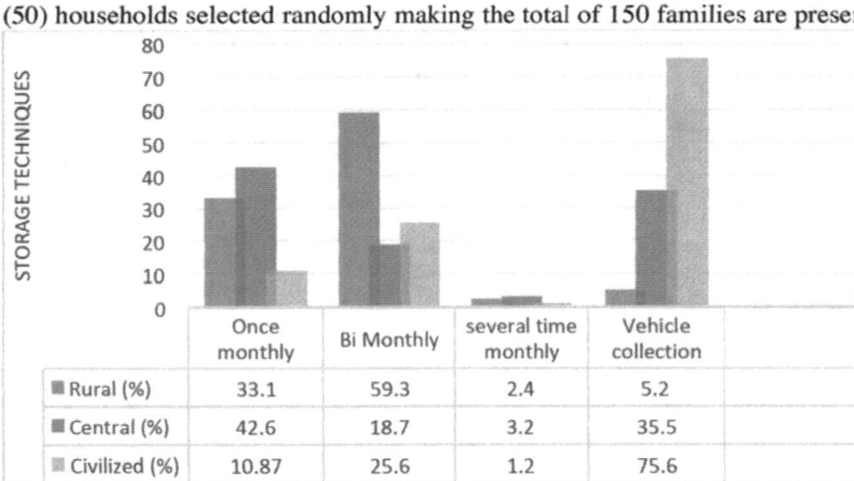

	Once monthly	Bi Monthly	several time monthly	Vehicle collection	
Rural (%)	33.1	59.3	2.4	5.2	
Central (%)	42.6	18.7	3.2	35.5	
Civilized (%)	10.87	25.6	1.2	75.6	

Figure 4: Burning Frequency by Residents of Gwagwalada Abuja

Figure 4 demonstrate that the bush burning practice happened in 83% of the households often once and twice monthly. Eighty-two (82.0%) contributors who practice bush burning burned garbage twice monthly and remain populace burned it once a month. The nonrural area of Gwagwalada local government accounted for 50 of 150 households (33.3%) that practice waste burning. One hundred fifty contributors stated that their neighbors burn waste habitually, with 67 (44.7%) of them from the Gwagwalada Local government (rural zone).The local waste experts gathered garbage from 97.0% of these contributors once per week in civilized zone and central but rural is once monthly or not at all.

Health effect of bush burning in Gwagwalada community

Lower airways signs

The lower health signs made using the three zones Questionnaire that was dispensed to fifty (50) families selected randomly making total of 150 people are presented in Figure 5.

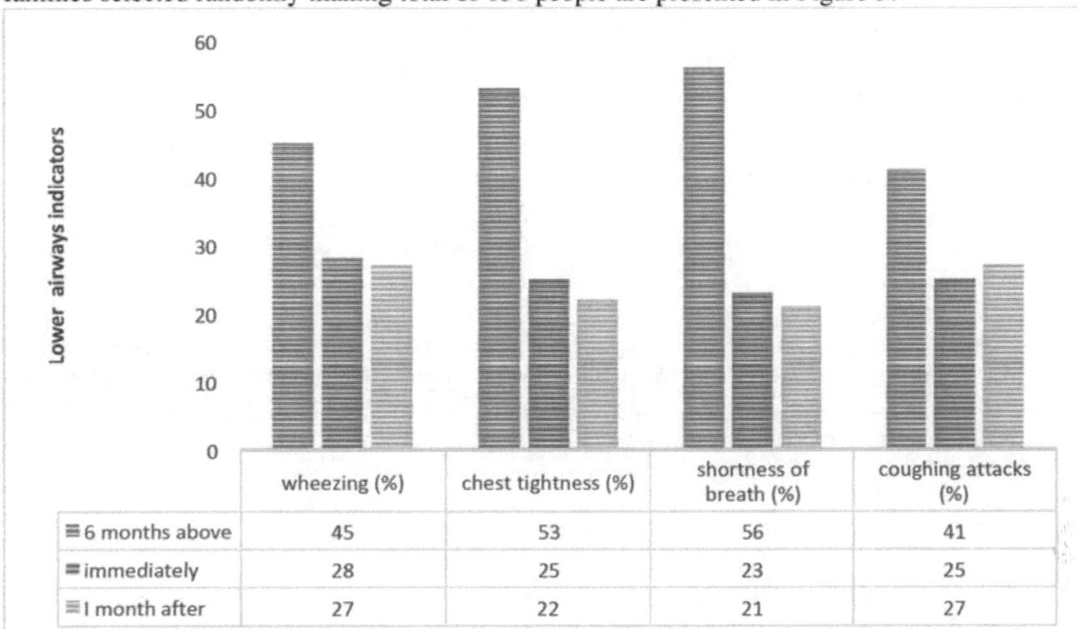

	wheezing (%)	chest tightness (%)	shortness of breath (%)	coughing attacks (%)
6 months above	45	53	56	41
immediately	28	25	23	25
I month after	27	22	21	27

Figure 5: Lower illness indicators of Gwagwalada Abuja Resident

Upper airways indications

The Upper health signs formed using the three zones Questionnaire that were dispensed to fifty (50) households selected randomly making total of 150 families are presented in Figure 6.

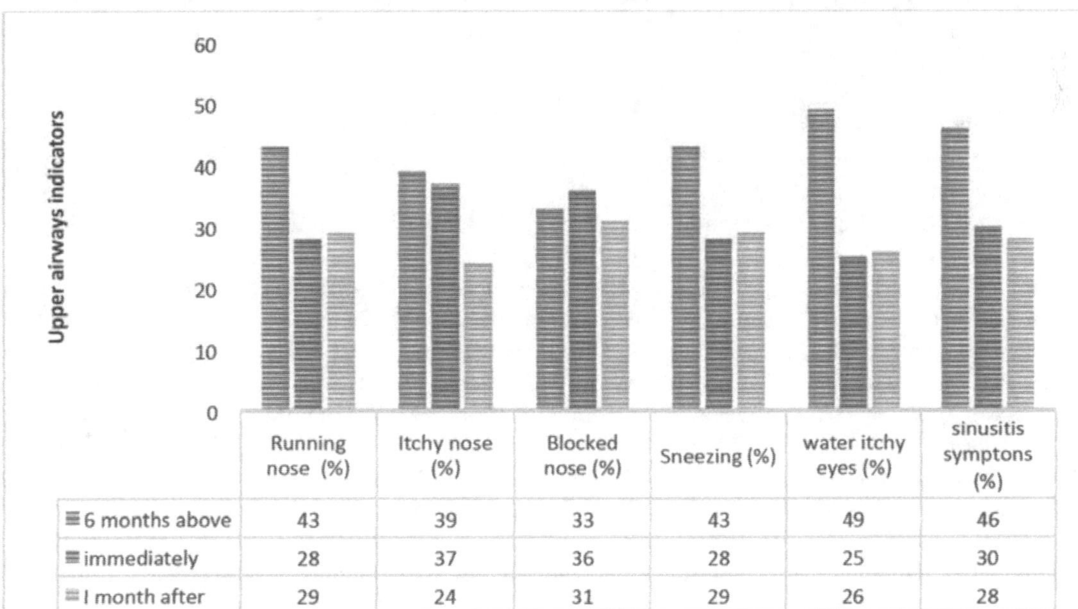

	Running nose (%)	Itchy nose (%)	Blocked nose (%)	Sneezing (%)	water itchy eyes (%)	sinusitis symptons (%)
6 months above	43	39	33	43	49	46
immediately	28	37	36	28	25	30
I month after	29	24	31	29	26	28

Figure 6: Upper illness indicators of Gwagwalada Abuja Resident

IOP Conf. Series: Earth and Environmental Science **331** (2019) 012010 doi:10.1088/1755-1315/331/1/012010

Figure 5 and 6 shows the trend for the lower and Upper health signs, the contributors who engaged in the waste burning practice had a faintly higher occurrence of breath shortness (56%) and water itchy eyes (49%) respectively. For each of the lower and upper health signs, the trend is such that 6 month above is greater than immediately, which is greater than those of one month after. These results imply that waste or bush burning effect on health is on the long – term basis.

Conclusion

In this study, total of 150 participants were enrolled and their retorts on household waste burning besides respiratory warning sign were investigated. The contributors who engaged in the waste burning practice had a faintly higher occurrence of cough, water itchy eyes and physician-diagnosed asthma. For lower and upper health signs, the participants who participated in the practice of waste burning had a faintly higher prevalence of breath shortness (56%) and Water itchy eyes (49%) respectively. For each of lower and upper health signs, the trend is such that 6 month above is greater than immediately, which is greater than those of a month after. Furthermore, there is probable disclosure misclassification, specifically those who do not burn waste but are still vulnerable to air pollutants from their neighbors' waste burning. Conclusively, these results imply that waste or bush burning effect on health is on the long – term basis.

References

[1] Akpinar-Elci, M., O. C. Elci, and A. Odabasi. 2011. United airways disease among florists. Chest Dis. Rep. 1(10):22–24. doi:10.4081/cdr.2011.e10.

[2] American Thoracic Society, Committee of the Environmental and Occupational Health. 1996. Health effects of outdoor pollution. Am. J. Respir. Crit. Care Med. 153:3–50.

[3] Arbex, M., L. Martins, R. De Olveira, L. Pereira, F. Arbex, J. Cancado, P. Saldiva, and A. Braga. 2007. Air pollution from biomass burning and asthma hospital admissions in a sugar cane plantation area in Brazil. J. Epidemiol. Commun. Health 61(5):395–400. doi:10.1136/jech.2005.044743.

[4] Estrellan, C., and F. Iino. 2010. Toxic emissions from open burning, Chemosphere 80:193–207. doi:10.1016/j.chemosphere.2010.03.057.

[5] Ferris, B. G. 1978. Epidemiology standardization project (American Thoracic Society). Am. Rev. Respir. Dis. 118:1–120.

[6] Fiedler, H., G. Solorzano Ochoa, G. Yu, T. Zhang, S. Marklund, and L. Lundin. 2010. Hazardous chemicals from open burning of waste in developing countries—Final report. United Nations Environment Programme, Division of Technology, Industry and Economics, Chemicals Branch.

[7] Kish, L. 1995. Survey Sampling. New York, NY: John Wiley & Sons.

[8] Lemieux, P. M., C. C. Lutes, J. A. Abbott, and K. M. Aldous. 2000. Emissions of polychlorinated dibenzo-p-dioxins and polychlorinated dibenzofurans from the open burning of household waste in barrels. Environ. Sci. Technol. 34(3):377–84. doi:10.1021/es990465t.

[9] Muge Akpinar-Elci, Kareem Coomansingh, James Blando & Larissa Mark (2015) Household bush burning practice and related respiratory symptoms in Grenada, the Caribbean, Journal of the Air & Waste Management Association, 65:9, 1148-1152, DOI: 10.1080/10962247.2015.1070773.

[10] Solorzano-Ochoa, G., D. de la Rosa, P. Maiz-Larralde, B. Gullett, D. Tabor, A. Touati, B. Wyrzykowska-Ceradini, H. Fiedler, T. Abel, and W. Carroll. 2012. Open burning of household waste: Effect of experimental condition on combustion quality and emission of PCDD, PCDF and PCB.Chemosphere 87:1003–8. doi:10.1016/j.chemosphere.2011.11.038.

[11] U.S. Environmental Protection Agency. 2013. United State Environmental Protection Agency. Wastes-Non Hazardous Waste-Municipal Solid Waste. http://www.epa.gov/osw/nonhaz/municipal/backyard (accessed June 17, 2013).

[12] U.S. Environmental Protection Agency. 2014. United States Environmental Protection Agency, Technology Transfer Network—Air toxics web site, health effects notebook for hazardous air pollutants. http://www.epa.gov/ ttn/atw/hlthef/hapindex.html (accessed July 7, 2014).

Influence of wastes management and viable improvement in Nigeria: case study of Kubwa, Abuja in Northern Nigeria.

Igibah, Ehizemhen. C; Agashua, Lucia. O; Sadiq, Abubakar. A

Department of Civil Engineering, university of Abuja, F.C.T, Nigeria.
*corresponding author: igibahchrist1@gmail.com; mobile: +234 7052275257

Abstract. The paper investigate waste management and viable improvement in Nigeria with specific reference to Kubwa, F.C.T, Abuja. The region was divided into three districts namely: District A (south-north), District B (west - east) and District C (central zone). Survey research stratagem was implemented with questionnaire as main tools applied for data gathering which were dispensed to twenty households carefully picked randomly making total of 60families that were studied. Illiteracy and income level was discovered has main cause of waste management difficult, equally city waste characterization for the Kubwa – Abuja reveals paper, food leftovers and water sachets as the highest waste created, while delivery bag and flexible bag are consistently utilized to dispose waste since bags are stress-free in disposing and low-priced. The waste storage frequency is between 12 – 24 hours whereas waste throwing frequency follows regularly once and bi-weekly gatherings. Some respondents have open dumps within their locality where waste is stock before thrown away. Nevertheless, there is great recycling event for metal and bottles since they are recycle at home and 15 % sell. Mutual solid waste management practice in Kubwa - Abuja are storage and highway or canal waste discarding whereas recycling and waste reduction which are waste minimization implements in the trio pyramid system which is the strength of most waste administration systems is weakly practiced.

1. Introduction

Solid wastes are the unfunctionable and undesirable goods in the solid form resulting from the society activities [8]. The term public solid discarded (PSD) regularly presumed to contain all discarded created in community with the eradication of manufacturing process and agricultural waste, even though main PSD sources encompasses of housing, construction, institutional and commercial [1,10]. Viable development have three features: social development, financial development and protection of the environmental [11]. Viable development guarantees an advanced world with safety and healthy environment for plants, animals as well human beings [2]. Main purpose of solid waste management is to guarantee public health protection alongside environmental quality and viable development thus, national and local authorities must implement viable solid waste administration systems in strong partnership with both the public and private sector. The structure and features of metropolitan solid waste is disposed thru certain matters, which comprises of the housing, commercial, the income level, etcetera [3]. Poor waste disposal practice of the populaces, weak government by-law, bad work approach, corruption, fund deficient, insufficient facilities for instance plants and equipment amongst others are factors stimulating against operative waste management towards viable development in developing nations. High- income districts usually produce more inorganic materials for instance plastics and paper whereas low - salary zones generate diminutive organic unwanted that were unsuitably located besides open dumpsites which introduce health exposures and mutilation of the aesthetic magnificence of many cities in Nigeria [4-6]. Hence, landfill availability, sufficient resources (Human, capital, plant, equipment and other tools) must be readily accessible for viable development in developing countries because there is need to safeguard future for the next generation through cleaning up of our environment from various kinds of waste and take into consideration physical and population growth of the nation [7,9,12]. The Public solid waste management problem in Nigeria metropolises has fascinating the researchers attention such as [8,9,13-15] and their research outcomes point to the need for allowable and a resourceful solution to solid waste nuisance in Nigerian cities. Additionally, it has been advocated that resourceful recycling and composting can save 15.5%waste management prices and 62.5% landfill costs, likewise applaud recycling approval as a footstep towards stimulating an unified solid waste management method besides, when diverse waste classifications are not considered during collecting and disposal, operative solid waste administration becomes problematical [8,12,15].

This research work will cover solid waste administration in three zones within Kubwa, Abuja Bwari Area Council, Abuja. The research boundaries would embraces scheme features of the unified solid waste administration method through joint solid waste management (JSWM) as a valuation device.

2.0 Material and Method

2.1 Study Area

In this study Kubwa town in Abuja were selected as countryside settlement illustration, growing to borough besides a major suburbia within the cosmopolitan area of Abuja. It is a residential region in Bwari one of the area council in the Federal Capital Territory (F.C.T), Nigeria, existence since 1990 as satellite city of Abuja. It is principally residential and compactly populated and approximately twenty six kilometres from the Wuse market, Abuja. Exact drivers of economic sustainability are ascribed to civil servants which are sixty percent (60%) of the residents who are staff of the Federal Government of Nigeria while the remaining forty percent (40%) is various categories of businesses.Being among the swiftest growing built-up centers in the FCT, the population over three point three million populaces, the locations are presented in Fig. 1.

Fig. 1: Kubwa - Abuja map

2.2 Research Methods and Sampling approaches

The sampling structure comprises of individuals living within case study areas, data was gathered through questionnaire and field analysis. The research equipment making is through JSWM which was exploited to framework the assessment areas survey content with closed housing questionnaire. The survey was circulated using simple random analyst within each status and distributed to people in open places and gardens in the cities. The questionnaire used was from previous studies but it was improved to suite this analysis purpose as well authenticated by three professionals and certified through trial study and reused after improved. Field survey comprises of visits to city and points collection within Kubwa, Abuja.

Figure 2: Open Garbage dump within a Housing Area

Table 1: Questionaire format

Section	Description	Scale Type	No. Of Items
A	Demographic Information	Categorical	12

B	Waste Management Practices	Binary	14

3. Results and Discussions

3.1 Demographic Analysis

A Sex and Occupation

The demographic investigation (Gender and Profession)formed using the three regions Questionnaire that were dispensed to twenty households selected randomly making total of 60 families are presented in Figure 3.

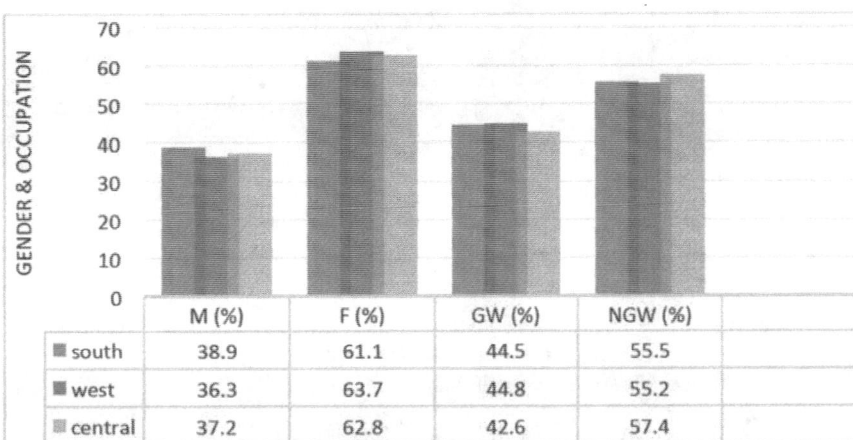

Government workers (GW), Non-government workers (NGW) and Male (M) Female (F)
Figure 3: Gender and Profession of respondent in Kubwa, Abuja.

B Age Group

The Age group formed using the three regions Questionnaire that were dispensed to twenty households selected randomly making total of 60 families are presented in Figure 4.

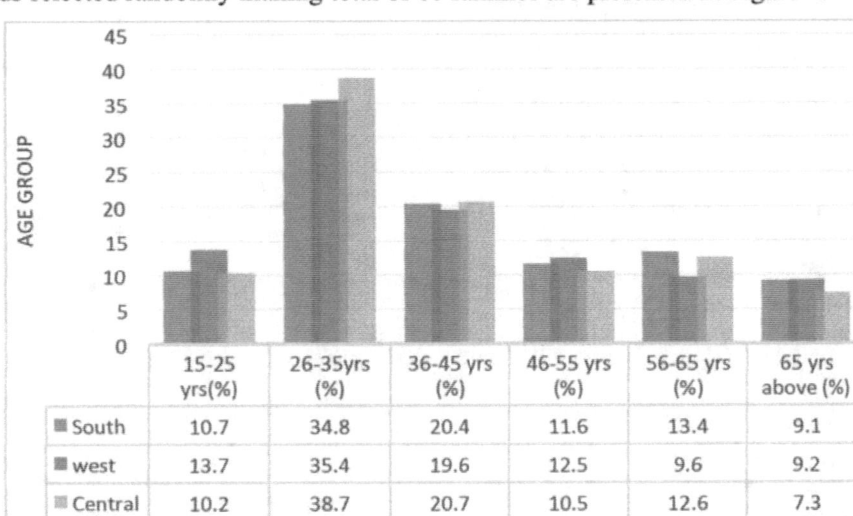

Figure 4: Age group for Kubwa Abuja Resident

IOP Conf. Series: Earth and Environmental Science **331** (2019) 012011 doi:10.1088/1755-1315/331/1/012011

C Education

The Education level formed using the three regions Questionnaire that were dispensed to twenty household's selected randomly making total of 60 families are presented in Figure 5.

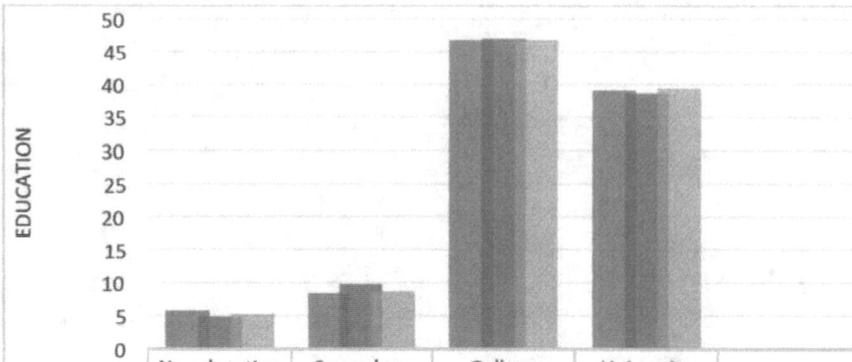

	No education	Secondary	College	University
South	5.7	8.4	46.8	39.1
West	4.8	9.7	46.9	38.6
Central	5.2	8.6	46.8	39.4

No education Qualification (NEQ), Secondary school education (SSE), College Education (CE) and University Education (UE)

Figure 5: Schooling Level of Kubwa Abuja Resident

D Salary

The salary formed using the three regions Questionnaire that were dispensed to twenty households selected randomly making total of 60 families are presented in Figure 6.

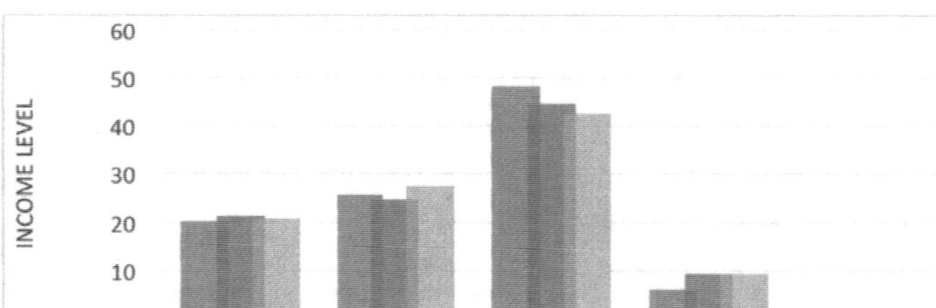

	0-100 USD (%)	101-200 USD (%)	201-500 USD (%)	501 above (%)
South	20.2	25.6	48.3	5.9
west	21.4	24.8	44.5	9.3
Central	20.8	27.4	42.5	9.3

Figure 6: Income level of Kubwa Abuja respondents

E Size of Families

Families sizes formed using the three regions Questionnaire that were dispensed to twenty households selected randomly making total of 60 families are presented in Figure 7.

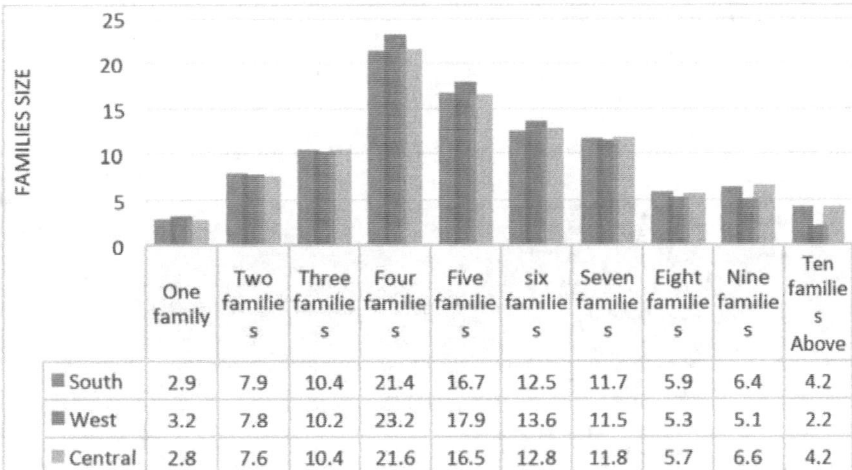

Figure 7: Household numbers of Kubwa Abuja Resident

Figure 1 – 7 shows that demographic figures (gender, profession, age, education, salary and households size) for each of the three zones, female discarded waste more than men, while Non-government employees discarded waste more than government staffs. These results indicate that income and education level of the populace regulate proper waste management.

3.2 Waste Administration

Municipal Waste Characterization

The civic waste characteristics quantities and percentages for the waste generated using the three regions in Kubwa, Abuja presented in Table 2 and graphically illustrated in Figure 8.

Table 2: Civic Waste Characterization for the Kubwa, Abuja

Civic Waste Composition	Quantities (tons)	%
Paper	20567.75	24.03
Textile	2134.65	2.49
Plastics	4678.56	5.47
Water sachets	11234.45	13.12
Glass	2743.43	3.21
Metals	2689.34	3.14
E-waste	2087.24	2.44

Organic materials; food &garden waste	35856.24	41.88
Other organic materials	3616.45	4.22
TOTAL	**85,608.11**	**100.0**

Source: (BAC, 2017)

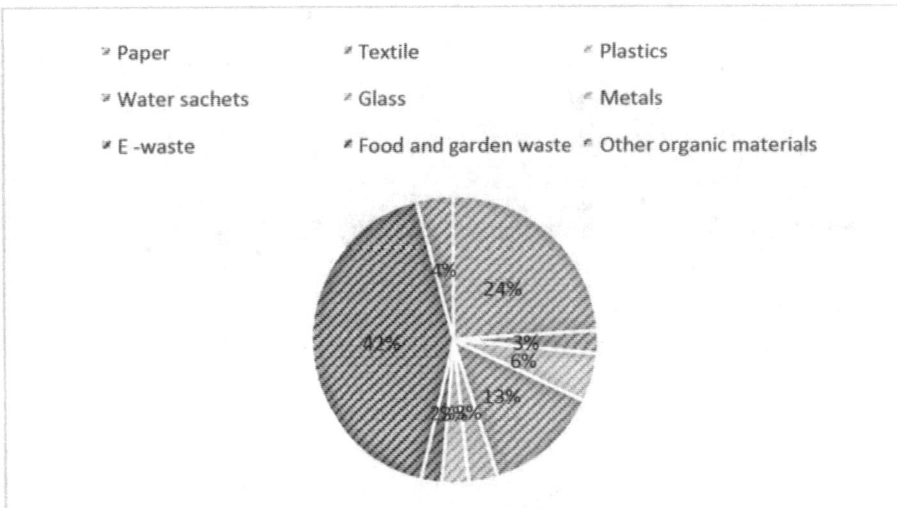

Figure 8: Civic Waste Characterization for the Kubwa-Abuja

Figure 8 reveals the frequency shape based on the group of waste created. It can be absorbed from the pie chart that paper, food waste and water sachets have greatest percentages which is in agreement with paper work by [16].

Waste Storage Techniques

The storage techniques gathered using the three provinces of Questionnaire that were assigned to twenty households selected randomly making total of 60 families are presented in Figure 9.

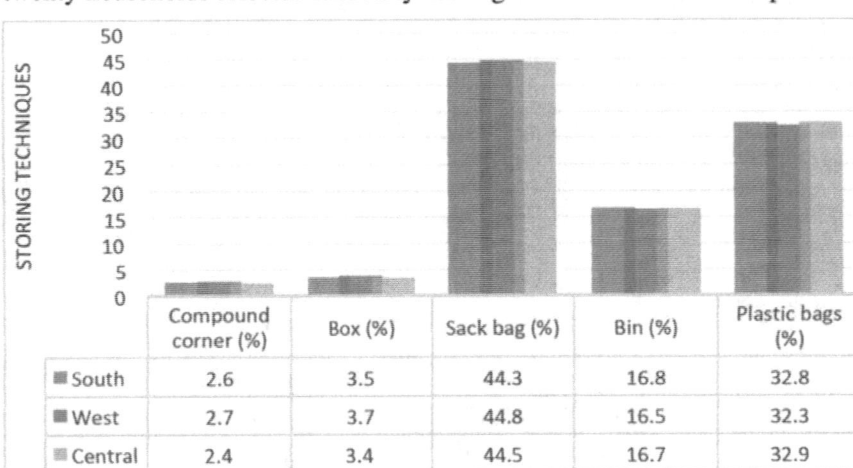

Figure 9: Types of Waste Storage by Residents of Kubwa

IOP Conf. Series: Earth and Environmental Science **331** (2019) 012011 doi:10.1088/1755-1315/331/1/012011

Waste storage style shows that storing by using sack and plastic bag are frequently since bags are easy to dispose and economical.

Storage Frequency

The storage frequency formed using the three regions Questionnaire that were dispensed to twenty households selected randomly making total of 60 families are presented in Figure 10.

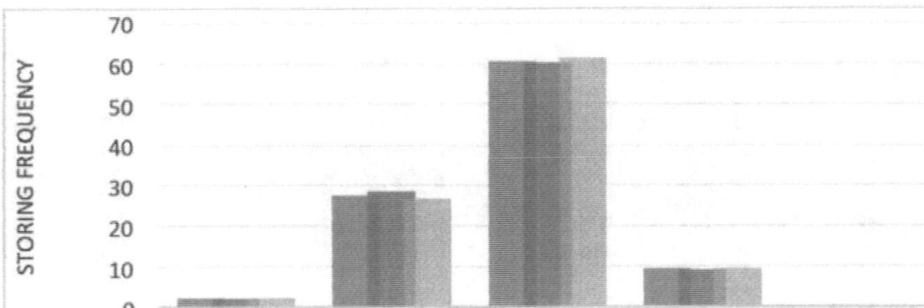

	Immediately (%)	Within 12 hrs (%)	12-24 hrs (%)	24-48 hrs (%)	
South	2.3	27.5	60.6	9.6	
West	2.1	28.4	60.3	9.2	
Central	2.4	26.7	61.4	9.5	

Figure 10: Waste Frequency by Residents of Kubwa Abuja

Figure 10 demonstrations the waste storage frequency style for the three regions in Kubwa Abuja, 12 – 24 hours storage is generally used.

3.3 Waste Disposal method and frequency

The Collecting and Disposal Frequency formed using the three regions Questionnaire that were dispensed to twenty households selected randomly making total of 60 families are presented in Figure 11.

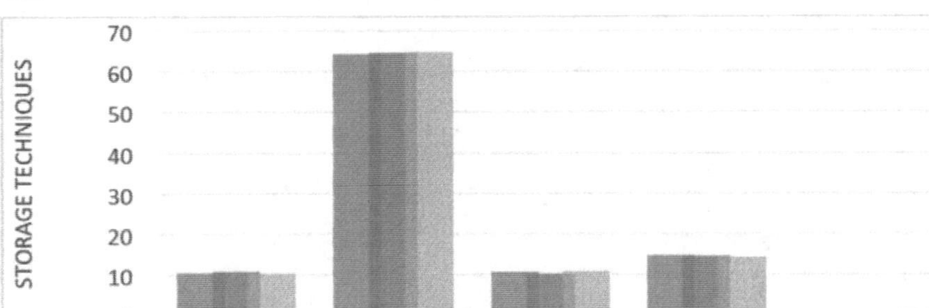

	Daily (%)	Once weekly (%)	several times weekly (%)	Longer interval (%)	
South	10.5	64.2	10.6	14.7	
West	10.8	64.5	10.2	14.5	
Central	10.4	64.6	10.8	14.2	

Figure 11: Waste Frequency by Residents of Kubwa Abuja

The waste disposing frequency style follows regularly once and twice weekly, some respondents have open garbage dump within their locality where waste is stock and thrown away through burning or recycling which is in agreement with paper work by (Ogwueleka, 2013).

3.4 Waste minimization (Frequency of Items Recycling)

The recycling frequency formed using the three regions Questionnaire that were dispensed to twenty households selected randomly making total of 60 families are presented in Figure 12.

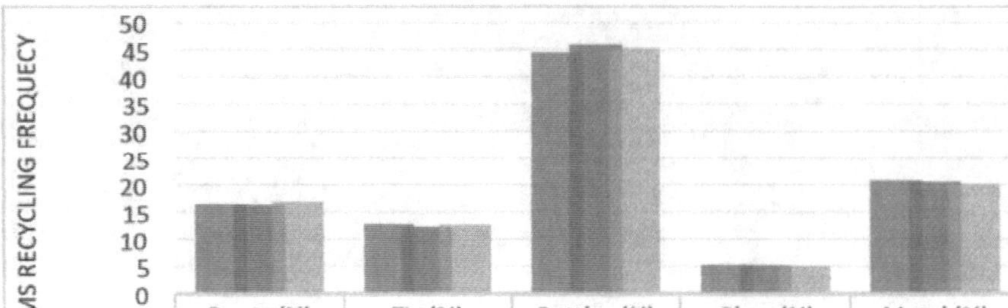

	Paper (%)	Tin (%)	Bottles (%)	Glass (%)	Metal (%)
South	16.5	12.8	44.5	5.4	20.8
West	16.3	12.2	45.8	5.2	20.5
Central	16.8	12.6	45.3	5.1	20.2

Figure 12: Items recycling frequency by Residents of Kubwa Abuja

There is a high recycling frequency for bottles and metal because they are reuse at home and 15 % sell which is in agreement with paper work by [17].

3.5 Solid Waste Management Practices (SWMP)

The solid waste practices among resident formed using the three regions Questionnaire that were dispensed to twenty households selected randomly making total of 60 families are presented in Figure 13.

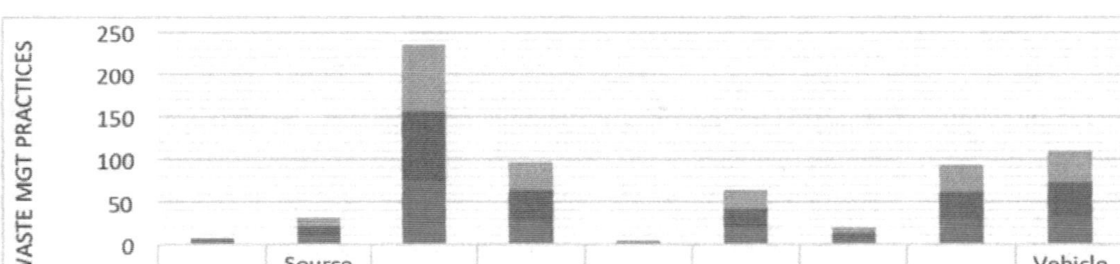

	Reduction (%)	Source separation (%)	Storage (%)	Recycling (%)	Composting	Burning (%)	Burying (%)	Dumping (%)	Vehicle Collection (%)
Central	2.1	10.2	78.6	31.7	1.2	21.6	6.5	30.5	35.7
West	2.7	10.5	80.6	34.7	1.5	21.1	6.2	31.2	38.5
South	2.5	10.8	75.6	29.7	1.2	21.5	6.8	30.8	34.8

Fig.13: Solid waste practices for Kubwa Abuja

IOP Conf. Series: Earth and Environmental Science **331** (2019) 012011 doi:10.1088/1755-1315/331/1/012011

The trend demonstrations high frequency of storage and waste dumping either highway side or canal as waste management normally practice in Kubwa, Abuja. Recycling and reduction which are waste minimization performs in the trio hierarchy structure, which also form waste administration backbone is faintly practiced which is in agreement with research work made by [19].

Conclusion

This study examined the effect of wastes management and viable improvement in kubwa, Abuja Northern Nigeria. Demographic figures of the three regions shows illiteracy and income level, has cause of waste management problems likewise metropolitan waste characterization for the Kubwa Abuja reveals paper, food waste and water sachets as the highest waste created, meanwhile sack and plastic bag are frequently used to throw away waste since bags are easy to dispose besides cost-effective. The waste storage frequency is between 12 – 24 hours whereas waste discarding frequency follows classically once and twice weekly gatherings. Some respondents have open dumps within their locality where waste is stock and discarded. Nevertheless, there is great recycling rate for bottles and metal because they are reused at home and 15 % sell. Public solid waste administration practice in Kubwa-Abuja are storage and waste disposal on highway or canal while recycling and reduction which are waste minimization implements in the trio hierarchy structure that also form backbone of most waste administration methods is slightly practiced.

For success to be achieve in this subject matter, constant and periodic environmental awareness programmes by means of miscellaneous communication mediums is obligatory so as to change the resident's waste management approach. Waste reduction need to be earnestly addressed in Kubwa- Abuja and stratagems need to be initiated to intensificate the recycling rate, waste minimizing and composting. Furthermore, Indiscriminate littering necessitates an upgrading collection points to miniscule transfer stations, intensifying municipal bin number in closely populous districts with daily evacuations during office hours to dodge traffic jams by haulage truck. Conclusively, institutional body must incorporate decrees and policies about environmental matters, perceptibly stated guideline for workforce and populaces to follow as well instigated, or formulated and enhanced for each portion of waste administration processes besides unified the institutions answerable for waste administration under singular body with the applicable tool put in place to expedite its appropriate function.

IOP Conf. Series: Earth and Environmental Science **331** (2019) 012011 doi:10.1088/1755-1315/331/1/012011

References

[1] Anantanatorn, A.; Yossomsakdi, S.; Wijaya, A.F.; Rochma, S. (2015). Public Service Management in Local Government, Thailand (Case Study of Solid Waste Management in Pattaya City). Int. J. Appl. Soc. 5, 5–15.

[2] Beatrice A and Kantola J (2013) Municipal Solid Waste Management Problems in Nigeria: Evolving Knowledge Management Solution, World Academy of Science, Engineering and Technology International Journal of Environmental and Ecological Engineering 7 (6), pp 303-308.

[3] Bwari Area Council (BAC) 2007. Field data.

[4] Eheliyagoda, D. (2016) SWOT analysis of urban waste management: A case study of Balangoda Suburb. J. Glob..Ecol. Environ. 5, 73–82.

[5] Ferronato, N.; Torretta, V.; Ragazzi, M.; Rada, E.C. (2017). Waste mismanagement in developing countries: A casestudy of environmental contamination. UPB Sci. Bull. 79, 185–196.

[6] Guerrero, L.A.; Maas, G.; Hogland,W. (2013) Solid waste management challenges for cities in developing countries.Waste Manag. 33, 220–232. [CrossRef] [PubMed].

[7] Hashemi, H.; Pourzamani, H.; Rahmani Samani, B. (2014). Comprehensive planning for classification and disposalof solid waste at the industrial parks regarding health and environmental impacts. J. Environ. Public Health.. 230163. [CrossRef] [PubMed].

[8] Kadafa A (2017). Solid Waste Management Practice of Residents in Abuja Municipalities (Nigeria). IOSR Journal of Environmental Science, Toxicology and Food Technology (IOSR-JESTFT). 11 (2), pp 87-106.

[9] Nabegu, A. B. (2012). Using Pearson product moment correlation to explore the relationship between different categories of Municipal solid waste in Kano metropolis, North-western Nigeria; Journal of Environment and earth Sciences, 2 (4).

[10] Nachalida Y, Beverley C and Kirstin R. Barriers to Effective Municipal Solid Waste Management in a Rapidly Urbanizing Area in Thailand. International Journal of Environmental Research and Public Health, 2017, 14 (1013), 1-23.

[11] Ndubuisi-Okolo P, Anekwe R, Attah E. (2016). Waste Management and Sustainable Development in Nigeria: A Study of Anambra State Waste Management Agency. European Journal of Business and Management. Vol.8, No.17

[12] Ogwueleka, T.C. (2013). Survey of Household Waste Composition and Quantities in Abuja, Nigeria. Resources,Conservation and Recycling, 77:52– 60.

[13] Rada E, Ragazzi M, Ionescu G, Merler G, Moedinger F, Raboni M, Torretta V. (2014). Municipal SolidWasteTreatment by Integrated Solutions: Energy and Environmental Balances. Energy Proced. 50, 1037–1044.[CrossRef].

[14] Rigamonti, L.; Sterpi, I.; Grosso, M. (2016). Integrated municipal waste management systems: An indicator to assesstheir environmental and economic sustainability. Ecol. Indic. 60, 1–7. [CrossRef].

[15] Siriratpiriya, O. (2014). Municipal Solid Waste Management in Thailand: Challenges and Strategic Solution.In Municipal Solid Waste Management in Asia and Pacific Island; Pariatamby, A., Tanaka, M., Eds.; Springer:Singapore, pp. 337–354.

[16] Topic M, Biedermann H. (2015). Planning of integrated/sustainable solid waste management (ISWM)—Model ofintegrated solid waste management in Republika Srpska/B&H. Serb. J. Manag. 10, 255–267.

[17] Yukalang, N.; Clarke Beverley, D.; Ross Kirstin, E. (2017). Solid waste management in Thailand: An overview andcase study (Tha Khon Yang sub-district). Rev. Environ. Health. [CrossRef] [PubMed].

[18] Zaman, A.U. Measuring waste management performance using the 'ZeroWaste Index': The case of Adelaide,Australia. J. Clean. Prod. **2014**, 66, 407–419. [CrossRef].

[19] Ziraba, A.K.; Haregu, T.N.; Mberu, B. A review and framework for understanding the potential impact ofpoor solid waste management on health in developing countries. Arch. Public Health **2016**, 74. [CrossRef][PubMed]

An Assessment of Energy Performance Index of the Nigerian Universities' Senate Building

Yahuza Abdulhamid[1], Aisha Suleiman Sani[2], Mudashir Gafar[3], Usman A.M.[1]
[1]Faculty of Engineering and Environmental Design; Usmanu Danfodiyo University, Sokoto
[2]Department of Architecture, Kano State Polytechnic, Kano
[3]Faculty of Environmental Design; Ahmadu Bello University Zaria, Nigeria
*corresponding author: yahuzatahirabdul@gmail.com

Abstract. Apparently, Nigeria is struggling with the developmental issues of the incessant power outage and lack of energy services. The primary sources of energy for office building are from national grid and gasoline generators. However, the former is not reliable and the latter is unsustainable; hence, the need to explore energy efficiency strategies to cut down the energy demand for cooling office building in Nigeria is unavoidable. The article aimed to assess the energy efficiency or performance index of the Nigerian Universities' senate buildings in the north-eastern region (dry sub-humid climate). The research employed a case study approach, construction of baseline model with the Revit architecture software and applied DesignBuilder building analysis software to evaluate the energy performance of the senate building as built. The research findings showed that energy efficiency indexes for both universities' senate buildings (57% and 63%) which were relatively below the recommended standard checklist. Also, the simulation results further showed that both case studies require 141.4wh/m2 and 127.3wh/m2 heating/cooling per annum compared to the stipulated LEED (Leadership in Energy and Environmental Design) standard of 15wh/m2. Therefore, research implication is that it is essential for the architects and engineers to incorporate more proactive energy efficiency strategies in the design of office buildings in Nigeria

IOP Conf. Series: Earth and Environmental Science **331** (2019) 012012 doi:10.1088/1755-1315/331/1/012012

1. Introduction

University Senate Building is a structure built to serve the central administrative function in university campuses, design to accommodate offices, exhibition hall, senate chamber, governing council chamber and other necessary facilities. The university senate is the agency for the articulation and representation of the views of the faculty [1].
Some buildings are said to consume more energy than others; office buildings consume a great deal of energy. Sadrzadehrafiei, Sopian, Mat, Lim, Hashim & Zaharim [2] asserted that great deal of energy is consumed by office buildings in Nigeria.
Therefore, reduction in the quantity of energy consumed is inevitable; reduction of energy consumption in buildings can be achieved by simple methods and techniques using an appropriate energy-efficient system such as passive cooling system [3].
On the other hand, if the existing institutional structures in Nigerian universities were designed with least focus on energy efficiency in the past, built environment professionals especially practicing architects, engineers and students can still change the direction of the practice to a more efficient way of energy conservation. However, despite the numerous benefits of passive cooling strategies practicing Architects still linger between theory and practical application. Thus, the article aimed to assess the energy efficiency or performance index of the Nigerian Universities' senate buildings in the north-eastern region (dry sub-humid climate).

2. Literature Review

2.1 Sources of energy supply in relation to energy efficiency in Nigeria

The energy crisis, which has engulfed Nigeria for almost two decades, has been enormous. Ref [4] explained the quantity of electricity generated, transmitted and distributed in Nigeria as summarized in Figure 2.1.

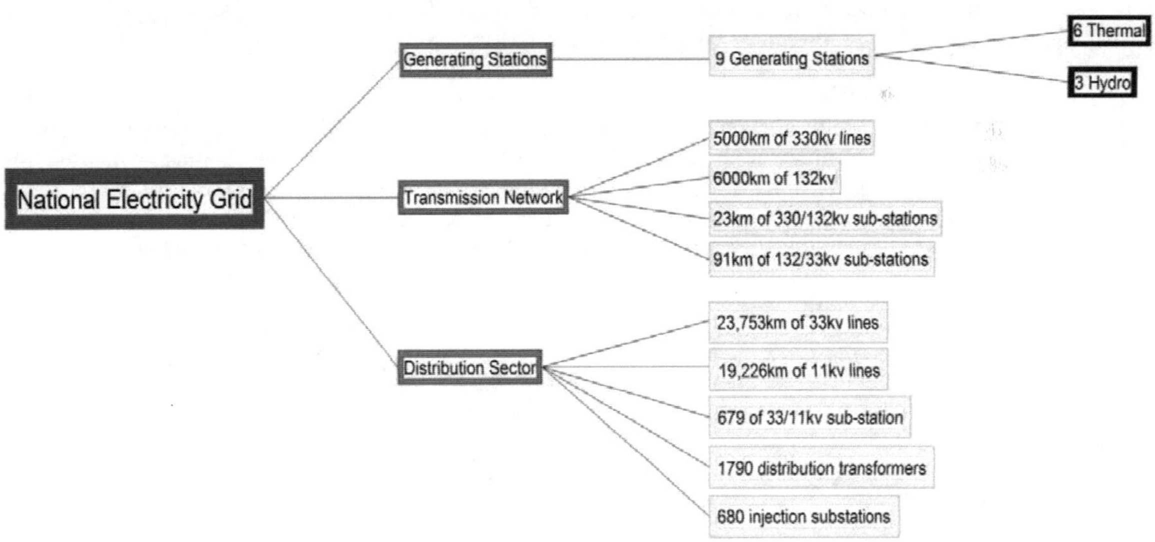

Figure 1 The structure of Nigerian National Electricity Grid [4]

The megawatts are certainly too low, to be dependent upon, it is in this regard that this research proposes energy efficient model for office building.

The triangle for low energy building design is shown in Figure 2 Energy savings in Nigeria will lead to personal income saving; reducing the building of more power stations allowing the funds to be spent on other sectors of the economy; improve access to energy; and decrease load shedding.

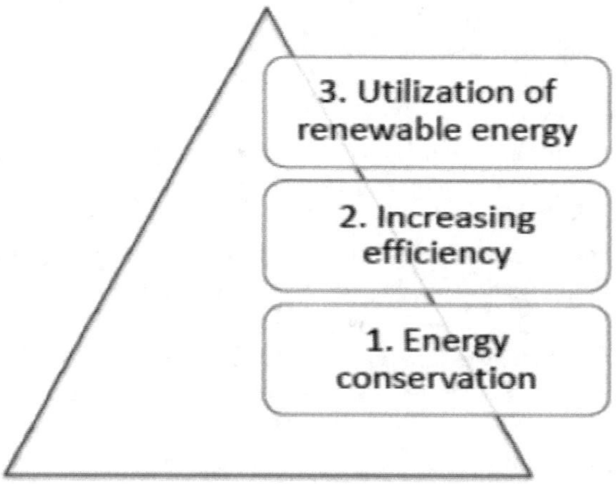

Figure 2 Triangle for low energy building design [5]

Buildings contribute more than one third of total global greenhouse gas emissions and consume more than 50% of energy in Nigeria; therefore, Architects should explore and apply the existing research findings that optimize passive cooling.

2.2 Contemporary Methods of Passive Cooling

Recent review papers [6] which collated individual studies from a broader spectrum of climatic regions agree that cooling strategies for buildings should be designed at three levels such as solar and heat control, heat modulation and heat dissipation using naturally driven phenomena

2.2.1 Solar and heat protection techniques

The most effective method to save energy and cool the building in is to keep the heat from building up in the first place [6]. Solar and heat can be avoided through two techniques namely: microclimate of an urban area can be modified by appropriate landscaping techniques, with the use of vegetation and water surfaces; and shading Among all other solar passive cooling techniques solar shading is the most relevant to thermal cooling of buildings especially in a developing country owing to their cost effectiveness and easy to implement [7].

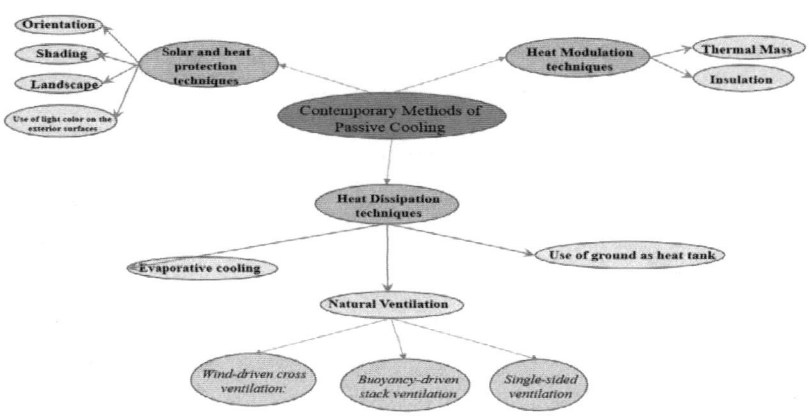

IOP Conf. Series: Earth and Environmental Science **331** (2019) 012012 doi:10.1088/1755-1315/331/1/012012

Figure 3 Contemporary Methods of Passive Cooling [6]

2.2.2 Modulation of heat gains

Heat modulation is associated with high thermal mass materials, like brick and concrete, in building structures that act as a storage for heat during daytime and cold at night. Modulation of heat gains is subdivided into two such as: Insulation whose effect is to reduce heat gain and heat loss [10] and thermal mass which stipulates that the time lag needs to be very long to delay the penetration of heat until the evening. However, insulation is typically a better option [6] because thermal mass requires very heavyweight construction.

2.2.3 Heat dissipation technique (*Remove internal heat)*

Effective dissipation of the excess heat depends on two main pre-conditions: (a) The availability of a proper environmental heat sink with sufficient temperature difference for the transfer of heat and (b) the efficient thermal coupling between the building and the sink [6]

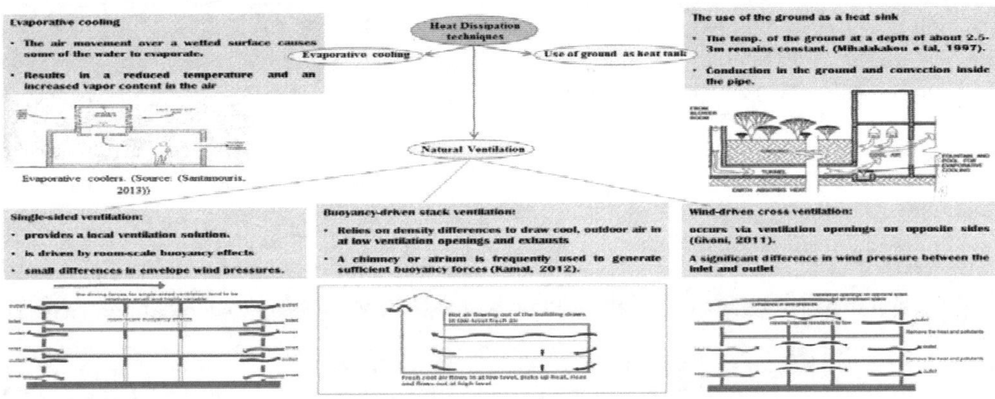

Figure 4 Heat Dissipation Technique

3. Research Methodology

3.1 Environment and Material Settings for the Experiment

The research employed a case study approach, construction of baseline model with the Revit architecture software and applied DesignBuilder building analysis software to evaluate the energy performance of the senate building as built and the interpretation and presentation of results in various tables and charts.

Table 1 Environment and Material Settings on Case Studies

General Information on Case Studies		
Location	Case A	Case B
Latitude &	10°18'57" N	9°13'48" N
Longitude	9°50'39"E	12°27'36"E
Activity Period	5 days a week from 08:00am to 05:00pm	5 days a week from 08:00am to 05:00pm
Occupants' Activities	Mostly paper based	Mostly paper based
Wall composition	sandcrete blocks (225mm); internal and external rendering (12.5mm); finished with medium	Ivory emulsion sandcrete blocks (225mm); internal and external rendering (12.5mm)

	Ocher Orange paint 3.11 U-Value; 3.97 Time Constant (Hours)	finished with vibrant ocher yellow; 3.11 U-Value; 3.97 Time Constant (Hours)
Roofing Material	0.75mm milk; no roof ventilation; 0.89 U-Value; 7-hour time lag.	0.75mm milk and ash; no roof ventilation; 0.89 U-Value; 7-hour time lag.
Window material	*single clear with blind* glazed windows	*single clear with blind* glazed windows

DesignBuilder is building analysis software that allows user to easily design and work in 3D; it offers a wide range of analysis options from acoustics to energy efficiency tools. The simulation runs on average hot season day (5th April) also, it is important to note that the general parameters for the simulation are summarized in Table 1

4. Results

4.1 Monthly heating/cooling loads

Heating Cooling loads of air conditioning system working with 95% efficiency and $14m^2$ per person at a state of office activity (55W) is considered for the calculations. The spaces observed are described in the following subsections

I. Senate buildings

Table 2 Monthly heating/cooling loads for cases A and B

MONTH	HEA A (kWh)	COOL. A (kWh)	HEA B (kWh)	COOL. B (kWh)
Jan	0	468.384	0	2299.477
Feb	0	7131.288	0	17730. 912
Mar	0	26365.46	0	46806.129
Apr	0	58443.12	0	84051.719
May	0	63080.76	0	89620.016
Jun	0	54036	0	77672.07
Jul	0	39352.02	0	58138.551
Aug	0	21149.6	0	45894.27
Sep	0	23829.49	0	43102.422
Oct	0	31048.23	0	59567.117
Nov	0	1381.722	0	11145.439
Dec	0	128.565	0	1291.624
Total		326414.64		537319.746

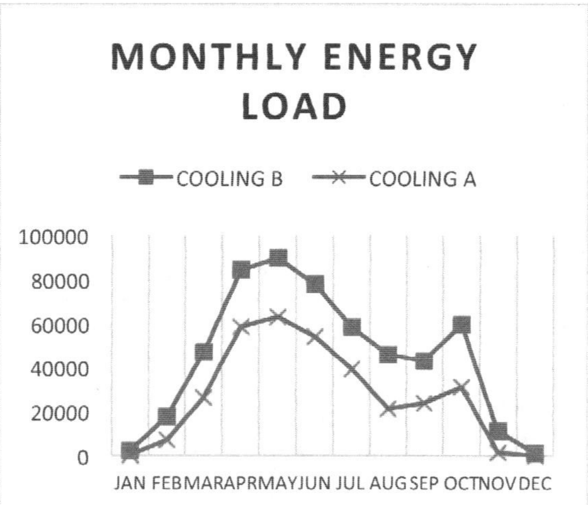

Figure 5 Monthly energy loads for case A and B

The thermal comfort zone is chosen as 18-28 0C for the temperature band that the calculation would take place according to the standards [8-9] Standard 55-2000R) applied to the Yozgat weather data, 2002. However, due to geographical location and climatic condition of dry sub-humid zone, the total annual energy required for heating is very small (which can be negligible) and it's only required for few hours in the two coldest months of the year which are January and

December. For both cases A and B, the coldest months are January with 468.384kwh and 2299.477kwh respectively; and December with 128.565kwh and 1291.624kwh respectively, while the hottest month is May with 63080.76kwh and 89620.016kwh respectively, see table 2.

The possible reasons for case B requiring a lot of energy for cooling may be connected to the fact that the senate building does not have enough shading devices, the landscapes are limited to the backyard of the building neglecting walkways, paving to receive and reflect excessive solar radiation and due to the presence of the hard landscapes from the eastern side of the building. conversely, case A building is compacted and receives cross ventilation. It comprises of four square blocks of offices connected by an octagonal atrium, as such the orientation of the building as a whole is relatively good because the longer side faces north and south, these may account for the reasons why case A requires lesser energy for cooling throughout the year.

II. Council Chambers

Table 3 Monthly heating/cooling loads for the council chambers of cases A and B

MONTH	HEA B (kWh)	COOL. B (kWh)	HEA A (kWh)	COOL. A (kWh)
Jan	0	32.36	0	0
Feb	0	300.896	0	125.403
Mar	0	912.891	0	486.853
Apr	0	1655.246	0	1019.876
May	0	1789.111	0	1111.967
Jun	0	1538.775	0	947.619
Jul	0	1206.649	0	729.012
Aug	0	947.293	0	420.783
Sep	0	892.097	0	430.406
Oct	0	1224.386	0	574.358
Nov	0	134.921	0	3.082
Dec	0	15.405	0	0
		10650.03		5849.359

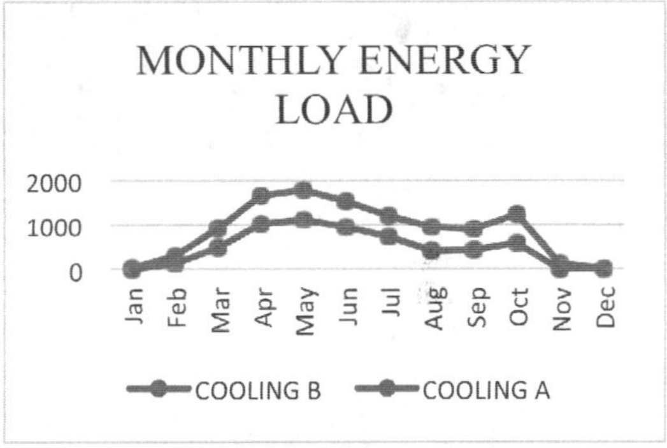

Figure 6 Monthly energy load for council chamber of cases A and B

Case B's council chamber requires more energy for cooling, see figure 6, then case A. This is because case B's apertures for council chamber face the east and does not have cross ventilation and shading devices. Case B's council chamber faces south and has balconies as shading devices as well as deciduous trees from the south elevation.

III. VC's Offices

Table 4 Monthly heating/cooling loads for the VC's offices of cases A and B

MONTH	HEA B (kWh)	COOL. B (kWh)	HEA A (kWh)	COOL. A (kWh)
Jan	0	34.683	0	0
Feb	0	282.702	0	67.624
Mar	0	819.425	0	282.359
Apr	0	1509.973	0	663.783
May	0	1622.735	0	715.997
Jun	0	1393.821	0	610.17
Jul	0	1065.179	0	409.508
Aug	0	834.885	0	165.472
Sep	0	797.043	0	208.454
Oct	0	1097.303	0	299.573
Nov	0	140.605	0	0
Dec	0	12.866	0	0
		9611.22		3422.94

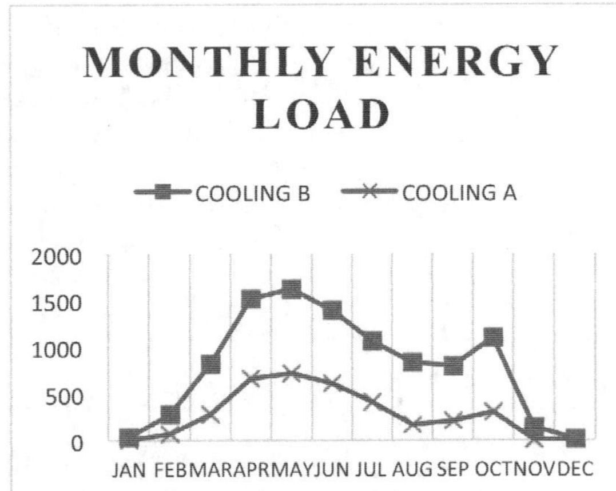

Figure 7 Monthly energy load for VC's office of case A and B

Case B's VC's office has more energy load followed by case A, see figure 4.6. This is because case A has better depth to span ratio and orientation.

4.1.1 input and output data for the case studies

The data input and output obtained for case Aare summarized in figure 8

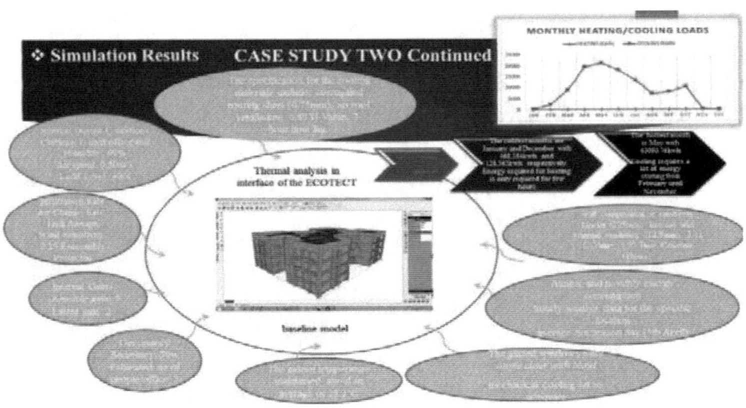

Figure 8 input and output for case A (source: field work)

The data input and output obtained for case B are summarized in figure 9

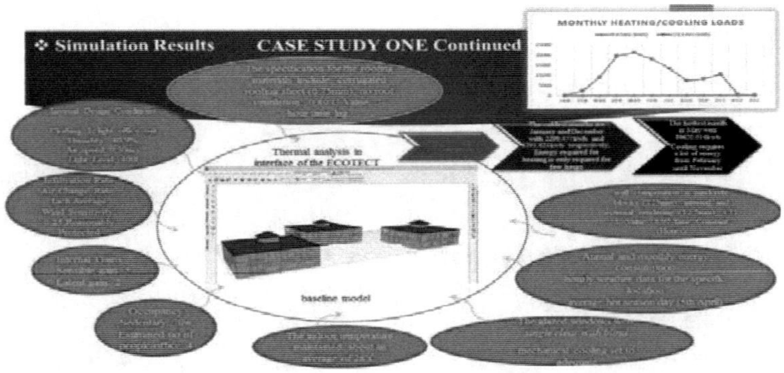

Figure 9 input and output of Case B (source: field work)

5. Conclusion

From the results presented above, it is observed that case A is more sustainable than case B. This may be connected to the fact that majority of the offices in case A are ensured with cross ventilation and vegetation.

1- Case A requires 127.3kwh/m2 annual energy
2- In case B, as high as 141.4kwh/m^2 of heating/cooling required per annum;
3- These results are far away from the British American standards of 15kwh/m2 or even the standards for Ireland of 120kwh/m^2 for energy efficient office buildings.

References

[1]. University, W. K. (2011). *Charter of University Senate* . Retrieved August 1, 2015 from http://www.wku.edu/senate

[2]. Emetere, M. E., Boyo, A.O., & Akinyemi, M.L., Statistical Analysis of the Thermal Comfort In The Urban Climate Of Ilorin-Nigeria: A Three Decade Event, Research journal of fisheries and hydrobiology. 10 (11), 178-185

[3]. ME Emetere (2019). A modified approach to estimating thermodynamic impact on buildings: A case study of poor urban setting in Lagos state, Journal of Engineering, Design and Technology 17 (1), 115-135

[4]. Dimoudi, A. (1996). Passive Cooling of Buildings . *In: Passive Cooling of Buildings [Santamouris, M. and D. Asimakopoulos (eds.)]. James & James (Science Publishers)* , pp. 35-55

[5]. Givoni, B. (2011). Solar Energy. *Indoor Temperature Reduction by Passive Cooling Systems* , 85, 1692-1726.

[6]. Burberry, P. (1997). *Environment and services.* Harlow: Longman.

[7]. Kamal, M. (2012). An overview of passive cooling techniques in building: Design concepts and Architectural interventions. *Acta Technica Napocensis: Civil Engineering & Architectural* , 55 (1) 84-97.

[8]. Al-Obaidi, K. M., Ismail, M., & Rahman, A. M. A. (2014). Passive cooling techniques through reflective and radiative roofs in tropical houses in Southeast Asia: A literature review. *Frontiers of Architectural Research, 3*(3), 283-297.

[9]. Leylian, M., Amirkhani, A., Bemanian, M., & Abedi, M. (2010). Design Principles in The Hot and Arid Climate of Iran, The Case of Kashan. *International Journal of Academic Research* , Vol. 2. No. 5, p: 11-17.

[10]. Nayak, J., & Prajapati, J. (2006). *Handbook on energy conscious buildings.* Indian institute of technology, Bombay and Solar energy center Ministry of non-conventional energy sources, Government of India.

IOP Conf. Series: Earth and Environmental Science **331** (2019) 012013 doi:10.1088/1755-1315/331/1/012013

The Level of awareness of automation technology in the construction industry.

O P Fadamiro[1] and A E Oke[2]
Federal University of Technology, School of Environment Technology, Department of Quantity Surveying, P.M.B. 704, Akure, Ondo-state, Nigeria.

E-mail: sefpatty@gmail.com and emayok@gmail.com

Abstract. The rapid development of infrastructure in a developing country like Nigeria may require the construction industry to totally embrace the innovative and advance technique of construction through the adoption of automation technology, so as to make construction project more valuable, enhance projects performance and stay competitive in the global and international industry. This study is strictly tailored towards investigating the level of awareness of the automation technique identified in the literature review section of this study. The study employed a cross-sectional survey design using a quantitative approach. A total of 191 construction professionals which include majorly the Architect, Builder, Engineer and Quantity Surveyor were the respondents for the study using a well-structured questionnaire. The questionnaire was designed on a five (5) point Likert-type scale having 5 sections of question in all. Data analysis were done using MIS, SD, Kruskal-Wallis and SNK Test. Twenty (25) automation techniques were pin-pointed from previous literature and materials that are linked with automation in construction industry. Study revealed that Computer Aided Design (CAD), cranes and Building Information Modelling (BIM) are the top three (3) techniques that are most aware of in the construction industry, while Concrete electric and infrared curing system, Humanoids and Contour crafting robots has the least rating. The study recommended seminars and workshop themes that are tailored towards the benefit of adopting automation in the construction industry. However, this research focused on level of awareness of automation techniques, future research can focus on cost implication of those techniques

1. Introduction

Science and innovative technology is rapidly developing and growing increasingly all over the universe at a level and speed that has certainly not been witness in the time past. The construction industry is well-structured, universal and a very large industry and contribute extensively to the nations international or global finances in areas like job creation, infrastructure, investing in the economy and massive contribution to Gross Domestic Product (GDP) [1]. The construction industry makes up about 25percent of the Gross Domestic Product (GDP) in developing countries and 10percent of the Gross Domestic Product (GDP) in countries that are developed [2]. The season of the countries national buoyancy are commonly linked to the periods where construction activity is at its peak in the country [3]. The level at which the construction industry affect the general environs is being consider through a wide range of its activities. As a result of the impact of construction industry on the environment, the industries have since embraced new and improved technology refer to as automation or industrialization in other to stay competitive and relevant in the international arcade [4]. Industrialization is described to be a broader modernization method of construction by transformation and the improvement of advanced techniques of manufacturing and innovative technology system, usually manufacturing works [1]. Mechanization and automation in the construction industry are viewed to deliver tasks faster than humans at job sites, construction service providers are also expected to complete tasks in time with high quality of construction product when using these innovative technologies [12]. In other to totally embrace automation in the construction industry. The researches that has been done so far on this innovative and advance technology focus more on the perceived impact and Barriers of automation in the construction industry where it has not been fully exploited. In addition to the previous researches, this study was conducted to fully investigate the level of awareness of construction automation techniques in Nigeria construction industry with an opinion of creating awareness among construction professionals and providing a template for enhancing project performance in the Nigerian Construction Industry through the adoption of automation principles.

2. Overview of Automation in the Construction Industry

Construction has been prevailing since the creation of man. Starting from the well-constructed great pyramids of Egypt and the magnificent Great wall in China to the most recent projects such as improved method of constructing subways, skyscraper, bridges and factory [13]. Construction has been a human effort for generations in all parts of the rondure [5]. The construction industry plays an essential part in ensuring the needs of the country's infrastructural and economic improvement is meet. Approximately seventy percent (70%) of the nation's funds goes into the construction industry year after year [6]. In many countries the construction industry has, however, attracted criticism for inefficiencies in outcomes such as inferior working conditions, occupational safety, labor shortage, decreasing quality and productivity, all these issues linked with the construction industry have paved way for the possibility of more revolutionary resolutions within the industry in terms of safety of construction workers, quality of construction product and productivity. The prospective competency to achieve higher output of products at a lesser unit cost, with improved quality of construction products could develop universal effectiveness, since the construction industry demands operative construction organizations, well-organized construction procedures and advanced construction procedures to efficiently play crucial role under increasing globalization market competition and technological innovation in the twenty-first century [7]. As a result of the problem faced by the construction industry, the industry has since embraced the innovative and technologically advanced method of construction which is "Automation" in other to remain relevant in the international arcade and in other to achieve better quality of projects. Adoption of automation technology are viewed as a means to solving the problem faced by the construction industry [14]. As a result of the rapid growth and development of computer hardware and software in the earlier years, vital improvement can be pin-pointed in areas like monitoring, transportation, localization, mapping, robotic control, sensing, vision and planning of modules. Furthermore, the application of automation in construction industry will result to lesser use of skilled and unskilled labor, timely completion of projects and cost incurred in executing a project will be minimal, thereby eliminating the high level of risk to the lives of the skilled and unskilled labor. The application of Automation has had a remarkable effect in a wide range of industries beyond automobile and manufacturing, where it all originate. According to Vaha [8] important reasons for the use of automation technology was highlighted by some industries applying the technology to their projects and some of the reasons given are: quality and reliability, standardization of components, safety of site workers, improvement of productivity, favorable working conditions for construction workers, reduction in labor costs, minimizing life cycle cost and simplification. The word "Construction Automation" (CA) is simply described as the application of electrical, computerized and machine-driven means in the construction industry in other to accomplish automatic control and smooth operation to ensure occupational safety and wellness of skilled and unskilled labor, increment in quality of construction output, productivity of construction workers and timely completion of projects is achieved [9]. Construction automation are modern type of innovative technology used in the construction industry, this technology involves the combination of electronics, mechanical and computer software to operate robots by using a special code to perform require functions [6]. In construction industry, the scope for these innovative technologies execution can be broad, cutting across through all the phases of the construction life-cycle, from the designing of the construction projects, through the eventual erection of the structure on construction site, even after the completion of the project, some technologies have been invented for maintaining and operating the edifice, and through to eventual disassembling or demolition of the structure [5]. However, the degree of application of this technologies differs considerably from one construction stage to another, for instance, automation of demolition stage of construction through the use of demolition Robots or the automation of design phase through the application of Computer Aided Design (CAD). It is therefore important to pin-point the automation techniques and different areas of the construction projects life-cycle that is applicable. Some of the automation techniques identify in the literature includes: Demolition Robots, 3D printing robots, Robotic Drones, Bricklaying Robots, Welding Robots, Exoskeletons (enhance a task or ability the human body lacks), Forklift Robots, Repaving robots, Humanoids, Mobile telepresence robots, Hydraulic elevators, Cranes, Repainting robots, Contour crafting robot, Transportation Drone, Surveying Drone, Monitoring Drone, Concrete steam curing system, Automated cutting Grinder, Concrete electric and infrared curing system, Inspection Laser sensor, Global Positioning System (GPS), Computer Aided Design (CAD), Computer Aided Manufacturing (CAM) and Building Information Modelling all this automation technologies and many others helps in smooth operation of construction process [5].

3. Research Methodology

A cross-sectional survey design was used for this study through the distribution of a well-structured questionnaire to construction professionals working with either construction firms, consulting firms, government establishment or client organization in Lagos state, Nigeria. These construction professionals include Engineer, Architect, Quantity Surveyor and Builders. Convenient sampling technique which is a non-probabilistic sampling technique was used to administer the questionnaires. A convenience sample is made up of construction professionals who are easy to reach and have adequate or broad knowledge of automation as well as good knowledge of the construction industry. The questionnaire was designed in two sections: the section A address the important background information of the respondent while the section B part was designed to get information from respondents on the level of awareness of automation technique in the construction industry. Identified automation techniques from literature were pointed-out in a table and 5-point Likert scale was used to seek the views of the respondents. Mean Item Score (MIS) and Standard Deviation (SD) were calculated for each of the identified variables based on the adopted Likert scale, this was used to rank and determine the level of importance of the factors. Kruskal-Wallis test was also employed in analyzing the difference in opinion of respondent based on professional group. Student-Newman-Keuls (SNK) Test was further conducted in this study to further group the sample means that are significantly different from each other.

4. Findings and Discussion

Table 1 shows various automation technique identified in the literature review of the study. Respondents were asked to rate the level at which they are aware of the listed automation techniques. A 5-point Likert scale of Very High, High, Neutral, Low and Very low was used to rate the level of the respondent awareness. Table 1 presents the overall mean and the overall standard deviation result of all the respondent that participated in the study based on their profession. "Cranes" and "Building Information Modelling (BIM) was rated the most by the architect as the automation techniques that is most know to the profession with the mean value of 3.93 and 3.80 respectively while quantity surveyor rated "Computer Aid Design" (CAD) and "Building Information Modelling" (BIM) as the most know techniques. This is partly in line with the opinion of the engineers that believe "Computer Aid Design" (CAD) and "Global Position System" (GPS) are the technique that is most aware of. The builders are of the opinion that "Transportation drone and Cranes" are the most know techniques. According to the overall mean and standard deviation based on professional grouping on level of awareness of automation, Computer Aided Design (CAD) with (MIS=4.13, SD=1.09, R=1) was rated highest as the techniques that is most aware of. In addition the results showed that Cranes (MIS=3.92, SD=0.99, R=2), Building Information Modelling (BIM) (MIS=3.88, SD=1.17, R=3), Computer Aided Manufacturing (CAM) (MIS=3.85, SD=1.05, R=4), Global Positioning System (GPS) (MIS=3.78,SD=1.20, R=5), Robotic Drones (MIS=3.74, SD=1.10, R=6), Hydraulic elevators (MIS=3.68, SD=1.09, R=7), Transportation Drone (MIS=3.54, SD=1.13, R=8), Demolition Robots (MIS=3.53, SD=0.98, R=9), Surveying Drone (MIS=3.47, SD=1.22, R=10), all these techniques were ranked amongst the top ten (10) techniques in Nigeria construction industry. Other important techniques include Monitoring Drone (MIS=3.45, SD=1.12, R=11), Forklift Robots (MIS=3.42, SD= 1.00, R=12), 3D printing Robots (MIS=3.38, SD=1.04, R=13), Concrete steam curing system (MIS=3.38, SD=0.99, R=14), Inspection laser sensor (MIS=3.31, SD=0.98, R=15), Repaving robots (MIS=3.26, SD=0.93, R=16), Automated cutting Grinder (MIS=3.24, SD=1.02, R=17), Repainting Robots (MIS=3.23, SD=1.07, R=18), Bricklaying Robots (MIS=3.22, SD=1.04, R=19), and Exoskeletons (MIS=3.19, SD=1.08, R=20). Furthermore, techniques such as Welding Robots (MIS=3.18, SD=1.11, R=21), Mobile telepresence robots (MIS=3.10, SD=1.16, R=22), Concrete electric and infrared curing system (MIS=3.09, SD=1.03, R=23), Humanoids (MIS=3.04, SD=1.08, R=24) and Contour Crafting Robot (MIS=2.86, SD=1.05, R=25) were ranked as the least technique that is been aware of, in Nigeria construction industry.

Kruskal- walis H test was carried out to determine the difference in the sample means of different groups of respondent relating to the level of awareness of automation techniques. The result of the test showed that there is statistically significant difference between the opinions of the construction professionals because the P-value is lesser than 0.05 (an alpha of 0.05 is used as the cutoff for significance). Chi-square value =9.847; Degree of freedom (DF) =3; P-value =0.020, with a mean rank for Architect 116.50, 87.33 for Quantity Surveyor, 96.81 for Engineer and 80.80 for Builders.

Newman-Keuls (sometimes called Student–Newman–Keuls or SNK) is a post hoc test for measuring differences in means. Once an ANOVA has given a statistically significant result, you can run a Newman-Keuls to see which specific pairs of means are different. The test is based on the studentized range distribution. The second column in the output for the S-N-K post hoc test contains the list of professionals in order from lowest to the highest mean (The mean of the Builder, Quantity Surveyor, Engineer and Architect are 3.2322, 3.3600, 3.4365 and 3.6771 respectively). The third column of the table, identifies the number of subjects in each group. The remaining columns identify the sets of groups that are statistically significantly different from each other. The result from Table 4.4 shows that Quantity Surveyor, Engineer and Architect opinion on awareness of automation in the construction industry are not significantly different from Subset 1 also Subset 2 shows that Quantity Surveyor, Engineer and Architect opinions are not significantly different. The only groups that are significantly different are the Builder and Architect opinion.

Table 1.1. SNK Post-Hoc Test on Level of Awareness of automation.

	Profession	**N**	**Subset 1**	**Subset 2**
Student-Newman-Keuls	Builder	33	3.2322	
	Quantity surveyor	52	3.3600	3.3600
	Engineer	62	3.4365	3.4365
	Architect	44		3.6771
	Sig.		0.3128	0.0633

Based on the ranking using the mean item score (MIS) and the standard deviation (SD), the finding indicated that the Architect are of the opinion that they are more aware of Cranes and Building Information Modelling (BIM) while the Quantity Surveyor are more aware of Computer Aided Design (CAD) and Building Information Modelling (BIM). The Engineer rated high Computer Aided Design and Global Position System (GPS) and the Builder are more aware of Transportation Drone and cranes. Form the overall mean Computer Aided Design (CAD) was rated first followed by Cranes and then Building Information Modelling. The findings are in line with the opinion of Mahbub [10] that described Computer Aided Design (CAD) has a tool that has increased from a mere tool used to communicate and collaborate or design function to encompass improvement in the management and control within all aspects of architectural practices. Computer Aided Design (CAD) gives an image of model towards the design, allows the designer and stakeholder to plan and communicate easily and Information can be kept in the data base to allow for modification by the designers during rectification of error or orders by the client, when change is recommended is easy to change on the computer than starting from the beginning when using paper drawings [11]. The idea of Computer Aided Design is not new in the construction industry, with on-going development constantly providing improvement in the tools used.

5. Conclusion and Recommendation

In order to investigate and examine the benefit of adopting automation in the construction industry in Lagos state, Nigeria, this study has investigated and examine the level of awareness of automation technique. The study on level of awareness of automation techniques based on over mean, standard deviation and ranking of the professional groups revealed that Computer Aided Design (CAD), Cranes, Building Information Modelling (BIM), Computer Aided Manufacturing (CAM), Global Positioning System (GPS), Robotic Drones, Hydraulic elevators, Transportation Drone, Demolition Robots, Surveying Drone, Monitoring Drone, Forklift Robots, Concrete steam curing system, 3D printing Robots and Inspection laser sensor are the top fifteen (15) techniques that are most aware of in the construction industry by the construction professionals while Welding Robots, Mobile telepresence robots, Concrete electric and infrared curing system, Humanoids and contour crafting robot are the least rated automation technique by the construction professionals. These indicate that the construction professionals are not really aware of the last five (5) automation technique. It is therefore necessary for construction professional body (NIQS, NSE, NIOB, NSE) to ensure that some of their seminars and workshop themes are tailored towards the benefit of adopting automation in the construction industry: enlighten them on areas in the construction activity that can be automated and enlighten them on construction automation techniques that are available in the construction industry, in a view to improve project quality and performance through the application of automation technique. The focus of this study was on level of awareness of automation technique in the construction industry. Further research can explain explicitly one of the automation techniques i.e. in terms of mode of application, cost implication and cost benefit. There can also be further research on benefit of the automation techniques in the construction industry.

IOP Conf. Series: Earth and Environmental Science **331** (2019) 012013 doi:10.1088/1755-1315/331/1/012013

Table 1.2. Level of awareness of automation techniques.

Factors	Architect		Quantity Surveyor		Engineer		Builder		Overall		Rank
	Mean	SD	Mean	SD	Mean	SD	Mean	SD	Mean	SD	
Computer Aid design (CAD)	3.66	1.01	4.88	0.32	4.33	0.98	3.13	1.20	4.13	1.09	1
Cranes	3.93	0.71	4.31	0.88	3.90	1.13	3.33	0.92	3.92	0.99	2
Building Information Modelling (BIM)	3.80	0.82	4.83	0.38	3.85	1.08	2.70	1.33	3.88	1.17	3
Computer Aid Manufacturing (CAM)	3.61	0.89	4.31	0.92	4.08	0.97	3.03	1.05	3.85	1.05	4
Global Position System (GPS)	3.52	0.82	4.00	1.49	4.15	1.08	3.15	1.06	3.78	1.20	5
Robotic Drones	3.39	0.99	4.10	1.00	3.92	1.14	3.30	1.10	3.74	1.10	6
Hydraulic elevators	3.58	0.93	3.96	1.24	3.84	0.87	3.09	1.21	3.68	1.09	7
Transportation Drone	3.38	1.03	3.35	1.34	3.87	0.93	3.42	1.15	3.54	1.13	8
Demolition Robots	3.36	0.99	3.79	0.70	3.76	0.95	2.94	1.14	3.53	0.98	9
Surveying Drone	3.59	1.00	3.42	1.53	3.55	1.13	3.24	1.12	3.47	1.22	10
Monitoring Drone	3.44	0.88	3.25	1.03	3.85	1.26	3.06	1.09	3.45	1.12	11
Forklift Robots	3.36	0.99	3.48	1.04	3.53	0.95	3.21	1.05	3.42	1.00	12
Concrete steam curing system	3.31	0.78	3.62	0.91	3.44	1.13	2.97	0.95	3.38	0.99	13
3D printing Robots	3.32	0.96	3.06	1.00	3.79	0.94	3.21	1.17	3.38	1.04	14
Inspection laser sensor	3.30	0.88	3.15	0.90	3.58	1.01	3.06	1.06	3.31	0.98	15
Repaving robot	3.30	0.93	3.35	0.86	3.24	0.92	3.12	1.05	3.26	0.93	16
Automated cutting Grinder	3.48	0.90	2.88	1.06	3.55	0.91	2.91	1.07	3.24	1.02	17
Repainting Robots	3.07	0.96	3.20	1.23	3.47	0.99	3.03	1.05	3.23	1.07	18
Bricklaying Robots	3.33	0.89	2.90	1.18	3.44	1.03	3.15	0.91	3.22	1.04	19
Exoskeletons (enhance a task or ability the human body lacks.	3.14	1.05	3.02	1.04	3.39	1.14	3.15	1.09	3.19	1.08	20
Welding Robots	3.30	0.98	2.92	1.17	3.34	1.03	3.12	1.29	3.18	1.11	21
Mobile telepresence robots	3.02	1.02	2.62	1.35	3.44	0.97	3.27	1.15	3.10	1.16	22
Concrete electric and infrared curing system	3.34	0.96	2.60	1.00	3.45	1.07	2.85	0.76	3.09	1.03	23
Humanoids	3.05	1.08	2.67	1.32	3.18	0.87	3.33	0.89	3.04	1.08	24
Contour crafting Robot	3.14	0.98	2.29	1.07	3.10	0.96	2.97	0.92	2.86	1.05	25

IOP Conf. Series: Earth and Environmental Science **331** (2019) 012013 doi:10.1088/1755-1315/331/1/012013

References

[1] Kamaruddin S S, Mohammad F M and Mahbub R 2015 Barriers and impact of mechanisation and automation in constructon to achieve better quality products *Procedia-Social and Behavioral Sciences* 111-120

[2] Kim J M, Chi H L, Wang X and Ding X 2015 Automation and robotics in construction and civil engineering *Journal of Intelligent Robot System.* **79** 347-350

[3] Clough R H, Sears G A and Sears S K 2000 *Construction project management* (Canada: John Wiley & Sons)

[4] Alshawi M and Ingirige B 2003 Web-enabled Project management: Emerging Paradigm in Construction *Automation in Construction.* **4** 349-64

[5] Alexander F 2016 Development of an unmanned autonomous concrete floor robotic trowelling system *Proc. Int. symp. Of robotics and automation in construction* vol 10 p 79-84

[6] Dantata S A 2008 *General Overview of the Nigeria Construction Industry* Msc Thesis in Massachusetts Institute of Technology

[7] Mahbub R 2008 *An Investigation into the Barriers to the Implementation of Automation and Robotics Technologies in the Construction Industry, PhD thesis* Queensland University of Technology Quantity Surveying (Australia)

[8] Vaha P, Heikkila T, Kilpelainen P, Jarviluoma M and Gambao E 2013 Extending automation of building construction- Survey on potential sensor technologies and robotic applications *Automation in construction.* **16** 168-178

[9] Hewitt M M and Gambatese J A 2002 Automation Consideration During Project Design *Int. symp. on Automation and Robotics in Construction (ISARC)* (Washington D.C.)

[10] Mahbub R 2006 Automation and robotics implementation in developing countries: opportunities for Malaysian construction industry *Proc. of Int. Conf. on Construction and Real Estate Management (ICCREM)* vol 12 p 261-77

[11] Boche F N and Haas C T 2008 Automated retrieval Of project three-dimensional CAD Object in range point cloud to support automatized dimensional QA/QC *Information technology in Construction.* **13** 71-85

[12] Oke A, Aigbavboa C and Mabena S 2017 Effect of automation on construction industry performance *Int. Conf. on Mechanics, Materials and Structural Engineering (ICMMSE)* vol 102 (Johannesburg: Atlantis Press) p 370-74

[13] Kamaruddin S S, Mohammad F M and Mahbub R 2013 Enhancing the quality of life by adopting IBS: an economic perspective on mechanisation and automation *Proc. Association of Malaysian Environment-Behaviour Researchers (AMER) Int. Conf. on Quality of Life (AicQol 2013). 'Quality of Life in the Built and Natural Environment (*Langkawi)

[14] Oke A, Aigbavboa C and Mabena S 2017 Drivers for the applcation of automation in construction activities *Commonwealth Association of Surveying and Land Economy (CASLE-Dar es Salaam)* (Johannesburg) p 80-87

Effect of digitalization on the performance of commercial banks in Nigeria

[1]Mayowa G Agboola, [2]Kiitan A Awobajo, [3]Stephen O Oluwatobi, [4]Mosunmola O Akinbode, [5]Michael O Fagbohun, [6]Ugwunwa C Esse, [7]Chidi D Segun-Adeniran, [8]Aderonke O Asaolu and [9]Chelsea M Betek

[1]Lecturer 1, Dept. of Business Management, Covenant University, Ota, Nigeria
[2]Relationship Manager, Guaranty Trust Bank, Nigeria
[3]Lecturer 2, Dept. of Economics, Covenant University, Ota, Nigeria
[4]Lecturer 1, Dept. of Business Management, Covenant University, Ota, Nigeria
[5]Librarian 2 Centre for Learning Resources, Covenant University, Ota, Nigeria
[6]Librarian 1, Centre for Learning Resources, Covenant University, Ota, Nigeria
[7]Librarian 1, Centre for Learning Resources, Covenant University, Ota, Nigeria
[8]Senior Librarian, Centre for Learning Resources, Covenant University, Ota, Nigeria
[9]PG Student, Dept. of Political Science and International Relations, Covenant University, Ota, Nigeria

Corresponding Author's Email – gbenga.agboola@covenantuniversity.edu.ng

Abstract. This study investigates how digitalization enhances the performance of commercial banks adopting the purposive method and simple random sampling selecting 370 non-managerial employees from a commercial bank. A self-structured questionnaire was used as the major instrument for data collection and was analysed using SPSS version 25. From the result, it was discovered that there was a mild significant and positive relationship between the digitalization process and commercial bank performance ($r = 0.114^*$; $p< .05$). Also, there is a positive significant relationship between product innovation and performance of commercial banks in Nigeria ($r = 0.186$; $p< 0.001$). The study recommends that digitalization processes if adequately and correctly implemented, will have a significant positive relationship on the performance of commercial banks in Nigeria ceteris paribus.
Keywords: Digitalization; Technological Change; Performance; Commercial Banks

1. Introduction

Digitalization has become paramount as industries and institutions including commercial banks strive to adopt better and more effective methods of service provisions to their clients and day to day customers. Despite the obvious shift to digitalized methods of service provision and delivery, it is pertinent to note that some financial institutions still see the entire process of digitalization as problematic and difficult (Carbo-Valverde, 2017). Abbasi and Weigand (2017) observed that information and communication technologies has brought about a boost in the financial sector specifically the support of banking services, risk management and increase in productivity and these factors have encouraged financial institutions to increase investments in the process of digitalization. A study carried out by Gareth et al. (2006) as quoted by Abbasi and Weigand (2017) revealed that 'banks in North America, Europe, Asia- Pacific and Latin America spent a total of $241billion on information technologies in the year 2016 and an overall increase of almost 4% when compared with the year 2015.' They also reiterated that the practice of digitalizing services carried out by financial institutions gives a new turn to service provision and make for better customer satisfaction as a result of diversity in service provision.

Technology has changed businesses and within-business connections. This has empowered reconfiguration of design, marketing, production, delivery and services through supply chains, independent structure, outsourced manufacturing and contract warehousing and delivery. The advent of technology has prompted advanced digital transformation, which is not simply restricted to tech-savvy businesses, also high-tech major established organizations have embarked on digital transformation venture too (Idiegbeyan-Ose, Ilo and Isiakpona, 2015; Weill and Woerner, 2018; Agboola, Asokomeh, Akinbode, Peter, and Deborah, 2017)

IOP Conf. Series: Earth and Environmental Science **331** (2019) 012014 doi:10.1088/1755-1315/331/1/012014

Digitalization represents a change in perspective that unavoidably influences the most traditional organizations and even effects the general society (Gimpel and Röglinger, 2015; Sambamurthy et al., 2003). It is important to state that digitalization has overturned large industries, retail businesses, media, transport etc. and this is currently sweeping across commercial banks in Nigeria.

The digitalization change has been around for a considerable length of time; but in recent times, its effect and the speed of change appear to be exceptional. Digitalization has changed the financial sector and its working condition. Albeit, it is important to say that financial services have been automated for quite a long time, but a more extreme change could be said to be delayed as a result of most financial organization trying to maintain their traditional financial mode of services (IFC, 2017).

As a result of the technological advances and digitalization, the collaboration between commercial banks and customers has likewise been improved and this has made better approaches for transactions. This quick change can be found in the banking sector, where digitalization has given the banks more innovative ways for reaching out to potential clients and in the meantime, this has helped them to enhance their services. Digitalization in the banking sector has brought about Internet banking and mobile banking, which has turned into the single greatest channel right now for reaching customers and for customers to be able to deal with their banking errands themselves (Deutsche Bank, 2016).

The impacts of digitalization on Nigerian's financial institutions can undoubtedly be associated to the measure of individuals utilizing the internet banking and mobile banking services to pay their bills, which is the dominating factor on an average person's banking errands. The breakthrough techniques for reaching customers are becoming less expensive and less demanding than the traditional methods (Coiera, 2000) (Magatef & Tomalieh, 2015) (Wright, 2017).

The effect of digitalization on the Nigerian financial sector and its services actually goes past retail and client applications and services, it also incorporates all components of the Nigerian financial sector's services and production processes. It also expands the need for promptness and customized services and products, which also put the Nigerian financial sector under the demand of meeting customers' requests with minimal effort, convenient approaches to transfer cash, investment and borrowings.

Digitalization of the Nigerian commercial bank has brought about new models in businesses, that is new ideas with regards to development and improvement in various areas, from mobile banking to financial transactions and internet banking. This has continued to grow and supplant the conventional banking services to the customers through innovative technologies to meet the developing complex necessities and challenges of globalization. The Nigerian banking and financial sector have experienced radical changes and improvement over the last few years and is in a consistent state of development (Gabor and Brooks, 2016; Abbasi and Weigand, 2017).

It is an obvious fact that the performance of an organisation is measured by how well customers are satisfied with products or services that are being offered as a result of their persistent patronage (Khadka and Maharjan, 2017). In the above premise, how can one justify the challenges that customers still face with banks in carrying out regular banking transactions? The Nigerian commercial banks may have digitalized their processes, but there are still several limitations being experienced by customers. Customers are still limited a great deal as they process their daily banking transactions such as still filling physical forms to sort out issues that ought to be have been attended to via a digital process. Due to this, one will still find lots of queues within the halls of commercial banks waiting to be attended to. With the effect of the cashless policy in 2017 by the Central Bank of Nigeria (CBN) and other similar policies to encourage digitalization, one will still wonder why this limitation is still lingering as most commercial banks have either adopted or adapted digitalizing their processes making the enormous investments on the required IT infrastructure and digital technologies to be in vain (Nigeria Today, 2017).

The main objective of this study is to investigate digitalization process and performance of commercial banks in Nigeria. More specifically, the study explores the extent to which digitalization process of banking services relate to the performance of commercial banks in Nigeria and investigates the relationship between product innovation and the performance of commercial banks in Nigeria.

IOP Conf. Series: Earth and Environmental Science **331** (2019) 012014 doi:10.1088/1755-1315/331/1/012014

2. Review of relevant literature

The process of digitalization in commercial banks is an activity of great importance; however, despite the importance, it is pertinent that hard copies of documents digitized should still be stored and there should be manual processes of carrying out services peradventure the automated means become faulty. Satyendra (2016) opined that the process of digitalization is a sequential procedure that is patterned in line with the activities of archiving, access and management. He further outlined some processes of digitalization which include the following:

The process of initiation and the start of the project: this involves the initial plans preceding the digitalization process such as acquisition of the devices to be used, training and hiring of the needed manpower to carry out the digital content creation.

Selection of documents and activities for digitalization: some hard copy documents are then sorted and selected in preparation for the digitalization. This step also involves setting time frames for achievements of certain task relating to the digitalization process. In a situation where manual methods have never been used, there would be no need for changing the physical formats of the documents

Conversion Process: this is the actual process of transforming the documents into machine readable format and putting in place all the gadgets needed for activity of digitalization.

Editing, Access and Maintenance: this involves checking through the already digitized documents to avoid errors, making the access points user friendly and putting in place proper procedure and plans for regular maintenance. It is important to note that maintenance is paramount when digitalization activities are put in place, to avoid sudden loss of data or total crash of the entire system.

There is no doubt that the digitalization of services offered by commercial banks has brought about a lot of positive influence on their respective clients or customers even the performance of the banks. Some of these influences include the following:

The process of inculcation of digital technologies into service provision in commercial banks has brought about a drastic reduction in the level of poverty (Mukherjee, 2017) (Naseem, 2017) (Maxima & Kim, 2018). Bank accounts can be opened without the prospective customer being physically in the bank simply because the process of digitalization has allowed for mobile communication in the financial system (Nyangosi, Nyangau, Nyariki, and Nyangau, 2014; Happiest Mind, 2014; Piirainen, 2016; Ortstad and Sonono, 2017). Customers are encouraged to open bank accounts because they are aware that with the click of a mouse or the use of mobile devices, they can carry out bank transactions; rather than having to go through the time consuming and stressful experiences at the banks and other financial institutions. In a survey conducted by Harigaya (2016), it was revealed that *"approximately 30% of time was saved during financial deposits and approximately 70% of time was saved during withdrawal as a result of the digital transformation in financial institutions."* The ease of access to commercial banks as a result of digitalization has in turn increased the amount of savings made and in the long run reduced the level of poverty (Gonzalez-Paramo, 2017). This point was further corroborated by Abbasi and Weigand (2017), noting that there was a gargantuan hike (700 million) in the number of people having accounts between the years 2011 and 2014.

The digitalization process of commercial banks has also led to better provision of banking services. With the utilization of information and communication technologies, the level of banking services offered to clients are more efficient and effective. Customers' details and transactions are easily created, stored and retrieved stored on a database. This helps to control the level of paper work that commercial banks have to handle. The process of opening accounts for new customers, prospective customers' evaluation process for new accounts and loan process has been made very easy, effective and efficient as a result of the digital transformation (Gonzalez-Paramo, 2017).

The level of comfort and convenience in the process of financial transactions derived as a result of digitalization is incontestable. The commercial banks tend to perform better with the process of digitalization. They make use of ICT devices and gadgets which offers better comfort and convenience to them without jeopardizing the quality of service offered (Harigaya, 2016). For instance, the use of computer systems to confirm payer and payee details (name, passport photograph, etc.) when transactions are made helps performance, reduces errors and in the long run heightens efficiency. Digitalization has also brought about innovative services such as online banking, online purchase, provision of Automated Teller Machine services, internet transfers, mobile banking, etc. (Forest and Rose, 2015) and evidently seen in (Akinbode, Adegbuyi, Kehinde, Agboola, & Olokundun, 2019). On the part of the customer, they obviously take advantage of the privilege of carrying out transactions from the comfort of their homes, offices and even in transit using mobile banking and internet banking. Harigaya (2016) also averred that the utilization of digital technologies has provided a stable platform for cheap and speedy purchase of goods and services thereby gradually transforming positively the payment landscape globally.

Customers and clients are seen as the core of the establishment of commercial banks and therefore the manner in which the provided financial services are communicated to customers cannot be taken with levity. The process of digitalization has brought about better level of efficiency in the customer relationship aspect of commercial banks in Nigeria and in the long run this boosts the level of sales and revenue of the banks. This was reiterated by Carletti (2017) whose research revealed that *"the institutions that adopted an effective digitalization strategy were able to increase their sales and revenue by 4% to 16% through the instrumentation of enhanced customer relationship."* Therefore, banks can now have better, faster, cheaper and more effective means of communicating with their clients. Some of these communication channels include using text messages, e-mails, phone calls and social media chats.

The digitalization process carried out by commercial banks also enhance the process of record keeping. Records are better organised and stored on computer systems and can easily be retrieved when needed. Records as old as possible can easily be retrieved with the click of a mouse unlike when activities in the banks were not digitized.

It is however important to note that despite the positive influence digitalization has on the performance of commercial banks some drawbacks are still being experienced. A common challenge is the increase in the level of cyber frauds and cybercrimes carried out in the commercial banks. Some individuals who are technology savvy sometime use mobile devices and online platforms to carry out theft and fraud. Some employees in commercial banks also take advantage of their access to information contained in the database of the banks and easily do some manipulations of the computer systems to carry put frauds which may not be detected.

Another effect the digitalization process of commercial banks has on performance is the negative effect of data protection and data loss. Most gathered data on transactions, customers' records, loan records and so on are stored mostly on computer systems or other digital storage devices which are prone to virus or harmful attacks and in the long run may lead to data loss if not solved quickly to remedy it (Gonzalez-Paramo, 2017). This is a negative effect that could affect performance of commercial banks

3. Methodology

The study adopted the purposive method and simple random sampling was used to select 370 non-managerial employees from a commercial bank in Nigeria. A self-structured questionnaire was used as the major instrument for data collection. Out of 370 copies of the questionnaire administered 344 copies were found useful. The data collected for this study was subjected to descriptive statistical analysis using frequency count, inferential statistics. Pearson correlation was used to test the hypotheses. The data was analysed using Statistical Package for Social Science (SPSS) version 25. In a similar study conducted by Mueni (2017), he used SPSS to analyse how digitalization influence Kenyan commercial banks.

4. Analysis and discussion

Research Question 1: What is the extent to which digitalization process of banking services relate to the performance of commercial banks in Nigeria?

Table 1 showed a highest proportion of the respondents with mean and standard deviation score $\bar{x}$=3.95; SD=0.735 indicated that Nigerian commercial banks' staff actively use their expertise in big data, for instance when developing new products or business models, As parts of digitalization process, majority ($\bar{x}$=3.85; SD=0.790) of the respondents also indicated that commercial banks in

Nigeria integrated digital channels or technologies such as mobile and social media, to communicate products and services to their users. A large proportion of the respondents $\bar{x}$=3.75; SD=0.764 agreed that that commercial bank extensively deployed the most recent digital options in order to automate their routine processes. This finding is in line with the report of Kamra (2015) who stated that digitalization of commercial bank is far beyond mere moving from old banking system to a digital system. It is a fundamental transformation of how commercial banks relate with and meet client's needs. An effective digitalization starts with recognizing digital patron behaviour, dislikes, likes, desires, choice, favourites, spoken and unspoken needs, etc.

Table 1: Frequency Distribution of Digitalization process of banking services of commercial banks in Nigeria

DIGITILIZATION PROCESS	SA		A		U		D		SD		N	Mean	Std. Dev.	Rank
	N	%	N	%	N	%	N	%	N	%				
Item 6	68	19.8	209	60.8	50	14.5	17	4.9	0	00.0	344	3.95	.735	1st
Item 4	85	24.7	122	35.5	137	39.8	0	00.0	0	00.0	344	3.85	.790	2nd
Item 5	34	9.9	225	65.4	51	14.8	34	9.9	0	00.0	344	3.75	.764	3rd

Source: Researchers' Field Survey

Research Question 2: What is the level of relationship between product innovation and the performance of commercial banks in Nigeria?

Table 2 sought to the level of product innovation in commercial bank surveyed. The findings revealed that employees of commercial bank surveyed regularly contribute ideas for digital products ($\bar{x}$=4.05;SD=0.690), majority indicated that the institution have expanded her products and services with digital services ($\bar{x}$=4.04;SD=0.741) while $\bar{x}$=3.99;SD=0.773 indicated actively inclusion of customers in the development of new ideas for digital products as parts of innovative method adopted. The finding of the study shows that some commercial banks in Nigeria are already embracing product innovation. This corroborates Marous (2019) who opined that today commercial banks are strategically positioning themselves for the future by changing their cultures and embracing technological innovations in order to deliver new innovative products and services which are considered as their top priority. Kamra (2015) parts of product innovation are meeting bank customers' needs on the move, anytime anywhere. This has become a mean of survival among commercial banks in this competitive marketplace place.

Table 2: Frequency Distribution of Product Innovation in the Nigerian Commercial Banks

PRODUCT INNOVATION	SA		A		U		D		SD		N	Mean	Std. Dev.	Rank
	N	%	N	%	N	%	N	%	N	%				
Item 2	86	25.0	188	54.6	70	20.3	0	00.0	0	00.0	344	4.05	.690	1st
Item 1	102	29.7	155	45.1	87	25.3	0	00.0	0	00.0	344	4.04	.741	2nd
Item 3	84	24.4	190	55.2	53	15.4	17	4.9	0	00.0	344	3.99	.773	3rd

Source: Researchers' Field Survey

Test of hypotheses

H_1: *Digitalization process of banking services does not relate to the performance of commercial banks in Nigeria.*

Table 3: Correlation Matrix between Digitalization Process and the Performance of Commercial Banks in Nigeria

Variable	Mean	Std. Dev.	N	R	P	Remark
Digitalization Process	3.8517	0.57958	344	.114*	.034	Sig. (2-tailed)
Commercial bank Performance	4.4874	0.42658				

*. Correlation is significant at the 0.05 level (2-tailed).

Results in Table 3 revealed that there was mild significant and positive relationship between digitalization process and commercial bank performance ($r = 0.114^*$; $p< .05$). This implies that effective digitalization process of commercial bank activities led to improved performance of commercial banks in Nigeria.

H_2: There is no relationship between product innovation and the performance of commercial banks in Nigeria.

The hypothesis sought to determine the significant relationship between product innovation and the performance of commercial banks in Nigeria. The findings revealed that there is a small positive significant relationship between product innovation and performance of commercial banks in Nigeria at $r = 0.186$; $p< 0.001$. This implies that the more commercial banks innovate their products the more the level of excellence performance as shown in Table 4.

Table 4: Correlation Matrix between Product innovation and the Performance of Commercial Banks in Nigeria

Variable	Mean	Std. Dev.	N	R	P	Remark
Digitalization Process	3.8517	0.57958	344	.186**	.001	Sig. (2-tailed)
Product Innovation	4.0271	0.62237				

**. Correlation is significant at the 0.01 level (2-tailed).

5. Conclusion and recommendation

Based on the findings of this study, it is therefore concluded that digitalization has a long way in affecting the performance of commercial banks in Nigeria. Even though the relationship is small, yet it is very significant indicating that commercial banks in Nigeria need to pay more attention on the digitalization processes of banking services to suit the demands of customers. If the digitalization processes of banking services are perfected, the issue of customers still visiting the banking halls and waiting on long queues to solve small issues will be a thing of the past. The essence of digitalizing the processes of banking services is to reduce waiting time, break bottlenecks and make transactions faster. If this is adequately and correctly implemented, then digitalization process should have a larger significant positive relationship with performance of commercial banks in Nigeria ceteris paribus.

Finally, still on the findings of this study, it is concluding to state that product innovation affects the performance of commercial banks in Nigeria. Though the result showed a small positive relationship between product innovation and performance of commercial banks, the management of Nigerian commercial banks can devote more to the enhancement of their various products and services. This is not to say that there are not enough banking products and services, but the very current ones should be worked upon and improved to meet the needs and demands of customers so as to avoid them to consistent visit to commercial banks and waiting in queues to be attended to.

Acknowledgements
This is to appreciate the Management of Covenant University and Covenant Centre for Research, Innovation and Discovery (CUCRID) for the support given to carry out and publish this study.

References

[1] Abbasi T and Weigand H 2017 The Impact of Digital Financial Services on Firm's Performance: *A Literature Review*. Retrieved from https://arxiv.org/pdf/1705.10294.pdf on 22/01/2019

[2] Abuka C A 2019 The Impact of digitalization on financial services. *New Vision: Uganda's Leading Daily*, January 2019. Retrieved from https://www.newvision.co.ug/new_vision/news/1468909/digitalisation-impact-financial-services on 16/01/2019

[3] Agboola M G, Asokomeh B, Akinbode M, Peter F and Deborah M 2017 Technology Acceptance and Usage: A Catalyst for Better Performance of the 21st Century SMEs in Nigeria. *Advanced Science Letters, 23*(9), 9329-9333. doi: https://doi.org/10.1166/asl.2017.10079

[4] Akinbode M, Adegbuyi O, Kehinde O, Agboola M and Olokundun M 2019 Percieved Value Dimensions on Online Shopping Intention: The Role of Trust and Culture. *Academy of Strategic Management Journal, 18*(1), 1-20. Retrieved from https://www.abacademies.org/articles/percieved-value-dimensions-on-online-shopping-intention-the-role-of-trust-and-culture-7895.html

[5] Bandi S, Angadi M and Shivarama J 2015 Best Practices in Digitisation: Planning and Workflow Processes. *Emerging Technologies and Future of Libraries: Issues and Challenges.* Pp. 332-339. Retrieved from http://eprints.rclis.org/24577/1/Digitalization%20ETFL-2015.pdf on 19/01/2019

[6] Carbo-Valverde S 2017 The Impact on Digitalization on Banking and Financial Stability. *Journal of Financial Management Markets and Institutions*, Vol 5 (1). Pp 133-140

[7] Carletti M 2017 *Whitepaper: FINTECH and the Digitalization of Financial Services*. Management and Advisory services Mas Ltd.

[8] Coiera E 2000 Information Economics and the Internet. *Journal of the American Medical Informatics Association*, 7(3): 215-221.

[9] Forest H and Rose D 2015 Delighting Customers and Democratising Finance: Digitalisation and the Future of Commercial Banking. *Global Transaction Banking*: Deutsche Bank

[10] Gareth L, Stephen G, Hua Z and Juan M 2016 *Global Tech Spending Forecast: Banking Edition*. Technical Report, Celent

[11] Garg A 2018, August 8. *What is the difference between digitization and digitalization?* Retrieved from Quora: https://www.quora.com/What-is-the-difference-between-digitization-and-digitalization

[12] Garg A 2018, August 8. *What is the difference between digitization and digitalization?* Retrieved from Quora: https://www.quora.com/What-is-the-difference-between-digitization-and-digitalization

[13] Happiest Mind 2014 *Banking Digitalization Revolution: A Way Forward.* Happiest Mind.

[14] Harigaya T 2016 *Effects of Digitalization on Financial Behaviours: Experimental Evidence from the Philippines*, pp. 1-50. Retrieved from http://scholar.harvard.edu/files/tharigaya/files/harigaya_jmp_11.23.16.pdf on 22/1/2019

[15] Idiegbeyan- ose, J, Ilo P and Isiakpona C 2015 The 21st Century Library and Information Services as a Veritable Strategy for the Enhancement of Teacher Education. Handbook of Research on Enhancing Teacher Education with Advanced Instructional Technologies, *IGI GLOBAL*, pp. 20-33

[16] Kamra S 2015 Digital Transformation in Banking – The Future of Banking. *Happiest Minds Technologies Report*. Retrieved from https://www.happiestminds.com/whitepapers/digital-transformation-in-banking.pdf

[17] Khadka K and Maharjan S 2017 *Customer Satisfaction and Customer Loyalty.* Jakobstad: Centria University of Applied Sciences.

[18] Magatef S G and Tomalieh E F 2015 The Impact of Customer Loyalty Programs on Customer Retention. *International Journal of Business and Social Science*, 8(1): 78-93.

[19] Marous J 2019 Five Innovation Trends That Will Define Banking in 2019. *Digital Banking Report*, retrieved from https://thefinancialbrand.com/77869/innovation-trends-banking-ai-api-personalization-payments/

[20] Maxima Q and Kim J Y 2018, June 06. *Achieving Financial Inclusion: Fintech, account usage, and innovation*. Retrieved from The World Bank: https://blogs.worldbank.org/voices/achieving-financial-inclusion-fintech-account-usage-and-innovation

[21] Mueni M B 2017. The influence of digitalization on the performance of digitalized commercial banks in Kenya. A Research Project Report Submitted to the Chandaria School of Business in Partial Fulfilment of the Requirements for the Degree of Masters in Business Administration (MBA). United States International University-Africa.

[22] Mukherjee A 2017, April 20. *Reducing Poverty in India with the Power of Digital Payments and UBI.* Retrieved from Center for Global Development: https://www.cgdev.org/blog/reducing-poverty-india-with-power-digital-payments-UBI

[23] Naseem K 2017, July 06. *Why digital banking services are vital to reduce poverty in the developing world.* Retrieved from Medium: https://medium.com/@IFC_org/why-digital-banking-services-are-vital-to-reduce-poverty-in-the-developing-world-35ac35758f04

[24] Nigeria Today 2017 *Nigeria Records Drop in Digital Technology, Smartphone Sales*. Retrieved May 11,

2017, from http://www.nigeriatoday.ng/2017/03/nigeria-records-drop-in-digital-technology-smartphone-sales/

[25] Nyangosi R Nyangau S N, Nyariki K O and Nyangau A S 2014 Digitizing Banking Services: An Empirical Analysis of Customer's Adoption and Usage. *Research Journal of Finance and Accounting*, 5(8): 47-53.

[26] Ortstad R and Sonono B 2017 *The Effects of the Digital Transformation Process on Banks' Relationship with Customers: Case Study of a Large Swedish Bank.*

[27] Piirainen L 2016 *Digitalization of the Financial Sector and Change Management.* Oulu University of Applied Sciences.

[28] Satyendra 2016 *Digitalization Process*. Retrieved from http://ispatguru.com/digitalization-process/ on 23/1/2019

[29] Wright K B 2017 Researching Internet-Based Populations: Advantages and Disadvantages of Online Survey Research, Online Questionnaire Authoring Software Packages, and Web Survey Services. *Journal of Computer-Mediated Communication*, https://doi.org/10.1111/j.1083-6101.2005.tb00259.x.

IOP Conf. Series: Earth and Environmental Science **331** (2019) 012015 doi:10.1088/1755-1315/331/1/012015

Twenty-first century technology of combating wildfire

D K De[a,b], O C Olawole[a], E S Joel[a], U I Ikono[a], S O Oyedepo[d], O F Olawole[e], O Obaseki[f] , I Oduniyi[c], M Omeje, A A Ayoola[g], M R Usikalu, T V Omotosho[a], E O Bolujo[h], and I P Ilo[i]

[a]Department of Physics, Covenant University, Canaan Land, Ota, Ogun State, Nigeria
[b]Sustainable Green Power Technologies, Mansfield, Texas 76063, USA.
[c]Research Engineer, Vicaland Global Ventures Limited, Ogun State, Nigeria
[d]Department of Mechanical Engineering, Covenant University, Canaan Land, Ota, Ogun State, Nigeria
[e]Department of Physics, Mountain Top University, Mowe, Ibafo, Ogun State, Nigeria
[f]Department of Physics, Landmark University, Omu Aran, Kwara State, Nigeria
[g]Department of Chemical Engineering, Covenant University, Canaan Land, Ota, Ogun State, Nigeria
[h]Department of Petroleum Engineering, Covenant University, Canaan Land, Ota, Ogun State, Nigeria
[i]Covenant University centre for Learning Resources

·E-mail: Dilip.de@covenantuniversity.edu.ng

Abstract. Wildfire causes havoc and serious destruction of forest resources and wildlife. It also causes occasionally serious damages to human lives and properties. Current technology of fighting wildfire employs fire retardant chemicals, which pollute the environment seriously, and damages both wild and aquatic lives. Based on the new novel technology, it is possible that in the near future the industrial emissions that pollute the environments and cause health problems would be completely captured in frozen solid forms of emissions products. One of these is the frozen carbon dioxide, which is known as dry ice, which would be available than in hundreds of millions of tons in any country. Cost analysis shows that the power and cement industries would significantly benefit by selling the dry ice captured using the technology, just at $0.07 per kg. The present paper discusses the new technology of applying dry ice that would thus be abundantly available in fighting a wildfire. The paper discusses how the new technology would be by far superior to current technology that employs fire retardant chemicals in terms of environmental protection and the cost-effectiveness over the current technology.

1. Introduction

Wildfire remains a deleterious threat to the animal existence, animal habitats, local weather pattern, organic flora and fauna. Researchers have found out that any affected habitats would experience a non-hygienic environment and atmosphere due to the high level of nuisance smoke and carbon dioxide from the wildfire [1]. The wildfire is a strange phenomenon that catches the firefighters unaware because it occurs in a secluded area that is filled with dry leaves, tree and wood logs as depicted in

IOP Conf. Series: Earth and Environmental Science **331** (2019) 012015 doi:10.1088/1755-1315/331/1/012015

Figure 1. The start-up of the fire known as surface fire and the out-of-control is called crown fire as shown in Figure 1 which might be caused by artificial agent such barbeque party and smoking or through natural action like an elevated temperature during summer and concentration of sunlight via lenses on a particular point for a longer time. Besides, wildfire destroys millions of hectare of forest on a yearly basis[2]. According to the fire rule of the game: one minute typifies one cup of water, two minutes typify a hundred litres of water and ten minutes stand one thousand litres of water. For decades, forest fires have been a major concern to the European Union (EU), and as a result, the European Forest Fire Information System (EFFIS) was established by the EU in 2000 [3] to provide a robust database on wildfires for Middle East, North Africa and European.

The annual report issued by the EFFIS in 2016 recorded over 54,000 wildfires across Europe, engulfing an area of almost 376 thousand hectares. The values reported in 2016 indicates a decrease in the various wildfire incidents by about 20% when compared to the average reported forest fires incidents between 2006-2015 [4]. This percentage reduction can be accredited in part to technological advancements in early forest fire detection. However, forest fires remain a major issue that is claiming lives and properties; and early detection and quick response are necessary to combat it. There are several types of forests fires such as crown or tree fires, ground fires and surface fires [5]; each of them requiring specific counteractions to successfully combat them. Forest fire detection has been carried out in several ways, from manned to the unmanned systems. In recent times, forest fire detection has witnessed tremendous improvement through the use of the latest technology; some of which involves the installation of Infrared cameras to indicate the presence of fire. However, much focus is on the potential detection of fires by examining broader smoke areas, or by recognizing environmental parameters prior to the spread of the forest fire.

Figure 1: Fire-brigadiers wrestling with wildfire[1]

2 Monitoring of Wildfire with Latest Technology

2.1 Unmanned Aerial Vehicles

The adoption of aerial vehicles in combatting forest fires began in the 1920s through the dropping water off aircraft to extinguish forest fires [6,7]. However, in these nascent years, using aircraft in the extinguishing of fire was not very successful. Therefore the aircraft were used for the detection of forest fires instead. After the Second World War, in late 1960, the use of aircraft to extinguish forest fires remerged with the use of aeroplanes and helicopters [6,7]. However, it was found to be both costly and risky as pilots are expected to fly really close to the fires. Also, the annual number of firefighting related aircraft accidents raised public concerns about its reliability and the merits of its use.

Developmental advancements of unmanned aerial vehicles (UAVs) have created a unique opportunity for the use of aircraft in forest firefighting as they can replace piloted aircraft. In recent years, various UAVs have been assessed as answers to the early awareness of wildfires [8]. An example of such was a sophisticated drone (enhanced with multi-rotor and a blimp) fabricated by Krüll *et al.* [9]; another is the project (project code: 2263) by Balkan-Med area as to monitor and forecast the trace of forest fire as shown in Figure 2 [10,11]. A rudimentary configuration of the system uses a network of ground cameras to provide steady surveillance of the forest region targeted. However, challenges involved in this implementation are the sites where the cameras ought to be installed which may include national parks and forest reserves. Also, it is a given that camera-based systems raise a lot of false positive signals. Therefore, to provide solutions to these concerns, a multi-rotor UAV is kept stand by to investigate signals given off by the ground cameras. This will solve the problems of false signals as well as put to efficient use the limited flight time of drones.

A more comprehensive configuration involves chiefly, the use of both a fixed winged and rotary winged UAVs which allows for close monitoring of difficult terrain that does not favour the use of ground cameras. The fixed winged high endurance UAVs which can last up to 8-10 hours are equipped with infrared cameras to provide long term surveillance of the region as depicted in Figure 2. As soon as increments in temperature levels are detected by the cameras, signals with the GPS coordinates of the region is sent to the controlled station. The UAV will, however, continue to patrol and monitor the region. When the signals are sent and received, two smaller rotary winged drones will be deployed to observe the region in close proximity to confirm whether or not a false signal was given off. Further advancement of the studies involves the use of surveillance data to explore the possibilities of implementing artificial intelligence to further improve forest fire monitoring systems using UAVs.

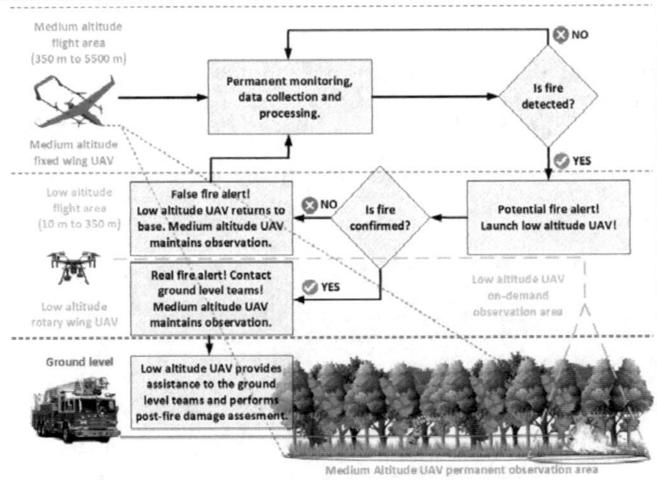

Figure 2: Network of unmanned aerial vehicles wildfire

2.1.1 *Wire-less Sensor-Network Technique*

The work of Alkhatib [2] has indicated that the wireless sensor network (WSN) is a suitable network to give a quick signal on the outbreak of a forest fire in the target area. This sensor is presently receiving global attention because of its usefulness in detecting forest fire, sensing its environment and computes the data. The wireless sensor network (WSN) has the potential to raise alarm, trip off during power surge, communicate distress messages and differentiate between the smoke of fake fires from real fires [28]. This WSN can also be adopted for early detection of a fire outbreak [28]. The wireless device tracks the presence of carbon (IV) oxide, carbon (II) oxide, nitrogen (II) oxide, pressure, relative humidity and temperature in the atmosphere. Consequentially, the sensors are economical self-organizing and self-healing wireless networking. A series of investigation has been conducted through WSN, specifically in a wood fire early detection. FireWxNet wireless device is seen to have the potential to supervise weather conditions of a harsh wildfire territory [13]. Also, in Figure 3, web cameras and wireless sensor-network have been used as a hybrid wood fire detection.

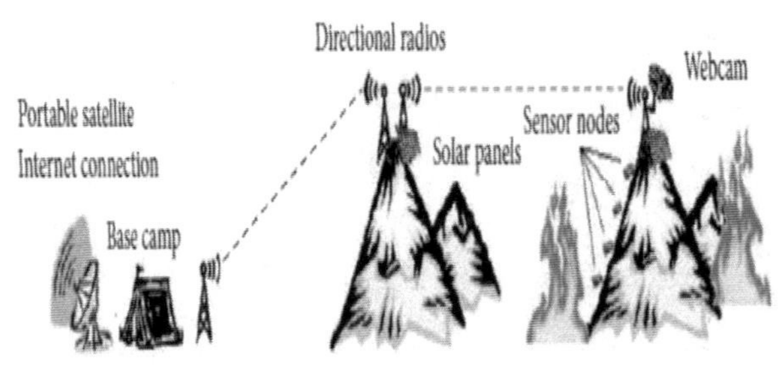

Figure 3: Schematic diagram of FireWxNet [13]

IOP Conf. Series: Earth and Environmental Science **331** (2019) 012015 doi:10.1088/1755-1315/331/1/012015

Interestingly, the adoption of GPS by Conrad et al. (2009) [14] showed that forest fires destroy nearly 2,554 acres yearly and this causes loss of socio-economic, human life and environment. Their study revealed that the use of GPS and fire-sensors devices as shown in Figure 4 are more suitable and effective than the beautiful look-out towers, aerial patrol and public phone lines.

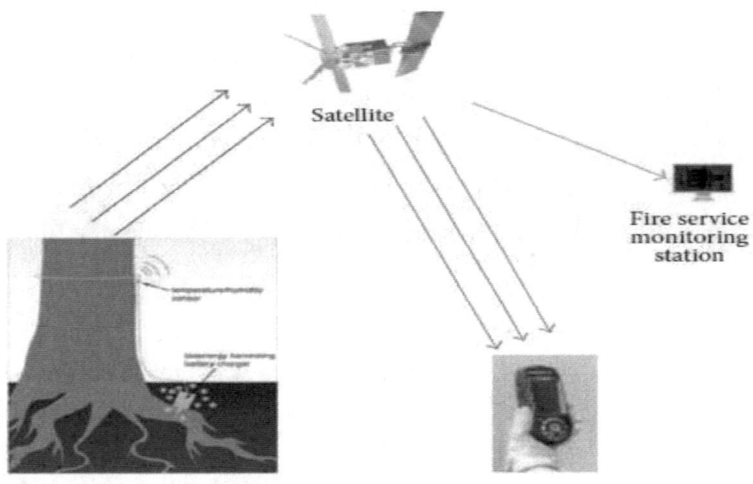

Figure 4: Outlook of GPS in detecting forest fire [14]

European Union has launched multi-sensor devices to monitor, protect and preserve their ecology from the activities of wildfire. Specifically, in the area volatile to the harsh weather condition as depicted in Figure 5 [15]

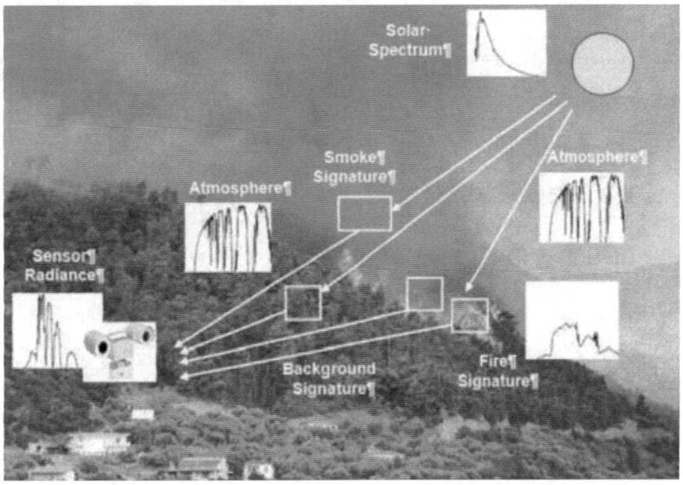

Figure 5: Wildfire monitoring with multi-sensor devices [15]

2.1.2 Spacecraft Technique

Spacecraft based technique has been adopted by researchers to have timely information and feedback on the occurrence of wildfire of such, are the advanced very high-resolution radiometer [4] and common resolution imaging spectroradiometer [5,16]. This method suffers limitless drawbacks in terms of accuracy, feedback and early detection due to the absence of antenna, frequency translation, downlink transmission, amplification reception and transponders in the mounted satellite for forecasting forest fire occurrence. In addition, the feedback of geostationary spacecraft (GEOS) and low earth orbit Spacecraft (LEOS) show that both satellites are not suitable for detecting wildfire because they are situated on orbits of 22,800 miles far apart from the earth. Substantially, inverse square law predicts that the GEOS and LEOS would become insensitive to the intensity of the infrared and optical radiation give off by wildfire flame at a far distance [Alkhatib Ahmad].

2.1.3 High-Tech Sensor and Camera Devices

In recent time, wildfires are supervised for early detection via cordless sensor and camera devices. Broadly speaking, the sensors are capable of tracking the nature of the smoke, thermal level of the fire, and backscattering of laser rays from the smoke of the wildfire [2]. These high-tech devices for monitoring wildfire operate on various algorithms according to their demands from manufactures;

Figure 6: Outlook of forest watch gadget [17]

however, they all monitor and send a signal to the control room on the amount of smoke and fire on the location of interest as represented in Figure 6.

Consequential, the Thailand government has adopted the use of camera devices (AlarmEYE) that can detect black, infrared and colour frequency of forest fires [18,19]. Also, Fire companies in Australia are using EYEFi equipped with GIS technology to visualize the smoke level and predict the time of forest fire [18–20]. France is utilizing UraFire with underlying science of clustering motion dependent on time input to observe smoke [18,21,22]. Additionally, forest fire finder, which uses the principle of how much chemical constituent of the atmosphere is absorbed by sunlight, is used to detect forest fire. It can also distinguish the occurrence of organic smoke from a forest fire and industrial smoke as far as 15 km away as depicted in Figure 7. This method is employed in Portugal to address wildfire menace [18,23].

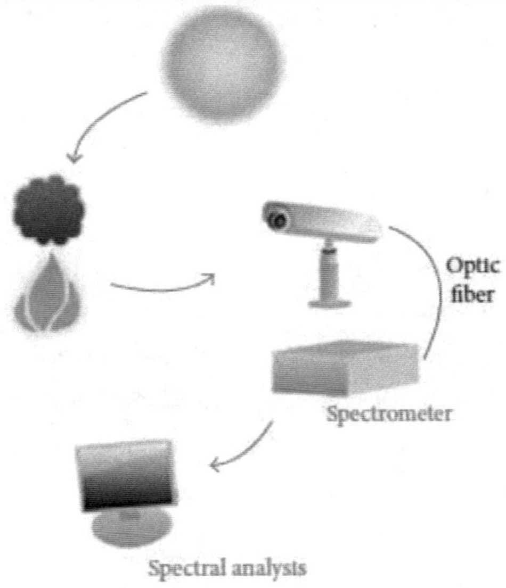

Figure 7: Forest fire detector [23]

Besides, forest watch is a semiautomatic device that can detect smoke of wildfire in the proximity of 16-20 km and deliver feedback in the range of 0.25 Mbps 3 G [19]. Fire agencies in South Africa, Greece, Swaziland, USA, Chile, Slovakia and Canada have adopted the forest watch technology to track the occurrence of a wildfire [2,19]. In recent time, fire watch with automated features that can detect and recognize smoke in the nearness of 10-40 km is being embraced by fire agencies in Cyprus, Czech Republic, USA, Spain, Greece, Germany, Estonia and Mexico [19]. Figure 8 portrays how the optical sensor system moves revolve 360^0 within 4 to 6 minutes during the day and rotate 8-12 minutes at night hours. Transfer of data from the site of interest to the central office is equally noted in Figure 8.

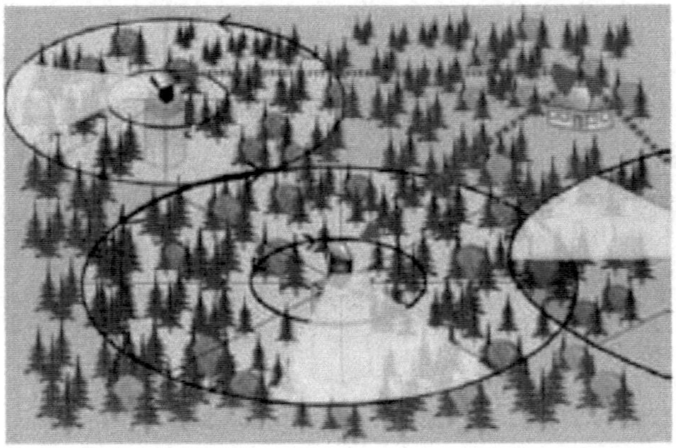

Figure 8: Schematic diagram of the fire watch [24]

Consequentially, three types of wildfire, research burning, agency burning and private burning were tested with EYEfi, Forest watch and fire watch methods are depicted in Table 1 [8]. In that, tower observer in Figure 9a gave precise and fast feedback to the control room ahead of EYEfi, Forest watch and Fire watch. This result shows that the detection effectiveness is dependent on the distance from the camera and fire magnitude. Matthews et al. (2010) observed that the manufacturers EYEfi, Forest watch and Fire watch failed to integrate the localization calculations of the land topography.

Figure 9a: Comparison of Tower observer with EYEfi, Fire watch and Forest fire [19]

IOP Conf. Series: Earth and Environmental Science **331** (2019) 012015 doi:10.1088/1755-1315/331/1/012015

Table 1: Performance of EYEfi, Forest watch, Fire watch and Tower observer on Research Fire [2]

Specimen	EYEfi	Forest watch	Fire watch	Tower observer
6-research fire	0	0	1	6
6- stationary plus two at night	0	3	3	nil

More importantly, Table 2 shows the comparison that exists among unmanned aerial vehicle, wireless sensor network, spacecraft technique and high-tech sensor camera devices.

Table 2: Summary of fire detectors technology [2]

Comparison	UAV	WSN	SCT	HTCD
Cost	Moderate	Moderate	Very High	High
Efficiency	High	High	Low	Moderate
False alarm(s)	Low	Moderate	Low	Moderate
Accuracy	High	High	Moderate	Moderate
Detection delay	long	Small	Very long	Long
Fire behaviour information	Yes	Yes	Yes	None
Other purposes	Yes	Yes	Yes	No

Unmanned Aerial Vehicle (UAV), Wireless Sensor Network (WSN), Space-Craft Technique (SCT) and High-Tech Camera Devices (HTCD)

2.1.4. Carbon (IV) oxide Technique

Wildfires of any size and ferocity can be quenched fairly rapidly if we could drop from helicopters a large number of small chunks of dry ice (solid CO_2) and liquefied CO_2, dry ice + water slush in vast quantities in and around the fireplaces. The immediate surroundings where the fire has not touched yet (say about 100 meters from the fire lines) should be completely cleared of any trees, plants, shrubs or anything that can catch fire. This clearing should be done as fast as possible. These clean areas should also be covered (up to 2- 3" thick) heavily with chunks of dry ice (-78 C) and liquefied CO_2 up to a width of 15 meters in the direction of approaching the fire. Liquefied CO_2 should be sprayed along with water from helicopters directly on the inferno and around and in places where the fire has not touched yet. Liquefied CO_2 sprayed from helicopters would create a thick layer where oxygen will be

IOP Conf. Series: Earth and Environmental Science **331** (2019) 012015 doi:10.1088/1755-1315/331/1/012015

mostly displaced. The water dropped from a helicopter will force the dry ice to accelerate the release of CO_2 and replenish the CO_2 that is diffusing out.

In a region where intense wildfire is approaching fast, the following should be done to protect the houses which are at least 200 meter from the fire lines: Spread dry ice chunks surrounding each house (outside up to a width of at least 1 m, 2-3" thick and parallel to the sides) and on the roof and inside, and outside verandah, after evacuating all the people. Spread liquefied CO_2 or CO_2 gas inside enough and close all windows and doors. Remove as many trees and dry combustible staff from yards as soon as possible if fire approaches the house through the ground. Then it will stop at the outside as the immediate area will be cold and filled with CO_2. Even if fire penetrates inside somehow, its intensity will be substantially less, as inside of the house is filled with CO_2. If fire touches the roof, it will not be intense if dry ice is spread there. Following the above methods, the inferno will be quenched, and the fire approaching line will retreat. If \$1.08 MN is spent for the liquefied CO_2 (50%) at a cost of \$50-55 per ton, see new emission capture technology below and dry ice (50%) at a cost of \$50-55 per ton, then these materials should be able to quench (when dropped from helicopters and trucks) complete wildfire (20 m high) over an area of $1\ km \times 1\ km$. It will save lives and properties. The wildfire in Carr [sees Fig. 9b], California can be extinguished in a far shorter time and with fewer personals and with less or no water when the above technique is used.

The above ingredients will be much more effective than any current fire retardant chemicals. Unfortunately, modern technologies cannot produce dry ice or liquefied CO_2 cost effectively.

To use dry ice in a vast amount to combat such huge wildfire fast we need different technology. The firefighter team must carry oxygen cylinders to breathe well during close combat with fire.

3. Sustainable Green Power Technologies

De and Oduniyi have developed a new technology to capture industrial emissions without uses of any chemicals/reagents [25]. The capture process produces very cost effectively a vast amount of liquefied CO_2 and dry ice. The cost of dry ice/liquefied CO_2 thus produced or liquefied can be as low as \$50 to \$70 per ton in markets (current market price is \$1000 to \$3000 per ton). Once implemented in industries, this new technology can ensure clean energy from coal and natural gas power plants and help mitigate global warming. The vast amount of dry ice that will be captured during the said process can quench wildfires like those in Carr, California. Wildfires also release a huge amount of CO_2 and other toxic gases that are bad for the environment and health. Such CO_2 can also contribute to global warming. Thus, the new technology will have a double blessing for us. With the vast amount of dry ice or liquefied CO_2 that is available at a very low price (\$27 per ton of dry ice at the coal/natural gas power plant), we can quench such wildfires easily, fast and very cost-effectively. The technology can be adapted by any industries such as cement, power, steel etc. that produce a lot of emissions. The cost includes the cost of separate capture of associated toxic gases (in small amounts).

Thus, the new technology when employed in power plants and cement industries will not only capture [25]. The unique feature of the new technology is that the capture of emissions is paid back by sales of just 30% of the captured products. Thus, it is profitable for industries (power plants, cement and steel plants etc.) to employ the technology unlike existing emission capture technologies [excluding government incentives]. This technology cannot only ensure a fully green planet for us in future but also could provide us with the means to quench wildfires rapidly. In addition, the investigation of Dirisu *et al.* [27] recommended that firefighter should not exceed 85 seconds in and out of a building under fire.

Despite all that said above, severe punishment is needed for anyone deliberately setting the wildfire. Punishment also should be there for negligence that can cause a wildfire. All dead and dry wood must be removed as precautionary measures to prevent wildfire.

Figure 9b: The sun sets over an area affected by the Carr Fire, which had burned more than 160,000 acres [26]

IOP Conf. Series: Earth and Environmental Science **331** (2019) 012015 doi:10.1088/1755-1315/331/1/012015

4. Strategy of Rapidly fighting wildfire

A scheme of rapid wildfire fighting is shown in Fig. 10. The central area A is the inferno where the fire is raging to a great height. The surrounding area B and C are two areas where the fire is gradually spreading and area D and E fires have not touched yet. With helicopter dry ice is dropped in areas B, C and D. The dry ice in area D should also be spread by firefighters and the dry ice should cover the entire circle up to a width of at least 25 m and 3" thick. The area E (width 25 m) also should be covered with dry ice after clearing all trees, debris and shrubs. With helicopter liquefied CO_2 and water or dry ice + water, slush should be dropped on the inferno area A.

Dry ice in area B and C will soon sublimate and fill the space with gaseous CO_2. Fire will be quenched in these areas soon. While the inferno in area A is being battled, it will not find easy to spread to area B and C as long as these areas are being sprayed with liquefied CO_2. Dry ice will displace ground air completely while liquefied CO_2 spread from a helicopter over B and C will displace air up to a height of 30 m. This will prevent the fire from inferno area A to spread to B and C. Then the area D & E should be also be sprayed with small chunks of dry ice. When a fire in areas B and C have subsided with the above procedure, area D should be cleared of trees, shrubs (grasses) and debris. Then again, some dry ice should be spread over area D. While area E is cleared of all debris and shrubs and covered with dry ice up to a width of 25 m and thick ness 3", the fire will not spread further. If dry ice and liquefied CO_2 is collected from coal and natural gas power plants where dry ice should be available at $27 to $30 per ton, then the cost of fighting wild fire(inferno) over a radius of 50 m and surrounding would not exceed $16000. Once thus quenched vast area can be freed of wildfires. This estimate assumes that while Inferno is raging over a radius of 50 m, an area with diameter 100 m (with centre coinciding with that of the inferno) will be covered with CO_2 gas up to a height of 10 m.

IOP Conf. Series: Earth and Environmental Science **331** (2019) 012015 doi:10.1088/1755-1315/331/1/012015

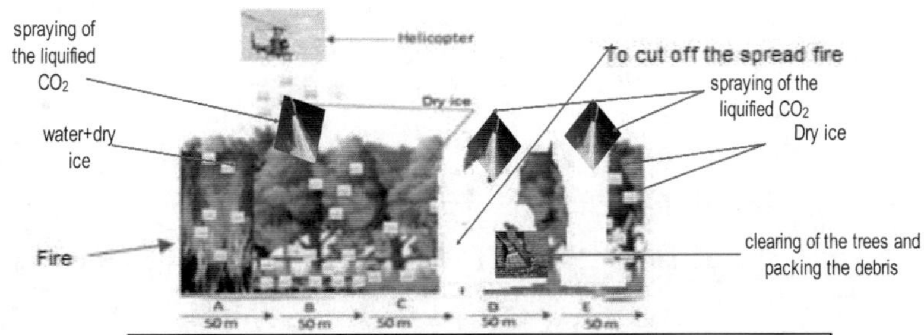

Figure 10: Schematic diagram of combating wildfire

Halting the spread of the fire in Figure 10, first the helicopter to spray liquefied CO_2, dry ice plus water slush in large quantity are positioned at area A. Second, the dropping of dry ice plus water slush is significantly available in area B. Third, a drop of dry ice chunks to cover the floor with up to 1.5 to 2" thick is seen in area C. Four, clearing of wood and debris and then dropping off dry ice chunks to cover up to 1" thick in conjunction with spray liquefied CO_2 are provided in area D. Fifth, clear wood and debris and spray liquefied CO_2.

Lastly, the firefighters on the ground who are clearing wood and debris must have an oxygen tank and mask. Once the fire is contained fully it will not spread further. The whole area of Figure 10 will be engulfed easily with a huge ball of CO_2 that will displace the air. Part of the water sprayed on the ground from the helicopter will dissolve CO_2 from dry ice and prevent the spread of fire further.

5. Conclusions

Conclusively, this study has compared various techniques for detecting and fighting wildfire. Wireless sensor network which is quite expensive looked suitable for the early detection and in the fight against the spread of wildfire. It would be seen that the method of spraying and spreading liquefied CO_2 and dry ice are cost-effective over other methods of putting out wildfire as stated in this research. More importantly, since the end product is CO_2 after the fire is quenched, then, the vegetation would grow back faster than other techniques of fighting the fire. We have proposed an innovative method of controlling a future wildfire in the sense that the emissions from coal and natural gas power plants can be captured cost-effectively and dry ice/liquefied CO_2 can be made available at the power plants at the rate $27-$30 per ton and at $50-$70 in the markets. This study recommends individual, private bodies, state and federal government of a nation to adopt the method in Figure 10 in addressing the imminent and future occurrence of wildfire.

Acknowledgement: The authors express profound gratitude to Covenant University Management via the Covenant University Centre for Research, Innovation and Discovery (CUCRID) for providing conference publication financial aid.

References

[1] Borenstein S 2018: Hotter weather turbocharges US West wildfires *Phys. Today* **71** 44–50
[2] Alkhatib A A A 2014 A Review on Forest Fire Detection Techniques *Int. J. Distrib. Sens. Networks* **10** 597368
[3] Anon 2004 Disclaimer System function
[4] Hernández Z, Sánchez D, Pecci J, Intrigiolo D S and Erena M *Use of remote sensing for the calculation of biophysical indicators*
[5] Anon 1999 NASA, MOIDS, National Aeronautics and space administration
[6] Anon FIRE-DEX: health and amp; Safety-focused firefighter
[7] Jendsch W 2008 *Aerial firefighting* (Schiffer Pub)
[8] Yuan C, Zhang Y and Liu Z 2015 A survey on technologies for automatic forest fire monitoring, detection, and fighting using unmanned aerial vehicles and remote sensing techniques *Can. J. For. Res.* **45** 783–92
[9] Krüll W, Tobera R, Willms I, Essen H and von Wahl N 2012 Early Forest Fire Detection and

Verification using Optical Smoke, Gas and Microwave Sensors *Procedia Eng.* **45** 584–94

[10] SFEDA Forest Monitoring System for Early Fire Detection and Assessment in the Balkan-Med Area

[11] Moulianitis V C, Thanellas G, Xanthopoulos N and Aspragathos N A 2019 Evaluation of UAV Based Schemes for Forest Fire Monitoring (Springer, Cham) pp 143–50

[12] Hristov G, Raychev J, Kinaneva D and Zahariev P 2018 Emerging Methods for Early Detection of Forest Fires Using Unmanned Aerial Vehicles and Lorawan Sensor Networks *2018 28th EAEEIE Annual Conference (EAEEIE)* (IEEE) pp 1–9

[13] Hartung C, Han R, Seielstad C and Holbrook S 2006 *FireWxNet: A Multi-Tiered Portable Wireless System for Monitoring Weather Conditions in Wildland Fire Environments,*p41.

[14] Conrad A, Liu Q, Russell J and Lalla J 2009 *Enhanced Forest Fire Detection System with GPS Pennsylvania.*

[15] Kose K, Tsalakanidou F, Besbes H and et al. 2010 FIRESENSE: Fire Detection and Management through a Multi-Sensor Network for the Protection of Cultural Heritage Areas from the Risk of Fire and Extreme Weather Conditions. CmpE WEB *Proceedings of the 7th Framework Programme for Research and Technological Development*

[16] Nakau K, Fukuda M, Kushida K, Hayasaka H, Kimura K and Tani H 2006 *Forest fire detection based on MODIS satellite imagery, and Comparison of NOA A satellite imagery with fire fighters' information,* 18-23.

[17] Son B, Her Y and Kim J-G 2006 *A Design and Implementation of Forest-Fires Surveillance System based on Wireless Sensor Networks for South Korea Mountains, International Journal of Computer Science and Network Security,* **6** (9), 124-130.

[18] EUFOFINET *Project Detection Synthesis of Good Practices,* (National Forest Centre, Zvolen, Slovakia)

[19] Mathews S, Ellis P and Hurle J H 2010 *Evaluation of Three Systems*, Bush fire Cooperative Research Centre, Australia,)

[20] French N H F, Goovaerts P and Kasischke E S 2004 Uncertainty in estimating carbon emissions from boreal forest fires *J. Geophys. Res* **109** 14–22

[21] Aslan Y E, Korpeoglu I and Ulusoy Ö 2012 A framework for use of wireless sensor networks in forest fire detection and monitoring *Comput. Environ. Urban Syst.* **36,** 614–25

[22] Guillemant P and Vicente J 2001 Real-time identification of smoke images by clustering motions on a fractal curve with a temporal embedding method *Opt. Eng.* **40** 554

[23] NGNS 2013 Forest fire finder http://www.ngns-is.com/ eng/FFF eng.html.

[24] FireWatch 2013 An early warning system for forest fires, successfully in the global use

[25] De K D and Oduniyi I A 2017 A novel and highly cost-effective technology for capture of industrial emissions without reagent for clean energy and clean environment applications-submitted (Nov. 20, 2017) to USA Patent Office (USPO). Patent Appl. No. 15915007. Under Examination.

[26] Https://pressfrom.info/us/news/offbeat/-173561-mendocino-fire-racing-at- unprecedented-speed-into-the-record-books.html

[27] Dirisu J O, Oyedepo S O, Fayomi O S and Okokpujie I P 2019 Ignition time of selected ceiling materials and escape temperature time prediction of fire fighter rescue mission. *Journal of Chemical Technology and Metallurgy*, **54** (1),1-8.

[28] Okokpujie, K, John, S N, Osaghae, E N, Okokpujie I P and Okonige R. 2019 A wireless sensor network based fire protection system with sms alerts. *International Journal of Mechanical Engineering and Technology*, **10** (2),1-10.

Noise Pollution and Waste Control Techniques in Building Construction in Nigeria: A Literature Review

Theme: Built Environment

A Sholanke, T Aina-Badejo, A Aina-Babajide and A Nara Jacob.

Department of Architecture, Covenant University, Ota, Ogun State, Nigeria.

E-mail: anthony.sholanke@cu.edu.ng, tolu.ainabadejo@gmail.com, ainababajide@yahoo.com, digitalbalanceltd@gmail.com

Abstract. The activities of the construction industry are often plagued with improper material control and operational measures that result in the production of noise and generation of waste in project sites. Waste and noise are generated in every step of the construction process, from the production of materials used, to the erection and finishing of structures. In this age of incorporating sustainability into the development of buildings and their environments, there is a need for the construction industry to substantially reduce the amount of noise and waste produced. This paper investigated various noise pollution and waste control techniques used in the construction industry in Nigeria, with a view to identify areas for further improvement. The paper is a systematic literature review which made use of document analysis to collect data and content analysis to analyse the data gathered. Descriptive approach was used to present the result. The paper examined the current state of waste and noise management and the recommendations provided. The study found that in Nigeria, the noise control technique employed is the use of noise barriers, while the waste control management strategies used are prefabricated elements, sorting waste generated, open dumping, open burning and composting disposal method. Some of the recommendations made include: discarding open dumping and open burning as waste control measures due to their negative environmental impact, developing and enforcing a legal framework for noise and construction waste. Furthermore, educating site workers on noise and waste control techniques and replacing machines that use combustion engines with those that use electrical motors. Students, educators, building industry practitioners and other construction industry stakeholders will find the study a useful material to work with, as well as build upon.

Keywords: Noise Pollution, Waste Control, Construction, Literature Review and Nigeria.

1. Introduction

The role of the construction industry is notably significant when it comes to a nation's economic development. The construction sector activities are considered the most important source of stimulus for development and the growth of the economy [1]. However, despite the benefits it provides, the impact on the society and the environment is tremendous when considering the relationship of the construction industry with other sectors such as the service sector and the various industries that deal in material production, distribution, how resources are being used and the quantity that is required [2, 3]. Nigeria as a developing country has a lot to catch up on. The country can to achieve a sustainable environment by learning from the mistakes of others. The activities of the construction processes generate noise and waste which pose a severe threat to the environment. Their effects happen over time, from crude materials being processed, to their utilization in construction, to the maintenance and demolition of the structure [4]. Therefore, there is a need to control the amount of noise and waste generated in order to limit the effect they have on the environment. Also, most of the techniques available for noise pollution and waste control in Nigeria are used sparingly, while some of the control strategies that are being used have been dropped by developed countries due to their unsustainability, negative impact on the environment and having a damaging effect on the wellbeing of the public and workers on construction sites. The design of a building and the materials that are procured strongly influences how waste on a construction project is generated [4]. Therefore, this paper identified the sources of noise pollution and waste generation in building construction and the various techniques of controlling them in Nigeria, in order to identify areas for further improvement. The paper also suggested additional noise and waste control techniques that can effectively minimise waste and noise, reducing the impact of their activities on the environment and eliminate additional cost on projects in Nigeria. The study's scope is limited to the activities that occur during construction. Students, educators, building industry practitioners and other stakeholders will find the study a useful material to build upon and work with. The following part of the paper is divided into 4 sections namely: research methodology, results, conclusion and acknowledgements.

2. Research Methodology

The paper is a systematic literature review. Data was gathered from secondary sources by contextual analysis. The data were mainly collected from internet sources. The internet sources used include: research gate and science direct sites, as well as google search engine and google scholar. The data gathered were analysed by content analysis. The result was grouped in themes and presented using descriptive approach with the aid of a table and diagram for better understanding.

3. Results

3.1. Noise Pollution and Construction Waste

Pollution refers to changes that are undesirable. These changes occur in the air, land and water affecting their biological, physical or chemical composition and may harmfully affect our living conditions, raw material resources, human life and other species [5]. Pollution in building construction is mainly in the form of environmental pollution which also includes air, dust and noise pollution. According to Dixon [6], the estimate of the pollution of the world that can be attributed to buildings are: air quality in cities (23%), drinking water pollution (40%), climate change gases (50%), ozone depletion (50%) and landfill waste (50%). Environments around construction sites are usually disturbed by the noise pollution generated from construction activities. This negatively affects neighbours and construction workers [7].

Several authors have varying definitions of what constitute waste in construction. Material waste in construction can mostly be considered as a by-product that is produced as a result of activities relating to construction, renovation and demolition of structures [8]. Ayegba [9] referred to construction material wastes as those materials supplied to the site for construction, but were not used. Ameh *et al* [8] viewed it as wastefulness that occurs from the utilization of assets in larger amounts than necessary in the construction of building. Construction waste also include debris removed from construction sites and disposals in landfills. From the above narratives noise pollution that emanates from construction activities are a source of disturbance to both neighbours and construction workers. While construction wastes are not only hazardous to people but the environment as well. It is therefore important that these two construction by-products are effectively controlled in order to minimize their adverse effects.

IOP Conf. Series: Earth and Environmental Science **331** (2019) 012016 doi:10.1088/1755-1315/331/1/012016

3.2. Sources of Noise Pollution and Waste in Building Construction

The source of noise pollution in building construction comes from the activities that take place at the construction site and the equipment that are used, such as bulldozers, air compressors, pneumatic hammers, loaders, pavement breakers and dump trucks [10]. The source of construction waste during building construction comes from the materials used, which are by-products of construction processes [11]. These by-products are caused by some factors shown in table 1 and ranked in order of their significance.

Table 1. Causes of construction waste [12]

Causes of Construction Waste	Mean Rank Value	Rank
Poor site management and supervision	7.17	1
Lack of experience	7.71	2
Inadequate planning and scheduling	7.78	3
Mistakes and errors in design	8.00	4
Mistakes during construction	8.34	5
Incompetent subcontractors	8.76	6
Rework	9.02	7
Frequent design changes	9.24	8
Labour productivity	9.35	9
Inadequate monitoring and control	9.40	10
Inaccurate quantity take-off	9.74	11
Shortage of site workers	10.09	12
Lack of coordination between parties	10.21	13
Slow information flow between parties	10.61	14
Shortage of technical personnel (skilled labour)	10.91	15
Changes in material specification and type	10.93	16
Equipment availability and failure	11.83	17
Effect of weather	11.90	18

Table 1 shows that the highest ranked sources in descending order are poor site management and supervision, lack of experience, inadequate planning and scheduling, mistakes and errors in design and mistakes during construction. The lowest ranked sources in descending order are slow information flow between parties, shortage of technical personnel (skilled labour), changes in material specification and type, equipment availability and failure and effect of weather. Other sources in between include, incompetent subcontractors, rework, frequent design changes, labour productivity, inadequate monitoring and control, shortage of site workers, inaccurate quantity take-off and lack of coordination between parties [12].

Waste can likewise be delegated as physical and non-physical waste as shown in Figure 1.

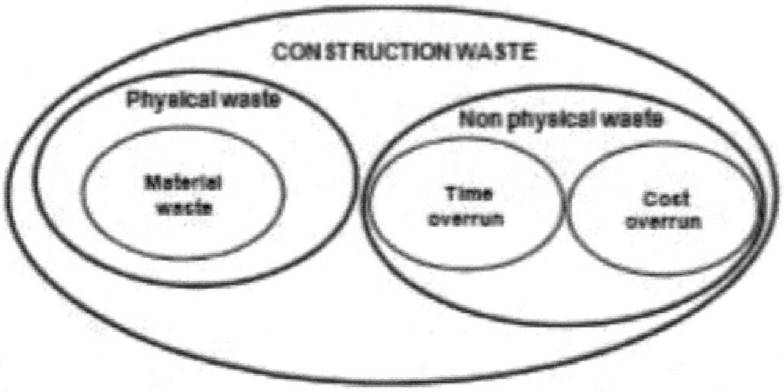

Figure 1. Categorization of waste types in construction [13]

Material waste falls under physical waste, while cost and time overrun are categorised under non-physical waste. Waste can also be grouped under six sources which include: residual, operation, handling, procurement, design and others [14]. According to the just-in-time concept, waste in construction can be grouped into seven categories as follows: waste from defects, delays, over-production, excess inventory, transportation, motion and unnecessary people processing [15].

Wahab *et al* [16] listed other sources of waste as: the design of building; material damage due to weather, inappropriate storage, mishandling or careless delivery; over consumption of resources; rework; vandalism; site office waste; lack of recording; materials supplied and their use on site. Construction waste are generally caused by slip-ups, working out of order, unnecessary repeated actions, untimely sources of information or not addressing issues on time.

3.3. Impact of Construction Waste on Cost

In construction projects, waste contributes up to 15% additional cost [14]. The cost implication of waste is due to the fact that materials become inadequate because of poor preparation, management and storage. The inadequate materials would then require replacement which increases the cost of the project [14]. Additional financial losses are incurred in a construction project due to purchases to replace squandered materials, delays on the project, revisions to address mistakes and managing produced waste, all of which can lead to heavy losses if not handled correctly [17]. Since materials represent around 30% to 70% of construction cost, material wastage results in the range of 20% to 30% of project cost overruns [17]. Materials are a significant component of a construction development venture. Therefore, poor management of materials in construction will lead to a variation in the cost of the project. Materials and installed equipment used in a construction project can account for 50% to 60% respectively, of the total project cost and affect 80% of its schedule [4]. This implies that unless the generation of waste is curtailed and effectively managed, the construction industry will continue to experience unnecessary financial losses. To this end, there is a need to find ways of improving the situation.

3.4. Impact of Noise Pollution and Construction Waste on the Environment

Exposure to noise pollution over time can lead to severe health risks to site workers and neighbours such as hearing impairment, stress, cardio vascular disease, hypertension, sleeping disturbance, loss of voice, loss of attention. It can also lead to lower productivity for site workers [7]. Therefore, the reduction of noise pollution is necessary to protect the health and wellbeing of site workers and neighbours.

Construction wastes that are not properly handled cause increasingly serious problems to the environment [18]. When there is no clear framework for waste control, they are disposed of illegally. This harms the environment and depletes landfills resources. Illegal practices include: illegal dumping in deserted and forested areas; dumping waste in waterways; mixing construction waste with domestic waste; burying waste in abandoned sites and illegal burning of waste [18]. Therefore, controlling

IOP Conf. Series: Earth and Environmental Science **331** (2019) 012016 doi:10.1088/1755-1315/331/1/012016

waste generation and noise pollution on construction sites is necessary in order to minimize the negative effect they have on the environment.

3.5. Noise Pollution and Waste Control Techniques in Building Construction

There are various methods of controlling and managing noise pollution and the generation of construction waste in a project site. Some of these control measures are examined in the following sections.

3.5.1. Noise Control Techniques. Noise control can be categorised into two main groups as follows: passive noise control and active noise control. These control measures aid in mitigating the transmission and emission of noise. Passive noise control focuses on reducing the transmission of noise, while active noise control involves reducing noise emission from its source [7].

Passive noise control includes the following: planning and organising the construction site layout of site in order to limit the exposure of workers to noise and to allow equipment and material move freely within the site without obstacles; organising construction activities in order to prevent unnecessary movement of equipment and materials around the site and organizing equipment to be able to move freely without obstacle. Others include: using noise isolation or sound proof barriers wherever possible; isolating noisy activities to noise zones where they would cause the least disturbance; marking noisy zones to prevent workers from entering unnecessarily; using hearing protection devices for site workers and adequately training personnel in order to avoid inadequate use of equipment and to appropriately protect themselves [7].

Active noise control techniques include: substituting equipment that use internal combustion engines with electric motors; use of power tools in place of traditional tools such as using nail guns instead of hammers; substituting noise generating activities with quieter methods and use of concrete mixers should be done at the construction plant preferably. Other control techniques are: substituting cranes with electric lifts if possible; use of damping to reduce impact noise; maintenance of machines periodically in order to reduce noise output and modifying machines to generate less noise. Also, the use of walkie-talkies for communicating instead of shouting commands and minimising the operating time for an equipment that generates noise, especially if it cannot be replaced by a less noisy alternative [7].

3.5.2. Waste Control Techniques. Reducing waste in construction can be achieved through the use of various techniques such as: site management, using standardized construction materials, adopting waste management systems, using waste rate estimation to improve the material handling, improve productivity and reduce the rate of waste. The use of alternative construction methods such as large panel systems, modularization, prefabricated components, off-site construction techniques and reducing the need for wet works can also be employed to reduce waste. Other methods include sorting construction waste as it is generated, stock control to minimize over ordering and educating site workers [4,14]. Waste management is necessary to deal with the already generated waste in an effective manner in order to reduce their effect on the environment. It involves using alternatives to waste disposal which are to: reduce, reuse, recycle and recover. These four techniques are known as the R's of construction waste management, while the fifth is disposal.

i. **Reduce:** This is the first process in waste management. It involves reducing the amount of material that is produced for building construction. It can be achieved by reviewing how projects are designed in order to produce less waste, coordinating a project properly, avoiding material over-ordering, using standard quantities and sizes of materials and ensuring that stores are secure, safe and weatherproof.
ii. **Reuse:** This method is effective for increasing cost savings on a construction project. This is achieved by reusing surplus materials in different projects, using cut-offs and selling off surplus materials. It helps to reduce the demand for new material and the impact of construction waste has on the environment.

iii. **Recycle:** This method begins by sorting the waste generated into different categories. The reusable materials are separated from solid and unsafe ones, treated and returned back to the economy as a constituent of other products. The benefits of recycling include the reduced environmental impact of waste, improved cost effectiveness of waste handling and the conservation of natural resources.

iv. **Recovery**: Recovery can be described as the removal of components or materials from the stream of waste such that it can be reused in the same form it was originally produced in. It reduces the volume of waste ending up in a landfill.

v. **Disposal:** Disposal is the least sustainable waste management method out of the above options. It is difficult achieving sustainability because it involves dumping the waste generated into the environment and does not contribute towards reducing generated waste. However, construction waste is disposed of in landfills by many countries [14].

Figure 2 shows waste reduction hierachy from the most favoured to the least favour in descending order which is reduce, reuse, recycle, recovery and landfill.

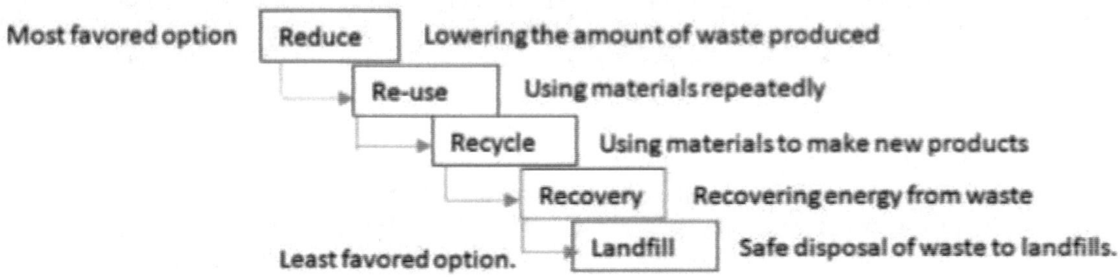

Figure 2. Waste Reduction Heirarchy[14].

Some other methods that can be used to minimise the quantity of waste generated on construction projects are itemised as follows:

i. Co-ordination between construction companies and designers;
ii. Encourage effective communication among project stakeholders in order to keep relevant parties abreast on steps taken;
iii. Planning the arrival of materials and products in order to stock them appropriately and in the order that they will be used;
iv. Storing gravel, sand, soil and other comparable materials effectively to prevent them from spilling;
v. Removing accident spills of materials before the day end;
vi. Rejecting faulty materials when delivered in order to avoid disposal;
vii. Materials delivered on site should be packed in order to reduce cracks during transportation and handling;
viii. The number and amount of packaging materials should be reduced, especially those that are difficult to recycling or reuse;
ix. Selecting products that leave the least amount of residue; and
x. Storing materials appropriately in safe areas to protect them from deterioration;

The just-in-time (JIT) concept is another technique used to reduce waste and increase productivity. It is guided by six principles. The principles are: eliminating the categories of waste; uninterrupted workflow; Kanban or pull system; total quality control which ensures that completed works are free from defects before going on to the following step; employee participation by providing training and responsibilities in several areas and supplier relations which ensure the delivery of continuous high-quality materials and components [15].

3.6. Noise Pollution and Waste Control Techniques in Nigeria

3.6.1. Noise Control Techniques. There are a number of ways by which noise pollution on construction sites are controlled in Nigeria. They include: modifying existing equipment; using quieter equipment; using barrier protection; maintenance of equipment and using partial enclosures. Others techniques are: total enclosures; work activity scheduling; noise perimeter zones; proper maintenance; modified operating procedures. Similarly, the relocation of noisy vents away from workers; replacement of equipment with a quieter version; modified room treatment to reduce noise transmitted; relocation of equipment; and running equipment at lower speed or less than its full power are other control techniques in Nigeria [19-21].

Reducing noise pollution by dampening at the source is one of the best methods used for noise reduction. Personal Protective Equipment (PPE) is necessary when noise exceed 85 dB(A). Using shock mounting assemblies, noise mapping, and smoothing the transport route to site are also noise reducing techniques employed [22].

3.6.2. Waste Control Techniques. Construction companies in Nigeria have used a number of techniques in order to minimize and manage the generation of waste on their sites, some of which include: use of prefabricated elements; sorting of generated waste; open dumping; open burning; and composting disposal method [16]. Although, the study carried out by Wahab *et al* [16] indicated that the percentage of respondents that used waste control techniques are as follows: use of prefabricated elements (9.10%); sorting of generated waste (14.29%); open dumping (70.5%); open burning (16%); composting disposal method (13.5%). Their study also indicated that: site space as a factor is highly ranked in affecting the choice of on-site sorting; during construction process most sites did not sort out waste generated; there were unplanned deposition of waste on sites and a 100% of the respondents had never calculated the volume of waste that could be generated per surface area using waste indices [16]. All of which has led to indiscriminate dumping of refuse in landfills without considering the negative impact their activities and materials have on the environmental and the public. Dania *et al* [23] showed that the reuse of materials and sale as scrap such as timber for firewood were methods that are widely adopted. Other methods are recycling, reducing the amount of materials used, controlled dumps and waste packaging. To effectively implement site waste management practices, a government legislative framework is required to make waste management a focus [11,24].

Some other important strategies used to reduce waste in descending order include: adequate site supervision and control; proper storage handling and usage of materials; effectual material planning and policy control; ensuring the comprehensive design detailing and specifications are correct; site security; education and training of site personnel; avoiding late changes at critical stages of the development and recycling of materials where economical [15].

4. Conclusion

The study set out to identify the noise pollution and waste control techniques in Nigeria. The result shows the sources of waste construction and noise pollution are from the equipment used and the activities employed on the construction site in addition to mistakes and unnecessary movements. The results also showed the negative impact construction waste has on cost and the damaging effects of construction waste and noise pollution on the environment and other control techniques. It is observed that though professionals in the construction industry are mindful of some of these waste control measures, they do not implement their control techniques adequately. There is a poor practice of waste management concepts and noise pollution techniques. Generally, awareness is needed for construction professionals to know the impact of noise pollution and construction waste on the environment and the public's wellbeing, as well as the benefits of using sustainable control strategies in their projects. The study recommends that outdated waste management practices such as open dumping and burning should be stopped because of their negative impact on the environment. Also, a legal framework for noise and construction waste should be developed and enforced. Furthermore, site workers should be educated on noise and waste control techniques and machines that use combustion engines should be replaced with those that use electrical motors. The authors recognise that the sources and the scope used for the paper are a limitation to the study. Hence, further studies could explore the use of more internet search engines to gather data. In addition, suggested areas for further research include: comparative analysis of noise control and waste management techniques employed before and after construction projects; noise pollution and waste management strategies used in construction projects in other countries; and to carry out similar studies using primary sources to collect data.

Acknowledgements
The authors are grateful to Covenant university for providing the funds, facilities and an enabling environment to carry out and publish this paper. The authors are also appreciative of researchers whose work were used and duly acknowledged in the study.

References

[1] Khan R A 2008 Role of construction sector in economic growth: empirical evidence from pakistan economy *1st Int. Conf. on Construction in Dev. Countries (ICCIDC–I)*

[2] Oladinrin T O, Ogunsemi D R and Aje I O 2012 Role of Construction Sector in Economic Growth: Empirical Evidence From Nigeria *FUTY Journal of the Environment* pp. 50-60

[3] Horsley A, France C and Quartermass B 2003 Delivering Energy Efficient Buildings: A Design Procedure to Demonstrate Environmental and Economic Benefits *Journal of Construction Management and Economics* pp. 246-345

[4] Gulghane A A and Khandve P V 2015 Management for Construction Materials and Control of Construction Waste in Construction Industry: A Review *Int. Journal of Engineering Research and Applications* pp. 59-64

[5] National Research Council (U.S.) 1966 Committee on Pollution, "Waste Management and Control," National Academies, Washington, D.C.

[6] Dixon W 2010 The Impacts of Construction and the Built Environment.

[7] Xin N, Jingyan Q, Wu C and Wang W 2019 Reducing noise Pollution by Planning Construction Site Layout via a Multi-Objective Optimization Model *Journal of Cleaner Production* pp. 218-30

[8] Ameh J O and Itodo D E 2013 Professionals' View of Material Wastage on Construction Sites and Cost Overruns *Int.l Journal of Organization, Tech. and in Construction* pp. 747-57

[9] Ayegba C 2013 An Assessment of Material Management on Building Construction Sites *Journal of Civil and Environmental Research* pp. 18-22

[10] Ijaiya H 2014 The Legal Regime of Noise Pollution in Nigeria *Beijing Law Review* pp. 1-6

[11] Adeagbo D O, Achuenu E and Oyemogun I M 2016 Construction Material Waste Managemtent Practices In Selected Construction Sites in Abuja, Nigeria *FUTA Journal of Management and Techn.y* pp. 96-104

[12] Sasitharan N, Ismail R A, Ade A, Aftab H and Rosli Z M 2012 Identifying Causes of Construction Waste - Case of Central Region of Peninsula Malaysia *Int.l Journal of Integrated Engineering* pp. 22-8

[13] Nagapan S, Rahman I A, Asmi A, Memon A H and Latif I 2012 Issues on Construction Waste: The Need for Sustainable Waste Management *Colloquium on Humanities, Science & Engineering Research* Sabah Malasia

[14] Nwachukwu L, Orji S, Agu E and Owoh C 2016 Reducing Material Wastes in Building Construction Sites: An Action for Sustainable Development *Civil and Environmental Research* pp. 50-7

[15] Odusami K, Oladiran O and Ibrahim S 2012 Evaluation of Materials Wastage and Control in Some Selected Building Sites in Nigeria *Emirates Journal for Engineering Research* pp. 53-65

[16] Wahab A B and Lawal A F 2011 An evaluation of waste control measures in construction industry in Nigeria *African Journal of Environmental Science and Tech.* pp. 246-54

[17] Adewumi T O and Otali M 2013 Evaluation of Causes of Construction Material Waste: Case of Rivers State, Nigeria *Ethiopian journal of Environmental Studies and Management* pp. 746-53

[18] Sapuay S. E. 2016 Construction Waste – Potentials and Constraints *Int.l Conf. on Solid Waste Management*

[19] Owhor S C, Abdul A I G, Ojo V K and Dan A D 2015 Innovative Control of Noise and Vibration of Industrial Equipments and Machines *American Journal of Engineering Research* pp. 193-205

[20] Oyedepo S O 2012 Noise Pollution in Urban Areas: The Neglected Dimensions *Environmental Research Journal* pp. 259-71

[21] Omubo-Pepple V B, Briggs-Kamara M A and Tamunobereton-ari I 2010 Noise Pollution in Port Harcourt Metropolis: Sources, Effects, and Control *The Pacific Journal of Science and Tech.* pp. 592-600

[22] Oyedepo S O 2013 Effective Noise Control Measures and Sustainable Development in Nigeria *World Journal of Environmental Engineering* pp. 5-15

[23] Dania A A, Kehinde J O and Bala K 2007 A Study of Construction Material Waste Management Practices by Construction Firms in Nigeria *The 3rd Scottish Conf. for Postgraduate Researchers of the Built and Natural Environment*, Glasgow, Scottland

[24] Amusan L, Akintaro H, Osawaru F and Makinde A 2018 Information on State of Challenges of Waste Management System in Nigeria Urban Housing System *Int. Journal of Mechanical and Production Engineering Research and Development* pp. 75-86

Neuro-fuzzy mid-term forecasting of electricity consumption using meteorological data

Paul A. Adedeji [1a], Stephen Akinlabi [b,c], Nkosinathi Madushele [a], and Obafemi Olatunji [a]
[a] Department of Mechanical Engineering Science, University of Johannesburg, South Africa
[b] Department of Mechanical and Industrial Engineering, University of Johannesburg, South Africa
[c] Department of Mechanical Engineering, Covenant University, Ota, Nigeria.

Email: pauladedeji2k5@gmail.com

Abstract. Forecasting energy consumption is highly essential for strategic and operational planning. This study uses the Adaptive-Neuro-Fuzzy Inference System (ANFIS) for a mid-term forecast of electricity consumption. The model comprises of three meteorological variables as inputs and electricity consumption as output. Two ANFIS models with two clustering techniques (Fuzzy c-Means (FCM) and Grid Partitioning (GP) were developed (ANFIS-FCM and ANFIS-GP) to forecast monthly energy consumption based on meteorological variables. The performance of each model was determined using known statistical metrics. This compares the predicted electricity consumption with the observed and a statistical significance between the two reported. ANFIS-FCM model recorded a better mean absolute deviation (MAD), root mean square (RMSE), and mean absolute percentage error (MAPE) values of 0.396, 0.738, and 8.613 respectively compared to the ANFIS-GP model, which has MAD, RMSE, and MAPE values of 0.450, 0.762, and 9.430 values respectively. The study established that FCM is a good clustering technique in ANFIS compared to GP and recommended a comparison between the two techniques on hybrid ANFIS model.

Keywords: ANFIS; Electricity Consumption; FCM; GP; Mid-term Forecasting

1. Introduction

Over the years, electricity has been observed to be central to a sustainable economy [1] and has been considered as the most important resource in the global economy [2]. It has also experienced global increased demand despite the annual increase in generation. Many developing countries still have low electricity access, most especially sub-Saharan Africa [3], [4]. Population increase, technological advancements, human migration, increase in energy related anthropogenic activities and many more have resulted in increased electricity demand. One of the groups of exogenous variables which influence electricity consumption is the meteorological variable. It has been established that air temperature affects electricity consumption [5]. It highly influences energy consumption during the winter seasons in temperate regions, where electricity is largely used for space heating [6]. Among other factors which influence electricity consumption is the relative humidity, wind speed, solar

IOP Conf. Series: Earth and Environmental Science **331** (2019) 012017 doi:10.1088/1755-1315/331/1/012017

irradiance, cloud cover and so on [7]–[9]. In order to efficiently manage the demand side of energy system, there is need for electricity consumption forecast relative to meteorological variables which can be used for the purpose of strategic planning.

The literature is replete with studies on electricity consumption forecast either in the form of an input-output model [10] or as an autoregressive model [11]. These forecasts has been divided based on the time scales to short term, mid-term and long-term forecast [12]. The short-term forecasts are hourly, daily, and weekly time scale forecasts. The mid-term forecast, on the other hand, suggests forecast on a monthly time scale while the long-term forecasts on the yearly time scale. Many of these studies are related to energy consumption in buildings. For example, an investigation of the a campus energy usage was performed by Huang et al. [13]. The authors applied support vector regression (SVR) for a short-term prediction of the next 24-hours consumption in the institution. The day of the week, weekend indicators, temperature factors, historical hourly energy consumption profile, and occupancy indicators were used as model inputs while the output remains the 24-hour- ahead energy consumption. Similarly, a study by Massana et al. [2] on the short-term load forecasting for non-residential buildings considered meteorological variables, occupancy, calendar, indoor environment, to forecast electric load of office buildings in a university environment. The study compared the efficiency of multiple linear regression (MLR), support vector regression (SVR), and multi-layer perceptron (MLP) for the forecast. Meteorological data are fuzzy in nature and this has a tendency of affecting the model output. Some studies have investigated the use of other models with fuzziness in the data accounted for. One of the commonest modelling techniques in this regard is the Adaptive Neuro-fuzzy Inference System (ANFIS).

ANFIS technique is a modeling technique which combines the functionality of both the artificial neural network and fuzzy logic in a data-driven model [14]. The adaptive nature of ANFIS makes a non-linear mapping between inputs and output possible within n-dimensional space. ANFIS models follow a supervised learning approach with inputs and outputs. The technique has been used in different studies some of which includes hydrology [15], [16], energy systems [17], [18], geosciences [19] and so on. It has also been used in electricity consumption forecast. For example, Cevik and Cunkas [20] experimented the use of fuzzy logic (FL) and ANFIS for load forecasting. While the model inputs were historical load, temperature difference and season the output is the load forecast. The study established the abilities of both models to be able to process large amount of data with accuracy of the forecast not compromised. Similarly, a one-step ahead forecast was performed by Cheng [21]. The study established the inability of statistical forecasting tools to perform efficiently in a non-linear data space. The study used a time series ANFIS model with a-step ahead integration. Also, Gholamreza et al. [10] applied ANFIS modelling to electricity demand using a combination of meteorological variables and socia-economic variables. Recently, the technique has been hybridized with optimization models like Particle Swarm Optimization (PSO) Genetic Algorithm(GA), Differential Evolution(DE) [22], [23], and so on towards achieving a global optimal parameter for effective predictions. Increased prediction accuracy has been recorded from these hybrids. Significant to the ANFIS model is the clustering technique. Clustering tools group multi-dimensional data into clusters based on density or distance functions according to specified radius of influence. Significant to this study is the study by Tiwari et al. [24]. This study investigated the efficiency of two clustering techniques: the subtractive clustering and the fuzzy-c-means clustering technique on water quality index of a river. However, the point of departure in this present study is the investigation of grid partitioning (GP) and fuzzy-c-means ANFIS integrated clustering technique on electricity consumption forecast. Several studies that utilizes ANFIS models for energy consumption forecast are

IOP Conf. Series: Earth and Environmental Science **331** (2019) 012017 doi:10.1088/1755-1315/331/1/012017

building related, however, this study applies ANFIS to forecast for a nexus of electricity consuming sectors which comprises the Gauteng Province of South Africa.

The world of fourth industrial revolution is tending towards not only smart manufacturing but smart buildings which include residential and non-residential buildings. These buildings are expected to be able to capture ambient variables to efficiently and intelligently optimize the energy consumption without user's interference. Such buildings require intelligent algorithms with fuzzy capabilities for intelligent forecast, which forms one of the relevance of this study. One of these intelligent algorithms is the ANFIS modelling tool selected in this study. Gauteng province used as a case study in this paper constitute 24.1 % South African Population [25]. The province constitutes one provinces in the country with high consumption of electricity [26]. These consumptions largely come from industrial, residential, and non-residential buildings. However, this study investigates the effect of the choice of clustering technique on ANFIS modeling in electricity consumption forecast relative to meteorological variables.

2. Methodology

Two clustering based ANFIS models were developed in this study: the ANFIS-FCM and the ANFIS-GP model. The two models were built with same inputs and output as shown in Figure 1. The performances of these models were evaluated using known statistical metrics and the results reported. The Gauteng province of South Africa was chosen as a case study.

2.1. Data Collection

Meteorological data from representative weather stations in the province were collected from the South African Weather Service within a period of 2007 to 2017 on monthly basis. These stations were selected based on proximity to residential and non-residential and buildings, knowing well that meteorological variables have a close relationship with electricity consumption [9], [27]. This was achieved using GoogleEarth Pro. The meteorological data used in this study include the average temperature, relative humidity, and wind speed as identified to influence electricity consumption. Monthly electricity consumption for the province within the same number of years was collected from the South African Monthly Bulletin.

2.2. Model description

The general ANFIS model follows the Type-II Takagi-Sugeno inference system [28]–[30]. The ANFIS model is a five-layered network consisting of the fuzzy layer, the product layer, the normalized layer, the de-fuzzification layer, and the total output layer. More information on these layers can be obtained in [28], [30]. ANFIS consists of adaptive nodes interconnected by directional links [31]. The ANFIS model used in this work comprises of the three meteorological variables as inputs and the electricity consumption within the period of the data collection as output (as shown in Figure 1).

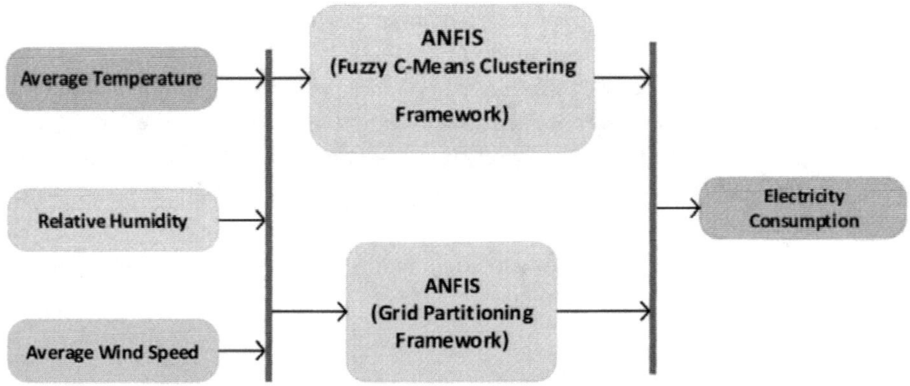

Figure 1. The forecasting model using ANFIS framework

Clustering forms an integral part of the ANFIS modeling technique. This technique groups individual members into fuzzy classes. It employs several properties to find the similarity groups in the data [32], [33]. Several clustering methods have been identified in the literature with each having its peculiarities. These are classified into partitioning methods, the hierarchical methods, the density-based methods and the model-based methods [34], [35]. In this study, two clustering techniques (FCM and GP) were integrated into the ANFIS model.

2.2.1. Fuzzy C-Mean Clustering

The FCM technique follows the conventional Euclidean distance function that includes hyperspherical clusters [36]. The technique is a fuzzified version of the k-means algorithm which minimizes the least square error function $J_m(U,v)$ also known as the c-means function as presented in eqn. (1). The FCM performs well when member classes are not pre-specified. It is very good at suggesting suitable classes for observations. In this study, the number of clusters, partitioning exponent, maximum number of iterations and minimum improvement used were 10, 2, 100, and 1×10^{-5} respectively.

$$J_m(U,v) = \sum_{i=1}^{N}\sum_{j=1}^{C} U_{ij}^{m} \left\|x_i - v_j\right\|_{A'}^{2}, \qquad 1 \leq m \leq \infty \tag{1}$$

where v_j = the centre of cluster j

C = number of clusters | $2 \leq C < n$

m = weighting exponent

x_i= vector data of observations

A = positive definite $(n\times n)$ weight matrix

$\|\ \ \|$ = n −dimensional Euclidean space wherein sample data belong.

2.2.2. Grid Partitioning

This is the most commonly used fuzzy partitioning technique to generate fuzzy members from observations within a dataset. Two other fuzzy partitioning techniques are the tree partitioning and the scatter partitioning. The GP technique divides the input space into different fuzzy slices with each specified by a membership function thereby forming a partition. The technique uses similar membership functions on the input space to generate equal partitions within the symmetric membership function [37]. It has been established that finer grids offers a better model performance [38]. It should be noted that one of the common setbacks of this method is what is referred to as the "curse of dimensionality". This is the exponential increase in the rule base as the problem size increase [39]. In this model, the Gaussian membership function was used for input and the linear membership function for output. The number of fuzzy rules is relative to the input-output.

3. Results and Discussion

3.1. Model result

The two models were trained and tested with 70 % (84 months) and 30 % (36 months) of the data collected respectively. Shown in Figure 2 is a comparison plot between the actual energy consumption and that produced in the use of the ANFIS-GP model. The predicted ANFIS-GP model fits the trend in the observed data. From Figure 2, a large deviation is observed between the actual electricity consumption and the forecast at June 2015, October 2015, September 2016. While there was significant over-prediction in June 2015 and October 2015, there was significant under-prediction in September 2016. This could be due to misfiring of the ANFIS node. The other time steps during the forecast appears closer to the actual electricity consumption.

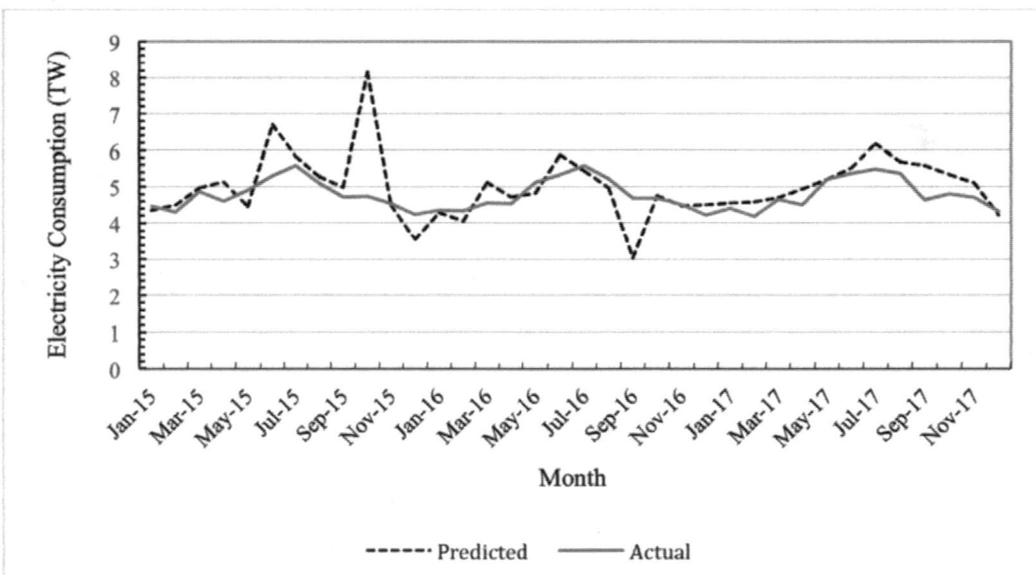

Figure 2. Prediction results for the ANFIS-GP model at testing phase between January 2015 to Dec 2017

Statistical analysis between the actual and predicted results obtained with ANFIS-GP was performed and presented in Table 1. At a confidence level of $\alpha = 0.05$, F (1.808) $< F_{crit}(3.978)$. The result shows that there is no statistically significant difference between the observed and the predicted electricity consumption.

Table 1. Statistical results between observed energy consumption and the predicted energy consumption using the ANFIS-GP model.

Source of Variation	*SS*	*df*	*MS*	*F*	*P-value*	*F crit*
Between Groups	0.867	1	0.867	1.808	0.183	3.978
Within Groups	33.573	70	0.480			
Total	34.440	71				

Similarly, a comparison between the actual and the predicted electricity consumption using the ANFIS-FCM model is reported (Figure 3). The result shows that the predicted electricity consumption follows a similar trend with the actual consumption as shown in Figure 3. The model under-predicted in the second month but predicted close to the observed electricity consumption in other months. The prediction in the ANFIS-FCM appears closer to the observed electricity consumption as compared to the ANFIS-GP model. After the significant underprediction in the second month (Figure 3), there was no significant underprediction in subsequent month as opposed to the ANFIS-GP model (Figure 2).

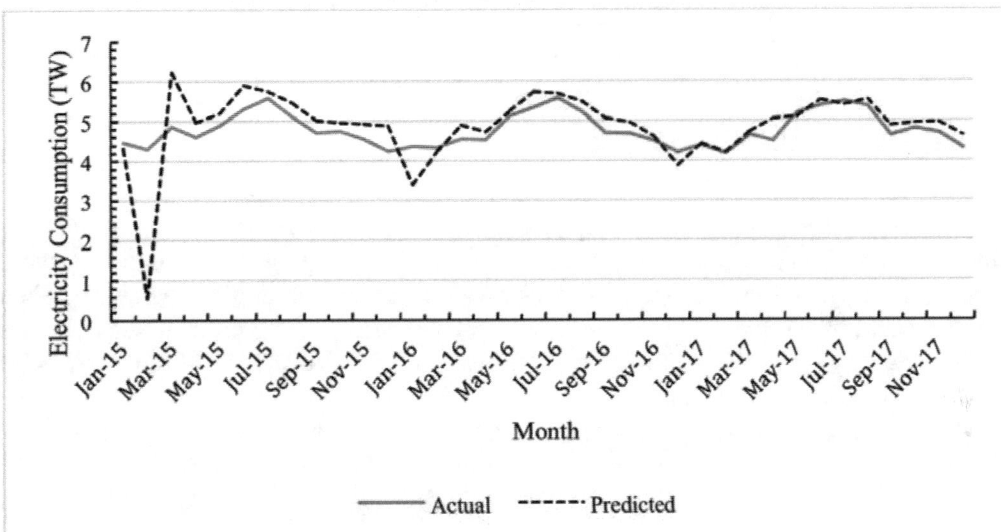

Figure 3. Prediction results for the ANFIS-FCM model at testing phase between January 2015 to Dec 2017

A statistical analysis to determine whether the difference between the observed and the predicted electricity consumption is statistically significant was performed and the results presented in Table 2. At a confidence level $\alpha = 0.05$, $F\ (0.285) < F_{crit}(3.978)$. The result obtained shows that the difference between the observed electricity consumption and the predicted is not statistically significant.

Table 2. Statistical results between observed energy consumption and the predicted energy consumption using the ANFIS-FCM model.

Source of Variation	*SS*	*df*	*MS*	*F*	*P-value*	*F crit*
Between Groups	0.149	1	0.149	0.285	0.595	3.978
Within Groups	36.586	70	0.523			
Total	36.736	71				

3.2. Performance Evaluation

Both clustering techniques show no statistically significant difference between their prediction and the actual energy consumption. In order to determine the model performance, the Mean Absolute Deviation (MAD), Root Mean Square Error (RMSE) and Mean Absolute Percentage Error (MAPE) were used on the predicted electricity consumptions and the results are presented in Table 3. The ANFIS-FCM clustering technique records a lower MAD, RMSE, and MAPE value compared to the ANFIS-GP technique. These lower indices basically signify that the two models can predict better energy consumption. ANFIS-AFCM model records a lesser MAPE value which shows a model accuracy of 91.387% while ANFIS-GP model is 90.57% accurate in prediction. FCM based ANFIS model offers a high prospect of model accuracy and is not associated with curse of dimensionality as known for GP-based models.

Table 3. Model Performance at the testing phase

Performance Metrics	ANFIS-GP	ANFIS-FCM
MAD	0.450	0.396
RMSE	0.762	0.738
MAPE	9.430	8.613

4. Conclusions

The effect of the clustering technique in forecasting electricity consumption using ANFIS modeling technique was investigated in this study. A case study of Gauteng Province, South Africa was used with three meteorological variables as model inputs and electricity consumption as output. The results obtained established that the FCM clustering technique performs better than the GP clustering technique. A study by [40] also establishes the effectiveness of the ANFIS-FCM model. GP clustering technique is also associated with curse of dimensionality, such that as the problem size increase, the rule base increases exponentially and thus resulting into increased computational time [39]. Investigating the effect of the clustering technique on a hybrid ANFIS model can be explored to establish the significance of clustering technique selection on either standalone ANFIS or hybrid ANFIS model.

Acknowledgements

The authors appreciate the South African Weather Service and the Statistics South Africa for providing the meteorological and the electricity consumption data respectively, used in this study.

References

[1] D. Chirambo, "Towards the achievement of SDG 7 in sub-Saharan Africa: Creating synergies between Power Africa, Sustainable Energy for All and climate finance in-order to achieve universal energy access before 2030," *Renew. Sustain. Energy Rev.*, vol. 94, no. July, pp. 600–608, 2018.

[2] J. Massana, C. Pous, J. Melendez, and J. Colomer, "Short-term load forecasting in a non-residential building contrasting models and attributes," *Energy Build.*, vol. 92, pp. 322–330, 2015.

[3] R. Best and P. J. Burke, "Electricity availability: A precondition for faster economic growth?," *Energy Econ.*, vol. 74, pp. 321–329, 2018.

[4] Abel Oyuke, P. H. Penar, and B. Howard, "Off-grid or 'off-on': lack of access, unreliable

electricity supply still plague majority of Africans," *Afrobarom. Dispatch*, no. 75, pp. 1–26, 2016.

[5] M. A. Lokoshchenko and N. A. Nikolaeva, "Dependence of energy consumption on air temperature in Moscow," *Russ. Meteorol. Hydrol.*, vol. 42, no. 12, pp. 783–791, 2017.

[6] S. C. Taylor, S. K. Firth, C. Wang, D. Allinson, M. Quddus, and P. Smith, "Spatial mapping of building energy demand in Great Britain," *GCB Bioenergy*, vol. 6, no. 2, pp. 123–135, 2014.

[7] D. Jose, M. Mathew, and A. Krishnan, "Weather dependency of electricity demand: A case study in warm humid tropical climate," *2016 3rd Int. Conf. Electr. Energy Syst. ICEES 2016*, pp. 102–105, 2016.

[8] F. Apadula, A. Bassini, A. Elli, and S. Scapin, "Relationships between meteorological variables and monthly electricity demand," *Appl. Energy*, vol. 98, pp. 346–356, 2012.

[9] L. Wang, X. Liu, and H. Brown, "Prediction of the impacts of climate change on energy consumption for a medium-size office building with two climate models," *Energy Build.*, vol. 157, pp. 218–226, 2017.

[10] G. Zahedi, S. Azizi, A. Bahadori, A. Elkamel, and S. R. Wan Alwi, "Electricity demand estimation using an adaptive neuro-fuzzy network: A case study from the Ontario province - Canada," *Energy*, vol. 49, no. 1, pp. 323–328, 2013.

[11] P. A. Adedeji, S. Akinlabi, O. Ajayi, and N. Madushele, "Non-linear autoregressive neural network (NARNET) with SSA filtering for a university enegy consumption forecast," in *16th Global Conference on Sustainable Manufacturing- Sustainable Manufacturing for Global Circular Economy*, 2019, pp. 176–183.

[12] H. Guo, Q. Chen, Q. Xia, C. Kang, and X. Zhang, "A monthly electricity consumption forecasting method based on vector error correction model and self-adaptive screening method," *Int. J. Electr. Power Energy Syst.*, vol. 95, pp. 427–439, 2018.

[13] Y. Huang, T. Lu, D. Xianfhua, and N. Gu, "Campus Building Energy Usage Analysis and Prediction: A SVR Approach Based on Muli-scale RDF Kernels," in *International Conference on Human Centered Computing*, 2015, vol. 8944, pp. 441–452.

[14] J. A. Awan and D. H. Bae, "Improving ANFIS Based Model for Long-term Dam Inflow Prediction by Incorporating Monthly Rainfall Forecasts," *Water Resour. Manag.*, vol. 28, no. 5, pp. 1185–1199, 2014.

[15] F. J. Chang and Y. T. Chang, "Adaptive neuro-fuzzy inference system for prediction of water level in reservoir," *Adv. Water Resour.*, vol. 29, no. 1, pp. 1–10, 2006.

[16] F. J. Chang and K. W. Wang, "A systematical water allocation scheme for drought mitigation," *J. Hydrol.*, vol. 507, pp. 124–133, 2013.

[17] S. R. Moreno and L. dos Santos Coelho, "Wind speed forecasting approach based on Singular Spectrum Analysis and Adaptive Neuro Fuzzy Inference System," *Renew. Energy*, vol. 126, pp. 736–754, 2018.

[18] D. Zheng, A. T. Eseye, J. Zhang, and H. Li, "Short-term wind power forecasting using a double-stage hierarchical ANFIS approach for energy management in microgrids," *Prot. Control Mod. Power Syst.*, vol. 2, no. 1, p. 13, 2017.

[19] A. Khashei-Siuki and M. Sarbazi, "Evaluation of ANFIS, ANN, and geostatistical models to spatial distribution of groundwater quality (case study: Mashhad plain in Iran)," *Arab. J. Geosci.*, vol. 8, no. 2, pp. 903–912, 2015.

[20] H. H. Cevik and M. Cunkas, "Short-term load forecasting using fuzzy logic and ANFIS," *Neural Comput. Appl.*, vol. 26, pp. 1355–1367, 2015.

[21] C.-H. Cheng and L.-Y. Wei, "One step-ahead ANFIS time series model for forecasting electricity loads," *Optim Eng*, vol. 11, pp. 303–317, 2010.

[22] L. M. Halabi, S. Mekhilef, and M. Hossain, "Performance evaluation of hybrid adaptive neuro-fuzzy inference system models for predicting monthly global solar radiation," *Appl. Energy*, vol. 213, no. January, pp. 247–261, 2018.

[23] A. Azad, H. Karami, S. Farzin, A. Saeedian, H. Kashi, and F. Sayyahi, "Prediction of water

quality parameters using ANFIS optimized by intelligence algorithms (Case study: Gorganrood River)," *KSCE J. Civ. Eng.*, vol. 00, no. 0000, pp. 1–8, 2017.

[24] S. Tiwari, R. Babbar, and G. Kaur, "Performance Evaluation of Two ANFIS Models for Predicting Water Quality Index of River Satluj (India)," *Adv. Civ. Eng.*, pp. 1–10, 2018.

[25] Z. J. Katima, O. I. Olukunle, O. loanna Kalantzi, A. P. Daso, and J. O. Okonkwo, "The occurrence of brominated flame retardants in the atmosphere of Gauteng Province, South Africa using polyurethane foam passive air samplers and assessment of human exposure," *Environ. Pollut.*, vol. 242, pp. 1894–1903, 2018.

[26] Department of Energy, "A survey of energy-related behaviour and perceptions in South Africa: The residential sector," 2012.

[27] L. O. Idahosa, N. N. Marwa, and J. O. Akotey, "Energy (electricity) consumption in South African hotels: A panel data analysis," *Energy Build.*, vol. 156, pp. 207–217, 2017.

[28] P. Adedeji, N. Madushele, and S. Akinlabi, "Adaptive Neuro-fuzzy Inference System (ANFIS) for a multi-campus institution energy consumption forecast in South Africa," in *Proceedings of the International Conference on Industrial Engineering and Operations Management Washington DC, USA, September 27-29, 2018*, 2018.

[29] D. Karaboga and E. Kaya, "Adaptive network based fuzzy inference system (ANFIS) training approaches: a comprehensive survey," *Artif. Intell. Rev.*, pp. 1–31, 2018.

[30] J. S. R. Jang, "ANFIS: adaptive-network-based fuzzy inference system," *IEEE Trans. Syst. Man. Cybern.*, vol. 23, no. 3, pp. 665–685, 1993.

[31] F. Chahkoutahi and M. Khashei, "A seasonal direct optimal hybrid model of computational intelligence and soft computing techniques for electricity load forecasting," *Energy*, vol. 140, pp. 988–1004, 2017.

[32] J. Bedi and D. Toshniwal, "Deep learning framework to forecast electricity demand," *Appl. Energy*, vol. 238, no. October 2018, pp. 1312–1326, 2019.

[33] R. Dubes and A. K. Jain, "Clustering techniques: The user's dilemma," *Pattern Recognit.*, vol. 8, no. 4, pp. 247–260, 1976.

[34] J. Han, J. Pei, and Kamber, *Data mining: concepts and techniques*. Elsevier, 2011.

[35] H. Motoda *et al.*, *Top 10 algorithms in data mining*, vol. 14, no. 1. 2007.

[36] T. Küçükdeniz, A. Baray, and K. Ecerkale, "Expert Systems with Applications Integrated use of fuzzy c-means and convex programming for capacitated multi-facility location problem," vol. 39, pp. 4306–4314, 2012.

[37] J. Galindo, *Fuzzy Information Processing in Databases*. 2008.

[38] Y. H. Joo, "Fuzzy Systems Modeling : An Introduction," vol. 1, pp. 2009–2011, 2009.

[39] T. Vasileva-Stojanovska, M. Vasileva, T. Malinovski, and V. Trajkovik, "An ANFIS model of quality of experience prediction in education," *Appl. Soft Comput. J.*, vol. 34, pp. 129–138, 2015.

[40] H. Fattahi, "Adaptive Neuro Fuzzy Inference System Based on Fuzzy C–Means Clustering Algorithm, a Technique for Estimation of Tbm Penetration Rate," *Int. J. Optim. Civ. Eng.*, vol. 6, no. 2, pp. 159–171, 2016.

Influence of nanoparticles deposition on surface roughness and heat transfer characteristics of nanofluids – A review

Mercy Ogbonnaya[1,2], Oluseyi O. Ajayi[2,4*], M.A Waheed[3], Sunday O. Oyedepo[2], [4]A.P.I. Popoola, [5]O.M. Popoola

[1]Department of Mechanical Engineering, University of Lagos, Akoka, Lagos State, Nigeria
[2]Department of Mechanical Engineering, Covenant University, P.M.B. 1023, Ota, Ogun State, Nigeria
[3]Department of Mechanical Engineering, Federal University of Agriculture, Abeokuta, Ogun State, Nigeria.
[4]Department of Chemical, Metallurgical and Materials Engineering, Tshwane University of Technology, P.M.B. X680, Pretoria, South Africa
[5] Department of Electrical Engineering, Tshwane University of Technology, P.M.B. X680, Pretoria, South Africa

[*]Corresponding author's email: mogbonnaya@unilag.edu.ng; (+234) 8025888540

ABSTRACTS

The inherent low thermal conductivity of working fluids used in transferring heat from energy system makes the performance of the system less efficient. The introduction of nanoparticles into the working fluid enhances the thermal conductivity of the fluid, thereby improving the overall performance of the energy system. Deposition of nanoparticles on the heat transfer surface of the heat exchanger occurs and this affects the mechanism of bubble formation, wettability, surface roughness of the heating surface and the overall heat transfer coefficient. In this review, the use of nanofluid showed improvement in the critical heat transfer flux (CHF) of the heat transfer fluid while the deposition of nanoparticles on the heater surface enhanced the wettability for pool boiling but caused deterioration in the heat transfer. The influence of nanoparticles deposition on surface characteristics and heat transfer coefficient during single and two phase flows in flow boiling needs to be further investigated.

Keywords: Nanofluids, bubble formation, wettability, heat transfer coefficient surface roughness,

1. Introduction

Enhancement of heat transfer is essential in energy systems such as nuclear reactors, refrigeration and air-conditioning, bio-engineering reactors, heat pumping systems and high power electronic components. The need to develop high performance and energy efficient thermal system, with reduced process time and high thermal rating has lead researchers to carryout various studies on potential methods of enhancing heat transfer processes in thermal systems. Several factors which govern the heat transfer performance of energy systems are [1-4]:

1. Thermophysical properties of the heat transfer fluid such as density, viscosity, thermal conductivity, surface tension and specific heat capacity
2. Properties of the material used for the heat transfer surface such as surface roughness, wettability, surface geometry and orientation
3. Working system properties such as pressure, temperature gradient between the surface and the fluid
4. Rate of bubble growth, departure and movement on the surface
5. Interaction between the fluid and the surface

Several approaches such as application of vibration to the surface, increase in flow rate, utilization of extended surfaces such as fins, use of micro channels, and altering the thermophysical properties and dynamics of the fluids are being used to enhance the heat transfer process and overall performance of industrial processes [5-7]. For heat exchangers, the extension of the surface and increase in the rate of fluid flow are mostly used in the improvement of system's performance but these methods have led to unacceptable enlargement in the size of the energy system and increase in pumping power respectively. Modification of the thermophysical properties especially the thermal conductivity of heat transfer fluid is important in the improvement of the energy systems performance. To achieve this, several researches have been conducted and one of which is the addition of metal and its oxides into the working fluid of an energy system. [8] dispersed micro-sized particles into the heat transfer fluid which increased the thermal conductivity of the fluid considerably but it was not sustainable due to solid particles settling rapidly and depositing on the heater surface hereby creating a layer of particles on the surface. Hence, this reduced the coefficient of heat transfer of the fluid and eroded the surface rapidly. The large size of the particles also caused clogging of the narrow cooling channels and considerably caused the pressure drop in the fluid to increase [9]. These observations hindered further research on the use of micro-sized solid particles in heat transfer fluids for heat transfer enhancement.

Nanofluids were introduced by [5] to overcome the challenges faced with addition of micro-sized particles in heat transfer fluid. Nanofluid is formed by the dispersion of nanoparticles of nominal diameter 1 nm to 100 nm into the base fluid. A considerably increase in the thermal conductivity of the nanofluid when compared to that of the base fluid was observed [5]. Therefore, it was proposed that the challenges encountered with deposition of micro sized particles on the heater surface will most likely not occur in the use of nanoparticles given that the nanoparticles are stable and uniformly distributed in the base fluid. Metals and its oxides in nanosized are mostly used in the preparation of nanofluid due to its high thermal conductivity and its application has been found to enhance the convective and conductive heat transfer and the overall system's performance [10]. According to [11], in a laminar developing region of constant heat flux, the average convective heat transfer coefficient and flow rate increased with the increase in nanoparticle concentration when aluminium oxide nanoparticles of 45 nm and 150 nm were dispersed in water. At constant wall temperature, [12] considered the dispersion of aluminum oxide (Al_2O_3) nanoparticles in water for a circular tube. Enhancement of heat transfer coefficient was found to be caused by the random movement of the nanoparticles in the base fluid and this caused a change in the structure of the flow field of the base fluid. [13] used similar Reynold number to study the convective heat transfer of Cu/water nanofluid at nanoparticles concentration of 2 vol%. The convective heat transfer improved by 60% while the change in the friction factor was negligible for low concentration of nanoparticles.

For proper analysis of the mechanism behind the enhanced heat transfer coefficient, [14] conducted an experimental analysis on the flow boiling in subcooled region using Zinc oxide dispersed in water. The heat transfer coefficient for the nanofluid increased by 126% when compared with water while an increase in the concentration of ZnO particles caused an increase in the surface roughness by 1367%. This increment was attributed to the deposition of nanoparticles on the heater's surface. [15] found out that the deposition of nanoparticle on the surface significantly caused deterioration in the boiling process and boiling characteristics of Al_2O_3/water nanofluid in pool boiling. It was further explained that the deposition caused alteration in the surface roughness hence changing the nucleation site density and caused impediment to bubble movement and stirring, hence affecting the heat transfer characteristics and decrease in boiling performance respectively.

From the above, it can be deduced that the dispersion of nanoparticles into heat transfer fluid has the tendency to enhance the coefficient of heat transfer of the base fluid. Despite this, the pressure drop of fluid flow in a channel and heat transfer are affected by the initial surface roughness and the deposition of nanoparticles on the surface [4,16-17]. With this, the need to properly understand the nucleate and flow boiling heat transfer characteristics of nanofluid, the relationship between the relative size of the nanoparticles and nanoparticles deposition with respect to the roughness of the heater surface becomes eminent. Hence, this review paper helps to achieve this by addressing the effect of nanoparticle deposition on surface micro-geometry, liquid – solid interaction, wettability, the heater surface roughness for pool and flow heat characteristics and the overall heat transfer mechanism.

2. Effect of nanoparticle deposition on wettability and critical heat flux

[18] studied the influence of nanoparticle concentration, boiling duration and initial surface roughness on surface wettability while varying one parameter at a time and controlling the others using nanoparticles. The study considered using aqueous Al_2O_3 nanofluid; Al_2O_3 nanoparticles were deposited on aluminum surface of varying roughness so as to investigate the enhancement of surface wettability due to topography change. With an increase in nanoparticles concentration, the treated surface roughness increased along with the wettability of the coated surface. It was observed that boiling and nanoparticle deposition treatment leads to a surface roughness higher than the uncoated ones. It was concluded that, the roughness of coated surface is dependent of the nanofluid concentration. To analysis the critical heat flux (CHF) of the nanofluid, [19] considered R-123 refrigerant on an uncoated and Al_2O_3 nanoparticle-coated heater. It was observed that the surface coated with nanoparticles improved the critical heat flux by 17% and caused little alteration in the heat transfer coefficient when compared with the uncoated surface. CHF enhancement was attributed to the high wettability of R-123 refrigerant, capillarity and porosity of the fluid. The deposited nanoparticles on the surface created a porous structure on it which enhanced the rewetting process caused by the increase in capillary action.

The improvement of critical heat flux through the variation of the surface energy using nanofluids along with pure fluids was investigated [20]. Nanofluid was prepared by dispersing Al_2O_3 nanoparticles of volume concentrations within 0.001 g/L to 10 g/L in water and ethanol. The heater surface was systematically changed with the deposition of metal on the surface and oxidization of the surface at various degrees. The boiling curves on the various surfaces for pure fluids and ethanol and water nanofluids were considered. The analysis showed that dilute suspensions of nanoparticles degraded or had negligible effect on the boiling performance but higher concentrations of 0.5 g/L caused approximately 37% increase in the CHF. Comparing the result obtained for the water and ethanol nanofluid, the ethanol exhibited more degradation in CHF because of its better wetting system while the water nanofluid which possesses poorly wetting systems on polished copper was enhanced through dispersion of nanoparticles into the water and fouling of the surface by nanoparticles.

For nucleate boiling, [21] considered the influence of nanoparticles deposition on surface wettability using zirconia, alumina and silica nanoparticles dispersed in water. The main focus was analysis the variations in surface wettability during pool boiling and its influence on CHF. From the result, there was significant enhancement in the CHF at nanoparticle concentration of 0.1%. Deposition of nanoparticles was observed on the surface during nucleate boiling. The layer of nanoparticle deposition considerably improved the surface wettability as indicated by the decrease in the static contact angle on the nanofluid surfaces when compared with the base fluid surface. A systematic analysis on surface modification during pool boiling using Al_2O_3/ nanofluid was carried out by [22]. The effect of surface roughness on wettability at different concentrations, heat fluxes and boiling rate using various aluminium substrates of different initial roughness was analysed. The analyses showed that an increase in wettability occurs when the boiling durations is extended and the nanoparticle concentration is increased. The increase in wettability was ascribed to nanoparticles deposition on the surface during the process and also the growth of hydroxides. The nanofluid caused topographical and chemical modification on the aluminium surface. [23] showed that the convective coefficient of heat transfer for water and water based nanofluid droplet decreased with the decrease in surface wettability but the droplet convective heat transfer coefficient for nanofluid was higher than that of water with the enhancement increasing with wettability but the heat transfer rate did not change as the nanoparticles concentration changed.

From the reviews, it is established that surface wettability variations caused by nanoparticle deposition signifies the foremost significant step towards identification and understanding the plausible mechanism for enhancement of heat transfer in nanofluids. Surfactant are known to be added to nanofluid to ensure stability of nanoparticles in the base fluid but most researches have not been able to separate the influence of surfactant and nanoparticles concentration on the surface wettability during the heat transfer analysis. Therefore further research will need to be conducted in this area.

3. Nanoparticles deposition on surface roughness characteristics and boiling heat transfer

The tendency of nanoparticles accumulating on the heater surface after a long period of time may cause fouling on the surface thereby causing a resistance to energy transfer. This causes reduction in heat transfer between the surface and the fluid but the CHF is likely to increase due to the increase in the surface wettability as observed in Section 2. The propensity of the increment or decrement of the surface roughness by the nanofluid depends on the size of the nanoparticle in relation to the initial surface condition. Considering the deposition of nanoparticles on the surface, the surface roughness will increase if the initial surface roughness is less than the size of the nanoparticles but when the initial surface roughness is greater than nanoparticles size, it will decrease due to nanoparticles depositing on the pores of the surface [24]. The surface structure is a function of the initial surface roughness scale and the nanoparticle size which affects the pool and flow boiling heat transfer.

3.1 Nucleate Pool Boiling

The nucleate boiling and overall heat transfer are affected by the surface geometry, surface material properties such as cleanness, orientation, thickness, finish, and type of heat transfer fluid and its wettability [2]. The commencement of nucleate boiling is governed by the size of cavities on the heater surface while nucleation characteristics are affected by the size distribution and amount of cavities existing on the heater surface [25]. Nucleate boiling thermal performance is also affected by the aging and roughening or addition of cavities on the heat transfer surface [26-28]. The influence of surface roughness and surface material on nucleate pool boiling was analysed using two different refrigerants namely, R-123 and R-134a at various reduced pressures and stainless steel, brass and copper cylindrical surfaces [2]. The rough surface exhibited better boiling thermal performance when compared to the smooth surface at low heat fluxes but this trend changed for the high heat flux. The slope between the heat transfer coefficient and heat transfer was largely dependent on the surface material and it was higher in copper and brass but lower in stainless steel. Using horizontal circular plates of copper, brass and aluminum [16] experimentally studied the influence of surface material on nucleate pool boiling at different heat fluxes ranging from 8 – 200 kW/m^2. In the analysis it was revealed that the surface material have significant effect on nucleate pool boiling. Copper showed the highest heat transfer coefficient while the least was aluminium. At low heat fluxes, there was negligible variance in the heat transfer of the brass, copper and aluminum. At high heat fluxes, copper exhibited 23% improvement in thermal performance than aluminum and 18% better than brass.

The deposition of nanoparticles on surface can affect nucleate pool boiling. The influence of nanoparticle layering on heat transfer surface using ZnO/water nanofluid on horizontal surface was investigated [29]. The result initially exhibited 24% heat transfer enhancement when compared to water on an unroughened surface. As the experiment proceeded, nanoparticles layering occurred on the surface thereby causing a steady decline in rate of heat transfer displaying a measured drift that is contrary to that of water. The suspension of nanoparticles on the surface caused a suppression of bubble formation and motion thereby causing a decrease in heat transfer rate. It was found that an increase in surface roughness caused by nanoparticle layering can significantly enhance boiling for the base fluid and to some extent decrease performance for the nanofluid. The analysis of the effect of various ranges of nanoparticle concentration, surface roughness and heat flux on the boiling heat transfer coefficient in pool boiling using Al_2O_3 and water showed that the surface was covered with tiny porous layer of Al_2O_3 deposit. This deposition modified the size of the surface cavities which affected the heat transfer coefficient of the nanofluid but the surface roughness, boiling heat transfer, nucleation site density were improved in the smooth surface because the deposited nanoparticles size was larger than the surface roughness. The variation in coefficient of heat transfer for Al_2O_3 nanofluid and water decreased with heat flux. At lower heat flux, the creation of nucleation was active in the larger cavities while at higher heat flux, the creation of nucleation was active at smaller cavities [30]. [15] investigated the effect of water/Al_2O_3 nanofluids with nanoparticle diameter of 20 nm on pool boiling in a horizontal tubular heater having different surface roughness at atmospheric pressure. From the analysis, it was observed that nanoparticles deposition on the surface modified the surface wettability and distribution of active nucleation sites, thereby causing deterioration in the boiling characteristics of the fluid. [24] also confirmed that the heat transfer coefficient of Al_2O_3/water nanofluid on a horizontal highly smooth flat surface is lower than that of water. To give a further explanation to the reason behind the deterioration of the boiling heat transfer coefficient, the surface roughness of the same samples were measured for water, 5% and 4% concentration of Al_2O_3 nanofluid. It was observed that the increase in nanoparticle concentration caused a decrease in the coefficient of heat transfer. This observation was attributed to the deposition of nanoparticles on the surface which caused reduction in the number of active nucleation sites with variation of surface roughness value. [31] conducted an experimental analysis on the transient nature of nanoparticle deposition and its influence on the heat transfer coefficient in pool boiling. Aluminium oxide of nanoparticle size of 40 – 50 nm was dispersed in water at concentration of 0.01, 0.1 and 0.5 vol. %. The pool boiling experiment was done on a horizontal flat copper surface. The same nanoparticle-deposited (NPD) surfaces were also considered for the pool boiling experiment for water. It was observed that the rate and uniformity of particles deposition depended on the concentration of nanoparticles dispersed in water. For low nanofluids concentration (0.01 vol. %), the deposition of particles occurred at a slower rate, thereby, increasing the heat transfer rate.

From the review, it is established that the deposition of nanoparticles on the surface affects the surface roughness and coefficient of heat transfer in nucleate boiling and the size of nanoparticles plays a significant role in the boiling process. The mechanism behind the enhancement or deterioration of coefficient of heat transfer with varying heater surface roughness still remains unclear. [32] conducted an experimental study to further understand the mechanism behind the boiling heat transfer enhancement/deterioration nanofluid over vertical tubes. Aluminum oxide nanoparticles of diameter 150 nm and 47 nm were used for the study while average surface roughness of 48 nm, 98 nm and 524 nm were considered. Due to the enhancement and deterioration in the boiling heat transfer, the surface particle interaction parameter (SPIP) was introduced. This parameter is the proportion of the average surface roughness to the average nanoparticle diameter. It was explained that the maximum deterioration in pool boiling performance occurred when the SPIP value is near unity.

[33] developed a dimensionless heat transfer coefficient for pure fluid relating to the surface roughness and heat flux shown in Equation (1)

$$\frac{h}{h_0} = C.F(P_r).\left(\frac{q}{q_0}\right)^n \quad (1)$$

Where h_0, q_0 and Ra_0 are the specific heat transfer coefficient the fixed reference heat flux and the reference value for roughness, respectively. Value for constant C is given in Equation (2).

$$C = \left(\frac{R_a}{R_{a0}}\right)^{0.1333} \tag{2}$$

$$R_{a0} = 0.4\mu m \tag{3}$$

[34] modified the Gorenflo correction considering nanoparticle size, pressure and heat flux for pool boiling nanofluid. The heat transfer enhancement was calculated directly using the simplified Equation (4).

$$\bar{h} = \frac{h_{nf}}{h_{bf}} = \left(\frac{1.73 \times \frac{Ra_b}{dp^{0.122}}}{Ra_b}\right) = \left(\frac{1.73}{dp^{0.122}}\right)^{0.57} \tag{4}$$

The result showed that the heat removal from the surface in the pool phenomena is enhanced. The nanoparticles cluster formation and agglomeration increased the surface roughness thereby improving the rate of heat removal.

3.2 Flow boiling

Convective and nucleate boiling are the major element in flow boiling heat transfer. The effect of surface roughness on the pressure drop and flow boiling heat transfer was investigated in a three microchannel heat sinks [35]. At a lower heat flux, the surface roughness minor impact on the saturation boiling heat transfer and the boiling. [36] also considered the effect of surface roughness on flow boing heat transfer, pressure drop and instability in microgap heat sink. From the result, it was concluded that the surface roughness in the microgap heat sink has no effect on the pressure drop. However, the inlet pressure instability and inlet pressure fluctuation increased with increase surface roughness at larger microgap heat sink but the bubble nucleation site density and heat transfer coefficient increase with surface roughness.

The effect of nano-sized particles deposition on the flow boiling heat transfer characteristics has not been fully explored in existing literatures. More investigations needs to be conducted to properly comprehend the influence of nanoparticles deposition on surface characteristics and the coefficient flow boiling heat transfer during single and two phase flows.

4. Conclusion and recommendation

The impact of nanoparticles deposition on surface roughness and heat transfer characteristics has been reviewed. It has been established that surface wettability variations caused by nanoparticle deposition signifies the foremost significant step towards identification and understanding the plausible mechanism for heat transfer enhancement in nanofluids. Further studies needs to be conducted to properly understand the nucleate and flow boiling heat transfer characteristics of nanofluid with respect to relative size of the nanoparticles and the roughness of the heater surface. For the nucleate boiling heat transfer analysis, the presence of nanoparticles in the heat transfer fluid affected the heat transfer characteristics of the fluid and surface roughness of the heat transfer surface but there was discrepancy on whether nanoparticles enhanced, deteriorated or caused the heat transfer coefficient to remain the same. Therefore, a full understanding on the mechanism behind the enhancement or deterioration of heat transfer coefficient with varying heater surface roughness and the influence of nanoparticles deposition on bubble growth and size and nucleation process still remains unclear. There are insufficient studies on the consequences of surface roughness and nanoparticle size on heat transfer coefficient during flow boiling. Hence, more research works need to be conducted to study the influence of deposition of nanoparticles on the change in surface roughness, wettability, flow boiling heat transfer characteristics and pressure drop.

REFERENCES

[1] Celen A., Çebi A., Aktas M., Mahian O., Dalkilic A. S., Wongwises S. (2014). A review of nanorefrigerants: Flow characteristics and applications. International Journal of Refrigeration. 44, 125–140.

[2] Jabardo J. M. S. (2010). An Overview of Surface Roughness Effects on Nucleate Boiling Heat Transfer. The Open Transport Phenomena Journal, 2, 24-34

[3] Incoprera F.P. and DeWitt D.P. (2007). Introduction to Heat Transfer. John Wiley & Sons.

[4] Kandlikar S. G., Joshi S. and Tian S. (2003). Effect of channel roughness on heat transfer and fluid flow characteristics at low Reynolds numbers in small diameter tubes. Heat transfer Engineering. 24(3), 4-16.

[5] Choi, S.U.S. (1995). Enhancing thermal conductivity of fluids with nanoparticles. ASME FED. 231, 99-103.

[6] Chou F.C., Lukes J. R and Tien C.L (1999). Heat Transfer Enhancement by Fins in the Microscale Regime. J. Heat Transfer, 121(4), 972-977.

[7] Gerken I., Brandner J. J., Dittmeyer R. (2014). Heat transfer enhancement with gas-to-gas micro heat exchangers. 4th Micro and Nano Flows Conference UCL, London, UK, 7-10 September 2014.

[8] Maxwell J. C. (1954). A Treatise on Electricity and Magnetism. Unabridged, Dover.

[9] Das S.K, Choi S.U.S., Patel H.E. (2006). Heat Transfer in Nanofluids-A Review. Heat Transf. Eng. 27 (10), 3-19.

[10] Sivashanmugam P. (2012). Application of Nanofluids in Heat Transfer. http://www.intechopen.com/books/an-overview-of-heat-transfer-phenomena http://dx.doi.org/10.5772/52496

[11] Anoop K.B., Sundararajan T., Das S.K., (2009). Effect of particle size on the convective heat transfer in nanofluid in the developing region. Int. J. Heat Mass Transf. 52, 2189-2195.

[12] Heris S.Z., S.G. Etemad, M.N. Esfahany, (2006) Experimental investigation of oxide nanofluids laminar flow convective heat transfer. Int. Commun. Heat Mass Transf. 33, 529- 535.

[13] Li Q., Xuan Y. (2002). Convective heat transfer and flow characteristics of Cu-water nanofluid. Sci. China. 45 (4 Series E), 408-416.

[14] Prajapati O. S. and Rohatgi N. (2014). Flow Boiling Heat Transfer Enhancement by Using ZnO/Water Nanofluids. Research Article. Hindawi Publishing Corporation Science and Technology of Nuclear Installations, 2014, Article ID 890316, 7 pages. http://dx.doi.org/10.1155/2014/890316

[15] Das S. K., Putra N. and Roetzel W. (2003). Pool Boiling Characteristics of Nano-Fluids. Int. J.Heat & Mass Transfer, 46, 851-862.

[16] Hosseini R., Gholaminejad A., and Nabil M. (2011). Concerning the Effect of Surface Material on Nucleate Boiling Heat Transfer of R-113. Journal of Electronics Cooling and Thermal Control, 1, 22-27.

[17] Pioro I. L., Rohsenow W., and Doerffer S. S. (2004). Nucleate Pool Boiling Heat Transfer. I: Review of Parametric Effects of Boiling Surface. Int. J. Heat Mass Transf., 47, 5033-5044.

[18] Zhang F., Jacobi A. M. (2014). Nanoparticle Deposition by Boiling on Aluminum Surfaces to Enhance Wettability. 15th International Refrigeration and Air Conditioning Conference at Purdue, July 14-17, 2014.

[19] Seo S. B. and Bang I. C. (2015). Effects of Al2O3 nanoparticles deposition on critical heat flux of R-123 in flow boiling heat transfer. Nucl. Eng. Technol, 47, 398-406.

[20] Coursey J. S. and Kim J. (2008). Nanofluid boiling: The effect of surface wettability. International Journal of Heat and Fluid Flow, 29 (6), 1577–1585.

[21] Kim S.J., Banga I.C., Buongiornoa J., Hub L.W. (2007). Surface wettability change during pool boiling of nanofluids and its effect on critical heat flux. International Journal of Heat and Mass Transfer, 50 (19–20), Pages 4105–4116

[22] Zhang F. and Jacobi A. M. (2016). Aluminum surface wettability changes by pool boiling of nanofluids. Colloids and Surfaces A: Physicochemical and Engineering Aspects, 06 (5), 438–444.

[23] Jackson R. G., Kahani M., Karwa N., Wu A., Lamb R., Taylor R., Rosengarten G. (2014). Effect of surface wettability on carbon nanotube water–based nanofluid droplet impingement heat transfer. Journal of Physics: Conference Series 525. doi:10.1088/1742-6596/525/1/012024.

[24] Bang I. C., Chang S. H. (2005). Boiling heat transfer performance and phenomena of Al2O3–water nano-fluids from a plain surface in a pool. International Journal of Heat and Mass Transfer, 48, 2407–2419

[25] Kandlikar S. G. and Spiesman P. H. (2000). Effect of surface finish on flow boiling heat transfer

[26] Stelute E. B. (2004). Effect of Surface Roughness on Nucleate Boiling Heat Transfer of Halocarbon Refrigerants on Horizontal Tubes. MSc Thesis, Escola de Engenharia de São Carlos, University of São Paulo, Brazil, 2004 (in Portuguese).

[27] Jabardo J. M. S. (2008). Boiling heat transfer in two-phase flow heat exchangers. First Brazilian Meeting on Boiling, Condensation, and Multi-phase Liquid-gas Flow, EBECEM, Florianópolis, SC, Brazil, April 28-29, 2008 (in Portuguese).

[28] Kang M.G. (2000). Effect of Surface roughness on Pool Boiling Heat Transfer. International Journal of Heat and mass Transfer, 43 (22) 4073-4085.

[29] White S. B., Shih A. J., Kevin P. (2000). Pipe Effects of nanoparticle layering on nanofluid and base fluid pool boiling heat transfer from a horizontal surface under atmospheric pressure. Journal of Applied Physics 107, 114302 (2010); doi: http://dx.doi.org/10.1063/1.3342584

[30] Vafaei S. (2015). Nanofluid pool boiling heat transfer phenomenon. Powder Technology, 277, 181–192

[31] Ahmed O., Hamed M.S. (2012) Experimental investigation of the effect of particle deposition on pool boiling of nanofluids. International Journal of Heat and Mass, Transfer, 55, 3423–3436

[32] Narayan, G. P., Anoop K. B., and Das S. K. (2007). Mechanism of enhancement/deterioration of boiling heat transfer using stable nanoparticle suspensions over vertical tubes. Journal of Applied Physics, 102(7), 074317-7.

[33] Gorenflo, D., (1993). Pool Boiling, VDI Heat Atlas (English Version), VDI-Verlag Dusseldorf.

[34] Niknam P. H., Kasiri N., Haghighi M., Khanof M. H. (2014). Surface roughness and particle size effects on nanofluids pool boiling heat transfer. 1st National Conference of Chemical and Petrochemical. https://www.researchgate.net/publication/265849643

[35] Jones B. J. and S. V. Garimella (2009). Surface Roughness Effects on Flow Boiling in Microchannels Journal of Thermal Science and Engineering Applications, 1 / 041007-1

[36] Alam T., Lee P.S, Yaap C.R (2013). Effect of surface roughness on flow boiling in silicon microgap heat sink. International Journal of Heat and Mass Transfer, 64, 28-41.

Wind energy generation from Nigeria continental shelf: A review.

A A Attabo, O O Ajayi[1] and S O Oyedepo
Mechanical Engineering Department, Covenant University, P.M.B. 1023, Ota, Ogun State, Nigeria.
E-mail: oluseyi.ajayi@covenantuniversity.edu.ng

Abstract. The menace of global warming and increase of green-house gases have motivated many countries to embrace renewable energy sources as an alternative to fossil fuels. Some of the means by which sustainable energy is tapped are solar, biofuel, and wind. Wind energy have been around for quite some time and the power in the wind have been harnessed on land. However, due to some constraints related to land use, visual impacts and turbulence induced by buildings and trees, offshore wind farms have been engaged to provide quality alternative to onshore wind energy. This study reviews various works that have been done in the offshore wind development and tries to make a case for the generation of electricity from this type of renewable energy source to boost the nation's electricity demand. Work done in offshore energy assessment, wake models, offshore turbine layout, future innovation in offshore wind turbine and Nigeria onshore energy scenario were reviewed. To speed up development of offshore wind energy in Nigeria, recommendation was made for government to invest in wind energy research, industries that operate in the offshore region should be willing to share meteorological data with researchers and tertiary institutions should encourage student to work on "issue-based" projects like proffering solution to the energy deficit crisis in the country.

Introduction

Unsteady oil prices, increase in energy demand and the reality of global warming has led to the push for cleaner sources of energy. One of such clean energy resources is wind energy and extracting energy from the wind is one of the oldest methods of tapping energy from nature. This can be traced back to pre-historic times i.e. about 3000 BC, where sailing boats were propelled using thrust from the wind [1]. The use of wind energy can also be found in olden day agriculture where wind energy was used to pump water from wells for irrigation, fill cattle water troughs and for grinding grain [2]. In the instances above, the wind energy was converted into torque by means of a spinning rotor and a shaft.

Experience from the oil and gas industry in offshore installations and foundations motivated the construction of wind turbine offshore, where the wind is stronger and steadier [3,4]. The desire to create more reliable and cheaper energy per unit price has led to the quest of adventuring offshore to install wind turbines [5]. Offshore wind turbines are wind Turbines installed in bodies of water for the purpose of electricity generation. According to [2,6] installation of wind farms offshore has gained momentum in Europe and it is expected to be the next big step in wind energy development. Some examples of notable offshore wind farms are the Ormode offshore wind farm shown as plate 1.1 and the Rev Horns and London array farm. As at February 2019, the total installed capacity for wind power was reported to be 600 GW [7], 4% of these wind power turbines where installed offshore [8]. With reference to the high capital cost that might be involved in foundation and logistics in moving materials offshore, installations of wind turbines have been on the increase in recent years [9].

Plate 1.1 Ormode Offshore Wind farm (FoundOcean Incorporation, 2013)

The exploration of offshore wind resources is quite novel and has yielded positive results in Europe. Other motivation for investing in offshore Wind farms includes the scarceness of suitable land that have sufficient wind speed, visual impact, environmental issues related to eddy noise from wind turbines, impact on bird life and flickering shadow from rotating turbine blade [10,11]. When compared to onshore wind resources in most parts of the world, offshore wind resources have been reported to possess higher potential, and the wind speed tend to increase as one move further into sea (i.e. further away from the shoreline). This therefore results to considerably higher energy yield, as the power output from a wind turbine is theoretically a function of the cube of the average wind speed [12].

1. Statement of problem

Power generation is the core of both technological and economic development of any nation as it is the bedrock of many crucial sectors of the economy for example, constant power is required in health sector for the preservation of drugs and vaccines, it is used to run machines in industries, schools need energy to run laboratories and multimedia learning equipment, streets lightings and state security gadgets all require electricity to work. The aforementioned make electricity generation and its distribution a very sensitive issue with several economic complexities therefore making the success of nations hinged on it. Sadly, Nigeria is still struggling with the generation of energy to meet her growing population of which a large percentage are youth who use electrical gadgets and equipment heavily [13,14].

Hydro and thermal plants are the two commercially available energy generating systems in the country. This is relatively low energy generation mix as it can easily drive the nation into energy crisis because the unavailability of one of the resources will strain the national grid. With the unpleasant security situation in Nigeria and youth restiveness in most south-south areas of Nigeria [15], frequent sabotage on gas supply pipelines that feed the thermal power plants which have led to unplanned shutdown of the thermal power plants in Nigeria [16]. Finally, the use of fossil fuel for electric generation have been a major culprit in the emission of greenhouse gases [17].

2. Background of offshore wind farms

The first known wind farm that was installed offshore was Vindeby offshore wind farm and it started operation in 1991 by DONG energy in Lolland, Denmark. The wind farm was made up of 11 units of wind turbines rated 450 kW each, located in a site that is 4 m deep and 2 km away from the shoreline. After 26 years of operation, the Vindeby wind farm was decommissioned in 2017 with gross electricity production of about 234 GWh. After the successful commissioning and operation of the Vindeby farm, many other offshore wind farms began to spring up, this led to an exponential growth in the development of offshore turbines with larger capacity and deeper water depths. At the end of

2014, about 3,230 turbines (in 84 offshore farms) had been commissioned in 11 European countries and grid-connected, with total capacity of about 11,027 MW [13].

Review of various studies on offshore wind energy

Works carried out on the development of wind energy were categorized into various sub topics reviewed as outlined below.

2.1. Review of offshore wind energy assessment

Oh et al. [14] investigated the wind potential around the south-western area of the Korean Peninsula. Wind potential in the projected site was measured using a meteorological mast constructed offshore in conjunction with Metrological and oceanographic special Research unit. Data obtained from the metrological mast and other adjourning areas where used to analyse the sea surface roughness, diurnal and seasonal variations in wind speed, wind direction pattern and estimation of wind speed using Measure-Correlate-Predict (MCP) model. The wind energy potential of the site was reported to be in class 3 regime but the effect of wake within the prospective wind farm was not considered.

Adaramola et al. [15] used the Weibull probability function to examine the wind speed characteristics in Ghana for the purpose of electricity generation. The six coastal locations that were studied are Warabeba, Oshiyie, Mankoadze, Aplaku, Anloga and Adafoah. Wind speed data was obtained from Ghana Energy commission at a height of 12 m. Results from the study indicates that wind resource along the coastal region can be classified into Class 2 or lower, thereby making the region suitable for marginal wind energy development. The recommendation was to use wind turbines having cut-in wind speed of < 3 m/s and 9-11 m/s rated wind speed to increase turbine availability factor. The study further proposed CF-100 to be the most suitable wind turbine model for the coastal region of Ghana and using that model of wind turbine, the estimated unit cost of electricity was reported to be between 0.01$/kWh and 0.05$/kWh. The effect of turbine induced wake was not considered in the paper.

Michael et al. [16] evaluated the wind resource available in California's offshore area using a mesoscale atmospheric model developed by the Pennsylvania State atmospheric research center called Mesoscale Model version 5 (MM5). Results from this mesoscale was then authenticated with results from measurement buoys with high resolution bathymetry image located in the ocean. The wind speeds were measured at 80 m turbine hub height for the project using monthly average for the month of January, April, July, and October of 2005 and 2006, the study also examined mean speed for the whole months in 2007. Resource assessment of the region was carried out using a 5 MW offshore turbine and key factors considered for the work were transmission access, water depth and legislation. In the study, offshore areas where wind speed at 80 m height were 7.5 m/s and higher were considered to be suitable for offshore wind development and their result showed that power generation could range between 1.4 and 64.9 GW (daily or annually). Based on the type of foundation type, power generation was categorized as follows: 1.4 - 2.3 GW could be generated using monopile, 4.4 - 8.3 GW could be generated using jacket type foundation and wind farms using floating turbine foundations could yield 52.8 - 64.9 GW of power annually. Diurnal variation for the area studied was found to be relatively stable hence power generation was possible and consistent both day and night for the whole year however reference was not made to the effect of turbine wake on wind energy potential.

Ahmmad [17] evaluated the wind energy potential in Bangladesh using a hybrid wind conversion system with a battery storage system. Data collected from the meteorological centres were analysed using the Rayleigh distribution functions. The results showed the hybrid system has an annual mean wind power profile and power density of 2351.76 K/hr and 194.2 W/m^2. respectively. The study used wind speed information retrieved from the Dhaka airport station.

To estimate the potential of wind energy available in the coastal area of Ceara, Brazil Lima et al. [18] used a computer based atmospheric model called Regional Atmospheric System (RAMS) that can predict mesoscale wind data at high resolution. The application was set at a 2 km horizontal resolution to estimate the offshore average wind speed for both dry and wet seasons, predominant wind direction and wind turbulence between 1996-2012. Numerical simulation results were then validated with measured data at 60.4 m and correlation between the two data set was satisfactory. Results from the study indicated that mean wind speed was about 8 m/s and calculated power density of the area was about 720 W/m2 for all seasons. Wind direction was predominantly from east to west for the dry season while the turbulence intensity was also calculated to be lower during this season compared to the wet season. Moreover, the validation of results was not done for some period between 1997 and 1998.

Satir et al. [19] evaluated the coastal areas of Turkey using the windPro software to analyse the suitability of the area for wind power generation. Bozcaaada area was recommended for offshore wind farm installation because it was observed to possess the highest wind potential. Technical analysis was conducted using Siemens SWT-3.6-120 and Vestas V90-3.0 wind turbines and annual energy production from the two turbines were given as 322,911 MWh/yr and 454,326 MWh/yr respectively. However, the work used 1 year of data which is not usually sufficient to carry out robust wind resource analysis

2.2. Review of studies that used Numerical Analysis for prediction of wind speeds

Monfared et al. [20] proposed a model for the prediction of wind speeds using fuzzy logic and artificial neural network. In the study, standard deviation, average and slope of the measured wind speeds were inputted in the fuzzy logic to improve the prediction of the model. The feed-forward neural network was used and the ANN was trained using a standard back propagation algorithm. The neural network was trained with 672 wind data samples from February to November 2007 and result from the work suggest that this method reduces computation time.

Li and Shi [21] carried out a comparative study of the application of three types of artificial neural networks as it applies to wind speed forecasting. It studied the following models: back propagation, adaptive linear element and radial basis function. Two observatory sites in North Dakota were used for the study and hourly mean wind speeds were collected. The three metrics used for the evaluation of the ANN are the root mean square error, mean absolute error and mean absolute percentage error. Results from showed that, no single artificial neural network model outclasses others even for the same data set. The work also recommended the use of robust post-processing method in forecasting wind speeds for better accuracy.

Fadare [22] modelled and predicted wind speed in Nigeria using a 3-layered, feed-forward, back-propagation artificial neural network (ANN) model. The ANN model was designed using the Neural Toolbox for MATLAB while the training and testing of the model was carried out using monthly and daily mean wind speeds data monitored at 10 m height obtained from the Nigeria Meteorological Services (NIMET). Result from the work showed that the correlation coefficient (r) between the predicted and the measured wind speed was obtained as 0.9380. Average monthly wind speed predicted for the 28 onshore metrological stations studied ranged from 0.9 to 13.1 m/s and annual mean speed was 4.7 m/s.

Koletsis et al. [23] researched further, to evaluate the effect of climate change on wind energy developments in the Black Sea and Mediterranean Sea using ENSEMBLES project software. The study covered the past climate of 1961 to 1990, predicted climate regime from 2021 to 2050) and 2061 to 2090. The estimated wind speeds and resultant wind power potential showed a reduction in average wind speed in areas around major parts of the Mediterranean Sea. However, increase in wind speed was noted in the Gulf of Lion, Alboran and Aegean seas. The prediction showed some areas where there are mutual future trend, but there was disagreement in the specific locations where these change will occur and the magnitude of the predicted change.

Sasaki [24] used numerical models to predict offshore wind and wave power globally considering 9 days look-ahead using various wave and wind models for the period 2008–2012. Wave power was calculated using a wave model termed "third-generation" model which was developed at the US National Oceanic and Atmospheric Administration – National Center for Environmental Prediction while the near-surface wind data from the ERA-Interim reanalysis was used to simulate wind power available offshore. To validate the accuracy of predicted wind and wave speeds, the results were statistically evaluated using the ensemble spread, mean prediction error and anomaly correlation coefficient (ACC). Results from their study showed that wave power can be predicted over vast zones of the global ocean with prediction error less than 20% for 3 days lead time but predictability of ocean wind speed showed lower correlation between the measure and predicted values using the ACC method.

2.3. Review of GIS methods for Offshore wind energy estimations

Ho et al. [25] evaluated the offshore wind potential in Malaysia with the aid of two GIS methods (the QuikSCAT and the WindSat) in estimating the potential energy inherent in offshore wind in the region. Annual average wind speed of 6-7 m/s at 50 m was observed throughout the year hence, the study made strong recommendation to the government and private investors to commence investment in power generation from offshore wind resources in Malaysia. However, the study did not validate GIS data with actual field data and cost analysis of intended wind project was not investigated.

Cavazzi and Dutton [26] developed an Offshore Wind Energy Geographic Information System (OWE-GIS) which was used in determining wind energy resource in United Kingdom's offshore sites. In the system, cost projections of offshore farms were estimated using; development costs based on water depth, potential annual energy production dependent on the locations' wind speed distribution, turbine availability, project lifetime and discount rate on capital expenditure used in executing the project. Results from the study showed that 150 GW of electricity can be produced form 10% of accessible UK waters at a production price lower than 280 $/MWh. Validation of GIS generated data with field data was not performed, as all cost estimation was done using only the novel model.

Garlapati et al. [27] used two years (2012-2013) OSCAT satellite data to determine the offshore wind energy resources in India. A Geographical Information System (GIS) was used to generate wind maps at 80 m taken to be the turbine height. The study showed that about 74% of the area of interest has class 1 wind speeds, annual average wind speed and wind power density were calculated to be 9.7-13.4 m/s and 1048.7-1632.5 W/m2 respectively. Using a 2.1 MW rated capacity wind turbine and matured offshore foundation type for analysis, the average annual output power excluding conflict areas was found to be approximately 1738.6 GW. However, the study used only two years' data for the analysis as against 10 years or more which would have been more suitable to cover for climate change variation.

Ulazia et al. [28] evaluated the suitability of various Weather Research and Forecasting (WRF) meteorological models in estimating wind speeds for energy conversion systems. The focus of the work was in the Bay of Biscay (i.e. around the Iberian northern coastline) between 1990 to 2001. WRF models that was used captured wind speeds at different time intervals of 24 hrs, 12 hrs and 6 hrs. Using surface wind data acquired from the Cross-Calibrated Multi-Platform (CCMP), a spatially distributed analysis was conducted for the region. The predictions from satellites data were validated with wind data obtained from observatory buoys. Result from the study revealed that the wind speed values obtained at data frequency of six hours yielded the best verification scores at a 95% confidence level, thereby being the most accurate at reproducing wind observations in the area. The study did not carry out econometrics of offshore wind turbines.

Zheng et al. [29] reviewed the various models that can be used for the estimation of offshore wind resources, noting the fact that it is not economically practicable to carry out continuous real time measurements on the sea for preliminary investigation of offshore wind energy potential. Some of the applications they reviewed were Quicksat, SatWind, Wind Atlas Analysis and Application Program (WAsP), SiteWind, Wind energy Simulating Toolkit, wind energy Power Prediction Tool (WPPT) and the use of Synthetic Aperture Radars (SAR). The study did not cover wind energy economics.

Scarcity of in-situ wind data at offshore locations in India motivated Gadad and Deka [30] to study the applicability of using Oceansat-2 scatterometer (OSCAT) to measure the speed of wind and wind power density in and around the Indian Ocean. The research assessed the power inherent in the offshore wind around Karnataka state using continuous measured satellite data that its precision was validated with INCOIS Real-time All-Weather Station (IRAWS) data. Power generation capacity evaluation was conducted using a Repower 5 MW wind turbine as the test turbine and two years (2011 and 2012) mean wind speeds obtained from the satellite. Results from the study showed that 9,091 MW of electricity could be generated using Monopile foundation (0–35 m water depth), about 12,000 MW with Jacket (35–50 m water depth), 24,000 MW using advanced Jacket (50–100 m water depth) and 118,000 MW using floating foundations (100–1000 m water depth). However, the levelized cost of energy from offshore wind turbines was not considered.

In a similar study, Nagababu et al. [31] used 14 years GIS information retrieved from European Centre for Medium-range Weather Forecasting to analyse available wind energy potential in the coast of India. Wind speed was simulated at a hub height of 80 m and wind turbine manufactured by

Gamesa (Model-G128- 5MW) was used for LCOE computations. For maximum energy yield, the study suggested coastal towns of Tamil-Nadu and Gujarat for the installation of offshore wind farms. The western coast of India was estimated to have wind power density between 13-294 W/m2 and while technical potential was about 38.7 TWh. on the other hand, the estimated values for wind energy potential and technical potential for the eastern coast of India are between and 63-393 W/m2 and 58.4 TWh respectively. This potential is about 42% of India's wind energy capacity however, validation of GIS data and field metrological data was not done.

Amirinia et al. [32] used GIS method to assess the potential of offshore wind in the Gulf of Persia. The study used uncertainty analysis method to estimate the wind potential in the area. The uncertainties considered in the study were sea surface roughness, wind power production and air density variation. The study evaluated 25-years (1984-2008) data acquired from ECMWF (European Center for Medium-Range Weather Forecasts) and then using the Monte Carlo simulation method, long term wind energy and speed were estimated. The study reports a 5.3% reduction in energy produced from the area when compared with conventional method, i.e. that the region can generate up to 2,980 GWh per year. A figure which is about 5.3% lower than outcomes from studies that used conventional analytical methods. Econometric study of the region reported that if the interest rate on capital was at 15%, operating wind energy turbines in the area cannot be competitive compared to the rate of energy production globally. However, if the interest rate is reduced by 10% (i.e. interest rate of 5%), some parts of the region will be able to generate electricity at a competitive price.

Carvalho et al. [33] carried out comparative experiment using offshore wind speeds acquired via remote measurements methods with the wind data obtained from equipment mounted on buoys located around the Atlantic coast (Iberian Peninsula). The study proved that GIS method is a credible alternative for measuring offshore wind speeds when physical measurement was not possible. It used Advanced-Scatterometer (ASCAT) of the European Space Agency, the Indian Space Research Organization's Oceansat-2 Scatterometer (OSCAT) and a numerical weather prediction model. Their result showed that ASCAT has the low variations and wind power flux estimation errors while the Weather Research and Forecast (WRF) model gave the best prediction for mean wind speed hence the study recommends using it for estimation of wind speed and weibull probability density functions offshore. The study however did not consider the levelised cost of energy produced from the wind turbines located offshore.

Schallenberg-Rodríguez and Montesdeoca [34] used the ArGIS method to investigate the energy potential in the wind around the Canary Island and reported the wind power potential to be 57 GW. The study used 515 units of 5 MW offshore turbines with fixed bottom foundations on an ocean area of 180 km2. The estimated energy production from the region was 3 times higher than the energy demand of the area. However, precise wake deficit as it affects the planned turbine array was not considered in the study.

2.4. Reviews on Turbine wake

This section presents some works that have been carried out to evaluate wake deficit in wind farms for better estimation of energy production. These studies include those of Kusiak and Song [35] which used generic model with few constraints considered while making assumption to estimate wake loss in wind farms. By transforming the constraints into second objective function, a multi facet algorithm was established to solve the turbine wake optimization problem. Solving this problem maximized output from the wind farm and reduces failure rate of the blades. The work focused only on one terrain type which was high speed terrains with large turbines.

Barthelmie and Jensen [36] developed a novel wake model which was for the wake prediction for Nysted offshore wind farm in Denmark. Comparing the novel model with WAsP, it reported that deep array effect (wake loss in the center) was underestimated by WAsP. One of the key recommendations from the study was that a wake decay coefficient of 0.03 is a better fit for the Nysted farm. The work was also carried out on an established wind farm hence wind resource assessment was not looked into.

Gaumond *et al.* [37] compared various wake models using the Horns Rev and Lillgrund farms located in Sweden and Denmark respectively. The study reported that the wake models evaluated presented large prediction error due to the fact that the distance between turbines are relatively small and the effect of turbine wake still affect the inner array turbines. The constant change in wind direction was one of the reasons for discrepancy between predicted wake deficit and actual deficit in energy produced from the wind farms. Work was carried out on existing wind farm hence consideration was not given to production cost and wind farm resource assessment.

Seim [38] evaluated three kinematic wake models which was validated on the WindSim software to test eight single wake cases in complex terrain condition. Result from the work showed that the Larsen's model predicted a wake whose width was overestimated. However, the model predicted a constant offset location thereby reducing the ambiguity involved in computing the overall power output from the wind farm. Jeon *et al.* [39] did a similar work in comparing various wake models head-to-head to check their accuracy. The study noted that the Jensen wake model gave a better prediction than both the Larsen and Eddy viscosity models. Albeit, the width of the wake was predicted with higher accuracy using the Eddy viscosity and Larsen models.

Niayifar and Porte-Agel [40] tested a proprietary analytical model which was developed for prediction of power loss as a result of turbine wakes in a wind farm. Field data obtained from the Horns Rev farm with turbine power curve from Vesta V-80 turbines was used in the study. Local wake growth was simulated by the superposition of velocity deficits from multiple wakes and validated using large Eddies Simulation (LES). Results from the studies suggest that the power prediction using application like WAsP is overestimated and their model showed better approximation of power produced from the farm.

Hou *et al.* [41] used the Mixed Integer Particle Swarm Optimization (MIPSO) algorithm in analyzing yield obtained from a wind farm and compared the result with a reference wind farm layout developed by the Norwegian Centre for Offshore Wind Energy (NORCOWE). Findings from the study established a 5% decrease in Levelized Energy cost and about 3.82% rise in energy yield. The study used the Horns Rev offshore farm as a case study. However, installation cost and foundation cost were not considered in the wind farm optimization algorithm.

2.5. Review on wind Turbine Layout

Attias and Ladany [42] formulated a discrete model to improve energy yields from wind farm with the resultant aim of increasing Net Present Value and the Internal Rate of Return of the wind farm. The study evaluated downwind turbines in a rectangular array, same turbine types assumed to have the same hub height. The study recommended that for maximum return on investment, turbine array of 24 units to be used with 4 row x 6 column will yield about 19.7% return on investment. Using the wind turbine layout optimization program (Multi-Population Genetic Algorithm).

Gao [43] optimized the arrangement of four theoretical offshore wind farms at the coast of Hong Kong. An area of about 357.78 km^2 was evaluated using 10 years' wind data obtained from government agencies. It calculated the cost of energy of the farms after optimization to range from 0.23 $/kWh to 1.52 $/kWh, payback time ranged from 14.44 years to 75.89 years. The effect of wake was considered with wind direction from all directions, the study developed a model named 'Wind Farm Power Generation Calculation Tools' (WFPGCTs) which the study recommends for calculation of wind power after the layout of the farm is established.

Shakoor *et al.* [44] reviewed different far wake models to evaluate the accuracy in prediction. It focused on the studies on wind farm layout optimization while paying attention to far wake models. The findings showed that 75% work on wind energy farm optimization used the generic algorithm. Hence, it was suggested that other optimization methods be explored to solve wind farm optimization issues.

As offshore wind turbines begin to increase Ajit *et al.* [45] carried out studies to optimize the layout of Middelgrunden wind farm located off the shores of Denmark. The work was carried out considering the standpoint of wind farm developers and profit maximization during the planning phase of future offshore wind farms. It designed new layouts for the farm using a generic algorithm (GA) and particle swarm optimization (PSO). These new layouts were then compared with the existing wind farm. Results from the study showed that both GA and PSO can improve yields from the layouts by a resultant 1-3.5% reduction in levelized cost of energy compared to the existing layout. Subsequently, notable recommendations were made for the use of advance algorithms during wind farm layouts because this will result in substantial savings over the lifetime of the wind farm. Ou et al. [46]

researched into development of practical zoning method for offshore wind farms in 10 coastal provinces in China. The Maritime Spatial Planning method was used to avoid the haphazard location of Offshore Wind Farm (OWF) in China waters. The study considered a time frame of 2011 to 2020 and zoning was based on the factors such as sea use status, demand for energy, environmental considerations and distance to shore. The results show that offshore wind farms sites can be such that it will co-exist with other economic activities like fishing, light-to-medium water transportation and industrial zone if it is well planned. The actual energy that can be produced form the region was however not considered in the study.

Amaral and Castro [47] used a novel approach they called the deterministic algorithm to predict on offshore wind park with 8 wind turbines arranged in two rows. The model was validated with the generic algorithm and particle swarm optimization models. The comparison was satisfactory as the optimization algorithms reached very good estimates of the annual energy produced, after 100 experiments the error of the generic algorithm was 0.03%.

2.6. Review of Offshore Turbine Logistics

Green and Vasilakos [48] made a research on the economies of offshore winds using the development of offshore wind turbine (OWT) in the European Union as a case study. It acknowledged the fact that installation cost for OWT is relatively high hence, there is need for government support in terms of grants, reduction in investors taxes and waivers where applicable. Some recommendation made include the development of hybrid foundation type (where OWT foundations can be used for other economic benefits), use of dual distribution lines for sales of electricity to various locations based on demand and the provision of green certificate for units of energy produced by the OWT. Moreover, it is worthy of note that improper planning during the installation of offshore wind turbine could lead to increased over-head cost which in turn will lead to increase in capital expenditure and reduced returns on investments. This assertion motivated Iris and Evrim [49] to carry out a study on the logistics of installing offshore wind farms. The study pointed that the installation of offshore wind farm (OWF) is highly complex, due to the high dependence on weather and the oversized components that might be complex to lift or transported. The study further analysed actual projects that were executed in the North Sea region and their findings revealed the major components to be pre-assembled from shore, the ideal vessel load to be used for transporting loads and the distance to shore. It therefore strongly recommended pre-assembling of turbine parts onshore thereby reducing the number of parts that will be installed on site.

He et al. [50] used the Strength Weakness Opportunity and Threat (SWOT) analysis method to evaluate both the external and internal factors affecting the development of the Chinese offshore power sector. The study shows that offshore wind technology has the following strengths: abundant wind resources, possibility of combining offshore wind generation with tidal turbines and environmental friendliness. Moreover, the drawbacks include high cost, lack of a local manufacturer (in China as at time of the research) and lack of coordination among departments. To enhance the strengths of offshore wind farms the study therefore proposed the need for more research to be carried out on OWT technology with the aim of cost reduction in foundation cost and increase reliability of the turbines.

Colmenar-Santos et al. [51] reviewed the status of offshore wind energy development in Spain considering the fact that steady growth have been made in the industry worldwide but Spain is yet to benefit from this technology. It referenced counties which have made giant strides in offshore wind energy technology, like Denmark, Holland, Sweden, Germany and United Kingdom. Notable hindrances to the development of offshore winds are listed to include social disapproval to installations near the coast, deep coastal seabed characteristics and absence of a stable regulatory policy for the establishment of offshore wind farm. Furthermore, it made recommendation for the establishment of a policy to encourage investment of offshore wind farm in Spain as preliminary resource assessment have shown that the region is viable for this type of renewable energy development. On the other hand, Sarker and Faiz [52] designed a model based on variables that affect installation of offshore wind farm. These variables include number of parts to be installed on site, turbine rated output, lifting capacity of crane, vessel deck and distance of port to farm. Numerical analysis was then performed to illustrate the model and to understand the general behaviour of

IOP Conf. Series: Earth and Environmental Science **331** (2019) 012019 doi:10.1088/1755-1315/331/1/012019

different system parameters. The aim of the work was to minimize cost of installation and a key recommendation was to encourage preassembly of turbine parts on land before moving them offshore as dynamic lifting operations requires more time and specialty to handle. It also noted that total cost is highly impacted by turbine size and routine/scheduled maintenance help reduce the cost of maintenance.

2.7. Reviews on novel approach to Offshore Wind Farms

Katsaprakakis and Christakis [53] investigated the technical and economic feasibility of using an offshore wind farm combined with Wind Powered Pumped Storage System (WP-PSS) at a location in Rhodes island. The concept is for the wind turbine to pump water into an elevated reservoir during period of high winds and the power can later be extracted from the water through a hydro turbine when the wind speed reduces. To increase the reliability of the wind park, it suggested that larger reservoirs and double penstocks should be use in designing the system as it will allow greater water reserve capacity and the simultaneous water fall from the penstocks will improve the hydro-turbine efficiency. Further to this, the study created an algorithm to simulate the action of the WP-PSS with two autonomous pipelines, one mainly for water fall and the other for pumping water, thereby making it possible for the simultaneous storage of energy and production of energy. Feasibility study was carried out using wind potential measurements captured from October 2009-March 2010 (six months) in conjunction with the WAsP software developed by Technical University of Denmark and a wind atlas of the proposed site. It also established the fact that a WP-PSS can be used in areas with marginal wind potential but have available land space. Result from the study shows that the WP-PSS model is feasible and could lead to an annual wind energy production that is 50% greater than the annual wind potential available in that site.

Michailides et al. [54] evaluated the operation of an offshore wind turbine combined with semi-submersible wind energy and Flap-type wave energy Converter (SFC) in extreme environmental conditions. The SFC is a combined equipment with a semi-submersible floating wind turbine and three fully submerged rotating flap-type Wave Energy Converters. The study developed a numerical model using the software Simo-Riflex (developed by MARINTEK). To validate their numerical results, a 1:50 scale model was built and load tests were conducted on it. Some of the load tested were static, quasi-static, regular waves and asymmetrical waves in windy conditions. The response of the SFC in regular waves (with absence of wind loading) was investigated considering fourteen different wave periods. Results showed satisfactory correlation between experimental data and predictions made with numerical models to validate the correctness of the numerical models.

Tsai et al. [55] studied the patents granted by the United States Patent and Trademark Office (USPTO) and the European Patent Office (EPO) in the offshore wind sector and results from the study showed that 59% of patents in 2014 were granted in this field. A review of patents collected indicated that technologies associated with engineering vessels, floating foundations, turbine installations, integration of multiple technologies, tower and mooring system were the researched areas for OWT development.

Qin et al. [56] proposed a novel method for storage of energy produced from offshore wind farms to compensate for wind variations. energy storage to compensate for loss of power due to wind variation. The aim of the study was to reduce the disparity between energy demand and energy generated from wind farms. Experiment was carried out using an offshore wind farm in Richmond Virginia by retrofitting the turbines with hydraulic power transmission (HPT) and Compressed Air Energy Storage (CAES) system. The HPT was installed in the nacelle of the turbine and it comprises a hydraulic pump with high compression ratio that can convert the wind energy into pressure energy; when the energy demand is low, the excess pressure is stored as compressed air in a receiver located at the base of the turbine tower. During low wind scenario and energy is demanded, the compressed air expands through the CAES system to generate power, hence meeting the energy demand. The study then used a cost and scaling model developed by the National Renewable Experimental laboratory for cost analysis and result showed that retrofitting conventional turbines with CAES and HPT can increase savings to about 21.6% if the turbine operates for 20-years.

2.8. Review of Nigeria Onshore Wind Energy Potential

Ajayi et al. [57] assessed the potential of wind energy in Shaki and Iseyin (two sites in south West Nigeria) using the Weibull two parameter distribution. 21 years' (i.e. 1987-2007) monthly average wind data was used to investigate monthly, seasonal and yearly wind characteristics and potential for the sites. The study recommended that wind turbine models with cut-in speed of 3 m/s, cut-out-speed of 25 m/s and rated wind speed of 11.6 m/s will be viable for small scale electricity generation in the sites. The study also showed that the generation of few Giga-Watt hours of electricity is possible at sites with average cost of electricity not higher than €0.049 per kWh.

Ohunakin [58] statistically analysed the characteristics of wind in five selected sites in some Nigeria North-central states, the locations investigated are Lokoja, Ilorin, Makurdi, Bida and Minna using wind speed data between 1971 to 2007 and the Weibull distribution function. Result from the study showed that the mean annual wind speeds are 3.158, 4.386, 4.570, 2.747 and 4.289 m/s for Lokoja, Ilorin, Makurdi, Bida and Minna respectively. Hence, none of the sites was recommended as a profitable site for large scale wind farm investment. Cost analysis for siting wind farm in the North central region of Nigeria was also evaluated by Ademola et al. [59] using a 37 years' wind data information. The results showed that Minna has the highest potential to be used as a site for wind power generation, while wind data for Bida showed it to have the least wind energy potential amongst the sites considered. The study also shows that if inflation rate increases from 0% to about 5%, the energy cost will drop by 29% and the return on investment will also reduce from about 12% to 6%.

In a similar study, Oyedepo et al. [60] examined wind potential for three eastern sites in Nigeria (Enugu, Owerri, and Onitsha) using wind data from NIMET that was captured at a height of 10 m. The annual mean speed for Enugu, Owerri and Onitsha were reported to be 5.4, 3.4 and 3.6 m/s, respectively, while the yearly wind speed carrying maximum energy for the site is given to be 6.5, 4.3 and 3.9 m/s. The study also simulated the performance of a windmill with rated power of 0.36 kW for pumping water and it reported a monthly mean water produced by the rotodynamic pump to be 3,442, 2,730 and 2,530 m3 for Enugu, Onitsha and Owerri respectively.

Ajayi et al. [61] used the Weibull probability density with two-parameter function to analyse electricity generation potential from wind in Kano. The result showed that the mean monthly wind speed ranged from 6.6 m/s to 9.5 m/s. When comparing seasonal variations in wind speeds between dry and wet season, the wet season had a higher mean wind speed of 9.5 m/s. Ajayi et al. [62] increased the scope of the study by evaluating the potential in wind at ten sites located in Nigeria's south-western region using twenty-four years' data (1987 to 2010) obtained from NIMET. Result from the study suggested that Lagos state (with mean wind speed of 2.9 m/s) and Oyo State (5.8 m/s mean wind speed) can be exploited for the generation of large-scale electricity using wind turbines while the other location could be used for standalone small to medium scale electricity generation. Five models of horizontal axis wind turbines where evaluated for the sites and GE 1.5 sle wind turbine model was recommended for use. Econometrics shows that electricity generation cost ranged from 0.04 $/kWh to 5.03 $/kWh, based on the model of turbine used.

Abdulkarim et al. [63] carried out statistical analysis to determine the frequency distribution method with the best goodness of fit using wind speed data collated from fourteen sites in Nigeria namely Abuja, Bauchi, Bida, Gombe, Gusau, Ibi, Jos, Kaduna, Kano, Katsina, Lafia, Lokoja, Makurdi, Minna and Nguru. The study evaluated Gamma, Weibull and Rayleigh distribution functions and the performances of the probability functions was rated based on the variation between the theoretical and measured wind power densities. The Weibull distribution frequency gave a better prediction than the other distribution functions, result from the study also stated that Jos, Kano and Minna can be used for both off grid and grid electricity generation because the sites were rated to be class 4 category wind site.

Fatigun et al. [64] assessed the energy potential available in the wind speed in Ikeja, south western Nigeria using two-parameter Weibull distribution function with data captured at 10 m. Results from the study shows that the wind turbulence in the area was higher during dry season compared to the raining season. Wind power density (WPD) extrapolated at 50 m hub height was reported to be between 116.3 W/m2 $\leq$ WPD $\leq$423.3 W/m2 with annual average of 257.9W/m2. These figures suggest that the sites have marginal wind energy potential. Therefore, it was recommended for small scale electricity generation.

IOP Conf. Series: Earth and Environmental Science **331** (2019) 012019 doi:10.1088/1755-1315/331/1/012019

3. Opportunity for further wind power researches in Nigeria

From the reviewed literature, it can be seen that many works have been done on wind farm but the following opportunity for research still exist.

- Extensive wind mapping has been done for the land locations (Onshore) in Nigeria but offshore wind data is yet to be captured and analysed. Offshore wind characteristics have been reported to be more consistent with higher speed in other parts of the world but wind energy mapping of Nigeria continental shelf is yet to be done.
- Research into appropriate GIS software/model to estimate wind speed and by extension wind power for Nigeria is required.
- There is need for further research into multilayer numerical models for the prediction of wind profile in the regions of Nigeria.
- Installation of an offshore metrological station at location targeted with relatively high speeds for real-time extended period for wind speed, direction, atmospheric pressure, relative humidity and other relevant data capturing will be very adequate.

4. Recommendations for Nigeria wind energy sector development

The development of offshore wind energy and other renewable in Nigeria have not made much progress in recent years because of some inherent problems that have stunted the growth of the industry. To achieve energy sustainability and increase the country's energy mix, there is need to consider the following recommendations:

4.1. Government incentive and policies framework

The lack of commitment from the government to execute the energy policies and framework have led to investors demoralization and withdrawal as there is no guarantee that the return on their investment will yield returns in the least possible time. It is expedient for government to show commitment and follow through with the renewable energy framework with evidence that the interest of the investors is protected even when there is a change in government. A robust regulatory and legal framework will increase the confidence of investors for renewable energies and that will translate into individuals willing to commit their time and resources into the development of renewable energy. The government should also consider granting import duty waivers, tax rebates, energy credits and other incentives to offshore wind energy investors since the project is capital intensive. Bureaucracy around energy permits and approval should be reduced so as to encourage some start-up companies to venture into renewable energy generation such as offshore wind turbine.

4.2. Institutions to emphasis R&D tailored towards wind energy projects

The role that tertiary institution plays in the development of a nation cannot be overemphasized as it serves as a hub for research and innovation. Hence, researchers and students from such institutions should be encouraged to focus on issued based projects like the development of wind power technology and work towards the development of indigenous technology that will suit the nation's peculiar climatic environment.

4.3. Information to aid R&D to be made available to public

Institutions with relevant information and data that can aid the growth of renewable energy should make such information available for free. For example, NIMET, NIMASA etc should make wind speed information free for research institutions to download, Oil and Gas company should also be willing to share water depth, wind speeds and cost of running, with research institutions. As it stands acquiring information from these establishments for research purposes is cumbersome and creates a bottleneck for speedy development of renewable technology in the country.

5. Conclusion

Nigeria has coastal line of about 853 km which has potential for the generation of electricity via the use of offshore wind turbines. However, technical investigation to evaluate the exact wind energy resources that abounds in the shores of the country has not been carried out. This paper therefore reviewed various work that have been carried out in other countries where offshore wind turbines gained some measure of success with the aim of sensitizing research and development of offshore wind energy and by extension reduce the country's energy deficit. This work reviewed work carried out in offshore wind resource assessment, the use of GIS methods to measure offshore wind speeds, offshore wind turbine layouts, Nigeria onshore wind assessment scenario, turbine wake deficit and future innovations in offshore wind development. For rapid development of offshore wind turbine in this region, it was recommended that the government should invest in wind energy R&D especially development of offshore wind technology, tertiary institutions to encourage students to carry out "issues based" researches like offshore wind energy development to alleviate the energy crisis in the coutry and marine and oil and gas industries should be willing to share wind energy data information with researchers to enhance ease of estimating wind energy potentials.

References

[1] Gary L J 2006 *Wind Energy Systems* (Manhattan, KS.USA)

[2] Breton S and Moe G 2009 Status, plans and technologies for offshore wind turbines in Europe and North America. *Renewable Energy* **34** 646–654

[3] Schneider J and Senders M 2010 Foundation Design: A comparison of oil and gas platforms with offshore wind turbines *J. Marine Technology Society* **44(1)** 32-51

[4] Chong W Z, Chong Y L, Jing P, Ming Y L and Lin L X 2016 An overview of global ocean wind energy resource evaluations. *Renewable and Sustainable Energy Reviews* **53** 1240–1251

[5] Walt M, Donna H, Philipp B, George S and Caroline D 2016 Offshore Wind Energy Resource Assessment for the United States National Renewable Energy Laboratory. 2-7

[6] Graziano M, Lecca P and Musso M 2017 Historic paths and future expectations: The macroeconomic impacts of the offshore wind technologies in the UK. *Energy Policy* **108** 715–730

[7] World Wind Energy Association. Press release February 25, 2019 https://wwindea.org/information-2/information/ assessed April 30th 2019

[8] International renewable Energy Agency 2018 IRENA, Renewable Energy Benefits; leveraging Local capacity for Offshore wind, *2018 report* 4

[9] Adesoji A, Charles M, Benjamin C and Yohannes H 2012 Assessing offshore wind potential *Energy Policy* **42** 191–200

[10] Esteban M D, Counago B, Lopez-Gutierrez J S, Negro V and Velliso F 2015 Gravity based support structures for offshore wind turbine generators: Review of the installation process. J. *Ocean Engineering* **4** 1-17. DOI: 10.1016/j.oceaneng.2015.10.033

[11] Kaldellis J K and Kapsali M 2013 Shifting towards offshore wind energy—Recent activity and future development. *Energy Policy* **53** 136–148

[12] Effiong S O, Nwankwojike B N and Abam F I 2016 Economic cost evaluation on the viability of offshore wind turbine farms in Nigeria. *Energy Reports* **2** 48–53

[13] Onakoya A B O, Onakoya A O, Jimi-Salami O A and Odedairo B 2013 Energy consumption and Nigerian economic growth: An empirical analysis European Journal of Scientific Research 9(4) 25-40

[14] Sama M C and Tah N R 2016 The effect of energy consumption on economic growth in Cameroon Asian Economic and Financial Review 6 (9) 510-521

[15] Akuhwa T P 2015 The Effects of Unemployment and Anti-Social Activities of Youth on Socioeconomic Development of Benue State, Nigeria J. of Humanities and Social Science 20 (8) 69-81

[16] Enyidah-OkeyOrdu G 2017 Violence, Aggression and Insecurity in the Niger Delta of Nigeria: An Exploratory Study of the Militancy in the Region J. Advances in Social Sciences Research 4 (2) 9-14

[17] Gadad S and Deka P C 2016 Offshore wind power resource assessment using Oceansat-2 scatterometer data at a regional scale J. Applied Energy 176 157–170

[18] International renewable Energy Agency (2016) IRENA floating foundations: a game changer for offshore wind power. *A supplement to innovation outlook: offshore wind.* 5-6

[19] Oh K, Kim J, Lee J, Ryu M and Lee J 2012 An assessment of wind energy potential at the demonstration offshore wind farm in Korea *J. Energy* **46** 555-563

[20] Adaramola M, Agelin-Chaab M, and Paul S S 2014 Assessment of wind power generation along the coast of Ghana *J. Energy Conversion and Management* **77** 61–69

[21] Michael J D, Cristina L A and Mark Z J 2010 California offshore wind energy potential. *J. Renewable Energy* **35** 1244–1254

[22] Ahmmad M R 2014 Statistical Analysis of the Wind Resources at the Importance for Energy Production in Bangladesh. *International Journal of Energy, Science and Technology* **7 (2)** 127-136 http://dx.doi.org/10.14257/ijunesst.2014.7.2.12

[23] Lima D K S, Leao R P S, Santos A C S, Melo F D C, Couto V M, Noronha A W T and Oliveira-Jr D S 2015 Estimating the offshore wind resources of the State of Ceara in Brazil. *J. Renewable Energy* **83** 203-221

[24] Satir M, Murphy F and McDonnell K 2017 Feasibility study of an offshore wind farm in the Aegean Sea, Turkey *Renewable and Sustainable Energy reviews* **34** 243-453 http://dx.doi.org/10.1016/j.rser.2017.06.063

[25] Monfared M, Rastegar H and Kojabadi H M 2009 A new strategy for wind speed forecasting using artificial intelligent methods *J. Renewable Energy* **34** 845–848

[26] Li G and Shi J 2010 On comparing three artificial neural networks for wind speed forecasting *J.Applied Energy* **87** 2313–2320

[27] Fadare D A 2010 The application of artificial neural networks to mapping of wind speed profile for energy application in Nigeria *J. Applied Energy* **87** 934-942

[28] Koletsis I, Kotroni V, Lagouvardos K and Soukissian T 2016 Assessment of offshore wind speed and power potential over the Mediterranean and the Black Seas under future climate changes *Renewable and Sustainable Energy Reviews* **60** 234–245

[29] Sasaki W 2017 Predictability of global offshore wind and wave power *International Journal of Marine Energy* **17** 98–109

[30] Ho L, Shaharin I, Sutarji K, Che-Musa C O and Ahmad M A 2012 Review of Offshore Wind Energy Assessment and Siting Methodologies for Offshore Wind Energy Planning in Malaysia *American International Journal of Contemporary Research* **2 (12)** 72-86

[31] Cavazzi S and Dutton AG 2016 An Offshore Wind Energy Geographic Information System (OWE-GIS) for assessment of the UK's offshore wind energy potential. *Renewable Energy* 87 212-228

[32] Garlapati N, Ravi S R, Natansh K N, Surendra S K and Vimal S 2016 Application of OSCAT satellite data for offshore wind power potential assessment of India. *5th International Conference on Advances in Energy Research, ICAER* 2015, 15-17 December 2015, Mumbai, India. Energy Procedia **90** 89 – 98

[33] Ulazia A, Saenz J and Ibarra-Berastegui G 2016 Sensitivity to the use of 3DVAR data assimilation in a mesoscale model for estimating offshore wind energy potential. A case study of the Iberian northern coastline. *Applied Energy* **180** 617–627

[34] Zheng C W, Li C Y, Pan J, Liu M Y and Xia L L 2016 An overview of global ocean wind energy resource evaluations *Renewable and Sustainable Energy Reviews* **53** 1240–1251

[35] Gadad S and Deka P C 2016 Offshore wind power resource assessment using Oceansat-2 scatterometer data at a regional scale *Applied Energy* **176** 157–170

[36] Nagababu G, Kachwaha S S and Savsani V 2017 Estimation of technical and economic potential of offshore wind along the coast of India *J. Energy* **138** 79-91

[37] Amirinia G, Mafi S and Mazaheri S 2017 Offshore wind resource assessment of Persian Gulf using uncertainty analysis and GIS *Renewable Energy* **113** 915-929

[38] Carvalho D, Rocha A, Gomez-Gestteira M and Santos S C 2017 Offshore winds and wind energy production estimates derived from ASCAT, OSCAT, numerical weather prediction

models and buoys-A comparative study for the Iberian Peninsula Atlantic coast *J. Renewable Energy* **102** 433-444

[39] Schallenberg-Rodríguez J and Montesdeoca N G 2018 Spatial planning to estimate the offshore wind energy potential in coastal regions and islands. Practical case: The Canary Islands *Energy* **143** 91-103

[40] Kusiak A and Song Z 2010 Design of wind farm layout for maximum wind energy capture *Renewable Energy* **35** 685–694

[41] Barthelmie R I and Jensen L E 2010 Evaluation of wind farm efficiency and wind turbine wakes at the Nysted offshore wind farm *J. Wind Energy* **13** 573–586

[42] Gaumond M, Réthoré P E, Bechmann A, Ott S, Larsen G C and Pena D A 2012 Benchmarking of wind turbine wake models in large offshore windfarms. In: Proceedings of the science of making torque from wind **34** 1-9

[43] Seim F 2015 Validating kinematic wake models in complex terrain using CFD (Master thesis). Department of Ecology and Natural Resource Management, Norwegian University of Life Sciences. Akershus, Norway.

[44] Jeon B K and Huh J 2015 Comparison and verification of wake models in an onshore wind farm considering single wake condition of the 2 MW wind turbine. *J. Energy* **38** 1769–1777

[45] Niayifar A and Porte-Angel F 2016 Analytical Modeling of Wind Farms: A New Approach for Power Prediction J. *Energies* **9** 1-13

[46] Hou P, Hu W, Soltani M, Chen C and Chen Z 2017 Combined optimization for offshore wind turbine micro siting *J. Applied Energy* **189** 271–282

[47] Attias K &\and Ladany S P 2011 Optimal Layout for Wind Turbine Farms. Proceedings World renewable energy Congress, 8-13 May 2011, Linkoping, sweden

[48] Gao X, Yang H and Lu L 2014 Study on offshore wind power potential and wind farm optimization in Hong Kong *J. Applied Energy* **130** 519–531 http://dx.doi.org/10.1016/j.apenergy.2014.02.070

[49] Shakoor R, Hassan M Y, Raheem A and Wu Y 2016 Wake effect modeling: A review of wind farm layout optimization using Jensen's model *Renewable and Sustainable Energy Reviews* **58** 1048–1059. http://dx.doi.org/10.1016/j.rser.2015.12.229

[50] Ajit C P, John C, Mahdi K, Sami B and Lars J 2017 Application of an offshore wind farm layout optimization methodology at Middelgrunden wind farm *J. Ocean Engineering* **139** 287-297

[51] Ou L, Xu W, Yue Q, Ma C L, Teng X and Dong Y E 2017 Offshore wind zoning in China: Method and experience *J. Ocean & Coastal Management* **10** 1-10

[52] Amaral L and Castro R 2017 Offshore wind farm layout optimization regarding wake effects and electrical losses. *Engineering Applications of Artificial Intelligence* **60** 26–34 http://dx.doi.org/10.1016/j.engappai.2017.01.010

[53] Green R and Vasilakos N 2011 The economics of offshore wind. *J. Energy Policy* **39** 496–502

[54] Iris F A and Evrim U 2016 Assessment approaches to logistics for offshore wind energy installation *J. Sustainable Energy Technologies and Assessments* **14** 80–91

[55] He Z, Xu S, Shen W, Zhang H, Long R, Yang H and Chen H 2016 Review of factors affecting China's offshore wind power industry *Renewable and Sustainable Energy Reviews* **56** 1372–1386

[56] Colmenar-Santos A, Perera-Perez J, Borge-Diez D and DePalacio-Rodríguez C 2016 Offshore wind energy: A review of the current status, challenges and future development in Spain *J. Renewable and Sustainable Energy Reviews* 64 1–18

[57] Sarker B R and Faiz T I 2017 Minimizing transportation and installation costs for turbines in offshore wind farms **J. Renewable Energy** 101 667-679

[58] Katsaprakakis D A and Christakis D G 2014 Seawater pumped storage systems and offshore wind parks in islands with low onshore wind potential. A fundamental case study *J. Energy* **66** 470-486. http://dx.doi.org/10.1016/j.energy.2014.01.021

[59] Michailides C, Goa Z and Moan T 2016 Experimental and numerical study of the response of the offshore combined wind/wave energy concept SFC in extreme environmental conditions *J. Marine Structures* **50** 35-54

[60] Tsai Y, Huang Y and Yang J 2016 Strategies for the development of offshore wind technology for far-east countries – A point of view from patent analysis *J. Renewable and Sustainable Energy Reviews* **60** 182–194 http://dx.doi.org/10.1016/j.rser.2016.01.102

[61] Qin C, Saunders G and Loth E 2017 Offshore wind energy storage concept for cost-of-rated-power savings *J. Applied Energy* **201** 148–157

[62] Ajayi O O, Fagbenle, R O and Katende J 2011 Assessment of Wind Power Potential and Wind Electricity Generation Using WECS of Two Sites in South West, Nigeria. *International Journal of Energy Science* ***1*** **(2)** 78-9

[63] Ohunakin S O 2011 Assessment of wind energy resources for electricity generation using WECS in North-Central region, Nigeria *J. Renewable and Sustainable Energy Reviews* **15** 1968–1976

[64] Adaramola M S, Paul S S and Oyedepo S O 2011 Assessment of electricity generation and energy cost of wind energy conversion systems in north-central Nigeria *J. Energy Conversion and Management* **52** 3363-3368.

[65] Oyedepo S O, Adaramola M S and Paul S S 2012 Analysis of wind speed data and wind energy potential in three selected locations in south-east Nigeria. *International Journal of Energy and Environmental Engineering,* **3 (7)** 1-11

[66] Ajayi O O, Ohijeagbon O D, Ogbonnaya M and Attabo A 2016 Wind power mapping and NPV of Embedded generation systems in Nigeria. *International Journal of Environmental, chemical, Ecological, Geological and Geophysical Engineering* **10 (5)** 394-405

[67] Ajayi O O, Fagbenle R O, Katende J, Ndambuki J M, Omole D O and Badejo A A 2014 Wind Energy Study and Energy Cost of Wind Electricity Generation in Nigeria: Past and Recent Results and a Case Study for South West Nigeria *J. Energies* **7** 8508-8534.

[68] Abdulkarim A, Abdelkader S M and Morrow D J, Falade A J and Adediran Y A 2017 Statistical analysis of wind speed for electrical power generation in some selected sites in northern Nigeria *Nigerian Journal of Technology,* **34 (4)** 1249-1257

[69] Fatigun A T, Akoshile C O, Ajibola T B, Salau R O, Faweya E B, Aduloju K A and Oyedele E A 2017 Evaluation of Wind Energy Potential of Ikeja, Southwest, Nigeria using Two-Parameter Weibull Distribution Function *J. of Engineering* **7(09)** 2278-2289

IOP Conf. Series: Earth and Environmental Science **331** (2019) 012020 doi:10.1088/1755-1315/331/1/012020

Criteria for measuring Sustainable Construction Project Performance in Nigeria

Olowosile Sanmi[1] and Oke Ayodeji[2]
[1]Department of Quantity Surveying, Federal University of Technology, Akure, Nigeria
[2]Department of Quantity Surveying, Federal University of Technology, Akure, Nigeria

Email: sanmijossy@yahoo.com

Abstract. Construction Project (CP) in Nigeria is currently still at a rear end as far as sustainable construction is concerned. Consequently, the objective of this study is to identify criteria necessary for measuring sustainable construction project performance and to elicit perceptions of their relative importance. Thorough review of related studies were deployed to generate a set of factors that makes construction project habitable, attractive and liveable, and also, factors for measuring construction project performance. A total of 87 questionnaires were distributed to seven (7) main groups of project participants namely; client, project manager, architect, quantity surveyor, electrical engineer, mechanical engineer and builder/civil/structural engineer who participated on nine (9) case study projects. The survey findings indicate that quality performance and productivity both with 81.6% of importance, innovation with 80.4% of importance, health and safety with 80.2% of importance and client's satisfaction with 80% of importance are the most important factors to consider while measuring sustainable construction project performance. For a construction project to be habitable, safety and incident risk, acoustic comfort and visual comfort are the most important factors to consider. When considering how attractive construction project should be, art in architecture, unique style, terrain choice and harmonious surrounding are the most important factors to consider. Also, when considering the liveability of a construction project, space efficiency, ventilation and public accessibility are the most important factors to consider. Finally, this research recommended that the factors listed above, are key factors to consider at the design stage of any construction project for both the designer and the regulatory agencies, and these factors are to be monitored throughout the construction process since they are vital in measuring the performance of any sustainable construction project to infer a more desirable and satisfactory sustainable building.

1. Introduction

The global and national search towards sustainable development has placed the construction industry in the spotlight since the industry is the major main consumer of natural resources. Construction projects has fundamental impacts on economic, environmental, social and technical sustainability. In considering these impacts, it is important to measure the performances of product of construction industry which is construction projects. In measuring the performance of construction projects, it is vital to identify criteria needed to assess the performance of the construction projects as this in turn simplify the process of performance measurement of construction projects.

Several construction project performance evaluation techniques or methods are presently in use all around the world; yet some of these methods neglect to incorporate the needed sustainable performance indicators and criteria, hence, the need to identify criteria to consider when measuring performance of construction project. Also, many criteria are to a great degree complex to make and are regularly scrutinized for disentangling the unpredictable issues of sustainability in CP performance. Today, great deal of effort is being placed worldwide in achieving sustainable performance in construction industry [4]. The sole objectives of these efforts are to regulate the energy consumption and to ensure construction project sustainability in extreme climatic condition and sustainable efficiency design in construction project.

2. Review of Related Studies

Performance measurement is utilized to enhance project execution by taking a gander at the criteria to be estimated and assessed to get the outcomes; this will be a guide for what associations need to design and do later on. Moves ought to be made in view of the after effects of assessment. Criteria measured in the assessment procedure are constituted by inside and outer parts of the organization (Hoque, 2004).

Thought of taking non-monetary measures, for example, consumer satisfaction and occupation fulfilment into account in estimating execution began in the 1990s. After numerous industries uncovered the restriction of money related measures, associations tended to take a gander at non-monetary viewpoints to be estimated and considered in assessing execution of CP. Business pioneers concur that a successful project will better serve its stakeholders by concentrating on the necessities of its clients, workers, providers and the more extensive network [1].

There is a developing mindfulness among organizations in the construction industry that measurement frameworks are vital for checking and controlling their execution and performance. [5] Study concede to a similar thought. They express that performance measurement is creating its way in the industry. There are three reasons. In the first place, the consistent fast improvement of performance measurement in different divisions amid the 1990s made a monstrous enthusiasm for its advancement in the construction industry. The second reason is the expanding intricacy of construction projects requiring proper measurement instruments and models to enhance execution and the third is the advancement and difficulties of project management and in addition building innovation as of late.

Some factors are essential to consider while measuring sustainable construction project performance in order to have a clearer picture of where these types of projects stands as far as sustainability is concerned and according to [8], these factors can be categorised under different phases in the construction process namely; Inception phase, Construction phase, Operational phase and Demolition phase. The following are the factors identified in his study as the most important factors to consider; finance plan, workers' health and safety assessment, Air pollution assessment, standardization, security consideration, energy consumption, public awareness, re-usable element, local economy, provision of services among others. [15] Study also recognises the impact construction chemical waste and waste disposal could cause throughout a construction project lifecycle. [11] Study emphasizes on occupational health and safety as one of the major benefits of applying sustainability measure in a construction project.

3. Research Methods

3.1. Research design. The design adopted for this study was survey method (quantitative research). Primary data were collected through questionnaire administered to professionals that participated in the case study projects for this study.

3.2. The Study Area. Nine (9) Construction projects were selected as case study for this research. These construction projects are project registered with a sustainability tool as at the time of construction in Nigeria.

3.3. Study Population, Sample Size and Sample Technique. The projects selected for this study are mainly construction projects constructed to meet sustainable development goals. They are listed below:

1. Nest Oil Tower Project, Lagos
2. The Heritage Place Tower Project, Lagos
3. Nox Building Project, Abuja
4. P & G Nigeria MDO Warehouse Project, Ogun State
5. No 4 Bourdilon Street Building Project, Lagos
6. Convention Centre Project, Uyo State
7. Wing Tower Project, Lagos
8. AfDB Nigeria Field Office Building Project, Abuja
9. Alliance Place Tower Project, Lagos

For One (1) Construction Project, respondent for the study are:

A. Client
 a. Client or Client representative = 1

B. Consultant
 a. Project Manager = 1
 b. Architect = 1
 c. Quantity Surveyor = 1
 d. Structural Engineer = 1
 e. Electrical Engineer = 1
 f. Mechanical Engineer = 1

C. Contractor
 a. Quantity Surveyor = 1
 b. Builder or Civil Engineer = 1
 c. Electrical Engineer = 1
 d. Mechanical Engineer = 1

 11

Therefore, for Nine (9) Construction Projects, number of respondent (R) will be

R = 11 x 9
 = 99

The number of questionnaire distributed for this study are Ninety Nine (99) but not all the questionnaires were returned, Eighty Seven (87) questionnaires were returned in which all were found suitable for the purpose of analysis, this represent 87.87% of the response rate which is adequate for the study.

The technique for sampling respondents for this study is a non-probabilistic sampling technique also termed purposive sampling technique. This is so because these type of projects are not so common in the country, therefore, careful selection of sustainable commercial construction projects in Nigeria was purposively done.

3.4. Data Collection Approach. The collection approach employed for this study includes a well-structured questionnaire self-administered to various professionals that are concerned through direct contact, e-mails, WhatsApp and telegram. The questionnaires were self-explanatory, clear and understandable.

3.5. Data Analysis Techniques. The data collected in this study was analysed with relative importance index and mean item score through SPSS package and Microsoft Excel software.

3.5.1 Relative Importance Index (RII). The criteria for measuring performance of sustainable construction was analysed using Relative Importance Index (RII) to rank the indicators accordingly to their order of importance. 5-point likert scale was employed for the collection of data, the research employ the formula below

$$\text{R.I.I} = \frac{\sum_1^5 W}{A \times N} \quad (1)$$

N = the total number of responses.
W = weighting given to each factor by the respondents and ranges from 1 to 5 where '1' is 'Not Important' and '5' is 'Most Important',
A = highest weight (i.e. 5 in this case).

3.5.2. Mean Item Score (MIS). Mean Item score was used to analyse criteria for sustainable construction. The sum of FX resulting from the multiplying frequencies of the respondents and ratio of the range of options between 1-5. FX was divided by the total number of questionnaires analysed to get the mean item score used to rank indicators from the highest to the lowest.

Mean score equation used for the calculation

$$\text{MIS} = \frac{5(FX5) + 4(FX4) + 3(FX3) + 2(FX2) + (FX1)}{X5 + X4 + X3 + X2 + X1} \quad (2)$$

Where MIS = mean item score

X = range 1 – 5 with 5 being the highest and 1 being the lowest

F = Frequency of respondent in each factor

4. Data Analysis and Result

Table 1. Demographic Information

Categories	Classification	Number	%
Profession	Client	9	10.34
	Project Manager	8	9.20
	Architect	9	10.34
	Quantity Surveyor	15	17.24
	Civil Engineer	6	6.90
	Structural Engineer	4	4.60
	Electrical Engineer	16	18.39
	Mechanical Engineer	15	17.24
	Builder	5	5.75
Year of Experience	6 – 10 years old	47	54.00
	11 – 15 years old	17	19.50
	Less than 5 years old	15	17.20
	16 – 20 years old	5	5.70
	More than 20 years old	3	3.40
Participation in SCP	Yes	78	89.70
	No	0	0.00
	Missing	9	10.30
	Total	87	100.00

IOP Conf. Series: Earth and Environmental Science **331** (2019) 012020 doi:10.1088/1755-1315/331/1/012020

From Table 4.1, respondents who were involved in filling of questionnaires and answering of interview questions are Clients or Client representative, Project Managers, Architects, Quantity Surveyors, Structural Engineers, Electrical Engineers, Mechanical Engineers, and Builders who actively participated in the construction projects selected as case study for the purpose of this study. respondents whose years of experience are between 6-10years are at the majority as they dominates 54% of the total population while respondents with 11-15 years of experience has 19.50% , those respondents who have less than 5years of experience have 17.20. The respondents with years of experience of over 20years are very limited as they are only 3.40% fraction of the entire populace, and thereafter followed by the number of respondents with 16-20 years of experience with a fraction of 5.7% of the entire populace. This reveals that as much as the idea gotten from this research is distributed among different age bracket of experience, there seem to be a larger turn out from the middle aged professional in the construction industry to participation in Sustainable Construction in Nigeria. The result gotten from the respondents shows that about 90% have well participated in Sustainable Construction project while none claim to have not participated in Sustainable Construction projects, at the same time, about 10% of the respondents left their opinion blank.

Table 2. Factors that makes Construction projects habitable.

Criteria for Habitable Projects	Descriptive Statistical Values	
	Mean	Ranking
Safety and Incident risks	3.56	1
Acoustic comfort	3.44	2
Visual comfort	3.09	3
Indoor air quality	2.94	4
Outdoor building related qualities	2.94	5
Influence of the user	2.86	6
Aesthetic comfort	2.67	7
Thermal comfort during dry season	2.48	8
Thermal comfort during rainy season	2.47	9

From Table 2 above, Safety and Incident risks with mean score of 3.56 has the 1st rank when comparing several criteria that makes habitable construction project. This rank was then followed by Acoustic comfort coming 2nd with a mean score of 3.44, thereafter Visual comfort with 3.09 mean and 3rd in ranking. This is not surprising as safety and comfortability of users of any construction project is maximally important which have been demonstrated in Table 2 above as the most important criteria to consider when measuring the habitability of a construction project. Thermal comfort during rainy season and during dry season has the lowest rank on the table with 9th and 8th ranking and mean value of 2.47 and 2.48 respectively. This also can as well be interpreted thus; that in considering how habitable a construction project can be, these factors are less important compared to the other factor mentioned earlier above.

Table 3. Factors that makes Construction Projects attractive

Criteria for Attractive Projects	Descriptive Statistical Values	
	Mean	Ranking
Art in Architecture	3.64	1
Unique style	3.62	2

Terrain choice	3.57	3
Harmonious surroundings	3.54	4
Landmark design	3.45	5
Colour or Finish choice	3.37	6
Acceptability by the public	3.10	7
Cultural identity	2.99	8
Size	2.06	9

In the attempt to identifying what makes Sustainable construction Project attractive, it was found that one of the most important criteria to consider is the Art in Architecture of the given project which ranked 1st and has a mean value of 3.64; thereafter, we can consider the uniqueness of the sustainable project's style as this ranked 2nd with a 3.62 mean value. Another criteria to consider when proposing an attractive sustainable construction project is the terrain choice where the sustainable construction project will be situated, this factors is ranked 3rd with a mean value of 3.57. The least of the criteria to consider for attractive projects have mean value of 2.06, 2.99 and 3.10 and they are size, cultural identity of the structure and acceptability by the public with 9th, 8th, and 7th ranking respectively. From the analysis, even though these three criteria happen to be the lowest ranked, it does not mean they really do not matter, as two out of these criteria has a mean rank well above average; which means that they are as well important even if not as important as other with higher ranking.

Table 4. Factors that makes Construction Projects Liveable

Criteria for Liveable Projects	Descriptive Statistical Values	
	Mean	Ranking
Space efficiency	3.94	1
Ventilation	3.92	2
Public accessibility	3.54	3
Shopping facilities	3.45	4
Vehicle comfort	3.37	5
Outdoor building related qualities	3.24	6
Capability of conversion	3.16	7
School facilities	3.09	8
Aesthetic	2.92	9

In the quest of finding out what makes construction project liveable, one of the most important reasons why construction building project are always embarked upon was found to be space efficiency with ranking 1st with a mean score of 3.94. This criterion happens to be very scarce to find in places like Lagos state Nigeria, where construction projects are being managed on the little land mass available due to scarcity of land in the area. Ventilation is also a very paramount criterion with 2nd rank and a mean rank of 3.92, as it is very essential for construction project especially building project to be well ventilated for the comfort and healthy living of the users. Public accessibility and shopping facilities are also a key criteria to consider in a construction project. In considering how liveable a construction project can be, all the criteria mentioned in Table 4 above should be well satisfied as they all ranked well above average which makes them all important in the life of a construction project.

Table 5. Criteria for measuring Sustainable Construction Project Performance

Criteria for Measuring SCPP	**RII**	**% of Importance**	**Ranking**

Quality performance	0.816	81.6%	1
Productivity	0.816	81.6%	1
Innovation	0.804	80.4%	3
Health and Safety	0.802	80.2%	4
Client's satisfaction	0.800	80.0%	5
Time performance	0.790	79.0%	6
Strategic planning and Management	0.782	78.2%	7
Supply chain management	0.770	77.0%	8

Table 6. Criteria for measuring Sustainable Construction Project Performance (Cont'd)

Criteria for Measuring SCPP	RII	% of Importance	Ranking
Profitability	0.766	76.6%	9
Cost performance	0.766	76.6%	9
Team work	0.752	75.2%	11
Risk Management	0.740	74.0%	12
Information Analysis	0.676	67.6%	13
Stakeholder's satisfaction	0.676	67.6%	13
Growth	0.674	67.4%	15
Training & development	0.646	64.6%	16
Leadership	0.646	64.6%	16
Partnering	0.626	62.6%	18

Table 6 shows that the two most important criteria for measuring sustainable construction project performance is quality performance and productivity of the project which both have a relative importance index of 0.816 and ranked 1st. Innovation with a relative importance index of 0.804 with 3rd ranking is also another important criteria for measuring sustainable construction project performance with a very high importance index. Health and Safety and Client's satisfaction are other cogent factors that needs to be taken seriously, as the relative importance index of these factors are 0.802 and 0.800 respectively with 4th and 5th ranking on the table. [11] presented health and safety as one of the major benefit of applying sustainability to construction project. The least of the factors listed in Table 5 above is partnering with 18th ranking and an importance index of 0.626, followed by Leadership with 16th ranking and an index of 0.646. Other factors which are low on the table includes but are not limited to Training and Development, Growth, Stakeholder's Satisfaction and Information Analysis which are all of indexes 0.646, 0.674 and 0.676, 0.676 with 16th, 15th 13th and 13th ranking respectively.

5. Discussion of Findings

Safety and Incident risks is the most important factor to consider that makes construction project habitable. Acoustic comfort and Visual comfort are as well considered important in making a construction project habitable. This is not surprising as safety and comfortability of users of any construction project is maximally important. Thermal comfort during rainy season and during dry season has the lowest negligible importance in considering how habitable a construction project can be, these factors are less important compared to the other factor mentioned earlier above. It was explained that to enhance safety practices, it is imperative to enhance professional's interest in safety management and also increased awareness [16]. Visual comfort is a relatively important criteria which includes provision of natural light, external views and reduction of glare among others. This is also used as a factor by Building Research Establishment Environmental Assessment Method (BREEAM) assessor to assess a sustainable construction; likewise acoustic comfort which includes overcoming the noise nuisance which may have negative effect on health and quality of life. An example by Design Building of the United Kingdom is the ability to hear the occupants of a neighbouring house through the walls of the building.

In the attempt to identifying what makes Sustainable construction Project attractive, it was found that the most important criteria to consider is the Art in Architecture of the given project, thereafter, we can consider the uniqueness of the sustainable project's style. Another criteria to consider when proposing an attractive sustainable construction project is the terrain choice where the sustainable construction project will be situated. The least of the criteria to consider for attractive projects are size, cultural identity of the structure and acceptability by the public respectively; even though these three criteria happens to be the lowest ranked, it does not mean they really do not matter. Art in architecture can be termed the expression or application of creative human imagination and skill in a visual form to be appreciated for their beauty and emotional power. Therefore whether the primary purpose of art in architecture of any construction project is to achieve beauty or to organize space in a strategic way is completely up for discussion [2].

In the quest of finding out what makes construction project liveable which is one of the most important reason why construction building project are always embarked upon, it was found that space efficiency is the most important criteria to consider. This factor happens to be very scarce to find in places like Lagos state Nigeria, where construction project are being managed on the little land mass available due to scarcity of land in the area. Ventilation is also a very paramount criterion, it is very essential for construction project especially building project to be well ventilated for the comfort of the users. Public accessibility and shopping facilities are also a key criteria to consider in a construction project. In considering how liveable a construction project can be, all the criteria mentioned above should be well satisfied as revealed in this study, as they are important in the life of a construction project. This result agrees with Physical planning division of the government of the commonwealth of Dominica who considers space, light and ventilation as the key criteria for a liveable and habitable construction project.

Two of the most important criteria for measuring sustainable construction project performance is quality performance and productivity of the project which both have a relative importance index of 0.816. It was opined that for quality performance in any construction project to be achieved, there is need for contractors to give a greater preference to the use of quality materials to improve project cost, time and quality [8]. (Patience, 2014; Liu, 2014) and Arora (2014) also justifies quality performance as a crucial measure of project performance. Innovation as a factor is also another important criteria for measuring sustainable construction projects. Safety and Client's satisfaction is another cogent factor that needs to be taken seriously. The least of the factors is Partnering, followed by Leadership. Other factors which are low on the table includes but are not limited to Training and Development, Growth and Stakeholder's Satisfaction [4][10].

6. Conclusion and Recommendation

Several criteria that makes habitable construction project among others are Safety and Incident risks, acoustic comfort and visual comfort. This is not surprising as safety and comfortability of users of any construction project is maximally important. In the attempt to identifying what makes Sustainable construction Project attractive, it was found that the most important criteria(s) to consider are the harmonious surrounding of the given project, the uniqueness of the sustainable project's style and design, and the terrain choice where the sustainable construction project will be situated. In the quest of finding out what makes construction project liveable, some of the most important reasons why construction building project are always embarked upon was found to be space efficiency, ventilation, and public accessibility and shopping facilities. Some of the most important criteria for measuring sustainable construction project performance is quality performance and productivity of the project. Also, there is need for contractors to give a greater preference to the use of quality materials to improve project cost, time and quality. Generally, it will be important to understand that the quality performance of the criteria necessary in achieving habitable, liveable and attractive construction project should be highly regarded and incorporated into other factors used when measuring the performance of sustainable construction project.

Finally, this research recommended that the factors listed above, are key factors to consider at the design stage of any construction project for both the designer and the regulatory agencies, and these factors are to be monitored throughout the construction process since they are vital in measuring the performance of any sustainable construction project to infer a more desirable and satisfactory sustainable building.

7. References

[1] Adams C A, Muir S and Hoque Z 2014 Measurement of sustainability performance in the public sector *Sustainability Accounting Management and Policy Journal* **51** 46-67
[2] Al Marashi H 2006 Facing the Environmental Challenge: Encouraging Sustainable Urban Development in the United Arab Emirates *Global Urban Development Magazine* **21**, 175-180
[3] Ariana Z 2017 Art in Architecture? 13 Projects that blur the boundary *Arch daily;* Retrieved April 13, 2019 from https://wwwarchdailycom/868877/art-or architecture-13-projects-that blur theboundary/
[4] Arora V 2015 A Balanced Scorecard Framework for Measuring Sustainability Performance of Business Organizations *Concordia University; Concordia Institute for Information Systems Engineering CIISE Library*
[5] Bassioni H, Price A and Hassan, T 2005 Building a Conceptual Framework for Measuring Business Performance in Construction: An Empirical Study *Journal of Construction Management and Economics* **235** 495-507
[6] Bawany N and Shamsi J 2015 Smart City Architecture: Vision and Challenges *International Journal of Advanced Computer Science and Applications* 247-255
[7] Edwards L 2015 Privacy, Security and Data Protection in Smart Cities: a Critical EU Law Perspective *RCUK funded Centre for Copyright and New Business Models in the Creative Economy CREATe* Retrieved June 13, 2018 from https://strathprintsstrathacuk/55917/1/Edwards_EDPLR_2016_Privacy_security_and_data protection_in_smart_cities_a_critical_EUpdf
[8] Enshassi A, Kochendoerfer B and Al Ghoul H 2016 Factors affecting Sustainable Performance of Construction Projects Life Cycle Phases *International Journal of Sustainable Construction Engineering and Technology* **71** 50-68
[9] Huovila P, and Koskela L 1998 Contribution of the Principles of Lean Construction to Meet the Challenges of Sustainable Development *Sixth international conference of the International Group for Lean Construction* Guarujá, Brazil
[10] Liu J 2014 *Developing a life cycle assessment model for measuring sustainable performance of building in China* Retrieved June 13, 2018 from University of Sydney Library: https://opuslibutseduau/bitstream/10453/30396/2/02wholepdf
[11] Jalilon L, and Poon S 2008 Sustainable Construction aspects of using prefabrication in dense urban environment *Journal of Construction Management and Economics* **269** 953 - 966
[12] Khalfan M and Bouchlaghem M, Dino A, Carrillo P 2002 Sustainable development and sustainable construction C-SanD Approach *e-Sm@rt conference*, pp 19-21 Salford, UK
[13] Nwokoro, I 2011 Sustainable or Green Construction in Lagos, Nigeria: Principles, Atribute and Framework *Journal of Sustainable Development* **44** 166-174
[14] Pew Research Internet Project 2014 *The Internet of Things will thrive by 2025* Retrieved March 14, 2018 from http://wwwpewinternetorg/2018/03/14/internet-of-thingspdf

[15] Shen L, Hao L, Tam V, and Yao H 2007 A checklist for assessing Sustainable Project of construction projects *Journal of Civil Engineering and Management* **134** 273-281
[16] Vitharana G, Subashi D, and Sudhira D 2015 Health Hazards, Risk and Safety Practices in Construction Sites *Journal of The Institution of Engineers* Sri Lanka **483** 35-44
[17] Wilkinson S, and Reed R 2007 The Structural and Behaviourial Barriers to Sustainable Real Estate Devt. *American Real Estate Society ARES Conference* San Francisco, USA

IOP Conf. Series: Earth and Environmental Science **331** (2019) 012021 doi:10.1088/1755-1315/331/1/012021

Review Of Heat Transfer Enhancement In Energy Conversion Systems; Nanotechnology

[1]Efemwenkiekie U. Kelvin, [1]Atiba E. Opemipo, [1]Oyedepo O. Sunday
[1]Department of Mechanical Engineering, Covenant University, Ota, Nigeria
E-mail: uyi.efemwenkiekie@covenantuniversity.edu.ng

Abstract. Different heat transfer fluids with very high thermal stability are highly desirable for heat exchangers. Inadequate understanding on the basic principle of nanofluid flow, unsatisfactory experimental data on nanofluid heat transfer and general context for heat transfer correlation is yet to be established. The work under focus will be on highlighting working principles, problems hindering practical application of unconventional heat exchangers devices, and the various means of utilizing nanotechnology in improved efficiency. The study investigates experiments on heat exchangers in the last decade, the gap in knowledge and future applications of nanotechnology in improving heat transfer. This review summarizes, the significant advances of various nanotechnology in heat transfer application of recent energy conversion system.

1. Introduction

It is almost impossible to go through any industry today without noticeably seeing the presence of heat transfer, this is cause wherever there is a difference in the temperature that exists in a medium or within medium there is bound to be that tendency that will cause the more energy medium to transfer some of its heat to the less energetic medium the main objective of this for the entire system to achieve equilibrium or uniformity. Whether its air conditioning, refrigeration, ventilation, heating or its just coolant trying to stabilize the temperature of a machine.

Almost a century since Maxwell (1973) started working on millimeter or micrometer particles but, ever since this concept(theory) of finding better development of material for engineering. There has been tremendous growth in this area especially with the development nanomaterials. With its characteristics of been fast economical and also finding new applications in various forms we never thought of, this margin is likely to grow in near future. Researchers all over the world are toiling all day and night to get new ways to better establish and give new meaning and application to nanofluid as engineering materials. (Fakehinde, 2019) Energy is a wealth generator as such the reason for interest the world shows on it. As at the time fossils fuel was first discovered we were only interested in in usage but as the years progressed more development came in with researchers pushing for the optimization of these system at later years prediction of expected depletion was also forecasted was also hinted and also continued threats of global warming and the resulting green as effects, experts have been on the look trying finds ways of mitigating the challenges posed to the world. Nanomaterial have been seen as an option in enhancing energy systems in conversion. Apart from the reduction of greenhouse gases effects, its cost-effective attribute and so many that will be further elaborated later.

The exchange of heat between physical systems is one of the most vital operations being utilized in industries, finished products and machines. Furthermore, large compactness and effectiveness of heat transfer systems are required to obtain an essential heat transfer of liquids with poor performance. The application of additives to liquids with solid particles less than 100nm constitute a major part towards improving of heat transfer and heat properties of devices. This technology comes either as nanocomposites or nanofluids.

Nanofluid with enhanced thermal Conductivity are of tremendous benefits in the application of heat exchange between physical systems since they do not clog the path of flow while inducing a little drop in pressure. The erratic movement of nanoparticles in base fluid makes nanofluids relatively stable, due to the continuous collision from molecules of the medium (Angayarkanni & Philip, 2015; Sarviya & Fuskele, 2017). The tendency of a material to allow heat to pass through it is known as heat conductivity. In addition, nano particles surface to volume ratio is higher than micro particles while nano fluids is the most preferred in thermal heat transfer applications due to its higher ability to conduct than and microfluid (Bhattacharya et al., 2004; Tawfik, 2017).

A heat transfer device is capable of transferring heat from one medium of fluid to another. They have great heat removal capability which makes them work over an extensive range of temperatures when being used. Heat exchangers have been discovered to be important in technologies. Like all other devices, performances of heat exchangers depend strongly on the properties of the material that it employs for heat transfer and energy storage (Dai et al., 2012). Environmentally benign devices are required to complement the new heat transfer systems, due to an intermittent feature on most heat exchangers, thus the need for nanofluids, nanoelectrodes, nanowires, nanocatalysts and other improved nanotechnology.

1.1 ***Classifications of Heat Exchangers***

Heat exchangers are used for adjusting system temperature by increasing or reducing the heat energy. Heat transfer occurs between two fluids due to the difference in their temperatures without application of external energy (Kumar et al., 2015).

In heat exchanger, energy is transferred primarily through a convection from the surface of the wall to the cold fluid. It should be noted that in no way are heat exchangers the same as mixers. In the latter, energy is transferred by partial or complete blending of the hot and cold fluid streams when the fluids combine in a control volume or separately so as to exit together as a single mixed stream.

Heat exchangers maybe categorized according to (a) the process of transfer, (b) fluids number, (3) fabrication, (4) principle of heat transfer, (5) surface compactness, (6) arrangement of flow, (7) the kind of surface. (Javadi et al., 2013). Other special type of heat exchangers are the recuperators and regenerators where the former are utilized when energy transfer exist between two fluid flowing at various level of temperature in a room that is parted by a thin strong wall (a separating sheet) while the latter are kind of heat exchanger where heat from the hot fluid is stored in a heat storage medium before it is transferred to the cold fluid alternately through the same surface at prescribed time intervals (Hussein et al., 2014).

2.0 ***Nanotechnology in Heat Transfer***

The blending of infinitesimal particles with a base fluid led to the improvement of convection and heat conductivity (Maxwell paper, 1987). However, some constraints caused by sedimentation, erosion, clogging and drop in high-pressure has kept the technology far from practice. The essence of thermal transfer improvement for fluid containing a small quantity of nanosized particles is to increase the heat conductivity and turbulence through the suspended nanoparticles (Godson et al., 2010). Functional materials made up of particulates with at least one dimension below 100 nanometers (nm) can be categorized into organics and inorganics where the former is made up of carbon nanotubes, graphite and nanofibers. (Scida et al., 2011). Nanofluids are classified as engineered colloids with possibility of replacing the upcoming generation medium of heat transfer (Sarkar, 2010) while the latter comprises of aluminum, zinc, copper, iron, aluminium oxide, iron oxide, titanium oxide (Dreizin, 2009).

2.1 Nanocomposites and Nanomaterials

The major driving force for nanotechnology researches lay in its vast range of applications and inert potentials. Nanocomposites can be inorganics or organics based such as paraffin blended nanocomposites, fatty acid blended nanocomposites, HDPE blended nanocomposites etc. They are also referred to as polymer composites (Wang et al., 2010; Socher et al., 2011). Nanofluids are nanoscale colloidal suspensions containing nanomaterial such as nanofibers, nanorods, nanosheets, nanoparticles, nanodroplets, nanowires or nanotubes in a conventional base fluid like water, bio-fluids and other common coolants. Nanofluids possess a higher specific surface area, dispersion stability, thermal conductivity, surface wettability, reduced particle clogging and pumping (Hussain et al., 2013)

2.1.1 Carbon Nanomaterials

Carbon is known to exist in three forms, namely amorphous carbon, graphite and diamond. The arrangement and bond of carbon atoms in the crystal lattice strongly affects the properties

2.1.2 Graphene

It contains two-dimensional (2D) shape of hexagonally lattice carbon atoms (Hu et al., 2015). The unique structure of graphene with large theoretical specific surface area (ca. 2600 m^2g^{-1}) (Luo et., 2012) and outstanding diffusiveness of Li^+ ranging from 10^{-7} to 10^{-6} cm^2s^{-1} further accounts for its uses in various high-power-density which includes lithium ion batteries, fuel cells, and solar cells (Yuan & Shi, 2013).

2.1.3 Other Nanomaterials

Thermophysical properties of nanomaterials includes heat conductivity, heat diffusivity and convection coefficients (Yu & Xie, 2011). (Behroyan et al., 2016). Below are some of the numerous properties of organic and inorganic nanomaterials in the operation of heat exchangers.

Table 1: Properties of organic and inorganic nanomaterials in heat exchangers

Organics	Inorganics
Merits	
Anti- corrosive	Higher enthalpy phase change
Chemical and thermal stability	
Low or non-undercooling	
Demerits	
Less enthalpy phase change	Low cooling
Lower heat conductivity	Corrosive
Non-flammable	Phase separation
	Absence of heat stability

3.0 Characteristics of Thermal Transfer in Nanofluids

Nanotechnology has been attractive to researchers in the past decades for reasons best attached to their superior thermophysical advantage. In terms of its application to energy conversion devices, there are little or no limitation in pressure fall (Raja et al., 2016). When describing the characteristics of heat transfer in nanofluids, some major properties which includes convective heat transfer, heat capacity and heat conductivity are to be considered during analysis.

Table 2: Summary of thermophysical properties of nanofluids

Investigators	Thermophysical Properties	Nanoparticle/ Base Fluid	Empirical Correlation / Relevant Information
Ghazvini et al. (2011)	Specific Heat Capacity	Nanodiamond: engine oil	$C_{p,nf} = 2.62 - 6 \times 10^{-3}T + 2 \times 10^{-5}T^2$ $278 \le T \le 373$
Sekhar and Sharma (2013)	Specific Heat Capacity	Al_2O_3: water	$C_{p,nf} = 0.8429\left(1 + \frac{T_{nf}}{50}\right)^{-0.3037}\left(1+\frac{d_p}{50}\right)^{-0.4167}\left(1+\frac{\emptyset}{100}\right)^{2.272}$ $20^oC < T_{nf} < 50^oC$ $0.01\% < \emptyset < 4.0\%$
Pakdaman et al. (2012)	Specific Heat Capacity	MWCNT: heat transfer oil	$\frac{C_{p,bf} - C_{p,nf}}{C_{p,bf}} = (0.0128 \times T + 1.8382) \times \emptyset^{0.4779}$ $0.1 < \emptyset < 0.4\ vol.\%$
Shin and Banerjee (2014)	Specific Heat Capacity	Al_2O_3: alkali carbonate salt eutectics	$C_{p,nf} = \frac{\rho_p \emptyset C_{p,p} + \rho_{salt}\emptyset_{salt}C_{p,salt} + \rho_{ns}\emptyset_{ns}C_{p,ns}}{\rho_p\emptyset + \rho_{salt}\emptyset_{salt} + \rho_{ns}\emptyset_{ns}}$
Khanafer and Vafai (2011)	Specific Heat Capacity	Al_2O_3: water	$C_{p,nf} = \frac{(1-\emptyset)\rho_{bf}C_{p,bf} + \emptyset\rho_p C_{p,p}}{\rho_{nf}}$
Kumaresan and Velraj (2012)	Specific Heat Capacity	MWCNT/EG: water	$C_{p,nf} = \left[(1-\emptyset)(\rho C_p)_{bf} + \emptyset(\rho C_p)_p\right]/[\emptyset\rho_p + (1-\emptyset)\rho_{bf}]$
Pantzali et al. (2009)	Convective Heat Transfer	CuO: water (6.0 vol.%)	$Nu = 0.248Re^{0.75}Pr^{0.4}$
Pandey and Nema (2012)	Convective Heat Transfe	Al_2O_3/EG: water	$Nu = (0.26 + 0.02\emptyset - 0.0051\emptyset^2)Pr^{0.27}$

Ray et al. (2014)	Convective Heat Transfer		$Nu = 0.3053Re^{0.75}Pr^{0.3}$ For $150 < Re < 1500$ and $Pr(4-27)$
Gulenoglu et al. (2014)	Convective Heat Transfer	Nanofluids	$Nu = \begin{cases} 0.32867Re^{0.68}Pr^{0.1/3}\left(\frac{\mu}{\mu_w}\right)^{0.14}, 300 \le Re \le 5000 \\ 0.3277Re^{0.675}Pr^{0.1/3}\left(\frac{\mu}{\mu_w}\right)^{0.14}, 300 \le Re \le 5000 \end{cases}$

3.1 Specific Heat of Nanofluid

Salman et al. (2012) Investigated the specific heat of single-phase and two-phase treatments. The specific heat of fluids containing small quantity of nanoparticles were determined based on the volume fractions determined as;

$$C_{p,nf} = \emptyset C_{p,p} + (1-\emptyset)C_{p,f} \quad \text{and}$$

$$(\rho C_p)_{nf} = \emptyset(\rho C_p)_p + (1-\rho C_p)_f$$

Where ρ = density and C = specific heat capacity. Elia et al. (2014) were able to deduce reduction in the specific heat of nanofluid with a rise in its volume concentration from experimental study where the specific heat energy was intensified by a rise in temperature

3.2 Convective Heat Transfer

The governing parameters for the convective heat transfer model are dimensionless variables. Convection coefficient in this study can be obtained from (Faizal et al., 2013)

$$h = \frac{Q_c}{A_p(T_p - T_a)}$$

According to Zoubida et al. (2011) assessed on the natural convection of nanofluids and arrived at a summary shown in the table below

Table 3: Experimental study on nanofluids with different heat transfer configuration

Authors	Configuration	Experiment Title / Method	Particle / Base Fluid	Observations
Das et al	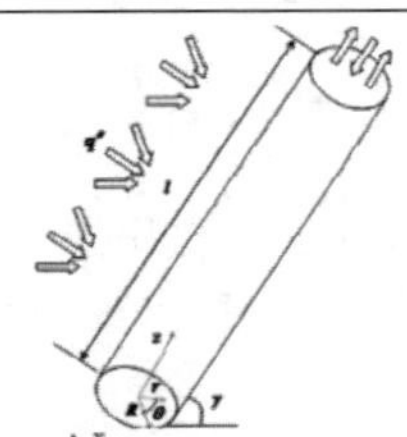	Nanofluids Science and Technology (Finite volume method)	Cu – water	Nusselt number and highest flow rate of mass at output are the increasing parameters of solid concentration.
Khanafer et al	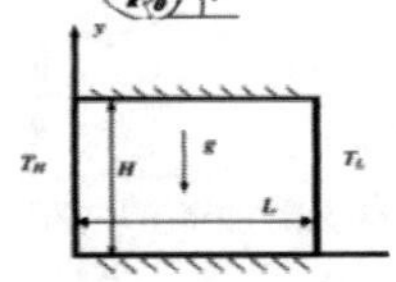	The utilization of nanofluids for the purpose of improving buoyancy-driven heat transfer in a two-dimensional enclosure (Finite-volume approach)	Cu – water	volume fraction of nanoparticles was noticed to rise with the rate of heat transfer of nanofluid
Jang and Choi	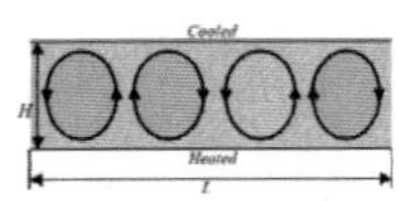	Improved heat conductivity of nanofluids through an irregular motion of particles (Theoretical study)	Cu/Diamond – water	Stability and greater heat transfer coefficient was observed from nanofluids
Hwang, Lee and Jang	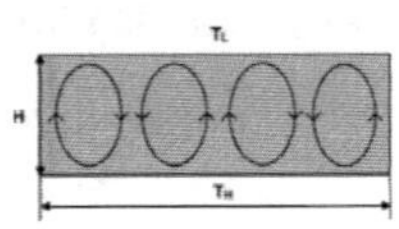	Buoyancy-driven heat transfer of water-based Al_2O_3 nanofluids in a rectangular cavity (Theoretical study)	Al_2O_3 – water	Reduced thermal transfer with an increasing nanoparticles
Ho et al.	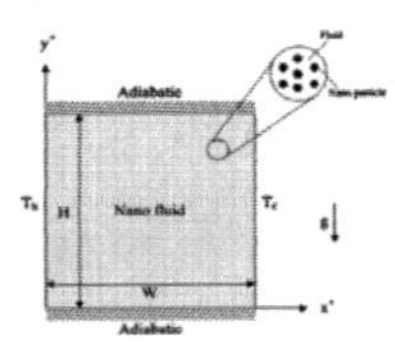	Mathematical Simulation of nanofluid natural convection in a square enclosure: impacts caused by the risk of viscosity and heat conductivity (Finite volume method)	Al_2O_3 – water	Dissimilarity between an improvement in the dynamic viscosity being evaluated from the two formulas led to contradictory heat transfer of nanofluid.
Oztop and Adu-nada	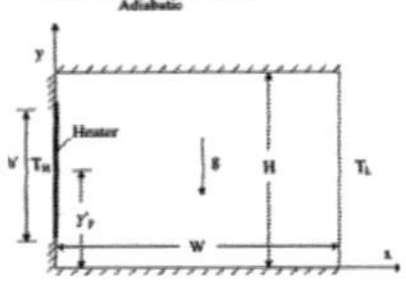	Mathematical study of natural convective heat transfer in a partially heated rectangular container being occupied with nanofluid (Finite volume method)	Al_2O_3/Cu/ TiO_2 – water	Thermal transfer improves with the volume of nanoparticles fraction
Abu-nada and Oztop	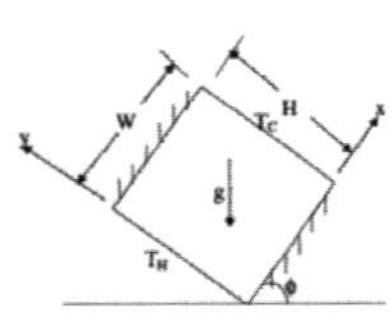	Impacts of tilt angle on natural convection in an enclosure occupied with Cu –water nanofluid (Finite volume method)	Cu – water	Impact of concentrating nanoparticle on Nusselt number was noticed at a large volume fraction
Abouali and Falahatpisheh	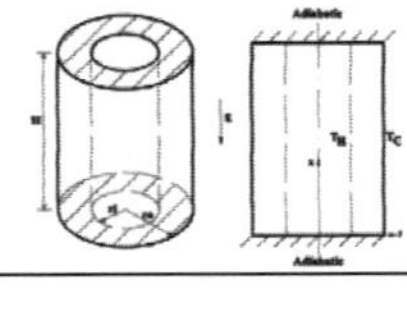	Impacts of unsteady viscosity and heat conductivity of Al_2O_3 – water on improving heat transfer in natural convection (Finite volume method)	Al_2O_3 – water	Thermal conductivity showed drop in natural convective heat transfer.

4.0 Models for Computing Thermal Conductivity

In nanotechnology applications in heat exchangers, properties of nanofluids being utilized are high thermal conductivities and absorption properties (Yu & Xie, 2011). Thermal conductivity constitutes a significant parameter when evaluating the energetic performance of nanofluids. When referring to nanotechnology, interest corresponding to heat transfer will include the nanomaterials consisting of nanoparticles or nanoarchitectured material for the device (Arico et al., 2015).

Several factors are responsible for heat conductivity of nanofluids, they include nominal temperature of the nanofluid (Ariana, Vaferi & Karimi, 2015; Ahmadi et al., 2018; Alirezaie et al., 2018; Alade et al., 2018), level of sonication of nanofluid (Shahsvar et al., 2017), size of the nanoparticles (Ren et al., 2005; Li & Peterson, 2007), concentration of solid phase (Akilu2017; Toghraie & Afran, 2018), aspect ratio (Kumar et al., 2017), nanoparticles size and shape (Sarviya & Fuskele, 2017). Numerous investigations have been done on heat conductivity of nanofluids so as to examine the functional parameters. Magnetic field strength of the nanoparticle is among the properties that affects heat conductivity of nanofluids.Moreover, thermal conductivity is directly proportional to the magnetic field strength of a nanoparticle which varies with operating temperature of the nanofluid (Nurdin, Yaacob & Johan, 2016)

4.1 Measurement of Heat Conductivity

The heat conductivity of hybrid nanofluids can be estimated in a number of ways. Various techniques that have being utilized to measure the heat conductivity of nanofluids in recent years, they include: thermal constant analyser, 3ω method, constant-state parallel plate technique and transient hot wire technique which is the commonest (Sarviya & Fuskele, 2017). All existing devices like the hybrid nanofluids employ one of the above methods, they include and not limited to LAMBDA, KD2-Pro thermal analyzer.

a. Transient Plane Source (TPS) method

TPS principle is applied by heat constant analyzer to evaluate the heat conductivity of nanofluids. The approach employs TPS element of an electrically transmitting thin foil as shown in Fig. 1. TPS works on the Fourier law of heat conduction as the basic principle for measuring heat conductvity. Heat conductivity is measured by a thermometer which also measures the resistance of the Transient Plane Source element immersed vertically in a container having steady temperature bath of nanofluid as shown in Fig. 2 (Kumar et al., 2015).

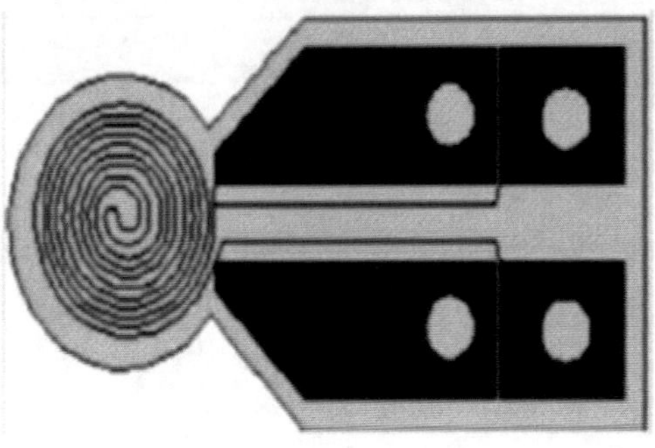

Fig. 1. TPS sensor (Paul et al., 2015)

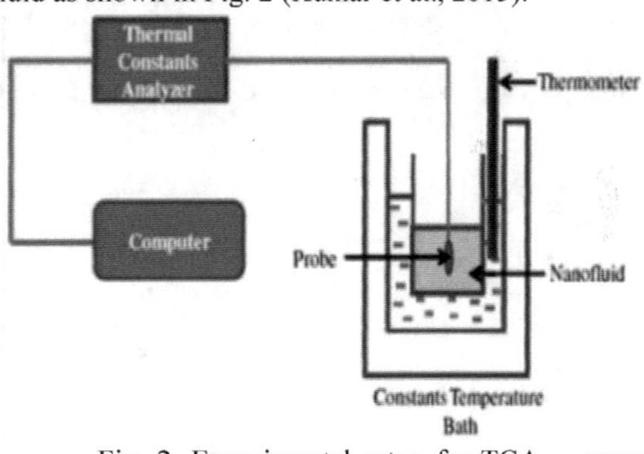

Fig. 2. Experimental setup for TCA measurement (Kumar et al., 2015)

b. Transient hot-wire (THW) method

This technique works on the concept of measuring the degree of hotness and time reaction of wire that are being subjected to an abrupt pulsating electrical current. In the THW approach, probe was refer to as a lengthy thin platinum wire. The wire is immersed into the nanofluid to measure the rise in temperature as constant current is supplied through resistive heating. According to Fourier's law of heat conduction , an increase in fluid temperature is proportional to the heat conductivity. Therefore the fluid heat conductivity (Eq. 1) was derived from the correlative variation in wire resistivity measured through four wire resistive measurement system shown in Fig. 3 (Kumar et al., 2015; Paul et al., 2015).

$$k = \left[\frac{q}{4\pi(T_2 - T_1)}\right] \ln\left(\frac{t_2}{t_1}\right)$$

Where k, = nanofluid heat conductivity, q = ratio of applied electric energy to wire length, and i= temperature diference between time t_1 and t_2.

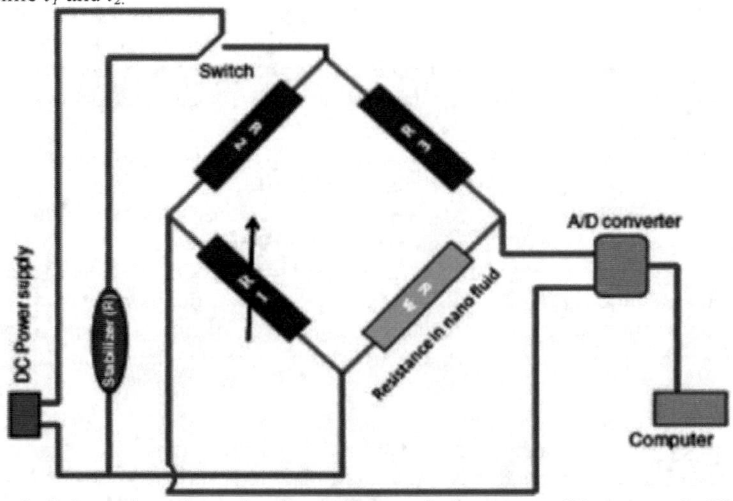

Fig. 3. Schematics of transient hot wire practical structure (Kumar et al., 2015)

Different models are represented to suggest heat conductivity of nanofluids. Each of these models are expensions to the four foundamental models - Maxwell model (Maxwell, 1873) which considers spherical particles, Hamilton & Crosser model for any particle with shape factor, Yu & Choi model which considers the nanolayer and Koo & Kleinstreuer model factors the size of particle, particle , temperature dependence, base properties of fluid and the particle subjected to an erratic random movement(Sarviya et al., 2017).

Table 4: Other models utilized to assertain the heat conductivity of nanofluids

Authors	Nanoparticle/Base Fluid	Empirical Correlation
Minsta et al. (2009)	Al_2O_3: water	$\frac{k_{nf}}{k_{bf}} = 1.72.\emptyset + 1$
Khanafer and Vafai (2011)	Al_2O_3: water	$k_{nf} = 1 + 1.0112.\emptyset + 2.4375.\emptyset\left(\frac{47}{d_p(nm)}\right) - 0.0248.\emptyset\frac{k_p}{0.613}$
Yiamsawasd et al. (2012)	Al_2O_3: water	$\frac{k_{nf}}{k_{bf}} = A\emptyset^B T^C \left(\frac{k_{nf}}{k_{bf}}\right)^D$ A = 0.946887, B = 0.073528, C = 0.03616, D = -0.016229
Patel et al. (2010)	Metal and metal oxides	$k_{nf} = \left[1 + 0.135\left(\frac{k_p}{k_{bf}}\right)^{0.273} \emptyset^{0.467}\left(\frac{T}{20}\right)^{0.547}\left(\frac{100}{d_p}\right)^{0.234}\right] k_{bf}$ A = 0.916205, B = 0.066817, C = 0.018789, D = 0.016229 Applicable to TiO_2: water

Hemmat Esfe et al. (2015)	Ag-MgO: water	$k_{nf} = \left(\frac{0.1747\times10^5 + \emptyset}{0.1747\times10^5 - 0.1498\times10^6\emptyset + 0.117\times10^7\emptyset^2 + 0.1997\times10^8\emptyset^3}\right)k_{bf}$ HNF $0<\phi<0.03$

4.2 Conventional Nanofluids

a. TiO_2 nanofluids

Maheshwary et al, (2017) measured the impacts of concentration, temperature and size of nanoparticles on the heat conductivity of TiO_2/water. The study included the use of cubic, rod and spherical nanoparticle of TiO_2. However, it was the cubic shaped nanoparticle that gave the highest thermal conductivity, with spherical nanoparticle performing better in thermal conductivity than the rod-shaped TiO_2 nanoparticle. Sonawane, et al, (2015) quantified the impact of ultrasonication time on heat conductivity of titanium dioxide. It was observed that the time application of sound energy to agitate particles would enhance an erratic random motion of particles in the suspension. Selah et al., (2014) found out an increases in heat conductivity; in addition to surfactants improving dispersion of nanofluids,. The result demonstrated the thermal conductivity to be higher when using SDS as stabilizer compared to CTAB; and CTAB has improved its tendency to conduct heat over Span-80. Duangthongsuk & Wongwises, (2009) discovered volume fraction of TiO2 deposited in water to increase by 1.8% while the improvement in thermal conductivity increased by 6%.

b. Al_2O_3 nanofluids

Sharifpur et al., (2017) examined the impacts of concentration, temperature and particle size on the heat conductivity of Al2O3/glycerol nanofluid. The sizes of particle involved in study was limited and the results indicated reduction in the ratio of heat conductivity of the nanofluid with an increasing particle size. Alawi et al., (2018) examined the heat conductivity behaviour of various nanofluids shape of Al_2O_3 using distilled water at different capacity. The best heat conductivity of Al_2O_3 nanoparticle in distilled water with increasing volume fraction was observed; platelets nanoparticle was discovered to be the least while blade nanoparticle, bricks nanoparticle, cylindrical nanoparticle and spherical nanoparticle are the best of the Al_2O_3 nanomaterial.

c. Cu/CuO nanofluids

Li and Peterson, (2006) found out the impact of volume fraction and temperature of CuO-water nanofluids on heat conductivity. Xuan et al., (2002) had carried out similar experiment on heat conductivity of Cu-water nanofluid and observed its enhancement to increase by 9% in 1% volume loading when temperature rises.

d. SiO_2/SiC nanofluids

In a natural square enclosure of a heat exchanger, Jahanshahi et al., (2010) investigated the impact of the heat conductivity of Silica–water nanofluid on the properties of heat transfer through an iterative experiment. They used model suggested by Hamilton et al., (1962) and derived a rise the Nusselt number and heat conductivity when using SiO_2-water nanofluid (Haddad et al., 2012).

e. CNTs and Graphene nanofluids

Ding et al. (2009) observed a significant improvement of heat conductivity of CNT nanofluids. The enhancement of it was noticed to increase by 69% for 0.49% nanofluid When there is a rise in temperature. In the work of Assael et al., (2005) they discovered that; heat conductivity increased by over 30% if 0.6% of CNTs is mixed with water.

f Other Nanofluids

Since nanofluids operate under varying experimental conditions, heat conductivity of other used nanofluids for heat transfer applications are presented in Fig. 4. Other group novel nanofluids called hybrid nanofluids are surfacing in the area of nanotechnology research. Hybrid nanofluids are fluids made from more than one nanoparticle that have been synthesized in either one base fluid or homogenized base fluids. Table 5 shows works and conclusions of some researcher using hybrid nanofluids in heat transfer applications.

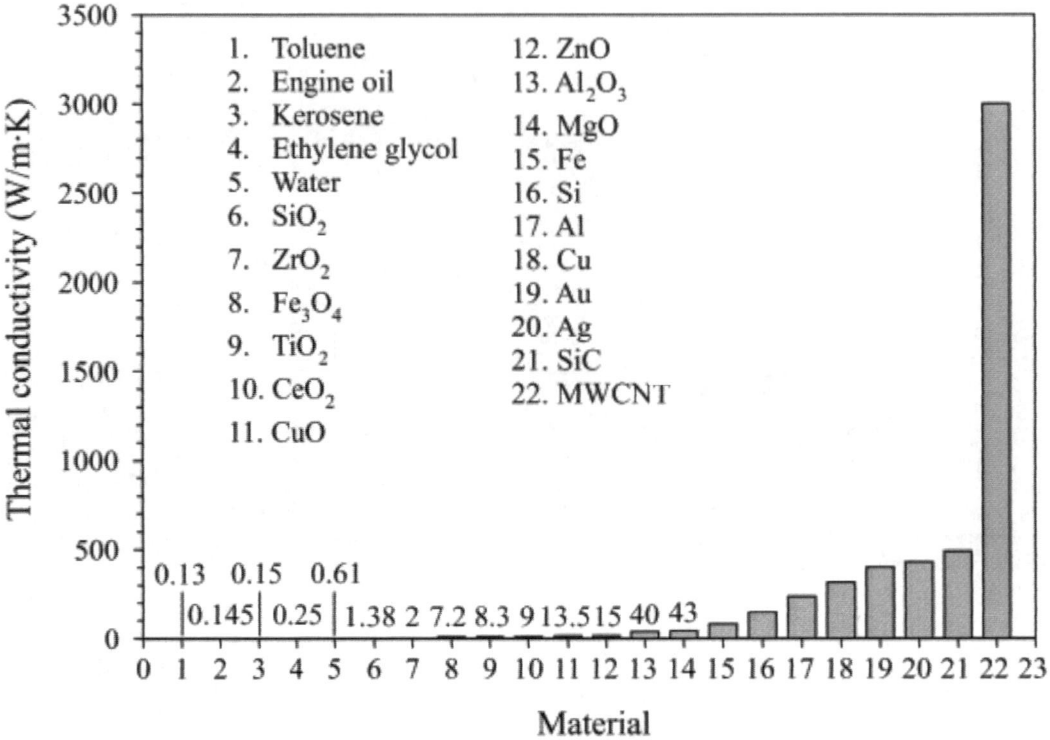

Fig. 4. Heat conductivity of the most frequently utilized fluid and materials at ambient temperature (Murshed et al., 2011)

4.3 Effects of heat Conductivity on Heat Exchangers

The need of fluids with better thermal conductivity is on the increase, many sectors like energy production, provision of clean water, increasing power of microprocessors and other heat application components are in search for a more efficient heat dissipating system. Improving performance of heat transfer enhancement methods or heat exchange systems can be of help to reduce the energy consumptions at industries and in a long run make things economical. Nguyen et al., (2007) observed the effects of nanoparticles on a heat block of a microprocessor and noticed an increasing volume concentration with temperature decrease. Rise in the flow rate of working fluid makes the coefficient of energy transfer for pure liquid and nanofluid better whereas the difference in temperature of heat transfer to the radiator was caused by changes in inlet temperature of the liquid (in the range tested (Peyghambarzadeh et al., 2011). This supports the observation made by Nguyen et al., (2007) that increase in mass flow rate can better improve the heat conductivity of nanofluids.

Table 5: Research summary on thermal performance of heat exchangers

Authors	Title of Experiment	Results /Knowledge
Huminic and Huminic (2011)	Heat exchange properties of a two-phase closed thermosyphons using nanofluid	The nanoparticles were discovered to have a crucial impact on the improvement of heat conductivity of two-phase closed thermosyphons using nanofluid (TPCT).
Alizad et al. (2012)	Heat performance and functional startup properties of flat-shaped heat pipes using nanofluid	They discovered better heat performance of heat pipes with flat shape using CuO, Al_2O_3 and TiO_2 with high volume concentrations.
Arun Kumar, Pradyumna & Jahar (2013)	plate heat exchanger performance comparism when several nanofluids are utilized	The improvement in the heat efficiency of heat pipe for nanofluid was limited to laminar fluid conditions and no pH was considered.
Vermalumoudi et al. (2014)	Practical examination on the performance of thermal transfer of image/water nanofluid in an air-finned heat exchanger	They noticed a peak improvement of few percentage in the total heat transfer coefficient
Hussein et al. (2014)	Improvement in thermal transfer by utilizing nanofluids in an automotive cooling system	Nanofluids works properly with automotive radiator. Also, the Nusselt number was noticed to rise with volume flow rate. The result was conducted for a crossed flow type heat exchanger and no consideration on pH.

5.0 Conclusion

The possibility of nanofluids enhancing heat transfer will lead to the advancement of more compact and effective heat transfer equipment in engineering. Researchers in the field, suggests that in improving the thermal efficiency of heat exchangers using nanotechnology attention should be placed on the following:

i. Similarity of nanomaterials and the base fluids by altering the boundary attributes of two phases might be among the paths to solving problem
ii. The benefits of additives shape in nanofluids for the properties
iii. Suspension stability for scientific findings and experimental use.

In conclusion, improvement of thermal transfer by nanofluids depends on different factors comprising of; thermal conductivity increase, random motion of nanoparticles, variations and interface. It is necessary to examine the products of nanofluid made from blending types with equal volume fraction for augmentation of heat transfer improvement and application.

Acknowledgements

The authors wish to acknowledge the financial support offered by Covenant University in actualization of this research work for publication

References

[1] Abu-Nada, E. & Oztop, H. (2010). Effects of inclination angle on natural convection in enclosures filled with Cu–water nanofluid. *International Journal of Heat and Fluid Flow, 30,* 669–678.
[2] Abu-Nada, E. (2010). Effects of variable viscosity and thermal conductivity of Al_2O_3-water nanofluid on heat transfer enhancement in natural convection. *International Journal of Heat and Fluid Flow, 30,* 679–690.
[3] Ahmadi, M.H., Nazari, M.A., Ghasempour, R., Madah, H., Shafii, M.B. & Ahmadi, M.A. (2018). Thermal conductivity ratio prediction of Al_2O_3/water nanofluid by applying connectionist methods. *Colloids Surf.* A Physicochem. Eng. Asp.
[4] Alade, L.O., Oyehan, T.A., Popoola, L.K., Olatunji, S.O. Bagudu, A. (2018). , Modeling thermal conductivity enhancement of metal and metallic oxide nanofluids using support vector regression. *Adv. Powder Technol.* 29, 157–167.

[5] Alawi, A., Sidik, N.A.C, Xian, H.W., Kean, T.H. & Kazi, S.N. (2018). Thermal conductivity and viscosity models of metallic oxides nanofluids. *Int. J. Heat Mass Transf.* 116, 1314–1325.
[6] Alirezaie, A., Hajmohammad, M.H., Hassani Ahangar, M.R., Hemmat Esfe, M. (2018). Price-performance evaluation of thermal conductivity enhancement of nanofluids with different particle sizes. *Appl. Therm. Eng.* 128, 373–380.
[7] Angayarkanni, S.A. & Philip, J. (2015). Review on thermal properties of nanofluids: Recent developments. *Advances in Colloid and Interface Science* 225, 146–176.
[8] Ariana, M.A., Vaferi, B. & Karimi, G. (2015). Prediction of thermal conductivity of alumina water-based nanofluids by artificial neural networks. *Powder Technol.* 278, 1–10.
[9] Arico, A., Bruce, P., Scrosati, B., Tarascon, J., & Van-Schalwkijk, W. (2015). Nanostructured materials for advanced energy conversion and storage devices. *Nature Materials, Vol. 4,* 366-377.
[10] Arun Kumar, T., Pradyumna, G. & Jahar, S. (2013). Performance comparison of the plate heat exchanger using different nanofluids. *Exp Therm Fluid Sci, 49,* 141–51.
[11] Assael, M.J., Metaxa, I.N., Arvanitidis, J., Christolilos, D. & Lioutas. (2005). Thermal conductivity enhancement in aqueous suspensions of carbon multi-walled and double-walled nanotubes in the presence of two differnet dispersants. Int. J. Thermophys., 26(3), 647-664.
[12] Botha, S.S., Ndungu, P. & Bladergroen, B.J. Physicochemical properties of oil-based nanofluids containing hybrid structures of silver nanoparticles supported on silica. *Ind. Eng. Chem. Res.* 50, 3071–3077.
[13] Dai, L., Chang, D., Baek, J. & Wen, L. (2012). Carbon nanomaterials for advanced energy conversion and storage. *Carbon Nanomaterials, 8,* 1130-1166.
[14] Das, S., Choi, S., Yu, W. & Pradeep, T (2008). Nanofluids Science and Technology. *Hoboken, NJ,* John Wiley & Sons.
[15] Ding, G., Peng, H., Jiang, W. & Gao, Y. (2009). The migration characteristics of nanoparticles in the pool boiling process of nanorefrigerant and nanorefrigerant–oil mixture. *Int'l J. Refrig.* 32 (1), 114-123.
[16] Dreizin, E. (2009). Metal-based reactive nanomaterials. *Prog. Energy Combust., 35,* 141-167.
[17] Duangthongsuk, W. & Wongwises, S. (2009). Measurement of temperature dependent thermal conductivity and viscosity of TiO_2-water nanofluids. *Exp. Thermal Fluid Sci.* 33 (4), 706-714.
[18] Faizal, M., Saidur, R., Mekhilef, S. & Alim, M. (2013). Energy, economic and environmental analysis of metallic oxides nanofluid for flat-plate solar collector. *Energy Conversion and Mgt.*, 2-6.
[19] Ghazvini, M., Akhavan-Behabadi, M., Rasouli, E. & Raisee M. (2011). Heat transfer properties of nanodiamond-engine oil nanofluid in laminar flow. *Heat Transfer Eng, 33,* 525–532.
[20] Godson, L., Raja, B., Mohan Lai, D. & Wongwises S. (2010). Enhancement of heat transfer using nanofluids - An overview. *Renewable and Sustainable Energy Reviews, 14,* 629-641.
[21] Gulenoglu, C., Akturk, F., Aradag, S., Uzol, N. & Kakac, S. (2014). Experimental comparison of performances of three different plates for gasketed plate heat exchangers. *Int J Therm Sci, 75,* 249–256.
[22] Haddad, Z., Oztop, H.F., Abu-Nada, E., Mataoui, A. (2012). A review on natural convective heat transfer of nanofluid. *Renewable and Sustainable Energy Reviews* 16, 5363-5378.
[23] Hafiz, A., Hassan, A., Hassan, L., Hafiz, M., Talha, B. & Ahmed, N. (2015). Experimental investigation of convective heat transfer augmentation for car radiator using ZnO-water nanofluids. *Energy 2015, 84,* 317–24.
[24] Hamilton, R.L., Crosser, O.K. (1962). Thermal conductivity of heterogeneous two component systems. *I&EC Fundamentals* 1, 187–191.
[25] Hemmat Esfe, M., Saedodin, S., Biglari, M. & Rostamian, H. (2015). Experimental investigation of thermal conductivity of CNTs-Al2O3/water: A statistical approach. *International Communications in Heat and Mass Transfer* 69, 29–33.
[26] Ho, B., Chen, M. & Li, Z. (2008). Numerical simulation of natural convection of nanofluid in a square enclosure: effects due to uncertainties of viscosity and thermal conductivity. *International Journal of Heat and Mass Transfer 2008, 51,* 4506–4516.
[27] Hu, C., Song, L., Zhang, Z., Chen, N., Feng, Z., & Qu, L. (2015). Tailored graphene systems for unconventional applications in energy conversion and storage devices. *Energy Environmental Science, 35,* 31-54.
[28] Huminic, G. & Huminic A. (2011). Heat transfer characteristics of a two-phase closed thermosyphons using nanofluids. *Exp Therm Fluid Sci, 35,* 550–557.
[29] Hussain, H., Saw, C. & Lukmon, O. (2013). Review on nanomaterials for thermal energy storage technologies. *Nanoscience and Nanotechnology, 13,* 59-71.

[30] Hussein, A., Bakar, R., Kadirgama, K. & Sharma, K. (2014). Heat transfer enhancement using nanofluids in an automotive cooling system. *Int'l Comm. in Heat and Mass Transfer, 53,* 195-202.

[31] Jahanshahi, M., Hosseinizadeh, S.F., Alipanah, M., Dehghani, A., Vakilinejad, G.R. (2010). Numerical Simulation of free convection based on experimental measured conductivity in a square cavity using water/SiO2 nanofluid. *International Communications in Heat and Mass Transfer* 37, 687–694.

[32] Javadi, F., Sadeghipour, S., Saidur, R., Boroumandjazi, G., Rahmati, B., Elias, M. M., & Sohel, M. R. (2013). The effects of nano fluid on thermophysical properties and heat transfer characteristics of a plate heat exchanger. *International Communications in Heat and Mass Transfer, 44,* 58–63.

[33] Khanafer, K. & Vafai, K. (2011). A critical synthesis of thermophysical characteristics of nanofluids. *Int J Heat Mass Transfer, 54,* 4410–4428.

[34] Kumar, P.M., Kumar, J., Tamilarasan, R., Sendhilnathan, S. & Suresh, S. (2015). Review on nanofluids theoretical thermal conductivity models. *Engineering Journal* 19(1), 67-83.

[35] Kumar, V., Tiwari, A. & Ghosh, S. (2015). Application of nanofluids in plate heat exchanger: A review. *Energy Conversion and Manangement, 105,* 1017-1036.

[36] Kumaresan, V. & Velraj, R. (2012). Experimental investigation of the thermophysical properties of water–ethylene glycol mixture based CNT nanofluids. *Thermochim Acta, 545,* 180–186.

[37] Luo, B., Wang, B., Li, X., Jia, Y., Liang, M. & Zhi, L. (2012). *Adv. Mater., 24,* 3538–3543.

[38] Maheshwary, B., Handa, C.C. & Nemade, K.R. (2017). A comprehensive study of effect of concentration, particle size and particle shape on thermal conductivity of titania/water based nanofluid. *Appl. Therm. Eng.* 119, 79–88.

[39] Maxwell, J. (1873). *Treatise on electricity and magnetism. Oxford,* Clarendon Press, 1873.

[40] Murshed, S.M., Nieto de Castro, C.A., Lourenco, M.J.V., Lopes, M.L.M. & Santos, F.J.V. (2011). A review of boiling and convective heat transfer with nanofluids. *Renewable and Sustainable Energy Reviews* 15, 2342-2354.

[41] Ngugen, C.T., Roy G., Gauthier, C. & Galanis, N. (2007) Heat transfer enhancement using Al2O3-water nanofluid for an electronic liquid cooling system. *Applied Thermal Engineering, 27*(8–9), 1501–1506.

[42] Nurdin, I., Yaacob, I.I. & Johan, M.F. Enhancement of thermal conductivity and kinematic viscosity in magnetically controllable maghemite (γ-Fe2O3) nanofluids. *Experimental Thermal and Fluid Science* 77, 265-271.

[43] Oztop, H. & Abu-Nada, E. (2008). Numerical study of natural convection in partially heated rectangular enclosures filled with nanofluids. *International Journal of Heat and Fluid Flow, 29,* 1326–1336.

[44] Pandey, S. & Nema, V. (2012). Experimental analysis of heat transfer and friction factor of nanofluid as a coolant in a corrugated plate heat exchanger. *Exp Therm Fluid Sci, 38,* 248–256.

[45] Pantzali, M., Mouza, A. & Paras, S. (2009). Investigating the efficacy of nanofluids as coolants in plate heat exchangers (PHE). *Chem Eng Sci, 64,* 3290–3300.

[46] Patel, H., Sundararajan T. & Das S. (2010). An experimental investigation into the thermal conductivity enhancement in oxide and metallic nanofluids. *J Nanopart Res, 12,* 1015–1031.

[47] Peyghambarzadeh, S. M., Hashemabadi, S. H., Jamnani, M. S., & Hoseini, S. M. (2011). Improving the cooling performance of automobile radiator with Al_2O_3/water nanofluid. *Applied Thermal Engineering, 31*(10), 1833–1838.

[48] Raja, M., Vijayan, R., Dinashukumar, P., & Venkatesan, M. (2016). Review on nanofluids characterization, heat transfer characteristics and applications. *Renewable and Sustainable Energy Reviews, 64,* 163-173.

[49] Ray, D., Das, D. & Vajjha, R. (2014). Experimental and numerical investigations of nanofluids performance in a compact minichannel plate heat exchanger. Int J Heat Mass Transfer, 71, 732–746.

[50] Salman, B., Mohammed, H. & Khaerbeet, A. (2012). Heat transfer enhancement of nanofluids flow in microtube with constant heat flux. *Int'l Communications in Heat and Mass Transfer, 39,* 1195-1204.

[51] Sarviya, R.M. & Fuskele, V. (2017). Review on Thermal Conductivity of Nanofluids. *Materials Today* 4, 4022-4031.

[52] Scida, K., Stege, P., Haby, G., Messina, G., & Garcia, C. (2011). Recent application of carbon-based nanomaterials in analytical chemistry: critical review. *Anal. Chim. Acta,* 691, 6-17.

[53] Sekhar, Y. & Sharma, K. (2013). Study of viscosity and specific heat capacity characteristics of water-based Al_2O_3 nanofluids at lowparticle concentrations. *J Exp Nanosci,* 1–17.

[54] Shahsavar, A., Salimpour, M.R., Saghafian, M., Shafii, M.B. (2017). An experimental study on the effect of ultrasonication on thermal conductivity of ferrofluid loaded with carbon nanotubes. *Thermochim. Acta* 617, 102–110.

[55] Sharifpur, M., Tshimanga, N., Meyer, J.P. & Manca, O. (2017). Experimental investigation and model development for thermal conductivity of α-Al_2O_3-glycerol nanofluids. *Int. Commun. Heat Mass Transfer* 85, 12–22.

[56] Shin, D. & Banerjee, D. (2014). Specific heat of nanofluids synthesized by dispersing alumina nanoparticles in alkali salt eutectic. *Int J Heat Mass Transfer 2014;74*, 210–214.

[57] Socher, R., Krause, B., Hermasch, S., Wursche, R. & Potschke, P. (2011). Electrical and thermal properties of polyamide 12 composites with hybrid fillers systems of multiwalled carbon nanotubes and carbon black. *Compos. Sci. Technol., 71,* 1053-1059.

[58] Sonawane, S.S., Khadler, R.S. & Wasewar, K.J. (2015). Effect of sonication time on enhancement of effective thermal conductivity of nano TiO2-water, ethylene glycol, and paraffin oil nanofluid and models comparisons. J. Exp. Nanosci, 10 (4), 310-322.

[59] Vermahmoudi, Y., Peyghambarzadeh, S., Hashemabadi, S. Naraki M. (2014). Experimental investigation on heat transfer performance of image/water nanofluid in an air-finned heat exchanger. Eur J Mech - B/Fluids, 44, 32–41.

[60] Wang, J., Xie, H., Xin, Z., Li, Y. & Chen, L. (2010). Enhancing thermal conductivity of palmitic acid based phase change materials with carbon nanotubes as fillers. *Sol. Energy, 84,* 339-344.

[61] William, C. (2003). Materials Science and Engineering - An Introduction. *John Wiley & Sons,* 757.

[62] Xuan, Y.M., Hu, W.F. & Li, Q. (2002). Simulations of structures and thermal conductivity of nanofluids. J. Eng. Thermophys. 23(2), 206-208.

[63] Yu, W. & Xie, H. (2011). A Review on Nanofluids: Preparation, Stability, Mechanisms, and Applications. *Journal of Nanomaterials, 12,* 1-17.

[64] Fakehinde, O. B., Fayomi, O. S., Efemwenkieki, U. K., Babaremu, K. O., Kolawole, D. O., & Oyedepo, S. O. (2019). Viability of Hydroelectricity in Nigeria and the Future Prospect. *Energy Procedia, 157,* 871-878.

[65] Oyedepo, S. O. (2012). Energy and sustainable development in Nigeria: the way forward. *Energy, Sustainability and Society*, *2*(1), 15.

Issues and Challenges Involved In Green Building Concept Innovations Adoption In Construction Practice.

Amusan Lekan M[1], Michael Oluwakorede[2], Oloke Clement[3], Adelakun Adegbenjo[4] Garba Rishi[5]
1-2 Building Tech. Dept. Covenant University. PMB 1023.Ota. Ogun State.Nigeria.
3 Estate Management Dept. Covenant University. PMB1023. Ota. Ogun State. Nigeria.
4 Building Tech. Dept. School of Vocational and Technical Education.Azare. Bauchi State. Nigeria.
5 Building Tech. Dept. Oke Ogun Polytechnic. Oyo State. Nigeria.
Email:1 lekan.amusan@covenantuniversity.edu.ng,

Abstract. Green concept is globally creating impact with diverse application at various facet of human endeavour. The concept involved the practice whereby energy is utilized in a way that it would not jeopardise continuity of energy usage in the environment. The aim of this study is to explore the issues and challenges that surround the adoption of green concept in Nigeria focusing on built environment with a view to enhancing the rate of adoption of the concept for an increased healthy environment. The study adopted convenience sampling method to pick the population for the study, which comprised of picking construction companies that engaged on a project that involved green building features. Population frame of 250 professionals that are registered with professional bodies that cut across 20 selected construction firms was used for the study. A structured questionnaire designed in Likert scale of 1 to 5 was distributed among 50 respondents for data collation. The data was later processed using content analysis, descriptive statistics and relative agreement index. Data was collated on parameters that include evaluation of existing issues and challenges involved in green building concept innovations adoption in construction practice, current green concept practice in construction, challenges involved in adoption of green concept in construction, critical issues involved in green concept adoption, drivers for success in the adoption of green concepts in construction, challenges involved in adoption of green concept in construction. Innovative illustrations of integration of green concepts in buildings. Mean Item Score was used with relative agreement index, this was used to rate the perception of respondents on parameters calibrated for respondents to answer. Categorical regression analysis was used to carry out comparative analysis of respondents' submission on success factors that influences the adoption of green building concept and challenges involved in the adoption of green building. Some other parameters were further explored, they include, innovative approach to sustainable construction practice and perspectives to green building practice. The analysis of the study brought to fore factors to be taken into consideration for effective adoption of green building concept. The challenges such as thought reengineering and philosophical and cultural adjustment need to be further harnessed for people to leave the built environment better than it was met.
Key words: Green-building, Innovations, Energy , Impact. Challenge

1. Introduction

These Green environment is highly desirable for every individual especially across the globe. The desire is often reflected in the choice they makes daily in the choice of type of habitation, internal and external decoration of such habitation and the interaction of the habitation with environment. The type of interaction with the environment is often dependent on the choice of input and output of activities to the immediate environment. However, the concept is not new to European and Asian countries. For instance in China the mediaeval construction which involves surrounding habitation with vegetation that replenishes the depleted oxygen due to human environmental interaction has been reenacted. Often times, human activities tend to overuse the environment resulting into environmental abuse such

IOP Conf. Series: Earth and Environmental Science **331** (2019) 012022 doi:10.1088/1755-1315/331/1/012022

as constructing non-ecofriendly houses, energy non efficient houses and non-green building compliant buildings. Similarly the low and moderate income population who can afford to construct low cost, sustainable, and eco-friendly houses are not always very keen in constructing such houses. Reasons for these kinds of attitudinal issues are to be tested using scientific and empirical studies. Thus, strategies for faster and equitable development of green buildings could be formulated. The study therefore assesses the level of development of Green building practice in the built environment of Nigeria. The aim of this study is to explore the issues and challenges that surround the adoption of green concept in Nigeria focusing on built environment with a view to enhancing the rate of adoption of the concept for an increased healthy environment.

1.1 Understanding Green Concepts In Built Environment

Green concepts in ecological parlance entails leaving an environment in a stable ecological condition without jeopardizing the biomass balance. In environmental protection parlance, green concept involve having zero emission to the environment in way that atmospheric air mass within such environment is maintained. In construction parlance, green concept starts from design stage, construction stage and up to the point of handling over to the client. In design, green concept involve the use of green material and green design concept. In construction it involve the use of eco-friendly materials with low emission of unusable energy [1] and [2].

However, some of the aspect of integration of green building concepts in building construction includes the following: Use of Solar system to heat water, use of glass panels for daylighting, harvesting rain water for storage, the use of environmentally friendly building materials and specifications. Also, waste minimization, healthy indoor air movement, minimizing and maximizing the use of energy in buildings and manufacturing and use of equipment and tools that are energy compliant and sufficient.

2. Review of Related Literature

Review of selected literatures was presented in this section. The selected presentation includes: [3],[4], [5] and [6].

In [6], a study was conducted on smart car parking system. The aim of the paper is developing a smart parking system. The intelligent features were incorporated in the car park system for prompt notification for opening and closing of the gates.

Also, in [5], an investigation was carried out on smart home security, issues, challenges and countermeasures involved. The study identified the significance of home security using green concepts approach. The study later recommended application of sensor based application to create home green home solutions

Similarly, a study was carried out by [4], on sustainable green smart buildings. The author submitted that the gateway to unlimited innovations adaptation in buildings construction and design is giving room to innovative approaches in design process and construction. This could be achieved by taking multi-perspective approach to the choice of design concept, choice of materials particularly eco-friendly materials. This approach would afford the building occupants access to luxury that accompanies green design concepts. The study access among other things The Gulf Energy Council Energy situation, Sustainable buildings solution and what should be a sustainable building objectives, high rated building systems and services.

Finally, [4] researched on Intelligence of Intelligent Building, the study was focussed on exploratory approach to green buildings. The study was carried out as a survey on projects at Technical research centre Finland VIT Building Technology in 1993 to 1997. The feasibility of integrating green component to building was explored. The study recommended inclusion of multi criteria concept in integration of functions in an intelligent building.

3. Methodology of Research

The study adopted convenience sampling method to pick the population for the study, which comprised of picking construction companies that engaged on a project that involved green building features. Population frame of 250 professionals that are registered with professional bodies that cut across 20 selected construction firms was used for the study. A structured questionnaire designed in Likert scale of 1 to 5 was distributed among 50 respondents for data collation. The data was later processed using content analysis, descriptive statistics and relative index. The data was analysed using Mean Item Score, Simple percentages and regression analysis. The study used questionnaire survey to collate information from respondents. The collated data was examined using content analysis method by classifying the information and categorised them under appropriate headings. Data was collated on parameters that include evaluation of existing issues and challenges involved in green building concept innovations adoption in construction practice, current green concept practice in construction, challenges involved in adoption of green concept in construction, critical issues involved in green concept adoption, drivers for success in the adoption of green concepts in construction, challenges involved in adoption of green concept in construction. Innovative illustrations of integration of green concepts in buildings. Mean Item Score was used with relative agreement index, this was used to rate the perception of respondents on parameters calibrated for respondents to answer.

3.1. Presentation of Data and Results of Data Analysis

Retrieved questionnaires were subjected to content analysis and revalidation, analysis was carried out on them and results presented in the Tables 1 to 5.

3.2. Current Green Concept Practice In Construction

Table 1 Existing Application of Green Concept Practice In Construction Field

S/N	Existing Application of Green Concept	R.A.I	Rank
1	Green Design and construction practice	0.87	1
2	Integration of Eco-friendly material in building design	0.87	1
3	Highly ergonomic design that reduces building operation cost	0.86	3
4	Highly ergonomic design that reduces energy consumption cost	0.83	4
5	Adoption of non-environmental pollution prone items	0.79	5
6	The use of system that reduce energy use in Buildings	0.78	6
7	Integration of features that improve indoor air quality	0.77	7
8	Recycling system for domestic generated building waste	0.76	8
9	Waste reduction in building	0.75	9
10	System that conserve natural resources	0.72	10

Legend: R.A.I. ...... Relative Agreement Index

In Table 1, Existing application of green concept practice in construction field was presented. Some of the current practice presented is as reflected in the Table 1. Green Design and construction practice top the list with RAI value 0.87 ranked 1st alongside Integration of eco-friendly material in building design with RAI value of 0.87. Similarly, highly ergonomic design that reduces building operation cost was ranked 3rd with RAI of 0.86. Highly ergonomic design that reduces energy consumption cost with RAI 0.83 was ranked 4th Adoption of non-environmental pollution prone items with RAI 0.79 was ranked 5th , The use of system that reduce energy use in Buildings RAI 0.78 ranked 6th, Integration of features that improve indoor air quality RAI 0.77 was ranked 7th Recycling system for domestic generated building waste was ranked 8th with RAI 0.76 while waste reduction in building RAI 0.75 and System that conserve natural resources with RAI 0.72 was ranked 9th and 10th respectively.

3.3. Critical Issues Involved In Green Concept Adoption

Table 2 Issues Involved In the Adoption of Green Concept

S/N	Critical Issues In Green Concept Adoption	R.A.I	Rank
1	Expensive nature of the eco-friendly materials	0.88	1
2	Expensive nature of the whole green building concept implementation.	0.88	2
3	Non-proliferation of Green building design codes	0.87	4
4	Lack of technical capability	0.83	5

5	Government reluctance in supporting of the spread of the concept	0.73	6
6	Need for professionals and artisan to upgrade their knowledge	0.72	7
7	Lack of green building projects that could serve as case study for referral purpose	0.71	8
8	Contractors' inadequate experience in the implementation of the green building practice.	0.70	9
9	Limited technology for green building practice	0.67	10
10	Challenge of availability of capacity to maintain green buildings	0.62	11

Legend: R.A.I. …… Relative Agreement Index

Table 2 presents critical issues involved in the adoption of green buildings which form one of the focal points of this study. Top on the list of the issues is Expensive nature of the eco-friendly materials ranked 1st high with RAI value of 0.88 alongside with Expensive nature of the whole green building concept implementation with RAI value of 0.88. Also, Non-proliferation of Green building design codes ranked 4th with RAI 0.87, Lack of technical capability RAI 0.83 ranked 5th, Government reluctance in supporting of the spread of the concept RAI 0.73 was ranked 6th, Need for professionals and artisan to upgrade their knowledge with RAI 0.72 was ranked 7th. The need to provide green project prototype as referral was ranked 8th with RAI of 0.71, while Contractors' inadequate experience in the implementation of the green building practice with RAI value of 0.70 and Limited technology for green building practice with RAI value of 0.67 were ranked 9th and 10th respectively.

3.3.1. Critical Success Drivers In Green Concept Adoption

Table 3a Drivers for Success In The Adoption of Green Concepts In Construction

S/N	Success Driver Parameters	R.A.I (Average)	Rank
1	Availability of standard codes for green building component implementation	0.79	1
2	Establishing training and skill acquisition centre for green building related empowerment	0.75	2
3	Green concept value reorientation	0.75	2
4	Developing sense of value of giving back to the ecosystem.	0.75	2
5	Provision of relevant technologies for green building practice	0.73	5
6	Provision of an affordable green building materials	0.72	6
7	Ergonomic design	0.70	7
8	Environmental regeneration conscious philosophy	0.70	8

Legend: R.A.I. …… Relative Agreement Index

In Table 3 critical success factors and drivers of effective adoption of green concepts was presented in Table 3. Availability of standard codes for green building component implementation was ranked 1st with RAI of 0.79, follow with Establishing training and skill acquisition centre for green building related empowerment which was ranked 2nd, Reorientation towards green concept value inculcation was ranked 3rd with RAI of 0.75, while developing sense of giving back value to the ecosystem with RAI 0.75 was also ranked 2nd . Similarly, Provision of relevant technologies for green building practice RAI of 0.72 was ranked 5th, Provision of an affordable green building materials with RAI 0.72 was ranked 6th , Ergonomic design with RAI 0.70 ranked 7th while Environmental regeneration conscious philosophy was ranked 8th with RAI value of 0.70.

Table 3b. ANOVA Test on Differences in Opinion of Professionals on Drivers for Success in the Adoption of Green Building Concepts.

Levene Statistics	Sum of Squares	df	Mean Square	F	Sig. (2 tailed)	Remark

Between Professional groups Censored	.680	135	.967	3.193	0.669	NSD

Levene's test for equality of variance was carried out on the data extracted from companies' professionals and the Levene Statistics analytical output was presented in the 1.5.3b. Categorical regression analysis was carried out on the tested parameters using an independent sample t-test in order to determine the level of significance difference in opinion among professionals sampled in the study area on the critical success factors that influence the adoption of green building concepts in construction work. Certain criterion were used they are criterion for effective size, .01 =small effect, .06 =moderate effect, .14 =large effect. S contained in a study carried out by [5] and [6].

The above table shows that all the variables that have their level of significant below 0.05 based on Levenes' Statistical values and ANOVA rule. There is no significant difference in opinion among the professionals on the identified factors since the Sig. Value of 0.669 was greater than Sig.Value 0.05, it shows no clear difference in the opinion of construction professionals as regards the adoption of green building concepts in building instruction.

3.3.2. Challenges Involved In Adoption of Green Concept In Construction.

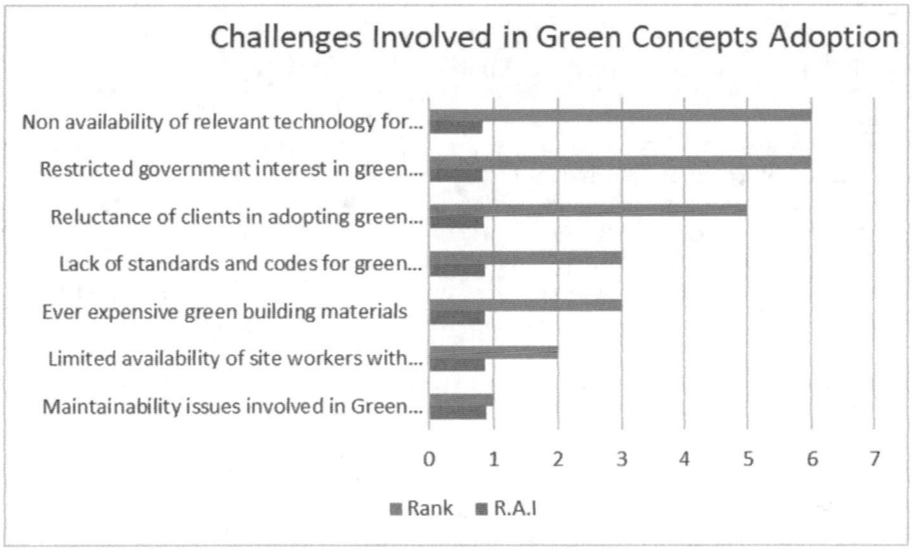

Fig 1. Challenges Involved in Green Concepts Adoption in Construction.

Legend: R.A.I. ...... Relative Agreement Inde

Challenges involved in green concept adoption in building was presented in Table 4. Some of the challenges identified include, Maintainability issues involved in Green building components and Limited availability of site workers with green building skills which was ranked 1st and 2nd with Relative Agreement Index Value (RAI) of 0.88 and 0.87. Also, Ever expensive green building materials RAI 0.86,ranked 3rd, Lack of standards and codes for green building practice RAI0.86 ranked 3rd , Reluctance of clients in adopting green building concepts with RAI 0.85 was ranked 5th , Restricted government interest in green building practice was ranked and Non availability of relevant technology for capacity development were ranked 6th with RAI value of 0.83 respectively

3.3.3. Inovative Approaches to Green Concept Integration In Building

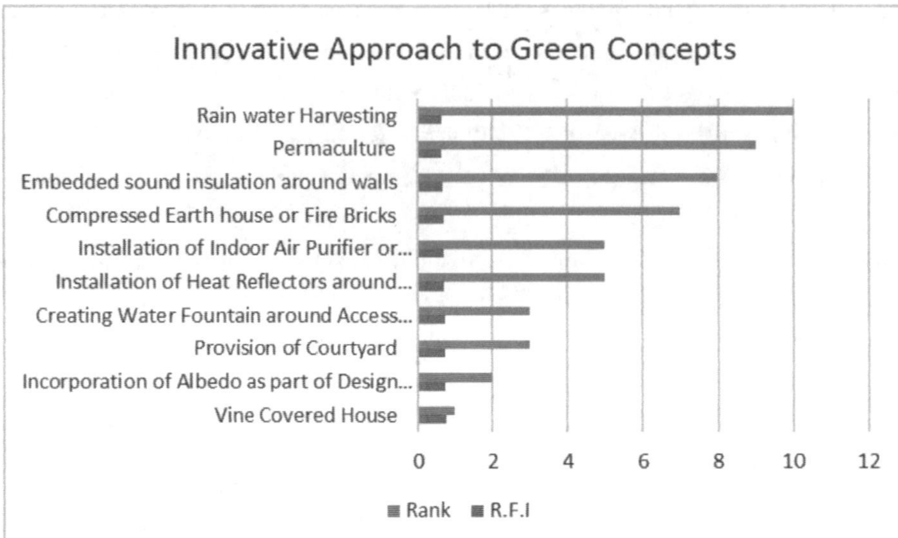

Fig. 2 Innovative Approach to Green Concepts
Legend: R.A.I. ...... Relative Agreement Index

Innovative approaches that could be used in green building construction was presented in Table 5. The innovations were subjected to familiarity test among the focused professionals, this assisted in determining the type of innovations that could be used to aid green concept compliance of buildings. Vine coverage and incorporation of Albedo as part of design in building was suggested, they were ranked 1st and 2nd with relative familiarity index (RFI) of 0.76 and 0.74 respectively. Courtyard provision, Inclusion of water fountain around courtyard or foyer were also suggested and ranked 3rd with RFI values 0.73 respectively. Others with their RFI values and ranks include installation of heat reflectors around building, installation of indoor air purifier or dehumidifiers, compressed earth house or fire bricks, embedded sound insulation around walls, permaculture and rain water harvesting.

3.4 Innovative Approaches That Could Be Adopted In Green Building

3.4. 1 Perspectives to Innovative Strategies in Green Building Management

In furtherance to presentation in Table 5, the issue of innovation in green project management is viewed in this study from the perspectives of technological innovation and non-technological innovations(Energy conserving design and innovations).

a. Technological-based Innovation: The configuration design and production process involved in green project execution is basically technological oriented. Therefore, management of the quality of end-product is largely determined by the interplay of actors and the process adopt. Actors refer to the professional involved in the project design and configuration. However, technological-based innovations in this context consist of Energy conserving design and innovations.

IOP Conf. Series: Earth and Environmental Science **331** (2019) 012022 doi:10.1088/1755-1315/331/1/012022

b. Energy conserving design and innovations: Traditionally, industrial pollution should be addressed right from the production process rather than the end-of-pipe approach which has proved to be ineffective, thus pollution and energy conservation process should form part of production process of organization. Environmental impact programme on the product of the organization should be carefully observed throughout the product lifecycle, through integrating environmental strategies and practices into the management of the production systems. Negative environmental impacts resulting from building elements are basically assumed to occur as a result of energy consumption in construction, renovations and maintenance works. The negative impacts are attributed to the emissions released due to the energy and material used for heating cooling and lighting of building and the renewal of the element at the end of its lifetime [5]. Waste and indoor air quality issues are also connected to the emissions released during use and renewal of building element. Therefore the negative environmental impacts of the emissions and 5 resource depletion can be grouped as: global warning, mineral depletion water depletion, acidification, fossil fuel depletion, human toxicity, smog, insolation, radiation effect and others. The negative impacts of energy consumption and activities in the built environment can however be taken care of, through the following suggested innovative approaches: vine covered-solar house, solar-paneled house, compressed-earth block-walls house, rainwater harvesting, albedo, permaculture, agri-aqua-culture.

i). Vine-covered House: This is an innovation will keep off poisonous gas from surrounding of a building through carbon dioxide gas emissions from house interior will be absorbed by the plant around the House. The idea behind this kind of house will be constructing a house that will be able to keep occupant free of any sort of heat stressor. Therefore the glass panes will be reflective type and solar reflective in nature. The house will as well have a waterfall, this is to absorb moisture from the air, dry air feels more comfortable, so there will be less need for air-condition. Therefore, this innovation will reduce energy consumption required in cooling house interior.

ii). Permaculture: This is the art and science that applies pattern found in nature to the design and construction of human and natural environments. Sustainable living system can truly be achieved when such patterns and principles are applied to the built environment. The elements should be designed in a way that will not disturb the natural environment. A balanced ecosystem can be created around the house to allow for exchange of resources from plant to human and the living things around.

iii). Compressed-Earth House: This is the situation whereby compressed earth block-wall which have been reinforced with wire and parched with plaster, with about 2 inches space between the facing walls are used in house construction. Recycled polystyrene materials like nylon or pure-water sachet, or even recycled styro-foam can be used to fill the 2inches gap.

iv). Solar Paneled House: Solar panel could be used as an alternative to power engine that combusts fossil fuels that result in environmental air pollution. Energy from the sun can be tapped by carefully installing the photo-voltaic system on the building for electricity. The panels can be installed at locations that are not in direct angle with isolation effect, since the photovoltaic panel functions best at fairly high temperature.

v). Rainwater Harvesting: This is another new option which can reduce urban draining problems which is creating serious challenge to conventional supply programme, however it is limited in that, it does not include rainwater harvesting, utilization of waste water or any programme for ground water recharge, thus the need for its inclusion

In ancient Japan, India and China and some other Asian countries, tanks, canals, stepwells and check-dam were often built to store excess rain water in flood prevention. The importance of this system lies in the charging of surrounding soil with water for later collection, water supply, storing of water in area where the groundwater is already toxified or with polluted surface water. The reviving of this dying old wisdom is necessary; it will help avoid over-flooding that tends to create heat aquifer when undergoing evaporation that often culminates in heat stress on the atmosphere and global warning.

4. Conclusions

The aim of the study that was stated at the beginning of the study has been achieved, that is, to explore the issues and challenges that surround the adoption of green concept in Nigeria focusing on built environment with a view to enhancing the rate of adoption of the concept for an increased healthy environment. The following has been achieved in the study, which includes: finding out the current application of green concept practice in construction field, issues involved in the adoption of green concept, critical success drivers in green concept adoption, challenges involved in green concept adoption in building and innovative illustrations of integration of green concepts in buildings. The study discovered among other things that Green Design and construction practice and Integration of eco-friendly materials and design during design and construction top the list. As one of the strong factors, integrating green practice while conceptualizing a building project should mark beginning of the implementation of green concept. As long as the practice is being initiated, it is a matter of time for it to be wholly adopted as the times rolls by. Similarly, the expensive nature of the whole green building concept implementation and expensive cost of eco-friendly application are considered as some of the critical factors to be considered in the adoption of the green building concept. The products and materials usually being used in green building application are imported and so the cost should be reduced or subsidized or better being produced locally. Also, there is a need for provision of codes and standards to guide the implementation of green concepts at design and up to post occupancy stage [4],[5] and [6].

However, the following challenges are very crucial if success implementation of green construction is desired, they include maintainability issues involved in Green building components, limited availability of site workers with green building skills, ever expensive nature of green building materials and non-availability of relevant technology for capacity development. The challenges need to be tackled and neutralized for an effective adoption of green building concept innovation in building.

5 Acknowledgement

Appreciation to Management of Covenant University and Covenant University Centre for Research Innovation and Discovery (CUCRID) for financial and moral support in providing fund for sponsorship of this study. Many thanks.

References

[1] Kepuska, V, Humaid, A (2016) International Journal of Science and Technology. Volume 5 (8).

[2] Himanon, M (2013) The Intelligence of Intelligent Buildings: The Feasibility of the Intelligent Building Concept in Office buildings.ESPOO. VIT Publications. 492-497.

[3] Abdul-Salam, Darwish, S.(2012) Sustainable Green Smart Buildings: Future Energy Survivor. ISESCO Journal of Science and Technology. 12(21).

[4] Sudhur C,Neha D,Arvind, S (2016) An Investigation Study for Smart Home Security: Issues, Challenges and Countermeasure.Wireless Sensor Networks. 8. 61-68

[5] Pallant, J. (2016) SPSS Survival Manual?: A step by step Guide to Data AnalysisUsing SPSS Program. 6th Edition. Mc Graw Hill Publishers London.

[6] Cohen, J. (1988) Statistical Power Analysis for the Behavioural Sciences. 2nd Edition Hillsdale LawrenceErlbaum Associate Publishers.

Protective Effect of *Cajanus cajan* in Hepatotoxic Rats

Evbakhavbokun, Winifred O. and Iweala E.J. Emeka

Department of Biochemistry, Covenant University, Ota. Nigeria.

Winifred.evbakhavbokun@stu.edu.ng

Abstract. Hepatotoxicity results from overload of chemicals and drugs including N-Nitrosodiethylamine (NDEA), a nitrosamine found in smoke, meat and food products. This study examined the hepatoprotective potential of *Cajanus cajan* in male Wistar rats. Hepatotoxicity was induced by administration of NDEA at 200mg/kg while *Cajanus cajan* was administered at 200mg/kg, 400mg/kg and 800mg/kg for 28 days. Body weight, liver weight and markers of hepatotoxicity including Alanine aminotransferase (ALT), Aspartate aminotransferase (AST) and Albumin (ALB) were evaluated. NDEA treated group showed a marginal increase in body weight and a significant ($p < 0.05$) increase in liver weight. The *Cajanus cajan* treated groups showed a significant ($p < 0.05$) increase and decrease respectively in body and liver weights. NDEA treated group significantly ($P<0.05$) increased ALT and AST and significantly reduced ALB. *Cajanus cajan* significantly ($P<0.05$) decreased ALT and AST and significantly ($P<0.05$) elevated ALB. The results indicate that *Cajanus cajan* ameliorates NDEA-induced hepatotoxicity.

1. Introduction

N-Nitrosodiethylamine (NDEA), a nitrosamine compound is a popular hepatic carcinogen causing liver damage [1]. They can be seen in industrial processes, food stuffs e.g. meat and milk, pharmaceuticals and tobacco smoke [2, 3]. NDEA leads to cellular injury and oxidative stress due to ROS [4]. Liver being the major site for metabolic biotransformation of NDEA, generation of ROS can lead to oxidative stress causing damage to the liver [5]. These ethyl radicals and other reactive oxygen species interact with DNA which leads to mutations, increase in blood markers such as ALT and AST, and also lead to neoplastic transformation in tissues of the liver [6, 7]. Various animal studies proved that increased intake of plant products are linked to decrease in the development of liver diseases and different types of cancers [8].

Cajanus cajan is commonly known as Pigeon pea. It's a therapeutic plant used for treating wounds, bedsore and malaria [9]. Chemical constituents' and antioxidant activities have shown *Cajanus cajan* leaves to be high in flavonoids, stilbenes, saponins and alkaloids [10]. Four essential compounds which include pinostrobin, vitexin, cajaninstilbene acid and orientin gotten from the leaves contain antioxidant properties [11]. It also contains various medicinal properties such as antioxidants [11], anthelmintic [7], protection against alcohol induced liver damage [12] and so on. Studies have shown the protective potential of *C.cajan* against CCl_4, hepatotoxicity in rats [13]. Also, anticancer activity of *C. cajan* was examined towards MCF-7 breast cancer and it showed that the plant inhibited MCF-7 cell growth [13]. The aim of this present work was to study the hepatoprotective effect of ethanol extract of *C. cajan* against NDEA hepatotoxicity so in future an efficient formulation could be produced or developed which will be specific for imparting hepatoprotection.

2. Materials and methods

1. Chemicals

N-Nitrosodiethylamine which was of analytical grade was obtained from Sigma chemical company, USA. Silymarin was purchased from Micro Labs and other chemicals/reagents were of analytical grade, supplied by Merck (India).

2. Animals

Male albino rats (n=35) which weighs 150-200g body weight were bought from the University College Hospital, Oyo State, Nigeria and then housed in cages and fed *ad libitum* with rat chows and distilled water under 12- hours light/dark cycle at room temperature. All experimental procedures carried out were approved by Covenant University Health Research Ethics Committee (CHREC/015/2019).

3. Preparation of plant extract

Cajanus cajan was obtained in June, 2018 from a market in Ogun State, Nigeria and properly identified and authenticated by a qualified taxonomist. It was plucked from the branches, washed and air dried at room temperature (26^{0}C) for four weeks. The leaves (349.1g) were then pulverized using electric blender and soaked in 80% ethanol (600g) as solvent for 72 hours using two successive extractions. It was then sieved out using muslin cloth and cotton wool. Rotary evaporator was used at 50^{0}C and water bath at 40^{0}C to obtain a yield of 32g semi-solid crude substance.

4. Research design

Male albino Wistar rats (thirty five) were separated into seven groups consisting having five animals each in different cages. Rats were acclimatized for two weeks. Group A served as the control and received no treatment. Group B was the negative control and was given intraperitoneal injection (i.p.) of only NDEA mixed in saline at 200mg/kg, Group C, D and E received i.p. of NDEA 200mg/kg followed by oral intubation of *Cajanus cajan* at doses of 200, 400 and 800 mg/kg respectively, Group F was the positive control and was given i.p. of NDEA followed with a standard drug Silymarin at 50 mg/kg, Group G was given by oral intubation of *Cajanus cajan* at 800mg/kg only. The experiment lasted for 28 days. The intraperitoneal injection of N-Nitrosodiethylamine was done following the procedure described by Mathur *et al.* [14].

5. Measurement of body weight

The initial body weights of the rats were weighed on the first day and the final body weight measured on the last day of the experiment using a weighing balance and the unit expressed in grams (g).

6. Collection of Blood

Blood obtained from the animals through cardiac puncture was put into heparinized tubes and spun for 15 minutes at 3500 rpm to get the plasma used for biochemical analyses [15].

7. Measurement of liver weight

After sacrificing the rats, their livers were excised and weighed using a weighing balance and units expressed in grams (g).

8. Measurement of ALT and AST activity

AST and ALT which are vital markers of liver function were measured using commercial Randox diagnostic kits according to the instructions of the manufacturer. A Spectrophotometer was used for measuring the absorbance at 340nm and unit expressed in U/L.

9. Measurement of Albumin

Albumin was measured using commercial Randox diagnostic kits according to the instructions of the manufacturer. A Spectrophotometer was used for measuring the absorbance at 578 nm and unit expressed in g/dl.

3. **Results**

The NDEA treated group showed a marginal increase in final body weight and a significant ($p < 0.05$) increase in liver weight in comparison to the control (Table 1). In the *Cajanus cajan* treated and Silymarin groups, there were significant ($p < 0.05$) increases in the body weights and significant ($p< 0.05$) decreases in the liver weights compared to the NDEA treated group.

Table1. Effect of *Cajanus cajan* on body and liver weights

Groups	Treatment	Initial body weight (g)	Final body weight (g)	Liver weight (g)	Relative Liver weight (liver/100g b.w)
A	Control	168 ± 12.90	198 ± 10.72	7.08 ± 0.24	3.58 ± 0.11
B	NDEA	160 ± 14.0	170 ± 11.60[a]	9.93 ± 0.44	5.98 ± 0.36[a]
C	NDEA + 200mg/kg	166 ± 11.49	182 ± 13.24[b]	8.16 ± 0.35	4.48 ± 0.25[b]
D	NDEA + 400mg/kg	164 ± 12.46	185 ± 13.26[b]	7.55 ± 0.26	4.19 ± 0.16[b]
E	NDEA + 800mg/kg	163 ± 11.47	189 ± 13.29[b]	7.04 ± 0.22	4.01 ± 0.13[b]
F	NDEA + Silymarin	161 ± 11.41	186 ± 15.08[b]	7.58 ± 0.11	3.91 ± 0.25[b]
G	800mg/kg Only	167 ± 11.06	195 ± 11.38	6.97± 0.35	3.57 ± 0.13

Values are expressed as mean ± SEM.
[a]Values significantly differ from Control group.
[b]Values significantly differ from NDEA treated group.

NDEA induction significantly ($p < 0.05$) increased AST and ALT activity when compared with the control (Table 2). Treatment with *C. cajan* significantly ($p < 0.05$) reduced ALT and AST when compared with NDEA treated group. This reduction was comparable to Silymarin.

Table2. Effect of *C. cajan* on AST and ALT activity

Groups	Treatment	ALT (U/L)	AST (U/L)
A	Control	69.73.38 ± 5.19	106.67 ± 5.47
B	NDEA	124.73 ± 22.06[a]	227.15 ± 25.5[a]
C	NDEA + 200mg/kg	109.23 ± 8.39[b]	180.34 ± 8.25[b]
D	NDEA + 400mg/kg	98.34 ± 6.81[b]	145.05 ± 6.75[b]
E	NDEA + 800mg/kg	74.86 ± 5.63[b]	112.87 ± 5.80[b]
F	NDEA + Silymarin	90.59 ± 6.77[b]	132.12 ± 5.96[b]
G	800mg/kg Only	70.33 ± 5.22	109 ± 5.76

Values are expressed as mean ± SEM.
[a]Values significantly differ from Control group.
[b]Values significantly differ from NDEA treated group.

The NDEA treated group had a significantly ($p < 0.05$) reduced albumin when compared with control (Figure1). Significantly ($p < 0.05$) increased albumin in the *Cajanus cajan* treated and Silymarin groups with 800mg/kg being the highest when compared with the NDEA treated groups was seen.

IOP Conf. Series: Earth and Environmental Science **331** (2019) 012023 doi:10.1088/1755-1315/331/1/012023

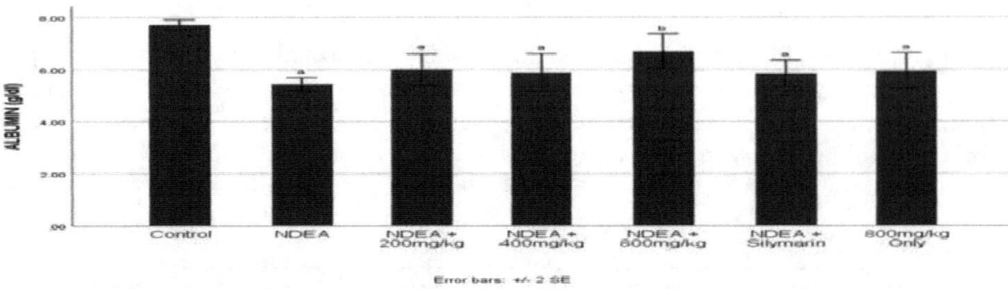

Values are expressed as mean ± SEM.
[a]Values significantly differ from Control group.
[b]Values significantly differ from NDEA treated group.

Figure1. Effect of *Cajanus cajan* on Albumin

4. Discussion

Hepatotoxicity means dysfunction of the liver due to overload of chemicals and drugs that are toxic to the body. NDEA is known to produce ROS that results to liver damage. It is known that DNA, lipids and proteins are the main targets of oxidative injury. Liver serves to filter out toxic substances from the bloodstream. When there are excessive chemicals filtering through the liver, it becomes overloaded and can lead to hepatotoxicity. The activities of particular enzymes are quickly changed in any sort of liver damage and transaminases are known to be vital markers of liver function. They sensitively reflect the status of liver damage [16]. This study showed that increase in marker enzymes were signs of cellular damage due to administration of NDEA. Treatment with ethanol extract of *Cajanus cajan* brought these enzymes back to near normal by protecting the functional integrity of the hepatocytes, showing its hepatoprotective property against NDEA induced liver damage. Also, Silymarin which is now widely used to treat hepatic damage also helped in protecting the liver.

This study showed that for AST, NDEA induction increased plasma AST activity compared to the control. This finding is similar to a previous study [17, 18]. Treatment with *Cajanus cajan* significantly ($p < 0.05$) reduced the activity to near normal levels compared with NDEA treated group. This reduction was comparable with Silymarin which confirms earlier observations [19, 20, 21].

This study showed that NDEA induction significantly increased plasma ALT activity in comparison with the control. There was a reduction in the *Cajanus cajan* treated and silymarin group when compared with NDEA treated group [17, 18]. This indicates protective activity of the *Cajanus cajan* extract revealing that the plant has the ability of protecting NDEA-damaged cells.

NDEA induction marginally increased the final body weight and elevated liver weight when compared to the control. This marginal increase in body weight in NDEA treated animals could predominantly be due to reduction of skeletal tissue following exposure to the chemical. This result is in line with a previous study which reported body weight loss after administration of N-nitrosodiethylamine [7, 22]

Albumins (ALB) are globular proteins made by the liver. From this study, the NDEA treated group recorded a significantly ($p < 0.05$) reduced albumin which may be associated with reduction of hepatocytes [23]. This observation is in agreement with a study which indicated that NDEA induced hepatic damage in experimental animals and reduced plasma albumin level [24, 25]. There was increased ALB in the *Cajanus cajan* and Silymarin treated groups indicating their protective role against liver cell damage.

5. Conclusion

The antioxidant and free radical scavenging properties of *C. cajan* on NDEA induced liver toxicity in rats was seen. Various reports have shown that 'alkaloids, flavonoids, steroids and also triterpenoids' have protective effects on the plasma and also liver as a result of their antioxidant properties. These phytochemicals are present in *Cajanus cajan* and may be responsible for the protective effect in NDEA induced hepatotoxicity.

IOP Conf. Series: Earth and Environmental Science **331** (2019) 012023 doi:10.1088/1755-1315/331/1/012023

References

1. Tricker, A., Pfundstein, B., Theobald, E., Preussman, R. and Spiegenhalder, B. (1989). Mean daily intake of volatile N-nitrosamines from foods and beverages in West Germany. Food and Chemical Toxicology, 29(2), 29–32.
2. Shank, C. (1975). Toxicology of N-nitroso compounds. Toxicology and Applied Pharmacology, 31(6), 729–732.
3. Bartch, H. and Montesano, R. (1984). Relevance of nitrosamines to human cancer. Carcinogenesis, 5(7), 81–93.
4. Bartsch, H., Hietanen, E. and Malaveille, C. (1989). Carcinogenic nitrosamines: free radical aspects of their action. Free Radical Biology and Medicine, 7, 637–644.
5. Gey, F. (1993). Prospects for the prevention of free radical disease, regarding cancer and cardiovascular disease. British Medical Bulletin, 49(6) 679–699.
6. Arul, D. and Subramanian, P. (2013). Inhibitory effect of naringenin (citrus flavonone) on N-nitrosodiethylamine induced hepatocarcinogenesis in rats. Biochemical and Biophysical Research Communications, 434, 203–209.
7. Singh, N., Singh, A., Sarma, B. and Singh, B. (2009). Potential chemoprevention of N-nitrosodiethylamine-induced hepatocarcinogenesis by polyphenolics from Acacia nilotica bark. Chemico-Biological Interactions, 181, 20–28.
8. Valko, M., Leibfritz, D., Moncol, J., Cronin, M., Mazur, M. and Telser, J. (2007). Free radicals and antioxidants in normal physiological functions and human disease. International Journal of Biochemistry and Cell Biology, 39, 44–84.
9. Aiyeloja, A. and Bello, A. (2006). Ethno botanical potential of common herbs in Nigeria. Educational Research Review, 3(1), 16-22.
10. Zu, Y., Fu, Y., Liu, W., Hou, L. and Kong, Y. (2006). Simultaneous determination of four flavonoids in pigeon pea (Cajanus cajan) leaves using RP-LC. Chromatographia, 63, 499-505.
11. Wu, N., Fu, K., Fu, Y., Zu, Y., Chang, F., Chen, H., Liu, X., Kong, Y., Liu, W. and Gu, B. (2009). Antioxidant activities of extracts and main components of Pigeonpea leaves. Molecules, 14(13), 10-44.
12. Kundu, R., Dasgupta, S., Biswas, A., Bhattacharya, A., Pal, B. and Bandyopadhyay, D. (2008). Cajanus cajan Linn. (Leguminosae) prevents alcohol-induced rat liver damage and augments cytoprotective function. Journal of Ethnopharmacology, 118(3), 440-447.
13. Ashan, R., Islam, M., Bulbul, J., Musaddik, A. and Haque, E. (2009). Hepatoprotective activity of methanol extract of some medicinal plants against carbontetrachloride-induced hepatoto-xicity in rats. European Journal of Scientific Research, 37(2), 302-310.
14. Mathur, N., Pande, K., Aeri, V., Kishore, A., Joshi, V., Madaan, A. and Verma, R. (2014). Determination of Antioxidants and Hepatoprotective ability of Flavonoids of Cichorium intybus. International Journal of Pharmacological and Toxicological Research, 6(4), 107-112.
15. Iweala, E. and Ogidigo, J. (2015). Prostate Specific Antigen, Antioxidant and Hematological Parameters in Prostatic Rats Fed Solanum macrocarpon Leaves. Asian Journal of Biological Sciences, 8, 30-41.
16. Zimmerman, J. (1978). The adverse effects of drugs and other chemicals on the liver. Separation and Purification Technology, 91, 121-133.
17. Ahmad, A., Maheshwari, V., Saleem, A. and Ahmad, R. (2012). Observation of esterase-like-albumin activity during N'-Nitrosodimethylamine induced hepatic fibrosis in a mammalian model. Macedonian Journal of Medical Sciences, 5, 55–61.
18. Rezai, A., Fazlara, A., Haghikaramolah, M., Shahriari, A., Zadeh, N. and Pashforosh, M. (2013). Effect of Echinacea purpurea on hepatic and renal toxicity induced by diethylnitrosamine in rats. Jundishapur Journal of Natural Pharmaceutical Products, 8(2), 60–64.
19. Moncada, C., Torres, V., Varghese, G., Albano, E. and Israel, Y. (1994). Ethanol derived immunoreactive species formed by free radical mechanisms. Molecular Pharmacology, 46, 786–799.

20. Albano, E., French, S. and Ingleman-Sundberg, M. (1999). Hydroxyethyl radicals in ethanol hepatotoxicity. Frontiers in Biosciences, 4, 533–540.
21. Zima, T., Fialova, L., Mestek, O., Janebova, M., Crkocska, J., Stipek, S., Mikulikova, L. and Popov, P. (2001). Oxidative stress, metabolism of ethanol and alcohol related diseases. Journal of Biomedical Sciences, 8, 59–70.
22. Sivaramakrishnan, V., Shilpa, P., Kumar, V. and Devaraj, S. (2008). Attenuation of NDEA-induced hepatocellular carcinogenesis by a novel flavonoloid. Chemico-Biological Interactions, 171, 79-88.
23. Wahid, A., Vijay, R. and Satish, B. (2009). Hepatoprotective activity of hydroalcoholic extract leaves of Alocassie indica. Indian Journal of Experimental Biology, 47, 816-821.
24. Zeashan, H., Amresh, B., Satyawan, Y. and Venkateswara, U. (2008). Hepatoprotective activity of Amaranthusspinosus in experimental animals. Food and Chemical Toxicology, 46(11), 34-71.
25. Khan, A. and Alzohairy, M. (2011). Hepatoprotective effects of camel milk against induced hepatotoxicity in Rats. Asian Journal of Biochemistry, 6(2), 171-180.

Evaluation Of Wind Energy Potentials In Ota, Ogun State, Nigeria

Akinpelu A[a], Emetere M,E [a,b] and Akinwumi S,A[a]
[a]Department of physics, covenant university, Ota, Nigeria
[b]Mechanical Engineering Science, University of Johannesburg, South Africa

Abstract— One of the fastest growing technologies in energy production business this time is wind energy. The unpredictable and epileptic state of power in Nigeria and the concerns on green house gas emission should is a huge concern for all and this has culminated into quest for strong demand for sustainable and renewable source of energy, wind energy generation. The absence of green house gasses emission, clean energy and unquenchable accessibility of wind that can be converted to electricity culminate in the main advantages of electricity generation from. Wind power prospect is estimated to be high or moderate in Nigeria, of which it has not connected this renewable resources to the grid, although, it is not sufficient for one to just bring to a close that wind turbines should be constructed and connected to the grid because there are sufficient wind speeds to drive the wind turbine. More often than not, the stability and reliability studies must be conducted whenever wind power is to be generated. This paper therefore describes the wind energy potential in Ogun, South-west, Nigeria and specifies the conditions to be met before the wind turbines can be made to generate electricity.
Keywords— Wind energy, Green house gas emission, Sustainability, Wind turbine, Wind power
Renewable energy

1. Introduction

Going by the historic designs, power has been gotten from the wind over hundreds of years, known as wind energy, constructed from stone and wood for the purpose of pumping water or for grinding. From history, typically large, heavy and inefficient, were replaced in the early centuries by fossil fuel engines. A better perceptive of aerodynamics and advances in materials science, particularly polymers, has led to the comeback of wind energy extraction towards the end of 20th. Electricity are now generated from wind power devices generally called wind turbines. Wind energy is a main sustaining column of tomorrow's renewable electricity generation. Since wind power was first used to generate electricity, its utilization and growth has been staggering [2]. Because of the concerns of climate change, nations have set aggressive targets to minimize CO_2 discharge by establishing renewable energy generation services. Of the obtainable renewable energy technologies, wind power has been the most rampant to date due to its lower maintenance costs, its established technological condition, the wealth of abundance of wind resources. In addition to reducing CO_2 emissions, renewable energy does not add to already existing pollution from particulate emissions and toxic emissions. Clean energy sources such as wind, hydro, geothermal and solar power have been the focal point in the transition to a healthy, efficient and less polluted ecosystem.

The wind manufacturing industries have set a new limestone for yearly installations in 2014 with wind power capacity on the rise by 44% according to the Global Wind Energy Council [3]. Figure 1 below indicates that nearly 70% of the world's 370GW of wind energy are located in China, USA, Germany, Spain and India [4].

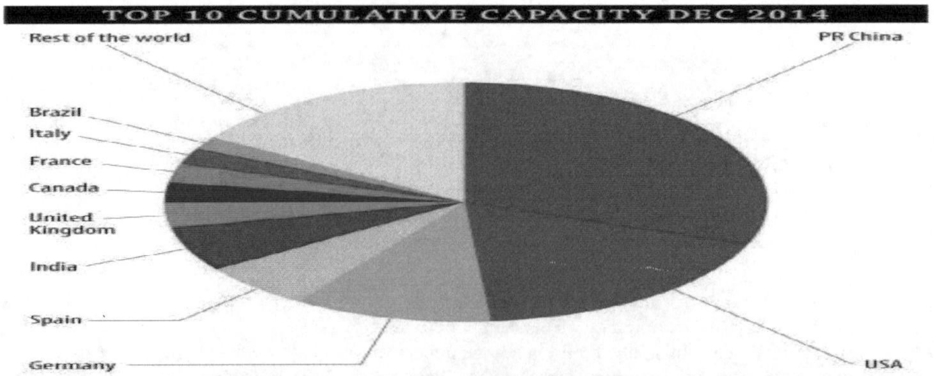

Country	MW	% SHARE
PR China*	114,763	31.0
USA	65,879	17.8
Germany	39,165	10.6
Spain	22,987	6.2
India	22,465	6.1
United Kingdom	12,440	3.4
Canada	9,694	2.6
France	9,285	2.5
Italy	8,663	2.3
Brazil**	5,939	1.6
Rest of the world	58,275	15.8
Total TOP 10	**311,279**	**84.2**
World Total	**369,553**	**100**

Source: GWEC

Figure 1- Total global capacity in 2014.

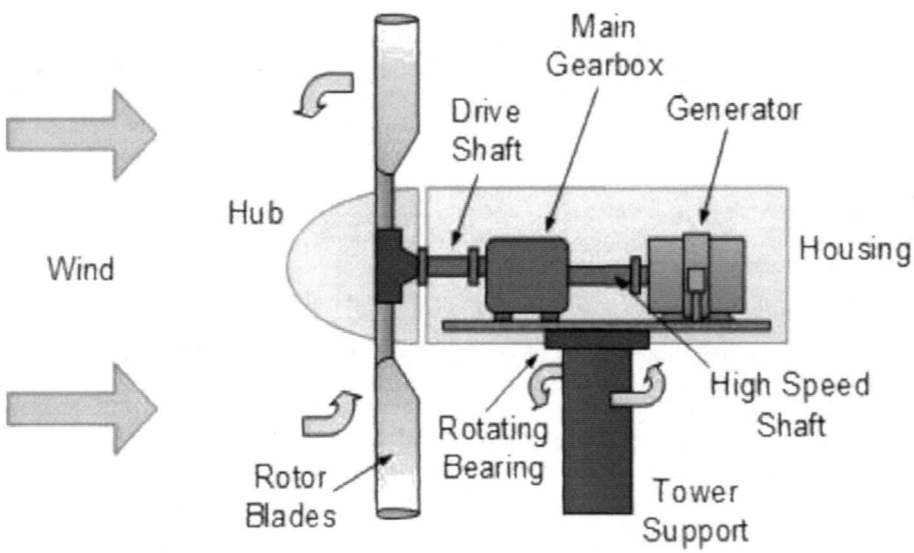

Figure 2- Inside view of a wind turbine
Source: http://www.alternative-energy-tutorials.com/wind-energy/wind-turbine-design.html

The essential components for large scale wind turbines include

I. Rotor: aerodynamic, brake and hub.
II. Drive train: rotor shaft, bearings, brake, gearbox and generator.
III. Supporting structure that is the tower and foundation.
IV. Electrical components which is for control and connection to the grid.

Wind turbines function on a straightforward principle. The energy in the wind turns propeller-like blades around the rotor. While the rotor is connected to the main shaft, which rotates within a stator surrounded by copper wire loops, then Electromagnetic induction is created and electricity is generated [6].

The first classification of the wind turbine is determined by the direction of the shaft and rotational axis. Horizontal Axis Wind Turbine (HAWT) is the turbine whose shaft mounted horizontally parallel to the ground. HAWT sits high atop towers to take advantage of the stronger and less turbulent wind at 100 feet (30 meters)[4].
Vertical axis wind turbine (VAWT) has its shaft at right angle to the ground (Figure 1).

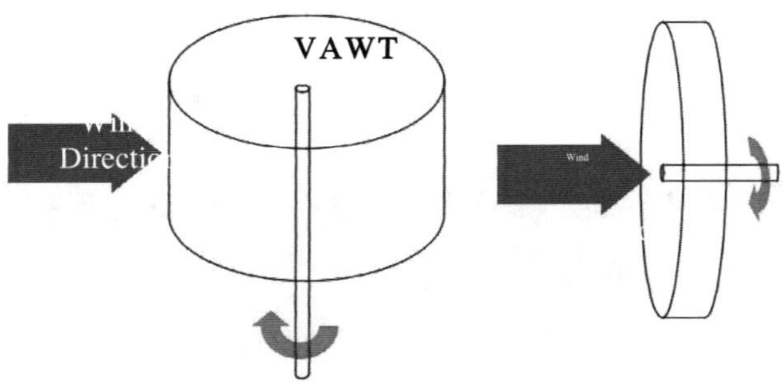

Figure 3. unusual configurations for shaft and rotor orientation.

Both HAWT and VAWT have totally differently rotor designs, each having its own unique configurations [1]. The low tip speed ratio and difficulty in controlling rotor speed is ascribed to a the discontinuity in the mainstream development of the VAWT. Difficulties in the starting of a VAWT have also disadvantaged progress [2]. However, the Vertical Axis Wind Turbine requires no extra mechanism to tackle the wind and weighty electromechanical equipment can be placed on the ground, therefore reducing tower loads. Consequently, the VAWT is not ignored completely for future prospects [3]. In addition to the problems associated with alternative designs, the popularity of the HAWT can be attributed to increased rotor control through pitch and yaw control. The Horizontal Axis Wind Turbine has therefore appeared as the leading design configuration, used by all of today's leading large scale turbine manufacturers.

1.1 Theoretical Efficiency

High theoretical efficiency is advantageous for increased wind energy extraction and should be exploited within the limits of affordable production. Energy (*E*) carried by moving air is expressed as a sum of its kinetic energy [Equation (1)]:

$$E = \frac{1}{2} * \ell A V^3 \qquad 1$$

Where,
ρ – Air density in kg/m³
A – Area of blade coverage in m²
v – Wind speed in m/sec

There is a limit to the amount of energy that can be gotten, which does not depend on the design.. The generation of energy is sustained in a flow process through the decrease of kinetic energy and consequent wind velocity. The enormity of energy exploited depends on the reduction in speed of air over the turbine. Hundred percent extractions simply means zero final velocity which is consequently zero flow. The zero flow situation cannot be accomplished therefore all the wind energy may not be made use of. This principle is extensively established [4,5] and specified that turbine efficiency cannot go beyond 59.3%. And this is generally referred to as the power coefficient *Cp*, where max *Cp* = 0.593 is called the Betz limit [6]. Losses in wind power efficiency are most of the time reduced by:

i. Not using low tip speed ratios.
ii. Choosing aerofoils which have a lift to drag ratio that is high.
And

iii. Dedicated tip geometries

Further explanation are established in the literature [4,6].

1.2 Practical Efficiency

Rotor designs suffer from the addition of minor losses practically, resulting from:

Tip losses

Wake effects

Losses in efficiency of drive train

Simplification losses of blade shapee

1.3 Aerodynamics

There is no perfect and efficient rotor design without aerodynamic performance [19]. The forc that is responsible for the power yield produced by turbine is aerodynamic lift and it is for that reason vital to make the most of this force using suitable design. A resistant drag force which resists the blade's motion is also produced by friction which of necessity, be minimized. So therefore, aerofoil section with a high lift to drag ratio, typically greater than 30 [20], be chosen for the design of rotor blade [19]:

$$LifttoDragRatio = \frac{CoefficientofLift}{CoefficientofDrag} = \frac{C_L}{C_D} \qquad 2$$

The co-efficient for the lift and drag of aerofoils is complex to mathematically predict it, even though there exists software that are freely available, for instance XFOIL [21] model results accurately [22,23]. Conventionally tables correlating lift and drag at given angles of attack and Reynolds numbers are used for testing aerofoils experimentally [24].

Previously aircraft technologies paved way for wind turbine aerofoil designs with similar number of Reynolds coupled with section thicknesses appropriate for conditions at the blade tip. Nevertheless, exceptional considerations must be made for the wind turbine design of specific aerofoil profiles due to the variation in operating conditions and mechanical loads.

2. Background Theoretical Wind Energy Information:

The wind energy turns two, three or more propeller-like blades about a rotor. The rotor is connected to the main shaft, which spins a generator to generate electric current [6].

The blade of the turbines behaves much similar to the wings of an airplane. The moment wind puffs, a low-pressure air coagulates on the downward part of blades. This low-pressure air afterwards drags the blade closer to it, causing th rotation of the rotoer. This is termed 'lift'. The force of the lift is obviously stronger than the force of the wind against the front side of the blade, which is described as 'drag'. The mixture of drag and lift makes the rotor to revolve like a propeller, then turning shaft spins generator to produce electricity [7].

The 3-phase power produced from a balanced system is gotten by

$$P_{3phase} = 3 * V_{phase} * I_{phase} \quad 3$$

The computation of frequency theoretically rely directly on the Revolution per minute and directly on the number of poles inside the generator

$$f = \frac{R.P.M * P}{120} \quad 4$$

where RPM is the generator revolutions per minute

P is number of generator poles.

The turbine pitch of the wind control system modifies the incidence of blades rotors in a wind energy production system that is dependent on real-time speed of the wind for the purpose of changing power output, reaching far more utilization efficiency of wind power and serving as shield for blade's rotor. When the rated speed is lower than the speed of the wind, the blade incidence stays at the highest power point, that is close to angle 0°. Otherwise, the pitch control mechanism changes blade incidence so that the output power of generator is within the allowed range. This serves both as a control feature and an optimisation plus a defense technique stopping runaway conditions of the generator [13].

The rotor shaft of the turbine forces a precision machined alloy planetary gearbox, which increases turbine rotor shaft revolution per minute by 4.

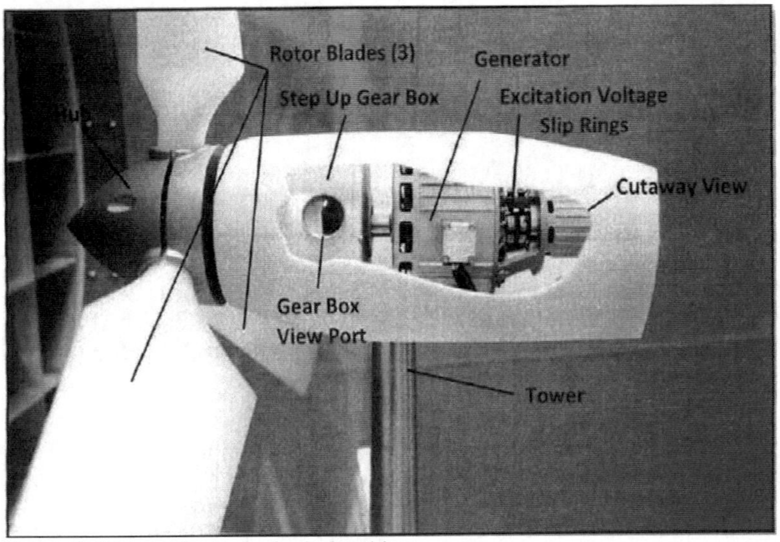

Figure 4: image of a typical wind turbine

IOP Conf. Series: Earth and Environmental Science **331** (2019) 012024 doi:10.1088/1755-1315/331/1/012024

3. Methodology

3.1 Data Collection & Importing To Excel From Davis Weather Station

Three months seconds by seconds wind data was imported to excel directly from 2012 Davis weather station, then with the wind speed, theoretical equation 1 was used to get theoretical power. We assumed air density to be 1.22g/m and the radius of blade to be 35.36m. MATLAB was used to get the Weibull Distribution Function while the data analysis of EXCEL was used for plotting the line graphs

3.2 THE WEIBULL PROBABILITY DENSITY FUNCTION

This is a function of a continuous random variable whose integral across an interval gives the probability that the variable lies within the same interval.According to Slootweg et al., 2001, it is given as:

$$f(t) = \frac{\beta}{\theta}\left(\frac{t}{\theta}\right)^{\beta-1} \exp\left[-\left(\frac{t}{\theta}\right)^{\beta}\right] \qquad 2$$

Where

β = shape parameter, or slope parameter
θ = scale parameter, or characteristic life parameter
t= time

4.0 Results and Discussion

Results are presented for time against the theoretical power for 8 cardinal directions that are; North, North-East, East, South-East, South, South-West, West and North-West directions which are figures 1-figure 8. Figure 9 indicates the wind speed of the station against frequency while figure 10 is the wind rose plot. From the plots, West directions have highest power values with West direction having the highest and relatively high theoretical values while other directions have power values that are not stable.

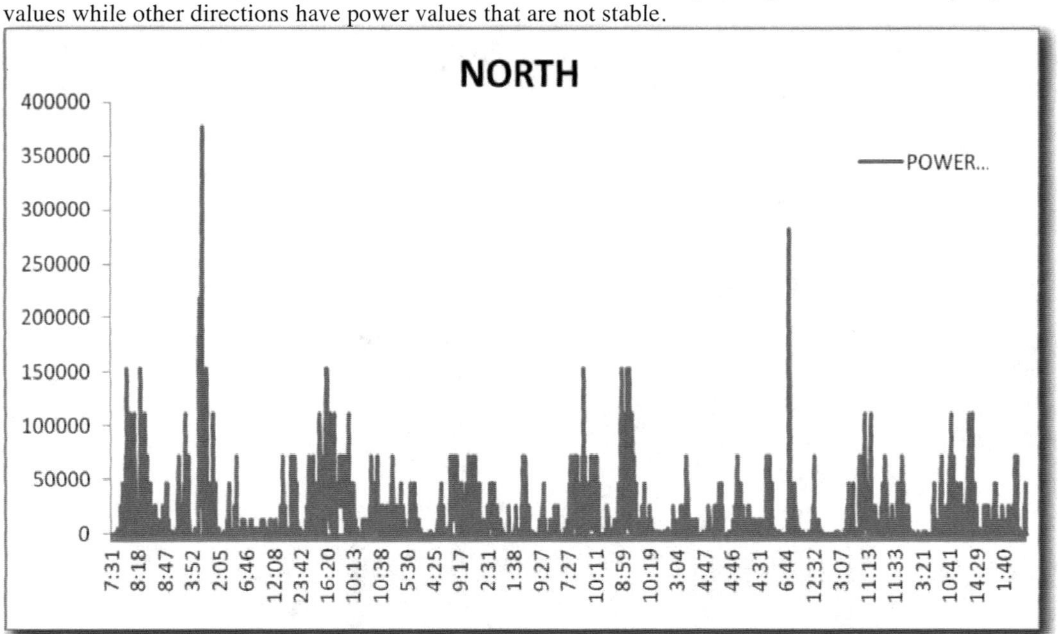

Figure 5: wind speed in north direction against time

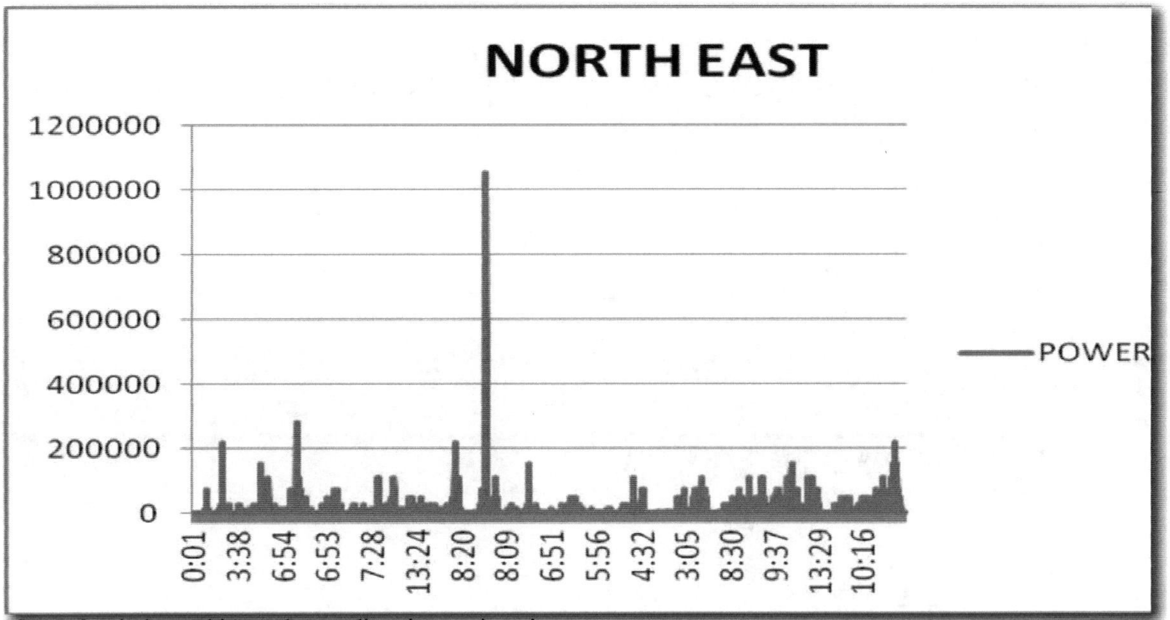

Figure 6: wind speed in north-east direction against time

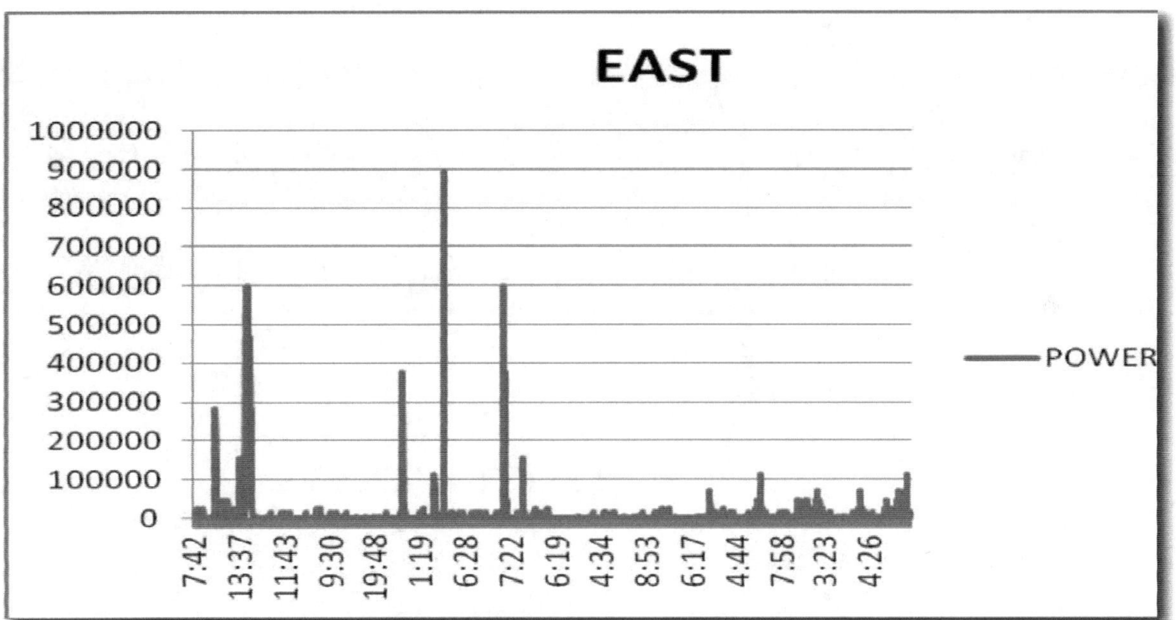

Figure 7: wind speed in east direction against time

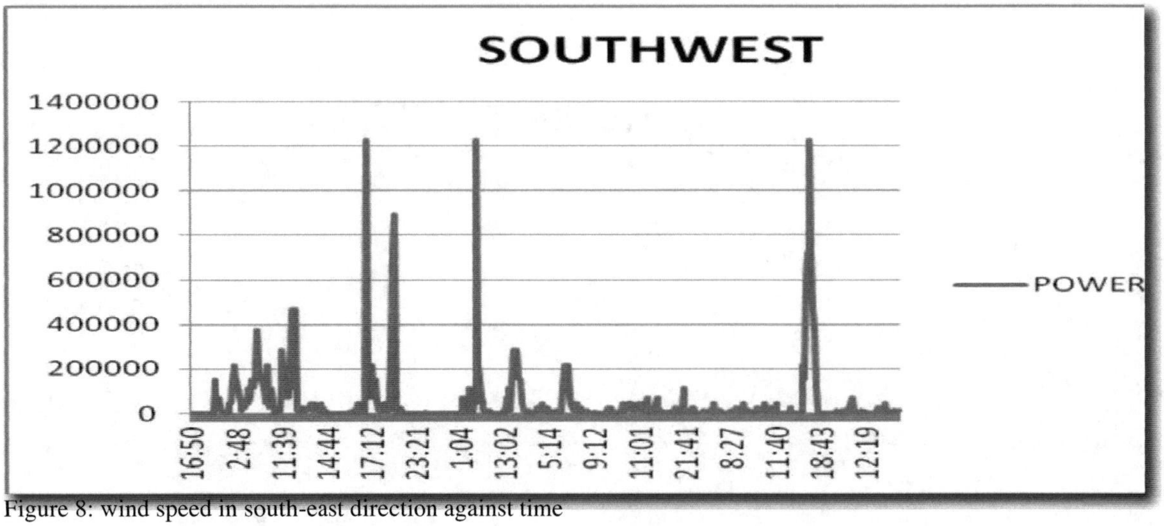

Figure 8: wind speed in south-east direction against time

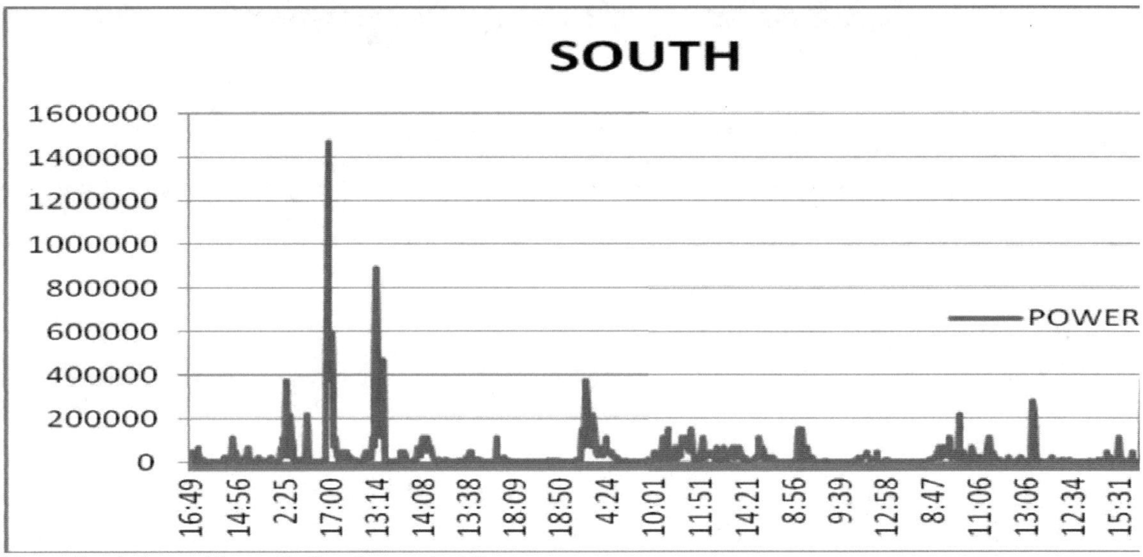

Figure 9: wind speed in south direction against time

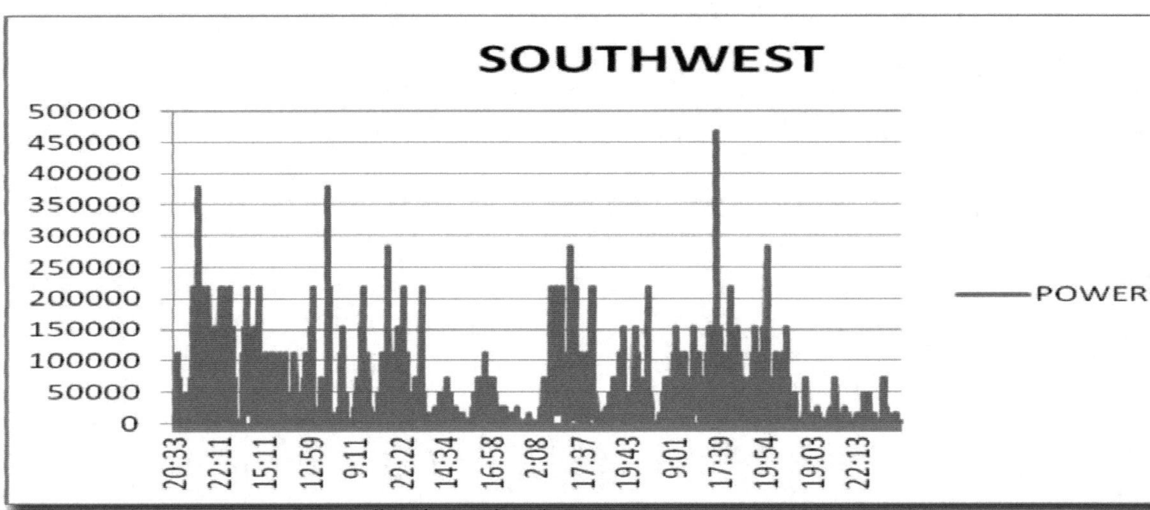

Figure 10: wind speed in south west direction against time

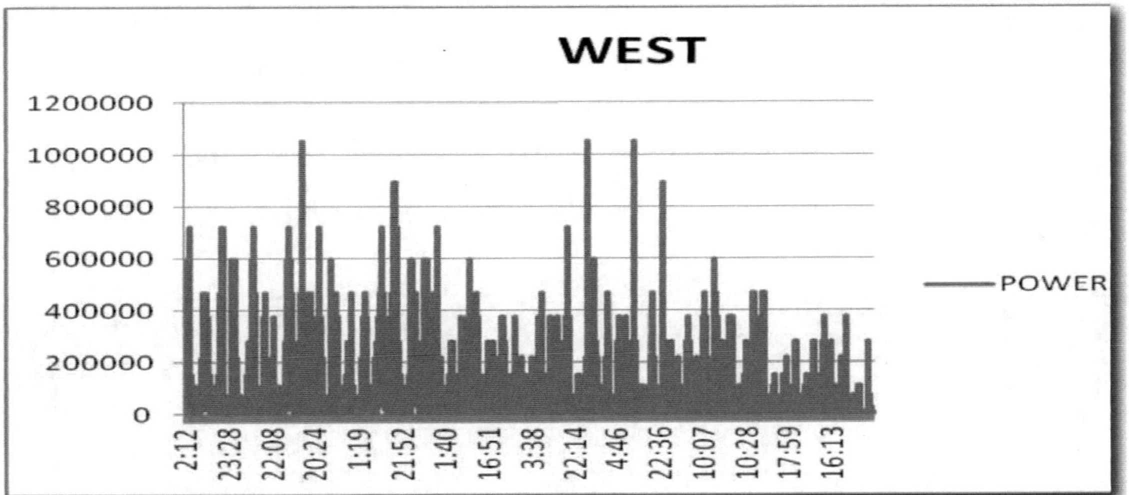

Figure 11: wind speed in west direction against time

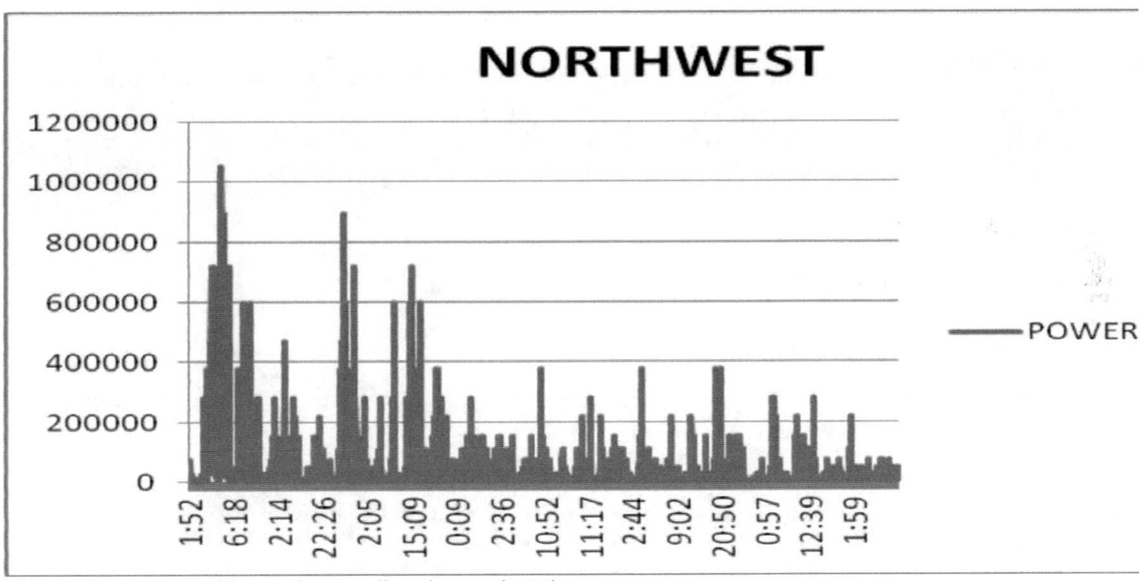

Figure 12: wind speed in north-west direction against time

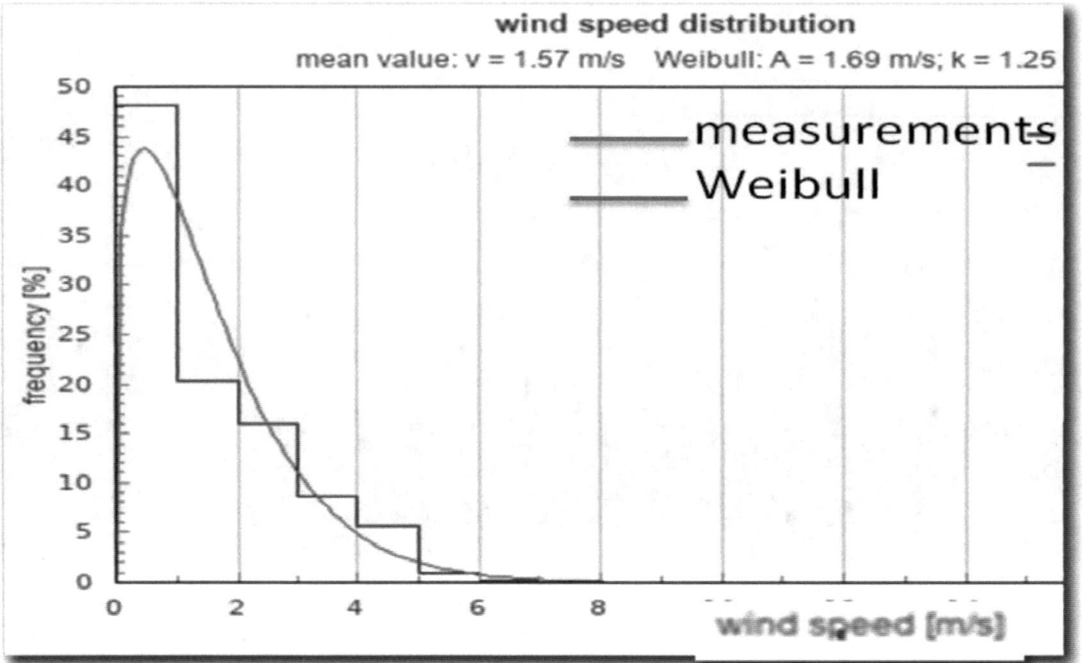

Figure 13: Weibull distribution of the data

5. Conclusions

In conclusion, the wind velocity, direction and characteristics in Ogun State, south-west Nigeria have been studied. The work evaluates wind energy potential in Ogun state and conclusively the wind speed in west direction is the most stable in Covenant University and has highest theoretical power compare to other directions. So therefore, having a wind turbine with above specifications will most likely work perfectly.

Acknowledgement

The authors appreciate the host institutions

References

[1]. Hau, E. *Wind Turbines, Fundamentals, Technologies, Application, Economics*, 2nd ed.; Springer: Berlin, Germany, 2006.
[2]. Dominy, R.; Lunt, P.; Bickerdyke, A.; Dominy, J. Self-starting capability of a darrieus turbine. *Proc. Inst. Mech. Eng. Part A J. Power Energy* **2007**, *221*, 111–120.
[3]. Holdsworth, B. *Green Light for Unique NOVA Offshore Wind Turbine*, 2009. Available online: http://www.reinforcedplastics.com (accessed on 8 May 2012).
[4]. Gasch, R.; Twele, J. *Wind Power Plants*; Solarpraxis: Berlin, Germany, 2002.
[5]. Gorban, A.N.; Gorlov, A.M.; Silantyev, V.M. Limits of the turbine efficiency for free fluid flow. *J. Energy Resour. Technol. Trans. ASME* **2001**, *123*, 311–317.
[6]. Burton, T. *Wind Energy Handbook*; John Wiley & Sons Ltd.: Chichester, UK, 2011.
[7]. Hull, D.G. *Fundamentals of Airplane Flight Mechanics*; Springer: Berlin, Germany, 2007

[8]. Anderson, D.; Eberhardt, S. *Understanding Flight*; McGraw-Hill: New York, NY, USA, 2001.
[9]. Yurdusev, M.A.; Ata, R.; Cetin, N.S. Assessment of optimum tip speed ratio in wind turbines using artificial neural networks. *Energy* **2006**, *31*, 2153–2161.
[10]. Duquette, M.M.; Visser, K.D. Numerical implications of solidity and blade number on rotor performance of horizontal-axis wind turbines. *J. Sol. Energy Eng.-Trans. ASME* **2003**, *125*, 425–432.
[11]. Oerlemans, S.; Sijtsma, P.; Lopez, B.M. Location and quantification of noise sources on a wind turbine. *J. Sound Vib.* **2006**, *299*, 869–883.
[12]. Chattot, J.J. Optimization of wind turbines using helicoidal vortex model. *J. Sol. Energy Eng. Trans. ASME* **2003**, *125*, 418–424.
[13]. Fuglsang, P.; Madsen, H.A. Optimization method for wind turbine rotors. *J. Wind Eng. Ind. Aerodyn.* **1999**, *80*, 191–206.
[14]. Jureczko, M.; Pawlak, M.; Mezyk, A. Optimisation of wind turbine blades. *J. Mater. Proc. Technol.* **2005**, *167*, 463–471.
[15]. Habali, S.M.; Saleh, I.A. Local design, testing and manufacturing of small mixed airfoil wind turbine blades of glass fiber reinforced plastics Part I: Design of the blade and root. *Energy Convers. Manag.* **2000**, *41*, 249–280.
[16]. Thresher, R.W.; Dodge, D.M. Trends in the evolution of wind turbine generator configurations and systems. *Wind Energy* **1998**, *1*, 70–86.
[17]. Gipe, P. The Wind Industrys Experience with Aesthetic Criticism. *Leonardo* **1993**. *26*, 243–248.
[18]. Chamberlain, D.E. The effect of avoidance rates on bird mortality predictions made by wind turbine collision risk models. *Ibis* **2006**, *148*, 198–202.
[19]. Maalawi, K.Y.; Badr, M.A. A practical approach for selecting optimum wind rotors. *Renew. Energy* **2003**, *28*, 803–822.
[20]. Griffiths, R.T. The effect of aerofoil charachteristics on windmill performance. *Aeronaut. J.* **1977**, *81*, 322–326.
[21]. Drela, M. *XFoil*; Massachusetts Institute of Technology: Cambridge, MA, USA, 2000.
[22]. Drela, M. *Xfoil User Primer*; Massachusetts Institute of Technology: Cambridge, MA, USA, 2001.
[23]. Timmer, W.A.; van Rooij, R.P.J.O.M. Summary of the Delft University wind turbine dedicated airfoils. *J. Sol. Energy Eng. Trans. ASME* **2003**, *125*, 488–496.
[24]. Abbott, I.H.; Doenhoff, A.V. *Theory of Wind Sections*; McGraw-Hill: London, UK, 1949.
[25]. Rooij, R.P.J.O.M.; Timmer, W. Roughness sensitivity considerations for thick rotor blade airfoils. *J. Solar Energy Eng. Trans. ASME* **2003**, *125*, 468–478.
[26]. Fuglsang, P.; Bak, C. Development of the Riso wind turbine airfoils. *Wind Energy* **2004**, *7*, 145–162.
[27]. Polinder, H. Comparison of direct-drive and geared generator concepts for wind turbines. *IEEE Trans. Energy Convers.* **2006**, *21*, 725–733.
[28]. Gupta, S.; Leishman, J.G. Dynamic stall modelling of the S809 aerofoil and comparison with experiments. *Wind Energy* **2006**, *9*, 521–547.
[29]. Stol, K.A.; Zhao, W.X.; Wright, A.D. Individual blade pitch control for the controls advanced research turbine (CART). *J. Sol. Energy Eng. Trans. ASME* **2006**, *128*, 498–505.
[30]. Bossanyi, E.A. Individual blade pitch control for load reduction. *Wind Energy* **2003**, *6*, 119–128.

[31]. Barlas T.K.; van Kuik, G.A.M. Review of state of the art in smart rotor control research for wind turbines. *Prog. Aerosp. Sci.* **2010**, *46*, 1–27.

[32]. Barlas, T.; Lackner, M. The Application of Smart Structures for Large Wind Turbine Rotor Blades. In *Proceedings of the Iea Topical Expert Meeting*; Delft University of Technology: Delft, The Netherlands, 2006.

[33]. Thor, S. The Application of Smart Structures for Large Wind Turbine Rotor Blades. In *Proceedings of the IEA Topical Expert Meeting*; Sandia National Labs: Alberquerque, NM, USA, 2008.

[34]. Kong, C.; Bang, J.; Sugiyama, Y. Structural investigation of composite wind turbine blade considering various load cases and fatigue life. *Energy* **2005**, *30*, 2101–2114.

[35]. Quarton, D.C. *The Evolution of Wind Turbine Design Analysis—A Twenty Year Progress Review*; Garrad Hassan and Partners Ltd.: Bristol, UK, 1998; pp. 5–24.

[36]. Ahlstrom, A. Emergency stop simulation using a finite element model developed for large blade deflections. *Wind Energy* **2006**, *9*, 193–210.

[37]. Kishinami, K. Theoretical and experimental study on the aerodynamic characteristics of a horizontal axis wind turbine. *Energy* **2005**, *30*, 2089–2100.

[38]. Brondsted, P.; Lilholt, H.; Lystrup, A. Composite materials for wind power turbine blades. *Ann. Rev. Mater. Res.* **2005**, *35*, 505–538.

[39]. Jensen, F.M. Structural testing and numerical simulation of a 34 m composite wind turbine blade. *Compos. Struct.* **2006**, *76*, 52–61.

[40]. Veritas, D.N. *Design and Manufacture of Wind Turbine Blades, Offshore and Onshore Turbines*; Standard DNV-DS-J102; Det Norske Veritas: Copenhagen, Denmark, 2010.

[41]. Case, J.; Chilver, A.H. *Strength Of Materials*; Edward Arnold Ltd.: London, UK, 1959.

[42]. Griffin, D.A.; Zuteck, M.D. Scaling of composite wind turbine blades for rotors of 80 to 120 meter diameter. *J. Sol. Energy Eng. Trans. ASME* **2001**, *123*, 310–318.

[43]. Shokrieh, M.M.; Rafiee, R. Simulation of fatigue failure in a full composite wind turbine blade. *Compos. Struct.* **2006**, *74*, 332–342.

[44]. *Wind Turbines. Part 1: Design Requirements*; BS EN 61400-1:2005; BSi British Standards: London, UK, January 2006.

[45]. Kensche, C.W. Fatigue of composites for wind turbines. *Int. J. Fatigue* **2006**, *28*, 1363–1374.

Performance enhancement using appropriate mass charge of R600a in a developed domestic refrigerator

Solomon O. Banjo[1*], Bukola O. Bolaji[2], Oluseyi O. Ajayi[1], Babalola P. Olufemi[1], Ibhadode Osagie[3], Anthony O. Onokwai[4]

[1] Mechanical Engineering Department, Covenant University, P.M.B. 1023, Ota, Nigeria
[2] Mechanical Engineering Department, Federal University, Oye-Ekiti, Nigeria
[3] Nigerian Building and Road Research Institute (NBRRI), Km 10 Idiroko Road, Ota, Ogun State, Nigeria
[4] Mechanical Engineering Department, Landmark University, Omu-aran, Kwara State, Nigeria
*Email Address: Solomon.banjo@covenantuniversity.edu.ng

Abstract: This paper presents performance analysis of a domestic refrigeration system using R600a as refrigerant. Investigation was conducted experimentally to select an appropriate refrigerant mass charge for the system and to compute the cooling system performance characteristics under ambient temperature of 29 °C. The choice of the hydrocarbon refrigerant (R600a) became so imperative due to notable influence of chlorofluorocarbons (CFCs), hydrochlorofluorocarbons (HCFCs) and hydrofluorocarbons (HFCs) on global warming and ozone depletion. UNFCCC and Montreal Protocol regulation have banned the use of halogenated refrigerants in cooling and heating systems. The paper reveals that the power consumed by the system using mass charge of 15 g refrigerant was reduced by 9.3 and 10.9% compared with the 10 and 25 g refrigerant mass charges respectively. In addition, the results show that the coefficient of performance (COP) of the refrigeration system while working with 15 g is 24.7 and 20.2% higher than when the system worked with 10 and 25 g refrigerants mass charge respectively. More so, the system attained an evaporator air temperature of -12 °C in 2 hours, which makes 15 g charge refrigerant appropriate for the running of the developed system that has a volume capacity of 68 liters.

1. Introduction

Refrigeration being a process of transferring heat from the refrigerating space to the surrounding required a working fluid which is referred to as refrigerant. The working fluid determines the rate of heat movement within a closed system. There are various classes of refrigerant, which are based on their chemical composition and safety [1]. Furthermore, the extensive effect of the greenhouse gas from the combustion of fossil fuel in generating power and emission of halocarbon refrigerants from the traditional refrigerator, mechanical heat pump and air-conditioning system has inflated the threat on ozone layers and in turns creating negative impact on the environment. In 1876, most of the cooling systems used inorganic refrigerants such as ammonia (R717), water (R718) and carbon dioxide (R744) as their working fluids but due to their toxicity and incompatibity with the materials used in the system, they were replaced with chloromethane (R40) and Sulphur dioxide (R764) [2-3]. More so, in 1931, halocarbon refrigerants (CFC_s and $HCFC_s$) were found to be better replacements for inorganic refrigerants [4]. Since the discovery of the harm that CFCs and HCFCs refrigerant produces on the environment and climate, the American Household Manufacturers have endorsed HFC134a as a possible alternative refrigerant for CFCs refrigerant in traditional refrigerators [5]. The rising of the hydrofluorocarbon (HFCs) refrigerants with suitable thermodynamic and thermo-physical properties have made them admissible both for the commercial and consumer set up in 90's [6]. They found applications in domestic vapour compression refrigerator, air-conditioning and heat pump and other heating and refrigerating appliances. However, HFC134a refrigerant has high global warming potential of 1430, which makes it non-environmental friendly [7-9]. A few years ago, the discovery of their immense contribution to the change in climate has led to the researching and investigation of various performance of refrigerant starting with class A1, class A2, and class A3 refrigerants. In 1997, Kyoto protocol of United Nations Framework convention on climate change (UNFCCC) worked on drop-in replacement refrigerants to abolish the predominant greenhouse gases [10]. To achieve reduction in the GHG, eco-friendly and energy saving refrigerants are needed to be introduced into the refrigeration system. Hydrocarbons (HCs) have been widely found to be alternative to halocarbon refrigerants in heating, ventilation and air-conditioning system (HAVC) and they are organic compounds with composition of hydrogen and carbon atoms [11]; [17]. These working fluids do not have potential to deplete the ozone layer because they have zero ozone depletion potential (ODP = 0), non-poisonous, low global warming potential (GWP<5), low cost, lower total equivalent warming impact (TEWI), miscible with mineral oil and high efficiency with better coefficient of performance COP [12-14]. The refrigerant is not only good for environmental preservation but it can step down energy consumption of the system and likewise, improve the cooling speed by 3-10%. Hydrocarbon miscibility in mineral oil enabled a smooth running of the compressor and its high critical temperature enhances the efficiency of the refrigeration system [15]. Combustion of hydrocarbon does not just occur by having contact with air. There are definite conditions that must be attained and these include: (i) a surface temperature should exceed 440 °C (ii) release of hydrocarbon is essential (iii) the substance has to mix with the true proportion of air [16]. The classes of refrigerant are shown in Table 1.

Table: 1 shows the classification of refrigerants with their thermo-physical properties [1]

S/N	Refrigerant	Chemical Composition	Safety Classification	ODP	GWP
1	R12	CCl_2F_2	A1	1	10200
2	R22	$CHClF_2$	A1	0.055	8100
3	R134a	CH_2FCF_3	A1	0	1430
4	R32	CH_2F_2	A2	0	675
5	R143a	CH_3CF_3	A2	0	4470
6	R152a	CH_3CHF_2	A2	0	124
7	R170	CH_3CH_3	A3	0	6
8	R290	$CH_3CH_2CH_3$	A3	0	3
9	R600a	$CH(CH_3)_2CH_3$	A3	0	3

The refrigerant used in this vapour compression refrigeration system is isobutane gas (R600a) and its hydrocarbon refrigerant. It has found its application in refrigeration and air-conditioning industry. Parts of European countries have switched to hydrocarbons from conventional refrigerants which increase the greenhouse gas and influenced climate negatively [17-18]. However, the purpose of this research work is to investigate the appropriate refrigerant charge level that will serve as specification for the developed domestic refrigeration system.

2. Experimental Procedure

A traditional refrigeration system was developed from locally sourced materials and used as experimental test rig. It contains four essential components, which are, evaporator, single hermetic sealed compressor, capillary tube and natural cooled condenser. The system operated under room temperature of 29°C in a refrigeration and air-conditioning laboratory as shown in Figure.1.

Figure.1: Experimental test rig with attached measuring devices.

Furthermore, temperature sensors were connected to each of the essential points of the refrigeration system to measure their temperatures at an interval of 15 minutes for a period of 5 hours and likewise, power meter was used in determined the compressor power input. Pressure gauge was also, used to measure the pressure at the suction and discharge of the compressor as shown in Fig. 1. In addition, the compressor of the refrigeration system was charged with a natural refrigerant known as isobutane (R600a). The mass charge of refrigerants was varied from 10, 15 and 25 g to determine the performance characteristics of the system such as Coefficient of Performance (COP), Refrigerating Effect (RE) and power consumption. REFPROP software (Version 9.1) was applied to capture the thermodynamic properties of the system. The performance parameters were computed using equation (1) to (5) and the results are shown in Fig. 2 to 7. More so, materials used in the fabrication of the domestic refrigerator and details of the application of temperature sensors on the refrigeration cycle are displayed in Table 2 and 3 respectively.

Table 2: Materials used for the construction of the refrigerator.

S/N	Component	Specification
1	Marine board	Plywood
2	Aluminum	Tin foil(<0.2mm)
3	Styrofoam	EPS-Thermal
4	Hermetic sealed compressor	1/12 (60-70W)
5	Condenser	Air cooled type
6	Evaporator	11/4 Copper pipe
7	Capillary tube	Copper tube type
8	Pressure gauge	No:536G-A
9	Refrigerant	R600a
10	Temperature Sensor	K-Type (-50-1300) °C

Table 3: Details of the temperature measurement in degree Celsius

Temperature Sensor	Fixed point
T_1	Temperature sensor fixed before compressor
T_2	Temperature sensor connected after compressor
T_3	Temperature sensor connected before capillary tube
T_4	Temperature sensor connected after capillary tube
T_5	Temperature sensor connected in the refrigerating space

IOP Conf. Series: Earth and Environmental Science **331** (2019) 012025 doi:10.1088/1755-1315/331/1/012025

The coefficient of performance was determined using the equation 1.

$$COP = \frac{Q_e}{W_c} = \frac{he_1 - he_4}{hc_2 - hc_1} \qquad (1)$$

Where, Q_e is the latent heat absorbed in the evaporator and Wc is the work done by the compressor

$$Q_e = \dot{m}(he_1 - he_4) \qquad (2)$$

Where Q_e is the latent heat absorbed in the evaporator, $\dot{m}$ is the mass flow rate of refrigerant and $he_1 - he_4$ is the differential in enthalpies at the evaporator (kJ/kg).

$$hcond = \dot{m}(hc_2 - hc_3) \qquad (3)$$

Where Q_{c}. is the heat rejected at the condenser, $\dot{m}$ is the mass flow rate of refrigerant and $hc_2 - hc_3$ is the differential in enthalpies at the condenser (kJ/kg)

$$Wc = \dot{m}(hc_2 - hc_1) \qquad (4)$$

Where, Wc is the compressor work, $\dot{m}$ is the mass flow rate of refrigerant and $hc_2 - hc_1$ is the differential in enthalpies at the compressor (kJ/kg).

$$Mass\ flow\ rate\ (\dot{m}) = \frac{cooling\ load}{he_1 - he_4}\ \text{in (kg/s)} \qquad (5)$$

Where $\dot{m}$ is the mass flow rate of the refrigerant, $R.E$ is the refrigerating effect and he_1-he_4 is the differential in enthalpies at the evaporator (kJ/kg).

3. Results and Discussion

Figure 2 shows that the coefficient of performance of the domestic refrigeration system when operating with 10, 15 and 25 g mass charge of refrigerant. However, the system performed better when working with 15 g refrigerant charge with 24.7 and 20.2% higher than when the system used 10 and 25 g refrigerants mass charge respectively. The refrigerating effect (R.E) of the domestic refrigerator was enhanced due to increase in the COP. Figure 3 revealed the energy saving trend of the cooling system, when working with different refrigerant mass charges of 10, 15 and 25 g. The energy consumed by the compressor when working with refrigerant charge of 15 g reduced below when the system was charged with 10 and 25 g by 9.3 an 14.0% respectively. This enables 15 g charge level to be appropriate refrigerants for the developed refrigerator as power consumption is one of the parameters to determine the performance characteristics of the system. The reduction in power input to the compressor, increases the COP and lowers the evaporator temperature of the system. Figures 4-6 displayed the variance of heat transfer within the natural cooled condenser and latent heat gained at the evaporator from the material to be cooled within the refrigerating space at an interval of 15 minutes for 5 hours. The thermal energy transfers in the condenser when the system was charged with 15 g of refrigerant was 7.4 and 14.8% higher than when the system was charged with 10 and 25 g and likewise the latent heat absorbed when the system was charged with 15 g of refrigerant was 11.5% and 3.8% higher than when the system was charged with 10 and 25 g of refrigerant respectively. More so, the rate of heat transfers within the evaporator and the condensing unit determine the overall performance of the refrigeration system.

The pull-down time is the time required for the system to attain its lowest temperature. Hence, the pull-down time for the 15 g mass charge achieved was 8.3 and 25.0% faster than when the system was charged with 10 and 25 g respectively. This enables the 15 g mass charge to have a better performance as displayed in Figure. 7.

IOP Conf. Series: Earth and Environmental Science **331** (2019) 012025 doi:10.1088/1755-1315/331/1/012025

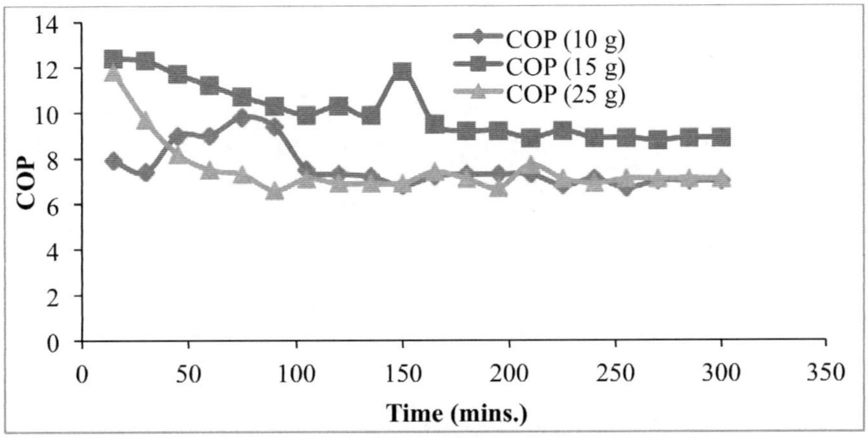

Figure. 2: Variation in COP with time when the system worked with 10 g, 15 g and 25 g refrigerant mass charge.

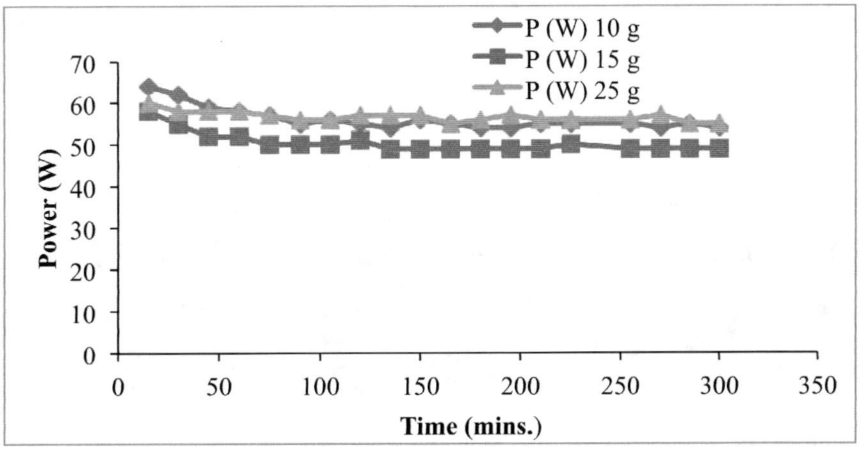

Figure. 3: Variation of power input to the compressor with time.

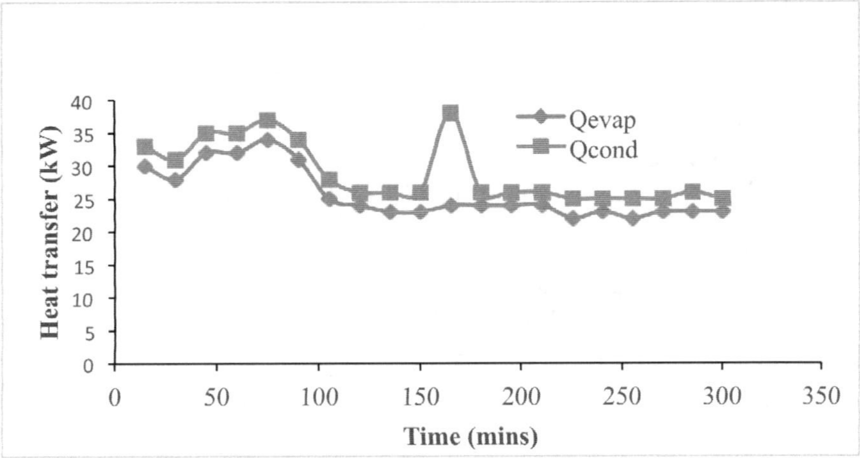

Figure. 4: Variation of heat transfer in the evaporator and condenser for 10 g of R600a.

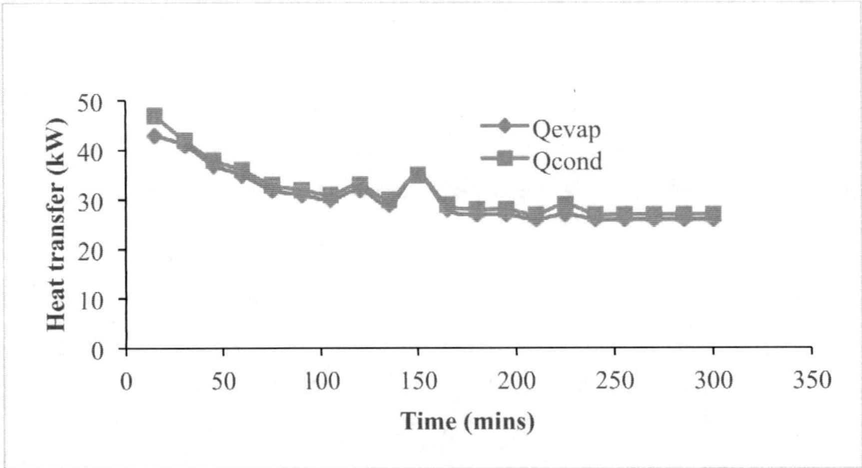

Figure. 5: Variation of heat transfer in the evaporator and condenser for 15 g of R600a.

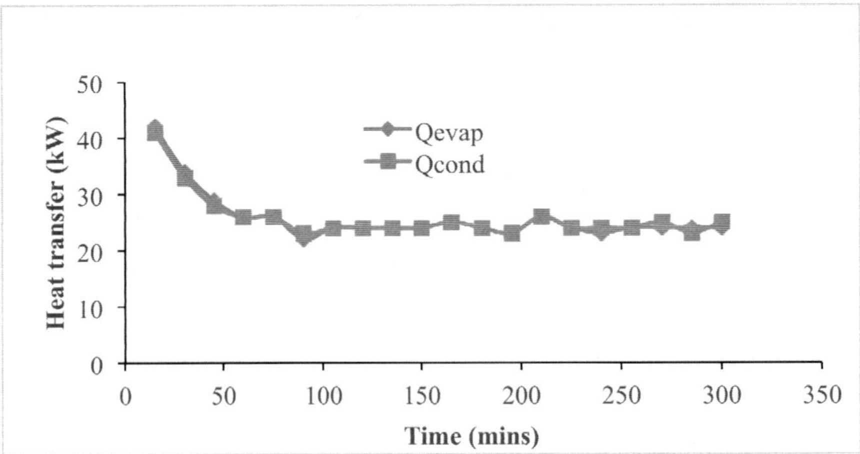

Figure. 6: Variation of heat transfer in the evaporator and condenser for 25 g of R600a.

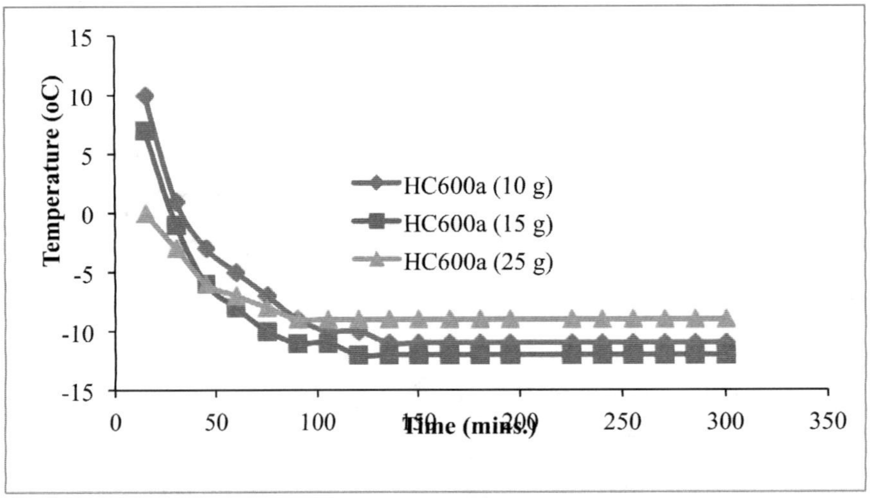

Figure. 7: Variation in pull-down time for the 10, 15 and 25 g of refrigerant mass charges.

4. Conclusion

From this experimental evaluation, effect of refrigerant mass charge was investigated using performance characteristics which include coefficient of performance, refrigerating effect and compressor work input. The selected hydrocarbon refrigerant (R600a) is an environmental friendly type with zero ozone depletion potential, miscible with mineral oil, negligible global warming potential and compatible with system materials. It was feasible from the experimental test results that refrigerant charge of 15 g was suitable to run the developed vapour compression refrigeration system (VCRS) efficiently under a normal room temperature in a tropical region.

Acknowledgement

The authors acknowledge the kind gesture of the management and proprietor base of the Covenant University for the Fund received in executing this project.

References

[1] ASHRAE (2001). Designation & Safety classification of refrigerant. American Society of Heating, Refrigerating and Air-conditioning Engineers, Atlanta, GA, USA, 34.
[2] McMullan J.T. (2002). Refrigeration & environmental issues & strategies for the future. International Journal of Refrigeration, 25, 89-99.
[3] Naeem A, Ali R.K, Nasrullah K, Aun H and Zahid S (2018). Natural and synthetic refrigerants, global warming: A review. Renewable and Sustainable Energy Science, Reviews, 90, 557-569.
[4] Bolaji B.O. (2010). Experimental study of R152a & R32 to replace R134a in a domestic refrigerator, Energy, 35, 3793-3798.
[5] Dalkilic A.S. and Wongwises (2010). A performance comparison of vapour-compression refrigeration system using various alternative refrigerants. International Communication in Heat and Mass Transfer, 37, 1340-1349.
[6] Ajayi O.O., Useh O.O., Banjo S.O., Owoeye F.T., Attabo A., Ogbonnaya M., Okokpujie I.P. and Salawu E.Y. (2018). Investigation of the heat transfer effect of Ni/R134a nanorefrigerant in a mobile hybrid Powered Vapour Compression Refrigerator. IOP Conference Series. Materials Science and Engineering, 391, 1-8.
[7] Tashtoush B. Tahat M. and Shudeifat M.A. (2002). Experimental study of new refrigerant mixtures to replace R-12 in domestic refrigerator. Applied thermal Engineering, 22, 495-506.
[8] Somchai W. and Nares C. (2005). Experimental study of hydrocarbon mixtures to replace HFC-134a in a domestic refrigerator. Energy conversion and management, 46, 85-100.
[9] Sekhar S.J. and Lal D.M., (2005). HFC134a/HC600a/HC290 mixture a result for CFC-12 system, International Journal of refrigeration, 28, 725-743.
[10] Aprea C, Creco A, and Maiorino A. (2017). An experimental investigation of the energetic performances of $HF0123_yf$ and its binary mixtures with 134a in a household refrigerator. International Journal of Refrigeration, 76, 109-117.
[11] Babarinde T. O., Akinlabi S. A. and Madyira D.M. (2018). Enhancing the Performance of Vapour Compression Refrigeration System using Nano Refrigerants: A review, IOP Conference Series: Materials Science and Engineering, 413, 1-8.
[12] Kundu A., Kumar R. and Gupta A. (2014). Performance comparison of zeotropic and azeotrpic refrigerants in evaporation through inclined tubes. 1ith International Conference on Mechanical Engineering, ICME, Procedia Engineering, 90, 452-458.
[13] Qureshi M.A. and Bhatt S. (2014). Comparative Analysis of COP using R134a & R600a Refrigerant in Domestic Refrigerator at steady state condition. International Journal of Science & Research (IJSR), Prestige Institute of Engineering & science, 3, 12.

[14] Sanket B. (2016). Substitution of non-ecofriendly refrigerants by hydrocarbon refrigerants: A review. International Journal of Advance in Science Engineering and Technology Aurangabad, 2, 122-125.
[15] Ahamed J.U, Saidur R, Masjuki H.H. and Sattar M.A. 2012. An analysis of energy, exergy and sustainable development of a vapor compression refrigeration system using hydrocarbon. International Journal of Green Energy, 9 (7), 702-717.
[16] Bolaji B.O. and Huan Z. (2013). Ozone depletion and global warming: Case for the use of natural refrigerant – a review. Renewable and Sustainable Energy Reviews, 18 49-54.
[17] Mohanraj M., Jayaraj S. and Muraleedharan C. (2009). Environment friendly alternatives to halogenated refrigerants---A review. International Journal of greenhouse gas control, 108-119.
[18] Peyyala A and Sudheer NVVS (2016). Experimental investigation of COP using Hydrocarbon Refrigerant in a Domestic Refrigerator. International conference on Advance Material Technology (DIET), Visikhapatnam Andhra Pradesh, India.

Rethinking growth and neo-liberalization development models in Africa: towards a sustainable environmental ethics

Daniel O Adekeye

Ondo State University of Science and Technology, Nigeria, deenofem@yahoo.com deenofem@gmail.com

Abstract: For a long period in the past, and even presently, many people think that if they can only get the economy right, everything else will fall into place. Soothing as this 'econo-mythical' presupposition seems, it has formed the bedrock of human predicament. It has constituted the perceptual model through which many individuals and societies see and understand the world, human activities and human relationship with the environment. This presupposition, which provides the intellectual foundation for neo-liberal conceptual framework compels individuals and societies to pursue development in strictly economic terms. This ideology amounts to a reductive definition of the idea of development. It implies a desperate pursuit of economic growth at the expense of the environment and other aspects of human existence. The paper undertakes a critical analysis of the neo-liberal growth model and its implications for development in Africa, especially in relation to the appropriation of development resources including economy and environment. It is discovered by this research that the neo-liberal growth model does not and cannot provide the proper normative principles for efficient energy and environmental management in Africa. Therefore, it is proposed that a new development paradigm which considers overall sustainability of present and future generations of Africans be embraced.

Keywords: Africa, Development, Energy, Environment, Ethics, Sustainability.

1. Introduction

The socio-economic situation in most African countries is pathetic in that, whereas sustainable development requires improved environmental management, these poor countries excessively exploit their environments in order to meet their basic needs. As a result of high level of poverty and poor technology, the reliance on environmental resource base is alarming. This is not to exonerate the industrialized nations from abuse and degradation of natural environment due to industrial activities. If the idea of sustainability involves maximizing the net benefits of economic development, subject to maintaining the services and quality of natural resources over time [1], then, Africa is far from achieving the minimum benchmark for sustainable development. The current level of individual and household poverty in Africa is blamed among other things on the prevailing development paradigm on the continent. Therefore, this article argues that a shift from the current neo-liberal development paradigm is a *conditio sine qua non* for environmental sustainability in Africa. As long as extreme poverty and unguarded consumption patterns persist, struggle for survival remains or becomes more demanding with its consequence being unhindered environmental abuse and degradation.

The paper has four main sections. The first sets the pace for the argument and establishes a common understanding between the author and the reader by clarifying salient concepts used in the paper. The scope of the second section is to examine the modern origins of mainstream development theories following the World War II, especially the growth theory and neo-liberalism. While the former constitutes a version of the Keynesian frameworks, the latter claims to be a reaction to Keynesianism. Also, it is expected that this section provides a causal relationship between praxis of these theories and the inevitable environmental abuses in Africa. The third section of the paper undertakes an evaluation of major ethical foundations for environmental sustainability discourse and provides us with the metaphysical underpinnings of some environmental ethical behaviour. In the fourth section, the article suggests the idea of eco-development as an alternative strategy for overall sustainability in Africa. This suggestion comes after a review of the idea had been done in the same section.

2. Concepts Clarification

It is pertinent to commence the argument by establishing a common understanding of the major concepts used in this paper, which are 'sustainability', 'development' and 'natural environment'. Sustainability simply refers to meeting our present needs without compromising the ability of future generations to meet their own needs. [2] The idea of sustainability involves three major and mutually inclusive aspects of human existence, namely: environmental conservation, social responsibility and economic development. Human existence as well as societies therefore becomes sustainable to the extent at which these components are enhanced and appropriated. Development is a phenomenon that pervades the essence of man. Development in human society is a many sided process involving the individual and the group. At the level of the individual, it implies increased skills and capacity, greater freedom, creativity, self-discipline, responsibility and material well-being. [3] As a result of the interdependence between the individual and society, the development of the individual is largely a product of the collective conventions and world-view of the community in which the person is accommodated. At least, every human being is socialized into a social group; and his/her development is the degree at which he/she is capable of internalizing these communicated norms. The individual is not an isolated island; therefore, the indices of development highlighted above are expressions of interpersonal relations whose language game is a structure put in place by the society.

At the level of society, development implies an increasing capacity to regulate both the internal and external relationships. [3] Development has often been used exclusively in economic sense. The justification, being that the state of economy is itself an index of other social features – a positioned influenced by the Marxist economic ideology. [4] Economic development is the overall increase in the capacity for relating with the environment, by members of the society. The quality of the capacity to appropriate the environment is a function of the extent at which man understands the laws of nature (science), on the extent at which the people put that understanding into practice, by devising tools (technology), and on the manner in which work is organized (institutions). [5] Going by this argument, one could infer that the economic status of a society is a reflection of developments in other critical areas of such a society.

However, this justification remains unconvincing for equating "development" with economic development. It amounts to logical reductionism when development is equated with the economy. It could even be untenable to argue that the end, to which human beings aspire, is a conducive and friendly economy which is a part of the development indices. Rather, one should consider the attainment of the desired economic condition as one of the requirement to enhance human capacity to exist. Economic development may be dependent on some other human dispositions and capacities such as a degree of moral knowledge and appreciation [6], the sense of justice and fairness that is higher than the brute. The individuals must have had a sense of justice which they draw upon in their daily lives and that the principles of justice are a public conception and a commonly acknowledged moral conviction. [7] Therefore, economic development becomes a part of the development equations and at best a reflection of the level of development in other aspects of the society. In other words, the "accounting system by which a nation measures its well-being is a

IOP Conf. Series: Earth and Environmental Science **331** (2019) 012026 doi:10.1088/1755-1315/331/1/012026

reflection of its basic values."[8]

The notion of development derives from a much older idea of progress, which has its roots in the ancient Greece. This idea presents "development" as a continuous process, as unending pathway toward better livelihood. The concept has been used to describe Western economic growth since the Middle Ages, as well as to explain the process through which all societies are expected to achieve certain economic, social, political, cultural, and other goals.

Development in human society is a conscious process which derives from man's desire to live a better life and his vehement pursuit of the desired status. Development constitutes the successive consequences of the relationship between human beings and the environment. In other words, the history of human development is an account of his relationship with the environment. It is an unfolding of his potentiality to study, understand and live by the realities of the environment as it was reported that much of human history has been a struggle for survival against natural hazards and against real and imagined enemies. [5] The first stage of human development as a social group is the capacity to make and use tools. Tools were made and used in order to control the hostility of the environment, and to maximally access the benefits thereof. Increasing efficiency in this respect could lead to better economic status, that is, the optimization of the resources available in the environment for the satisfaction of human needs.

The use of tools also had political implications and benefits. Hence, the Stone Age was regarded as the first human civilization. Development in the past has always meant the increase in the ability to defend the independence and sovereignty of the social group and indeed to infringe upon the freedom of others. The colonial experience of Africa by the imperial colonialists corroborates this notion. The colonial masters were mainly Europeans who were sophisticated in technology compared to the colonized populations. They invaded the societies with little or no advancement in technology; impoverished them and used their persons and resources to enrich their home economies. Even in the contemporary period, nations with unprecedented advancement in technology possess better economies and at the same time, they are the major global political actors. Therefore, the interconnection between knowledge, economy and politics cannot be over-emphasised.

Material welfare of the individual human person and of all the people has come to dominate the development discourse. This makes economic concerns occupy the central position in most development strategies and policies. Consideration for other spheres of human existence is to the extent to which it promotes the economy. Hence, development has come to mean the elimination of poverty, disease and ignorance; the increase of national wealth in such a way that every person has enough to consume. The chief aspect of development, according to this perspective, is the degree of access to the material wealth and factors of production. It is a transformation of the conditions of life of the majority of people in a beneficial manner to enhance their attainment of individual and collective material well-being.

Consequently, the global system is classified on the basis of the economic status of each component nation and region. The present global order is an interaction of three major economic worlds namely: the First World, the Second World and the Third World. On the basis of this global economic classification, the world is further stratified as 'developed', 'developing' and 'underdeveloped' with many African countries falling within the group of underdeveloped nations. Despite its enviable economic stature, the First World continues to strive for better economic positions. In what could be regarded as the survival of the fittest, the global economic space has been historically characterized by competition, exploitation and conspiracy, which calls for caution on the part of Africa and Third World nations in the execution of bilateral and multilateral

economic pacts.
Allan Gilpin (1995) defines the environment as the components of the earth and includes:

- Land and air including all layers of the atmosphere.
- All organic and inorganic matter and living organisms.
- The interacting natural systems that include components referred to in components above.

To corroborate the broadness of the above definitions, Gilpin argues that the word environment embraces the conditions of influences under which any individual or thing exists, lives or develops. For him, these conditions can be categorised into three.

1. The combination of physical conditions that affect and influence the growth and development of an individual or community.
2. The social and cultural condition that affects the nature of an individual or community.
3. The surroundings of an inanimate object of intrinsic social value.

These definitions imply interdependency and interaction between existents that inhabit the environment. The European Commission defines the environment as the "combination, the settings, the surroundings and the conditions of life of the individual and of the society, as they are or as they are felt". [9]

The natural environment constitutes a vital element in the process of development. It is a unique element in itself. It is immobile, and therefore incapable of any replacement, it can only be managed. The recurrent agitation for environmental responsibility is highly prudential and essentially obligatory. It is prudential in the sense that without improving environmental management, development may continually be elusive in poor countries, and the environment may continue to degrade. This conviction is encapsulated in the concept of environmentally sustainable development which presents Sub-Sahara Africa with the opportunity to view her development challenges in a new and different perspective, as well as to work out new solutions that are better adapted to the region.

The choice of this paper focusing on development and environmental sustainability is premised on the assumption that, all human activities are pursued within time and space. The environment constitutes the space within which other activities are executed. It could be regarded as the primary resource for human and natural existence. According to the Global Forum on Environment and Economic Growth (GFEEG), the links between the economy and the environment are manifold: the environment provides resources to the economy, and acts as a sink for emissions and waste. Natural resources are essential inputs for production in many sectors, while production and consumption also lead to pollution and other pressures on the environment. Poor environmental quality in turn affects economic growth and wellbeing by lowering the quantity and quality of resources or due to health impacts, etc. [10] Consequently, there is a relationship between the development model that is prevalent in the society and people's attitudes towards the environment. In other words, socio-economic factors determine the possibility of environmental sustainability.

To this extent, the main achievement of the World Commission on Environment and Development [2] was not so much that it revealed new insights; rather it placed environmental problems, issues and concerns on the agenda on the local, national, regional and international levels more forcefully than ever. In his article "Environmental Sustainability with Development: What Prospects for a Research Agenda", Harold Brookfield writes

> Without apology, I concentrate almost entirely on 'environmental sustainability', on the premise that the practice of development, especially but not only in the Third World, depends ultimately on the continued viability of the resource base. [11]

In the same vein, Brohman emphasised the centrality of environmental concerns in development and sustainability discourse in the following words:

> Worsening environmental problems have prompted a rethinking of the development agenda linking issues of sustainability to more traditional concerns for growth and equity. [12]

Environmental concerns are presently addressed at the local, national and international levels. In other words, frameworks for sustainable development are moving away from their prior technical fixation toward a more holistic focus which stresses the contextual specificity of environmental problems and includes a people-oriented agenda based on the needs and rights of local people. [12] The effects of rapid environmental change are being felt at all levels of human groupings. Nevertheless, concern for environmental change has until recently been confined to the margins. Hence, among other conferences, the Rio Earth Summit in 1992 enunciated that:

> Without improving environmental management, development will be undermined, and without accelerated development in poor countries, the environment will continue to degrade. [13]

3. Growth, Neo-Liberalization and Environmental Abuse in Africa

At the Bretton Woods conference, the intellectual shadow of the leading economic thinker of the age, John Maynard Keynes, had a great impact, and he made important contributions to its proceedings. Keynes' thought had come to exercise a profound impact on a generation of political leaders. His postulation for economic development was absorbed for both international and domestic economies. His idea of a smoothly running capitalist economy rested on a much greater role of the state than had been tolerated in classical and neo-classical methods of development, which had been more concerned with the free market. [14] To deal with recession, Keynes advocated the use of fiscal policy-government spending. By building roads and dams, for example, a government could create jobs, which in turn would create more demand for goods and services, which would cause factories to increase their output and thereby employ more labour, and so on in an upward spiral. Once the economy is revived, government could regulate it by taking money out of it. Put simply, Keynes' prescription for improving the capitalist economy was for "governments to save in good times, spend in bad." [14] He allowed governments to borrow, if necessary, to inject money into the economy. He reasoned that the loans would be repaid later from the earnings generated by a newly robust economy.

In the late 1940s, governments in Western Europe and North America started adopting Keynesian model. By then, the division of Europe into two blocs was becoming more consolidated. Capitalism only firmly reestablished its hold on Western Europe when United States instituted the Marshall Plan, whereby it injected billions of dollars into the reconstruction of Western Europe's ravaged infrastructure- a reflection of Keynesian thesis. It was a recipe for social peace like none seen before. Scholars at the time became so optimistic and proclaimed that correct economic management would prevent there ever being another Depression.

The idea that Third World development was a process and goal for which the First World especially the United States, should provide assistance grew out of the auspicious experience of the Marshall Plan in Europe, as well as Cold War fears and ambitions. [15] The mainstream framework of growth and modernization theories both arose in the context of Cold War competition between the superpowers for influences in the South (Africa). However, it was later realized that much of the mainstream development theory had served as instrument for American foreign policy and that it had ever been for her economic interest, more than anything else.

Growth theory as a school within development studies was popularized from the late 1940s to mid-1950s, after which it was gradually absorbed into the broader framework of modernization theory. It was based on the transfer of a series of Keynesian models for analyzing economic growth that had been developed in the United States and Europe to the Third World. The idea of development was conceived rather narrowly as economic growth. Social and cultural growth received attention only in their role of either facilitating or militating against appropriate societal changes that would accompany economic growth. Its intellectual roots being extensively based in Keynesian economics, growth theory was sometimes termed International Keynesianism. The Keynesian program of growth took a strongly interventionist position toward Third World development, which stressed comprehensive development planning by reformist states in collaboration with foreign donors.

Although growth theory rejected some aspects of neoclassical economics as inappropriate for Third World development, its models and theoretical framework nevertheless remained firmly rooted in Western economic history. This implied that growth theory was structured by a Eurocentric vision of development based on a Keynesian interpretation of the unique, albeit historically important, experience of core industrial capitalism. The 'new nations' of the post-colonial world, mostly in Africa were required to adopt the Western model in order to develop; this represented a 'modernization imperative' whereby countries pass through a sequence of stages on a unilinear path toward higher Western-style development. Most mainstream theorists in the 1950s and 1960s rejected the classical liberal emphasis on market forces and outward oriented growth as inadequate for the special Third World needs of rapid development, on the account of its inability to address the essentially dynamic problems of the Third World.

However, by the end of the 1960s, there was a shift to a more orthodox neo-liberal approach in major development sub-fields. By the late 1970s, the rise of anti-Keynesian conservatism in Europe and the US, and the seeming inability of growth theory to proffer solutions for the mounting Third World problems prompted a major change within mainstream theory as a whole. The political conditions at the same time favoured liberal theorists, and brought liberalism to the fore after a prolonged period at the margins of development discourse. The economic crisis experienced by many Third World nations in the late 1970s also provided additional impetus for neo-liberal counter-revolution.

Liberalism has evolved as a prominent ideology which determines and influences the contemporary global politics and economy. Hence, key to understanding the prominent economic features of the current global arrangement is the policy of neo-liberalization, which became prominent in the 1980s during the Reagan administration in the United States and the Thatcher government in the United Kingdom. [16] Neo-liberal economics mandate massive reduction in social spending and investment by the state, streamlining of government duties through job cuts, divestitures in public enterprises and the promotion of the private sector. The neo-liberal economic policy compels the state to divest public investment and participation in commerce. States' regulatory functions are often considered as "interference", thereby subjecting citizens to the unbridled power of the so called free and fair market. Multilateral institutions such as the World Bank and the International Monetary Fund (IMF) have been saddled with the responsibility of ensuring the implementation of the neo-liberal economic policy across the globe. Over the past three decades, neo-liberalism has moved from the fringes to the center of development studies, thereby displacing Keynesianism as the dominant development orthodoxy. To ascertain the influence of liberalism as an ideology and America as major economic actor on the global economic order, globalization has been described in some quarters as 'liberalization' and 'westernization'.

A major feature of the traditional neo-liberal development model is that it tends to focus singularly on the economic value of monetary transactions, i.e. Gross Domestic Product (GDP) at the exclusion of other progressive indicators such as social and environmental values. Growth, usually measured in terms of aggregate levels of production and consumption- typical of the neoliberal capitalist mindset- has been the prevailing measure of development. This Western vision of progress and modernization has commonly produced a pattern of development that ignores African traditional human-nature relations and inadequately addresses issues of social equity, ecological balance and overall sustainability.

From growth to modernization, and to neo-liberalization, mainstream development initiatives have applied 'nature-conquering' technologies on a massive scale, with devastating effects on many ecologically vulnerable Third World areas. Tropical ecosystems around the world are being destroyed at a rate of 25 million acres per year. [17] The relentless spread of large scale infrastructure projects has caused especially widespread damage to fragile environments and indigenous cultures in many countries. Many of these projects are typically initiated with the caption 'development needs', and are usually sponsored by foreign national governments. In some cases, the covert interests of some of these projects are sometimes elite groups (politicians, contractors, businessmen) linked to multinational capitals, whose interests are furthered through concerted lobbying within national and multinational development agencies. The poor indigenes of the affected communities are mainly excluded from the decision-making process while at the same time bearing the burden of the environmental cost of these projects. Moreover, economic growth . . . does not trickle down to the poor. An inequity aggravated by the fact that the economic wealth in the world so far is accomplished through unsustainable subsidies from the planet, while the cost for these subsidies are largely taken up by others, in general to poorest communities. [18] A case in point to illustrate environmental abuse by business corporations is the oil exploration activities by International Oil Companies (IOCs) in the Niger-Delta region of Nigeria.

The Niger Delta region is an epitome of oil wealth. It holds at least 20 billion barrel of oil reserves, while the Shell led-petroleum activities pumps about 2 million barrels of oil daily from the region. The region represents the core area where exhaustive production and exploratory activities take place. Activities of the IOCs in the Niger Delta region include extractions , drilling, earth-moving activities, deep-sea explorations, opening of roads, bush clearing, deployment of heavy equipment, digging of borrow pits, massive construction works which include petroleum terminals. There are pipelines network that criss-cross parts of the region for easy crude oil transportation.

The Niger Delta today inhabits about 31 million people of more than 40 ethnic groups with about 250 dialects. [19] The region is inhabited by Nigerian minority groups such as Itsekiris, Urhobos, Ijaws, Ilajes, Ogonis, Kalabaris', Efiks, Ikweres. Various reports have identified ways by which the IOCs' operational activities impact negatively on their host communities. First, there is serious implication for human displacements. As the IOCs expand their exploration activities to capture new oil opportunities, they embark on human displacements with tragic interferences and disruptions of human settlement equilibrium. Farmlands, houses and other settlements were made to be supplanted by erected oil facilities. Second, operations of the IOCs often result in massive oil spills across lands, forests and sea in the Niger-Delta region. Oil-related chemicals spill on available surfaces across the lands and water in the operating areas. When this happens, the spillage disrupts the eco-balance. In addition, water is polluted and the earth is de-fertilized for good planting. Third, oil companies flare about 1.7 billion standard cubic meters of gas.

The negative effects of gas flaring are unimaginable. The recurring gas flares close to Niger Delta host communities precipitate air pollution. [20] [21] It contributes to global warming. It also leads to many destructive air hazards. Fourth, exploratory activities also pose serious health implications to its host communities in the Niger Delta region. Apart from the fact that host communities are made to contend with polluted water and other related health risks; there is strong challenge of hydro-carbon contamination that over-hangs every inhabitant in Niger Delta region. Contrary to the enormous wealth that accrues to the national government of Nigeria and to the capitalist Multinationals, the majority of the population in the Niger-Delta region are forced to live with wanton poverty. Encroachments on farmlands and water ways due to oil spillage and massive water pollution have denied Niger Delta host communities major sources of livelihood. Oil spills make farm lands unfit for cultivation, available plants are easily exterminated while fishes are poisoned to death. By implications, exploratory activities have engendered and perpetuated personal and community poverty across Niger-Delta Region of Nigeria.

Apart from the large-scale economic and developmental projects, many other forms of economic practices and activities are responsible for environmental destruction especially in the poor communities of the Third World nations. Such is the case, for example with the Green Revolution in agriculture, which has been associated with social inequalities, groundwater depletion, soil erosion and degradation, chemical contamination and the reduction in biogenetic diversity in many countries. Among the most serious environmental problems arising from unsustainable agricultural practices has been the gradual deterioration of the soil base in many Third World areas- a phenomenon the Food and Agriculture Organization (FAO) has termed the ‘quiet crisis’. [22]

Either in the industrialised North or the developing countries of the South, the pursuit of economic development in line with the modern and neo-liberal growth models constitutes a major threat to development and environmental sustainability. It is reported that “over the last forty years in the United States, the genuine progress indicator (GPI) has been considerably lower than the Gross Domestic Product (GDP). This disparity indicates that while personal consumption may be going up, the inequalities, social malaise, and degradation of nature are sending the GPI lower”. [23] The Brundtland report traced how ecological devastation in many developing countries is amplified by poverty-induced behaviour and by the need to secure foreign exchange. Production of primary export commodities is often linked with environmental damage. The current debt situation and adjustment policies orchestrated by the neo-liberal capitalist multinational development agencies and creditors coerce countries to increase their production of export commodities with concomitant further pressure on the bio-physical environment. The pressure on the developing countries to survive economic hardships and the pressure on the rich industrialised countries to sustain and expand their economic power have resulted in the global scope of environmental abuses.

It has become increasingly clear that the effects of environmental degradation do not respect national borders. With the current knowledge about a number of global environmental challenges - a weakening ‘green lung’ capacity of tropical rain forests, perforation of the ozone layer and global warming, shrinking of bio-diversity base, and pollution of oceans with potential implications for oxygen production by plankton – it is pertinent to note that global cooperation and international collaborations are urgently needed to salvage the situation holistically. The following section of the paper attempts an evaluation of pertinent environmental ethical theories which serve as the basis for prescribing right conducts and behaviour for environmental sustainability, especially as it relates to African realities.

4. Environmental Sustainability: Ethical Evaluation

According to Mark Sagoff, “the environment is what nature becomes when we view it as a life-support system and as a collection of materials." [24] He explained that one can either think of the system in a mechanistic fashion, as a collection of materials with various physical and chemical interactions or in a more organic way, giving attention to the many ways in which the individuals are interdependent in their very nature. From the latter viewpoint, the environment stands related to the beings within it intrinsically. As such, the environment as an organic system takes on an even greater significance than from a mechanistic perspective. These viewpoints further influence the value people place on the environment.

One distinction that plays a significant role in environmental ethics is the distinction between intrinsic value and instrumental value. Things have intrinsic value when they have worth in themselves for some reasons. Things have instrumental value if their worth is dependent on their usefulness for some purpose and for people. Some environmentalists believe that plants and animals for example, have only instrumental value. They believe that trees are valuable because of their usefulness to man. Other environmentalists insist that plants and ecosystems have value in themselves in some sense. Another term being used in environmental ethics is "*prima facie*" value. This literally means something like "at first glance". Something has *prima facie* value if it has the kind of value that can be overridden by other interests or values. For example, the financial interests of a community could be valued and thus be considered as a *prima facie* value if it could be waived for stronger interests or values such as human health. These considerations regarding the nature and kinds of value play a crucial role in judging ethical issues and conducts relating to the environment. This is exemplified by the following perspectives in environmental ethics.

4.1. *Anthropocentrism:* As ethics of environmental sustainability, anthropocentrism prescribes a minimization of environmental abuse, ills and degradation as long as human good is effected in the process. For example,

> It is difficult to think of an action which would do irreparable damage to the environment or ecosystem but which would not also threaten human well-being . . . if a polluter dumps toxic waste in a river, this action could be said to be wrong . . . because the interests of the river are violated, but also . . . because there are human interests in having clean water. [25]

The basic assumptions in the above position are: first, that only human beings are morally significant persons. Second, that the duty humans owe the environment is an indirect one, since keeping the environment safe and healthy will guarantee human survival.

According to critics, anthropocentrism is the mistake of giving exclusive or arbitrarily preferential consideration to human interests as opposed to the interests of others. [26] It asserts that only humans have intrinsic value or worth. According to this ethics, those things are good that promote the interests of human beings. Thus, it is believed that animals are valuable in so far as they are useful to man in a variety of ways such as providing emotional, aesthetic, nutritional, clothing, entertainment and medical benefits. Anthropocentrism according to Midgley is simple human chauvinism, narrowness of sympathy comparable to national or race or gender chauvinism. [27] Barbara Mackinnon pointed out another difficulty with anthropocentric view in that sometimes, anthropocentric values conflict. [28] The reality of value conflicts has brought about the Cost-Benefit Analysis (CBA) idea which helps in guiding and directing assessment of interests and values.

Anthropocentrism in ethics derives its negative and misleading normative appellation when used as analogous of egocentrism. Anthropocentrism cannot simply be condemned for being human-centred if it is to perform the critical functions envisaged for it; since there are also respects in which human-centredness is unavoidable, unobjectionable or even desirable. [26] Human-centredness is unavoidable for the following reasons:

i. Every being's worldview is shaped and limited by the position of the being within the world. That is, all beings and species posit the kind of perspective in which they are at the centre of the scheme of things. Therefore, humans can only think as humans. In other words, 'perspectival anthropocentrism' would appear to be inescapable. [29]
ii. Humans, like other beings, have legitimate interests which they are justified to pursue.

Moreover, the desirability of human-centredness consists in that self-love, properly understood, can be considered a precondition for loving others, so, by analogy, it could be maintained that only if humans know how to treat their fellow humans decently will they begin to be able to treat species decently. [30] Human-centredness, in Hayward's analysis does not necessarily involve the exclusion of the well-being of other species, rather, serve to promote it.

4.2. *Biocentrism:* This is also known as extensionism of life-principle ethic, which by implication means that moral patiency should be extended beyond human interest to include other sentient and living things that can suffer pain and experience pleasure. The basis of this theory is that all existents have life as the common origin and characteristic. The advocates of this ethics are of two categories: those who argue that moral considerability be extended to all animals and plants; and those who hold that it must be restricted to animals only.

Albert Schweitzer in his *Reverence for Life* initiated the biocentric ethics. He emphasized that value be accorded to non-human forms of life including plants. According to him, life is the basic characteristic which all living things have in common and since life is sacred, then, it must be treated as such. It implies that humans have moral duty not just towards fellow human beings, but also to other forms of life. The following remarks succinctly capture Schweitzer's position:

> A man is ethical only when he obeys the compulsion to help all life that he's able to assist and shrinks from injuring any that lives. He also not asks how far this or that life deserves one's sympathy . . . and to what degree it is capable of feeling. Life as such is sacred to him. He tears no leaf from a tree, plucks no flower and takes care not to crush an insect. [31]

Taylor, in his *Respect for Nature*, as shown by Singer, advocated moral extension to all beings since they possess intrinsic value, therefore, they are purposeful. He reasoned thus,

> . . . Every living thing is pursuing its own good in its own unique way, once we see this, we can see all living things as we see ourselves and therefore we are ready to place the same value on their existence as we do on our own. [32]

Both Schweitzer and Taylor based their arguments on the principle of the value of life which commands one to respect life as the ultimate good. However, biocentrism commits the error it sets out to correct. It is an extreme position, since it could imply that both human and animals may not feed on plants and other existents in the ecosystem. This position may be too rigid and unrealistic for by this, the natural food chain will be disrupted. It negates the sustenance of life it is committed to preserve. Because of the apparent problems with this position, some of its proponents have attempted some sort of modifications. They argue that moral considerability should only be extended to animals or sentient beings since they have the capacity to experience pain and pleasure. This argument relies on the assumption that there is neither a distinct nor essential morally relevant differences between mankind and mankind's closest kin in the phylogenic scale. [33]

This modification seems to have failed; for the basis of the argument is unfounded and untenable. The restriction of experience of pain and pleasure to sentient existents is devoid of logical rigour. Experiencing pain and pleasure is a response to stimuli, which may be physical, psychological or physiological or sometimes combined. Plants also respond to stimuli. For instance, when a tree is exposed to excessive heat or burning at a close distance, it begins to respond to that experience while the leave begin to shrink and drop from the trunk. It is therefore evident that the condition is hostile to that plant. The point is that there is no morally relevant difference between a beast and the *Iroko* tree. The discrimination between animals and plants is morally unjustifiable just like racism. The question is to what extent can biocentrism address environmental challenges?

From Schweitzer's and Taylor's perspective, animals and plants will only be good for nothing. They will exist, die and rot. The logic is that if humans forbid their use in order to preserve their intrinsic value, then human beings will be passive and other existents will be redundant. It is obvious then that no problem will be solved. Although biocentrism claims to rely on the principle of life-ethic, it at the same time contradicts the natural process of life forms. It should be noted that decomposition, tear and wear, are natural realities that necessarily occur as natural processes. Human activities can only alter the rate or speed of these processes. Human activities are therefore encouraged to strive at attaining and maintaining natural equilibrium. Going by this version of biocentric environmental ethics, man would have perpetually remained in primitivity - no science, medicine, technology, education and development. There would not have been any difference between human and brute. The responsibility of man to the natural environment lies in his unparalleled ability to coordinate the affairs of other life forms and to curtail the inconveniences imposed by nature. Hence, a situation in which man is absolutely inactive and totally constrained from the use of various forms and levels of life for his immediate, remote, direct and indirect benefits can only remain at the level of hypothesis.

From Singer's position, biocentrism may also be inadequate because naturally there are some carnivorous animals (predators) that inflict pains and feed on other less powerful animals (prey). Should all animals who feed on other animals be exterminated so that pain will not be inflicted on other animals? Would this sort of extermination not lead to self-contradiction on the part of Singer? This version of biocentrism is inconsistent and inadequate, even as shown in Singer's argument where he admitted that the existence of carnivorous animals do pose one problem for the ethics of animal liberation. [34] Goodpaster stresses this inadequacy by rejecting the sentiency criterion put forward by Singer. He alleges that it is guilty of the same limitation and arbitrariness because if rationality or any other criteria for moral considerability proposed by the humanist is arbitrary and irrelevant to the status that it would confer, how is sentiency any less arbitrary and irrelevant. [35]

4.3. *Ecocentrism:* Ethicist of this tradition stress that it is whole systems or ecosystems that has intrinsic value and not humans only. This ethic is grounded on a metaphysics that is radically different from that of the preceding ethics. The traditional metaphysical bifurcation of man and nature has been the basis for moral conclusions in anthropocentrism and biocentrism. On the other hand, the basic presupposition in ecocentrism is that "no metaphysical cleavage can be made between man and animals [nature]". [36] An ecosystem is a whole of interacting and interdependent parts within a circumscribed locale. Ecosystems are a continuum of variations, a patchy mosaic with fuzzy edges. Some interactions are persistent, others occasional. [37] The ecosystems are loosely structured wholes. The boundary changes and some members come and go. There are varieties of relationship, sometimes there is competition within the whole; sometimes there is symbiosis. The need to survive pushes various creatures to be creative in their struggle for an adaptive fit. Although it is loose and decentralized, there is a unity to the whole.

IOP Conf. Series: Earth and Environmental Science **331** (2019) 012026 doi:10.1088/1755-1315/331/1/012026

Every section of life on human planet consists of a system of life, intricately interwoven and interdependent elements that function as a whole. It forms a basic pyramid with myriad smaller organisms at the base and gradually fewer and more complex organisms at the top. Plants depend on the soil, insects depend on plants, and other animals depend on insects. In this system, individual organism feeds off one another. Some elements come, others go, it is the whole that continues. The health or otherwise of the ecosystem is attributable to the whole system. Hence, "a thing is right when it tends to preserve the integrity, stability, and beauty of the biotic community. It is wrong when it tends to do otherwise". [38] The system has integrity because it is a unity of interdependent elements that combine to make a whole with a unique character. It has stability, not in that it does not change, but that it changes only gradually. Finally, it has specific beauty which reflects as a matter of harmony, well-ordered form, or unity in diversity. [39]

Morality becomes a matter of preserving this system or doing only what fits it. Hence, ecocentrism is an ethic that deals with humans' relations to land and to the animals and plants that grow upon it. [40] The standard of measuring morality becomes the effects human actions have on the ecosystem. An action is morally worthy if it produces good effects, guarantees the preservation, stability, integrity and the beauty of the biotic community. It is bad if it hinders the sustenance of the same. It therefore means that ecocentrism dictates caution on freedom of action in the struggle for survival, such that human freedom is not devoid of responsibility.

However, some criticisms have been raised against ecocentrism despite its inspirational tendency in advocating environmental responsibility. It is alleged of being consequentialist, since the rightness or otherwise of an action depends on how its consequence benefits or harms the environment. Hence, it fails in providing genuine traditional moral principle. But this criticism is not capable of invoking any major upset in ecocentric ethics because there is no basis for regarding traditional moral principle as a better standard than consequentialism. Moreover, the existence of a single moral principle *per se* is still the object of a hot debate in the field of ethics. Hence, the above criticism is a reflection of a philosophical prejudice.

Another objection states that ecocentric ethic cannot establish the fact that all things or beings in the ecosystem possess equal intrinsic value, and if it does, it fails to tell us the value of individual micro-organism or plant. Therefore, some scholars believe that it fails as a normative theory. They assert that it only fosters emotional inclination and awareness of environmental danger awaiting humankind and the need to change. If this is the case, then ecocentrism has made a laudable contribution by evoking a tremendous shift in human attitude from the selfish approach, such that humans are now conscious of their responsibility and obligation to the biotic community.

Those who claim that ecocentrism fails as a normative theory are not explicit enough, because the observations and objections they raised are not sound enough for such a conclusion. A theory is normative when it advocates a certain conviction that is considered morally praise worthy and fit for practice. The "ought to be" is always advocated in any normative theory, and the principle of universalizability is a key condition that makes a theory normative. Persuasions could be one of the efficient ways of changing the status-quo. Raising awareness is also an expedient effort in an effective campaign against environmental abuse. Hence, ecocentrism remains an ethical theory for environmental sustainability.

4.4. African Environmental Ethics: Africa, as a continent, has multiplicity of cultural practices and belief systems that make it difficult to generalize or sum up the entire African moral dispositions into a consistent whole without a sweeping generalization. To this extent, one of the Africans renowned scholars, Kwasi Wiredu once argued that there is no basis for an "African culture", except for the common oppression suffered under the Westerners and present indigenous misgovernment. [41] However, there are basic constitutive socio-cultural elements that cut across the continent of Africa. Some of these cultural practices and beliefs define the place and role of humans in natural community and prescribe norms that help to regulate human conducts in the society. Regarding humans and the natural environment, it is widely believed by African cultures that there is a sense of sacredness that characterizes the link and relationship between human beings and other beings and things in the world. These relationships are governed and monitored by some metaphysical forces, which ensure the natural order of things by allocating rewards and punishment. This belief has helped, among other things, to preserve social order and environmental values from generations to generations across many African societies.

To a large extent, the traditional normative principle upon which African land ethic is based, and which in fact may help to address the current ecological crises in Africa, is found in their belief in communalism and supernaturalism. Communalism and supernaturalism are fundamental world-views which provide the basis for almost all aspects of African life. Communalism, in this regard simply refers to the African social organization that sets community interests prior to individual interests. The traditional African social system is such that the diverse individual aspirations must align with the common good as defined by the community. Therefore, the achievement of communal objectives takes precedence over personal interests. In this regard, communal property is treated with respect and care by every member of the community. The traditional African idea of the community is a relationship of kinship that comprises of the unborn, the living and the dead. For instance, the Yoruba, a socio-cultural group in the South-West of Nigeria, Africa, view community as an integrated whole in which human and non-human forces and powers- physical and quasi-physical- interact in a mutually reinforcing manner. To corroborate this belief, Omari argued that "the important thing which united all African societies with regard to ownership of land was that land was considered a communal property belonging to both the living and the dead. Those ancestors who had lived on the land belonged to the social unit which owned and controlled the land and each individual who used the land felt a communal obligation for its care and administration before passing it to the next generation". [42] This metaphysical conception of the community by many African cultural groups is extended to their conception of communal assets which include their natural environment.

By supernaturalism, the Africans believe that there is a particular force and spirit in control of every existent and its position in nature. For example, it is the Yoruba belief that the world is not one, over which a person or group of persons can claim to have total control. It is not a place where an individual can assert his unbridled and unguarded freedom. Supernaturalism is the basis for the practice by many traditional African societies to reserve places such as forests, rivers and mountains as sacred abode of spiritual and quasi-physical existents. These secluded portions of their natural environment were accorded great respect and care as a matter of communal and individual obligations. The spiritual explanations provided for the obligation to respect and care for the environment was a communal element that the individuals have imbibed as basis for personal environmental responsibility. The ownership, allocation and control of land, forests and water resources all fell within the spiritual realm. Several forest phenomena; trees, rocks, mountains, pools, mermaids, snakes and large trees were made holy and conserved by cultural and spiritual design. [43] The affinity between the traditional African and his environment is captured in the statement "the forest is our skin and if one removes the skin of the human being, the end result is death". [44]

Theoretically, the African environmental ethical orientation is similar to the ecocentric thesis. The closest affinity between the two orientations is in the conception of the place of man within the biotic community and man's responsibility towards others in the natural environment. The traditional African environmental ethics commits human beings to the obligation to sustain the integrity, balance and beauty of the ecosystem. This commitment is reinforced by and reflected in the African agricultural practices, festivals and rituals, totems and taboos, industry, housing, etc. There were several hunting and farming rules that were generally observed by the hunters and farmers to ensure the sustainability of the natural environment. Although these laws were occasionally violated by few individuals, the laws were largely and easily obeyed because they appealed to people's cultural sensibilities. Therefore, it becomes very important to maximally engage cultural and indigenous resources in addressing environmental concerns both at the theoretical and practical realms; a proposition which is emphasized in the 'eco-development' approach to development.

5. Appraising Eco-development as Alternative Strategy for Development and Environmental Sustainability in Africa

Since the early postwar period, the central focus of the mainstream development strategies in Africa has been economic growth and the top-down diffusion of development impulses. The pattern of development was such that important decision-making was controlled by major international institutions in collaboration with local Third World (African) elites. In the execution of development projects, participation of the concerned people or communities was restricted to some hastily organized meetings in which foreign experts updated the people about the objectives and implementation of such projects. This approach to development had crippled the principles of self-help and community development in most part of Africa thereby inhibiting popular development.

In view of the above, alternative theory of development was born out of the realization that economic growth was not necessarily correlated with other development objectives such as rapid employment creation, eradication of poverty and inequalities, provision of basic needs and enhancement of human freedoms. The experience of the 1950s and 1960s suggested that, while growth was important, it was by no means a sufficient condition to induce broad-based development. It was noted that more than a decade of rapid growth in underdeveloped countries has been of little or no benefits to perhaps a third of their population. [45] Successful development should be measured not in abstract, aggregate growth indices, but according to other people-oriented criteria, such as the universal provision of basic needs, promotion of social equity, enhancement of human productive and creative capabilities through education, and most importantly, the capacity of communities to set and meet their own development goals. Thus, the themes that alternative theorists of development grapple with include: refocusing on human needs; new concepts of planning; participation and power; women and gender; environment and sustainability; popular development; among others.

Environmental considerations were given particular prominence within the alternative concept of 'eco-development', which was introduced by Maurice Strong at the 1972 Stockholm conference and was further elaborated and popularized by Ignacy Sachs. [46] The eco-development approach did not necessarily seek to halt the growth processes entirely, but called for industrialization and other development processes to be made more compatible with environmental sustainability through means such as the adoption of appropriate technologies, the encouragement of conservationist life-styles, and the use of bottom-up participatory planning approaches. Eco-development includes other elements such as harmonizing consumption patterns and lifestyles with environmental needs; using ecologically based productive systems; maintaining low energy profiles and promoting renewable energy bases; limiting depletion of non-renewable resources through recycling and other means; finding more socially and environmentally sustainable uses for existing resources and utilizing decentralized planning methods to encourage local participation. [47]

The alternative concept of eco-development is also devoid of any universal or abstract meaning of

IOP Conf. Series: Earth and Environmental Science **331** (2019) 012026 doi:10.1088/1755-1315/331/1/012026

development. As a strategy, it consists of specific elements- a particular people with their own values and needs, living in a distinct region with culturally specific resource. Eco-development contends that, in the real sense, development does not occur in a generalized form. It can only be of something specific, such as a certain eco-region with its own special conditions and needs. [47] Eco-development appreciates the reality of cultural differences and the contextual specificity of the various human groups. However, it is important to note the danger of implying the problematic cultural 'relativism' from this realistic cultural 'difference'. The former involves windowless ideas that cannot engage in dialogue and communication. [48] It could also be difficult to strip development of any universal character for this would mean denying it of any global relevance and discourse. It also implies its negation as a human phenomenon. Likewise, to affirm it only as a phenomenon of specific culture- ecoregion- involves a contradiction because it is possible to logically group the entire human race as a single social group. In this sense, we shall be talking of humanity as a single human culture.

Despite the possible misrepresentations of eco-development, it remains a strategically viable idea, calling on countries, especially in Africa, to be more self-reliant- to create strategies appropriate to their own ecological and cultural contexts rather than looking to the North and West for development solutions. This does not rule out collaborations and occasional foreign interventions. It advocates that development strategies should make use of the resources in a given region in ways that both sustain ecological system and provide for basic human needs. This strong normative component gives the eco-development approach a paradigmatic quality. [49]

To be truly sustainable, development must meet the needs of local people, because otherwise, many people will be compelled by necessity to take more from the environment than is advisable. Struggles over the environment often involve survival or the struggle to meet basic needs; the cost of individuals pursuing their own self-interests is often borne by the group. The appeal to idealism or altruism to protect the environment is questionable and unrealistic when households are forced to behave 'selfishly' in their struggle to survive. Hence, sustainable development should be linked with goals of distributional equity and social justice – within and between countries as well as generations. This underscores its inherently political nature. As the WCED (1987) argues, achieving sustainability implies basic political change in line with an alternative development agenda.

> The pursuit of sustainable development requires a political system that secures effective citizen participation in decision making. . . This is best secured by decentralizing the management of resources upon which local communities depend, and giving these communities an effective say over the use of these resources. It will also require promoting citizen's initiatives, empowering people's organizations, and strengthening local democracy. [50]

As it is contained in "Our Common Future", sustainability, particularly in the context of African continent, has a strong political element linked to the needs and interests of the people. This concern for meeting popular needs also inevitably gives sustainability an important livelihood component. This connection is most explicit in the concept of "sustainable livelihood security" developed by Chambers and Conway:

> [S]ustainable livelihood security [is] an integrating concept . . . livelihood is defined as adequate stocks and flows of food and cash to meet basic needs. Security refers to secure ownership of, and access to, resources and income – earning activates, including reserves and assists

> to offset risk, ease shakos and meet contingencies. Sustainable refers to the maintenance or enhancement of resource productivity on a long-term basis. [51]

The concept implies conserving natural resources and other environmental factors that are essential to people's livelihood. It includes a broader analysis of links between environmental sustainability, basic-needs provisions, and issues of power at the national and international levels. For all these reasons, eco-development represents a significant advance in our ability to conceptualize sustainability especially in Africa.

6. Conclusion

Arising from the arguments put forward in this article, one could safely submit that traditional mainstream development models are largely responsible for the massive global abuses that have brought about the various ecological challenges on humanity in general and Africans in particular. Under the neo-liberal model of development, the vast majority of the African population is alienated and enslaved to material greed. Human crave for consumption has been expanded with resultant pressure on limited natural resources. The high rate of environmental violations and non-conservationist tendencies could be blamed on excessive crave for material acquisition which is typical of neo-liberal economic system. In this regard, sustainable consumption pattern becomes desirable and inevitable for healthy environment.

For Africans, the traditional pattern of consumption used to align with the necessity to preserve the ecosystem. However, cultural dislocation which is endemic on the continent of Africa has led to the neglect of those traditional values that have hitherto sustained prosperous economy and healthy environments on the continent. Although, absolute reliance on appeal to traditional pattern of consumption may not be applicable in this age, national policies and legal frameworks should be formulated and implemented to ensure sustainable household, corporate and public consumptions patterns. Some of these policies which could be reviewed and adopted for regulations include: the Energy Labelling Directive [52], the Ecolabel Regulation [53], the Green Public Procurement Communication [54], etc.

Therefore, meaningful improvement in human attitude towards the environment, especially in Africa, is possible when there is a paradigm shift in dominant definition of and approach to development. The redefinition of the concept of development will imply a reassessment of development priorities in a way that the neo-liberal economic error is not perpetuated. The error which proceeds from reducing development to economic progress, translates to fundamental human problems as captured below:

> In the pursuit of economic development, human beings have abused the biosphere in their way to economic ascendancy; they have violated the peaceful cohabitation of their fellows, as in slavery and the slave trade. The blind pursuit of economic growth has brought the current calamitous conditions upon them. Environmental degradation, social breakdown, moral decadence, psychological turmoil, cultural dislocation, insecurity, and economic depression are some of the problems that threaten their [human] very existence. [55]

In a nutshell, green and innovative energy solutions are necessary requirement for the realization of the desired sustainable development across Africa but they are not sufficient without corresponding fundamental changes in the African socio-economic and political systems.

References

[1] Pearce, D.W. and R.K. Turner. 1990. *Economics of Natural Resources and the Environment*. (United Kingdom: Harvester Wheatsheaf). 24

[2] World Commission on Environment and Development 1987. *Our Common Future*. (Oxford: Oxford University Press).

[3] Walter, Rodney. 1982. *How Europe Underdeveloped Africa.* (Washington D.C.: Howard University Press). 3

[4] Presbey, M. Gail. 2000. *The Philosophical Quest: A Cross Cultural Reader.* 2nd edition. (Boston: McGrawHill Higher Education. 163

[5] Walter, Rodney. 1982. *How Europe Underdeveloped Africa.* (Washington D.C.: Howard University Press). 4

[6] Omoregbe, Joseph. 1993. *Ethics*. (Lagos: Joja Educational Research Publishers). 26

[7] Adeigbo, F.A. 1994. *Readings in Social and Political Philosophy*, **2** (Ibadan: Claverianum Press). 5-6

[8] Cobb, Clifford W., John B. Cobb, and Human Economy Center. 1994. *The green national product: A proposed index of sustainable economic welfare*. (Lanham, MD: University Press of America). 9

[9] Allan, Gilpin. 1995. *Environmental Impact Assessment Cutting Edge for the Twenty First Century.* (Cambridge: Cambridge University Press). 1

[10] OECD. 2016. Towards quantifying the links between environment and economic growth. *Global Forum on Environment.* https://www.oecd.org/economy/greeneco/global-forum-on-environment-2016.htm accessed on 30, May 2019.

[11] Stokke, Olav. (ed.) 1991. *Sustainable Development.* (London: Frank Cass & Co.). 42

[12] Brohman, John. 1996. *Popular Development: Rethinking the Theory and Practice of Development.* (Cambridge: Blackwell Publishers Inc.). 305

[13] World Bank. 1996. Towards Environmentally Sustainable Development in Sub-Sahara Africa: A World Bank Agenda. Washington D.C. 1

[14] Rapley, John. 1997. *Understanding Development: Theory and Practice in the Third World.* London: University College Press). 7

[15] Black, J. 1991. *Development in Theory and Practice: Bridging the Gap*. (Boulder: Westview). 48

[16] Ajayi, O. Oluwaseun. 2005. “Globalization and the Politics of Marginality”. *Globalization and Marginalization*, edited by Olufemi Vaughan, Marcheta Wright, and Charles Small. (Ibadan: Sefer Books). 204

[17] Black, J. 1991. *Development in Theory and Practice: Bridging the Gap.* (Boulder: Westview). 120

[18]Johan Rockstrom. 2017. 5 reasons why the economy is failing the environment, and humanity. *World Economic Forum.* https://www.weforum.org/agenda/2017/01/5-reasons-why-the-economy-is-failing-the-environment-and-humanity/ accessed on 30, May 2019.

[19] Adekeye, D.O. 2014. "Emerging Identity, Social Values and the New Social Order in the Niger Delta". *Africana.* African Studies Center, Boston University, USA. **7** 61-95

[20] Ndubuisi, O.L & Aisha, I.O. 2007. Environment Pollution in Oil producing area of the Niger Delta Basin, Nigeria. Empirical Assessment of Trends and Peoples' Perception. *Environment Research Journal.* **1**18 – 26

[21] Aniefiok, E. Ite and Udo, J.I. 2013. Gas Flaring and Venting Associated with Petroleum Exploration in the Niger Delta Region. *America Journal of Environmental Protection* **1**(4)70 – 77

[22] Mackenzie, F. 1993. *Exploring the Connection: Structural Adjustment, Gender and the Environment.* Geoforum. 71

[23] Jason, Venetoulis. 2010. "If the GDP Is Up, Why Is the GPI Down? From Normative Debate to Progressive Democratic Economics". *Sustainable Development: Principles, Frameworks and Case Studies.* Edited by Okechukwu Ukaga, Chris Maser and Mike Reichenbach. (New York: Taylor and Francis Group). 2

[24] Sagoff, Mark. 1993. *Population, Nature and the Environment.* Report from the Institute for Philosophy and Public Policy, **3**(4). 10

[25] Shrader-Frechette, K.S. 1981. *Environmental Ethics.* (California: Pacific Grove). 17

[26] Hayward, Tim. 1997. "Anthropocentrism: A Misunderstood Problem." *Environmental Values.* **6**. 49-63 (Cambridge: The White Horse Press). 51

[27] Midgley, Mary. 1994. "The End of Anthropocentric?" in *Supplement to Philosophy.* Royal Institute of Philosophy. **36** 11

[28] Mackinnon, Barbara. 1998. *Ethics: Theory and Contemporary Issues.* 2nd edition. (Belmont: Wadsworth Publishing Company). 323

[29] Ferre, Frederick. 1994 "Personalistic Organicism: Paradox or Paradigm? *Philosophy and Natural the Environment.* Edited by Robin Attfield and Andrew Belsey. (Cambridge: Cambridge University Press).

[30] Hayward, Tim. 1997. "Anthropocentrism: A Misunderstood Problem." *Environmental Values.* **6** 49-63 (Cambridge: The White Horse Press). 52

[31] Schweitzer, Albert. 1965. *The Philosophy of Civilization.* (New York: Wiston). 49

[32] Singer, Peter. 1979. *Practical Ethics.* (Cambridge: Cambridge University Press). 279

[33] Callicott, J. Baird. 1986. *The Search for an Environmental Ethic in Matters of Life and Death: New Introductory Essay in Moral Philosophy.* 2nd edition, (New York: Random House Inc.). 398

[34] Singer, Peter. 1975. *Animal liberation: a New Ethics for our Treatment of Animals*. (New York: Avon). 238

[35] Goodpaster, Kenneth. 1978. "On Being Morally Considerable". *Journal of Philosophy*, **75** 308-325

[36] Rollin, B.E. 1981. "The Metaphysics of Anthropocentrism: A Review of Paul Ehrenfield's "Arrogance of Humanism" and Mary Midgley's "Beast and Man". *International Journal for the Study of Animal Problems*, **2**(2) 75-80

[37] Rolston, Holmes. 1988. *Environmental Ethics: Duties to and Values in the Natural World*. (Philadelphia: Temple University Press). 169

[38] Aldo, Leopold. 1949. "The Land Ethic". *Sand County Almanac*. (New York: Oxford University Press). 262

[39] Hospers, John. 1982. *Understanding the Arts*. (Englewood Cliffs: Prentice-Hall).

[40] Singer, Peter. 1975. *Animal liberation: a New Ethics for our Treatment of Animals*. (New York: Avon). 280

[41] Wiredu, Kwasi. 1995. *Conceptual Decolonization in Africa: Four Essays*. Selected and Introduced by Olusegun Oladipo. (Ibadan: Hope Publications). 13

[42] Omari, C.K. 1990. "Traditional African Land Ethics". *Ethics of Environment and Development: Global Challenge, International* Response. J.R. Engel and J.G. Engel (eds.). (London: Belhaven Press). 162-175

[43] Kwashirai, V. 2010. *Green Colonialism in Zimbabwe: 1890-1980*. (New York: Cambria Press). 398

[44] Bujo, Benezet. 1998. *The Ethical Dimension of Community*. (Nairobi: Paulines Publications). 217

[45] Chenery, H. et al. 1974. *Redistribution with Growth. (*New York: Oxford University Press). xiii

[46] Sachs, I. 1974. "Ecodevelopment". *Ceres,* **17(**4) 9

[47] Glaeser, B. and V. Vyasulu. 1984. "The Obsolenence of Ecodevelopment?" *Ecodevelopment: Concepts, Prospects and Strategies*, edited by (Glaeser B. Oxford: Pergamon). 25

[48] Wiredu, Kwasi. 1993. "Canons of Conceptualization." *The Monist: An International Journal of General Philosophical Inquiry*, **76**(4) 450-476

[49] Hettne, B. 1990. *Development Theory and the Three Worlds*. (Essex: Longman). 187

[50] World Commission on Environment and Development 1987. *Our Common Future*. (Oxford: Oxford University Press). 63

[51] Chambers, R. and Conway G. 1992. *Sustainable Rural Livelihood: Practical Concepts for the 21st Century Institute for Development Studies Discussion*. **296** 7

[52] EU 2010, *Directive 2010/30/EU* of the European Parliament and of the Council of 19 May 2010 on the indication by labelling and standard product information of the consumption of energy and other resources by energy-related products (recast), OJ L 153, 18.6.2010, pp. 1–12

[53] EU 2009, *Regulation (EC) No 66/2010* of the European Parliament and of the Council of 25 November 2009 on the EU Ecolabel, OJ L 27, 30.1.2010, pp. 1–19

[54] EC 2008, *Communication from the Commission to the European Parliament*, the Council, the European Economic and Social Committee and the Committee of the Regions Public procurement for a better environment, COM/2008/0400 final.

[55] Adekeye, D.O. 2014. "An Alternative Platform for Sustainable Development: A Social Reform Agenda for Nigerian Intellectuals". *Africa Today,* African Studies Program, Indiana University, USA. **60**(3) 102-117

Renewal and Upgrading System For A Sustainable Urban-Rural Housing System Development: Panacea To Accommodation, Employment And Healthcare Issues.

Amusan, L. M,[1] Akanya C.N[2], Adeyemo, K. A,[3] Ayo, C. K.[4] Owolabi J.D[5], Dabo Baba Mohammed[6]
1,2,4&5. Building Tech. Dept. Covenant University, Ota, Nigeria. lekan.amusan@covenantuniversity.edu.ng
2.cinwon@yahoo.com
3. Lead City University, Ibadan. Nigeria vc@lcu.edu.ng
4. Computer Science Dept. Covenant University. Ota. Nigeria. charles.ayo@covenantuniversity.edu.ng
5. Building Tech. Dept. Covenant University. faithy012@yahoo.com
6. Technical Education Dept. Aminu Saleh College of Education. Azare Bauchi state. Nigeria.dbhammad@gmail.com

Abstract: Developments in the Urban communities has become issue of great concern. Urban development has been rapid with urban centres springing up from rural communities and sometimes creating an imbalance in environmental systems with attendant consequences and advantages. The study therefore explored the urban system housing development within the context of formulating framework for development, renewal and upgrading in order to achieve sustainable development in urban and rural communities. The aim of the study was to carry out a longitudinal survey of selected areas where there is occurrence of concentration of rural-urban, urban-rural settlement within the selected study areas. Survey was conducted on some parameters as part of the calibrated questionnaire that was designed in Likert Scale 1-5 used for the study. The parameters include evaluation of existing renewal and upgrading pattern, upgrading system, renewal strategies, features of rural-urban growth, challenges of urban renewal and regenerations, influence of urban and rural upgrading system on employment availability, health challenges associated with urban-rural settlement and post occupancy study of satisfaction level of residents of urban and rural settlement housing facilities.

The study engaged population of 5 different locations with 100 residents drafted from various selected locations classified as urban and rural settlement as respondents. The data was analysed using Mean Item Score, Simple percentages and regression analysis. Some of the striking contribution in this study includes: System to formulate good renewal and resuscitating, it also include resuscitating decayed component of rural setting at urban location. The study also contribute to knowledge in advocating strategies for urban and renewal strategy like compilation of plans for physical development of the area, development of identified locations, re-accommodation and resettlement of displaced settlements, re-accommodation and resettlement of displaced settlements and reconstruction of the areas devastated by development programmes. Similarly, rural and urban development would bring employment opportunity for all classes of trade either skill labour, un-skill labour and semi-skill labour. Construction activities would attract and provide fortune to different workers category there creating prosperity for all. Finally, the study developed a system framework that could be used for urban and rural settlement regeneration, revitalization, redevelopment and reconstruction. The developed framework would assist in eliminating accommodation challenges, employment problems and health care issues often associated with disparity in settlement growth when rural settlement metamorphosed into urban settlement.

Key words: Settlement, System, Sustainability, Transition.

1.0 Introduction

Modern day urbanism has led to encroachment on existing development on affected areas. Modernization of community is essential for some attendant desirable effects that come with it. Community modernization however in recent times has brought about introduction of modern houses in the midst of poor communities and traditional communities. This phenomenon has led to urban development and resuscitation. Urban development has been notably adjudged as catalyst of rural-urban metamorphosis because it is often lead to economic enlightenment and expansion. In rural-urban metamorphosis, slums are often found in the midst of modern development with no particular development pattern. Sometimes, there is a need for urgent intervention to be adopted. Some of urgent intervention that could be deployed to correct the imbalance include regeneration of rural section, restoration of destroyed structure and revitalization for setting up formal structure in such environment. It is on this note that this study explored the urban system housing development with a view to propose a framework for creating a sustainable urban-rural and rural-urban development. However there is a need to formulate adequate strategy to be able to have a successful rural urban development. Some of important parameters that could be used include: formulating adequate urban and rural developmental plan, discovering challenges and barrier to effective development and most importantly developing a robust framework that could guide in successful rural-urban and urban-rural development. Similarly, rural and urban development would bring employment opportunity for all classes of trade either skill labour, un-skill labour and semi-skill labour. Construction activities would attract and provide fortune to different workers category thereby creating prosperity for all.

1.1 Urban Renewal And Upgrading Theory

There are number of theories that underpins urban-rural and rural-urban renewal and upgrading. According to [1], some of the theories includes economic theory, behavioural theory and spatial theory among others. Spatial theory borders on allocation of space in an ergonomic ways during design and configuration stage of construction projects. The theory stipulates regulation as regards allocation of land spaces to owner and occupiers. The occupier is restricted in term of percentage allocation while the right of displaced land owners remaining on the claimed land is protected. Also, Behavioural theory delineates the scope of owner right or resident's right. The appropriate behaviour is to allow residents of area to be upgraded or renew to have input indecision made in relation to the land. The behaviour expected should be sympathetic and apathy in nature towards displaced or those about to be displaced. The shock of the relocation by the resident should be shared under behavioural theory.

Furthermore, Economic theory that impacts micro and macro-economic aspect of communities. The micro economic theories includes the influence of forces of demand and supply, mobility of labour, price movements and factors of productivity. The macro on the other hand has to do with interplay of forces of Inflation, deflation, balance of trade, balance of payment and gross domestic operation among others. There tendency for demand for certain commodity to rise at locations where rehabilitation, reconstruction and regeneration just occurred. Also, the cost implication of rising house demands ay be daunting in scope to the extent that local resident of the displaced community on account of developmental programme may not be able to concede. Therefore, immediate need of local residents should be proactively taken into consideration while planning renewal and upgrading of a location.

1.2 Urban and rural housing development systems

Urban and rural development initiative could be viewed in two perspectives, the demand for housing vis a vis dynamics of urban space vacancy and dynamics of regeneration and revitalization. Urbanization is usually associated with space demand, and the demand sometimes could be urgent, the urgency of the demand often necessitates evaluation of current residents at the time of initiating the urban development, for the purpose of setting up necessary framework. The framework is necessary in order to accommodate all necessary requirements and needs that would be residents oriented in term of their compensation package. In rural housing development type of space usually in demand is Commercial accommodation that is, office and business related spaces than residential accommodation. However, regeneration is the concept that is usually being used to describe emergence of civilized environment in the midst of slum and less privileged society. According to [2], in a study titled ''Urban regeneration and crisis, described the concept of regeneration in another form, as urban renaissance, redevelopment, reconstitution of society profile, resurgence of decay spaces, urban renewal and urban revitalization. Also, [3a]described urban regeneration succinctly using 7 concepts, economics, sociological content, spatial allocation, economic orientation, environmental approach, economic approach, socio-economic content and social integration. Holistic interpretation of the concepts would undoubtedly provide an access to decoding the myth surrounding failure in developmental programme implementation in the developing world.

2.0 Review of related literatures

Some selected works that treated issues that are related to Urban renewal and regeneration were reviewed and presented in this section. Some of the issued covered in the works include urban regeneration, urban renewal, security issues involved in urban and rural upgrading among other issues addressed.

For instance, Silvia Saccomani [3b], carried out a study on Urban regeneration and crisis. The study used urban regeneration concept in developing some section of selected locations in Turin Italy. The study identified consequences that could occur on account of renewal. The study advocated greater commitment from the public administration so as to deal with attendant social exclusion and cohesion challenges.

In another related study,[4] identified significant urban features that could be used for urban renewal initiative. The study adopted questionnaire survey on 30 respondents that are building professionals. The study recommended that opinion of residents should be duly censored before embarking on any development programme in the location of residents, this would guide against developmental error of judgement which could jeopardize future related development.

Similarly, [1] conducted a research on Urban renewal and Security issues involved. The study recognize that urban renewal can bring about redevelopment, rejuvenation, revitalization and resurgence. The study recommend that packages should be in place to mitigate the attendant negative outcome of the developmental programme, they include; insurance package on account of sudden relocation, eviction pension, eviction grant and provision of social welfare fund among others.

Finally, [5] worked on Urban regeneration of deprived neighbourhood in a selected metropolitan region of Essen in Finland. The aim of the study was to bring to the fore details about conversion of deprived neighbourhood into a healthy communities. The study recommended that urban planners should focus more on supporting healthy community and sustainable communities through their policy planning and formulation.

3.0 Methodology of research

Survey was conducted on some parameters as part of the calibrated questionnaire that was designed in Likert Scale 1-5 used for the study. The parameters include evaluation of existing renewal and upgrading pattern, upgrading system, renewal strategies, features of rural-urban growth, challenges of urban renewal and regenerations, influence of urban and rural upgrading system on employment availability, health challenges associated with urban-rural settlement and post occupancy study of satisfaction level of residents of urban and rural settlement housing facilities.

The study engaged population of 5 different locations with 100 residents drafted from various selected locations classified as urban and rural settlement as respondents. The data was analysed using Mean Item Score, Simple percentages and regression analysis. The study used questionnaire survey to collate information from respondents. The collated data was examined using content analysis method by classifying the information and categorised them under appropriate headings. Data was collated on parameters that include evaluation of existing renewal and upgrading pattern, upgrading system, renewal strategies, features of rural-urban growth, challenges of urban renewal and regenerations, influence of urban and rural upgrading system on employment availability, health challenges associated with urban-rural settlement and post occupancy study of satisfaction level of residents of urban and rural settlement housing facilities. Mean Item Score was used with relative agreement index, this was used to rate the perception of respondents on parameters calibrated for respondents to answer. Finally, the parameters were collated and presented in tables and charts, while factor analysis was used to reduce the data to sizeable number that could be represented by group of seven factors. The seven factors formed the component of framework develop in this study for urban and rural regeneration and development.

3.1 Presentation of data and analysis

3.1.1 Evaluation of Existing Renewal and Upgrading Agenda,

Table 1 Existing Renewal and Upgrading Agenda

S/N	Existing Renewal and Upgrading Pattern	R.A.I	Rank
1	Resuscitating Decayed Component of rural setting at urban location	0.872	1
2	Relocation of displaced residents of rural component upgraded to Urban standard	0.872	1
3	Clearing of deteriorated components of urban settlement	0.863	3
4	Remodification of obsolete structures in areas to be renewed	0.862	4
5	Compensating residents of displaced component of area under renewal and upgrading	0.752	5
6	Replacing old structures with modern structures	0.859	5
7	Removal of structures that cannot be remedied from renewal and upgrading sites.	0.856	6

In Table 1 existing and renewal and upgrading pattern was presented, it contained issues on evaluation of existing renewal and upgrading agenda was presented. Some of the agenda of renewal and upgrading agenda include resuscitating decayed component of rural setting at urban location, relocation of displaced residents of rural component upgraded to urban standard, clearing of deteriorated components of urban settlement, remodification of obsolete structures in areas to be renewed and compensating residents of displaced component of area under renewal and upgrading.

Renewal and upgrading agenda include resuscitating decayed component of rural setting at urban location and relocation of displaced residents of rural component upgraded to urban standard were ranked 1st with RAI value of 0.872, clearing of deteriorated components of urban settlement was ranked 3rd with RAI of 0.863, remodification of obsolete structures in areas to be renewed was ranked third with RAI value of 0.862 while

IOP Conf. Series: Earth and Environmental Science **331** (2019) 012027 doi:10.1088/1755-1315/331/1/012027

compensating residents of displaced component of area under renewal and upgrading with RAI value of 0.752 was ranked 5th. Relocation of residents and upgrade of a location should be priority of developmental programme of a community, also clearing of deteriorated component of a community should also be a centre of focus of a community. This would ensure continuity of developmental programme in any community.

3.1.2 Upgrading System Renewal Strategies,

Table 2 Upgrading System Renewal Strategies

S/N	Upgrading and renewal Parameters	R.A.I	Rank
1	Compilation of plans for physical development of the area	0.875	4
2	Profiling and classification of features to identify urban-rural and rural-urban location	0.882	2
3	Presentation of adequate design and codes for residents relocation from area to be upgraded	0.881	3
4	Feasibility survey of economic, socioeconomic and demographic requirements of the area under upgrading and renewal	0.887	1
5	Development of identified locations	0.790	5
6	Reconstruction of the areas devastated by development programmes	0.754	7
7	Re-accommodation and resettlement of displaced settlements	0.763	6

In Tables 2 Upgrading systems and strategies that could be used in urban and renewal system was presented. Some of the strategies includes. Feasibility survey of economic, socioeconomic and demographic requirements of the area under upgrading and renewal, Profiling and classification of features to identify urban-rural and rural-urban location, Presentation of adequate design and codes for residents relocation from area to be upgraded, Compilation of plans for physical development of the area, Development of identified locations, Re-accommodation and resettlement of displaced settlements, Re-accommodation and resettlement of displaced settlements and Reconstruction of the areas devastated by development programmes.

Feasibility survey of economic, socio-economic and demographic requirements of the area under upgrading and renewal was ranked 1st with RAI of 0.887, Profiling and classification of features to identify urban-rural and rural-urban location was ranked with RAI of 0.882, Presentation of adequate design and codes for residents relocation from area to be upgraded with RAI 0.881 was ranked 3rd, Compilation of plans for physical development of the area with RAI value of 0.875 was ranked 4th Also, Development of identified locations was ranked 5th with RAI value of 0.790, Re-accommodation and resettlement of displaced settlements ranked 6th with RAI value of 0.763, Re-accommodation and resettlement of displaced settlements was ranked 6th with RAI value of 0.763 while Reconstruction of the areas devastated by development programmes was ranked 7th with RAI value of 0.754.

3.1.3 Features of Rural-Urban Growth

Table 3 Characteristics of Rural-Urban Growth

S/N	Features of Rural-Urban Growth	R.A.I	Rank
2	Upgrading of slum settlements to urban standard	0.889	1st
1	Instituting Renewal development activities	0.873	2nd
3	Afforestation of overcrowded urban settlement for air mass control and ecosystem balance	0.765	3rd
4	Revitalization of desertified and eroded locations of occupied settlements in urban-rural and rural-urban location	0.689	4th
6	Reconstruction of dilapidated section of rural settlement within areas upgraded areas	0.667	5th
5	Redevelopment of components of rural-urban and urban-rural locations	0.657	6th
7	Reinvigoration of amenities and services in slum locations of urban components to urban standards	0.579	7th

Features of Rural and urban growth was presented in Table 3. It involves items that needs to be taken into consideration while upgrading settlements, it forms the crux of items that stimulates the planning and organization of policy for upgrading and development. In Table 3 the following order was presented. Upgrading of slum settlements to urban standard was ranked as 1st with RAI value of 0.889, Instituting Renewal development activities ranked first with RAI value of 0.873, Afforestation of overcrowded urban settlement for air mass control and ecosystem balance ranked 3rd with RAI value of 0.765, while Revitalization of desertified and eroded locations of occupied settlements in urban-rural and rural-urban location with RAI value of 0.689 was ranked 4th. Similarly, Reconstruction of dilapidated section of rural settlement within areas upgraded areas was ranked 5th with RAI value of 0.667, Redevelopment of components of rural-urban and urban-rural locations was ranked 6th with RAI values of 0.657 while Reinvigoration of amenities and services in slum locations of urban components to urban standards was ranked 7th with RAI value of 0.579. The main aim of any developmental programme is to renew the environment there upgrading the environment would be a matter of priority. Also, for any government or local authority that worth its onion would take instituting renewal and developmental programme. Also, afforestation of community landscape would reduce or eliminates the erodibility effect on the landscape of community land mass. The afforestation would ensure creating balance in ecosystem component of such community [6], [7] and [8]

3.1.4 Challenges of Urban Renewal and Regeneration

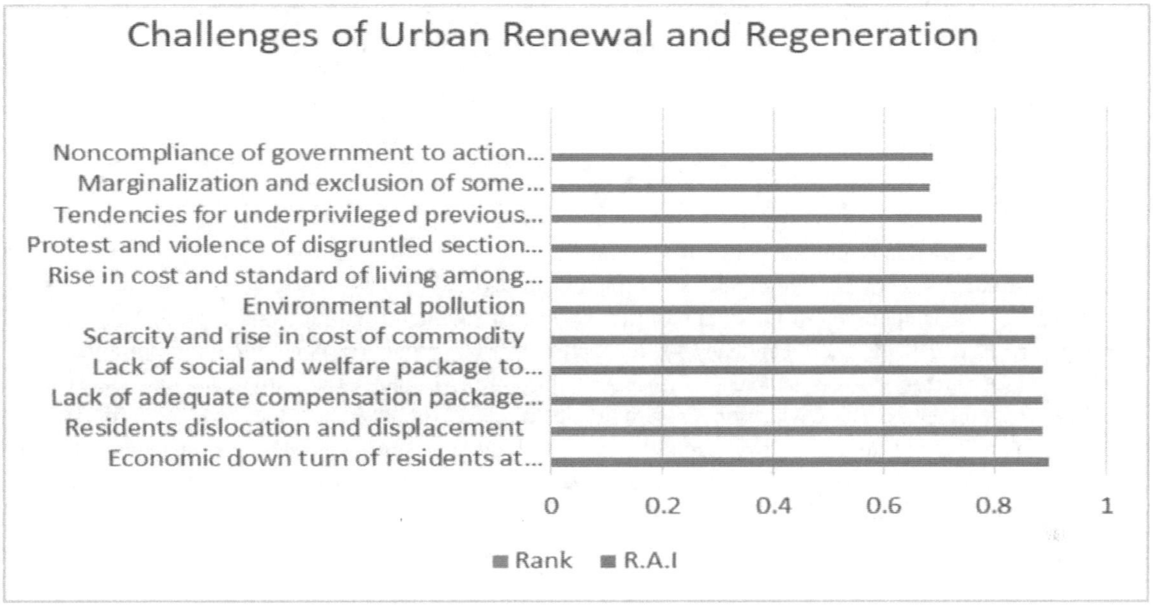

Fig 1. Challenges of Urban Renewal and Regeneration.

Figure 1 contains challenges that could confront development that involves Urban renewal and regeneration. The challenges were evaluated and summarized in Figure 1. Economic down turn of residents at displaced locations with RAI value 0.897 was ranked 1st , Residents dislocation and displacement, Lack of adequate compensation package and insurance for displace residents, and Lack of social and welfare package to cushion the effect of eviction with RAI values 0.886 were ranked 2nd respectively in the order presented. Also, Scarcity and rise in cost of commodity withRAI value 0.873 was ranked 4th while Environmental pollution RAI 0.872 and Rise in cost and standard of living among residents with RAI 0.786 were ranked 5th respectively. Also, Protest and violence of disgruntled section of unsettled residents of the displaced locations with RAI 0.786 was ranked 6th , Tendencies for underprivileged previous land owners in displaced locations to become tenants with RAI of 0.778 ranked 7th, Marginalization and exclusion of some residents of developed location with RAI value 0.685 was ranked 8th while Noncompliance of government to action plan for resettlement and compensation plan for likely displace residents of developed location with RAI value 0.689 was ranked 9th.

3.1.5 Influence of Urban and Rural Upgrading System on Environment

Table 4 Influence of Urban and Rural Upgrading System on Environment

S/N	Influence of Urban and Rural Upgrading System on Environment	R.A.I	Rank
1	Tendency for resources not to be made available in natural but unnatural forms on account of urbanism	0.889	1st

2	Enhanced employment opportunity	0.779	2nd
3	Reinvigoration of settlement structures of upgraded locations	0.768	3rd
4	Disruption of ecosystem and environmental biomass	0.768	3rd
5	Tendency for erosion attack on vulnerable section of environment	0.759	5th
6	Health associated challenges may occur in term of pollution	0.734	6th
7	Job or employment scarcity	0.645	7th

Influence of Urban and Rural Upgrading System on was presented in Table 5. The influence of urban and rural upgrading system on communities sampled was used as litmus test to the expected as regards the influence of areas upgraded in the communities. Some of the identified factors are presented with rankings using their RAI. For instance, Tendency for resources not to be made available in natural but unnatural forms on account of urbanism was ranked 1st with RAI value 0.889, Enhanced employment opportunity with RAI 0.779 was ranked 2nd, Reinvigoration of settlement structures of upgraded locations with RAI 0.768 and Disruption of ecosystem and environmental biomass and Disruption of ecosystem and environmental biomass were ranked 3rd with RAI of 0.768 respectively. Also, Tendency for erosion attack on vulnerable section of environment with RAI 0.734 was ranked 6th while Job or employment scarcity with RAI value 0.645 was ranked 7th i.e. employment availability, health challenges associated with urban-rural settlement

3.1.6 Post Occupancy Study of Satisfaction Level of Residents of Urban and Rural Settlement Housing Facilities.

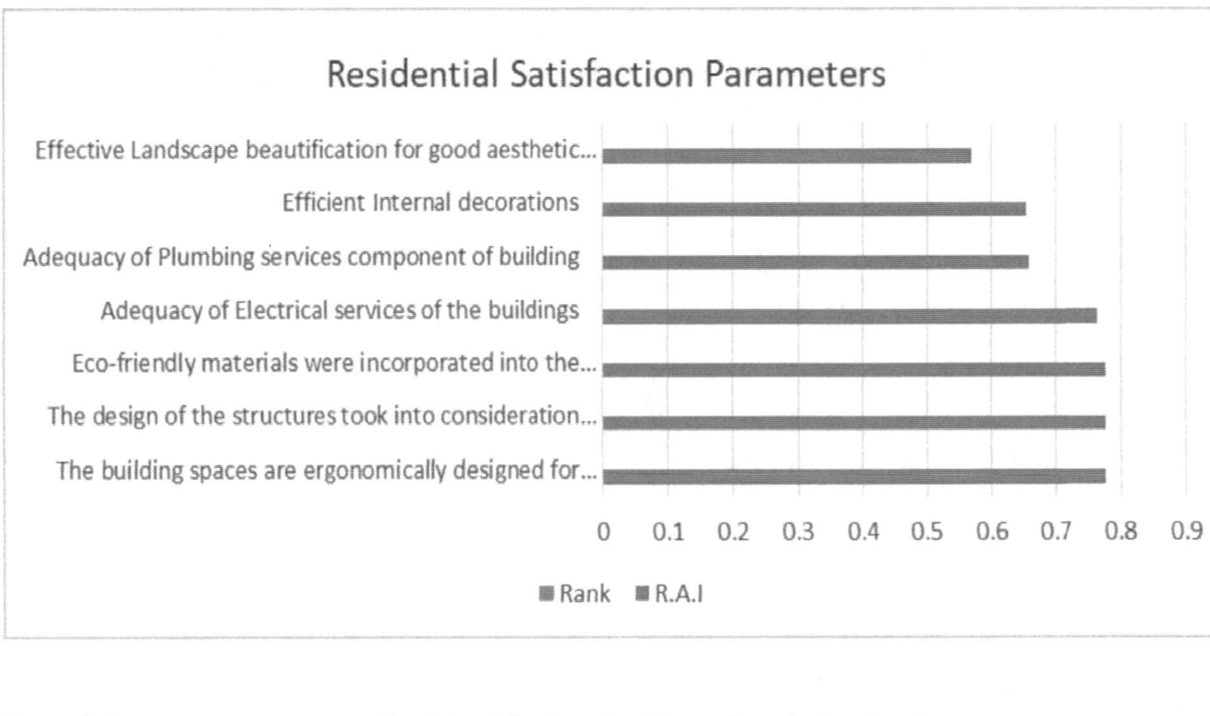

Figure 2: Parameters to measure residential satisfaction of residents of rural-urban housing.

Figure 2 contains presentation on post occupancy study of satisfaction level of residents of urban and rural settlement housing facilities. The occupants expressed satisfaction and their dissatisfaction about the components of facilities that are built on developed spaces. Parameters were developed to capture details about the related parameters. The parameters covers design, material, methodology, services and landscaping of environment. The building spaces are ergonomically designed for intended function with RAI 0.766, Eco-friendly materials were incorporated into the housing facilities design and The design of the structures took into consideration relevant urban renewal theory with RAI 0.766 were ranked 1st. Also, As regards the service component of the building was evaluated, for instance, Adequacy of Electrical services of the buildings with RAI value 0.764 was ranked 4th and Adequacy of plumbing services component of building was ranked 5th with RAI 0.658. There was consensus as to the adequacy of services in the constructed buildings. Similarly, Internal decorations was reported to be efficient in the buildings, this factor was ranked 6th while landscape on the terrain around the buildings was ranked 7th with 0.569.

Table 5 Framework Development

S/N	Factor Codes	Factor1	Factor2	Factor3	Factor4	Factor 5	Factor6
1	CDCU04	1.000					
2	UPRP01		1.000				
3	UPRP02			1.000			
4	UPRP03				1.000		
5	UPRP04					1.000	
6	UPRP05					1.000	1.000
7	UPRP06					.999	.999
8	UPRP07		.			.999	.998
9	FRUG01	.997	.992				
10	FRUG02	.997					
11	FRUG03					.992	.993
12	FRUG04	.991		.990		.990	.992
13	FRUG05		.993	.998			
14	FRUG06	.999	.	.			
15	FRUG07	.992		.			.991
		Factor7	Factor8	Factor9	Factor10	Factor 11	Factor12
16	CDCU01	1.000					
17	FRUG01		1.000				
18	FRUG02		1.000	1.000			
19	FRUG03	.990			1.000		
20	FRUG04	.991	.993	.993	.999	1.000	
21	FRUG05			.	.990	.990	1.000
22	FRUG06		.999	.999	.989	.995	
23	FRUG07	.993	.990	.989	.992	.997	

IOP Conf. Series: Earth and Environmental Science **331** (2019) 012027 doi:10.1088/1755-1315/331/1/012027

Eighteen factors were reduced to 9 representative factor. Data reduction method was used and rotated with Varimax and the factor were selected setting parameters of upper and lower value of Eigen Value. The Eigen value of between 0 and 1 was set. The factors with Eigen values of 0.9 to 1.0 were extracted and used in modelling. The matrix of the 9 factors was presented in Table 7. While the interpretation of the codes is as presented in Table 8.

Table 6 Legend for Coded factors and their Titles.

A.	Clearing of deteriorated components of urban settlement CDCU
1	Re-modification of obsolete structures in areas to be renewed CDCU01
2	Compensating residents of displaced component of area under renewal and upgrading CDCU02
3	Replacing old structures with modern structures CDCU03
4	Removal of structures that cannot be remedied from renewal and upgrading sites.CDCU04
B.	**Upgrading and renewal Parameters UPRP**
5	Compilation of plans for physical development of the areaUPRP01
6	Profiling and classification of features to identify urban-rural and rural-urban locationUPRP02
7	Presentation of adequate design and codes for residents relocation from area to be upgradedUPRP03
8	Feasibility survey of economic, socioeconomic and demographic requirements of the area under upgrading and renewalUPRP04
9	Development of identified locations UPRP05
10	Reconstruction of the areas devastated by development programmesUPRP06
11	Re-accommodation and resettlement of displaced settlementsUPRP07
C.	**Features of Rural-Urban Growth FRUG**
12	Renewal development activitiesFRUG01
13	Upgrading of slum settlements to urban standard FRUG02
14	Afforestation of overcrowded urban settlement for air mass control and ecosystem balance FRUG03
15	Revitalization of desertified and eroded locations of occupied settlements in urban-rural and rural-urban location FRUG04
16	Redevelopment of components of rural-urban and urban-rural locationsFRUG05
17	Reconstruction of dilapidated section of rural settlement within areas upgraded areas FRUG06
18	Reinvigoration of amenities and services in slum locations of urban components to urban standardsFRUG07

The Legend for that could assist in interpreting the factors generated is presented in Table 8 above. For instance CDCU04 represents removal of structures that cannot be remedied from renewal and upgrading sites represents a factor under CDCU which could be describe as Clearing of Deteriorated Components of Urban settlement CDCU. Also, UPRP represents Upgrading and renewal Parameters UPRP while UPRP1 indicates Compilation of plans for physical development of the area. Similarly, FRUG stands for Features of Rural-Urban Growth FRUG while FRUG01 represents Renewal Development activities which is a factor under Features of Rural-Urban Growth.

4.0 Summary of the Developed Framework

One of the objectives of the study was to develop a framework of application for Urban renewal and upgrading. The factors were selected based on the values of Eigen values of 9 parameters set while processing the data. Nine[9] factors emerged with Eigen values of between 0.99 and 1.00. The nine factors are summarized below, that is Factor F1 to Factor F9.

1. F1 ---- 1.000CDCU04, 0.997FRUG01,0.997FRUG02, 0.991FRUG04,0.999FRUG06,0.992FRUG07.
2. F2 ---- 1.000UPRP04, 1.000UPRP05,0.999UPRP06, 0.999UPRP07, 0.992FRUG03,0.990FRUG04.

3. F3 ---- 1.000UPRP05,0.999UPRP06, 0.998UPRP07,0.993FRUG03, 0.992FRUG04,0.991FRUG07.
4. F4 ---- 1.000FRUG01, 1.000FRUG02, 0.993FRUG04, 0.999FRUG06, 0.990FRUG07.
5. F5 ---- 1.000FRUG03, 0.999FRUG04, 0.990FRUG05, 0.989FRUG06,0.992FRUG07.
6. F6 --- 1.000FRUG 02, 0.993FRUG04, 0.993FRUG06, 0.989FRUG 07
7. F7 ---- 1.000CDCU01, 0.990FRUG03,0.991FRUG04,0.993FRUG07.
8. F8 ----1.000FRUG04,0.990FRUG05, 0.995FRUG 06, 0.997FRUG07
9. F9----- 1.000UPRP02, 0.990FRUG 04, 0.998FRUG05

From the above Each of the factors incorporates parameters that could be adopted independently of other factors and could be combined for further effectiveness. Also, the factors could be combined with other factors for maximum possible effect, the various combinations that could be adopted is as presented in Equations 1 to 9 below.

F1=[F2+F3+F4+F5+F6+F7+F8+F9] ------- Equation 1

F2=[F1+ F3+F4+F5+F6+F7+F8+F9] -------Equation 2

F3 =[F1+F2+F4+F5+F6+F7+F8+F9] ---------Equation 3

F4 =[F1+F2+F3+ F5+F6+F7+F8+F9] ---------Equation 4

F5 = [F1+F2+F3+F4+F6+F7+F8+F9] ---------Equation 5

F6 =[F1+F2+F3+F4+F5+F7+F8+F9] ----------Equation 6

F7=[F1+F2+F3+F4+F5+F6+F8+F9] ----------Equation 7

F8=[F1+F2+F3+F4+F5+F6+F7+F9] -----------Equation 8

F9=[F1+F2+F3+F4+F5+F6+F7+F8] ------------Equation 9

In term of application, an individual may decide to adopt Factor one, the action points when Facor F1 is picked could be described as follows:

F1=[F2+F3+F4+F5+F6+F7+F8+F9] ------- Equation 1

1.000CDCU04, 0.997FRUG01,0.997FRUG02, 0.991FRUG04,0.999FRUG06,0.992FRUG07.

Removal of structures that cannot be remedied from renewal and upgrading sites.CDCU04; Renewal development activitiesFRUG01; Upgrading of slum settlements to urban standard FRUG02; Feasibility survey of economic, socioeconomic and demographic requirements of the area under upgrading and renewalUPRP04; Reconstruction of the areas devastated by development programmes UPRP06 and Reinvigoration of amenities and services in slum locations of urban components to urban standards FRUG07. Other factors F2 to F10 could be selected and describe in similar ways [9] and [10].

4.1 Recommendation and Conclusion

CONCLUSION: URBAN RENEWAL AND UPGRADING AS PANACEA TO ACCOMMODATION, EMPLOYMENT AND HEALTHCARE ISSUES.

The analysis carried out in this study has indicated the potency of Urban-rural or Rural urban development in providing civilization and poverty Vis-a-vis unemployment eradication if properly managed.

In urban regeneration, slums gives way to modernized building units with accompany luxury. Most of the slums in parts of the world are always identified with certain inadequacies, e.g. lack of employment opportunities, poor medical services, scarcity of accommodation and the like. However rural upgrading enables rural community to receive face lift. Social amenities like pipe borne water, electricity supply, modern housing, and rural road upgrade are always being introduced. Upgrading will lead to people from less privileged community to flock into the upgraded areas thereby stimulating economy in the direction of more provision of accommodation and this in turn would lift up economic activities. Influx of people would also translate to more customer and more income as indicated in Table 5 of this study.

Similarly, rural and urban development would bring employment opportunity for all classes of trade either skill labour, un-skill labour and semi-skill labour. Construction activities would attract and provide fortune to different workers category there creating prosperity for all.

Moreover, there is environmental pollution that often come with urbanization, for instance upgrading often leads to balancing of the ecosystem and environmental bio mass which would tend to replenish environmental elements that have been depleted. However, urban theory tend to provide for wellbeing of residents therefore provision should be made for any developmental programme so as to ensure continuity. In the study area consulted, many health care centres, convalescence homes, maternity centres, drug dispensing centres, pharmacy and medical care centre both private and public were introduced when the communities were upgraded from slums to urban centres. Medical facilities are easily available in upgraded community as compared to those not yet engaged. This is an indication that people in those upgraded communities would have access to health care facility advantage over others.

Therefore, the study recommends adoption of workable framework in the upgrading of slums and rural components of a community into a more civilized form where all organs of environmental system would be present. For instance, a framework that could guide in policy formulation and implementation rooted in the Urban theory, System theory, regeneration theory, economic theory and spatial theory was presented in the study, it is a believe in this study study that proper application of the theory is a panacea to accommodation, employment and healthcare issues that characterizes underdeveloped community in our time.

Acknowledgement

The support of Center for Innovation and Research Covenant University is appreciated in the support of this study with research grant. Many Thanks.

References

[1]. Uwadiegwu Ben (2015) Urban Renewal and Security Issues. British Journal of Environmental Sciences. Vol. 3(2). Pg. 21-23.

[2]. Saccomani, S. (2011) "Reflecting critically on Turin's Strategic Planning Experience." https://www.researchgate.net/profile/Silvia_Saccomani/publications, 10.13140/RG.2:13185.9362/1.

[3]. Silvia Saccomani(2016a) Urban Regeneration and Crisis. EUPA Conference CityLights, Cities Beyond notwithstanding the Crisis. Torino. Italy.

[4]. Saccomani S. and Caruso N. (2016b). "Turin Metropolitan Region: From path-dependency dynamics to nowadays challenges", Balducci A., Curci F. and V. Fedeli (Ed.), *Post-Metropolitan Territories: Looking for a new urbanity*, London, Routledge.

[5]. Lee Grace and Chan Edwin(2016) Effective Approach to Achieve Sustainable Urban Renewal in Densely Populated Cities. 1st International CIB Endorsed PhD Graduate Conference. Ankara. Turkey.

[6]. Skodra Julita, Alexander Scmdit and Susanne Moebus (2018) Towards the Healthy Neighbourhood: Urban Regeneration of Deprived Neighbourhood in Metropolitan Regions. Universitat Duisburg Essen.

[7]. Artesio, E. (2002). "Strumenti di intervento e pratiche di rigenerazione urbana: il progetto speciale periferie di Torino", Governa, F., Saccomani and S. (Ed.), *Periferie fra riqualificazione e sviluppo locale*,Firenze, Alinea: 49-53.

[8]. Atkinson, R. and C. Rossignolo (2010). "Cities and the 'soft side' of Europeanization.The role of Urban networks", a. Hamedinger and A. Wolfdhardt (Ed.), *The Europeanization of cities*,Amsterdam, Techne Press: 197-210.

[9]. Dansero, E., A. Mela (2015). "Turin 2006's Legacy after eight years: Theories on territorialisation in the aftermath of the Olympic Games", V. Viehoff and G. Poynter (Ed.), *Mega-event Cities: Urban Legacies of Global Sports Events*,london, Ashgate: 99-108.

[10]. Zhiyong Yi, Guiwen Liu, Wei Lang, Asheem Shrestha and Igor Martek(2017) Strategic Approaches to Sustainable Urban Renewal in Developing Countries: A Case Study of Shenzhen, China

[11]. Couch, C. (1990). *Urban Renewal: Theory and Practice.* London, Macmillan.

IOP Conf. Series: Earth and Environmental Science **331** (2019) 012028 doi:10.1088/1755-1315/331/1/012028

Fabrication of P3HT: PCBM bulk heterojunction organic solar cell

R. O. Kesinro[1]., A.O. Boyo[2]., M. L. Akinyemi[1]., G. T. Mola[3].
[1]Covenant University, Ota, Ogun State. [2]Lagos State University, Ojo, Lagos State. [3]University of KwaZulu-Natal, Pietermartizburg, South Africa.
E-mail: olakesinro02@gmail.com

Abstract. The fabrication of organic solar cells (OSCs) employing solution processing has proven to be a convenient method to guarantee the fabrication of OSCs at low cost and large-scale production. This manuscript demonstrates the successful fabrication of TFOSC with device structure: poly(3,4-ethylenedioxythiophene) polystyrene sulfonate (PEDOT: PSS)/poly (3-hexylthiophene) (P3HT): [6, 6]-phenyl-C61-butyric acid methyl ester (PCBM)/lithium fluoride (LiF)/aluminium (Al) under ambient conditions. An overall power conversion efficiency of 3.25% and charge carrier mobility of $1.22e^{-2}$ $cm^2v^{-1}s^{-1}$ was obtained and PEDOT: PSS was beneficial in improving the charge transport processes in the preparation of organic solar cells.

1. Introduction

The increase in the world population, energy consumption per capita and more industrial usage has necessitated a search for alternative sources of energy. The discovery of petroleum in the 19th century marked a giant evolution in human history (Steele and Heinzel, 2001). It is agreed generally that the increasing use of fossil fuels, which represent the main part of the world energy production (Grossiord, Kroon, Andriessen, and Blom, 2012), is responsible for greenhouse gas emissions and the increasing ecosystem and climate instabilities (Tong and Soboyejo, 2012) which has resulted in an increase of natural disasters occurring repeatedly in different parts of the world. However, solar energy has emerged as a form of alternative energy source to traditional sources such as coal and fossil fuels. Solar energy is reliable is abundant and free. Also, the total energy from the sun reaching the earth per year is estimated to be about 6700 times the annual world primary energy consumption.

Organic solar cells (OSCs) are promising new devices for the conversion of light into electricity. The technology currently demonstrates moderate power conversion efficiencies (PCEs) and operational stabilities that do not qualify OSCs for rapid, large-scale commercialization. Nonetheless, academic curiosity and the potential of the technology has motivated intense research into this field. It's still relatively new, but it involves photoactive organic layers, comprised of chains and bucky-balls of p-type and n-type semiconducting polymers (Yuan, Ouyang, Cimrová, Leclerc, Najari, and Zou, 2017; Chabinyc and Yueh-lin, 2006) and oligomer (Murphy, Jean, and Fréchet, 2007) materials. NREL, (2019) established that the power conversion efficiency has continually shown increase, with laboratory solar cell efficiency reaching 10% in 2012 (Tong, Asare, Rwenyagila, Anye, Oyewole, Fashina, and Soboyejo,

IOP Conf. Series: Earth and Environmental Science **331** (2019) 012028 doi:10.1088/1755-1315/331/1/012028

2015) and recently increased to ~15.6% (NREL, 2019) and ~12% in well-optimized OSC systems with non-fullerene acceptor materials (Chen and Zhang, 2017), while more efforts are being explored to improve semiconducting polymeric materials (Vyprachticky, Demirtas, Dzhabarov, and Cimrova, 2017).

Presently, crystalline silicon solar cells are the majorly used solar cells but the cost of fabrication of such solar cells is expensive and there has not been any obvious improvement in the power conversion efficiency. The high cost of production of the solar cells has encouraged research into other alternatives of fabrication and types of solar cells (OSCs) at a low cost. In this study, we are reporting the photovoltaic parameters obtained using P3HT: PCBM based bulk heterojunction organic solar cell. Also, the charge transport property of the device was investigated.

2. Materials and Methods.

The OSC was fabricated using the solution-processing method. First, the ITO substrate was covered with polystyrene sulfonate (PEDOT: PSS) after it has been sonicated for 10 minutes. A thin layer of PEDOT: PSS (40nm thick) was deposited on the substrate by spin coating at 3500rpm for 60s and annealed at 100°C for 30 minutes under ambient condition. The active layer was prepared by dissolving P3HT and PCBM at a ratio of 1:1 in chloroform at a concentration of 20 mg/ml. The active layer solution was spin coated at 1200rpm for 40s on the dried PEDOT: PSS. The sample was then transferred to the furnace for annealing at 100°C for 10 minutes in nitrogen atmosphere (Oseni and Mola, 2017). Finally, 0.4 nm and 60 nm thick lithium fluoride (LiF) electron transport layer and aluminum (Al) back contact were deposited thermally under vacuum respectively using Edward Auto 306 deposition unit at a pressure of 1e-6 mbar. After few minutes of post deposition annealing in the nitrogen furnace for 10 minutes, the devices were characterized in ambient environment using Keithley HP2420 source meter under AM 1.5 solar simulator (Model SS50AAA) operating at a light intensity of 100mW/cm^2. The effective area of the solar cells was estimated to be 0.05 cm^2. The thin film absorption characteristics of the device was studied using the ultraviolet-Visible (UV-Vis) spectrometer (T80-PG-Instrument limited).

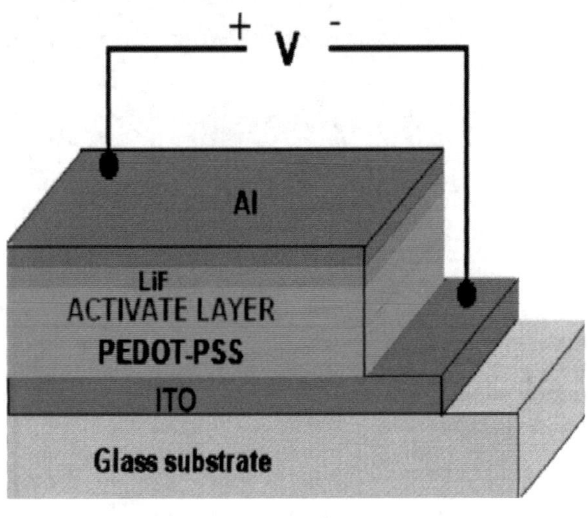

Figure 1: Schematic for the device

3. Results and discussion

IOP Conf. Series: Earth and Environmental Science **331** (2019) 012028 doi:10.1088/1755-1315/331/1/012028

The optical absorption spectra taken for the fabricated OSC was obtained using the ultraviolet-Visible (UV-Vis) spectrometer (T80-PG-Instrument limited). Figure 2 shows the absorption spectra with two distinct intensity peaks. The first peak is observed between 350-400 nm which can be attributed to the presence of the PEDOT: PSS and PCBM. Also, the second intensity peak which is more pronounced and observed between 450-600 nm indicates the presence of the active layer blend (P3HT: PCBM). These peaks indicate the ability of the device to absorb solar radiation within visible range (400-700 nm).

The photovoltaic parameters of the fabricated OSC was studied using the Keithley HP2420 source meter under AM 1.5 solar simulator (Model SS50AAA) operating at a light intensity of 100 mW/cm^2. A current density (J_{sc}), open circuit voltage (V_{oc}), fill factor (FF) and power conversion efficiency (PCE) of 13.37 mA/cm^2, 0.57 V, 42.60 and 3.25 % was obtained respectively from the current density – voltage curve in Figure 3. The open circuit voltage of the device is similar to the values reported in literatures (Hoppe and Sariciftci, 2006; Steim *et al.*, 2010). The high Jsc obtained can be attributed to improved charge carrier transport value determined in figure 5 which leads to a PCE of 3.25 %.

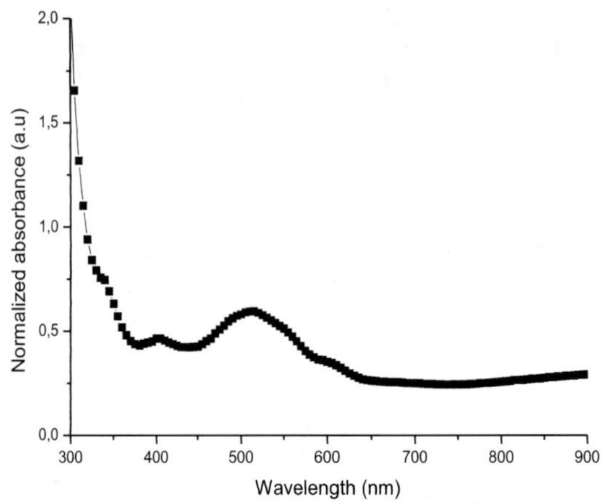

Figure 2: UV-Vis spectra of the fabricated OSC.

IOP Conf. Series: Earth and Environmental Science **331** (2019) 012028 doi:10.1088/1755-1315/331/1/012028

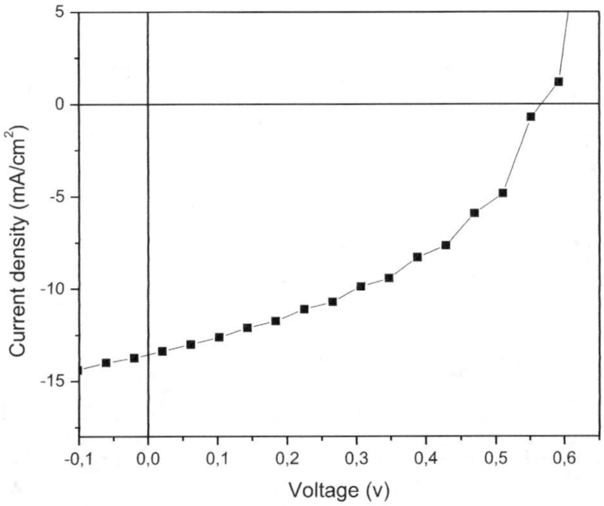

Figure 3: Current density-voltage curve of the device

The charge transport property of the device was studied using the measured space charge limited current (SCLC). The experimental SCLC data taken from the device dark current density -voltage curve (Figure 4) was compared with the Mott-Gurneys law given by the equation (1):

$$J = \frac{9}{8}\varepsilon\varepsilon_o\mu_o exp\left(0.89\gamma\sqrt{\frac{V}{L}}\right)\frac{V^2}{L^3} \quad (1)$$

Where ε the relative dielectric permittivity of the material, ε_o is the permittivity of free space; μ_o is the zero-field mobility; L is the thickness of the active layer, and V is the voltage dropped across the sample which is corrected for built-in potential V_{bi}. The space charge transport mobility determined in Figure 5 are compared with current density equation 1. The dark boxes in Figure 5 are the experimental data while the solid line is the computer fit according to equation 1. The zero-field mobility derived from the fit is μ_o $1.22*10^{-2}$ $cm^2V^{-1}s^{-1}$ is comparable to the values reported in literature (Coropceanu, Cornil, da S. Filho, Olivier, Silbey, and Bredas, 2007).

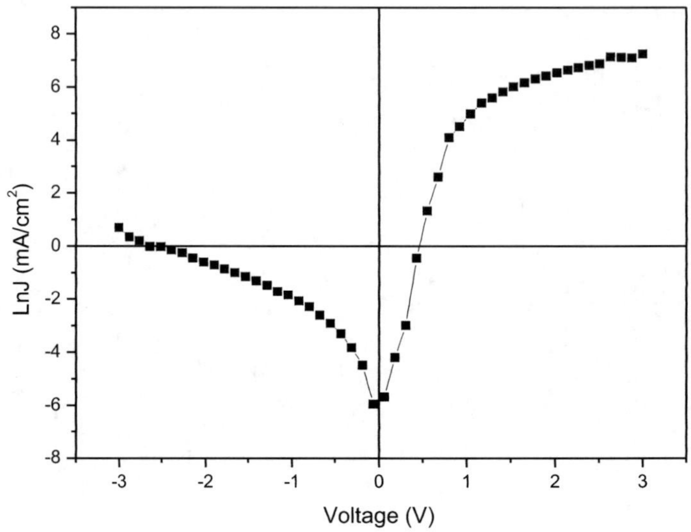

Figure 4: current density – voltage curve under dark illumination.

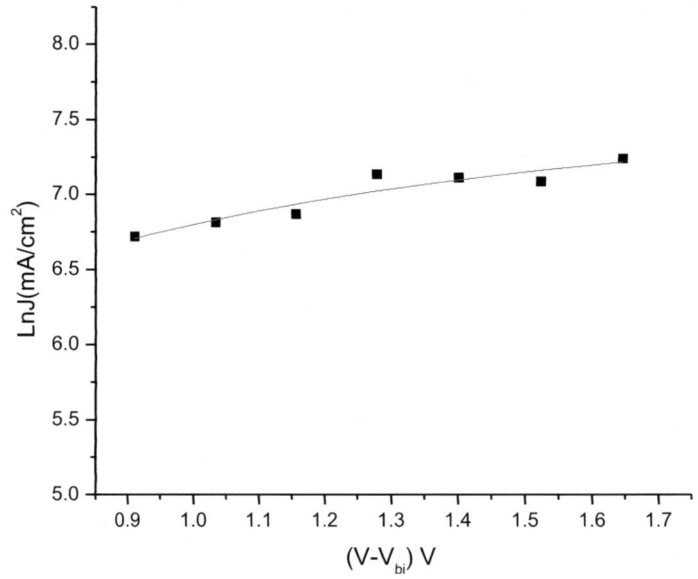

Figure 5: Charge carrier transport for the device

4. Conclusion

High current density is obtained from P3HT: PCBM bulk heterojunction organic solar cell fabricated using solution process method in ambient laboratory condition. The observed high current density can be attributed to the high PCE of 3.25% obtained. Also, the optical property of the device was found to be suitable for light transmission and as well as absorption of short wavelength emission within the visible region of the electromagnetic spectrum. The observed results are attributed mainly to less trap states in the active layer and smooth interfacial contacts between the active layer and PEDOT: PSS. Also, the power conversion can be improved by the introduction of nanoparticles and solvent additives which will improve the morphology and photovoltaics properties of the OSCs.

5. References

[1] Chabinyc, M. L. and Yueh-lin, L. (2006). Semiconducting polymers for thin-film electronics, *Journal of Macromolecular Science, Part C: Polymer Reviews*, 46(**1**), 1–5.

[2] Chen, W. and Zhang, Q. (2017). Recent progress in non-fullerene small molecule acceptors in organic solar cells (OSCs), *Journal of Materials Chemistry C*, **5,** 1275–1302.

[3] Coropceanu, V., Cornil, J., da S. Filho, D. A., Olivier, Y., Silbey, R. and Bredas, J-L. (2007). Charge Transport in Organic Semiconductors, *Chemical Reviews*, **107**, 926-952.

[4] Grossiord, N., Kroon, M. J., Andriessen, R. and Blom, P. W. M. (2012). Degradation mechanisms in organic photovoltaic devices, *Organic Electronics,* **13**, 432–456.

[5] Hoppe, H. and Sariciftci, N.S. (2008). Polymer Solar Cells. *Advances in Polymer Science*, 214, 1.

[6] Murphy, R. A., Jean, M. J. and Fréchet, M. J. (2007). "Organic semiconducting oligomers for use in thin film transistors," *Chemical Reviews,* 107(**4**), 1066–1096.

[7] Tong, T. M., Asare, J., Rwenyagila, E. R., Anye, V., Oyewole, O. K., Fashina, A. A and Soboyejo, W. O. (2015). A study of factors that influence the adoption of solar powered lanterns in a rural village in Kenya, *Perspectives on Global Development and Technology*, 14(**4**), 448–491.

[8] Tong T. M. and Soboyejo, W. O. (2012). Adhesion and Interfacial Fracture: from Organic Light Emitting Devices and Photovoltaic Cells to Solar Lanterns for Developing Regions, Dissertation, Princeton University, NY.

[9] Vyprachticky, D., Demirtas, I., Dzhabarov, V. and Cimrova, V. (2017). "New copolymers with thieno[3,2-b] thiophene or dithieno[3,2-b:2′,3′-d] thiophene units possessing electronwithdrawing 4-cyanophenyl groups: synthesis and photophysical, electrochemical, and electroluminescent properties," *Journal of Polymer Science Part A: Polymer Chemistry,* 55(**16**) 2629–2638.

[10] Yuan, J., Ouyang, J., Cimrová, V., Leclerc, M., Najari, A and Zou, Y. (2017). "Development of quinoxaline based polymers for photovoltaic applications," *Journal of Materials Chemistry C*, **5**, 1858–1879.

Antibacterial studies on fruit-skin and leaf extracts of *Annona muricata* in Ota, Nigeria

***[1]Wisdom O. Iyanda-Joel, [2]Emmanuel A. Omonigbehin, [1]Emeka E. J. Iweala and [1]Shalom N. Chinedu**

[1]Department of Biochemistry, College of Science and Technology, Covenant University, Ota, Nigeria

[2]Department of Biological Sciences, College of Science and Technology, Covenant University, Ota, Nigeria

wisdom.iyanda-joel@covenantuniversity.edu.ng

Abstract. The current study examined the phytochemical constituents and antibacterial activity of leaf and fruit-skin extracts of *Annona muricata* Linn. The extracts were screened qualitatively for their phytochemical constituents following established protocols; disc diffusion method was employed for the antibacterial study on *Klebsiella pneumoniae*, *Staphylococcus aureus* and *Pseudomonas aeruginosa*, after which their minimum inhibitory concentrations were determined via double dilution method. Furthermore, selected extracts were screened against 50 multidrug-resistant clinical isolates from immunocompomised patients who are already down with secondary acute respiratory infections. The phytochemical screening of the extracts collectively revealed the presence of tannins, saponins, terpenoids, steroids, phenols, flavonoids, coumarin, alkaloids, antocyanins and betacyanins. Only aqueous leaf (ALA) and fruit-skin ethanol (ESA) extracts showed remarkable activity with 15 – 17 mm inhibition diameter (DIH) against the test organisms. The minimum inhibitory concentrations (MIC) for the ALA and ESA were 25 and 6.25 $mgmL^{-1}$ respectively. The two extracts further showed activity (6-18 mm DIH) against 80% of the multidrug resistant isolates tested. It can therefore be inferred from this study that *Annona muricata* possibly contains potent bioactive antibacterials that can be developed into nature-friendly drugs to counter the effects of pathogenic bacteria, thereby preserving lives as well as sustaining a viable environment.

.Keywords: *Annona muricata*, antibacterials, minimum inhibitory concentration, multidrug-resistant isolates, phytochemicals.

1. Introduction

Provided the indispensable activity of a class of supramolecular complexes called viruses were not in existence, bacteria would have overrun the entire planet [1-3]! Bacteria are a set of highly prolific microscopic organisms that occupy any and everywhere called space. They fill up both habitable and non-habitable zones of the planet Earth. It is quite alarming to note that the total bacterial soil carbon in existence worldwide today, is estimated at the entire weight of the United Kingdom [4-7]. Be that as it may, bacteria can be as much beneficial as they are detrimental [8-10]. In the past few decades, Scientists have harnessed the use of bacteria through diverse technologies to better the lot of mankind, other living systems as well as the environment. In a bid to further ameliorate the impact of deleterious bacteria, natural herbs with ancient folklore claims are also being investigated worldwide to establish their ethnopharmacological effect [11-12]. Anyanwu and Okoye [13] more recently published a well detailed review on the antimicrobial activity of Nigerian medicinal plants and effects of local herbs found in Nigeria against several pathogenic bacteria and viruses such as *Salmonella typhi, Pseudomonas aeruginosa, Candida albicans, Escherichia coli, Staphylococcus aureus*, *Klebsiella pneumoniae*, *Streptococcus pyogenes*, Neisseria gonorrhoeae, *Mycobacterium tuberculosis, Streptococcus pneumoniae*, *Aspergillus niger*, *Vibrio cholerae* to mention a few; with as low as 4.9 µg mL⁻¹ minimum inhibitory concentration (MIC). Some genus that were reviewed with remarkable activity are as follows: *Allium spp. Amaranthus spp., Cassia spp., Eucalyptus spp., Euphorbia spp., Ocimum spp., Pterocarpus spp., Stachytarpheta spp., Terminalia spp.* Quite a number of specific species also studied include *Vernonia amygdalina*, *Psidium guajava*, *Jatropha curcas*, *Crinum jagus, Moringa oleifera,, Anogeissus leiocarpus, Gongronema latifolium, and Fagara (Zanthoxylum) zanthoxyloides*, *Terminalia avicennioides*, *Stachytarpheta angustifolia and Struchium sparganophora*. Deploying state-of-the-art technologies into exploring these medicinal plants are most likely the way out of the currently emerging threats of antimicrobial resistance that is rendering modern medicine ineffective in so many ways. One of such herbs that has been designated as a crop for the future to combat bacterial infections amongst other disease conditions is *Annona muricata* Linn. *Annona muricata* also commonly called soursop is an evergreen low-branching, slender and bushy tree towering between 5 and 9 metres and 0.15-0.83 metres wide. The soursop tree is extravagantly distributed in the tropics largely because of the favourable climatic conditions such as annual rainfall above 1500 mm, altitudes below 1200 m above sea level, 60 to 80% relative humidity between and temperatures between 25 and 28 °C. A remarkable feature of the tree that makes it of utmost economic importance is its high and continuous productive profile [14-17]. For many decades now, the antibacterial activity of *A. muricata* have been explored on several gram-negative and gram-positive organism with interesting results [18]. A review report showed that *A. muricata* extract demonstrated antibacterial activity comparable to streptomycin which is a potent standard antibiotic [19]. One very key observation in the antibacterial activity studies of *A. muricata* is that extract activity is largely dependent on extraction solvent; denoting the fact that extracts prepared with organic solvents for instance, showed higher activity over aqueous extracts. Viera *et al.* [20] highlighted and confirmed this phenomenon, where ethanolic and methanolic *A. muricata* extracts were active against *Staphylococus aureus,* whereas no such activity was observed in the aqueous extract of the fruit skin. As a result, the polarity of the organic solvent used in extraction further significantly influenced antibacterial activity. Amidst several studies, the current examination of the fruit-skin and leaf extracts of this unique crop is hereby presented.

2. Materials and Methods

Collection and identification of plant samples

Unripen mature fruits of *Annona muricata* were bought from local fruit markets in Lusada and Agbara, Ogun state, Nigeria; and the leaves were collected from trees on farmlands also in Ota, Ogun state, Nigeria. Fresh samples of the healthy fruits and leaves were deposited in the Forestry herbarium of the Forestry Research Institute of Nigeria (FRIN), Ibadan, Oyo State, Nigeria. The plant samples were identified and authenticated with Voucher referencing Number FHI. 110177.

Test Organisms

Clinical isolates of *Klebsiella pneumoniae, Staphylococcus aureus and Pseudomonas aeruginosa* coupled with those listed in Table 4 which were isolated from HIV/AIDS patients already down with secondary lower respiratory infections.

Culture media and plates

Mueller-Hinton agar (HiMedia Laboratories Ltd, Mumbai, India - ISO 13485-2003) and MacConkey agar (without salt: Rapid Labs), petri dishes and universal bottles were all purchased from standard scientific stores.

Methods

Preparation and extraction of plant samples

The plant samples of *A. muricata* were washed properly in clean running water. The fruits were peeled manually with knife, a portion of the peels (fruit skin) were blended and frozen while another portion was frozen directly for subsequent extractions. Also, a portion of the leaves were shade-dried and blended, while the remaining portion was frozen for respective extraction. Both fruit skin and leaves of *A. muricata* were extracted with five different categories of solvents following stipulated protocols itemized as follows. The extracts were concentrated using a rotary evaporator and the percentage (%) yield of the dry residue calculated as described by [21].

Aqueous extraction

Weighed amount of powdered leaves of *A. muricata* were soaked in distilled water (1:10) for 12 hours and sieved with muslin cloth to remove fiber particles; then filtered with a vacuum membrane filter device as well as filter paper into conical flasks. The filtrate was concentrated in a rotary evaporator and dried to powder in a water bath. The powder was weighed and percentage yield calculated. The aqueous leaf extract is labeled as subsequently referred to as ALA. The blended and frozen fruit skin were thawed and extracted with distilled water (1:5) for 12 hours, sieved and filtered. The filtrate was frozen and dried in a freeze dryer; it is labelled ASA [22]

Sodium chloride extraction

A specified volume of 1% sodium chloride was poured into weighed powdered leaves of *A. muricata* in a glass jar and left for 12 hours. The mixture was sieved, filtered and concentrated in a rotary evaporator and subsequently freeze-dried. The resulting powder was weighed and percentage yield calculated. Sodium chloride leaf extract is labelled as SLA. Frozen peels of *A. muricata* are weighed and blended into fine extractible particles; this was then soaked in a known volume of 1% sodium chloride for 12 hours. The mixture was also sieved, filtered and concentrated in a rotary evaporator and freeze-dried. Sodium chloride fruit skin extract is labelled as SSA [23-24].

Methyl-tert-butyl ether (MTBE) extraction

Fresh leaves and fruit skin of *A. muricata* were extracted with Methyl-*tert*-butyl ether (MTBE) amongst other solvents (MTBE/methanol/water: 10/3/2.5; volume by volume) for an accurate lipidome profiling. Methanol (970 mL) was added to 900g of blended fresh leaves of *A. muricata* in a cylindrical glass jar with lid; the content was shaken together to mix well. This was followed by adding 3.2 L of MTBE and incubating the mixture at room temperature for 55 minutes. After the incubation phase separation was induced with the addition of 810 mL of distilled water. Following 10 minutes of incubation at room temperature, the mixture was sieved with 8-fold muslin cloth and the resultan filtrate was poured into a separating funnel. The upper organic phase was collected, concentrated in a rotary evaporator and dried in a weighed beaker on a heat block set at 55 ◦C. Extracted lipids were kept in the refrigerator. Methyl-*tert*-butyl ether leaf and fruit skin extracts of *A. muricata* were labelled as MLA and MSA respectively [25].

Ethanol extraction

Powdered leaves of *A. muricata* were extracted with 95 % ethanol using soxhlet apparatus. Likewise the blended fresh peels of *A. muricata* were soaked in 95 % ethanol overnight (8-12 hours), sieved with an 8-fold muslin cloth to take out fiber particle and then filtered a vacuum membrane filter. Both the extract from the leaves and the filtrate from the fruit skin were concentrated in a rotary evaporator and dried in a weighed beaker on a heat block at 65 ◦C. The ethanolic leaf and fruit skin extracts of *A. muricata* were labelled as ELA and ESA respectively [26].

Isopropanol extraction
Frozen leaves and fruit skin of *A. muricata* were blended with a mixture of isopropanol/hexane/water in the ratio 55:20:25 (volume by volume) and incubated at 55.5 °C for 15 min with occasional shaking. After incubation, the mixture was sieved with an 8-fold muslin cloth and transferred to a separating funnel. The upper organic phase was collected with a micropipette into a weighed beaker and dried on a heat block set at 55 °C. isopropanol/hexane/water leaf and fruit skin extracts of *A. muricata* were labelled as ILA and ISA respectively [27].

Phytochemical screening
Selected qualitative phytochemical screening tests were carried out according to methods previously described on the extracts for the detection of the following compounds: alkaloids, anthocyanins & betacyanins, coumarins, flavonoids, glycosides, phenols, quinones, saponins, steroids, tannins, terpenoids and triterpenoids. Prior to the screening, the extracts were dissolved in their respective solvent [28-30].

Antibacterial studies
In vitro screening for antibacterial activity
The agar well diffusion method of sensitivity testing were used to determine the growth inhibition of bacteria by the plant extracts [31-33].

Determination of minimum inhibitory concentrations
Minimum inhibitory concentrations (MIC) of the extracts were determined by agar well dilution method as described by Russell and Furr (1972). Different dilutions of the extracts were prepared to give final concentration in the range of 100.0, 50.0, 25.0, 12.5, 6.25, 3.12 and 1.56 mg mL^{-1}. Two milliliter (2 mLs) of each dilution was mixed with 18 mL of Mueller Hinton agar (HiMedia Laboratories Ltd, Mumbai, India - ISO 13485-2003), poured into petri-dishes and allowed to set. The agar was streaked with an overnight broth culture of the bacterial isolates and incubated overnight. The plates were then examined for the presence or absence of growth. The minimum concentration that completely inhibited macroscopic growth was regarded as the minimum inhibittory concentration of the respective extracts [18].

Figure 1: *Annona muricata* fruit on the tree

3. Results and discussion
Qualitative phytochemical study
The qualitative phytochemical screening of the leaf and fruit-skin extract pairs of *A. muricata*, as shown in table 1 below indicated the presence of alkaloids, anthocyanins and betacyanins, coumarins, flavonoids, glycosides, phenols, saponins, steroids, quinones, tannins and terpenoids.

Table 1 Phytochemical screening results of extracts from the fruit-skin and leaves of *Annona muricata*

Test/Extract	ALA	ASA	SLA	SSA	MLA	MSA	ELA	ESA	ILA	ISA
Alkaloids	_	_	_	_	_	+	+	+	+	_
Antocyanins-betacyanins	_	_	+	_	_	_	+	_	_	+
Coumarins	_	_	+	+	_	_	_	_	_	_
Flavonoids	+	_	+	_	+	_	+	+	_	+
Glycosides	_	_	_	_	_	+	_	_	_	_
Phenols	+	_	_	_	_	+	_	+	+	_
Quinones	_	_	+	_	+	+	_	+	_	_
Saponins	+	+	+	+	+	+	_	+	_	+
Steroids	_	_	+	_	+	_	+	+	+	_
Tannins	+	_	_	+	_	_	_	+	+	_
Terpenoids	+	+	+	+	+	+	+	+	+	+
Triterpenoids	_	_	_	_	_	_	_	_	_	_

(+) denotes presence of phytochemical; (-) denotes absence or no trace of phytochemical in the extract tested. ALA: Aqueous leaf extract of *Annona muricata,* ASA: Aqueous skin extract of *Annona muricata*, SLA: 1 % Sodium chloride leaf extract of *Annona muricata*, SSA: 1 % Sodium chloride skin extract of *Annona muricata*, MLA: methyl tert-butyl ether leaf extract of *Annona muricata*, MSA: methyl tert-butyl ether skin extract of *Annona muricata*, ELA: ethanolic leaf extract of *Annona muricata*, ESA: ethanolic skin extract of *Annona muricata*, ILA: isopropanol leaf extract of *Annona muricata*, ISA: isopropanol skin extract of *Annona muricata*

Antibacterial sensitivity test for fruit-skin and leaf extracts of Annona muricata against Staphylococcus aureus, Klebsiella pneumoniae and Pseudomonas aeruginosa

Sensitivity testing on the five extract pairs of *A. muricata* against *Klebsiella pneumonia*, *Pseudomonas aeruginosa*, *Staphylococcus aureus* revealed antibacterial activity (Plates 4.14-4.16) for two of the extracts: ALA (16, 14 and 16 mm) and ESA (16, 16 and 17 mm) with zones of inhibition as shown in Table 2

Table 2: Activity of five extract pairs of *Annona muricata* against *Staphylococcus aureus*, *Klebsiella pneumoniae* and *Pseudomonas aeruginosa*

SN	Organism Extract	*Staphylococcus* aureus Zone of inhibition (mm)	*Klebsiella pneumoniae*	*Pseudomonas aeruginosa*
1.	ALA_1	16 ± 0.58 (+)	16 ± 1.15 (+)	14 ± 0.58 (+)
2.	ALA_2	-	-	-
3.	SLA	-	-	-
4.	ELA	-	-	-
5.	MLA	-	-	-
6.	ILA	-	-	-
7.	ESA	17 ± 0.58 (+)	16 ± 0.00 (+)	16 ± 0.58 (+)
8.	MSA	-	-	-
9.	ISA	-	-	-
10.	ASA	-	-	-
11.	SSA	-	-	-

ALA: Aqueous leaf extract of *Annona muricata*; SLA: 1% sodium chloride leaf extract of *Annona muricata*; ELA: ethanol leaf extract of *Annona muricata*; MLA: methyl tert-butyl ether leaf extract of *Annona muricata*; ILA: isopropanol/hexane/water leaf extract of *Annona muricata*; ESA: ethanol fruit-skin extract of *Annona muricata*; MSA: methyl tert-butyl ether fruit-skin extract of *Annona muricata*; ISA: isopropanol/hexane/water fruit-skin extract of *Annona muricata*; ASA: Aqueous fruit-skin extract of *Annona muricata*; SSA: Aqueous fruit-skin extract of *Annona muricata*

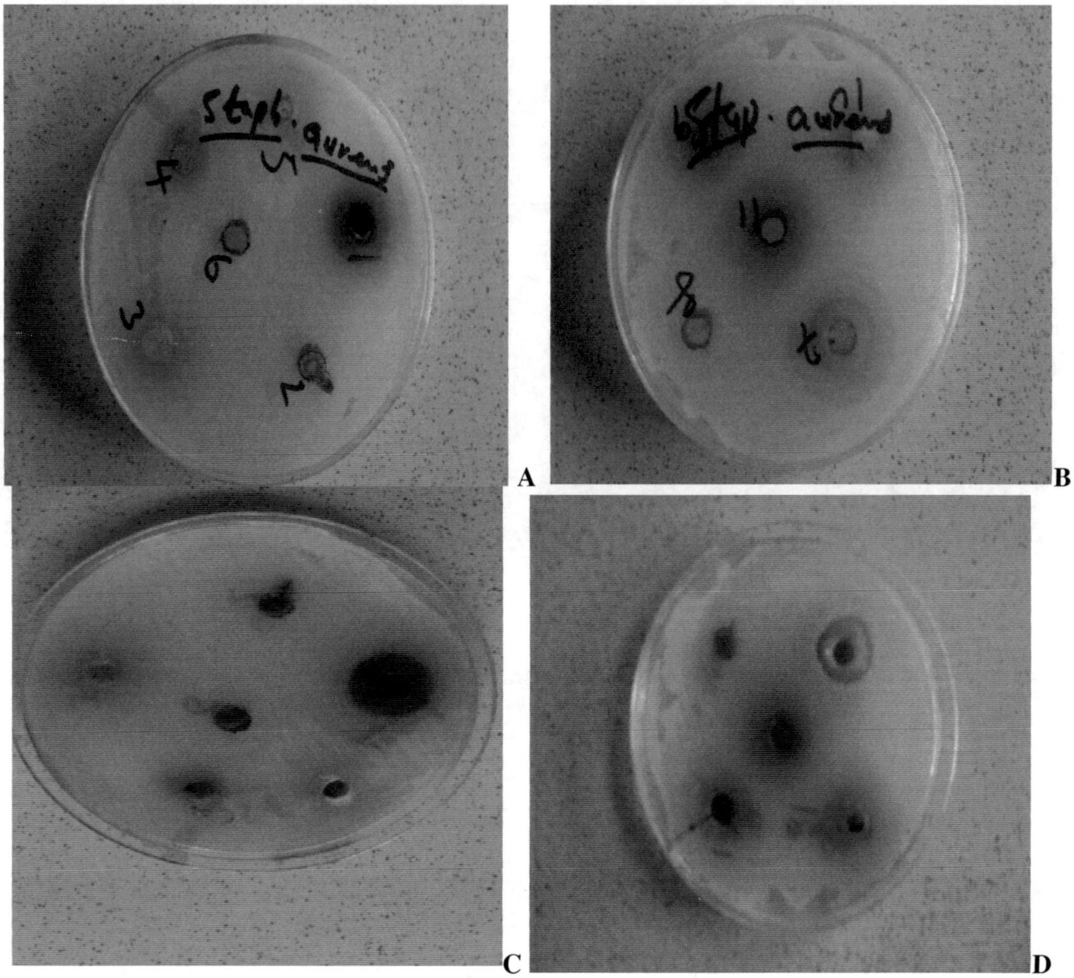

Plate 1 A, B, C and D: Agar diffusion plates showing the activities of *Annona muricata* extracts (100 $mgmL^{-1}$) against *Staphylococcus aureus*

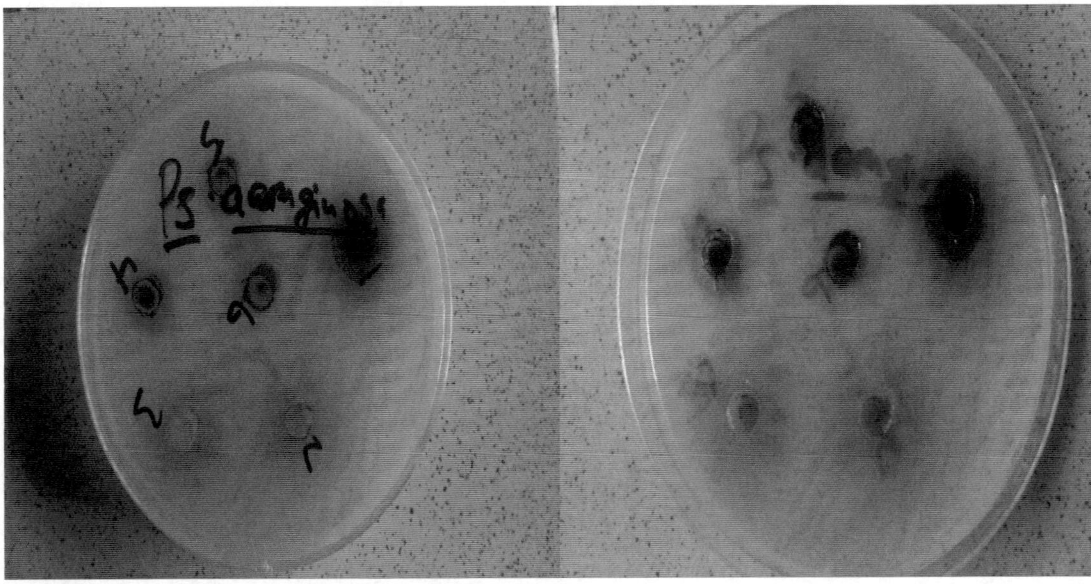

A↑ and B↑

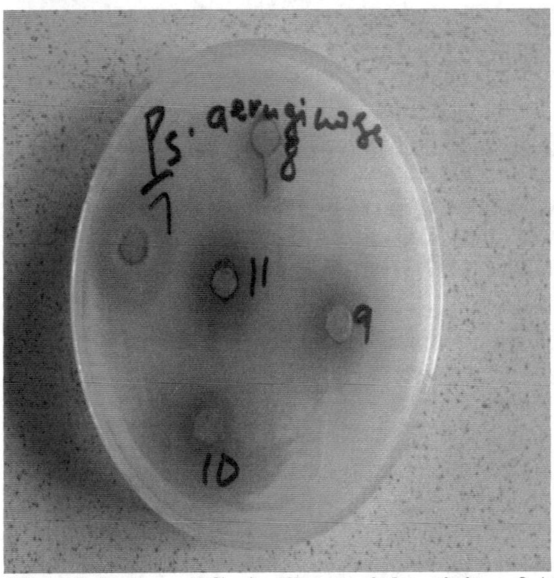

C

Plate 2 A, B and C: Antibacterial activity of *Annona muricata* extracts against *Pseudomonas aeruginosa*

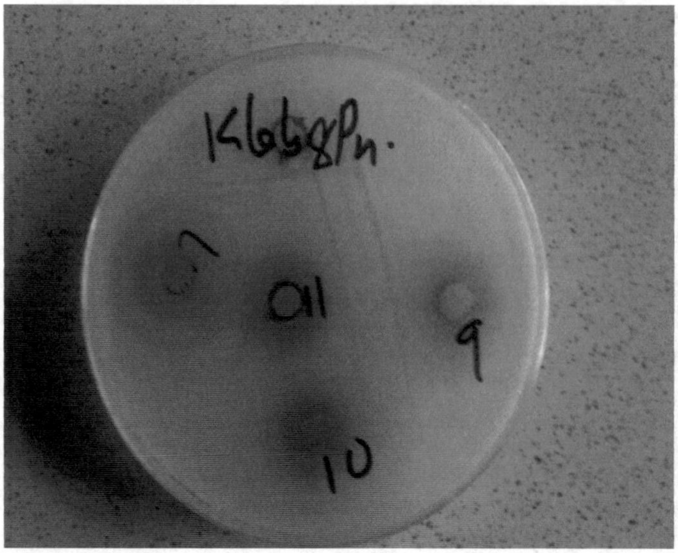

Plate 3: Antibacterial activity of *Annona muricata* extracts against *Klebsiella pneumoniae*

Minimum inhibitory concentration (MIC) of aqueous leaf and ethanol skin extracts of Annona muricata against Staphylococcus aureus

The minimum inhibitory concentration (MIC) of aqueous leaf (ALA) and ethanol fruit-skin (ESA) extract of *Annona muricata* against *Staphylococcus aureus* are 25 $mgmL^{-1}$ and 6.25 $mgmL^{-1}$ respectively as shown in table 3.

Table 3: Minimum inhibitory concentration of aqueous leaf and ethanol skin extracts of *A. muricata* against *Staphylococcus aureus*

SN	Dilution in mg mL^{-1} Extract	Zone of Inhibition (mm)						
		100	**50**	**25**	**12.5**	**6.25**	**3.12**	**1.56**
1.	ALA	16	14	12	-	-	-	-
2.	ESA	17	13	11	9	7	-	-

Antibacterial activity of aqueous and ethanol leaf extracts of Annona muricata against multidrug resistant (MDR) organisms isolated from immunocompromised (HIV/AIDS) patients with lower respiratory infections.

The two active extracts of *Annona muricata* (ALA and ESA) further screened against multidrug-resistant organisms isolated from immunocompromised patients co-infected with acute respiratory infections demonstrated antibacterial activity; and most of the organisms were sensitive to the extracts with zones of inhibition ranging from 7-18 mm (Table 4.6). Furthermore, a remarkable phenomenon was observed in the activity of ESA apart from showing a zone of inhibition such that the organisms appeared morphologically altered. This alteration occurs immediately after the zones of inhibition forming a purported zone of reversion which can be as wide as 28 mm in diameter.

Table 4 Antibacterial activity of aqueous leaf and ethanol skin extracts of *Annona muricata* against multidrug-resistant organisms isolated from immunocompromised (HIV/AIDS) patients with lower respiratory infections.

SN	Organism	MAR index	ALA Zone of Inhibition (mm)	ESA Zone of Inhibition (mm)
1.	*Klebsiella pneumoniae ozaenae*	0.2	14	15
2.	*Klebsiella pneumoniae pneumoniae*	0.5	10	7
3.	*Klebsiella pneumoniae pneumoniae*	0.7	14	9
4.	*Klebsiella pneumoniae ozaenae*	0.7	17	7
5.	*Klebsiella pneumoniae pneumoniae*	0.3	12	6
6.	*Klebsiella pneumoniae pneumoniae*	0.4	12	16
7.	*Klebsiella pneumoniae pneumoniae*	0.4	0	0
8.	*Klebsiella pneumoniae pneumoniae*	0.7	16	12
9.	*Klebsiella pneumoniae pneumoniae*	0.3	12	0
10.	*Klebsiella pneumoniae pneumoniae*	0.4	12	0
11.	*Klebsiella pneumoniae pneumoniae*	0.3	0	0
12.	*Klebsiella pneumoniae oxytoca*	0.1	0	0
13.	*Klebsiella pneumoniae ozaenae*	0.5	10	0
14.	*Klebsiella pneumoniae oxytoca*	0.7	0	0
15.	*Klebsiella pneumoniae pneumoniae*	0.3	11	0
16.	*Pseudomonas aeruginosa*	0.6	0	0
17.	*Klebsiella pneumoniae pneumoniae*	0.2	0	0
18.	*Klebsiella pneumoniae pneumoniae*	0.3	16	14
19.	*Klebsiella pneumoniae pneumoniae*	0.2	16	14
20.	*Klebsiella pneumoniae pneumoniae*	0.1	12	0

21.	*Pseudomonas putida*	0.6	15	13
22.	*Proteus mirabilis*	0.7	16	13
23.	*Burkholderia cepacia*	0.8	17	15
24.	*Burkholderia cepacia*	0.9	17	8
25.	*Shewanella putrefaciens*	0.8	16	12
26.	*Acinetobacter haemolyticus*	0.7	13	6
27.	*Enterobacter cloacae*	0.7	14	6
28.	*Burkholderia cepacia*	0.7	14	12
29.	*Burkholderia cepacia*	0.8	14	8
30.	*Acinetobacter baumannii*	0.7	13	15
31.	*Serratia marcescens*	0.9	12	0
32.	*Providentia stuartii*	0.8	12	15
33.	*Pseudomonas putida*	0.7	13	12
34.	*Acinetobacter baumannii*	0.7	15	11
35.	*Citrobacter diversus*	0.7	16	0
36.	*Escherichia coli*	0.7	16	10
37.	*Escherichia coli*	0.7	16	12
38.	*Burkholderia cepacia*	0.7	0	0
39.	*Acinetobacter iwofii*	0.7	0	0
40.	*Pseudomonas aeruginosa*	0.9	14	16
41.	*Burkholderia cepacia*	0.7	14	8
42.	*Pseudomonas aeruginosa*	0.9	14	16
43.	*Burkholderia cepacia*	0.7	14	12

44.	*Pseudomonas aeruginosa*	0.9	14	16
45.	*Pseudomonas aeruginosa*	0.9	14	16
46.	*Pseudomonas aeruginosa*	0.9	14	16
47.	*Proteus mirabilis*	0.7	18	14
48.	*Proteus mirabilis*	0.7	14	13
49.	*Pseudomonas putida*	0.7	13	12
50.	*Staphylococcus aureus*	0.8	14	14

MAR: multiple antibiotic resistance; ALA: Aqueous leaf extract of *A. muricata;* ESA: Ethanol fruit-skin extract of *A. muricata*.

Antibacterial screening of extracts of *A. muricata* against test organisms responsible for specific acute respiratory infections, particularly pneumonia and tuberculosis; revealed bioactivity as earlier reported by Solomon-Wisdom *et al.* [34] and Radji *et al.* [36] respectively. Bussmann *et al.* [37] recorded huge antibacterial activities in screening aqueous and ethanol extracts of approximately a gross population of plant species including *Annona muricata* against selected gram-negative and gram-positive bacteria. They observed that 81% of the ethanol extracts inhibited *Staphylococcus* aureus, while 36% inhibited *Escherichia coli* with minimum inhibitory concentration range between 0.008 - 256 $mgmL^{-1}$. This result confirms the activity of extracts tested in this study, particularly the two extracts: ESA and ALA which showed the most remarkable antibacterial activity with MIC of 6.25 and 25 $mgmL^{-1}$ respectively. In the same vein, the diameter of inhibition for fruit-skin ethanol extract of *Annona muricata* against *Staphylococcus aureus*, *Klebsiella pneumonia* and *Pseudomonas aeruginosa*: 17 ± 0.58, 16 ± 0.00 and 16 ± 0.58 mm respectively, and for aqueous leaf extract: 16 ± 0.58, 16 ± 1.15 and 14 ± 0.58 mm; all agree with the findings of Solomon-Wisdom *et al.* (34] and Yakubu *et al.* [35] in inhibition zone and that of Martínez *et al.* [38] with Meléndez and Capriles [39]; who reported *Staphylococcus aureus* as one the most susceptible bacteria to plant extracts. The non-significant difference in antibacterial activities observed between the aqueous leaf and fruit-skin ethanol extracts of *A. muricata* implies that the antibacterial activities of both extracts against selected organisms are closely related; although several studies have proved that the extent of antibacterial activity for *Annona muricata* is solvent-dependent; affirming that ethanolic and methanolic extracts showed higher activity compared to aqueous extract which often times show low or no antibacterial activity [40]. The antibacterial activity demonstrated by ALA and ESA against 80% of the multidrug-resistant organisms isolated from immunocompromised patients co-infected with acute respiratory infections, with zones of inhibition ranging from 7-18 mm agree well with the findings of [40].

Conclusion

Extracts from the fruit-skin and leaves of *Annona muricata* are composed of diverse phytochemicals that possess both antibacterial and possible antimultidrug-resistant bioactive principles that may serve as potential antidote against deadly bacterial infections. There is need for more research on the fruit-skin extracts of *Annona muricata* Linn., as future findings from this may further enhance life quality via ameliorating targeted infections. Bioactive-directed fractionation of the fruit-skin extract to isolate pure bioactive compounds is also highly recommended.

Acknowledgements

We duly acknowledge the entire management of Covenant University for making requisite facilities available for scientific research and selected Technical staff in the Biochemistry for collaborative support and assistance in getting the project executed.

Funding

This research did not receive any specific grant from funding agencies in the public, commercial, or not-for-profit sectors.

Conflict of Interest

We declare no conflict of interest whatsoever.

References

[1]. Siddharth R. Krishnamurthy, Andrew B. Janowski, Zhao G, Barouch D, Wang D (2016). Hyperexpansion of RNA Bacteriophage Diversity. PLOS Biology, **14 (3)**: e1002409.

[2]. Vitorino LC, Bessa LA (2017). Technological Microbiology: Development and Applications. Frontiers in Microbiology, **8**: 827.

[3]. Forterre P (2010). Defining Life: The Virus Viewpoint. Origins of Life and Evolution of Biospheres **40(2)**: 151–160.

[4]. Brown JH (2014). Why are there so many species in the tropics? *Journal of Biogeography*, **41(1)**: 8–22.

[5]. University of Georgia (1998). First-Ever Scientific Estimate of Total Bacteria on Earth Shows Far Greater Numbers Than Ever Known Before. *Science Daily*. Retrieved 10 November 2014.

[6]. Vaisberg HG, Eschweiler U, Reitz G, Wehner J, Willimek R, Strauch K (1995). Biological responses to space: results of the experiment "Exobiological Unit" of ERA on EURECA I. *Advances in Space Research*, **16(8)**: 105–118.

[7]. Zhang KD, Bieger-Dose A, Dillmann R, Gill M, Klein A, Meinert H, Nawroth T, Risi S, Stride C. (1995). ERA-experiment "space biochemistry". *Advances in Space Research*, **16(8)**: 119–129.

[8]. Eckburg PB, Bik EM, Bernstein CN, Purdom E, Dethlefsen L, Sargent M, Gill SR, Nelson KE, Relman DA (2005). Diversity of the human intestinal microbial flora. Science, 308(5728):1635-1638.

[9]. Bry L, Falk PG, Midtvedt T, Gordon JI (1996). A model of host-microbial interactions in an open mammalian ecosystem. Science. 273(5280):1380-1383.

[10]. Tlaskalova-Hogenova H, Stěpánková R, Kozáková H, Hudcovic T, Vannucci L, Tučková L, Rossmann P, Hrnčíř T, Kverka M, Zákostelská Z, Klimešová K, Přibylová J, Bártová J, Sanchez D, Fundová P, Borovská D, Srůtková D, Zídek Z, Schwarzer M, Drastich P, Funda DP. (2011). The role of gut microbiota (commensal bacteria) and the mucosal barrier in the pathogenesis of inflammatory and autoimmune diseases and cancer: contribution of germ-free and gnotobiotic animal models of human diseases. Cellular and Molecular Immunology. 8:110-120.

[11]. Moreira D, López-García P. (2002). The molecular ecology of microbial eukaryotes unveils a hidden world (PDF). Trends in Microbiology 10 (1): 31–38.

[12]. Kumamoto CA, Vinces, M.D. (2005). Contributions of hyphae and hypha-co-regulated genes to Candida albicans virulence. Cellular Microbiology 7 (11): 1546–1554.

[13]. Anyanwu MU, Okoye R.C. (2017). Antimicrobial activity of Nigerian medicinal plants. *Journal of Intercultural Ethnopharmacology*, **6(2)**: 240–259.

[14]. Iyanda-Joel WO, Adegbite OS, Ajetunmobi OB, Rotimi SO, Iweala EEJ, Chinedu SN. 2019. Phytochemical, antioxidant and mitochondrial permeability transition studies on fruit-skin ethanol extract of *Annona muricata*. Conference Proceedings of the 3rd International Conference on Science and Sustainable Development; May 6-8, 2019 at the Covenant University Center for Research, Innovation and Development, Covenant University, Ota, Nigeria.

[15]. Badrie N, Schauss A. 2010. Soursop (*Annona muricata* L.): Composition, nutritional value, medicinal uses and toxicology. In: Watson R, Preedy V, editors. Bioactive foods in promoting health. Oxford: Academic Press; pp. 621–643.

[16]. Pinto AC de Q, Silva EM. 1996. Graviola for Export. Technical Aspects of Production. Embrapa / SPI, Brasilia, Brazil.

[17]. National Research Council. 1975. Underexploited tropical plants with promising economic value: report of an ad hoc panel of the Advisory Committee on Technology Innovation, Board on Science and Technology for International Development, Commission on International Relations. National Academy of Sciences. Washington DC., USA; Libraries Australia ID: 5664342. https://trove.nla.gov.au/version/16206736.

[18]. Nwinyi OC, Chinedu N.S. and Ajani OO (2008). Evaluation of antibacterial activity of Pisidium guajava and Gongronema latifolium. Journal of Medicinal Plants Research, **2(8)**: 189-192.

[19]. Coria-Téllez AV, Montalvo-Gónzalez E, Yahia EM, Obledo-Vazquez EN. *Annona muricata*: A comprehensive review on its traditional medicinal uses, Phytochemicals, pharmacological activities, mechanisms of action and toxicity. Arabian J. Chem., 2018; 11(5):662–691.

[20]. Viera GH, Mourao JA, Angelo AM, Costa RA,Vieira RH. Antibacterial effect (*in vitro*) of Moringa *oleifera* and *Annona muricata* against gram positive and gram negative bacteria. Rev Inst Med Trop Sao Paulo, 2010; 52:129–132.

[21]. Lawal, I.O., Grierson, D.S. and Afolayan, A. J. (2015). Phytochemical and antioxidant investigations of a *Clausena anisata* Hook, a South African medicinal plant. *African Journal of Traditional, Complementary and Alternative Medicine*, **12(1)**: 28-37.

[22]. Larbie, C., Arthur, F.K.N., Terlabi, E.O. and Woode, E. (2011). Evaluation of acute and subchronic toxicity of Annona muricata (Linn.) aqueous extract in animals. European Journal of Experimental Biology, 1(4): 115–124.

[23]. Igbinoba, V.I. (2014). Comparative analysis of the phytochemical and in vitro antioxidant properties of Annona muricata leaf extracts. Unpublished Dissertation. pp 31–38.

[24]. Sarver, J.P., Blumer, L.S. and Denton, M.K. (2008). Methanol-free method of extracting secondary chemicals. In: Tested Studies for Laboratory Teaching, K.L. Clase (Editor). Proceedings of the 29th Workshop/Conference of the Association for Biology Laboratory Education (ABLE), 29: 376–388.Edeoga, H.O., Okwu D.E. and B.O. Mbaebie, B.O. (2005). Phytochemical constituents of some Nigerian medicinal plants. *African Journal of Biotechnology* **4(7)**:685-688.

[25]. Matyash, V., Liebisch, G., Kurzchalia, T.V., Shevchenko, A. and Schwudke, D. (2008). Lipid extraction by methyl-tert-butyl ether fir high throughput lipidomics. Journal of Lipid Research, 49(5): 1137–1146.Kokate, C.K. (2000). In: Practical Pharmacognosy, Vallabh Prakashan, Delhi, India, p. 218

[26]. Ronpirin, C., Charueksereesakul, T., Thongrakard, V. and Tencomnao T. (2014). Effects of ethanolic extract of Annona squamosa L. leaves on the expression of EGFR. Journal of Chemical and Pharmaceutical Research, 6(4): 791–797.

[27]. Matyash, V., Liebisch, G., Kurzchalia, T.V., Shevchenko, A. and Schwudke, D. (2008). Lipid extraction by methyl-tert-butyl ether fir high throughput lipidomics. Journal of Lipid Research, 49(5): 1137–1146.

[28]. Harborne, J.B. (1998). Phytochemical methods: A guide to modern techniques of plant analysis. 3rd edition. Chapman and Hall, London: p 235. ISBN-13: 978-0-412-23050-9 e-ISBN-13: 978-94-009-5921-7 doi: 10.1007/978-94-009-5921-7

[29]. Sofowora, A. (1993). Phytochemical screening of medicinal plants and traditional medicine in Africa. 2nd Edition Spectrum Books Limited, Nigeria: pp 150 – 156.

[30]. Trease, G.E. and Evans, W.C. (1989). Pharmacognosy, 12th Edition, Balliere-Tindall, London: pp 241-260.

[31]. Basu, S., Ghosh, A. and Hazra, B. (2005). Evaluation of the antibacterial activity of *Ventilago madraspatana* Gaertn., *Rubia cordifolia* Linn., and *Lantana camara* Linn.: isolation of emodin and physcion as active antibacterial agents. *Phytotherapy Research*, **19**: 888-894.

[32]. Sateesh, V. and Naikpatil R.J.L (2011). Selective isolation and antimicrobial activity of rare actinomycetes from mangrove sediment of Karwar. Journal of Ecobiotechnology, 3(10): 48-53.

[33]. Bauer, A.W., Kirby, W.M.M., Sherris J.C. and Turck, M. (1966). Antibiotic susceptibility testing by a standardized single disk method. American Journal of Clinical Patholology, 45: 493–496.

[34]. Solomon-Wisdom, G.O., Ugoh, S.C. and Mohammed, B. (2014). Phytochemical screening and antimicrobial activities of Annona muricata (L.) leaf extract. American Journal of Biological, Chemical and Pharmaceutical Sciences 2: 1–7.

[35]. Yakubu, O.F., A.H. Adebayo, T.O. Famakinwa, O.S. Adegbite, T.A. Ishola, L.O. Imonikhe, O.A. Adeyemi, O.A. Awotoye, and E.J. Iweala (2018). Antimicrobial and toxicological

studies of ricinodendron heudelotii (Baill.). Asian Journal of Pharmaceutical and Clinical Research, 11(1): 299-305.

[36]. Radji, M., Kurniati, M. and Kiranasari, A. (2015). Comparative antimycobacterial activity of some Indonesian medicinal plants against multi-drug resistant Mycobacterium tuberculosis. Journal of Applied Pharmaceutical Science, 5(1): 19–22.

[37]. Bussmann, R.W., Malca, G., Glenn, A., Sharon, D., Nilsen, B., Parris, B. and Townesmith, A. (2010). Minimum inhibitory concentrations of medicinal plants used in Northern Peru as antibacterial remedies. Journal of Ethnopharmacology, 132: 101–108.

[38]. Martínez, M.J., Betancourt, J., Alonso-González, N. and Jauregui, A. (1996). Screening of some Cuban medicinal plants for antimicrobial activity. Journal of Ethnopharmacology, 52: 171–174.

[39]. Meléndez, P.A. and Capriles, V.A. (2006). Antibacterial properties of tropical plants from Puerto Rico. Phytomedicine, 13(4): 272–276.

[40]. Viera, G.H., Mourao, J.A., Angelo, A.M., Costa, R.A. and Vieira, R.H. (2010). Antibacterial effect (*in vitro*) of Moringa oleifera and Annona muricata against gram positive and gram negative bacteria. Journal of the Institute of Tropical Medicine São Paulo, 52: 129–132.

The Impact of Traffic Gridlock on Values of Residential Properties in Apapa, Lagos Nigeria

C O.Iroham,[1], T Okubanjo,[1] O Joshua[2], H I Okagbue,[3] M E Emetere,[4] O A Akinjare,[1] J N Peter,[1] and O M Akinwale[1]
[1]Department of Estate Management, Covenant University, Ota, Ogun State Nigeria
[2]Department of Building Technology, Covenant University, Ota, Ogun State Nigeria
[3]Department of Mathematics, Covenant University, Ota, Ogun State Nigeria
[4]Department of Physics, Covenant University, Ota, Ogun State Nigeria

Abstract: Traffic gridlock has been a concern in research particularly in determining its effect on property values. Resultant findings have either been a negative or positive impact. For the case of Apapa in Lagos State which is the focus of this present study, traffic gridlock has been an experience of over ten years, hence the need to discover its impact on various property value types as against the collective findings in most previous research. Student-T test of significance at a 95% level of confidence was adopted with results derived for all property types of the calculated T result far outweighing that of the stipulated Tabulated T result of 1.645. These findings for all property types indicate that the increase in property values for various property types have not been distorted by the traffic gridlock and as such investors in residential properties in the study area will always make their profit notwithstanding the gridlock. The researchers hereby advocated further research in determining the choice of residency in Apapa notwithstanding the traffic gridlock.

Keywords: Traffic Gridlock, Residential Property, Property Values, Apapa

1. Introduction

One vital way by which cities develop has been linked to free movement to enhance economic activities which can easily be boosted by a good traffic flow [1]. However, one major hindrance to the free flow of movement is traffic gridlock which the World Bank has attributed to create about 54.5% of all noticeable urban transport externalities. Traffic gridlock which is also regarded as traffic congestion can be defined as a form of traffic blockage whereby incessant queue of various automobiles particularly vehicles obstruct a whole network of interconnected streets resulting to slower speeds, longer trip times and even a complete standstill of vehicular movement. Traffic gridlock is broadly caused by two principal categories. They are micro-level and macro-level factors. The micro-level factors relate to the road while macro-level factors relate to the overall demand for road use. Examples of the micro-level factors include struggle for people and freight to move at the same time, too many vehicle competing for limited road space, accident, vehicle breakdown, poorly timed traffic signal and other special occurring events on the roads such as political rallies, mass social gatherings amongst others. Examples of the macro-level factors include land use patterns, car ownership trends, regional economic dynamics, infrastructure investment, income level, and employment patterns amongst others [2]. In the long run traffic gridlock results to increase cost and loss of work hours amongst others [3]. Others are lost worker productivity, lost business earnings, higher transportation cost, increase in time spent on travelling, additional cost and less consistent times for pick-up and delivery for those that operate trucks. In fact from study conducted by the Portland Business Alliance, Port of Portland and Metro it was discovered that traffic gridlock will result to the loss of 6, 500 jobs in the local economy and $844 million a year by 2025[4]. With the increase in ownership of vehicles to support transportation particularly in the developing economies to a tune of about 15 to 20% annually, traffic gridlock is not farfetched particularly in Lagos State, Nigeria [5]. This traffic gridlock has however been established in virtually every capital city in Nigeria [6]. However, for the case of Lagos State which is believed to be one of the fastest growing cities of the world, traffic gridlock is very much evident. This is to buttress the assertion of [7-10] that traffic gridlock is a function of the population of any area. Other reasons attributed to traffic gridlock in Nigeria particularly in Lagos State include, the inertia of spatial planning in metropolitan Lagos[11]; Inability of road network to contain the volume of traffic that make use of it [12]; insufficient road infrastructure, dilapidated road condition; inadequate traffic planning, absence of integrated transport systems, behavior of road users particularly the drivers and accidents [13,14]. According to Aderamo & Atomode [15] road intersection, traffic wardens and parking problems are major causes of traffic gridlocks on the roads. Ukpata and Etika [16] disclosed that reckless driving is the most important cause of traffic congestion while poor road networks, lack of parking facilities, poor drainage, inadequate road capacity, poor parking habits, poorly designed junctions/roundabouts, lack of efficient mass transport system, poor traffic control/management and presence of heavy vehicles were discovered to be other trigging factors. Olaogbebikan, Ikpechukwu, Akinsulire and Enosko [17] attributed traffic gridlock in the state to much dependency on small occupation vehicles, indiscriminate parking, on-street trading, narrow road, incapacity of the traffic management agencies to evacuate crashed or breakdown vehicles on time, loading and off-loading of goods and passengers on the road, and ineffective traffic control measures. Results from research conducted by Popoola, Abiola and Adeniji [18] reveals inadequate road capacity, poor drainage system, poor parking habit, poor traffic management, presence of heavy trucks, poor road pavement, poor design junctions/round-about, lack of road furniture and lack of pedestrian and parking facilities as causes of traffic gridlock in the state. Impatience by road users in the state has also been discovered as another triggering factor of traffic gridlock in the state [19]. Notwithstanding efforts by scholars in transport management and various road traffic management strategies employed by the state such as the one way approach, construction of new routes and flyovers, park and ride system, application of odd and even numbers light rail schemes, use of water ways, Bus Rapid Transit (BRT) and para-mass transit system amongst others [20-27], the traffic gridlock is yet to be eliminated. While Lagos is synonymous with traffic gridlock [1, 26, 28-30] however that of Apapa, which lies around the mouth of Lagos lagoon, has become unbearable due to the location of the Tincan Island Port, the second largest in West Africa after that in Abijan. In Apapa, there are noticeable tankers packed on the road causing traffic gridlock in the area substantiating the findings of [26, 32]. This traffic gridlock has manifested in various forms describing the different types of gridlock propounded by Brownfield, Graham, Eveleigh, Ward, Robertson and Allsop [33]. These are recurrent congestion occurring at regular peak times usually mornings and evenings. Non-recurrent congestion, occurring at non regular times usually due to vehicle breakdowns and pre-congestion also known as borderline congestion which occurs where there is a free-flow conditions breakdown while full congestion is yet to occur. Due to the slow pace of movement which this traffic gridlock brings about, its out-turn on economic development can always been linked. One of such is the linkage of gridlock with property values.

2.0 Effect of road, accessibility, traffic gridlock on property values

Road can be described as the greatest investment of any nation where its citizens are shareholders. This statement buttresses the import of road, its network and accessibility because in every given society people are bound to abide. The accessibility of any given road thereby relates to location which has been attributed as the principal determinant of property values[34]. This explains why according to Aderamo [23], road network constitutes a significant constituent in urban development as roads provide accessibility to the different land uses in the urban area. In fact even the mere announcement of proposed road projects is believed to have effect on property values which was discovered to be a function on the size of the proposed project [35]. This is in line with the classic urban location theory which states that lower transport costs resulting from accessibility will bring about higher land and property values. Notwithstanding, academic researches needed to be conducted to situate peculiarities of various places. In light of this, various researches have tried to link road, its improvement, accessibility with economic development of which lies research in values of properties[36]. Right from 1956 when the Highway revenue Act came up, studies have sprung up on the effect of highways on property values. One of such remarkable earlier studies is that carried out by Adkins [37] in Dallas, Houston and San Antonio. An appreciable rise in value of about 450% was discovered particularly in Dallas and Houston while the appreciable rise in value was more in manufacturing land use to a tune of about 200% while single-family residential land use had an insignificant rise in value in San Antonio. Bone and Wohl [38] carried out a similar study in Massachusetts. They had divergent results of which in one community the highway has resulted in the appreciation of property values while in the other community the highway had little effect on property values. Ever since then other researchers have also come up with various divergent findings on highway improvement on property values. One of such recent studies is that carried out in Netherlands where it was discovered that changes in accessibility resulting from highway development project has a significant positive effect on the price of housing in close by municipalities, even though the ensuing enhancement in noise pollution and traffic intensity levels brought about a decline in prices. However, the combination of all the effects of all externalities, the effect of highway development on house prices is positive and the effect is prominent even before the completion of the project as a result of public anticipation effect [39]. Some other researchers have identified road construction to have a positive impact on property values [40-44] discovered varied effects of both a positive and negative nature while Aigbe, Ogundele and Aliu [45] discovered a negative effect on property values from road construction. The varied results gotten from the researches are not farfetched considering the fact that apart from boost in economic activities, certain externalities such as noise and air pollution, compulsory purchase and acquisition by government results to the negative effects on property values [46-49]. It is obvious then that whatsoever will inhibit the free flow of movement of people on the way will attract attention for research. This has even led to the study of effect of speed bumps on residential property values where divergent results were attained. To most of the respondents speed bumps is related with decline in values of residential property while for the elderly and homeowners with children, that enhances the value of residential properties [50]. Much more will be the likelihood of an envisaged attention by researchers on greater impediment to movement such as traffic gridlock even as it relates to property values. However, there is dearth of literature on impact of gridlock on property values.

Research on the effect of traffic gridlock on property values as recorded in the early eighties was conducted in Grand Rapids Michigan. It was discovered that residential property values displayed a surprising high elasticity with respect to decrease in traffic flow [51]. This is against findings gotten from some latter researches where decrease in traffic flow results to decrease in housing values [52-54]. There was a varied result gotten from a research conducted in Portsmouth city, England where it turned out to be that the effect of traffic gridlock on property (terrace houses) values varies according to economic conditions. During the period of high sales demand and low availability of property the impact is less compared to periods of low demand and higher supply

of property [55]. Hou [56] likewise discovered that house prices tend to increase more with vicinities of higher free flow that those of congested flow. However, this is more pronounced within the middle income neighbourhood compared to their counterparts in lowest and highest income neighbourhoods. In a more recent research relating traffic gridlock with house prices it was discovered that while there no significant relationship exist between congestion growth and house prices in rural areas and non-metro areas, the situation is different from the urban areas with an obvious adverse effect on house prices [57].
With all these varied results gotten it is imperative that a study of traffic gridlock and its impact on property values be studied in Apapa, Lagos State, Nigeria. This will enable findings of actual outcome of this gridlock.

3.0 Methods

This present study being a cross-sectional survey involves the distribution of seventy (70) questionnaires and Eight (8) questionnaires to practicing estate surveying and valuation firms in Lagos Island and Apapa respectively. The choice of Lagos Island is due to the much concentration of estate surveying and valuation firms in Lagos Island while the entire estate surveying and valuation firms in Apapa, being the study area were surveyed. The field work was carried out within a space of three months (February –May, 2019). The data gotten was principally on rental values of all types of residential properties in Apapa within a space of 18 years (year 2000 and year 2018). This was intended to capture both the period of free traffic flow and traffic gridlock in the study area. Data collected was analyzed using the Student T Test of Significance at the 95% confidence level hypothesized that the traffic gridlock had a negative effect on the various residential; property values in the study area. The choice of student T was due to its applicability in testing hypothesized significant difference between two pairs of sample means. In this study property values before and after the traffic gridlock in the study area were compared.

4.0 Results and Discussion

Out of the seventy (70) questionnaires distributed to estate firms in Ikeja and eight (8) questionnaires in Apapa, Lagos state, a total of sixty-eight (63) questionnaires were retrieved from the estate firms in Lagos Island and 5 questionnaires from Apapa. This represents a response rate of 87%. The researchers considered this substantial enough for a conclusive result. Moreover all estate surveying and valuation firm studied have dealings in the study area and have been in professional practice for between seven (7) and twenty-six (26) years.
For the various property types the values of such properties were analyzed before and after the gridlock using the Student-T Test of Significance at the 95% confidence level with the assumption that all other determinants of property values are kept constant. The analysis of the various property types are as follows:

Tenement building

Before Traffic Gridlock

$X_1 = 62.85$

$S^2_1 = 18250.84$

`N_1 .506

After Traffic Gridlock

$X_2 = 102.87$

$S^2_2 = 73574.71$

N_2 .619

Calculating F ratio:

$\frac{73574.71}{18250.84} = 4.0313$

Tab f = 1.00

Since Calf >tabf, we reject the null

Hence $S^2_1 . S^2_2$

Also $N_1 . N_2$

We deduce using Separate variance model with $_d f = (N_1 . N_2) \div 2$

$T = \frac{X_1 - X_2}{\sqrt{((S^2_1 . N_1) + (S^2_2 . N_2))}}$

$\frac{62.85 - 102.87}{\sqrt{((18250.84 \div 506) + (73574.71 \div 619))}}$

$\frac{40.02}{\sqrt{((36.0689) + (118.8606))}}$

$\frac{40.02}{\sqrt{(154.9295)}}$

$\frac{40.02}{12.4471}$

Cal T = 3.2152

Tab T = $_d f = (N_1 . N_2) \div 2$ (506 +619) ÷ 2 = 562.5

1 tail test directional of 562.5 @ 0.05 = 1.645

Decision Rule

Since Cal T > Tab T we reject null hypothesis meaning

Property Values of Tenement Building after the gridlock in Apapa is greater than Property Values of Tenement Building before the gridlock. Hence the gridlock had no impact on the increase of property values of tenement building in Apapa

Self-Contain
Before Traffic Gridlock
X_1= 124.78
S^2_1=2906.75
N_1 .574

After Traffic Gridlock
X_2= 194.73
S^2_2=6938.08
N_2 .513

Calculating F ratio:
6938.08
2906.75 = 2.3869
Tab f = 1.00
Since Calf >tabf, we reject the null
Hence S^2_1.S^2_2
Also N_1.N_2
Wededuce using Separate variance model with $_d$f= (N_1.N_2)÷2

T= X_1.X_2
√((S^2_1.N_1) + (S^2_2.N_2))

124.78 – 194.73
√ ((2906.75 ÷ 574) + (6938.08÷ 513))

69.95
√ ((5.0640) + (13.5245))

69.95
√ (18.5885)
69.95
4.3114
Cal T = 16.2244
Tab T = $_d$f= (N_1.N_2)÷2 (574 +513) ÷ 2 = 543.5
1 tail test directional of 543.5 @ 0.05 = 1.645
Decision Rule
Since Cal T > Tab T we reject null hypothesis meaning
Property Values of Self Contain after the gridlock in Apapa is greater than Property Values of Self Contain before the gridlock. Hence the gridlock had no impact on the increase of property values of Self Contain in Apapa

Two bedroom Flat
Before Traffic Gridlock
X_1= 180.03
S^2_1= 8612.10
N_1 .454

After Traffic Gridlock
X_2= 353.98
S^2_2=29803.76

N_2 .452

Calculating F ratio:
29803.76
8612.10 = 3.4607
Tab f = 1.00
Since Calf >tabf, we reject the null
Hence S^2_1.S^2_2
Also N_1.N_2
We deduce using Separate variance model with $_df$= (N_1.N_2)÷2

T= $\frac{X_1 . X_2}{\sqrt{((S^2_1.N_1) + (S^2_2.N_2))}}$

$\frac{180.03 - 353.98}{\sqrt{((8612.10 \div 454) + (29803.76 \div 452))}}$

$\frac{173.95}{\sqrt{((18.9694) + (65.9375))}}$

$\frac{173.95}{\sqrt{(84.9069)}}$
$\frac{173.95}{18.8779}$
Cal T = 18.8779
Tab T = $_df$= (N_1.N_2)÷2 (454 +452) ÷ 2 = 453
1 tail test directional of 453 @ 0.05 = 1.645

Decision Rule
Since Cal T > Tab T we reject null hypothesis meaning
Property Values of Two Bedroom Flat after the gridlock in Apapa is greater than Property Values of Two Bedroom flat before the gridlock. Hence the gridlock had no impact on the increase of property values of Two Bedroom Flat in Apapa

Three Bedroom Flat
Before Traffic Gridlock
X_1= 524.42
S^2_1=72479.42
N_1 .495

After Traffic Gridlock
X_2= 1033.96
S^2_2= 45160.04
N_2 .490

Calculating F ratio:
72479.42
45160.04 = 1.6049
Tab f = 1.00
Since Calf >tabf, we reject the null

IOP Conf. Series: Earth and Environmental Science **331** (2019) 012030 doi:10.1088/1755-1315/331/1/012030

Hence $S^2_1 \cdot S^2_2$
Also $N_1 \cdot N_2$
We deduce using Separate variance model with $_df= (N_1 \cdot N_2) \div 2$

T= $\underline{X_1 \cdot X_2}$
$\sqrt{((S^2_1 \cdot N_1) + (S^2_2 \cdot N_2))}$

$\underline{524.42 - 1033.96}$
$\sqrt{((72479.42 \div 495) + (45160.04 \div 490))}$

$\underline{509.54}$
$\sqrt{((146.4231) + (92.1633))}$

$\underline{509.54}$
$\sqrt{(238.5864)}$
$\underline{509.54}$
15.4462
Cal T = 32.9880
Tab T = $_df= (N_1 \cdot N_2) \div 2$ (495 +490) ÷ 2 = 492.5
1 tail test directional of 492.5 @ 0.05 = 1.645

Decision Rule

Since Cal T > Tab T we reject null hypothesis meaning
Property Values of Three Bedroom Bungalow after the gridlock in Apapa is greater than Property Values of Three Bedroom Bungalow before the gridlock. Hence the gridlock had no impact on the increase of property values of Three Bedroom Bungalow in Apapa

Four Bedroom House

Before Traffic Gridlock
X_1= 1234.71
S^2_1=442192.31
N_1 .507

After Traffic Gridlock
X_2= 1942.06
S^2_2=1099820.10
N_2 .452

Calculating F ratio:
$\underline{1099820.10}$
442192.31 = 2.4872
Tab f = 1.00
Since Calf >tabf, we reject the null
Hence $S^2_1 \cdot S^2_2$
Also $N_1 \cdot N_2$
We deduce using Separate variance model with $_df= (N_1 \cdot N_2) \div 2$

T= $\underline{X_1 \cdot X_2}$
$\sqrt{((S^2_1 \cdot N_1) + (S^2_2 \cdot N_2))}$

$\underline{1234.71 - 1942.06}$
$\sqrt{((442192.31 \div 507) + (1099820.10 \div 452))}$

$$\frac{707.35}{\sqrt{((872.1742) + (2433.2303))}}$$

$$\frac{707.35}{\sqrt{(3305.4045)}}$$

$$\frac{707.35}{57.4926}$$

Cal T = 12.3033

Tab T = $_d$f= $(N_{1+}N_2)\div 2$ (507 +452) ÷ 2 = 479.5

1 tail test directional of 479.5 @ 0.05 = 1.645

Decision Rule

Since Cal T > Tab T we reject null hypothesis meaning

Property Values of Four Bedroom House after the gridlock in Apapa is greater than Property Values of Four Bedroom House before the gridlock. Hence the gridlock had no impact on the increase of property values of Four Bedroom House in Apapa

5.0 Recommendation/Concluding Remarks

This research work is novel as it did not just study traffic gridlock with respect to property values which is sparse in literature but focused on Lagos, Apapa known for its huge traffic issues. More so Lagos is one of the vibrant property markets in Nigeria together with Portharcout and Abuja. It has also been established that majority of the estate surveyors and valuers in Nigeria have their head offices in Lagos State. An earlier related study in Nigeria focused on commercial properties in Bauchi State while this present study is hinged on residential properties. Other earlier researches carried out outside the country studied residential properties holistically but this present study focused on the individual residential properties found in the study area. Thus, this present research has been able to prove that beyond all negative externalities associated with traffic gridlock, Apapa still commands increase in the various property values. It is therefore obvious that over the years that the traffic gridlock in Apapa had no effect on the values of all residential property types. Hence, the researchers opine that investors in residential properties can always make such investment which will invariably turn out profitable in the study area notwithstanding the traffic gridlock. The question then is what are the features in Apapa that makes residents to opt for the area notwithstanding the traffic gridlock? Are there other property types such as commercial, recreational, industrial that are affected by this gridlock? These are questions which further research related to this work in the study area will find out.

Acknowledgement

The researchers hereby appreciate Covenant University, Ota, Ogun State Nigeria for the sponsorship of this work

References:

[1]. Adaramo, A. (2003). A Graph Theoretic Analysis of Intra-Urban Road Network in Ilorin, Nigeria, Journal of the Nigerian Institute of Social and Economic Research, 17, 18, 19, (1), (2), 221-240

[2]. Adejare , Q. A., Olusina, J. O., and Olaleye, J. B. (2017). Developing Bimodal Choice Functions for Lagos Urban Transportation System using Hybrid Technique. Journal of Engineering Research, 22(1), 11-22

[3]. Adekanmbi, S. T. (2015). The impact of traffic congestion on Lagos – Ibadan Road (A case study of Lagos end toll gate, Lagos State. Unpublished B.Sc Dissertation Department of Transport Management Faculty of Management Sciences, Ladoke Akintola University of Technology, Ogbomosho Oyo State Nigeria

[4]. Aderamo, A. J. (1990). Road Development and Urban Expansion: The Case of Ilorin, Nigeria. Ph.D. Thesis submitted to the Department of Geography, University of Ilorin, Ilorin. Kwara state.

[5]. Aderamo, A. J & Atomode. T. I. (2011). Traffic Congestion at Road Intersection in Ilorin, Nigeria. Australian Journal of Basics and Applied Sciences, 5 (9), 1439-1448.

[6]. Adkins, W. G. (1959). Land value impacts of expressways in Dallas, Houston, and San Antonio, Texas, Highway Research Board Bulletin, Number 227, 1959

[7]. Aigbe, G.O, Ogundele, F.O and Aliu, I.R (2012). Road Facility and Maintenance in Lagos State Nigeria. British Journal of Arts and Social Sciences, 4(2), http//www.bjournal.co.uk/BJASS. Aspx

[8]. Al-Mumaiz M, and Evdorides H. (2017) Modeling the Impact of Road Construction on Land Value. A PhD Thesis submitted to the University of Birmingham UK. Ministry of Higher Education and Scientific Research, Iraq (MOHE)

[9]. Alimi R.K, Ayedun C.A and Oni A.S (2014). An Appraisal of the Relationship between Road Improvements and Immediate Neighborhood Residential Properties Values in Metropolitan Lagos. American International Journal of Contemporary Research, 4(6); 215-222

[10]. Aliyu, A. H., Abubakar, S. I. and Adamu, H. (2015). Impact of traffic congestion on commercial property rental values in Bauchi metropolis. Book of Proceedings - Academic Conference of Cambridge Publications & Research International on Sub-Sahara African Potentials in the New Millennium, 3(2). 30th July, 2015- FUTA, Hilltop Conference Hall, Akure, Ondo State

[11]. Ameyan, O. (1996). Environmental and Energy Issue in Urban Transport System Development Paper Presented at the FUMTA Seminar on Urban Transport Policy for Nigeria.Atubi, A. (2008). Journey to work pattern in the Niger Delta: An Empirical Analysis of Warri and Environs Department of Geography and Regional Planning, Delta State University, Abraka.

[12]. Aworemi, J. R., Abdul-Azeez, I. A., Oyedokun, A. J. and Adewoye, J. O (2009). A study of the causes, effects and Ameliorative Measures of Road Traffic Congestion in Lagos Metropolis. European Journal of Social Sciences. II (1), 2009

[13]. Bagby, D. G. (1980) The effect of traffic flow on residential property values. Journal of the American Planning Association, 46(1), 88-94

[14]. Bello, S. A. (1993). Urban Public Transport in a Growing City: The Case of Ilorin, Nigeria. Ph.D. Thesis Submitted to the Department of Geography, University of Ilorin, Ilorin. Kwara state.

[15]. Bolade, T. (1989). Transport in Metropolitan Lagos, Transport Review, 16(2).

[16]. Bone, A. J. and Wohl, M. (1959). Massachusetts route 128 impact study, Highway Research Board Bulletin, Number 227

[17]. Bowhill, P. (2008). The effect of traffic congestion on property values in Portsmouth. (unpublished BSc dissertation), School of Environmental Design and Management, University of Portsmouth, Portsmouth

[18]. Brownfield, J., Graham, A., Eveleigh, H., Ward, H., Robertson, S., and Allsop, R. (2003). Congestion and accident risk. Road safety research report No.44, Department of Transport UK

[19]. Carey, J. (2001). Impact of highways on property values: Case study of the superstition freeway corridor. (A. D. o. Transportation, ed.), Arizona, USA.

[20]. Chay, K. Y. and Greenstone, M. (2005). Does Air Quality Matter? Evidence from the Housing Market. Journal of Political Economy, 113(2), 376–424

[21]. Downie, A. (2008). The World Worst Traffic Jams time. Available at: http://www.time/world/article/0,8599,1733872,00.html. (Retrieve on 28th May, 2015.

[22]. Fadare, S. O. and Ayantoyinbo, B. B. (2010). A study of the effects of road traffic congestion on freight movement in Lagos Metropolis. European Journal of Social Sciences, 16(3), 429- 437

[23]. Gatauwa, J. M., and Murungi, M. (2015). Infrastructure Development and Real Estate Values in Meru County, Kenya. Research Journal of Finance and Accounting, 6(8), 212-221

[24]. Graham, J. E. and Jones, A. T. (2019). A bump in the road: speed bumps' impact on property values, International Journal of Housing Markets and Analysis, 12(1), 43-58

[25]. Hanley, N., Shogren, J. F., and White, B. (2001). "Introduction to Environmental Economics," 1/Ed. Oxford University Press

[26]. Hou, Y. (2016). Traffic congestion, accessibility to employment, and housing prices: A study of single-family housing market in Los Angeles County. Urban Studies, 54(15), 3423–3445

[27]. Iroham, C. O., Durodola, O. D., Oluwatobi, A. O. and Peter, N. J. (2015). Proceedings of the 25th International Business Information Management Association Conference- Innovation Vision 2020: From Regional Development Sustainability to Global Economic Growth, IBIMA 2015, 3756-3763

[28]. Jimi-Oni, M. and Oluwatobi, A. O. (2017). Impact of Road Rehabilitation on Property Value: A Case of Lasu-Isheri Road, Igando, Lagos State. Covenant Journal of Research in the Built Environment, 5(2), 75-88

[29]. Jin, J. and Rafferty, P. (2018). Externalities of auto traffic congestion growth: Evidence from the residential property values in the US Great Lakes. Journal of Transport Geography, 70(c), 131-140

[30]. Knaap, G. (1998). The Determinants of Residential Property Values: Implications for Metropolitan Planning, Journal of Planning Literature, 12(3), 267-282

[31]. Kutz, M., ed. (2008). "Environmentally conscious transportation," Vol. 1, pp. 1-350. John Wiley and Sons, Inc., New Jersey.

[32]. Lagos Metropolitan Area Transport Authority (LAMATA) (2014): Description of transportation projects currently happening in Lagos. http:///www.lamata.ng.com. Accessed 12/20/2016

[33]. Levkovich, O., Rouwendal, J., and Van Marwijk, R. V. (2016). The effects of highway development on housing prices. Transportation, 43(2), 379 – 405

[34]. Moses, S. O. (2011). Information technology applications in transportation system. Proceedings of the National Conference of Nigerian Society of Engineers in Calabar.

[35]. Neelawala, P., Wilson, C., and Robinson, T. (2009). Impacts of Major Roads on Property Values: An Analysis of an Existing and a Proposed Road Corridor Project,

School of Economics and Finance, Queensland University of Technology, Brisbane Australia (2009)

[36]. Obadina, E. O., and Akinyemi, O. Y. (2018). Analysis of Traffic Congestion on Lagos/Abeokuta Expressway-Agege Motorway in Lagos Metropolis. Journal of Environment and Earth Science, 8(1), 7-17

[37]. Odeleye, J. A. and Oni, S. I (2007). A study of road traffic congestion in selected corridors of Metropolis Lagos. Nigeria. Proceedings of 11th World Conference on Transport Research, Berkeley, California. 24/6/2007

[38]. OECD (2007). (Organisation for Economic Cooperation and Development) and ECMT (European Conference of Ministers of Transport). Managing Urban Traffic Congestion. Joint Transport Research Centre

[39]. Ogunbodede, E. F. (2009). Assessments of Traffic Congestions in Akure (Nigeria) using GIS approach: Lessons and Challenges for Urban Sustenance; at http://download.sue- mot.org/Conference007/Papers/Ogunbodede.pdf. Retrieved on 28th May, 2015

[40]. Ogunleye O.S. (2011). Spatial Structure of Road Infrastructure in Ekiti State, Nigeria: Options for Transformation. Advances in Natural Science, 4(2), 138-142

[41]. Ogunsanya, A. A. (1987). Traffic Congestion in an Urban Centre: The Case of Ilorin, Nigeria. Nigerian Geographical Journal, 27(1& 2), 84-95.

[42]. Olagunju, K. (2015). Evaluating traffic congestion in developing countries – a case study of Nigeria. A paper presented at the 2015 chartered institute of logistics and transport (CILT) Africa forum held at Mount Meru Hotel Arusha, Tanzania

[43]. Olaogbebikan, J. E, Ikpechukwu, N., Akinsulire, E. S. & Enosko, O. (2013). Traffic Management Problems in Lagos: A Focus on Alaba International Market Road, Ojo, Lagos State Nigeria. Journal of Economics and Sustainable Development, 4(4), 144-153

[44]. Omiunu, F. G. I. (1988). Flooding and traffic management in Benin City Region in Sada, P.O. and Odemerho, F.O. (eds) Environmental Issues and Management Development Lagos: Evans Brothers (Nig.) Ltd.

[45]. Oruonye, E. D. (2014). An assessment of the impact of road construction on land use pattern in urban centres in Nigeria, A case study of Jalingo LGA, Taraba State Nigeria. Mediterranean Journal of Social Sciences, 5(10), 82-88

[46]. Palmquist, R. B. (1992). Valuing localized externalities. Journal of Urban Economics, 31(1), 59–68.

[47]. Peter, N. J., Ayedun, C. A., and Iroham, C. O. (2018). Registration into associate membership status of the Nigerian institution of estate surveyors and valuers (NIESV): The challenges. International Journal of Civil Engineering and Technology, 9(10), 1239-1251

[48]. Popoola M. O., Abiola S. O., and Adeniji, W. A. (2013). Traffic congestion on highways in Nigeria causes, effects and remedies. International Journal of Civil and Environmental Engineering, 7(11), 858-863

[49]. Rao, A. M., and Rao, K. R. (2012). Measuring Urban Traffic Congestion – A Review. International Journal for Traffic and Transport Engineering, 2(4), 286 – 305

[50]. Ukpata, J.O. and Etika, A.A. (2012). Traffic congestion in major cities of Nigeria. International Journal of Engineering and Technology, 2, 1433-1438.

[51]. Wilhelmsson, M. (2000). The impact of traffic noise on the values of single-family houses. Journal of Environmental Planning and Management, 43(6), 799–815

IOP Conf. Series: Earth and Environmental Science **331** (2019) 012031 doi:10.1088/1755-1315/331/1/012031

Technology usage and employee behaviour: controversies, complications, and implications in the Nigerian business environment

[1]Mayowa G Agboola, [2]Mosunmola Akinbode, [3]Chidi Segun-Adeniran [4]Peter Ifeanyichukwu Dibia, [5]Anuoluwapo Oloruntoba, [6]Tamarapreye Dinyain, [7]Tolulope Johnson Fabunmi, [8]Temitope Funsho, [9]Oluwaseun Abayomi Akinde and [10] Chelsea M Betek

[1-2]Lecturer 1, Dept. of Business Management, Covenant University, Ota, Nigeria
[3]Librarian 1, Centre for Learning Resources, Covenant University, Ota, Nigeria
[4-8]PG Student, Dept. of Business Management, Covenant University, Ota, Nigeria
[9]Researcher, Dept. of Economics, Federal University, Ilorin, Kwara State, Nigeria
[10]PGStudent, Dept. of Political Science and International Relations, Covenant University, Ota, Nigeria.

Corresponding Author's Email – gbenga.agboola@covenantuniversity.edu.ng

Abstract:
Organizations are among the key units of the society. An organization comprises of several people including employees. The behaviour of employees determines the performance and efficiency of the organization. Technology is an important factor that distinguishes one organization from the other. Employee's acceptance, rejection, and adaptability to technology being used determine to a great extent how organizations will thrive in a business environment. The drive of technology usage is to ensure that work processes are made easier, faster and to improve the organization's economic efficiency. In addition, technology usage greatly influences employee behaviour. The aim of this paper is to find out how technology usage affects employee behaviour and vice versa. It is important to understand that in order to improve employee's usage of technology incorporated into the organization, individual behaviour of employees must be examined cum their attitude. The results of the study indicate that employee behaviour mainly affects technology usage, promotes individual learning, increases efficiency and effectiveness, and improves organizational performance. The paper therefore recommends periodic review of organizational policies and in addition, procure hardware and software to guide against hackers and scammers to avoid losing those important information.

Key Words:
Technology acceptance model, organisational performance, behavioural intentions

1. Introduction

Technology is one of the most relevant components of organization's business environment which offers countless opportunities for organisations to succeed in their business environment [1]. Technology and the workplace are one of the trending research areas that researchers are showing keen interest. The presence of technology in the workplace is becoming obvious and the fear that, it could replace some jobs in the next decade or less cannot be farfetched [2]. Then how are organisations responding to this? Why so much crave for latest ultra-modern and state of the art technology in business processes [3].

It was reiterated that the reason for quest for the adoption or adaptation of the latest technology was because it helps both employees and organisations to be more effective, efficient, proficient, knowledgeable and profitable [4].

Aside these direct influences of technology on both employees and organisations, the use of technology has helped to break geographical barriers, language barriers and cultural barriers, thereby allowing employees to be more effective and efficient, which are evident in organisational work processes, productivity, competitiveness and profitability [5]. Consequently, technological changes cause business practices and processes to evolve, which makes organizations to invest more in it [6]. In fact, organizations have always sought and adopted technologies that enhance the effort of their manpower in production and management in the management of organizational operations.

However, the relationship between technology usage and organisational behaviour had been controversial [7]. It has been argued that the use of technology in the workplace leads to high employee productivity as other variables interfere such as employee's high morale [8]. Another research showed also that there is loss of productivity when employees use technology in the workplace [9]. A report from Nucleus Research [10] stated that the use of Facebook at work had cost companies 1.5 percent of their total productivity. But, on the other hand, the ability to use Facebook was an added advantage to be retained in the workplace as this will allow employees to connect with their loved ones and colleagues in the workplace. To buttress this further, it was stated that AT&T discovered that the use of social networking sites at work had helped improved employees' productivity in the workplace by 65 percent [11]. Part of the advantage of introducing technology effectively in the workplace is that employees become more motivated, work processes and managerial processes in the workplace become better organised and structured [12]. Technology usage always lead to handling large volumes of work at an increased speed, while using less resources [13].

Furthermore, empirical studies conducted to assess the influence of technologies in different areas of human endeavour such as social sciences and the sciences [14], medical sciences[17], engineering (Kühnle, 2010), and in business practices [15]. However, most of the studies in the area of business focused on the influence of technological usage on employee productivity [16] or organisational productivity [18-20] without accounting for its effect on employees' behaviour at work. It is worthy of note that employees' behaviour is a crucial factor for any organization to optimise the advantages inherent in the usage of any technology. Therefore, monitoring of employees' usage of technologies is now generating debate on how to strike a balance between employer's security rights and employee's privacy issues [21].

Nigeria business environment today depends on technological innovation and the impact of its usage on employee behaviour as well as organization's business decision making process are important research area for organisational behaviour scholars, since, employees will drive any technological innovation in any organization. Therefore, this study provides systematic review of empirical studies on the influence of technology usage on employee behaviour while highlighting the controversies and complications inherent in the relationship. Implications for the Nigerian business environment were provided based on findings from the review.

2. Review of Relevant Literatures

2.1 Conceptual Framework

Technology usage is all about uptake and continuance [22] while employee behaviour is the manner in which employees react to certain situations at the workplace [23]. It is also interesting to note that the behaviour of employees is shaped and nurtured by both the culture of the organisation they work and their own culture [24]. Every organizational culture emphasised that employees must be well behaved so that they can remain in the organization and earn promotion [24].

The conceptual framework in figure 1 shows that when a new technology is deployed for a task or an activity, that new technology may influence employee's performance if the technology is adopted (employee behaviour) which could improve the productivity of the organisation justifying management investment in the technology.

IOP Conf. Series: Earth and Environmental Science **331** (2019) 012031 doi:10.1088/1755-1315/331/1/012031

Conversely, if an employee rejects the new technology (employee behaviour) they will either refuse or abuse it resulting in shortfall in management's expectations. The introduction of new technology creates a perception of job insecurity in the mind of the employees due to the changes that will be made to their job description which could result in their redundancy or retrenchment [25]. The introduction of robotics and artificial intelligence, for example, is enough to cause psychological trauma for employees that work in an organisation.

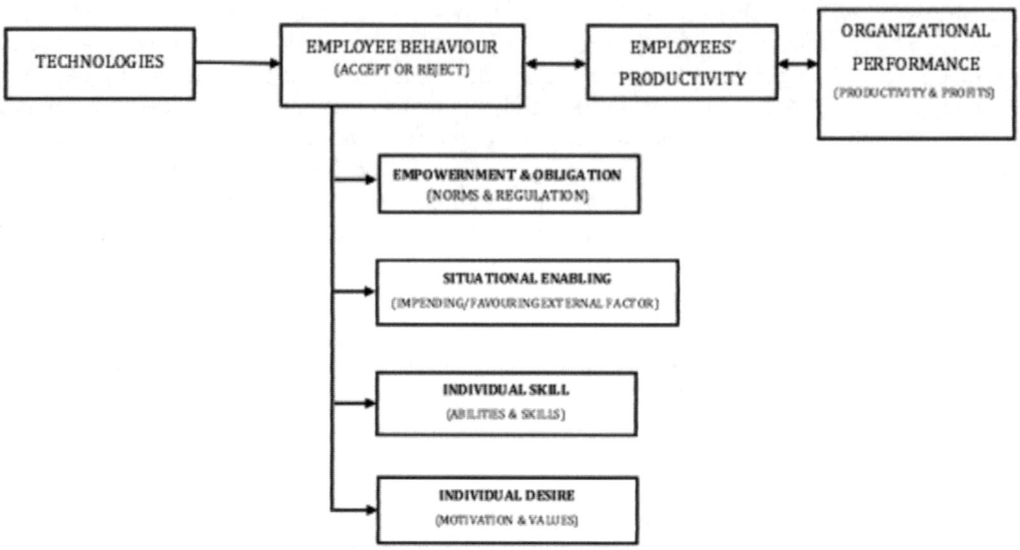

Figure 1: Conceptual Framework
Source: Adapted from [26]

2.2 Theoretical Background

This paper reviewed four related theories namely the theory of planned behaviour (TPB), the theory of reasoned action (TRA), the technology usage model (TUM), the unified theory of acceptance and use of technology 2 (UTAUT2) and the technology acceptance model (TAM). Each of these theories or models is described briefly below.

2.2.1 The Technology Usage Model (TUM). Nakarin Pinpathomrat propounded the technology usage model in 2015 at a conference tagged International Conference on Science and Technology (TICST). The model was fallout of three other theories of technology, which are: technology acceptance model (TAM), the adaptation level theory (ALT) and cognitive dissonance theory (CDT). It was in this theory that [27] explained that technology is used in two ways. These two forms, as earlier mentioned, are uptake and continuance. The uptake explains the first-time experience of the user of Information Technology (IT), which is expressed as t_0 to t_1. On the other hand, the continuance was explained as the period the technology is being or was used, which is expressed as t_1 to t_2. The major reason for this approach was to examine the explanatory power of the model of technology, assess the factors that affect the uptake and continuance of technology. From the study of the technology usage model, it was opined that the use of technology had a moderating explanatory power over the uptake and continuance of technology.

It was concluded that the expectations of the technology users were major elements affecting the uptake and continuance of technology usage in the two time periods expressed earlier (Pinpathomrat, 2015).

2.2.2 Theory of Planned Behaviour (TPB). The theory of planned behaviour (TPB) is an advancement of the theory of reasoned action (TRA) based on the limitations that were identified with respect to the behaviour of people studied. The theory of planned behaviour introduced a new element, known as the perceived behavioural control (PBC) [28]. Perceived behavioural control has to do with determining whether an individual possess the ability to behave in a certain way [29]. The theory of planned behaviour is a model that forecasts an intentional behaviour because it is believed that behaviour may be planned or deliberates [30]. According to the model, the best predictor of behaviour is the behavioural intention. The elements that predict the behavioural intention of an individual are attitude, subjective norms and the perceived behavioural control. Furthermore, there are three beliefs that guide the behavioural intention of an individual, which include: behavioural beliefs, normative beliefs and control beliefs [31]. The behavioural beliefs have to do with the outcome of behaviour and the evaluation of behaviour. On the other hand, normative beliefs have to do with perceived behavioural expectations from important referral persons around the person, while control beliefs have to do with beliefs on certain factors present that can facilitate the performance of behaviour and the power behind these factors.

2.2.3 Extended Technology Acceptance Model (TAM2). The extended technology acceptance model (TAM2) described the intention of users of IT and their behaviour towards the use of IT as propagated by Davis, Bagozzi, and Warshaw [32] and extended Chen, *et al.* [33]. Two very vital elements were identified in the model, which were perceived ease of use and perceived usefulness to predict users' behaviour. Perceived usefulness has to do with the extent that a user thinks that an IT will improve his or her performance in a workplace, while perceived ease of use has to do with the extent to which a user thinks that using an IT will require little or no effort. It was believed that the attitude of a user will likely influence the behavioural intention to use IT as it in turn influence the actual behaviour to use IT. From the model, it was assumed that perceived usefulness has a straight impact on behavioural intention, aside its indirect impact through users' attitude [32]. Other researches have shown a strong relationship between perceived ease of use and perceived usefulness. Also, researches have shown that perceived usefulness, over time, has been the major predictor of a user's attitude in the use of IT (Chen, *et al*., 2017). There is still a controversy on whether attitude should be part of the model or not as some researcher had remove attitude completely from technological models.

2.2.4 Unified Theory of Acceptance and Use of Technology 2 (UTAUT2). UTAUT2 is the extended version of the UTAUT model looking at it from a consumer point of view, while the first version of the UTAUT model was the extended version TAM2 from a worker's point of view. It was introduced by Venkatesh, Morris, Davis and Davis in 2003. The idea of the UTAUT was to examine the already existing technological acceptance models, to do an empirical comparison of eight technological models, to express the UTAUT model and validate the UTAUT model (Agboola, 2014). On the other hand, the UTAUT2 model was formulated as an extended version of the main UTAUT model. The UTAUT2 model presented three new concepts, which are habit, price value and hedonic motivation. Other moderating variables such as age, gender and experience were assumed to moderate the effect of the three new concepts and others on technology usage and behavioural intention. Another study supported the introduction of the new concepts, which had further improved the explanation of the variance that the three concepts had on technology usage and behavioural intention [34]. The authors posited that with the three concepts, one can create a strategy of market segmentation to arouse consumers to use technology. In the context of this paper, consumers, which are users can be seen as workers or employees who actually use the technology in the workplace. Finally, the authors suggested that the producers of technology should make the designs of technologies much better and categorise their technologies based on the demographic nature of the users at different stages of use.

2.3 Empirical Background

The systemic review of literature relating to technology usage, job and organizational performance revealed confirmed that the introduction of technology influence both enhances both job and organizational performance but adversely affect employee behaviour.

In fact, two studies found that successful implementation of technology were adversely affected by the attitude of employee towards it. Hence, the controversy that how could technological usage boost job and organizational performance without appreciable improvement in employee behaviour.

Year	Country	Author(s)	Objective(s)	Methods	Findings
2018	Nigeria	Adeniji, C., Adeyeye, O., Iyiola, O., Olokundun, M., Borishade, T., Falola, H., & Salau, O.	Data on impact of technological change on employees' cognitive attitude and organizational performance	Primary data 600 questionnaires to 3 tiers management employees of 3 manufacturing industry in Lagos ANOVA, Hierarchical regression	technological change significantly increase employees' cognitive attitude and organizational performance
2015	Nigeria	Onu, C.A., Olabode, I. O & Fakunkunmoju, S.K.	Effect of Information Technology Investment on Organisational Productivity and Growth of Small and Medium Scale Enterprises in Developing Countries	Primary data 100 questionnaires Frequency tables, percentages and simple regression	Technological usage boost productivity
2014	South Africa	Binuyo, A. O. & Aregbeshole, R.A..	Effect of ICT investment and cost efficiency on bank performance	Secondary data sourced from 4 big banks in S/Africa (Absa, FirstRand, NedBank & standard bank Ordinary Least Square and Dynamic Panel Regression	ICT investment and ICT cost efficiency improve bank performance
2013	USA	Moqbel, M., Nevo, S. & Kock, N.	Organization member use of social networking sites and job performance	Primary data Questionnaires to 193 employees Variable-based structural equations modelling	Social network usage significant increase job performance
2011	Nigeria	Dauda & Akingbade	Effect of technological change on employee performance	Primary data 1256 questionnaires to 30 manufacturing industries in Lagos (beverages, textile, steel, cement and chemical industries) ANOVA	Technological change significantly raise employee skills and performance
2008	Nigeria	Osabuohien, E.	Assessed anticipated role of ICT and operation of selected banks in Nigeria	Primary data 3 commercial banks in Lagos 180 questionnaires	ICT raise the speed of bank operations, productivity and profitability
2003	Nigeria	Agboola, A. A.	Assess attitude of workers on the application of IT in banking industry	Primary data 6 commercial banks in Lagos 60 questionnaires Descriptive; frequency & percentage	Usage of ICT complicate work and alter the skills required
2003	Nigeria	Ehikhanenor, F. A.	Assess application of ICT in Nigeria banks	Primary data 56 banks in Lagos	54.6% of them achieve successful implementation of ICT

Source: Author's Review (2019)

3 Controversies, Complications and Implications of Technology Usage in the Nigerian Business Environment

3.1 Controversies of Technology Usage in the Nigerian Business Environment

Despite the advantages of technology usage within and among organisations in the Nigerian business environment, there is one major controversy militating against the use of technology in the workplace among employees, and in turn influences employee behaviours to work, which is known as employee monitoring. This is a major point of conflict between employees and management [35]. It is worthy to note, even though the advent of technology is not intended to replace workers, those without basic technological skills and are unwilling to acquire it will soon be replaced [36]. The question is how can balance be established when an employee is tech savvy and uses it more for personal gains instead of the organisation?

A scenario was illustrated by Perchstone and Graeys [37] as quoted below:

"It was about noon. Samuel had concluded some routine workplace assignments but had a few more to attend to. He heaved a sigh, and thought- 'I'll resume to those, after this well-deserved time-out'. Yes, time-out to look up the social media messages piling up, and the intermittent flashing notifications which never ceased to teasingly announce their arrival (into his smartphone). He was soon engrossed with this very absorbing act...far too busy to notice his Line Manager standing a few steps away, making a hand gesture in an apparent bid to get Samuel's attention. This continued for almost a minute, until he (the Line Manager) had to call out sharply, visibly irritated. Samuel was visibly startled. He had been animatedly responding to a thread message on a recent Facebook posting. The Line Manager left angry, only to return issuing Samuel with a query. Later that week, the office churned out a new policy on the use of social media in the workplace" (para. 1).

The above scenario illustrates what typically happens in the workplace and the reason why most organisations monitor their employees. The Nigerian business environment is a very dynamic one and organizations are poised to ensure continuous productivity and performance to maintain competitive advantage over other organisations. One major way they do this, as earlier mentioned, is to have a close monitoring of every employee within the organization from when they resume to when they close for the day and assess their contribution to the organization each day, week and month. Some other employers go as far as knowing their employees' family and the relationships they keep. This is gradually becoming typical of the Nigerian business environment as advanced technologies are employed to gather all the required information all in the name of knowing their employees.

Whereas, it has been argued that the data gathered from employees is way too much and sensitive that it has encroached into the privacy of employees [38]. Another question that bothers one is that, what is being done to protect employees after gathering all their required data? In some indications, there are justified reasons for using technology to monitor employees such as to alleviate the misuse of resources, to ensure proper usage of organizational time, to avoid gossip, to monitor the illicit visit of inappropriate webpages and to protect both employee and customers' profile or information from cybercrimes[39].

Yerby [40] conducted a study which tressed the need to monitor employees in the workplace. It was made clear in the study that aside the fact that organisations desire increases in productivity, there are other challenges, such as, mis-use of email clients, internet and potential legal liability, which gives reason to employee monitoring. It was further supported that the major reason for employee monitoring in the Nigerian business environment has to do with attack of viruses due to employees visiting pornographic sites and online gaming sites [41]. This was earlier mentioned that there was a study carried out for two years, which showed that the word "sex" was the number one most searched word on the internet, while the word "porn" was the fourth most searched word on the internet.

Other words that were mostly searched were, "nude", "erotic", "playboy" and "xxx", which were also among the top twenty most searched words on the internet [42]. The authors also posited that 70% of these searches occurred during working days (Mondays to Fridays) and during working hours (8 am to 5 pm) leading organisations to discipline such employees found using the internet for that purpose.

Employee monitoring within organisations or at workplace is made possible through several software and hardware which are easy to use but costly depending on its sophistications and specifications. The extent of employee monitoring in an organisation will determine the hardware or software to be deployed, which ranges from technological solutions for monitoring website visit, social media chats, monitoring of email conversations and many others [40]. These technological solutions can provide daily reports of technology or internet usage and more sophisticated ones sends alerts once employees visit unapproved websites and monitor conversations via the phone. In this case, some organisations create the awareness that phone calls are being monitored for better feedback purposes.

In more advanced and developed countries, the government have set policies, rules and regulations for organisations to protect sensitive information going in and out. According to Yerby [40], it was opined that there is no law against monitoring employees since organisations had raised evidences as reasons to monitor employees. In the Nigerian business environment, it has been observed that depending on the organisation, the legal part of employee monitoring varies depending on the relationship the organisation has with the labour union and the nature of the organisation in terms of what it does.

The subject of employee monitoring raises some ethical issues, such as privacy rights on the part of employees. What an organisation may consider unethical may be ethical for the employee and vice versa. Technology solutions deployed for employee monitoring do not have sensors to determine when an employee is feeling ill and perhaps needs rest, thereby giving a wrong or incomplete data about the employee, which may lead to suspension because it would be assumed that the employee was taking rest instead of working. This is where there must be a balance. There are other cases where employees feel brain or mind fatigue due to information overloading and they need to just relax with an online game or social media chat to ease off the stress. When organisations are strict and rely on their technology for employee monitoring, then they are likely to cause more harm than good to the entire organisation. On the other hand, while it seems that there are no provisions from the law in Nigeria to govern employee monitoring in the workplace, there are constitutional provisions for right protections [35]. The Nigerian constitution of 1999 as amended states *"...the right to privacy of citizens, their homes, correspondence, telephone conversations and telegraphic communications is hereby guaranteed and protected"*. Based on this premise, it seems the law is in favour of employees, but that is when the organisation is found misusing their employees' information.

It is therefore important to note that employee monitoring in various aspects could decrease the morale of employees to work and the sense of trust that ought to be established between the employee and organisation could be eroded adversely affecting his behaviour and the attainment of organisational goals [43].

3.2 Complications of Technology Usage in the Nigerian Business Environment

Whether we like it or not, from what we are experiencing around the world, technology is a major factor that is determining how things are done in any environment. Technology serves and plays a major role in every works of life ranging from government, military, manufacturing, education, entertainment, commerce, banking and any entire industry at large [44]. Despite the role technology plays in the Nigerian business environment, there are difficulties it poses and negative effects it has already on the human behaviour. As earlier explained, technology usage has caused major ethical issues resulting to employee behavioural feedbacks.

IOP Conf. Series: Earth and Environmental Science **331** (2019) 012031 doi:10.1088/1755-1315/331/1/012031

There are three ethical issues of concern, which are access rights, personal privacy and harmful actions [44].Unknowing to several people is the fact that they grant many hardware and software application access right to their personal information during installation of some software. Access rights simply means giving a user or application the permission to read, store, write, transfer and erase files or information on a device [45]. Before now, applications, when installed, automatically have access to personal stuffs on devices, but now, the way it goes is that, when applications are installed, they further ask for permissions to grant them access rights. However, in most cases when access is not fully granted usability of the software of device is not optimised. Now that international commerce and online shopping has become very popular on the internet, access right has moved from low priority to high priority for most organisations [44]. The issue of access right has led to lots of cybercrimes committed by hackers thereby creating lots of illegal access rights [46]. With this situation, employees within organisations have put up some sort of negative behaviour towards the use of devices and applications by offering fake and untrue personal information [47].

The second ethical issue is personal privacy. Total dependence on the use of technology has resulted to the high transfer and exchange of large data and information on a large scale among people or organisations around the world from one location to another resulting to big data [48]. It has been made clear that since large data are transferred around the world, there is tendency that there will be disclosure or leakage of personal information, thereby resulting to violation of the privacy of individuals or groups around the world [44]. This becomes a very important ethical issue since sensitive information are involved. This is another reason why employee behaviours are negative towards the use of technology at the workplace in the Nigerian business environment.

Over the years, there had been several cyber-attacks that had resulted to loss and destruction of vital information with respect to various individuals [49]. Various governments are investing a lot in dealing with cyber wars, where a simple and small device can be used as a weapon to cause damages that the entire world will feel. The virtual space is the battlefield and there are moves to intentionally destroy and modify files or programmes, which could lead to loss of information [44]. With such awareness, individuals and employees are always showing fear leading to negative behaviours towards the use of technology in the workplace. These harmful actions mostly come in the form of malicious codes usually through emails and employees are not always aware of such strategies, for example, phishing, password attack, birthday attack, malware attack and so on [50]. When organization suffers these cyber-attacks the private and sensitive information of the employee is exposed. Thus, organisation and the employee should constantly be sensitized of these trends; Based on the above scenarios, employee behaviour towards the usage of some technologies in the workplace becomes complicated and a very critical issue or challenge that an organisation must need to deal with especially in a very volatile business environment such as the Nigerian business environment. Other ethical issues regarding the use of technology that may affect the behaviour of employees in the workplace include plagiarism, piracy, hacking, files sharing, education [44].

According to Jegede and Olowookere [51], there had been quite some complications experienced with the use of technology that has caused trauma to users in the Nigerian business environment which include areas like e-business where money and products are usually transferred through the internet. Based on the study carried by the authors, it asserted that e-commerce is the most prominent platform that scammers or "opportunists" take advantage of to use the internet for their illicit benefits. In view of the related threats, there had been cases where intrusion of privacy was associated with local institutions like commercial/deposit banks, telecommunication companies and e-commerce firms, thus, establishing the basis of negative exposures [52]. Other complications that were identified relating to the use of technology in the workplace within the Nigerian business environment are issues regarding health of the employees, job insecurity and pressure on users of technology [44].

3.3 Implications of Technology Usage in the Nigerian Business Environment

The permeability of the use of internet have been widely study in research [37, 451-52,53-56].

Hence, how do these controversies and complications affect employee behaviour and use of technology in the Nigerian business environment as well as its implications? Organisations around the world have different policies that govern the use of technology at the workplace. Ironically, these policies are not at the reach of employees, which makes employers continue to monitor their employees [37]. Use of technology policies are one of the commonest organisational policies that govern the manner in which employees use an organisation's technology in terms of what is expected of them and what they should expect, such as privacy issues, ethical issues and so on in developed countries [44].

Despite the fact that the use of technology may allow for fraud and other forms of cyber security breaches, it also results in generating threats to personal (employee) details or information and banking details too, while using the internet to transact, which negatively affect employee behaviours to use technology in the work place. Securing the internet from cyberwars and hackers is not cheap especially for private individuals. Currently, it is very expensive to manufacture goods and services as there are many challenges in the Nigerian business environment ranging from unstable power supply, bad roads, obsolete infrastructures and economic instability [57-58]. Therefore, there is no one organisation that may be facing these challenges within the Nigerian business environment that would want to spend so much on the state-of-the-art technological software and hardware to fight scammers and hackers, which on the other hand leaves their employees and customers at large to take precautions by themselves. Instead, organisations would prefer to go for insurance policies that will cover them in the event of loss of data or information [59]. Operating a business in Nigeria comes with a lot of cost in production of goods and services, which affects the interest to purchase software to tackle and enhance cyber security of their online services or even take an insurance policy to cover in the event of loss of employee data or information. This can sufficiently affect the behaviour of employees towards technology usage in their organisations.

4. Conclusion and Recommendations

Technology has played a vital role in most facets of organizations in the Nigerian business environment, which has contributed to the behaviour of employees as well as global interactions among business subsidiaries, customers and all stakeholders. Despite all the benefits and advantages of technology, there are a lot of negative experiences in the Nigerian business environment resulting to several complications and controversies, which brings about the implications discussed. The law governing the Nigerian business environment is clear on the right to privacy for employees in the workplace but with a few exceptions. The law may have permitted employers to monitor employees, but they must be duly informed, while their personal information and data must not in any way be misused. If this is properly implemented, a positive behaviour from employees towards the use of technology could be achieved, thereby improving their productivity and performance of the organisation at large.

Every employee has a right to privacy, and this is essential (Perchstone & Graeys, 2018). This means that every employee needs their own personal space to communicate and share information as they will. No matter the rate of advancement of technology being experienced, it has not taken the right of privacy from employees. Even if employers want to monitor the employees' usage of technology, it has not changed the fundamental rights of privacy that the employees have. It is therefore needed that an "acceptable use policy' be crafted to tackle the issue of monitoring employees and the issue of privacy, which should not only be drafted by an IT department alone, but getting contributions from other departments in an organisation. Other departments such as human resource, legal, finance and so on need to come together to determine what activities, how, where and when should employees should be monitored, also, what level of data or information should be available in reports and who should monitor.

When this is done and it is in place, employees would be motivated having known their extent of privacy rights and to what extent they are being monitored, thereby not prompting a negative behaviour to work. It is therefore recommended that organisations should review their policies from time to time as there are constant and sporadic changes experienced in and use of technology. Policies in the Nigerian business environment should be well explicit and understandable for all employees within an organisation to avoid misinterpretation, misrepresentation and to make it consistent with the best practices of the world business environment. The sensitiveness of data and information with respect to employees, customers and the organisation cannot be overemphasized and overlooked. Therefore, the procurement of hardware and software to guide against hackers and scammers should not be considered expensive because loosing those data and information into the wrong hands would cause an organisation much more than the cost of procurement.

Acknowledgements
This is to appreciate Covenant University Centre for Research, Innovation and Discovery (CUCRID) and Covenant University as a whole for the support and sponsor of this study.

References

[1] Adeniji C, Adeyeye O, Iyiola O, Olokundun M, Borishade T, Falola H and Salau O 2018 Data on impact of technological change on employees' cognitive attitude and organizational performance. Data in Brief, **18**, 1360–1364. doi:10.1016/j.dib.2018.04.024

[2] Agboola A A 2003 Information Technology, Bank Automation, and Attitude of Workers in Nigerian Banks. Journal of Social Sciences, **7** 215-22. doi:10.1080/09718923.2003.11892383

[3] Agboola M G 2014 Employee Attitude towards Attributes of Technological Change as a Determinant of Organizational Performance in Nigerian Commercial Banks. Ota: Covenant University.

[4] Ajzen I 2005 Icek Ajzen, Homepage. Milton-Keynes, England: Open University Press/McGraw-Hill. Retrieved from Behaviour Intention Based on the Theory of Planned Behaviour: https://people.umass.edu/~aizen/

[5] Altraide D 2019 Big Brother: China Edition! Australia: YouTube. Retrieved January 15, 2019, from https://www.youtube.com/watch?v=0jVq04poU9I

[6] Anadon L D, Chan G, Harley A G, Matus K, Moon S, Murthy S L and Clark W C 2016 Making technological innovation work for sustainable development. Proceedings of the National Academy of Sciences of the United States of America, **113** 9682–90. doi:10.1073/pnas.1525004113

[7] Andersen E 2013 9 Ways To Deal With Difficult Employees. Retrieved from Forbes: https://www.forbes.com/sites/erikaandersen/2013/11/21/9-ways-to-deal-with-difficult-employees/#604d41ae40c4

[8] Barley S R 2015 Why the Internet Makes Buying a Car Less Loathsome: How Technologies Change Role Relations. Academy of Management Discoveries, **1** 31–60. doi:10.5465/amd.2013.0016

[9] Bernal P 2016 Data gathering, surveillance and human rights: recasting the debate. Journal of Cyber Policy, **1** 243–64. doi:10.1080/23738871.2016.1228990

[10] Binuyo A O and Aregbeshole R A 2014 The impact of information and communication technology (ICT) on commercial bank performance: evidence from South Africa. Problems and Perspectives in Management, **12** 59-68.

[11] Boop G 2018 Cyber Liability Insurance - Coverage for Data Breaches. Retrieved from The Balance Small Business: https://www.thebalancesmb.com/cyber-liability-insurance-coverage-for-data-breaches-462582

[12] Brandimonte M A and Ferrante D 2015 Effects of Material and Non-Material Rewards on Remembering to Do Things for Others. FrontiersinHumanNeuroscience, **9** 1-8. doi:10.3389/fnhum.2015.00647

[13] Cascio W F and Montealegre R 2016 How Technology Is Changing Work and Organizations. The Annual Review of Organizational Psychology and Organizational Behavior, **3**, 349-375. doi:10.1146/annurev-orgpsych-041015-062352

[14] Cayford M and Pieters W 2018 The effectiveness of surveillance technology: What intelligence officials are saying. The Information Society, **34** 88–103. doi:10.1080/01972243.2017.1414721

[15] Chau P Y and Hu P J 2014 Examining a Model of Information Technology Acceptance by Individual Professionals: An Exploratory Study. Journal of Management Information Systems, **18** 191-229. doi:10.1080/07421222.2002.11045699

[16] Chen H, Rong W, Ma X, Qu Y and Xiong Z 2017 An Extended Technology Acceptance Model for Mobile Social Gaming Service Popularity Analysis. Mobile Information Systems, 1-12. doi:10.1155/2017/3906953

[17] Cherry K 2018, October 4 Motivation: Psychological Factors That Guide Behavior. Retrieved from Very Well Mind: https://www.verywellmind.com/what-is-motivation-2795378

[18] Computer Hope 2017, April 4 Unauthorized access. Retrieved from Computer Hope: https://www.computerhope.com/jargon/u/unauacce.htm

[19] Court L and Warmington C 2004 The workplace privacy myth: why electronic monitoring is here to stay. Employment and Labor Law, **1** 1-20.

[20] D'Angelo M 2018 December 19. 6 Small Business Tech Trends. Retrieved from Business News Daily: https://www.businessnewsdaily.com/10469-business-technology-trends.html

[21] Dauda Y A and Akingbade W A 2011 Technological Change and Employee Performance in selected Manufacturing Industry in Lagos State of Nigeria. Australian Journal of Business and Management Research, **1** 32-43.

[22] Davis F D, Bagozzi R P and Warshaw P R 1989 User Acceptance of Computer Technology: A Comparison of Two Theoretical Models. Management Science, **35** 903-1028. doi:10.1287/mnsc.35.8.982

[23] Demaerschalk B M, Vargas J E, Channer D D, Noble B N, Kiernan T E, Gleason E A, Bobrow B J 2012 Smartphone teleradiology application is successfully incorporated into a telestroke network environment. Stroke, **43** 3098-101. doi:10.1161/STROKEAHA.112.669325

[24] Ehikhamenor F A 2003 Information technology in Nigerian banks: The limits of expectations. Information Technology for Development, **10** 13-24. doi:10.1002/itdj.1590100103

[25] Fan Z and Chiffelle C R 2018, June 6 These 5 technologies have the potential to change global trade forever. Retrieved from World Economic Forum: https://www.weforum.org/agenda/2018/06/from-blockchain-to-mobile-payments-these-technologies-will-disrupt-global-trade/

[26] Forson J E 2012 Impact of Motivation on the Productivity of Employees at GTBank Ghana. Kumasi: Kwame Nkrumah University of Science and Technology.

[27] Gallagher S 2014 Ethical considerations in monitoring employee Internet usage. Ethikos, 5-6.

[28] Hardin-Fanning F and Ricks J M 2017 Attitudes, social norms, and perceived behavioral control factors influencing participation in a cooking skills program in rural Central Appalachia. Glob Health Promot, **24** 43-52. doi:10.1177/1757975916636792

[29] Haythornthwaite C and Wellman H 2008 The Internet in Everyday Life: An Introduction. Blackwell. doi:10.1002/9780470774298

[30] Hughes B 2017 January 20. How Technology Is Rapidly Changing the Way Things Get Done Across Industries. Retrieved from Entrepreneur: https://www.entrepreneur.com/article/288025

[31] Human Rights Watch. 2018 World Report 2018: Rights Trends in China. Retrieved from Human Rights Watch: https://www.hrw.org/world-report/2018/country-chapters/china-and-tibet

[32] Imran M, Maqbool N and Shafique H 2014 Impact of Technological Advancement on Employee Performance in Banking Sector. International Journal of Human Resource Studies, **4** 57-70. doi:10.5296/ijhrs.v4i1.5229

[33] Jegede A E and Olowookere I E 2014 Cyber Risks and Fraud in the Nigeria's Business Environment: A Postmortem of Youth Crime. Journal of Social and Development Sciences, **5** 258-65.

[34] Joel S A, Ojo A G and Adeola A A 2015 Monitoring Employees' Use of the Internet in Ladoke Akintola University of Technology Ogbomoso, Oyo State, Nigeria. Greener Journal of Social Sciences, **5** 42-50. doi:10.15580/GJSS.2015.2.291114398

[35] Joyce C, Fisher J, Guszcza J and Hogan S K 2018, April 16. Positive technology: Designing work environments for digital well-being. Retrieved from Deloitte Insights: https://www2.deloitte.com/insights/us/en/focus/behavioral-economics/negative-impact-technology-business.html

[36] Katzenbach J, Oelschlegel C and Thomas J 2016, February 15. 10 Principles of Organizational Culture. Retrieved from Strategy-Business: https://www.strategy-business.com/feature/10-Principles-of-Organizational-Culture?gko=3e299

[37] Katzenbach J, Oelschlegel C and Thomas J 2016, February 15. 10 Principles of Organizational Culture. Retrieved from Strategy-Business: https://www.strategy-business.com/article/10-Principles-of-Organizational-Culture?gko=71d2f

[38] Kh R 2016, July 28. The Importance of Security When it Comes to Data Accuracy. Retrieved from Datafloq: https://datafloq.com/read/importance-security-when-it-comes-to-data-accuracy/2196

[39] KPMG 2017 Leading Through Digital: Business Models that Exploit the Digital Opportunity. Nigeria: KPMG International.

[40] Kühnle H 2010 Distributed Manufacturing: Paradigm, Concepts, Solutions and Examples. New York: Springer. doi:DOI 10.1007/978-1-84882-707-3

[41] Kumar S 2017 Technological and Business Perspective of Wearable Technology. Centria University of Applied Sciences, Kokkola, Finland.

[42] Kuti F, Obi U and Azubuike S 2017. The Privacy, Data Protection and Cybersecurity Law Review (4th ed.). (A. C. Raul, Ed.) London: Gideon Roberton.

[43] Landeta J M and Cortés C B 2013 Effect of competitive strategies and resources and market-oriented capabilities on the growth of organizations. Contaduría y Administración, **58** 169-97. doi:http://dx.doi.org/10.1016/S0186-1042(13)71202-6

[44] Long J 2017 December 7. 5 Factors That Contribute to the Success of Your Business. Retrieved from Entrepreneur: https://www.entrepreneur.com/article/305688

[45] Lord N 2019 January 3. 101 Data Protection Tips: How to Keep Your Passwords, Financial and Personal Information Safe in 2019. Retrieved from Digital Guardian: https://digitalguardian.com/blog/101-data-protection-tips-how-keep-your-passwords-financial-personal-information-safe

[46] Lord N 2019 January 17. Phishing Attack Prevention: How to Identify and Avoid Phishing Scams in 2019. Retrieved from Digital Guardian: https://digitalguardian.com/blog/phishing-attack-prevention-how-identify-avoid-phishing-scams

[47] Manokha I 2018 Surveillance, Panopticism, and Self-Discipline in the Digital Age. Surveillance and, **16** 219-237.

[48] Mardonova M and Choi Y 2018 Review of Wearable Device Technology and Its Applications to
the Mining Industry. Energies, 1-14.

[49] Marković M 2018 March 28. Information security and privacy protection aspects of CCTV systems. Retrieved from Government Europa: https://www.governmenteuropa.eu/information-security-cctv-systems/85930/

[50] Marr B 2018 April 27. What Does GDPR Really Mean For HR Teams? Retrieved from Forbes: https://www.forbes.com/sites/bernardmarr/2018/04/27/what-does-gdpr-really-mean-for-hr-teams/#29beb26a30e0

[51] Melnick J 2018, May 15 Top 10 Most Common Types of Cyber Attacks. Retrieved from NETWRIX Blog: https://blog.netwrix.com/2018/05/15/top-10-most-common-types-of-cyber-attacks/

[52] Moqbel M, Nevo S and Kock N 2013 Organizational members' use of social networking sites and job performance: An exploratory study. Information Technology and People, **26** 240-264. doi:10.1108/ITP-10-2012-0110

[53] Murray A 2015, April 22. Fortune's tech coverage: Your guide for tumultuous times. Retrieved from Fortune: http://fortune.com/2015/04/22/fortune-new-gigaom-writers/

[54] Nam T 2019 Technology usage, expected job sustainability, and perceived job insecurity. Technological Forecasting and Social Change, 138, 155-165. doi:10.1016/j.techfore.2018.08.017

[55] Nucleus Research 2009, July 21. Facebook Costs Companies 1.5 Percent of Total Productivity. Retrieved from Nucleus Research: https://nucleusresearch.com/news/facebook-costs-companies-1-5-percent-of-total-productivity/

[56] Ogbaudu D 2016, July 20. Technology transfer: The change we need in the technology sector. Retrieved from The Cable: https://www.thecable.ng/technology-transfer-change-need-technology-sector

[57] Olawale J 2018, September 19. Entrepreneurship development in Nigeria: problems and prospect.
Retrieved from Legit: https://www.legit.ng/1125968-entrepreneurship-development-nigeria-problems-prospect.html

[58] Ugbogbo, S. N., and Michael, A. O. 2016. Ethical Issues in Information Technology- A Conceptual
Approach. Igbinedion University Journal of Accounting, **2**, 174-243.

[59] Yerby, J. 2013 Legal and Ethical Issues of Employee Monitoring. Online Journal of Applied Knowledge Management, **1**

IOP Conf. Series: Earth and Environmental Science **331** (2019) 012032 doi:10.1088/1755-1315/331/1/012032

Thermal Degradation and Spectroscopic study of Neat Palm Oil, Waste Palm Oil, and Waste Palm Oil Methyl Ester

Omojola Awogbemi[1], Emmanuel I Onuh[2], and Clement A Komolafe[3]
[1]Discipline of Mechanical Engineering, University of KwaZulu-Natal, Durban 4041, South Africa. jolawogbemi2015@gmail.com
[2]Discipline of Mechanical Engineering, University of KwaZulu-Natal, Durban 4041, South Africa. onuhe@ukzn.ac.za
[3]Department of Mechanical Engineering, College of Engineering, Landmark University, Omu Aran, Nigeria. komolafe.adekunle@lmu.edu.ng

Abstract. Research into finding an affordable, renewable, biodegradable and environmentally friendly alternative fuel to power compression ignition engine has been intensified in the past few decades. The application of waste palm oil (WPO) as feedstock for biodiesel has gained traction and has led to a decline in the production cost of biodiesel, and prevents the contamination of aquatic and terrestrial habitats. The present effort is geared towards characterization of neat palm oil (NPO), WPO and waste palm oil methyl ester (WPOME) through Thermogravimetric analysis (TGA), Derivative thermogravimetric (DTG), and Fourier Transform Infrared Spectroscopy (FTIR). Samples of NPO and WPO were collected from takeaway outlets. The WPO was converted into WPOME by transesterification. The outcome of DTG analysis showed that the vaporization of NPO, WPO, and WPOME occurred between 320 °C and 480 °C; 320 °C and 470 °C, and 27 °C and 103 °C respectively. The TGA analysis revealed that NPO, WPO, and WPOME exhibited similar thermal degradation trend and lost 98 %, 29 %, and 63 % of their weight respectively, during the one stage of thermal degradation. The spectra curve for NPO and WPO followed the same trend and achieve peaks at nearly the same wavelength of 1743 cm^{-1} while the peculiar peak of the spectra curve for WPOME occurred at 1022 cm^{-1}. In conclusion, this work reinforces that usage and transesterification alters the thermal degradation and spectroscopic pattern of palm oil.

1. Introduction

The use of petroleum-based diesel fuel (PBDF) to power compression ignition (CI) engines has come with an attendant cost, engine performance, and environmental challenges. This scary scenario has necessitated the need for more interest and research in biodiesel, an important member of the biofuel family, as an alternative to PBDF. Despite some shortcomings, including the high cost of feedstock, low energy conversion, degradation during transportation and storage, biodiesel offers obvious advantages in the world's quest to meet its energy needs [1, 2]. Biodiesel, commonly referred to as fatty acid methyl ester (FAME), is generated by transesterification of feedstocks, mainly vegetable oils (edible or inedible), animal fats, etc. using alcohol (methanol or ethanol) with or without a catalyst [3]. Despite its renewability, low sulfur content, safer handling, higher cetane number, etc., large-scale production and application of FAME has been hampered by the high cost of feedstock, food versus fuel debate and the long time required to cultivate inedible vegetable feedstocks e.g 3- 4 years for palm tree to bear fruit [4], and 2 - 3 years for moringa tree to bear fruit [5]. When compared to PBDF, application, and combustion of FAME in CI engines are more environmental friendly as it generates less carbon monoxide (CO), between 75 % and 90 % reduction in unburnt hydrocarbon (UHC) and fewer polycyclic aromatic hydrocarbons (PAHs) [6]. Economically, it costs about US$0.35 to produce a liter of PBDF from fossil fuel compared to about US$0.5 to produce a liter of FAME, with raw materials accounting for most of the cost [7]. The feedstock is believed to account for between 70 % and 95 % of the cost of FAME production [8, 9]. One of the strategies to make the commercial production of FAME attainable and affordable is the adoption of waste cooking oil (WCO) as feedstock. The initial hurdles in collection logistics and infrastructure pointed out by Janauna and Ellis [10] and Atadashi et al. [11] are being overcome by partnering with operators of fast food, takeaway, and restaurants.

Palm oil is gotten from an oil palm tree (*Elaeis guineensis*). It is generally regarded as the world's most versatile and most widely used vegetable oil. Global production of palm oil rose from about 15.2 million tonnes in 1995 to an excess of 62.6 million tonnes in 2015 and has been projected to attain 240 million tonnes by 2050. Malaysia, Indonesia, and Nigeria top the 43 palm oil growing countries globally. Production and domestic use of palm oil by some African countries between 2013 and 2018 are shown in figure 1 [12]. Application of palm oil includes excellent frying properties at a cheaper cost compared with cottonseed or sunflower oils, a good material for packaging baked foods, raw materials for shampoo, liquid soap, detergent, paints, cosmetics, inks, lacquers, as preservatives in processed foods and an adhesive for binding fibreboards [13].

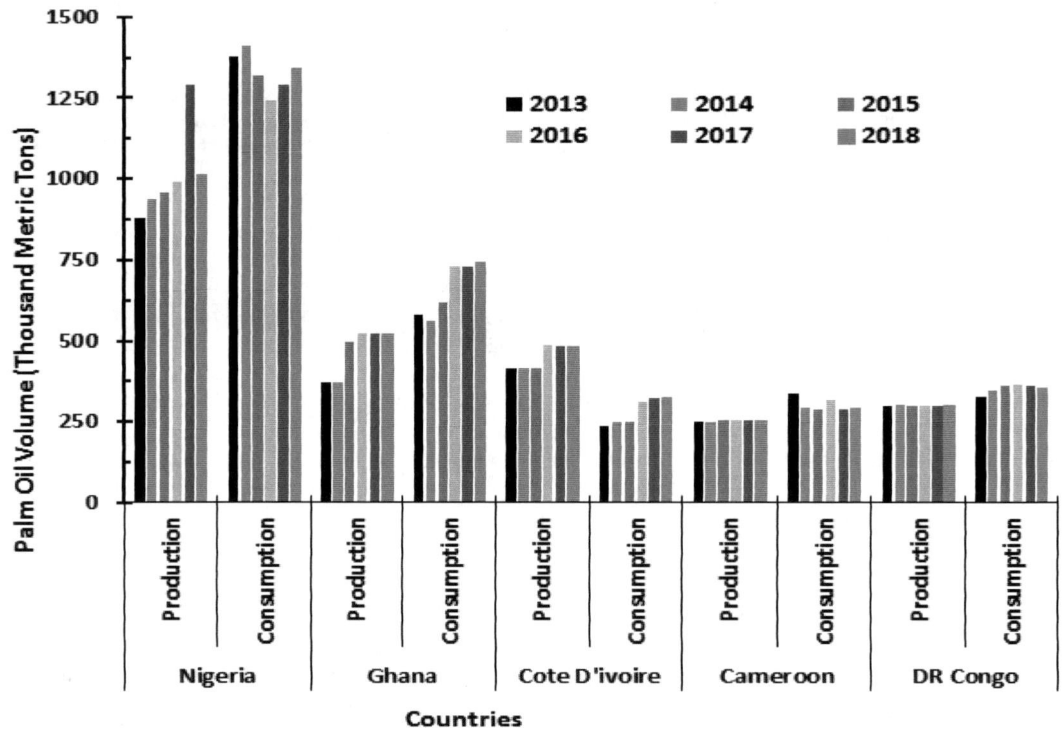

Figure 1. Production and domestic consumption of palm oil in some African countries (Thousand metric tonnes). Adapted from [12]

Due to population explosion, urbanization, and change in lifestyles, the consumption of palm oil, particularly in households, restaurants and fast food outlets for cooking and frying, has continued to rise in recent years with palm oil accounting for the greatest percentage of 34 % and 33 % in production and consumption of all the edible oils respectively. Application of palm oil is, however, not limited to cooking and frying but include lubrication, grease, cosmetics, animal feed, glycerol, polyurethane and alkyl resins which are raw materials for production of paints, inks, lacquers, varnishes and enamels as well as other socio-cultural purposes thereby providing income and employment in the entire value chain. Sadly, growing oil palm is credited to contribute to deforestation with damaging effects on wildlife and forests. Inappropriate disposal of the waste palm oil (WPO) has created problems by blocking the sewer pipes, corrode metals and structural problems by weakening concrete structures [14, 15].

During frying, the oil is heated repeatedly to between 170 ^{O}C and 220 ^{O}C in the presence of oxygen, and sometimes moisture, which causes the palm oil to be exposed to physical, thermal and chemical degradations. Thermal degradation also predisposes the palm oil as a suitable candidate for FAME production by various techniques including pyrolysis, dilution, microemulsion, and transesterification [16, 17]. In an attempt to gain more insight into this feedstock, a lot of investigation are being conducted for its characterization using various non-destructive techniques.

Thermogravimetric analysis (TGA) is a technique for comparing the mass of a sample against temperature or time when the sample is exposed to thermal decomposition through a change in temperature or heated. The analytical technique measures the change in mass of samples during oxidation, reduction, sublimation, decomposition, or vaporization. TGA is usually conducted in a controlled environment. However, outcomes of TGA can be affected by heating rate, buoyancy, and thermal expansion, sample-furnace-sensor arrangement, condensation and reaction, the thermal conductivity of the sample, atmospheric turbulence, etc. [18]. Fourier Transform Infrared (FTIR) spectroscopy is a non-destructive analytical method that uses ultraviolet light to identify the chemical properties of the sample. FTIR has been widely used for the detection, classification, authentication, and categorization of contaminations, additives, fats and oils, fibers, powders, solids, films, polymers, etc. The major advantage of using FTIR include its wavelength precision and competency to quantify spectra with signal-to-noise ratio [19, 20]. The combination of TGA)/DTG and FTIR has been found useful in the evaluation of thermal degradation and oxidative stability characteristics of the physicochemical properties of the samples. TGA/DTG shows weight loss and derivative weight change against temperature.

Over the last decade, a few researchers have characterized either unused vegetable oil or used vegetable oil with a view at comparing the effect of usage on their structure, properties, and degree of saturation. Ullah et al. [21] carried out the characterization of used and un-used palm cooking oil using TGA/DTG, and FTIR to unravel the effect of usage on saturation and properties of the oil. They reported no significant spectroscopic and structural changes in the sample, though there was a remarkable difference in the TGA spectra. Li et al [22] investigated thermal degradation of peanut oil and peanut biodiesel using TGA and FTIR. They observed similarities in their spectra which imply the existence of matching chemical groups in the chemical composition, though the discovery of CH_3 axial deformation in O–CH_3 at 1435.6 cm^{-1} in peanut biodiesel was not noticed in peanut oil. The TGA revealed similar thermal degradation narratives throughout the process though, the two samples were found to be thermally stable at different temperature and varying rate.

Similarly, Elkady et al. [23] researched the characterization of waste vegetable oil and its biodiesel using TGA and FTIR techniques. The differences in the FTIR spectra were attributed to the alteration, during transesterification, of the ester groups in the used vegetable oil into methyl esters in the biodiesel [24]. The TGA showed different commencement and end of temperatures for the thermal degradation, different degradation temperature ranges and percentage weight loss for the samples. Rabelo et al. [25] employed FTIR for the characterization of neat soybean oil and FAME produced from soybean oil. The slight divergences in the position of the peaks in their profile were attributed to the effect of the transesterification process. The difference in their spectrum was noticed in the fingerprint section between 1500 cm^{-1} and 900 cm^{-1}. Also, Matwijczuk et al. [26] engaged the use of FTIR to characterize sunflower oil used to fry French fries, sunflower oil used to fry fish and commercial unblended FAME. The FTIR spectrum revealed a clear difference in the band shapes of the profile of the waste sunflower oil within range of 1770-1670 cm^{-1} due to the effect of the items fried with the oil. In the case of the spectrum for commercial B100 FAME, it was found to be least intensive as a result of the effect of the carbonyl C=C vibration.

With the advancement in research in the production and use of FAME in the last few decades, there is a need to investigate thermal degradation trajectories along the value chain of biodiesel from neat vegetable oil, used vegetable oil to biodiesel produced from the used vegetable oil with a view to engendering better understanding on the quality, behavior, and storage of FAME and its various feedstocks. This current effort aims at the characterization of the neat palm oil (NPO), waste palm oil (WPO) samples that were used to fry different food items and the biodiesel generated from the waste palm oil samples to deepen the information available in the literature. Relevant questions to be answered through this investigation include; what are the effects of thermal degradation on neat palm oil (NPO), waste palm oil (WPO), and waste palm oil methyl ester (WPOME). This research is limited to the comparison of TGA, DTG and FTIR analysis of NPO, WPO and WPOME with a view for better understanding of this aspect of FAME and its feedstock.

2. Materials and Experimental work

The materials used for this research include neat palm oil, waste palm oil samples, biodiesel generated from the waste palm oil samples, chemicals, reagents, and glassware.

2.1. Materials collection and treatment

Neat palm oil and waste palm oil samples were collected from the local owner-operated restaurant at Durban central, KwaZulu Natal province, Republic of South Africa. The waste palm oil has been used for repeatedly for 14 days to fry various food items including fish, and chips at the point of disposal. The cycle of usage and the frying temperature could, however, not be ascertained. Waste chicken eggshells were obtained from the University of KwaZulu Natal, Durban cafeteria. The waste chicken eggshells were converted to CaO catalyst through high-temperature calcination at 900 °C as earlier reported [27].

The WPO was heated to 110 °C to remove moisture, allowed to cool to about 70 °C, where the sample is still in liquid form, and filtered by using laboratory sieve to remove any food debris and other particles trapped in the samples. The acid value of the clean WPO sample was determined to know its suitability for the transesterification process while the remaining sample was kept in glass bottles to prevent contamination.

Methanol (analytical grade, 99.5 %; Merck, South Africa, Univar) was used as alcohol. Magnesol® (analytical grade, 60-100 mesh, the molar weight of 100.39 g/mol, Sigma-Aldrich, Germany) was used as an adsorbent. Magnesol® becomes is hygroscopic once the package is opened. It was therefore carefully re-sealed as tight as possible and contact with eyes was prevented.

2.2. Experimental work

The experimental work consists of the conversion of the WPO to biodiesel via transesterification and the characterization of the samples, namely NPO, WPO, and WPOME, by TGA, DTG, and FTIR.

2.2.1. Transesterification of waste palm oil. The acid value of WPO was determined to ascertain their suitability for conversion to WPOME through one stage transesterification process. The clean feedstock was separately mixed with methanol and calcined CaO catalyst into a 500 ml flat-bottom flask in a required quantity and heated on an electric cooker with a magnetic stirrer. A reaction temperature of 60 °C was selected because it is close but lower than the boiling point of methanol [28]. Other process parameters for the transesterification reaction include methanol to oil ratio of 6:1, the catalyst particle size of 75 μm and 1 %w/w catalyst: oil ratio. The homogeneous mixture was maintained by a magnetic stirrer at 1200 rpm while a digital thermocouple was employed to verify the reaction temperature throughout the process.

The resulting mixture was thereafter filtered in a Buchner funnel filtration system set up to recover the powder catalyst. The filtered solution was conveyed to a separating funnel and exposed to 12 h sedimentation process for the glycerol to coagulate and settle at the base of the separating funnel. The glycerol is tapped out and the remaining crude biodiesel is decanted ready for the purification process.

1 %w/w Magnesol: crude biodiesel was added to the crude biodiesel and heated to 60 °C for 30 mins and mixed at 60 rpm by a magnetic stirrer. The resulting solution was filtered using vacuum filtration, heated to 110 °C to vaporize excess moisture and methanol trapped in the biodiesel and further polished by using 0.45 μm PTFE membrane syringe filter. The purified WPOME was stored in an airtight glass vial and labeled accordingly for characterization.

2.2.2. Characterization of samples. Three samples, namely NPO, WP, and WPOME were subjected to TGA, DTG and FT-IR analyses. The FTIR was analyzed using a Perkin-Elmer Spectrum 100 spectrometer from 380 cm^{-1} to 4000 cm^{-1}. The TGA-DTG were determined using a Perkin-Elmer Simultaneous Thermal Analyzer STA 6000 at a temperature from 30 °C to 600 °C, heated at 10 °C/min with nitrogen as the carrier gas at a gas flow rate of 20 ml/min for the analysis.

3. Results and Discussion

3.1. DTG analysis of NPO, WPO, and WPOME

The outcomes of the DTG analysis of NPO, WPO, and WPOME is shown in figure 2. The DTG curves of NPO and WPO samples showed a similar trend during the entire thermal degradation process which leads to the vaporization of the samples. The evaporation starts at 320 °C for NPO and WPO and ends at 480 °C for and 470 °C respectively until the samples are fully vaporized. On the other, the evaporation of WPOME commences and ends at 27.63 °C and 103 °C, and derivative weight of 3.42 %/m and 0.48 %/m respectively, after which the curve becomes flat, signifying the sample has been fully vaporized. The curves show a single peak for all the samples but at different temperature ranges showing one weight loss stage of the evaporation. The peak in the curves are found as 18.63 $\%m^{-1}$ at 418.37 °C for NPO, 18.5 $\%m^{-1}$ at 418.37 °C for WPO, and 15.25 $\%m^{-1}$ at 63.19 °C for WPOME. This shows that WPOME is thermally degraded in an endothermic reaction at a lower temperature compared with NPO and WPO samples under nitrogen atmosphere. This accounts for why FAME is a better CI fuel than neat vegetable oils because it easily combusted at a lower temperature. These results agree with outcomes of similar researches by Li et al. [22], Niu et al. [29], and Li et al. [30].

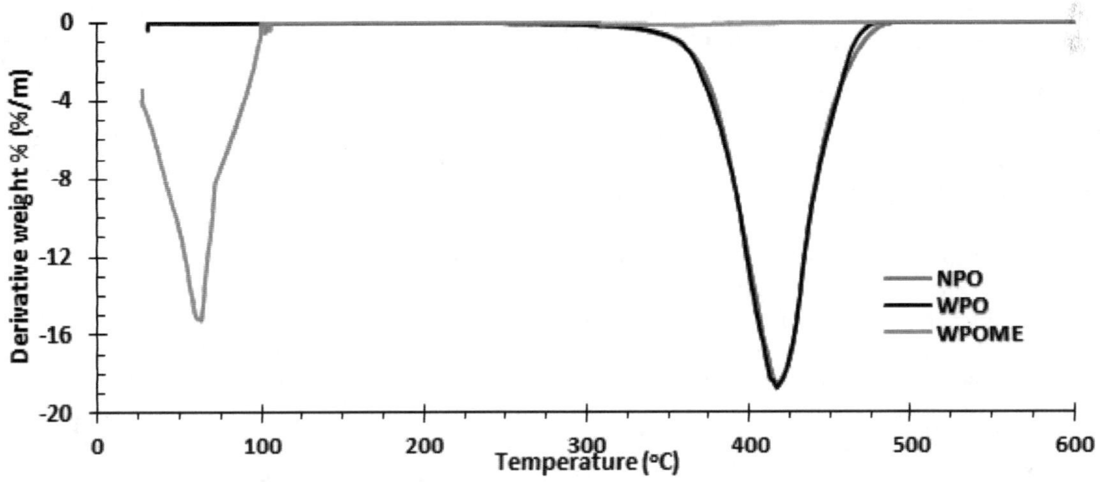

Figure 2. DTG curves for NPO, WPO, and WPOME

3.2. TGA analysis of NPO, WPO, and WPOME

The outcomes of the TGA analysis of NPO, WPO, and WPOME is shown in figure 3. NPO, WPO, and WPOME present similar trends during the entire thermal decomposition process with a single stage weight loss visibly identified. The TGA curves show a distinct step from 30 °C to 600 °C. The thermal degradation range of NPO varies from 330 °C to 480 °C with overall percentage weight loss of 98% while the overall percentage weight change within degradation scope of 350 °C to 470 °C for WPO is 29 %. For the WPOME sample, thermal degradation starts at 350 °C and ends at 470 °C and approximately lose 63 % of its weight. The NPO is degraded at a higher temperature than WPOME due to the fact that NPO is more viscous than WPOME and therefore requires a higher temperature to be degraded. The Lower degradation temperature of WPO compared to WPOME may be attributed to the effect of repeated heating and contamination during usage which has lowered its viscosity and make the bonds to be degraded at a lower temperature. The outcomes of the TGA characterization in this study are in consonance with those in the published and well-referred journals [22, 23, 31, 32].

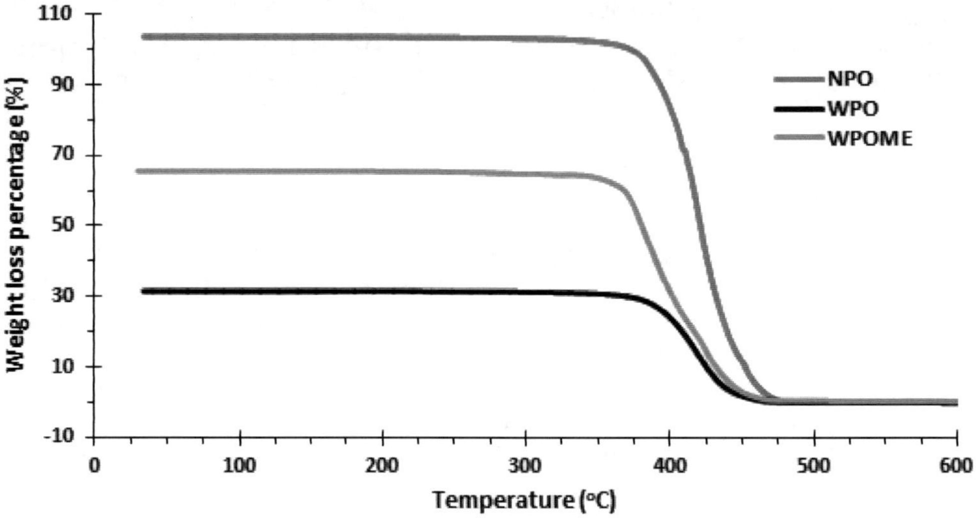

Figure 3. TGA curves of NPO, WPO, and WPOME

3.3. FTIR analysis of NPO, WPO, and WPOME

Four identifiable regions are noticed with characteristics peaks in the FTIR spectrum. The functional groups in the first region with the peaks from 4,000 cm^{-1} to 2,500 cm^{-1} represents N-H, C-H, and O-H single bonds. Peaks from 2,500 cm^{-1} to 2,000 cm^{-1}, showing triple bonds represent the second region while the third region which ranges between 2,000 cm^{-1} and 1,500 cm^{-1}, and the peak corresponds to absorption instigated by double bonds like C=O, C=N, and C=C. From 1,500 cm^{-1} to 400 cm^{-1}, the fingerprint region has many single bonds peaks represents the fourth region [33, 34]. The IR spectrum of the present samples does not contain any peak in the second region as shown in figure 4.

The FTIR spectra showing the functional groups of NPO, WPO, and WPOME are illustrated in figure 4. The resemblances in the FTIR spectra of NPO and WPO 3012 cm^{-1}, 2922 cm^{-1}, 2855 cm^{-1}, 1744 cm^{-1}, 1457 cm^{-1}, 1379 cm^{-1}, 1233 cm^{-1}, 1159 cm^{-1}, and 722 cm^{-1} is an indication of the presence of similar chemical groups in their compositions [35, 36]. This is ascribed to =C-H asymmetrical stretching in alkenes, C-H asymmetrical stretching in alkanes, C-H asymmetrical stretching in methylene, C=O bending in ester carbonyl functional group in FAME, =C-H and –CH_2 bending in methylene groups of cis distributed alkenes and aromatic, respectively [22, 29], as shown in table 1. The peaks for the spectra of WPOME are located at 627 cm^{-1}, 1023 cm^{-1}, 1450 cm^{-1}, 2832 cm^{-1}, 2941 cm^{-1} and 3327 cm^{-1}. These peaks correspond to C-Cl axial deformation in alkyl halide, C-O axial deformation in alcohol, C=C axial deformation in aromatic, O-H axial deformation in axis, C-H axial deformation in alkane, and O-H free axial deformation in alcohol, respectively. The detection of spectra peak at unique wavelengths in WPOME but not in NPO and WPO can be attributed to the distinctive variation between FAME and triacylglycerides [22, 37, 38], and a confirmation of conversion of WPO into WPOME [39].

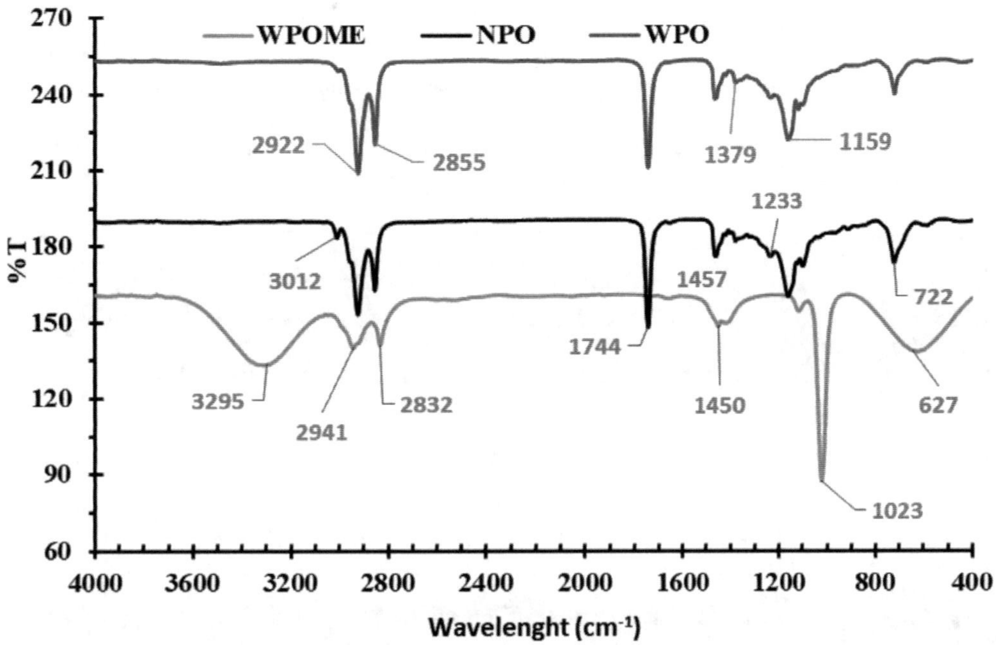

Figure 4. FTIR spectra of NPO, WPO, and WPOME samples

Table 1. Functional group frequencies of NPO, WPO, and WPOME samples [40-44]

Wave number (cm^{-1})	Types of vibration	Functional group	Intensity
3012	Stretching	=C-H, C-H	Strong, broad
3008	Asymmetrical stretching	=C-H of alkenes	
2922	Asymmetrical stretching	C-H of alkanes	
2853	Asymmetrical stretching	C-H of methylene	Medium
1745	Stretching	C=O ester band	
1462	Bending	=C-H (cis)	Variable
1381	Bending in plane	-C-H(CH_3)	Variable
1159	Stretching, Bending	-C-O, -CH_2-	
1024	Stretching	C-O	
722	Bending	=C-H and –(CH_2)n	
627	Out of plane bending	=C-H cis in alkene and aromatics	Strong

4. Conclusions

The DTG, TGA, and FTIR of NPO, WPO, and WPOME are presented. The WPO was taken from the NPO while the WPOME was from the catalytic transesterification of WPO. Neat palm oil was exposed to repeated heating and contamination when used to fry food items. These activities alter the form, and degree of saturation of the palm oil, leading to a change in properties and behavior. The thermal and spectroscopic state of the palm oil is altered during frying, and conversion to biodiesel through transesterification process. The DTG and TGA of NPO, WPO, and WPOME followed similar trend through the thermal degradation and weight loss occurred at different temperatures, owing to the influence of usage and transesterification process. The DTG of the three samples occurred between 330 °C and 480 °C in a one stage weight loss. The NPO exhibited the lowest commencement of thermal degradation temperature followed by WPOME and NPO in that order. The spectra curve of NPO and WPO followed a similar pattern and have their peaks at nearly the same wavelength while WPOME presented a peculiar peak wavelength.

Arising from the outcome of this research, the authors are investigating WPO samples used in frying varieties of food at varying temperature and cycle of frying with a view to establishing the effects of temperature and contamination from food items on its thermal decomposition. Since the outcome of GCMS of fuels has been used extensively to predict the properties, engine performance, combustion, and emission characteristic of internal combustion engines, the use of FTIR should be employed to predict or estimate the fatty acid composition and properties of fuels.

5. References

[1] J. C. Kurnia, S. V. Jangam, S. Akhtar, A. P. Sasmito, and A. S. Mujumdar, "Advances in biofuel production from oil palm and palm oil processing wastes&58; A review," *Biofuel Research Journal,* vol. 3, no. 1, pp. 332-346, 2016.

[2] M. Saifuddin, P. Goh, W. Ho, K. Moneruzzaman, and A. Fatima, "Biodiesel production from waste cooking palm oil and environmental impact analysis," *Bulgarian Journal of Agricultural Science,* vol. 20, no. 1, pp. 186-192, 2014.

[3] Biofuel Fact Sheet: Fatty Acid Methyl Esters (FAME). Available on http://www.etipbioenergy.eu/images/fame-fact-sheet.pdf 2011 [Online].

[4] M. F. Awalludin, O. Sulaiman, R. Hashim, and W. N. A. W. Nadhari, "An overview of the oil palm industry in Malaysia and its waste utilization through thermochemical conversion, specifically via liquefaction," *Renewable and Sustainable Energy Reviews,* vol. 50, pp. 1469-1484, 2015/10/01/ 2015.

[5] A. K. Azad, M. G. Rasul, M. M. K. Khan, S. C. Sharma, and R. Islam, "Prospect of Moringa Seed Oil as a Sustainable Biodiesel Fuel in Australia: A Review," *Procedia Engineering,* vol. 105, pp. 601-606, 2015/01/01/ 2015.

[6] A. Demirbas, "Importance of biodiesel as transportation fuel," *Energy policy,* vol. 35, no. 9, pp. 4661-4670, 2007.

[7] A. E. Atabani, A. S. Silitonga, I. A. Badruddin, T. Mahlia, H. Masjuki, and S. Mekhilef, "A comprehensive review on biodiesel as an alternative energy resource and its characteristics," *Renewable and sustainable energy reviews,* vol. 16, no. 4, pp. 2070-2093, 2012.

[8] I. B. Banković-Ilić, I. J. Stojković, O. S. Stamenković, V. B. Veljkovic, and Y.-T. Hung, "Waste animal fats as feedstocks for biodiesel production," *Renewable and Sustainable Energy Reviews,* vol. 32, pp. 238-254, 2014/04/01/ 2014.

[9] H. H. Mardhiah, H. C. Ong, H. H. Masjuki, S. Lim, and H. V. Lee, "A review on latest developments and future prospects of heterogeneous catalyst in biodiesel production from non-edible oils," *Renewable and Sustainable Energy Reviews,* vol. 67, pp. 1225-1236, 2017/01/01/ 2017.

[10] J. Janaun and N. Ellis, "Perspectives on biodiesel as a sustainable fuel," *Renewable and Sustainable Energy Reviews,* vol. 14, no. 4, pp. 1312-1320, 2010.

[11] I. Atadashi, M. Aroua, A. A. Aziz, and N. Sulaiman, "Production of biodiesel using high free fatty acid feedstocks," *Renewable and sustainable energy reviews,* vol. 16, no. 5, pp. 3275-3285, 2012.

[12] M. Barrientos. Palm oil. Available on https://www.indexmundi.com/about.html [Online].

[13] P. Tullis, "How the world got hooked on palm oil. ," in *The Guardian. Available on https://www.theguardian.com/news/2019/feb/19/palm-oil-ingredient-biscuits-shampoo-environmental*, ed. London, N1 9GU, United Kingdom., 2019.

[14] D. Panadare and V. Rathod, "Applications of Waste Cooking Oil Other Than Biodiesel: A Review," *Iranian Journal of Chemical Engineering,* vol. 12, no. 3, 2015.

[15] R. Pool, *The Nexus of biofuels, climate change, and human health: workshop summary.* National Academies Press (US), 2014.

[16] K. Ganesan, K. Sukalingam, and B. Xu, "Impact of consumption of repeatedly heated cooking oils on the incidence of various cancers-A critical review," *Critical reviews in food science and nutrition,* pp. 1-18, 2017.

[17] A. O. Falade, G. Oboh, and A. I. Okoh, "Potential health implications of the consumption of thermally-oxidized cooking oils–a review," *Polish journal of food and nutrition sciences,* vol. 67, no. 2, pp. 95-106, 2017.
[18] J. D. Menczel and R. B. Prime, *Thermal analysis of polymers: fundamentals and applications.* John Wiley & Sons, 2014.
[19] A. E. Atabani *et al.*, "Integrated valorization of waste cooking oil and spent coffee grounds for biodiesel production: Blending with higher alcohols, FT–IR, TGA, DSC and NMR characterizations," *Fuel,* vol. 244, pp. 419-430, 2019/05/15/ 2019.
[20] B. C. Smith, *Fundamentals of Fourier transform infrared spectroscopy.* CRC press, 2011.
[21] Z. Ullah, M. A. Bustam, and Z. Man, "Characterization of waste palm cooking oil for biodiesel production," *International Journal of Chemical Engineering and Applications,* vol. 5, no. 2, p. 134, 2014.
[22] H. Li, S. Niu, C. Lu, and Y. Wang, "Comprehensive investigation of the thermal degradation characteristics of biodiesel and its feedstock oil through TGA–FTIR," *Energy & Fuels,* vol. 29, no. 8, pp. 5145-5153, 2015.
[23] M. Elkady, A. Zaatout, and O. Balbaa, "Production of biodiesel from waste vegetable oil via KM micromixer," *Journal of Chemistry,* vol. 2015, 2015.
[24] A. Tiwari, V. Rajesh, and S. Yadav, "Biodiesel production in micro-reactors: A review," *Energy for sustainable development,* vol. 43, pp. 143-161, 2018.
[25] S. N. Rabelo, V. P. Ferraz, L. S. Oliveira, and A. S. Franca, "FTIR analysis for quantification of fatty acid methyl esters in biodiesel produced by microwave-assisted transesterification," *International Journal of Environmental Science and Development,* vol. 6, no. 12, p. 964, 2015.
[26] A. Matwijczuk *et al.*, "Spectroscopic studies of the quality of WCO (Waste Cooking Oil) fatty acid methyl esters," in *BIO Web of Conferences*, 2018, vol. 10, p. 02019: EDP Sciences.
[27] O. Awogbemi, F. L. Inambao, and E. I. Onuh, "Development and Characterization of Chicken Eggshell waste as Potential Catalyst For Biodiesel Production," *International Journal of Mechanical Engineering and Technology,,* vol. 9, no. 12, pp. 1329 - 1346, 2018.
[28] T. Ahmad *et al.*, "Optimization of process variables for biodiesel production by transesterification of flaxseed oil and produced biodiesel characterizations," *Renewable Energy,* 2019.
[29] S. Niu, Y. Zhou, H. Yu, C. Lu, and K. Han, "Investigation on thermal degradation properties of oleic acid and its methyl and ethyl esters through TG-FTIR," *Energy Conversion and Management,* vol. 149, pp. 495-504, 2017/10/01/ 2017.
[30] H. Li, S.-l. Niu, C.-m. Lu, and S.-q. Cheng, "Comparative evaluation of thermal degradation for biodiesels derived from various feedstocks through transesterification," *Energy Conversion and Management,* vol. 98, pp. 81-88, 2015/07/01/ 2015.
[31] M. Dantas *et al.*, "Thermal and kinetic study of corn biodiesel obtained by the methanol and ethanol routes," *Journal of Thermal Analysis and Calorimetry,* vol. 87, no. 3, pp. 835-839, 2007.
[32] M. Todaka, W. Kowhakul, H. Masamoto, M. Shigematsu, and S. Onwona-Agyeman, "Thermal decomposition of biodiesel fuels produced from rapeseed, jatropha, and coffee oils with different alcohols," *Journal of thermal analysis and calorimetry,* vol. 113, no. 3, pp. 1355-1361, 2013.
[33] D. Michael. (2017) "How to Read IR Spectrums." Sciencing, https://sciencing.com/read-ir-spectrums-6809192.html. 25 April 2017.
[34] J. McMurry, *Organic chemistry.* Belmont, CA: Thomson Brooks/Cole, 2008.
[35] P. Nautiyal, K. Subramanian, and M. Dastidar, "Experimental investigation on performance and emission characteristics of a compression ignition engine fueled with biodiesel from waste tallow," *Clean Technologies and Environmental Policy,* vol. 19, no. 6, pp. 1667-1677, 2017.
[36] Y. Isah, A. Yousif, K. Feroz, Y. Suzana, A. Ibraheem, and A. Soh, "Comprehensive Characterization of Napier Grass as Feedstock for Thermochemical Conversion. Open Access," *Energies Journal. www. mdpi. com/journal/energies,* vol. 8, pp. 3403-3417, 2015.
[37] F. Zhang, D. Adachi, S. Tamalampudi, A. Kondo, and K. Tominaga, "Real-time monitoring of the transesterification of soybean oil and methanol by Fourier-transform infrared spectroscopy," *Energy & Fuels,* vol. 27, no. 10, pp. 5957-5961, 2013.
[38] J. A. Melero, M. M. Clavero, G. Calleja, A. García, R. Miravalles, and T. Galindo, "Production of biofuels via the catalytic cracking of mixtures of crude vegetable oils and nonedible animal fats with vacuum gas oil," *Energy & Fuels,* vol. 24, no. 1, pp. 707-717, 2009.

[39] J. Nisar *et al.*, "Enhanced biodiesel production from Jatropha oil using calcined waste animal bones as catalyst," *Renewable Energy,* vol. 101, pp. 111-119, 2017/02/01/ 2017.
[40] A. Winter, *Organic Chemistry I For Dummies*, 2nd Edition ed. Wiley, 2016.
[41] A. Oyerinde and E. Bello, "Use of fourier transformation infrared (FTIR) spectroscopy for analysis of functional groups in peanut oil biodiesel and its blends," *British Journal of Applied Science & Technology,* vol. 13, no. 3, pp. 1-14, 2016.
[42] M. Y. Talpur *et al.*, "Application of multivariate chemometric techniques for simultaneous determination of five parameters of cottonseed oil by single bounce attenuated total reflectance Fourier transform infrared spectroscopy," *Talanta,* vol. 129, pp. 473-480, 2014.
[43] S. Y. Lim, M. S. Abdul Mutalib, H. Khaza'ai, and S. K. Chang, "Detection of fresh palm oil adulteration with recycled cooking oil using fatty acid composition and FTIR spectral analysis," *International Journal of Food Properties,* vol. 21, no. 1, pp. 2428-2451, 2018.
[44] Silverstein R.M., Bassler G.C., and M. T.C., *Spectrometric Identification of Organic Compounds*, 5th edition ed. New York, United States. : Wiley, 1991.

IOP Conf. Series: Earth and Environmental Science **331** (2019) 012033 doi:10.1088/1755-1315/331/1/012033

Energy poverty and environmental sustainability in Nigeria: an exploratory assessment

Agaptus Nwozor[1], Segun Oshewolo[1] and Oluwaseun Ogundele[1]

[1]Department of Political Science and International Relations, Landmark University, Omu-Aran, Kwara State, Nigeria.

Abstract. Energy constitutes an indispensable catalyst for sustainable socio-economic development. The imperative of energy security is highlighted by the pervasive energy poverty that characterizes developing countries, especially Nigeria. Despite being a major oil producer and having four refineries, Nigeria imports 90 percent of its domestically needed refined petroleum products. Similarly, Nigeria requires about 98,000 megawatts (MW) to achieve national electricity security. However, its thermal and hydropower stations have a combined installed capacity of only 12,522MW but generate between 3,000MW and 4,000MW and sometimes less than 3,000MW. The electricity challenge is deepened by fluctuations. The implications of this suboptimal power generation include: only 40 percent of Nigeria's 193 million population have access to electricity, erratic nature of electricity supply and the long hours of power outages. Thus, Nigeria can be categorized as an energy poor country. The major goal of this paper is to assess the link between energy poverty and environmental sustainability given that households must take alternative steps to bridge energy gaps that impact them. It further evaluates the opportunities and barriers to energy security and holds that the untapped potentials, especially among renewable energy resources, hold the key to sustainable energy development to tackle energy poverty in Nigeria.

1. Introduction

Nigeria is not only the most populous black country but also the largest economy in sub-Saharan Africa. Additionally, it is well endowed with enormous resources, which it could harness to provide energy for its teeming population, but its energy situation is deplorable and presents a serious paradox. Nigeria is a major oil producer with enormous hydrocarbon reserves. It has four refineries with a total installed capacity to refine 445,000 barrels per day. It has the resources to even expand this refining capacity in order to achieve self-sufficiency in refined petroleum products. However, none of these refineries has ever refined up to 20 percent of its capacity resulting in the country importing about 90 percent of its required refined petroleum products [1,2]. This scenario has led to several distortions and contradictions in the economy of Nigeria ranging from high cost of these products, uncertainty in their availability, encouragement of corrupt practices due to subsidies, and unpredictability in supply often resulting in the unavailability of these products.

Similarly, Nigeria has the potential to harness its abundant renewable and non-renewable resources to generate electricity but its electricity situation is more than deplorable. Nigeria requires an estimated 98,000MW to achieve energy security. By energy security, we mean the minimization of a country's vulnerability to the various energy systems and the possession of the capacity to provide energy services within its jurisdiction [3,4]. Nigeria has an installed capacity to generate 12,522 megawatts (MW) of electricity from its existing thermal and hydropower stations. However, it has only managed to generate between 3,000MW and 4,000MW and most times much lower than this range. The result is

[1] ORCID id: http://orcid.org/0000-0002-9782-6604; nwozor.agaptus@lmu.edu.ng

that only 40 percent of Nigerian households have access to electricity with the rural areas having marginal access as only approximately 10 percent of rural households are connected to electricity [5].

All the foregoing indicators suggest energy insecurity which invariably underpins Nigeria's energy poverty. In other words, Nigeria's energy profile suggests that it is an energy poor country which also means that its people are energy poor. The goal of this paper is to assess the relationship between energy poverty and environmental sustainability, that is, whether energy poverty faced by households has led them to engage in certain practices that have constituted threats to the environment. Relying on the traditions of exploratory research design, the paper utilizes secondary data sourced from statistical reports, government publications and journals to examine and analyze the inter-linkages between domestic energy poverty and environmental sustainability. This is against the backdrop that households must evolve survival strategies to augment shortfalls in their energy needs whether it is because they are unavailable or because they lack the resources to procure them. The paper finds linkage between energy poverty and the depletion of forestry resources which invariably leads to the deepening of household poverty. It further reviews and evaluates the opportunities and barriers to energy security and holds that the country must seek for effective ways to exploit its untapped renewable and non-renewable energy potentials.

2. Potentials and Paradoxes: Nigeria's Energy Profile

Nigeria has abundant renewable and non-renewable energy resources that could be harnessed to meet its energy needs, both present and future. But despite these enormous ensembles of energy resources, Nigeria is an energy poor country as majority of its 193 million people both lack access to energy resources and also the financial capacity to pay for them in order to meet their household needs. Paradoxically, Nigeria possesses the capability to lift its teeming population out of energy poverty but a combination of factors including among others weak policy framework and policy summersaults, corruption and lack of vision undermine efforts to bridge the energy poverty gap between the country and global averages in energy production and consumption.

2.1. Renewable energy resources

The global emphasis on renewable energy is a product of established relationship between non-renewable energy sources and climate change. Thus, the key emphasis of the international community centers on gradual transition to renewable energy sources as a panacea to the climate change menace. In other words, the world is evolving, adopting and emphasizing renewable energy options as veritable means to achieving energy sustainability and by extension, environmental sustainability.

Nigeria is blessed with abundant renewable energy resources. By renewable energy, we refer to all energy sources that are permanently available or renew themselves within a short frame of time. Nigeria is endowed with rivers, waterfalls, dams and river basins from which it could generate hydropower, adequate sunshine across the country for solar energy, strong wind for wind energy, adequately endowed environment with diverse biological mass to develop biomass energy and coastal corridor to generate wave and tidal energy. However, the challenge that Nigeria currently faces is suboptimal development of these energy sources due, largely in part, to inchoate policies and weak incentives to attract investors to the sector [6].

2.1.1. Hydropower energy: this is the major renewable energy source for the generation of electricity in Nigeria. This is mainly as a result of several large rivers, waterfalls, dams and river basins in the country. However, the major water sources that account for Nigeria's rich hydropower potentials are rivers Niger and Benue as well as the Lake Chad basin [7]. Despite the centrality of hydropower and its potential high capacity to generate Nigeria's electricity needs, it is grossly underexploited. This is in great contrast with the picture at the global arena.

Previous estimates of the potentials of Nigeria's seven out of eleven river basins, namely Chad, Cross River, Sokoto, Hadejia-Jama're, River Niger, Upper Benue, Lower Benue, showed a combined small-scale hydropower (SHP) capacity of about 734.2MW [8], but current adjusted estimates of river basins in Nigeria indicate a potential energy output of 3,500MW [9]. Notwithstanding this huge potential, Nigeria's installed capacity for SHP is approximately 45MW, which means that it has only developed about 1.3 per cent of its SHP potentials [10]. In the same vein, the total estimate of Nigeria's hydropower potential is about 14,750MW but, until recently, it only generated about 15 percent of this capacity or 1,930MW [11]. As at 2017, the total percentage contribution of Nigeria's hydropower stations (Jebba, Kainji and Shiroro) to the country's energy mix rose to 26 per cent [12]. This was essentially the result of the US$100 million maintenance, repairs and investment undertaken by the African Development Bank (AfDB) [7].

Nigeria has projected expansion in the contribution of hydropower to its energy mix as its economy grows towards being among the world's 20 largest economies by 2020. Thus, it is envisioned that hydropower's contribution (both large and small) would rise from 2,030MW in the short run to 6,664MW in the medium-run and ultimately 67,000 MW in the long-run [7, 11]. Efforts appear to be at various stages to strengthen the capacity of several existing hydropower plants, including the 30MW Gurara 1, the 10MW Tiga, 10MW Oyan, the 8MW Challawa and the 6MW Ikere plants as well as construct new ones like the 700MW Zungeru and the 40MW Kashimbila hydropower stations [7]. While these efforts will contribute in alleviating Nigeria's energy poverty, the commitment for their full completion remains to be seen.

2.1.2. Solar energy*:* the potential to develop solar energy due to the vast supply of sunshine all year round is very high in Nigeria. Solar energy involves the conversion of sunlight into electricity through photovoltaic or concentrated solar power systems. The prospects of solar energy to boost Nigeria's energy production and thus eliminate energy poverty lies in a combination of factors, namely, its capacity to deliver the country's annual energy requirements in just an hour; its status as a clean energy source, free of pollutants and greenhouse gases; and its non-susceptibility to geo-political constraints and conflicts [13].

Although solar resources are readily available around world, the amount available to each region differs. The same applies to Nigeria as it has various climatic regions. Thus, despite its location on the equator with attendant advantage of being placed within a high sunshine belt with fairly well-distributed solar radiation, there are still variations across the country. Osueke, Uzendu and Ogbonna [14] have pointed out that the amount of solar energy that is available in any geographical location is indicative of the measure of solar irradiance that falls on that location. The solar potential of Nigeria is very high. This implies that all the regions of the country have viable potentials for solar energy production, but in terms of hierarchizing solar energy potentials, the northern region has more potential than the southern region. Figure 1 below shows the radiation range of different parts of Nigeria.

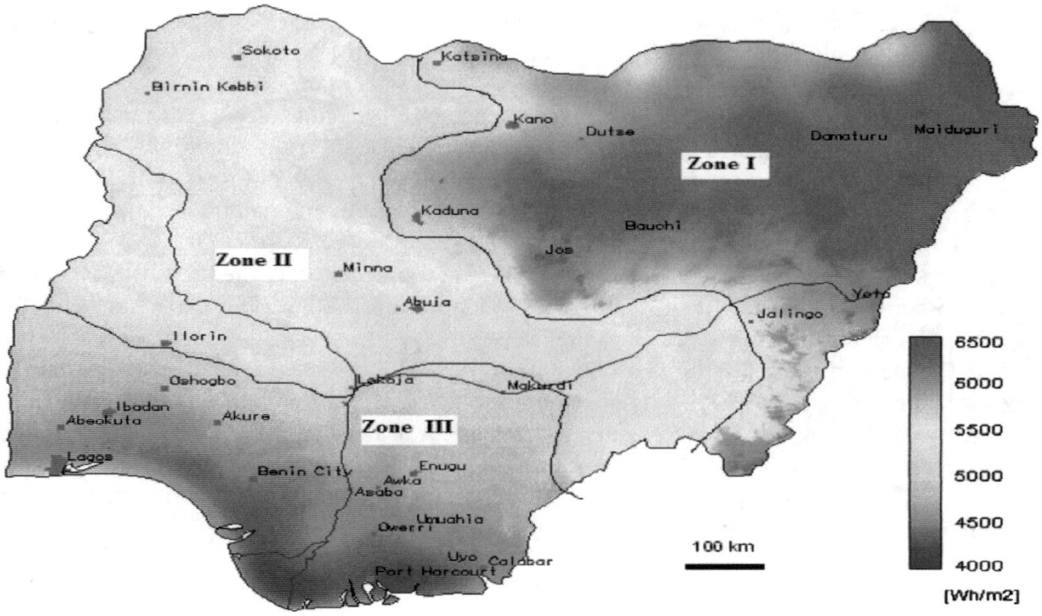

Figure 1: Map of Nigeria showing solar radiation zones [13]

Nigeria's sunshine profile is quite impressive. The sunshine radiates about 17,459,215.2 million megajoule (MJ) of solar energy per day on Nigeria. When this is broken down, it shows that the average daily solar radiation varies from around 12.6 MJ/m2/day (equivalent to 3.5 kWh/m2/day) in the coastal region to about 25.2 MJ/m2/day (equivalent of 7.0 kWh/m2/day) in the northern region [6, 11]. What this implies is that Nigeria has utilizable solar energy to lift its population out of energy poverty and ensure energy security.

Scholars have painted various scenarios to demonstrate the viability of solar energy in Nigeria. Maren, Agontu and Mangai [8] contend that the country could generate electricity that is equivalent to 4.66 million barrels of oil per day from 1 percent of its surface area if solar energy appliances with just 5 percent efficiency are used. In the same vein, Shaaban and Petinrin [11] project that if only 0.1 percent of the estimated 17,459,215.2 million MJ/day of solar energy that falls within the country's territory were to be harnessed at an efficiency of 1 percent, Nigeria's total energy demand would be met. In spite of the potentials painted by the foregoing scenarios, solar energy is yet to be included in the country's energy mix.

2.1.3. Wind energy: the wind is another untapped energy resource. Currently, the wind does not contribute to Nigeria's energy mix. In fact, the only large-scale harnessing of wind energy is a 10MW facility that is currently under construction in Katsina State. And yet every part of Nigeria has varying wind speed and strength that can be harnessed for various purposes, especially to generate power. According to estimates by scholars, the variation in the annual average wind speed at 10 meters heights could range from approximately 2 m/s in the coastal regions to approximately 4 m/s in the northern region with a peak wind speed recorded in the months of April through August for most sites and at 50m, the range is 2 - 8 m/s. There are both small and large modern wind turbines that can be customized and localized to generate electricity for utilities and other small energy needs [15].

Essentially, wind energy is produced by channeling the wind towards the blades of windmills or wind turbines, which in turn rotates them and drives electrical generators to produce power [15, 16]. Several studies have characterized wind speed and pattern in various parts of the country with a positive verdict on the viability of wind as a source of power generation [16-21]. However, studies also showed that across locations and even within a very close proximity, there is a wide variation in wind speed [22]. The variations in wind speed is normal around the globe. What is required based on the various conflicting data generated by scholars on wind speed from different parts of the country is, as Brimmo et al. [22]) have argued, to create a platform to validate all available data in order to ensure minimum error in mathematical models to predict wind speed.

Wind energy is very important if Nigeria must meet its target of clean energy as well as attain energy security. A major step towards actualizing it is to develop a clear policy direction in this regard, which of course will deal with the challenges of low financing, lack of awareness and attraction of investors who will encourage the development of appropriate technical capacities and wind technologies.

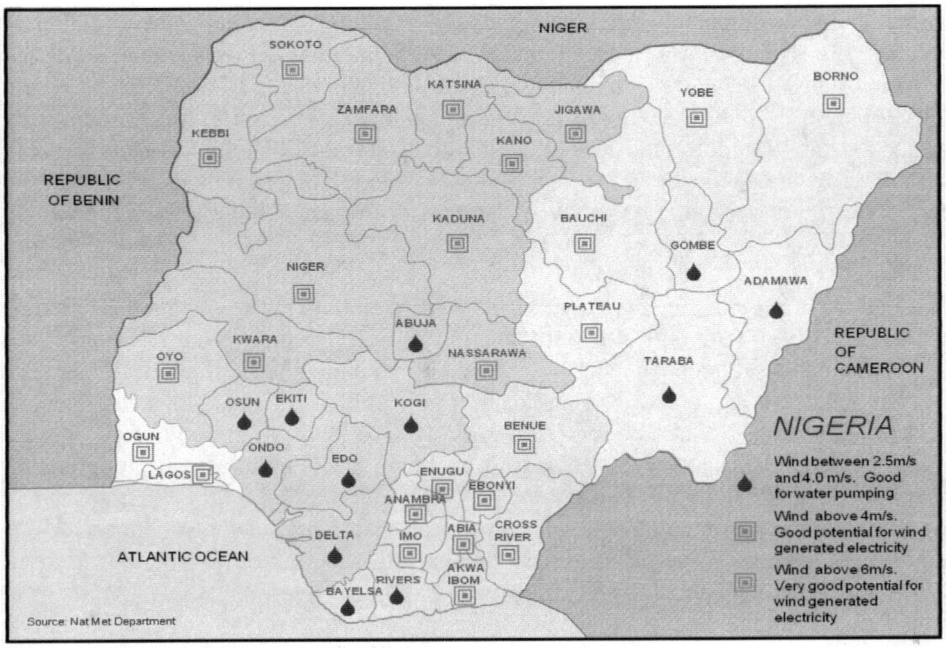

Figure 2: Wind strength across the country
(Source: http://www.neenigeria.com/Nigeria_wind_NEW.png)

2.1.4. Biomass energy: this is the oldest source of energy ever used by mankind. Its attraction lies in its abundance, cheapness and accessibility. Biomass energy is generated directly or indirectly from biological sources ranging from plants, algae, micro-organisms to animals. Thus, Nigeria has abundant biomass resources from which biomass energy could be generated. These biomass resources include: grasses and shrubs, water hyacinth, agricultural residues, livestock manure and animal wastes, urban refuse and industrial wastes. Generally, both food and non-food biomass could be converted into biofuels and they may either be solid, gas or in liquid forms [20]. Biomass is converted into electric power through several methods. These methods range from direct combustion, gasification, pyrolysis, fermentation of wet wastes to anaerobic digestion.

The various biomass conversion methods are specifically suitable for different biomass resources. In other words, different types of biomass require different forms of processes based on the type of energy that is being targeted to be produced. For instance, woody biomass, namely pellets, sawdust and wood chips as well as other categories of wastes like household refuse, dry organic wastes, industrial and agricultural wastes are combusted or gasified to generate electricity [23]. Similarly, both anaerobic digesters and pyrolysis are used to convert wheat straw residues, corn stover, and special fast growing trees into biogas, bio-oil or biofuel. Wet wastes such as those from animals and humans are also converted into energy through anaerobic digesters [15].

Nigeria has a great potential to generate bio-energy as a result of the availability of biomass resources. Although available data do not appear to converge, anecdotal evidence based on Nigeria's

IOP Conf. Series: Earth and Environmental Science **331** (2019) 012033 doi:10.1088/1755-1315/331/1/012033

population and the volume of domestic and industrial refuse generated by households and industries respectively, which often overwhelm waste management authorities across the country, it could safely be said that Nigeria can generate the necessary biomass resources to generate electric power. While Shaaban and Petinrin [11] assert that Nigeria can obtain 200 million tonnes of dry biomass from forage grasses and shrubs to generate energy, Aliyu, Dada & Adam [20] aver that Nigeria generates about 227,500 tonnes of animal waste on daily basis, which could potentially produce about 6.8 million m^3 of biogas per day. This is based on the calculation that a kilogramme of fresh animal waste yields about 0.03 m^3 of biogas [24].

The use of firewood or fuel wood is also an aspect of biomass energy. As a result of energy poverty, it is estimated that more than 70 percent of the people in the rural areas depend on fuel wood and associated charcoal for their household energy needs [25]. What this indicates is that although fuel wood is critically important to many poor households since it contributes to their energy provision, it also hurts the environment by putting pressure on forest resources and contributing to deforestation and degradation [26. Notwithstanding that the global charcoal trade in 2015 amounted to 2.6 million tonnes and Nigeria was ranked alongside Indonesia, Myanmar, Namibia, and Poland as a top charcoal-exporting country [26], the current energy policy of Nigeria deemphasizes the use of fuel wood for energy [25, 27].

2.1.5. Geothermal energy: Nigeria has the resources through which it can derive geothermal energy. Across the Northern and Southern regions, there are major river basins, river alluviums, deltaic sediments, sandstones and basalt deposits that signpost potential geothermal resources for energy generation[22]. Thus, the geology of Nigeria indicates the viability of generating thermal energy considering the availability of such geothermal resources as lavas, hot springs, geysers, fumaroles, volcanoes and mud pots [28]. According to Brimmo et al [22], "geothermal energy is derived from heat generated by Earth's formation, and subsequent radioactive decay of the earth's minerals. This heat is driven from the core to surrounding rocks which superheats ground water, pressurizing it to the surface." The current situation with geothermal energy is that it is uncommon and not in the forefront of energy mix being seriously considered for electricity generation in Nigeria.

2.1.6. Wave and tidal energy: Nigeria's massive coastline of 853 kilometers provides prospects for the harnessing of energy from the wave and tides of the ocean. Wave and tidal energy is derived from the oceans and seas. As Amoo [29] has explained, "tides are caused by the gravitational pull of the moon and the sun on the rotating earth. The associated motion of these bodies leads the surfaces of the oceans to be raised and lowered periodically, according to some interacting cycles." Thus, wave and tidal energy could be seen as a form of hydropower that depends on the conversion of the rising and falling of the ocean and sea levels to generate electricity. Waves are prominent in the oceanic areas of Nigeria and thus could be harnessed to produce electricity. However, this is not the case as wave and tidal energy is yet to be exploited at all in Nigeria [27, 29]. Thus, what is required is to mainstream wave and tidal energy into Nigeria's prospective scheme of low-carbon, clean energy and environmentally-friendly energy mix as well as initiate dedicated research in this sector.

2.2. Non-renewable energy resources

Nigeria is rich in non-renewable energy resources, which mainly consist of hydrocarbons. Not only is Nigeria richly endowed, it is classified as a major oil producing country. Thus, hydrocarbons occupy a very important place in its economic development. Except for the period between 2004 and 2009 when rebellion which was orchestrated by the Niger Delta militants disrupted oil production activities, Nigeria's oil production outputs average between 2.1 and 2.5 million barrels per day [30]. Additionally, in terms of the outlay of oil production activities, Nigeria had cumulatively produced some 75.9 billion barrels of crude oil between 1960 and 2016 [31].

Notwithstanding this volume of crude oil production, the proven reserves of Nigeria's crude oil stock was estimated at 37.453 billion barrels in 2016, thus placing it on the 11th position among oil producing countries [31]. Furthermore, the Nigerian government projects to expand the volume of proven oil reserves to 40 billion barrels by 2020 [32]. Nigeria is also enormously endowed with natural gas. Its enormous natural gas holdings in terms of proven reserves which were 5,475.2 billion standard cubic meters as at 2016 placed it among the top 10 in the world [31]. In addition to these enormous reserves, the cumulative marketed natural gas between 1970 and 2016 was put at 88,312.4 million standard cubic meters [31]. Nigeria's thermal stations which have total installed capacity of 10,142MW depend on natural gas to power them. However, the thermal stations barely meet the target as they operate at less than 25 percent of installed capacity. A major challenge often cited as being responsible for poor electric power generation is shortfall in gas supply as a result of either mechanical fault or sabotage [33].

Nigeria also has significant deposits of high quality lignite coals. Coal deposits are found mainly in the Anambra basin and the Benue trough with some minor deposits in the stretch of the Bida and Sokoto basins [34]. Coal was the first conventional energy resource that was discovered and utilized in Nigeria. Coal reserves are estimated at 2.8 billion tonnes in 17 identified coalfields with proven reserves of 639 million tonnes [6,35]. Although coal contributed to the economic development of Nigeria from the time it was discovered in Enugu, Nigeria in 1909 to the 1960s, it has since been abandoned. In other words, Nigeria's coal resources have remained unexploited following their abandonment. Before its abandonment, coal production in Nigeria peaked at 790,030 tonnes in 1956 and about 70 percent of this amount went into the country's energy generation [6].

Coal played an important role in Nigeria's early quest for industrialization before it was displaced by the combination of crude oil, which was discovered in commercial quantity in the country and the global evolution from coal-powered to oil-powered industrialization. According to Odesola, Samuel and Olugasa [36], coal production facilitated the expansion and consolidation of such firms as the defunct National Electric Power Authority (NEPA), the Nigerian Cement Factory, the Nigerian Railway Corporation (NRC), and other industrial ventures. The major interrelated reasons for the abandonment of coal by Nigeria included: its diminishing lucrativeness due to oil, the modernization processes by those firms which had hitherto relied on coal for their operations, dwindling export profile due to declining global demands and the effects of the Nigerian civil war [6,36].

But coal is still relevant in augmenting global energy needs as exemplified by its contemporary resurgence in terms of usage despite concerns about climate change. It was estimated that 6.6 billion tonnes of coals were consumed worldwide in 2012 especially to augment shortfalls in energy requirements [37]. The strongest demand for coal as well as its biggest market is Asia as countries like Japan, Chinese Taipei and Korea use it to supplement their energy requirements to generate electricity and steel production due to the shortfall of natural energy resources [37]. In addition to electricity generation and steel production, coal has many important uses, especially in the manufacture of cement and as a liquid fuel. Furthermore, the high quality of Nigerian coal, due to its low sulphur and ash content as well as low thermoplastic properties, makes it ideal for "boiler fuel, production of high calorific gas, domestic heating, briquettes, formed coke and the manufacture of a wide range of chemicals including waxes, resins, adhesives and dyes".

The paradox of Nigeria's coal sector is that the country imports coal even with its abundance and high quality as it is mostly lignite and sub-bituminous. This is despite the announced intention of the Nigerian government to generate 30 percent of its electricity from coal [38-39]. There appears to be efforts in this regard going by the various coal electricity projects that have been flagged off in Nigeria since 2013 which target to generate and supply coal-derived electricity of 4,800 MW [6] but as at now, coal is yet to make contributions to Nigeria's energy mix. The total result of this effort remains to be seen as some of the projects expected to be completed in 2015 are still on-going.

3. Linking Environmental Degradation to Energy Insecurity and Energy Poverty

Nigeria has several energy resources to guarantee energy security. Like the concepts that have the term "security" attached to them, energy security is multidimensional and highly context-dependent as it is variously interpreted in terms of the perceptions of different stakeholders [40]. The term "energy

IOP Conf. Series: Earth and Environmental Science **331** (2019) 012033 doi:10.1088/1755-1315/331/1/012033

security" has metamorphosed over time to acquire a wider meaning. It has transformed from preoccupation with uninterrupted or stable supply of, and access to, oil to encompass a wider range of issues relating to energy policy problems such as providing equitable access to modern energy, issues of pollution and environmental degradation, energy efficiency and mitigation of climate change [40-43].

Although Månsson, Johansson and Nilsson [40] have noted the difficulty of a single definition that could offer an acceptable interpretation of energy security due to differences in perspectives and priorities, this paper is inclined to conceptualize energy security in terms of having "access to clean, reliable and affordable energy services for cooking, heating, lighting, communications and other productive uses" (cited in [44]). Therefore at the epicenter of energy security are the considerations about whether energy resources are available in the required quantities and if so, whether they are accessible in terms of easy procurement; whether they are affordable, which is concerned with their cost and ease of payment by various segments of the society; and the consideration of acceptability, which is preoccupied with the question of whether or not these energy resources are harming the environment or indeed contributing to environmental sustainability [41]. Energy security, therefore, underpins the formulation of energy policy strategies. Energy security plays some roles in the economies of countries considering the centrality of energy resources in driving contemporary global development. Thus, there is a link between energy insecurity and energy poverty.

Nigeria possesses real potentials to ensure energy security on account of its endowments. In other words, it can produce more than enough energy from diverse energy sources, with the exception of nuclear sources, to meet national energy requirements. Paradoxically, despite Nigeria's endowments with energy resources, it is confronted with energy poverty such that majority of its population lacks access to modern sources of energy. Energy poverty manifests differently in various regions of the world but it essentially depicts energy vulnerability and deprivation arising from non-availability of energy resources or the inability of people to access adequate energy for their needs due to unsustainable costs. Non-availability of energy resources has a tendency to affect affordability by households. Thus, energy poverty results from unsustainable increase in the costs of energy in the face of stagnation in the average income of households [45] or national incapacity to produce energy resources in such magnitude as to satisfy the needs of the population. The non-availability and expansionary trend in energy costs without commensurate increase in household incomes have increased household expenditure on energy thus deepening the dilemma of households, particularly those with lower incomes.

The different circumstances of developed and developing countries underpin their divergent thresholds benchmarking energy poverty. For instance, although there is a wide distinction at the country level in the European Union about the benchmarks for energy poverty, they oscillate around struggling to achieve adequate warmth by households; inability to promptly pay utility bills; incommensurate energy bills in relation to household incomes and general negative effects on people's health and well-being due to insufficient energy services in the home or inadequate access to energy resources [45, 46]. Specifically, the UK has benchmarked fuel poverty through its low income high cost (LIHC) indicator. Within the framework of this indicator, a household is classified as fuel poor if it is operating a "relatively energy inefficient home, and stands to be left in relative income poverty as a result of paying fuel bills assuming they heat their home to the recommended regime" [47]. Notwithstanding the nuanced benchmarking of energy poverty, it generally denotes both non-availability and a lack of financial capacity to afford the type of energy which households require for normal functionality.

In Nigeria, energy poverty encompasses non-availability of energy resources in the diversity necessary for households to sufficiently make choices, financial constraints to access energy services, and general unavailability of modern energy resources that are clean, safe to the environment and support economic and human development in terms of meeting the basic needs of households and economic activities in the productive sector. A major sustaining factor of energy poverty in Nigeria is lack of diversification. Thus, national energy development has traditionally depended on Nigeria's oil sector to power domestic energy needs. This has thrown up a host of prominent challenges like resource constraints, inadequate generation and distribution of energy and environmental pollution with serious implications for human security, national development and environmental sustainability [48].

4. The Implications of Energy Poverty on Environmental sustainability

It may be difficult to sustain a green environment in the face of energy poverty as households must find alternative energy sources to meet shortfalls in their domestic requirements. And most times these alternatives have negative impacts on the environment. Nigerians are grossly energy poor on many fronts, especially in the provision of both refined petroleum products and electricity. Thus, households have evolved survival strategies that undermine environmental sustainability. For instance, only about 40 percent of Nigeria's 193 million people have access to electricity [27].

Nigeria is a major oil producer and exporter as well as a member of OPEC. Despite OPEC's quotas to deal with market fundamentals in the international oil market, Nigeria's oil production outputs have averaged between 2.1 million barrels per day. However, in extreme cases of sabotage and vandalization of oil production pipelines, the outputs record declines [48]. Despite Nigeria's status as an oil producer, it is unable to satisfy domestic demand for refined petroleum products. As a result of its incapacity to domestically refine sufficient quantities of petroleum products to satisfy the populace, Nigeria imports these products. For a long time now, Nigeria has been subsidizing the importation of refined petroleum products by up to US$ 3.85 billion yearly [49].

Nigeria operates four refineries whose combined nameplate capacity is 445,000bpd, yet the country imports more than 90 percent of refined petroleum products needed for domestic usage. *Table 1* below shows a 5-year capacity utilization of the four refineries. The nature of the underperformance of these refineries is underscored by the level of their capacity utilization. Since 2014, none of them has produced up to 20 percent of its nameplate capacity [50] with the effect that Nigeria is now a net-importer of refined petroleum products (*See Table 2 below*). Notwithstanding the subsidies on refined petroleum products, it is generally believed that these petroleum products are priced out of the reach of the average citizen.

Table 1: 5-Year Capacity Utilization of Refineries (%)

Refinery	Total Installed Nameplate Capacity	2012	2013	2014	2015	2016
Kaduna Refining & Petrochemical Co. Ltd	**110,000bpsd***	**29.12**	**29.33**	**9.24**	**2.98**	**9.24**
Port-Harcourt Refining Co. Ltd**	**210,000bpsd**	**11.95**	**9.18**	**17.28**	**4.66**	**17.28**
Warri Refining & Petrochemical Co. Ltd	**125,000bpsd**	**27.88**	**35.99**	**12.03**	**7.07**	**12.03**

Source: NNPC 2016, 20. ***bpsd:** barrels per stream day. **PHRC Ltd is made up of two refineries: The first refinery nameplate capacity of 60,000bpsd commissioned in 1965 and the second refinery with an installed capacity of 150,000 bpsd commissioned in 1989.

Table 2: Importation/Consumption Profile of Refined Petroleum Products, 2014-2017 (In Litres)

Product	Description	2014	2015	2016	2017
Premium Motor Spirit (PMS)	Imported	18,572,793,515	18,794,323,749	18,660,108,114	17,313,512,018
	Distributed/ Consumed	17,591,021,628	17,779,846,057	17,382,007,474	18,335,739,889
	ADC*	48,194,580	48,711,907	47,491,824	50,234,904
Automotive	Imported	3,483,665,839	3,875,483,242	4,665,966,344	4,277,630,139

IOP Conf. Series: Earth and Environmental Science **331** (2019) 012033 doi:10.1088/1755-1315/331/1/012033

Gas Oil (AGO)	Distributed/ Consumed	3,314,583,036	3,249,662,707	3,896,774,724	4,748,102,354
	ADC*	9,081,049	8,903,185	10,646,925	13,008,500
Household Kerosene (HHK)	Imported	2,939,640,169	1,824,619,750	548,651,307	340,333,700
	Distributed/ Consumed	2,956,144,224	1,714,053,716	951,571,008	944,390,851
	ADC*	8,099,025	4,696,038	2,599,921	2,587,372

Source NBS, 2018; *ADC: Average Daily Consumption

In the electricity sector, Nigeria's total installed electricity capacity is 12,522MW consisting of 10, 142MW from thermal sources and 2,380MW from hydropower sources. Despite the gross inadequacy of this installed power generation capacity, the actual output has never been anywhere near that mark. As a matter of fact, power generation has normally hovered between 3,000 and 4,000MW, and on some occasions even less. If this is juxtaposed with the projected national demand of 98,000MW to achieve energy security, then it shows how energy poor Nigerians are. This also explains why only about 40 percent of Nigerians have access to power supply and the high incidence of power outages. The implication is that those without access to energy find alternative sources of energy with implications for the environment.

The Nigerian environment is rich due to the diversity of forests and wildlife. It is estimated that Nigeria's forests support wildlife consisting of at least 899 species of birds, 154 reptiles, 53 amphibians, 274 mammals and 4,715 species of higher plants. Besides timber and fuel wood, there are abundant non-timber forest products (NTFPs) ranging from medicines, pharmaceuticals, bio-chemicals, aromatics and toxins, food and forage, fiber to ornamentals which also constitute means of livelihood to households, especially in the rural areas [51]. However, Nigeria has been losing its forest resources at a pace this is unsustainable. For instance, between 1990 and 2005 it lost nearly 80 percent of its old-growth forests which was unprecedented. It was qualified as the highest rate of deforestation of natural forest in the world during that period. In a similar vein, FAO sources indicated that while Nigeria's forest cover was 17.5 million hectares in 1990, it declined to 13.5 million hectares in 2000 (cited in [51]). The implication of this depletion is that Nigeria recorded a forest cover loss of 2.6 percent or approximately 400,000 hectares per annum. At this rate of consumption of forestry resources, experts have warned that Nigeria might not have forests by 2047 [52]. Recently, the Nigerian Conservation Foundation estimated that the country had lost about 96 per cent of its original forest[53]. It is imperative that the Nigerian government boosts other energy sources, in terms of availability and affordability, as a countermeasure to the continued use of fuel wood as a source of energy. However, this problem could be solved by converting the ubiquitous mountains of refuse across Nigeria into much-needed energy. An example can be found in Ethiopia where Africa's first waste-to-energy plant has been installed to incinerate 1,400 tons of waste per day. According to Mourdoukoutas [54], the Ethiopia Reppie thermal plant has capacity to handle and convert 80 percent of Addis Ababa's waste into electricity and thus provide 30 percent of the energy needs of its about 3 million inhabitants.

5. Conclusion

The relevance of energy in the matrix of national development is already established. Energy availability and consumption per capita constitute an index to evaluate the quality of life and development of a country or region. Nigeria is an energy poor country even though it has the resources to ensure reasonable energy sufficiency. A major reason for Nigeria's seeming energy poverty is its sole reliance on non-renewable energy sources and non-utilization of its abundantly available renewable resources.

Undoubtedly, Nigeria needs to boost its energy generation considering the huge gap between what it needs (about 98,000MW) and what it is currently producing (between 3,000MW and 4,000MW). The impact of energy poverty is felt on the environment both directly and indirectly. Nigeria's forestry resources are being lost at a very fast rate. Not only are trees lost through logging for use as fuel wood as well as for other purposes but also non-timber forest products (NTFPs). All these have negative implications for the environment, economic activities and the people, including health challenges due the effect of inefficient combustion of solid fuels (The Secretary-General's Advisory Group on Energy and Climate Change, 2010). Currently, all the various sources of renewable energy are not exploited by Nigeria. In other words, renewable energy does not play any role in Nigeria's energy mix. Sustainable energy must necessarily rely on a multiplicity of energy sources.

Therefore, this paper recommends that Nigeria scales up its quest for energy security by expanding its energy mix to include renewable energy. Thus, what it needs to do is to pursue the implementation of its renewable energy master plan (REMP), attract investors in the various areas of renewable energy development, invest in further research for evidence-based data and evolve plans to develop domestic capacity to drive the entire process.

Acknowledgement
The authors would like to thank the management of Landmark University for providing the financial support that made it possible for the lead author to attend the conference.

References

[1]. Cherp A and Jewell J 2014 The concept of energy security: Beyond the four As. *Energy Policy* 75, 415–421. https://doi.org/10.1016/j.enpol.2014.09.005
[2]. Adekoya L O and Adewale A A 1992 Wind energy potential of Nigeria. *Renewable Energy*, 2(1), 35-39. https://doi.org/10.1016/0960-1481(92)90057-A
[3]. Ajayi O O, Fagbenle R O, Katende J, Ndambuki J M, Omole D O and Badejo A A 2014 Wind energy study and energy cost of wind electricity generation in Nigeria: Past and recent results and a case study for South West Nigeria. *Energies*, 7, 8508-8534. https://doi.org/10.3390/en7128508
[4]. Akuru U B, Onukwube I E, Okoro O I, and Obe E S 2017 Towards 100% renewable energy in Nigeria. *Renewable and Sustainable Energy Reviews* 71, 943–953. http://dx.doi.org/10.1016/j.rser.2016.12.123.
[5]. Aliyu A S, Dada, J O and Adam I K 2015 Current status and future prospects of renewable energy in Nigeria. *Renewable and Sustainable Energy Reviews* 48, 336–346. https://doi.org/10.1016/j.rser.2015.03.098
[6]. Amoo L M. 2018 Techno-economic assessment of energy production potential from tidal streams in Nigeria. *International Journal of Energy and Environmental Engineering* 9, 81–98. https://doi.org/10.1007/s40095-017-0256-2
[7]. Ang B W, Choong W L and Ng T S 2015 Energy security: Definitions, dimensions and indexes. *Renewable and Sustainable Energy Reviews* 42, 1077–1093. http://dx.doi.org/10.1016/j.rser.2014.10.064
[8]. Brimmo A T, Sodiq A, Sofela, S and Kolo I 2017 Sustainable energy development in Nigeria: Wind, hydropower, geothermal and nuclear (Vol. 1). *Renewable and Sustainable Energy Reviews* 74, 474–490. http://dx.doi.org/10.1016/j.rser.2016.11.162
[9]. Dahunsi S O, Adesulu-Dahunsi, A T and Izebere J O 2019 Cleaner energy through liquefaction of Cocoa (Theobroma cacao) pod husk: Pretreatment and process optimization, *Journal of Cleaner Production* (in press), https://doi.org/10.1016/j.jclepro.2019.04.112.
[10]. Daily Trust 2018 Deforestation: Nigeria has lost 96 % of its forest – NCF. 2 September. https://www.dailytrust.com.ng/deforestation-nigeria-has-lost-96-of-its-forest-ncf-268655.html (accessed 15 May 2019)

[11]. Day R, Walker G and Simcock N 2016 Conceptualising energy use and energy poverty using a capabilities framework. *Energy Policy* 93, 255–264. http://dx.doi.org/10.1016/j.enpol.2016.03.019.

[12]. Ebhota W S and Tabakov, P Y 2018 The place of small hydropower electrification scheme in socioeconomic stimulation of Nigeria. *International Journal of Low-Carbon Technologies* 13(4), 311–319 https://doi.org/10.1093/ijlct/cty038.

[13]. Ejoh E 2017 NNPC lists steps to achieve 40bbl reserve by 2020. *Vanguard,* February 28. https://www.vanguardngr.com/2017/02/nnpc-lists-steps-achieve-40-bbl-reserve-2020/ (accessed 24th April 2019).

[14]. Emodi N V and Boo K 2015 Sustainable energy development in Nigeria: Current status and policy options. *Renewable and Sustainable Energy Reviews* 51 (2015) 356–381. http://dx.doi.org/10.1016/j.rser.2015.06.016.

[15]. Ewepu G 2017 Nigeria yet to transform 2.8 bn metric tonnes of coal in 17 fields into wealth. *Vanguard*, 15 August. https://www.vanguardngr.com/2017/08/nigeria-yet-transform-2-8-bn-metric-tonnes-coal-17-fields-wealth/ (accessed 15 May 2019).

[16]. Fadare D A 2010 The application of artificial neural networks to mapping of wind speed profile. *Applied Energy*, 87(3), 934-942. https://doi.org/10.1016/j.apenergy.2009.09.005.

[17]. FAO (Food and Agriculture Organization) 2018 *The state of the world's forests 2018 - Forest pathways to sustainable development.* https://reliefweb.int/sites/reliefweb.int/files/resources/I9535EN.pdf (accessed 15 May 2019).

[18]. FGN (Federal Government of Nigeria) 2015a. *National Renewable Energy and Energy Efficiency Policy, 2015.* https://nerc.gov.ng/index.php/library/documents/func-download/560/chk,29956501ae6207539706fa8e714631f4/no_html,1/ (accessed 15 May 2019)

[19]. FGN (Federal Government of Nigeria) 2015b. *Nigeria: Fifth national biodiversity report.* https://www.cbd.int/doc/world/ng/ng-nr-05-en.pdf/ (accessed 15 May 2019)

[20]. FGN (Federal Government of Nigeria). (2015c). Nigeria power baseline report. https://mypower.ng/wp-content/uploads/2018/01/Baseline-Report.pdf(accessed 18 May 2019)

[21]. Gouveia J P, Palma P and Simoes S G 2019. Energy poverty vulnerability index: A multidimensional tool to identify hotspots for local action. *Energy Reports* 5, 187–201. https://doi.org/10.1016/j.egyr.2018.12.004.

[22]. Hettiarachchi H and Kshourad C 2019 Promoting Waste-to-Energy. *Current developments in biotechnology and bioengineering, 163–184.* https://doi.org/10.1016/b978-0-444-64083-3.00009-

[23]. International Hydropower Association 2018 Hydropower status report: Sector trends and insights. https://www.hydropower.org/sites/default/files/publications-docs/iha_2018_hydropower_status_report_4.pdf (accessed 15 May 2019).

[24]. Månsson A, Johansson B and Nilsson L J 2014 Assessing energy security: An overview of commonly used methodologies. *Energy, 73, 1–14.* https://doi.org/10.1016/j.energy.2014.06.073.

[25]. Maren I B, Agontu J A and Mangai M M 2013 Energy security in Nigeria: Challenges and way forward. *International Journal of Engineering Science Invention* 2(11), 1-6.

[26]. Monyei C G, Adewumi A O, Obolo M O and Sajou B 2018 Nigeria's energy poverty: Insights and implications for smart policies and framework towards a smart Nigeria electricity network. *Renewable and Sustainable Energy Reviews*, 81, 1582–1601. https://doi.org/10.1016/j.rser.2017.05.237.

[27]. Mourdoukoutas E 2018 Africa's bumpy road to sustainable energy. *Africa Renewal,* 16 October. https://www.un.org/africarenewal/web-features/africa%E2%80%99s-bumpy-road-sustainable-energy (accessed 15 May 2019).

[28]. Mshandete A M and Parawira W 2009 Biogas technology research in selected sub-Saharan African countries – A review. *African Journal of Biotechnology* 8 (2), 116-125.

[29]. Munshi N 2018 Nigeria's fuel subsidies bill set to soar on rising oil price. *Financial Times,* 8 October. https://www.ft.com/content/8094bfd4-ca29-11e8-9fe5-24ad351828ab

[30]. Nwozor A Audu, J and Adama, J. I. 2019 The political economy of hydrocarbon pollution: Assessing socio-ecological sustainability of Nigeria's Niger Delta region. *International Journal of Energy Economics and Policy* 9(1):7-14. https://doi.org/10.32479/ijeep.7058.

[31]. Nwozor A 2019 Depoliticizing environmental degradation: Revisiting the UNEP environmental assessment of Ogoniland in Nigeria's Niger Delta region. *Geojournal* (in press). https://doi.org/10.1007/s10708-019-09997-x

[32]. Nzeh E, Eboh E and Nweze N J 2015 Status and trends of deforestation: An insight and lessons from Enugu State, Nigeria. *Net Journal of Agricultural Science* 3(1), 23-31.

[33]. Obaje N G, Amadi A N, Aweda A K, Umar U M and Shuaibu I 2018 Processing Nigerian coal deposits for energy source. *Environmental Earth Sciences* 77:176. https://doi.org/10.1007/s12665-018-7362-1

[34]. Odesola I F, Samuel E and Olugasa T 2013 Coal development in Nigeria: Prospects and challenges. *International Journal of Engineering and Applied Sciences* 4(1), 64-73.

[35]. Ohimain, E I 2014 Can Nigeria Generate 30% of her Electricity from Coal by 2015? *International Journal of Energy and Power Engineering* 3(1), 28-37. https://doi.org/10.11648/j.ijepe.20140301.15.

[36]. Ohunakin O S 2011 Wind resource evaluation in six selected high altitude locations in Nigeria. *Renewable Energy* 36, 3273–81.

[37]. Ohunakin O S, Adaramola M S, Oyewola O. M and Fagbenle R O 2014 Solar energy applications and development in Nigeria: Drivers and barriers. *Renewable and Sustainable Energy Reviews* 32, 294–301. http://dx.doi.org/10.1016/j.rser.2014.01.014.

[38]. Okolie S T A, Ozuor O, Fakehinde O, Ongbali S O, Fayomi O S I and Agu F A 2019 Study of Nigeria geothermal energy resources' viability, brief production techniques and transportation. *Energy Procedia* 157, 1475–1485. https://doi.org/10.1016/j.egypro.2018.11.312

[39]. OPEC (Organization of the Petroleum Exporting Countries) 2017 *OPEC annual statistical bulletin.* http://www.opec.org/opec_web/static_files_project/media/downloads/publications/ASB2017_13062017.pdf (accessed 18 May 2019).

[40]. Osueke C O, Uzendu, P and Ogbonna I D 2013 Study and evaluation of solar energy variation in Nigeria. *International Journal of Emerging Technology and Advanced Engineering* 3(6), 501-505.

[41]. Premium Times 2013 Nigeria could generate 30% of its electricity from coal – Jonathan. https://www.premiumtimesng.com/business/143045-nigeria-could-generate-30-of-its-electricity-from-coal-jonathan.html (accessed 18 May 2019).

[42]. PWC (PricewaterhouseCoopers) 2017 Nigeria's refining revolution. https://www.pwc.com/ng/en/assets/pdf/nigerias-refining-revolution.pdf (accessed 15 May 2019).

[43]. Robinson C, Bouzarovski S and Lindley S 2018 Getting the measure of fuel poverty: The geography of fuel poverty indicators in England. *Energy Research & Social Science* 36, 79–93. http://dx.doi.org/10.1016/j.erss.2017.09.035.

[44]. Shaaban M and Petinrin J O 2014 Renewable energy potentials in Nigeria: Meeting rural energy needs. *Renewable and Sustainable Energy Reviews* 29, 72–84. http://dx.doi.org/10.1016/j.rser.2013.08.078.

[45]. Speight J G 2019 Energy security and the environment. *Natural Gas, 361–390.* https://doi.org/10.1016/b978-0-12-809570-6.00010-2.

[46]. The Secretary-General's Advisory Group on Energy and Climate Change 2010 *Energy for a sustainable future: Report and recommendations,* 28 April.

https://www.un.org/millenniumgoals/pdf/AGECCsummaryreport[1].pdf (accessed 18 May 2019).

[47]. ThisDay 2017 Fashola: Hydro now contributes 26% off Nigeria's power generation. 18 October. https://www.thisdaylive.com/index.php/2017/10/18/fashola-hydro-now-contributes-26-of-nigerias-power-generation/ (accessed 19 May 2019).

[48]. UNIDO (United Nations Industrial Development Organization) and ICSHP (International Center on Small Hydro Power) 2016 *World Small Hydropower Development Report 2016*. https://www.unido.org/sites/default/files/2016-11/WSHPDR_Executive_Summary_2016_0.pdf (accessed 15 May 2019).

[49]. Vanguard 2013 Coal: Origin, types, mining and uses. 5 November. https://www.vanguardngr.com/2013/11/coal-origin-types-mining-uses/ (accessed 15 May 2019).

[50]. Vanguard 2017 Nigeria'll be without forest in 30 years – FUTA Don. 27 June. https://www.vanguardngr.com/2017/06/nigeria-ll-without-forest-30-years-futa-don/ (accessed 19 May 2019).

IOP Conf. Series: Earth and Environmental Science **331** (2019) 012034 doi:10.1088/1755-1315/331/1/012034

Spectral filtering of photovoltaic cells using novel bio-filter: Silver coated Hibiscus extract using methanol solution

Emetere M.E.[1,2], and Adeyemo Nehemiah[1]

[1]Department of Physics, Covenant University, Ota, Nigeria.
2Department of Mechanical Engineering & Science, Univ. of Johannesburg, APK, South Africa
emetere@yahoo.com

Abstract. Spectra filtering is a fairly new concept to control the damaging contributions of solar IR radiations to photovoltaic (PV) cells or panel. In this research, the bio-filter was synthesized and tested. The bio-filter was proposed as the panacea to filtering the solar IR radiations. The bio-filter performance was adjudged to show little success, as it was 3% lower than the normal performance in the monocrystalline solar PV panel and 50% higher than the normal performance in the polycrystalline solar PV panel. The varying results in the polycrystalline and monocrystalline PV panels may arise as a result of unequal coating of the panel. Hence, the coating technique is recommended to be uniform and precise in further research work.

1. Introduction

It has been reported that solar infrared spectrum is dangerous to photovoltaic panel because it destroys PV cells and reduce its lifespan [1-3]. The infrared spectrum refers to the invisible part of the electromagnetic spectrum with short frequency than those of visible light but less than that of microwaves, wavelengths of infrared ranges from about 700nm to 1mm [4-5]. Hence the energy of infrared is less than that of visible light. This wave spectrum is commonly associated with heat radiated by bodies. On the other hand, the visible spectrum is unique because it is the segment of the electromagnetic spectrum that the human eye can view. Furthermore, this range of wavelengths is called visible light. Surprisingly, the human eye can detect wavelengths from 380 to 700 nm. Violet is the shortest wavelength of the visible spectrum (at around 380 nm) and red has the longest wavelength, at around 700 nm [6]. Of the light that reaches Earth's surface, infrared radiation makes up 49.4% of while visible light provides 42.3% [7]. Scientists have been working hard on the PV technology to solve this challenge [8-10]. In this research, we present the concept of bio-filter that can be sprayed on the PV panel to filter the solar IR radiations. This bio-filter is not responsible to make solar irradiance uniform on the PV panel, in the contrary; it tries to normalize seemingly fluctuations. The research seeks to screen plant extract to elongate the life span or improve the efficiency of solar cells by investigating the different wavelength of light incident on the PV panel and how they affect the cell, thereby devising a means to filter out different spectral using silver coated plant extract which will reflect heat from the cells and allow light to pass through it [11-12]. This experiment would improve on the spectra selection incident on the solar PV cell, thereby aiding maximum performance of solar panels [13-14].

The plant extract used for this study is the hibiscus flower. Hibiscus is a genus of flowering plants that belongs to the Malvaceae family [15]. Hibiscus has over three hundred species and it is believed to have salient chemical properties to trap solar IR radiation. The bio-filter is expected to optimize and stabilize the performance of the PV panel. The polycrystalline and monocrystalline PV panels are used to test the efficiency of the synthesized bio-filter.

2. Experimental Design, Materials and Methods

The materials used for the experiment includes: monocrystalline panel (3watt), solarimeter, silver nitrate (Ag_2NO_3), polycrystalline panel (4 watt), methanol solution, multimter, weighing balance, beaker and data logger.

The bio-filter was synthesized in the laboratory using hibiscus flower extract and silver nitrate. The hibiscus flower were gotten around the Ota environ, Nigeria. The flowers were divided into four parts. Each portion of the flower was grinded (using electronic blender) in 20ml of methanol. After blending, extracts were filtered out and mixed with 1.84g/mole of silver oxide (Ag_2NO_3) and then left for a day so that the silver compound can dissolve with the mixture. The filtrate was not heated to conserve the vital chemical components of the hibiscus flower (saturated hydrocarbons from 25 to 33 carbon atoms, esters, and carboxylic acids [15]). The two-polycrystalline PV panels (i.e. one panel sprayed with the bio-filter and the second panel not sprayed) and four monocrystalline panels (i.e. two sprayed with the bio-filter and the other two panels not sprayed) were connected to the data logger as shown in Figure 2.

Figure 1: Bio-filter composition

The data logger was used to measure desired parameters such as radiation, current, voltage and power for two types of PV panels i.e. monocrystalline and polycrystalline [16]. A data logger records and saves all readings gotten from the PV panels in an SD card.

Figure 2: Experimental set-up for field measurement

3. Results and Discussion

IOP Conf. Series: Earth and Environmental Science **331** (2019) 012034 doi:10.1088/1755-1315/331/1/012034

The graph below (Figure 3) shows the UV radiation (w/m^2) against time (sec) plot of the silver coated hibiscus extract in methanol solution for the polycrystalline panel. The UV over the period showed peaks at 700 sec and 4600 sec indicating high UV radiation.

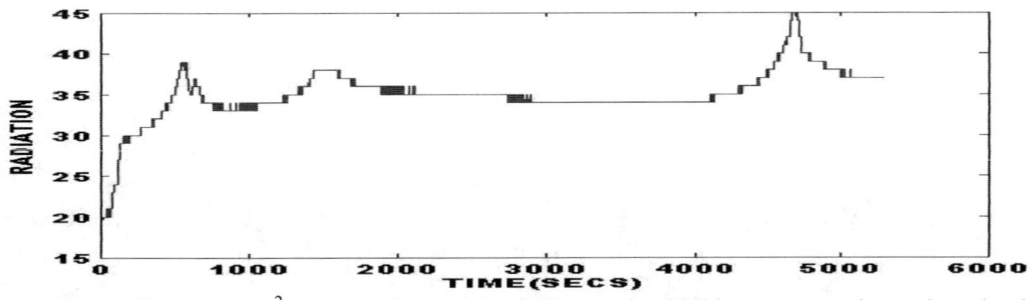

Figure 3: UV radiation (w/m^2) against time (sec) of silver coated hibiscus extract in methanol solution for the polycrystalline panel

The graph below (Figure 4) shows the voltage (V) against time (sec) plot of the silver coated hibiscus extract in methanol solution for the polycrystalline panel. The blue legend shows the coated panel while the green legend shows the uncoated polycrystalline panel. It is observed from that the coated panel produced a higher voltage relative to time than the uncoated panel. It is also noticed that there was a steady rise and fall in the voltage as time increased.

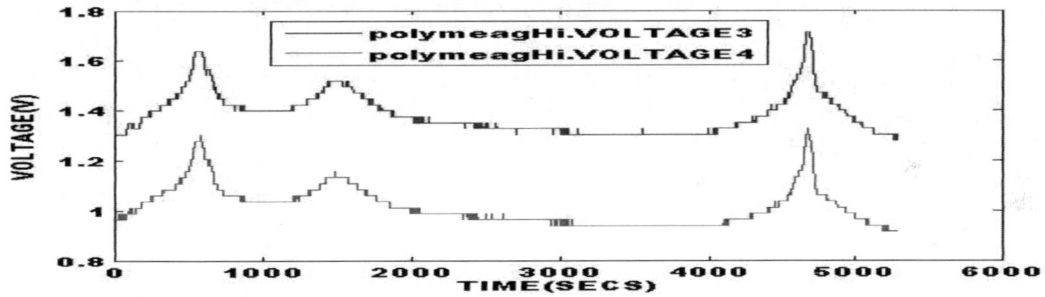

Figure 4: Voltage (V) against Time (Sec) of the Silver coated hibiscus extract in methanol solution for the polycrystalline panel.

The graph below (Figure 5) shows the current (A) against time (sec) plot of the silver coated hibiscus extract in methanol solution for the polycrystalline panel. The blue legend shows the coated panel while the green legend shows the uncoated polycrystalline panel. It is observed from that graph that the coated panel produced a higher current relative to time than the uncoated panel. It is also noticed that (like the voltage trend) there was a steady rise and fall in the current as time increased.

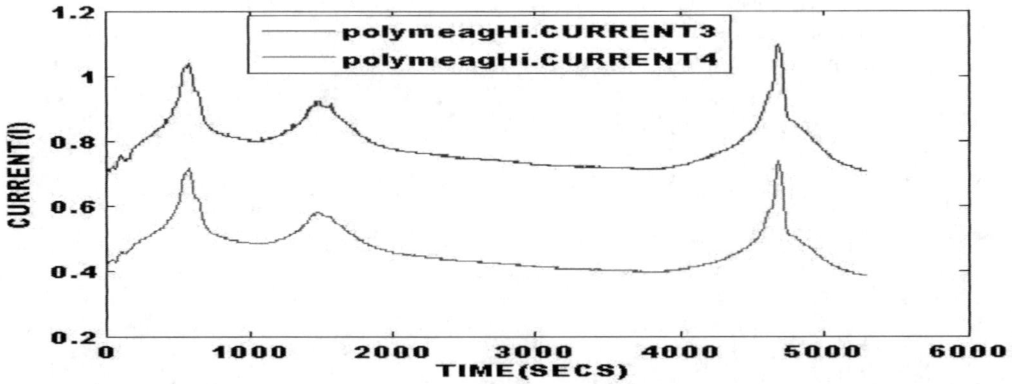

Figure 5: Current (A) against Time (Sec) of the Silver coated hibiscus extract in methanol solution for the polycrystalline panel.

The graph below (Figure 6) shows the power (W) against time (sec) plot of the silver coated hibiscus extract in methanol solution for the polycrystalline panel. The blue legend shows the coated panel while the green legend shows the uncoated polycrystalline panel. It is observed from that the coated panel produced a higher power output relative to time than the uncoated panel. It was also noticed that there was a steady rise and fall in the power. Hence, in Figures 4-6, the bio-filter improved the performance of the polycrystalline panel by 50%.

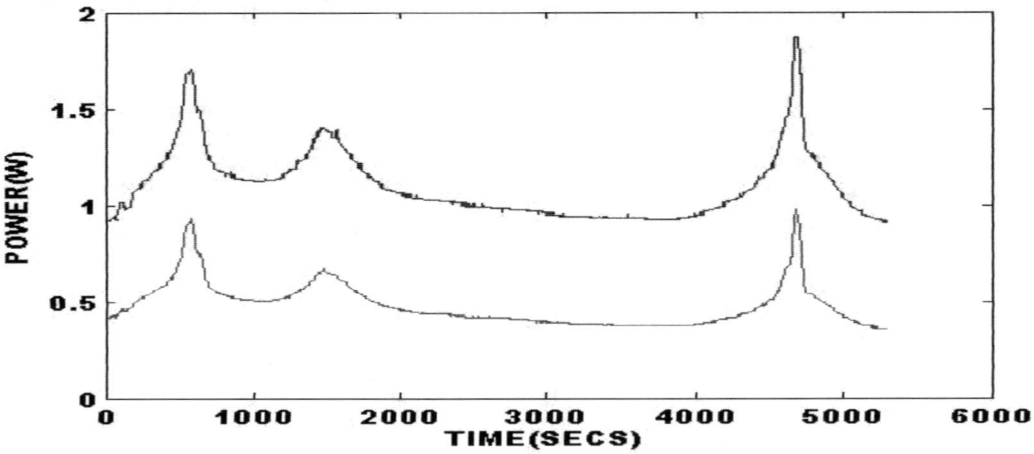

Figure 6: Power (W) against Time (Sec) of the Silver coated hibiscus extract in methanol solution for the polycrystalline panel.

The graph below (Figure 7) shows the UV radiation (w/m^2) against time (sec) plot of the silver coated hibiscus extract in methanol solution for the monocrystalline panel. The UV was very low because it was a cloudy day.

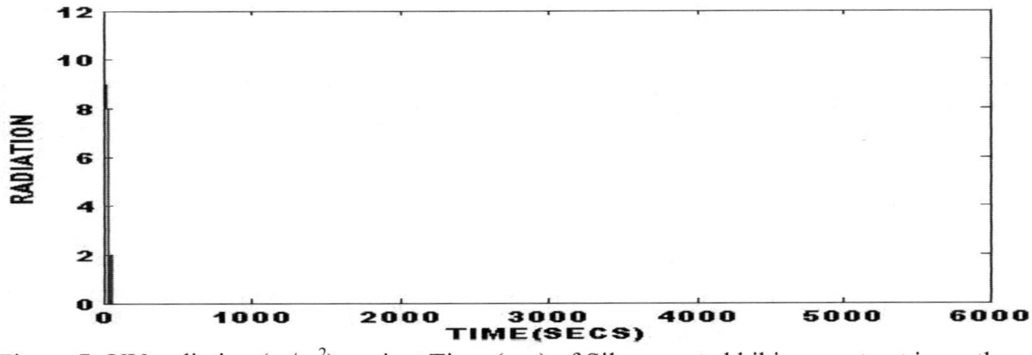

Figure 7: UV radiation (w/m^2) against Time (sec) of Silver coated hibiscus extract in methanol solution for the monocrystalline panel.

The graph below (Figure 8) shows the voltage (V) against time (sec) plot of the silver coated hibiscus extract in methanol solution for the monocrystalline panel. The blue legend shows the coated panel while the green legend shows the uncoated monocrystalline panel. It is observed from the graph that the uncoated panel produced a high voltage output with respect to time than the coated panel.

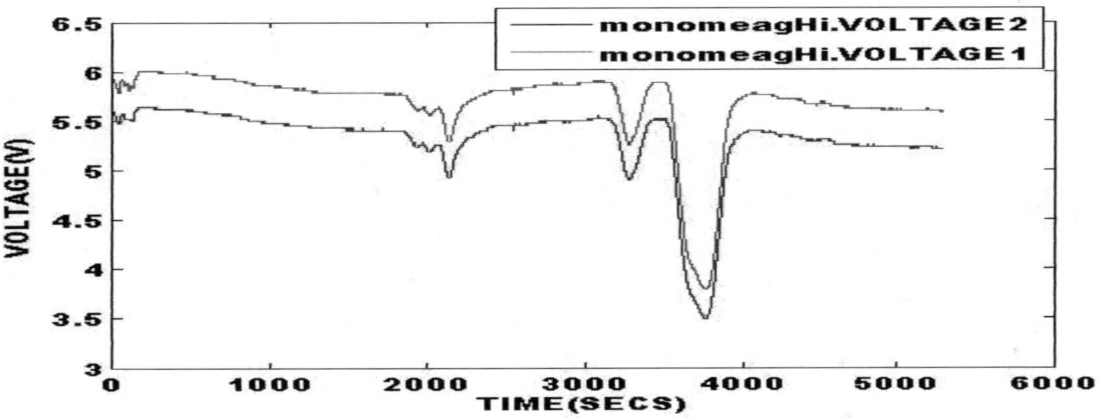

Figure 8: Voltage (V) against Time (Sec) of the Silver coated hibiscus extract in methanol solution for the monocrystalline panel.

The graph below (Figure 9) shows the current (A) against time (sec) plot of the silver coated hibiscus extract in methanol solution for the monocrystalline panel. The blue legend shows the coated panel while the green legend shows the uncoated panel for the monocrystalline panel. It is observed from the graph that the uncoated panel produced a high current output with respect to time than the coated panel.

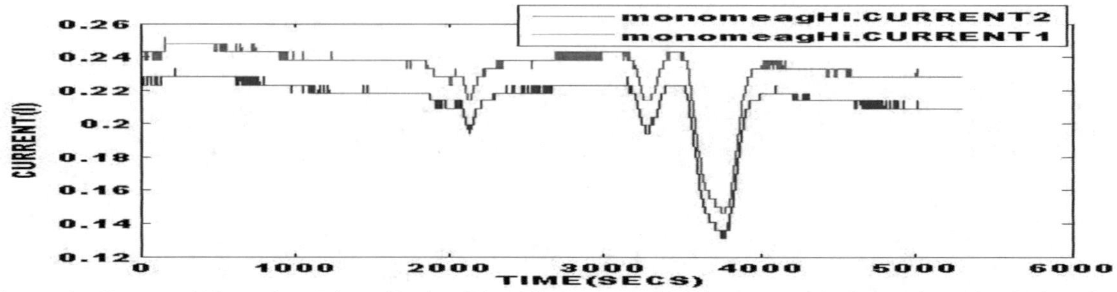

Figure 9: Current (A) against Time (Sec) of the Silver coated hibiscus extract in methanol solution for the monocrystalline panel.

The graph below (Figure 10) shows the power (W) against time (sec) plot of the silver coated hibiscus extract in methanol solution for the monocrystalline panel. The blue legend shows the coated panel while the green legend shows the uncoated panel for the monocrystalline panel. It is observed from the graph that the uncoated panel produced a high-power output with respect to time than the coated panel. Hence, in Figures 8-10, the effect of the bio-filter on the monocrystalline panel was as low as 8%. It was because there was not uniform coating of the monocrystalline panel.

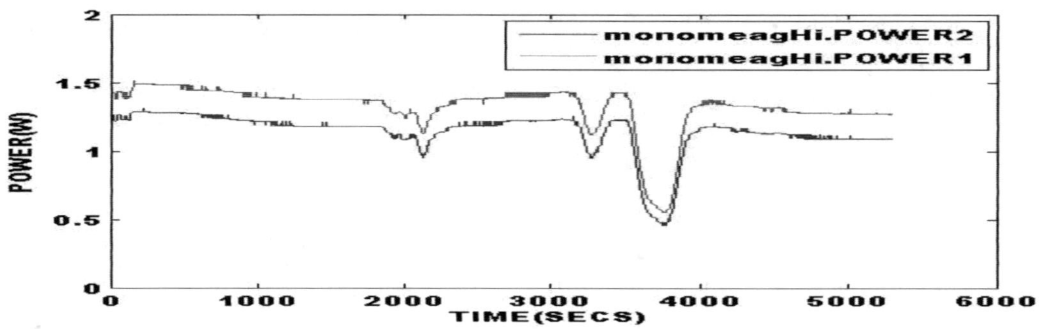

Figure 10: Power (W) against Time (Sec) of the Silver coated hibiscus extract in methanol solution for the monocrystalline panel.

4. Conclusion

The bio-filter performance was adjudged to show little success, as it was 3% lower than the normal performance in the monocrystalline solar PV panel and 50% higher than the normal performance in the polycrystalline solar PV panel. This experimentation supports bio-filter as an external material that will optimize and stabilize the performance of the PV panel. The varying results in the polycrystalline and monocrystalline PV panels may arise as a result of unequal coating of the panel. Hence, the coating technique is recommended to be uniform and precise in further research work.

.

Acknowledgement
This paper is partly funded by Covenant University.

References

[1]. Ariyo, F.; Famutimi, B.; Olowu, T.; Akintade, S.; Abbas, A. Web-Based Application for the Sizing of a Photovolatic (PV) Solar Power System. American J. Eng. Res. (AJER) 2016, 5, 219–222.

[2]. Mekonnen, Y.; Sarwat, A.I. Renewable energy supported microgrid in rural electrification of Sub-Saharan Africa. In Proceedings of the 2017 IEEE PES PowerAfrica, Accra, Ghana, 27–30 June 2017; pp. 595–599.
[3]. Ferrara, C.; Philipp, D. Why do PV modules fail? Energy Procedia 2011, 15, 379–387, doi:10.1016/j.egypro.2012.02.046.
[4]. Wen, L.; Chen, Q.; Song, S.; Yu, Y.; Jin, L.; Hu, X. Photon harvesting, legending and polarizing in photovoltaic cell integrated legend filters: Efficient energy routing strategies for power-saving displays. Nanotechnology 2015, 26, 265203.
[5]. Sinha, A.; Sastry, O.S.; Gupta, R. (2016), Detection and characterization of delamination in PV modules by active infrared thermography. Nondestruct. Test. Eval. , 31: 1–16.
[6]. Hui S.K., Chang R.L., Jeong H.I., Ki B.L., Thomas M., Ariana M., Soo J.M., Ronin H.B., Jun H.Y., Jacques E.M., Micheal G., and Nam G.P. (2012). Lead iodide perovskite sensitized all solid- state submicron thin film mesoscopic solar cell with efficiency exceeding. International journal of scientific research,485: 486-489
[7]. Fondriest Environmental, Inc. "Solar Radiation and Photosynethically Active Radiation." Fundamentals of Environmental Measurements. https://www.fondriest.com/environmental-measurements/parameters/weather/solar-radiation/ (Accessed 12/08/2019).
[8]. Dijk, L.; Paetzold, U.W.; Blab, G.A.; Schropp, R.E.I.; Vece, M.D. 3D-printed external light trap for solar cells. Prog. Photovolt. 2016, 24, 623–633.
[9]. Liu, S.; Niu, X.; Shan, W.; Lu, W.; Zheng, J.; Li, Y.; Duan, H.; Quan, W.; Han, W.; Wronsky, C.R.; Yang, D., Improvement of conversion efficiency of multicrystalline silicon solar cells by incorporation a reactive ion etching texturing. Sol. Energy Mater. Sol. Cells 2014, 127, 21–26.
[10]. Apostoluk, A.; Zhu, Y.; Masenelli, B.; Delaunay, J.J.; Sibin´ski, M.; Znajdek, K.; Focsa, A.; Kaliszewska, I. Improvement of the solar cell efficiency by the ZnO nanoparticle layer via the down-shifting effect. Microelectron. Eng. 2014, 127, 51–56
[11]. Mohd Fairuz Shahidan, Dr. Dato' Elias Salleh, Dr. Mustapha Kamal Mohd Shariff, (2006), The Influence of Tree Canopy on the Thermal Environment in a Tropical Climate: A Preliminary Study, iNTA Conference 2006 – Harmony in Culture and Nature, pp. 1-10
[12]. IOWA, Thermal Energy: Plant-Radiation Relationships, http://agron-www.agron.iastate.edu/courses/Agron541/classes/541/lesson09a/9a.5.html (accessed on 13 August, 2019)
[13]. Nofuentes, G.; García-Domingo, B.; Muñoz, J.V.; Chenlo, F. Analysis of the dependence of the spectral factor of some PV technologies on the solar spectrum distribution. Appl. Energy 2014, 113, 302–309.
[14]. Fernandez, E.F.; Soria-Moya, A.; Almonacid, F.; Aguilera, J. Comparative assessment of the spectral impact on the energy yield of high concentrator and conventional photovoltaic technology. Sol. Energy Mater. Sol. Cells 2016, 147, 185–197
[15]. Melecchi, M. I. S.; Martinez, M. M.; Abad, F. C.; Zini, P. P.; Filho, I. N.; Caramão, E. B.; Chemical composition of Hibiscus tiliaceus L. flowers: A study of extraction methods, J. Sep. Sci. 2002, 25, 86.
[16]. Lynn, P.A. Electricity from Sunlight: An Introduction to Photovoltaics; Wiley: Singapore, 2010; ISBN: 978-0-470- 74560-1

IOP Conf. Series: Earth and Environmental Science **331** (2019) 012035 doi:10.1088/1755-1315/331/1/012035

Perspectives on environmental CO_2 emission and energy factor in Cement Industry

G U Fayomi [1, 2], S E Mini[1], O S I Fayomi[3, 4*], A A Ayoola[5]
[1]Strategic Business Services, Covenant University, Ota, Nigeria
[2]Department of Geography, University of South Africa, Johannesburg, South Africa
[3]Department of Mechanical Engineering, Covenant University, P.M.B 1023, Ota, Ogun State, Nigeria
[4]Department of Chemical, Metallurgical and Materials Engineering, Tshwane University of Technology, Pretoria, South Africa
[5]Department of Chemical Engineering, Covenant University, P.M.B 1023, Ota, Ogun State, Nigeria
*Corresponding author: ojosundayfayomi3@gmail.com, +2348036886783
Ayodeji.ayoola@covenantuniversity.edu.ng

Abstract. The global challenges of pollution emission in cement industry in relation to energy consumption factor are overwhelming. No doubt it has been established that economic development has an impact on the environmental pollution and ecological system. With major constituent of environmental affluence been CO_2, the consideration has to be on the type of manufacturing process, the consumption mix and the additive ratio. This paper focuses on the cement industry and aims to provide a systematic review of the specific operations, its trend and its impact on environmental pollution in terms of energy consumption and emissions evolved.

Keywords: CO_2, Pollutions; energy consumption; environmental pollution

1. Introduction

Cement is widely regarded as one of the most important building materials available to man, and concrete, which is primarily composed of cement, is classified as one of the most universally utilized materials on earth [1, 2]. The buildings sector is classified as the third largest CO_2 emitting institution worldwide [3, 4], with cement production figures stand at between 3.0 - 3.6Gt/year, and it has been estimated that per 1kg of cement produced, 0.5-0.9kg of CO_2 emissions are evolved, and this equates to about 3.24 billion tons of CO_2/year for 3.6 billion tons of cement produced annually [5] Gartner 2004). Over 90% of the energy required for cement manufacture is obtained from fossil fuels, while the remaining 10% of energy is obtained via electricity, the statistical figure of energy consumption for the cement industry stands at 2% of global primary energy consumption or about 5% of total global industrial energy consumption [1]. Ali et al., 2011 affirmed to this, that the cement industry is one of the highest consumers of fossil fuel energy, it approximately consumes 12-15% of total industrial energy use, with an estimated 1.75±0.1 MJ of energy required to produce 1kg of cement, and China being the largest manufacturer of cement in the world with 57.3% of total world production [4, 6]. In terms of mean energy consumption and CO_2 emissions, the cement industry ranks as one of the highest, and this still comes after radical measures were enforced on the industry in the 1970s due to an embargo on oil and gas, that led to innovative solutions and necessitated a need for increased efficiency of production processes, this led to a reduction in consumption of fossil fuel due to the embargo [7].

Having established the fact that industrialization directly influences environmental pollution, the cement industry historically, has affirmed to this trend, and has been observed to generate a high volume of particulate emissions (CO_2), particularly in the production of Portland cement clinker, due to the usage of high carbon fuels and emissions from the cement production process [7, 8]. [9], argued that research trends estimates CO_2 emissions from the cement industry to stand at 5% of the overall global CO_2 emissions due in part to two reasons: the Portland cement make, which is the highest percentage of cement variety in production, having a highly pollution production process, and two, the use of outdated industrial equipment (vertical kilns) in some plants, that consume high amounts of energy as well as evolve a high percentage of particulate matter. While it has been established that economic development has an impact on the environmental pollution and ecological damage. This paper focuses and narrows down on the cement industry and aims to provide a systematic review of the specific operations of the cement industry, and its impact on environmental pollution in terms of energy consumption and emissions evolved.

2. Raw Materials Utilized In Cement Industry

Pollution in the cement industry in part stems from the raw materials used in the making of Portland cement, ordinary Portland cement (OPC), is the major cement product, and is produced from a mixture of limestone and clay, pyrolysed in a kiln at temperatures around 1,450°C, and then blended with some additives [2, 8]. According to [12], the average worldwide production figure for clinker stood at 3.6 billion, with worldwide production figures projected to reach over 5 billion by 2030. Typical reactions for the production of clinker include:

$$3CaCO_3 + SiO_2 \rightarrow Ca_3SiO_5 + 3CO_2\text{: with clay } (SiO_2)$$
$$CaCO_3 + \text{heat} \rightarrow CaO + CO_2$$

As can be seen from the process reaction all clinker production processes evolve CO_2, the major constituent of Portland Cement is CaO, and the amount of CaO found is clinker is estimated at 65%, the CO_2 emissions produced are estimated at about 0.5-0.9kg per 1kg of clinker produced [10-12] .Therefore a subsequent increase in the amount of clinker produced indirectly leads to an increase in worldwide CO_2 pollution as seen in Table 1.

Table 1. Constituents of modern Portland cement clinker [12]

Constituents of a finished modern Portland cement clinker		
Tricalcium silicate	50%	Ca_3SiO_5 or $3Ca.SiO_2$
Dicalcium silicate	25%	Ca_2SiO_4 or $2Ca.SiO_2$
Tricalcium aluminate	10%	$Ca_3Al_2O_6$ or $3CaO.Al_2O_3$
Tetracalcium aluminoferrite	10%	$Ca_4Al_2Fe_2O_{10}$ or $4CaO.Al_2O_3.Fe_2O_3$
Gypsium	5%	$CaSO_4.H_20$

The hydration (mixture) of these clinker compounds with water is what produces cement, in which the constituents undergo a chemical reaction and produce an amorphous solid hydrate.

$$2CaSiO_4 + 5H_20 \rightarrow 3CaO_2SiO_2 + 4H_20 + Ca(OH)_2$$
(Phase 2 cement)

2.1. Pollution in Cement Industry

The major industrial pollutant emanating from the manufacture of cement is the evolution of CO_2, an estimated 40% of the total CO_2 generated from the industry, emanates from fossil fuel burning which is used in the production process, and another 50%, from the raw materials utilized and the manufacturing process, and 10% from indirect emissions by transportation of finished goods [3, 12, 13]. According to [1], for every 1kg of cement produced, 0.9kg of CO_2 is evolved, and this equates to the evolution of about 3.24 billion tons of CO_2/year for the current projected 3.6 billion produced annually, and these figures don't take into account the emissions from the quarrying and transportation of raw materials and the transport and delivery of produced cement.

The emission sources at different stages of the product life cycle process are:

- The combustion of fossil fuel in the clinkering process to heat the raw material of limestone ($CaCO_3$), produces CO_2 at temperatures exceeding 1450°C.
- The calcination process (raw material conversion) in cement production process, also generates a significant amount of CO_2.
- Indirect emission from transportation and delivery of raw materials and finished goods.
- CO_2 generated from fossil fuel based electricity generation means, for running plant and equipment [14].

It should however be observed that the amount of CO_2 evolved in the manufacturing process also depends on:

- The type of manufacturing process adopted i.e. type of kiln used
- The type of fuel used (petroleum, natural gas, coal etc.)
- The clinker/cement ratio i.e. percentage of additives [14, 15].

Thus these three areas provides a wide scope for improvement and innovation in the production process for the both the management of energy consumption and reduction of CO_2 emissions. Figure 1 provides an estimate of CO_2 emissions produced at 0.38MJe/kg of electricity/clinker, emission factor of electricity production 0.22kg/MJe, fuel use: 3.35MJ/kg of clinker for dry process and 5.4MJ/kg for wet process.

Table 2. CO_2 emissions in kg per 1kg cement produced for dry and wet cement production process for various fuels and clinker/cement ratios [1]

	Process emissions	Process and fuel-related emissions (CO_2)							
		Dry process				Wet process			
Clinker/cement ratio	Clinker	Coal	Fuel oil	Natural gas	Waste	Coal	Fuel oil	Natural gas	Waste
55%	0.28	0.55	0.5	0.47	0.36	0.67	0.59	0.53	0.36
75%	0.38	0.72	0.66	0.61	0.47	0.88	0.77	0.69	0.47
(Portland) 95%	0.49	0.89	0.81	0.75	0.57	1.09	0.95	0.9	0.57

Cement production is largely an energy-dependent process, which begins with: raw materials preparation, fuel preparation and finish grinding. The world cement production has been observed to be increasing from the 1990s and is projected to keep increasing till 2050 as can be seen in the figure below. Cement is observed to be the second most consumed resource after water, as demand increases annually due to infrastructural development in developing and developed nations [15].

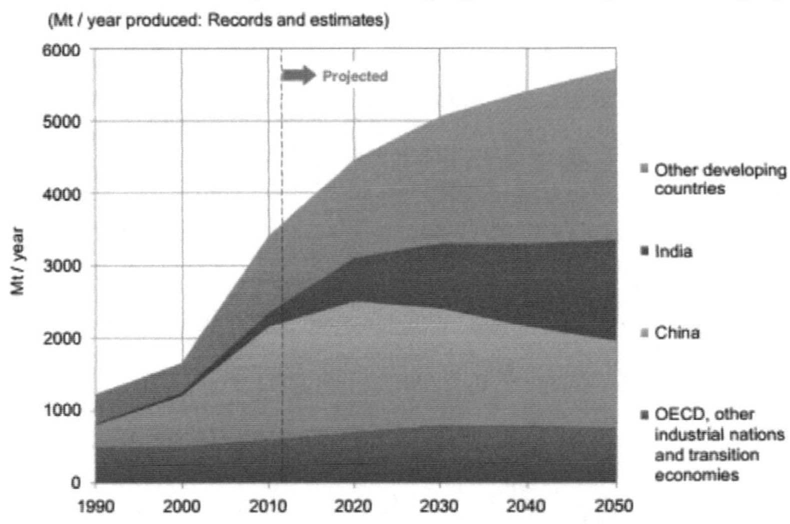

Figure 1. Current and estimated global cement production figures [12].

2.2. Historical Review of Cement Industry

Statistical estimates show that global trend for cement production, emission rates and energy consumption have been consistently increasing [16] in 1994, an estimated 1.381Gt of cement was produced, accompanied by an estimated 1.126Gt of CO_2 emissions. Average primary energy consumption for production was estimated at 4.8MJ/kg of cement, while consuming 6.6EJ of primary energy overall, and evolution of CO_2 was estimated at 1126Tg (=1.126Gt) of CO_2, these estimates account for ≥5% of the world carbon emission [1]. It should be noted that the amount of energy consumed and quantity of CO_2 emitted during these periods came after OPEC enforced an embargo on oil and gas in the 1970s, which forced the cement industry to innovate, become less fuel dependent and more fuel efficient, turning to cokes, coal and other forms of fuels, hinting that prior to this period, the emissions generated and energy consumed were even much higher. The table 3 and 4 presents the primary cement production data and growth rates, production figures have grown from 594Mt in the 1970 to 1453Mt in 1995, at an estimated rate of 3.6%

Table 3. Cement production data between 1970-1995 for major world regions [1]

	Cement Production						Average Annual Growth	
	1970	1975	1980	1985	1990	1995	1970-1995	1990-1995
Region/Country	Tg	Tg	Tg	Tg	Tg	Tg	%	%
China (incl. Hong Kong)	27	47	81	148	211	477	12.2%	17.7%
Europe	185	194	223	178	196	181	-0.1%	-1.7%
OECD-Pacific	69	83	113	100	126	154	3.3%	4.1%
Rest of Asia	20	31	49	57	89	130	7.8%	8.0%
Middle East	19	29	44	75	93	116	7.4%	4.6%
Latin America	36	52	76	71	82	97	4.1%	3.4%
Eastern Europe/ former Soviet Union	134	177	190	190	190	96	-1.3%	-12.7%
North America	76	73	79	81	81	88	0.5%	1.5%
India	14	16	18	31	49	70	6.6%	7.3%
Africa	15	20	28	35	38	44	4.5%	2.7%
World	594	722	901	965	1156	1453	3.6%	4.7%

Global estimates have predicted the mean carbon emission intensity of CO_2 emissions from cement production for this period as 0.81kg of CO_2 per kg of cement, with China being the largest emitter of CO_2. Table 4 provides the key statistics for carbon emission for cement production as at 1994.

Table 4. Global emission of CO_2 from cement production [1]

	Cement Production	Clinker/Cement Ratio	Primary Intensity	Primary Energy	Process Carbon Emissions	Carbon Emissions. Energy Use	Total Carbon Emissions	Share of World Total
Region/Country	**Tg**	**%**	**MJ/\kg**	**PJ**	**Tg CO_2**	**Tg CO_2**	**Tg CO_2**	**%**
China	423	83%	5.0	2117	175	197	372	33.0%
Europe	182		4.1	749	73	56	129	11.5%
OECD Pacific	151		3.5	533	65	41	105	9.3%
Other ASIA	124		4.9	613	56	179	105	9.3%
Middle East	111		5.1	563	51	44	95	8.4%
North America	88		5.4	480	39	40	78	7.0%
EE/FSU	101		5.5	558	42	38	80	7.1%
Latin America	97		4.7	462	41	30	71	6.3%
India	62	89%	5.0	309	28	30	60	5.1%
Africa	41		4.9	201	18	15	33	2.9%
World Total	1381		4.8	6585	587	830	1126	100.0%

3. Conclusion

- As observed from the statistical data, cement production and energy consumption has increased steadily between 1994 and 2010, and this has also characterized an increase in the emission of CO_2.
- It was observed that the cement industry has been one of the highest energy consuming and CO_2 emitting institutions worldwide (rated third worldwide), and while improved techniques and modern technologies have brought about an increase in efficiency of production, reduced energy consumption and reduced CO_2 emissions through a variety of techniques.
- However, despite the efforts of the industry and government to increase efficiency of production and regulate the emission of CO_2 associated with cement production, a rise in CO_2 continues particularly as a consequence of increased demand for infrastructural development.

Acknowledgement
The author will like to acknowledge the financial support from Covenant University CUCRID toward the publication of this manuscript.

References

[1] Hendriks, C. A., Worrell, E., Jager, D. De, Blok, K., & Riemer, P. (2004) (n.d.). Emission Reduction of Greenhouse Gases from the Cement Industry, 1–11.

[2] Cai, B., Wang, J., He, J., & Geng, Y. (2016). Evaluating CO 2 emission performance in China's cement industry : An enterprise perspective. *Applied Energy*, *166*, 191–200. https://doi.org/10.1016/j.apenergy.2015.11.006

[3] Habert, G., Billard, C., Rossi, P., Chen, C., & Roussel, N. (2010). Cement and Concrete Research Cement production technology improvement compared to factor 4 objectives. *Cement and Concrete Research*, *40*(5), 820–826. https://doi.org/10.1016/j.cemconres.2009.09.031

[4] Lei, Y., Zhang, Q., Nielsen, C., & He, K. (2011). An inventory of primary air pollutants and CO_2 emissions from cement production in China, 1990-2020. *Atmospheric Environment*, *45*(1), 147–154. https://doi.org/10.1016/j.atmosenv.2010.09.034

[5] Ali, M. B., Saidur, R., & Hossain, M. S. (2011). A review on emission analysis in cement industries. *Renewable and Sustainable Energy Reviews*, *15*(5), 2252–2261. https://doi.org/10.1016/j.rser.2011.02.014

[6] Madlool, N. A., Saidur, R., Hossain, M. S., & Rahim, N. A. (2011). A critical review on energy use and savings in the cement industries. *Renewable and Sustainable Energy Reviews*, *15*(4), 2042–2060. https://doi.org/10.1016/j.rser.2011.01.005

[7] Gartner, E. (2004). Industrially interesting approaches to "'low-CO_2"' cements $, *34*(January 2004), 1489–1498. https://doi.org/10.1016/j.cemconres.2004.01.021

[8] Zhang, C., Han, R., Yu, B., & Wei, Y. (2018). Accounting process-related CO_2 emissions from global cement production under Shared Socioeconomic Pathways. *Journal of Cleaner Production*, *184*, 451–465. https://doi.org/10.1016/j.jclepro.2018.02.284

[9] Benhelal, E., Zahedi, G., Shamsaei, E., & Bahadori, A. (2013). Global strategies and potentials to curb CO 2 emissions in cement industry. *Journal of Cleaner Production*, *51*, 142–161. https://doi.org/10.1016/j.jclepro.2012.10.049

[10] Du, G., Sun, C., Ouyang, X., & Zhang, C. (2018). A decomposition analysis of energy-related CO 2 emissions in Chinese six high-energy intensive industries. *Journal of Cleaner Production*, *184*, 1102–1112. https://doi.org/10.1016/j.jclepro.2018.02.304

[11] Ke, J., Zheng, N., Fridley, D., Price, L., & Zhou, N. (2012). Potential energy savings and CO_2 emissions reduction of China's cement industry. *Energy Policy*, *45*(2012), 739–751. https://doi.org/10.1016/j.enpol.2012.03.036

[12] Imbabi, M. S., Carrigan, C., & Mckenna, S. (2013). Trends and developments in green cement and concrete technology. *International Journal of Sustainable Built Environment*, *1*(2), 194–216. https://doi.org/10.1016/j.ijsbe.2013.05.001

[13] Chen, C., Habert, G., Bouzidi, Y., & Jullien, A. (2010). Environmental impact of cement production : detail of the different processes and cement plant variability evaluation. *Journal of Cleaner Production*, *18*(5), 478–485. https://doi.org/10.1016/j.jclepro.2009.12.014

[14] Hasanbeigi, A., Price, L., & Lin, E. (2012). Emerging energy-efficiency and CO 2 emission-reduction technologies for cement and concrete production : A technical review. *Renewable and Sustainable Energy Reviews*, *16*(8), 6220–6238. https://doi.org/10.1016/j.rser.2012.07.019

[15] Schneider, M. (2015). Cement and Concrete Research Process technology for efficient and sustainable cement production. *Cement and Concrete Research*, *78*, 14–23. https://doi.org/10.1016/j.cemconres.2015.05.014

[16] Hasanbeigi, A., Price, L., Lu, H., & Lan, W. (2010). Analysis of energy-efficiency opportunities for the cement industry in Shandong Province, China : A case study of 16 cement plants. *Energy*, *35*(8), 3461–3473. https://doi.org/10.1016/j.energy.2010.04.046

IOP Conf. Series: Earth and Environmental Science **331** (2019) 012036 doi:10.1088/1755-1315/331/1/012036

Climate Variability and Renewable Energy Planning

Emetere M.E.[1,2], and Akinyemi M.L.[1]
[1]Department of Physics, Covenant University Canaan land, P.M.B 1023, Ota, Nigeria.
[2]Department of Mechanical Engineering and Science, University of Johannesburg, APK, South Africa

E-mail: emetere@yahoo.com

Abstract. The impact of climate change is a global event. Its significant role in the climate system altered meteorological parameters over geographical locations. MISR satellite data have shown that climatic variability has impact on rainfall and surface vegetation; and increased aerosol loading and solar radiation. This research comparatively compared the meteorological parameters as well as its potentially accruable energy. The dataset was obtained from the automatic weather station (AWS). The location of research is Ota, southwest region of Nigeria. The research recommends that a more sophisticated technology may be required to boost wind and solar energy generation in the area.

1. Introduction

The reality of climate change and its effect on rainfall, surface temperature, solar radiation, surface vegetation, aerosol retention etc. are quite alarming in recent times [1]. In the West African tropics, there is reduced thermal comfort as a result of increased solar radiation [2]. Satellite measurement has shown that renewable energy parameters (such as wind and solar activities) are significantly altered. For example, Hall and Peyrille [3] reported that there are sharp seasonal differences in low-level winds and variability in monsoon winds. In other words, climate variability has intense impact on the meteorological parameters required for renewable energy planning. Also, solar irradiance dataset have been very dynamic in recent time [1], hence solar energy budgeting is becoming difficult by the day. More worrisome is the fact that the lifespan of photovoltaic (PV) module is grossly affected in the tropical region of West Africa.

In this research, two salient meteorological parameters were considered in reviewing renewable energy planning i.e. solar radiation and wind speed. Also, the meteorological parameters and its accruable renewable energy were examined to guide energy specialist on the choice of relevant technologies - required to acquire maximum usage of natural resources in Ota, Nigeria.

Ota is located on latitude 7.9452° N and longitude 4.7888° E of southwest Nigeria. The research location is a fast growing town with a population of over 300,000 [4]. The climate in Ota is classified as Aw by the Köppen-Geiger system [5].

2. Methodology

The dataset was obtained from the automatic weather station (AWS) located at the Covenant University, Ota (Figure 1). The dataset coverage was for the year 2014. The solar radiation and the solar energy were obtained as primary parameters in the automatic weather station. The interpretation of the graph was done using the CERN Root open source application. The optimized version of the solar energy was calculated using parameters such as solar panel dimension (of 1.2m by 0.5m), solar panel yield (of 0.45), average solar radiation (obtained from dataset) and performance ratio of 0.75. The optimized energy from the wind was obtained-using the air density (1.225), length of the wind blade (1 m, 2 m, 3 m, 12 m, 13 m, 14m and 15 m) to calculate the swept area, and the wind speed

IOP Conf. Series: Earth and Environmental Science **331** (2019) 012036 doi:10.1088/1755-1315/331/1/012036

obtained from the dataset. The distribution plot, conventional plot and 3D plot were used to illustrate salient observation in the dataset.

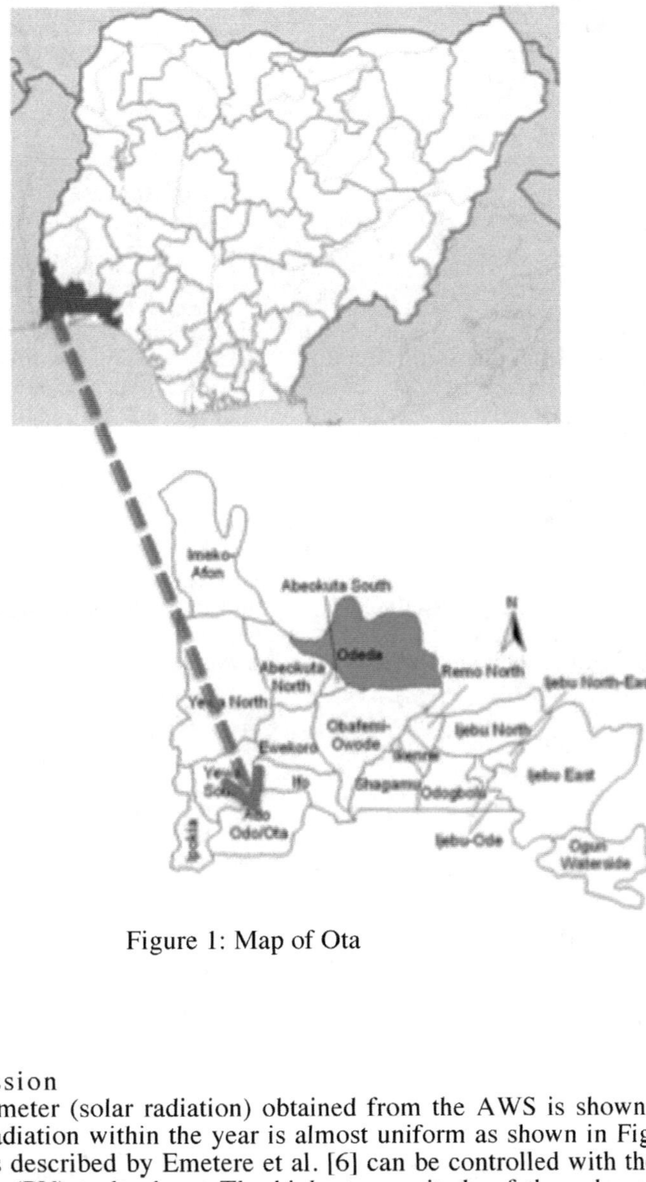

Figure 1: Map of Ota

3. Results and Discussion

The meteorological parameter (solar radiation) obtained from the AWS is shown in Figure 2a. The magnitude of the solar radiation within the year is almost uniform as shown in Figure 2a. Hence, the effect of solar shading as described by Emetere et al. [6] can be controlled with the help of emerging materials in photovoltaic (PV) technology. The highest magnitude of the solar radiation within the year is given as 267 W/m2. The solar energy as recorded by the AWS show very scanty and low magnitude of solar energy – at the beginning of the year (Figure 2b). A comparative assessment between the solar radiation and the solar energy generated on the AWS may explicitly explain the climate variability on solar energy production. The distribution of the dataset is shown in Figure 2c & d.

The distribution plot shows a number of information. For example, it shows the spot analysis of a dataset distribution. Spot analysis show the concentration of data over a spot or it shows the mode of the dataset values on each spot. For example, the raw dataset in Figure 2b shows scanty data. However, the spot analysis shows that there is a large concentration of solar energy of magnitudes

below 0.4 J. The second information that can be drawn from a distribution plot is the continuity of the dataset. The continuity of a dataset shows the degree of consistency of a dataset. Hence, the possibility of obtaining high error margin is low when the dataset is used to perform mathematical operations or modeling. The third information that can be drawn from distribution plot is the trend features. The trend feature may either describe a mathematical function or relation. It can also describe the area of a shape. The area of the shape can be calculated and may also be used to describe the dataset either in differential or integral form.
Figure 2c show that the solar radiation over the study area is consistent with significant concentration of dataset (yellow color) below 80 W/m2. The trend of the distribution plot is almost rectangular. This further corroborates the information in Figure 2a. The distribution plot in Figure 2d obviously shows the significance of climate change on solar energy production. In other words, climate variability is one of the salient factors that must be considered during renewable energy planning. In Figure 1e, a scenario where PV panel of specific parameter described in the research methodology was applied (Figure 2e). Figure 2f, illustrates a 3D plot showing the optimized solar energy that was calculated based on the solar radiation over the study region at a given period of time. The wind analysis is shown in Figures 2a-f.

The wind speed dataset from the AWS is shown in Figure 3a while the wind energy production is shown in Figure 3b. The wind energy production dwindles as shown in Figure 3b. This observation may be worrisome based on the dataset. The wind climate over the study area has been studied exclusively by Akinyemi et al. [7]. Hence, the energy generation would be somewhat unstable. However, adequate deployment of recent technology may stabilize energy production in actual sense [8]. The distribution plot shown in Figures 3c & d corroborates the effect of climate variability on renewable energy production. In Figure 3 e, a scenario of a windmill with wind blade of 15 m was considered. Though the energy production increased (i.e. compared to Figure 3b), it is observed that the trend of the meteorological parameter (wind speed) affects the power output in the system. Also, the use of storage system may stabilize power transmission to end-users. The 3D plot shows the outcome of the optimized wind energy that is calculated from the wind speed over a period of time. Unlike the information on Figure 1g, the outcome in Figure 3g is unpredictable because of the extended influence of climate change on the wind convections.

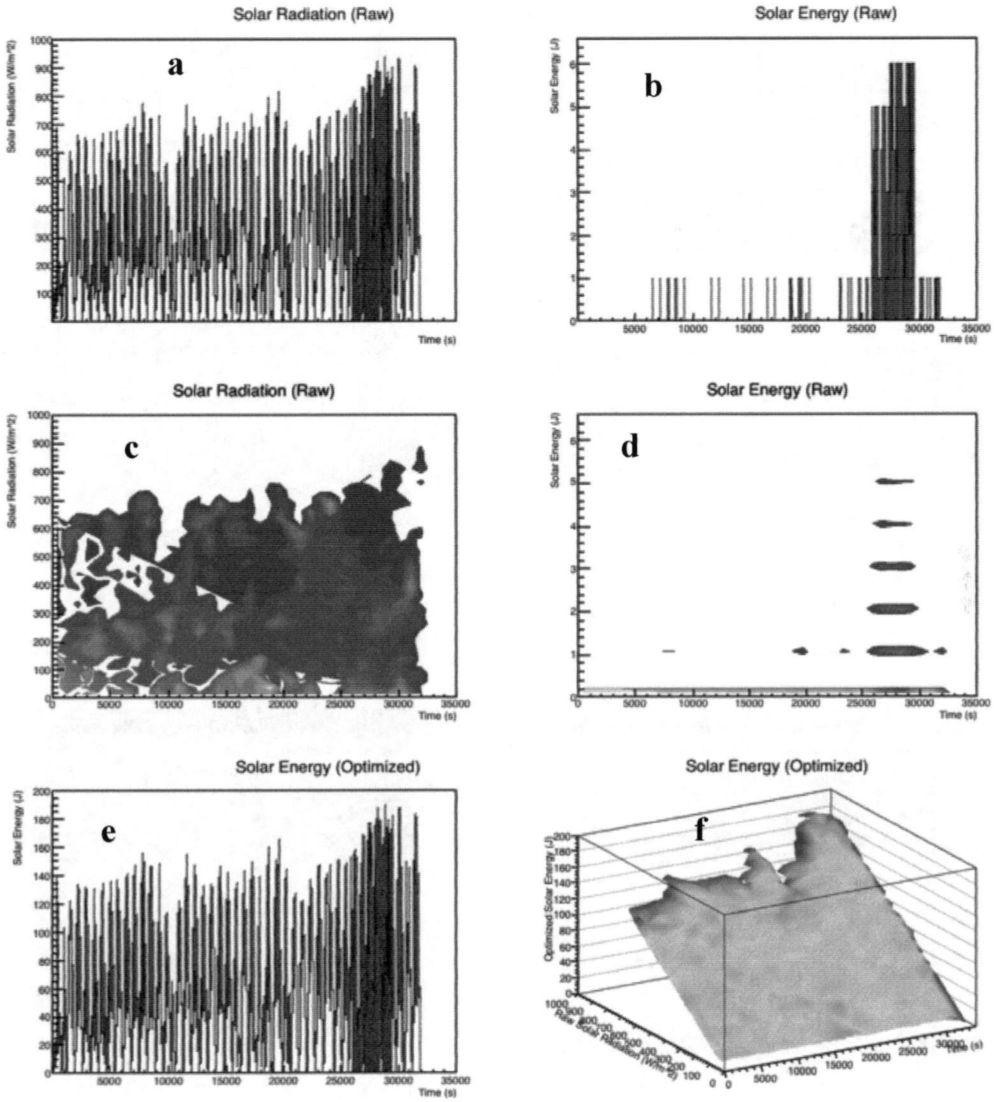

Figure 2: Comparative analysis between meteorological and solar energy parameter

IOP Conf. Series: Earth and Environmental Science **331** (2019) 012036 doi:10.1088/1755-1315/331/1/012036

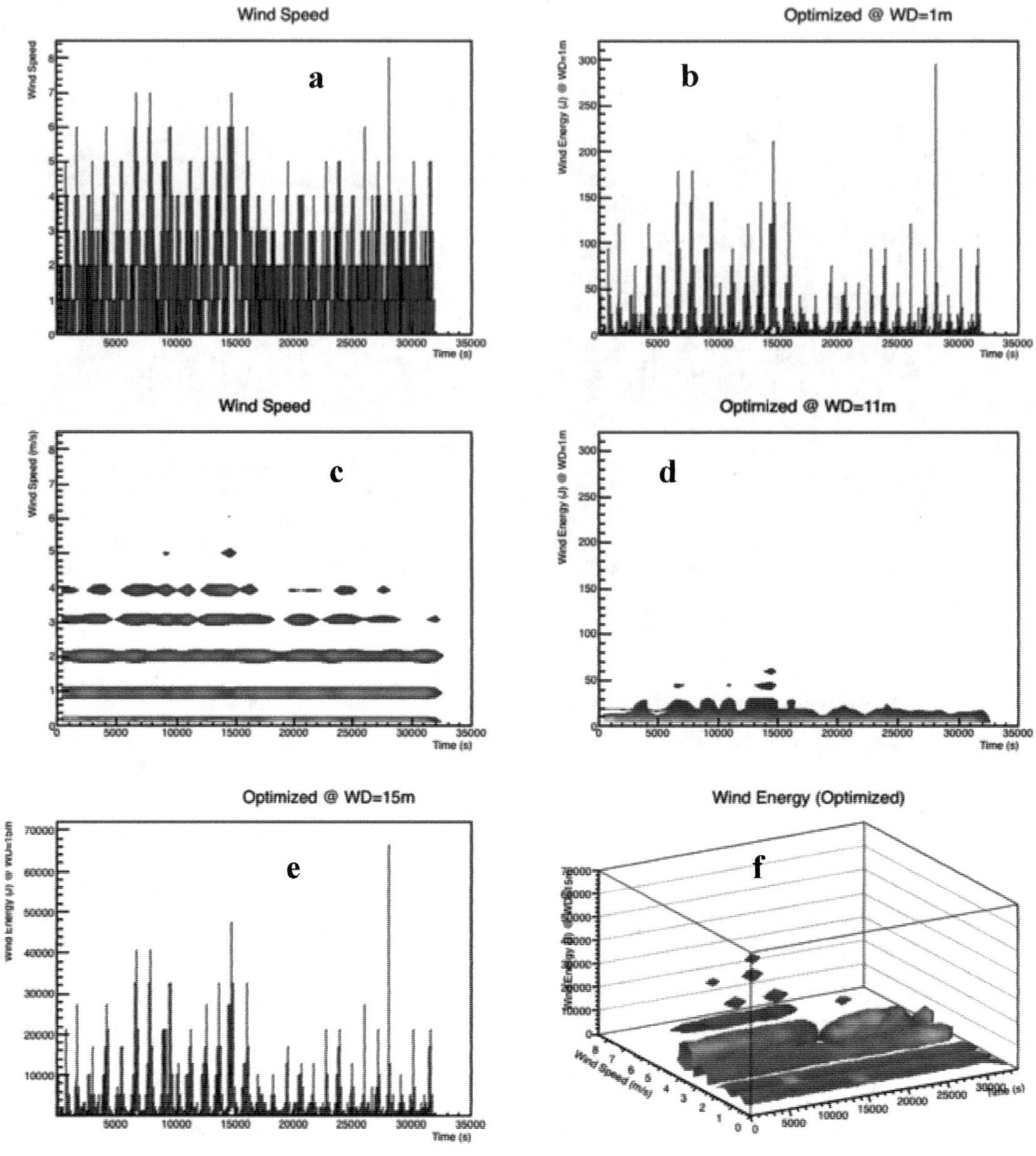

Figure 3: Comparative analysis between meteorological and wind energy parameter

4. Conclusion

It is observed that climate variability had significant influence on energy production i.e. considering solar and wind energy generation. However, climate variability was more significant on wind energy production. In other words, it is important for energy experts to consider the climate variability over a geographical location to understand the adequate technology to be used to acquire maximum energy production.

Acknowledgement
The authors appreciate the partial sponsorship of Covenant University.

References

[1]. Emetere M.E., Akinyemi M.L., and Edeghe E.B., (2016). A Simple Technique for Sustaining Solar Energy Production in Active Convective Coastal Regions, International Journal of Photoenergy 2016, 3567502, 1-11

[2]. Morice, C. P., J. J. Kennedy, N. A. Rayner, and P. D. Jones, 2012: Quantifying uncertainties in global and regional temperature change using an ensemble of observational estimates: The HadCRUT4 data set. J. Geophys. Res., 117, D08101, doi:10.1029/2011JD017187. http://doi.wiley.com/10.1029/2011JD017187 (Accessed September 25, 2015).

[3]. Hall, N.M.J and P. Peyrille, 2006: Dynamics of the West African monsoon. J. Physique, 139, 81-99.

[4]. Funnso Ogunlade, (2016). http://thenationonlineng.net/atan-overdue-for-a-bank/ (accesed 22-03-2018)

[5]. Climate (2018). https://en.climate-data.org/location/406205/ (accesed 22-03-2018)

[6]. Akinyemi M. L., Emetere M. E. and Akinwumi S. A., (2016). Dynamics of Wind Strength and Wind Direction on Air Pollution Dispersion, Asian Journal of Applied Sciences, 4(2): 422-429

[7]. Emetere Moses E., Akinyemi M.L. & Oladimeji T.E. (2016) Statistical Examination Of The Aerosols Loading Over Kano-Nigeria: The Satellite Observation Analysis, Scientific Review Engineering and Environmental Sciences, 25 (2), 72

[8]. Shejal B. D. and Jamge S. B., (2010), Wind power stabilization to achieve proper grid connection using power convertor & DSP Controller, AIP Conference Proceedings 1324, 147, https://doi.org/10.1063/1.3526179

IOP Conf. Series: Earth and Environmental Science **331** (2019) 012037 doi:10.1088/1755-1315/331/1/012037

Optimization of solar photovoltaic efficiency using angular inclination: Effect on surface temperature distribution

Emetere M.E.[1,2], **and Adeyemo N..**[1]
[1]Department of Physics, Covenant University Canaan land, P.M.B 1023, Ota, Nigeria.
[2]Department of Mechanical Engineering and Science, University of Johannesburg, APK, South Africa

E-mail: emetere@yahoo.com

Abstract. Temperature distribution over the surface of a solar photovoltaic (PV) panel is essential as over heating lowers the efficiency of the PV module. There have been multiple solutions towards cooling the PV surface during extremely hot sunshine. The mechanism is envisaged to be expensive to construct and maintain. In this research the PV panel was inclined at various degree to determine the best inclination that would allow convective cooling. It was observed that at an inclination of 45°, the convective cooling from surrounding wind was able to regulate the surface temperature. Hence, the sinusoidal trend of thermal heating was distorted significantly-signifying that the convective wind was able to mitigate the PV surface heating.

Keywords: energy, solar, temperature, solar energy, renewable energy

1. Introduction

Renewable energy serves as a source of clean and environmentally friendly energy to reduce the level of climate change and combat the anticipated scarcity of conventional energy. Solar energy offers one of the best alternative energy sources. The effect of energy as one of the key factors for economic growth and social development cannot be overemphasized. As a result, renewable energies that have the potential of providing the needed energy at little or no adverse effect to the environment are constantly being developed. Photovoltaic (PV) system (Figure 1) is one of the means of obtaining renewable energy [4]. Solar photovoltaic (PV) modules, a common type of solar energy utilization technology, which converts radiations from the sun into useable electricity, is widely in use due to its potential of providing clean and green energy continuously [3]. In 1839, Edmond Becquerel a French scientist discovered photovoltaic power. This has evolved through the years with a wide range of PV cell technologies and applications. The three generations of PV cell technologies based on the basic materials used are; crystalline silicon, thin film and concentrated photovoltaic and organic material [1]. Teo et al., [5] and Rahman et al., [6] reported that incident solar radiation on PV systems generates about 15 to 20% electrical energy and 75 to 80% thermal energy resulting in overheating of the PV system and a reduction in its efficiency. This is due to the fact that operating temperature affects the performance of a PV module even when other factors are controlled. Van et al., [7] reported that PV modules lose up to 7% power at higher temperatures due to decrease in voltage. Though increased solar intensity increases PV power, it decreases it's efficiency due to the change in the temperature of the cells. Renewable energy target, that requires electricity retailers to source a certain proportion of the total electricity sale from renewable energy sources is currently 0% in

Nigeria and the country aims for a 7% target by 2025 [2]. This research could provide a possible means of actualizing this target.

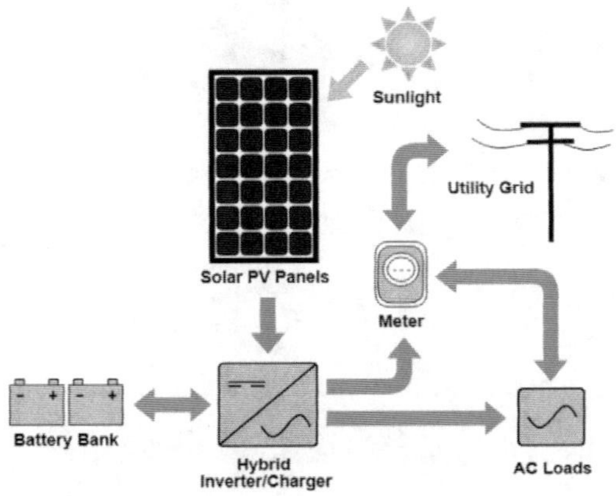

Figure 1: Typical photovoltaic system

Researchers have proven that photovoltaic (PV) efficiency of solar cells is inversely proportional to their operating temperature [8-11]. Hence, the temperature distribution in a PV module will also give rise to thermal stresses within the module. In this research, we examine the influence of PV inclination on the temperature distribution on the PV module.

2. Experimental Design, Materials and Methods

A manual logger was designed to measure the temperature and current simultaneously [12]. A polycrystalline 10V solar panel was used for the experiment. The PV panels were connected to the logger facing the south-east direction. The direction was determined using a GPS hand-held device. The angle of inclination was 0°, 30° and 45° as presented in figures 2-4 below. The dataset was processed using the MATLAB software.

Figure 2: Condition 1 (0°)

Figure 3: Condition 2 (30°)

Figure 4: Condition 3 (45°)

3. Results and Discussion

The current versus time plot for 0° is presented in Figure 5. It is observed that the current generation was optimal at great part of the day. However, the current drop suggests that the solar radiation dropped. It is observed that the sinusoidal pattern which depicts high concentration of heat on the surface of the PV panel (Figure 6) allows the current production drop.

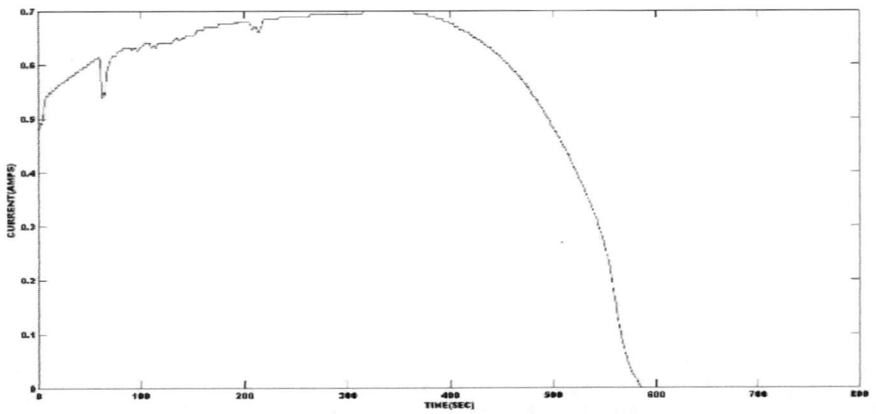

Figure 5: graph of current against time for 0°

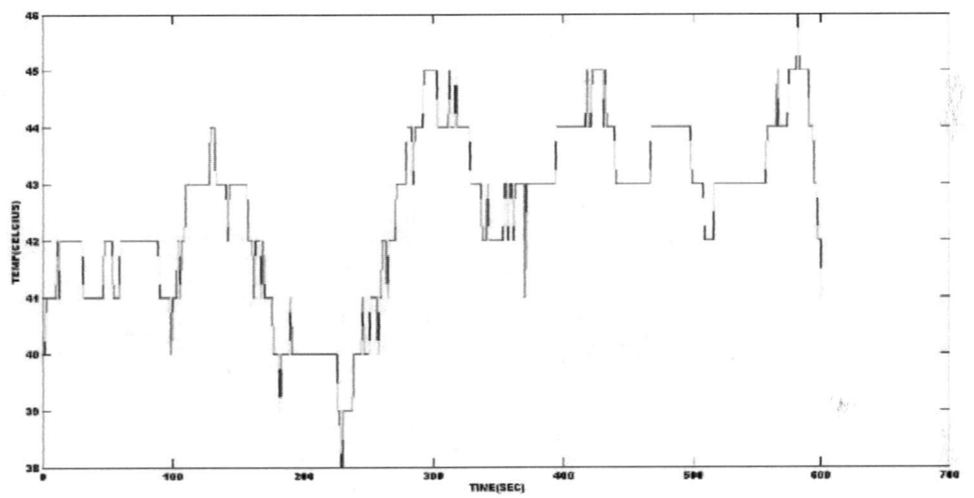

Figure 6: graph of temperature against time for 0°

The graph in figure 6 shows a step sinusoidal plot of time against temperature. Temperature changes with time but is constant at certain point for a little e period of time before it changes again hence the step form. The current production of the PV module describes a stepwise negative parabola (Figure 7) that makes it clear that the temperature distribution (Figure 8) produces high heat energy. The graph in figure 7 above shows the graph of time against current that exhibits a step sinusoidal form. The form taken by this graph indicates that at certain points in time, current was constant. The step sinusoidal form of the graph in figure 8 indicates the variation in temperature and points in time at which the temperature remained constant for a short period of time. Figure 9 shows that the PV inclination supports the convective wind system to cool the surface of the PV panel (Figure 10).

IOP Conf. Series: Earth and Environmental Science **331** (2019) 012037 doi:10.1088/1755-1315/331/1/012037

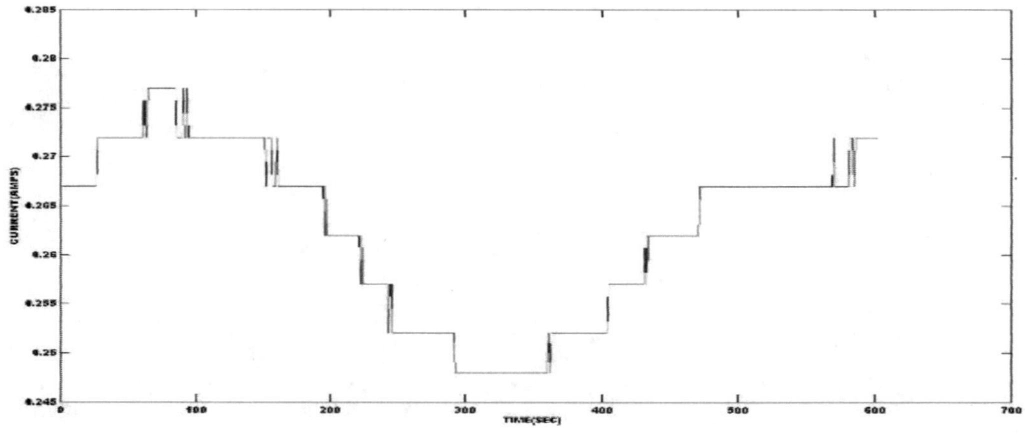

Figure 7: graph of current against time for 30◦

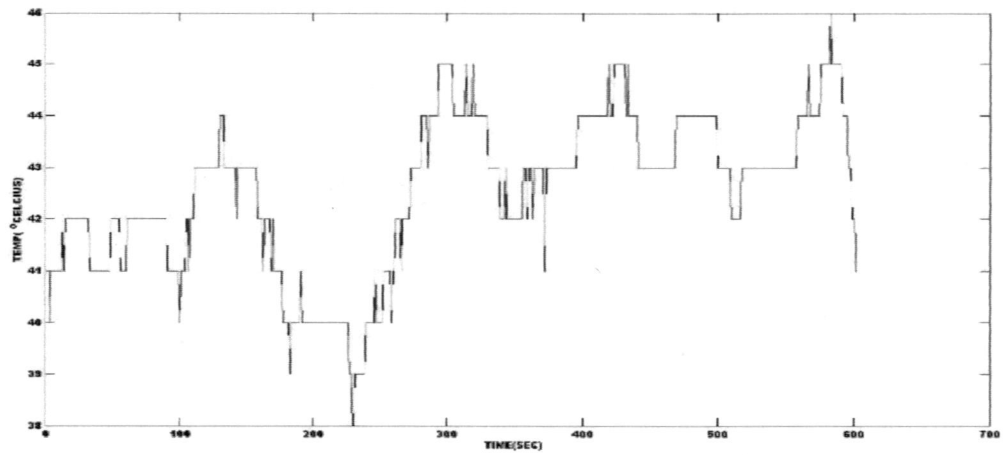

Figure 8: graph of temperature against time for 30◦

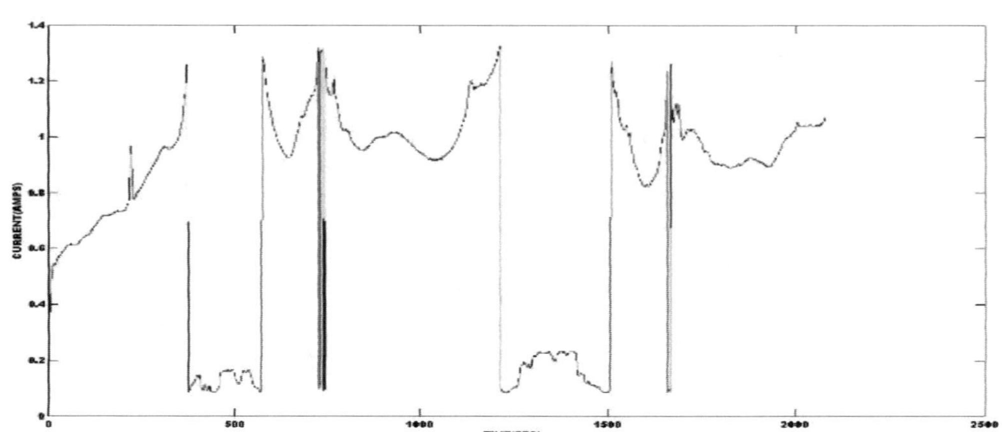

Figure 9: graph of current against time for 45◦

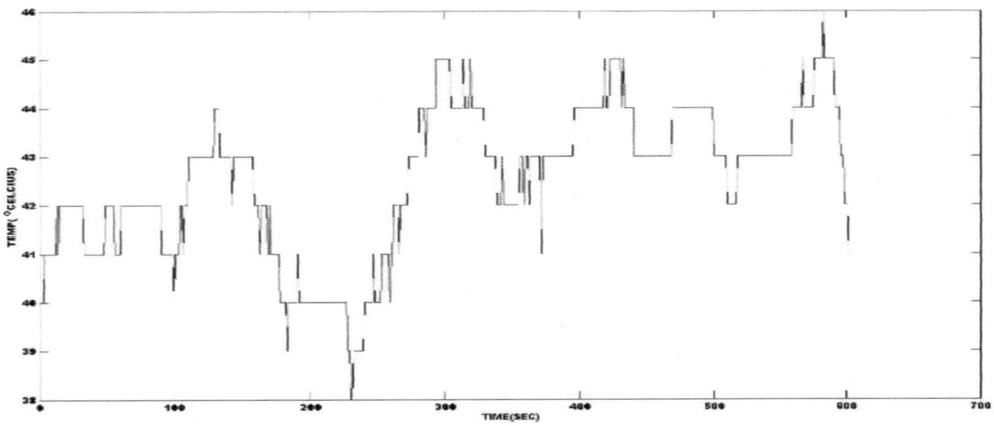

Figure 10: graph of temperature against time for 45◦

4. Conclusion

From the experimentation, it was proven that temperature distribution had significant influence on the current production of the solar PV panel. The solar collapse at high temperatures hence the need for PV cooling in extreme weather. It was observed that at an inclination of 45◦, the convective cooling from surrounding wind was able to regulate the surface temperature. Hence, the sinusoidal trend of thermal heating was distorted significantly-signifying that the convective wind was able to mitigate the PV surface heating.

Acknowledgement

The authors appreciate Covenant University for partial sponsorship. The authors acknowledge NASA for primary dataset.

5. References

[1]. Eldin A H, Refaey M and Farghly A (2015). A review on photovoltaic solar energy technology and its efficiency. Conference paper.
[2]. Lawal NS. Renewable energy as a solution to Nigerian energy crisis. Vaasan Ammattimkorkeakoulu, Vasa Yrkeshogskola, University of Applied Sciences, 2010.
[3]. Yu J and Lin L (2016). Experimentally investigating the effect of temperature differences in particle deposition process on solar photovoltaic (PV) modules. *Sustainability*. 2016:8, 1091.
[4]. Nasrin R, Hasanuzzaman M and Rahim NA (2017). Effect of high irradiation on photovoltaic power and energy. *International Journal of Energy Research*. 42:1115-1131.
[5]. Teo HG, Lee PS and Hawlader MNA (2012). An active cooling system for photovoltaic modules. *Appl. Energy*. 90(1):309-315.
[6]. Rahman MM, Hasanuzzamana M and Rahim NA (2017). Effects of operational conditions on the energy efficiency of photovoltaic modules operating in Malaysia. *Journal of Cleaner Production*. 143:912-914.
[7]. Van Dyk, EE, Scott BJ, Meyer EL, Leitch AWR (2000). Temperature dependence of performance of crystalline silicon modules. *S Afr J Sci*. 96: 198-200.
[8]. H.A. Zondag, D.W. De Vries, W.G.J. Van Helden, R.J.C. Van Zolengen, A.A. Van Steenhoven, The thermal and electrical yield of a PV-thermal collector, Sol. Energy, 72 (2) (2002), pp. 113-128
[9]. A. Tiwari, M.S. Sodha, Performance evaluation of solar PV/T system: an experimental validation, Sol. Energy, 80 (7) (2006), pp. 751-759

IOP Conf. Series: Earth and Environmental Science **331** (2019) 012037 doi:10.1088/1755-1315/331/1/012037

[10]. Y. Lee, A.A.O. Tay, Finite element thermal analysis of a solar photovoltaic module, Energy Proc., 15 (2012), pp. 413-420
[11]. Servant JM. Calculation of the cell temperature for photovoltaic modules from climatic data. In: Bilgen E, Hollands KGT, editors. Proceedings of the 9th biennial congress of ISES – Intersol 85, Montreal, Canada, extended abstracts; 1985. p. 370.
[12]. K. N. Akposionu, and A. O. C Nwokoye (2012), Design and Fabrication of A Low-Cost Data Logger for Solar Energy Parameters, Journal of Energy Technologies and Policy, Vol.2, No.6, ISSN 2224-3232

Monitoring of Background Radiation in Selected Schools in Ota, Ogun State Nigeria by Direct Measurement of Terrestrial Radiation Dose Rate

C. A. Onumejor, A. Akinpelu, T. E. Arijaje, M. R. Usikalu, O. F. Oladapo, M. E. Emetere, M. Omeje, J. A. Achuka
Covenant University, Ota, Ogun State Nigeria

E-mail: charity.onumejor@covenantuniversity.edu.ng

Abstract. Terrestrial radiations are natural types from radionuclides found in rocks, building materials and soils in an environment. Monitoring of background radiation energy levels in selected school in Ota, Ogun State Nigeria by direct measurement of terrestrial radiation dose rate was done. The mean dose rate ranged from 35.4 to 62.34 $nGyh^{-1}$. The measured mean values are within world average recommended limits of 59 $nGyh^{-1}$ except for study location SC7 with mean dose rate value of 62.34 $nGyh^{-1}$. On the over all, it can be concluded that the concentration of NORM (naturally occurring radioactive materials) and radiation dose rate in an environment is at safe level. Further environmental radiation assessment is recommended to ascertain the possible sources of background radiation dose rate and exposure, beside terrestrial radiation sources.

Keywords: Dose Rate, Terrestrial Radiation, Background Radiation Monitoring, Ota.

1. Introduction

Terrestrial radiations are natural types from radionuclides found in rocks, building materials and soils in an environment. Natural radioactivity requires concentration monitoring for sustainable health and environmental safety for general public. Radiation energy from naturally occurring radionuclides in the immediate environment contributes greatly to the background radiation level of that environment. Naturally, occurring radioactive materials (NORMs) are constituents of the earth crust; they continuously emit ionizing radiation as natural cause from the disintegrations of the radioactive materials. The geological formation of an area, makeup the earth crust, determines the radionuclide concentration that influences the background radiation level of the area. Research findings have it that radiation dose from NORMs makeup over 85% of the total dose that gets to man and 15% comes from artificial sources [1, 2, 3, 4]. Many individual has little or no reason to be exposed to manmade radiation, thus natural background radiation remains their total exposure. Natural background radiations are subdivided into cosmic radiation, radon inhalation, internal exposure and terrestrial external radiation [5, 6]. Of concern in this study is the monitoring of background radiation concentration levels in selected school by direct measurement of terrestrial radiation dose rate and radionuclide heads such as Thorium-232, Uranium-238 and Potassium-40. The environmental rocks and soils in an area naturally emits gamma radiation from the series Thorium-232, and Uranium-238 and none series Potassium-40. The emitted radiation gets to the general public both outdoor and indoor in an environment. It is therefore essential to measure terrestrial background radiation for assessment of exposure level to the public. Usikalu et al, 2018, worked on background radiation monitoring in industrial area of Ota town, the mean dose rate of the finding is presented in figure 1, the study concluded that the industrial areas of Ota environ is safe for dwellers [5].

IOP Conf. Series: Earth and Environmental Science **331** (2019) 012038 doi:10.1088/1755-1315/331/1/012038

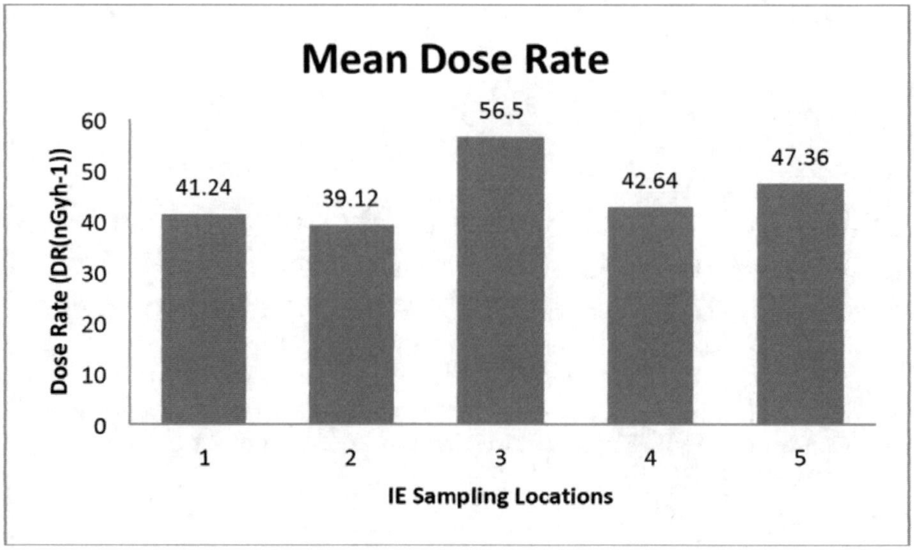

Figure 1. Mean dose rate for IE sampling locations [5]

2. Material and Method

The study area is Ota, a town in Ogun State, Nigeria with coordinates 07°57'N 04°47'E, it is the capital town of the Ado-Odo Local Government Area. There are many private and government (public) owned schools in Ota. Ten (10) out of the schools that are located along Idiroko- Owode road Ota were selected as shown in the figure 2, both private and public schools alike. A total of ten (10) schools were monitored for terrestrial background radiation assessment using RS-230 BGO Super-SPEC Handheld Gamma-Ray Spectrometer. The direct measurement of background radiation dose rates DR ($nGyh^{-1}$) were measured and reported as shown in table 1. RS-230 BGO device used was set on assay mode to obtain the dose rate concentrations at sampling time of 90 seconds. Sample locations ID are SC1, SC2, SC3, SC4, SC5, SC6, SC7, SC8, SC9 and SC10 for the 10 selected schools. Five (5) sampling points were taken for each sample location and the mean dose rate was obtained and used for assessment. The RS-230 BGO was held one (1) meter above the ground level during active sampling.

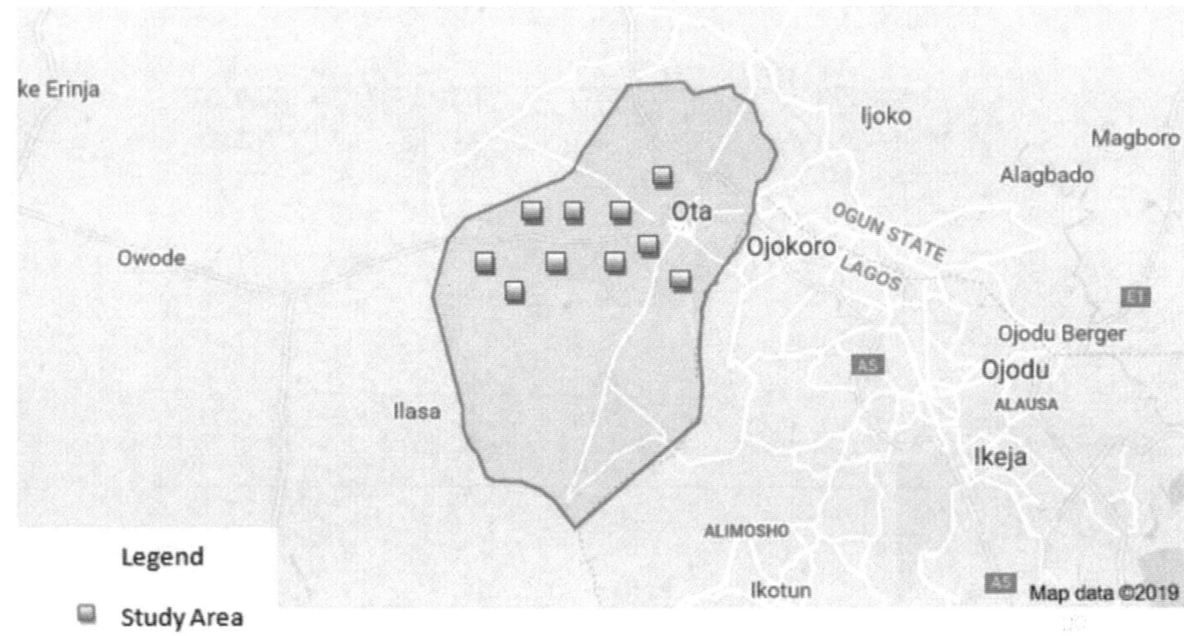

Figure 2. Map of Study Area

3. Results and Discussion

The terrestrial background radiation dose rate DR ($nGyh^{-1}$) of 10 selected schools in Ota was measured and the obtained results are presented in table 1. The plot of the mean dose rate for sample SC1 to SC10 is shown in figure 3 for better mean dose rate comparison of the different sample locations. The obtained mean dose rate ranges from 35.4 to 62.34 $nGyh^{-1}$. Sample location SC7 recorded the highest mean dose rate value of 62.34$nGyh^{-1}$ while the least value of 35.4 $nGyh^{-1}$ was recorded in location SC2. The respective mean dose rate for locations SC1, SC2, SC3, SC4, SC5, SC6, SC7, SC8, SC9 and SC10 are 49.44, 35.40, 52.10, 45.86, 51.90, 38.90, 62.34, 44.76, 58.04 and 39.48 $nGyh^{-1}$. Dose rate from terrestrial background radiation is usually assessed with the international unit of nanoSieverts per hour ($nSvh^{-1}$) for ease of comparison with world standards. Basically, 1 Gray (1 Gy) is equal to 1 Sievert (1 Sv) for gamma radiation, hence 1nGyh-1 will equal 1$nSvh^{-1}$. The US EPA report of 2017 presented dose rate limit for terrestrial radiation (background radiation from rocks, building materials and soils in an environment) to be less than 6 to more than 83 nSvh-1 in comparison with other possible exposure units such as milliRoentgen/hour (mRh^{-1}) for monitoring stations and this is comparable to world mean dose rate limit of 59 $nGyh^{-1}$ [8]. A comparison of the obtained results with the world mean value limit for dose rate (DR) shows that all the obtained dose rate from all study is well within set limit except location SC7 that went above limit with the value by 3.34 $nGyh^{-1}$. A closer look at constituent dose rate readings that gave the mean value of 62.34 $nGyh^{-1}$ for SC7, it was observed that there was an offshoot value of 75 $nGyh^{-1}$. This high offshoot value influenced the high mean dose rate obtained for location SC7. A comparison of the current work with the world standard and previous study in industrial area of Ota, as shown in figure 4, it can be seen also that SC7 went above the world standard. Man made factor or experimental error could be attributed to the value offshoot, but none the less, a further assessment is recommend for location SC7.

Table 1. Measured Terrestrial Background Radiation for Assessment of Public Exposure Radiation

Sample ID	DR (nGyh⁻¹)					Mean
SC1	47.1	47.6	53.9	51.4	47.2	49.44
SC2	36.1	31.9	33.8	33.9	41.3	35.40
SC3	55.9	41.1	66.3	50.3	46.9	52.10
SC4	49.3	39.7	44.7	45.7	49.9	45.86
SC5	55.0	52.1	52.4	60.0	40.0	51.90
SC6	49.0	39.0	31.1	38.2	37.2	38.90
SC7	58.3	61.6	56.5	60.0	75.3	62.34
SC8	45.1	48.7	48.1	42.9	39.0	44.76
SC9	60.3	65.1	53.4	54.9	56.5	58.04
SC10	46.9	39.3	34.0	38.9	38.3	39.48

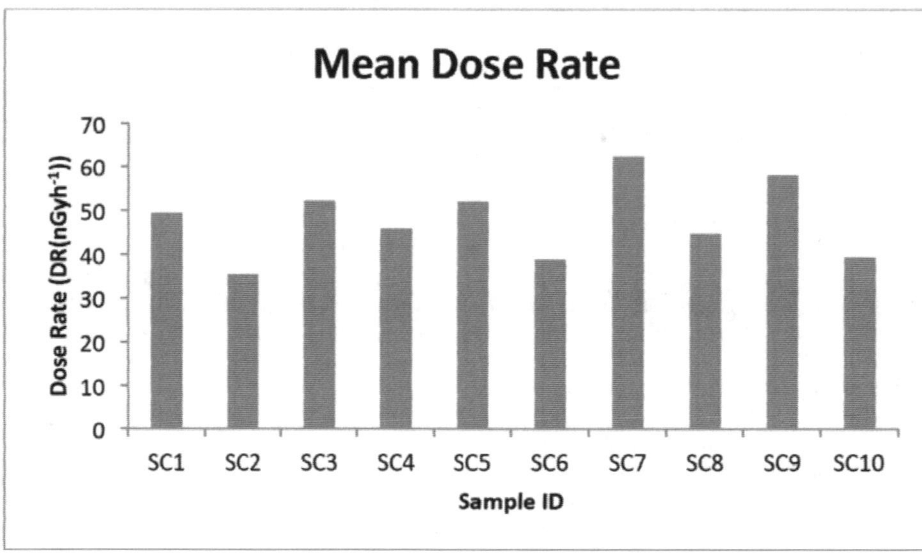

Figure 3. Mean Dose Rate for sample SC1 to SC10

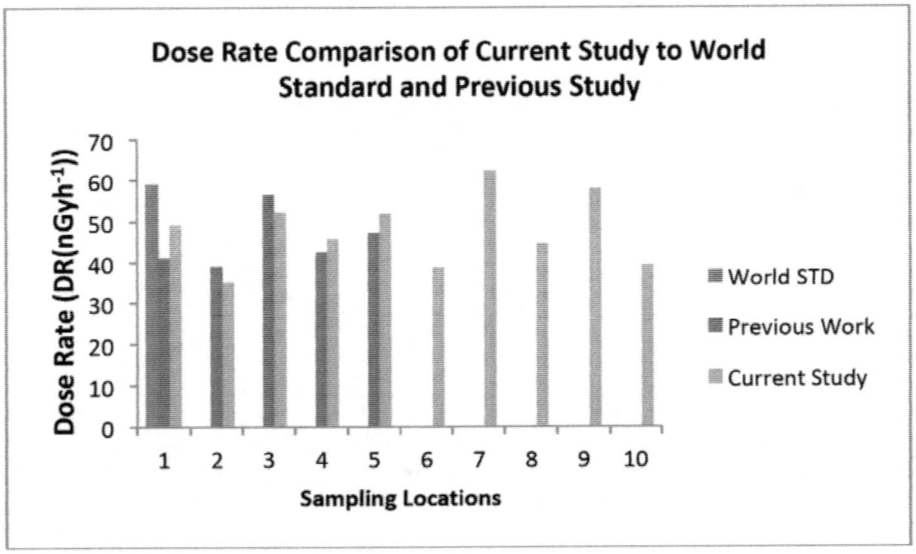

Figure 4. Comparison of Current study with World Standard and Previous Study

4. Conclusion

The dose rate of 10 selected schools in Ota environ were measured using RS-230 Super Spec gamma ray spectrometer. The measured mean values are within world average recommended limits of 59 nGyh^{-1} except for study location SC7 with mean dose rate value of 62.34 nGyh^{-1}. On the over all, it can be concluded that the concentration of NORM (naturally occurring radioactive materials) and radiation dose rate in an environment is at safe level.

5. Recommendation

For study location SC7, a further environmental radiation assessment is recommended to ascertain the possible sources of background radiation dose rate and exposure, apart from the terrestrial radiation sources.

Acknowledgment

The authors appreciate Covenant University for conference the support.

Reference

[1] IAEA (1986) International Atomic Energy Agency. Facts about low level Radiation Exposure. Publications No: A.A. E985-06482

[2] UNSCEAR (2000): United Nation Scientific Committee on Effect of Atomic Radiation-Ionizing Radiation Sources and Biological Effect. Report to the General Assembly (New York)

[3] Onumejor C. A., Balogun F. A., Akinpelu A., Arijaje T. E., Usikalu M. R.(2018) Rutherford Backscattering Spectrometry (RBS) method for the Determination of Elemental Constituent of Tropical Wood Matrices from Western Nigeria. IOP Conf. Series: Earth and Environmental Science 173 (2018) 012005 doi :10.1088/1755-1315/173/1/012005

[4] Usikalu M. R, Onumejor C. A, Akinpelu A and Ayara W. A, (2018) Improvement on Indoor Radon Accumulation Rate in CST Laboratories at Covenant University, Ota, Nigeria, International Journal of Mechanical Engineering and Technology, 9(10), 2018, pp. 135–148. http://www.iaeme.com/IJMET/issues.asp?JType=IJMET&VType=9&IType=10

[5] M. R. Usikalu, C. A. Onumejor, A. Akinpelu, J. A. Achuka, M. Omeje and O. F. Oladapo (2018) Natural Radioactivity Concentration and Its Health Implication on Dwellers in Selected Locations of Ota. IOP Conf. Series: Earth and Environmental Science 173 (2018) 012005 doi :10.1088/1755-1315/173/1/012005

[6] The United Nations Scientific Committee on the Effects of Atomic Radiation (UNSCEAR) 2008 report Vol. I: sources of ionizing radiation (New York: United Nations Publications) (2008)
[7] US EPA (2017) Data from U.S. Geological Survey, summarized by Health Physics Society. Environmental Radiation Fact Sheet: About Exposure and Dose Rates |RadNet | 6 pp, 635.47 K
[8] UNSCEAR "Source, effect and risk of ionizing radiation," (2000) New York, United Nations.

The Future of Renewable Energy for Electricity Generation in sub-Saharan Africa

Paul A. Adedeji [a], **Stephen Akinlabi** [b,c], **Nkosinathi Madushele** [a] **and Obafemi Olatunji** [a]

[a] Department of Mechanical Engineering Science, University of Johannesburg, South Africa.
[b] Department of Mechanical and Industrial Engineering, University of Johannesburg, South Africa
[c] Department of Mechanical Engineering, Covenant University, Ota, Nigeria.

Email: pauladedeji2k5@gmail.com

Abstract: Energy transition in the last decade has experienced increased quota of renewable energy in the global energy mix. In sub-Saharan Africa (SSA), the transition from the fossil fuel to the renewable energy source has been gradual. The state of renewable energy in the region in the next decade is the focus of this study. This study uses a single-layer perceptron artificial neural network (SLP-ANN) to backcast from 2015 to 2006 and forecast from 2016 to 2020 the percentage of renewable energy for electricity generation, exempting the hydropower in the energy mix of the SSA based on historical data. The backcast percentage renewable energy mix was evaluated using known statistical metrics for accuracy measures. The root mean square error (RMSE), mean absolute deviation (MAD) and mean absolute percentage error (MAPE) obtained were 0.29, 0.18, and 14.69 respectively. The result shows possibility of an increase in the percentage of renewable energy in the electricity sector in the region. In 2020, the percentage of renewable energy in sub-Saharan region is expected to rise to 4.13% with exclusion of the hydropower. With government policies encouraging the growth of the renewable energy as a means of power generation in the region, the predicted percentage and even more can be realized.

Keywords: Non-linear autoregressive ANN; Renewable energy; sub-Saharan Africa

1. Introduction

Renewable energy is fast becoming a global mantra aimed at mitigating climate change and global warming associated with energy sector. The African continent is among the continents endowed with high potentials for renewable energy sources with their abundance being a geospatial variant. While the solar energy is ubiquitous, the biomass, and the hydropower are more abundant in the wet, southern, and central regions of the continent. Wind energy on the other hand has its large presence in east, north the southern regions of the continent [1]. It is alarming that about 56% of the countries in Africa still experience periodic power outages and load shedding [1]. Quantitatively, about 2% of her

IOP Conf. Series: Earth and Environmental Science **331** (2019) 012039 doi:10.1088/1755-1315/331/1/012039

gross domestic products (GDP) is being lost to these [1]. The SSA is notable for high potential of renewable energy for power generation [2]–[6]. However, the SSA region is still referred to have the least energy access rate on the globe [7]. The renewable energy sources among other available energy sources in SSA have been identified to possess a very high prospect of transforming Africa's industrial sector with increased energy access either at mini-grid, micro-grid or nano-grid levels, thus naturally creating competitiveness among small and medium-sized enterprises, and modular industries.

The sub-Saharan region constitutes a part of the African continent where about 80% of the continent's population resides [8]. The region accounts for 13% of the world population, but, provides only 4% of its energy demand [1]. Despite the proliferation of the renewable energy sources as alternative and clean source of power generation, statistics showed that in 2010, about 95% of people in the sub-Saharan region still depends on the solid biomass for cooking [9] and other unclean sources for power generation [10]. However, recent statistics show that about 600 million people in the region lack access to electricity and 890 million still depend on conventional types of fuels for cooking [11]. Even though, this statistic has reduced in compared to a decade ago, the percentage reduction is not laudable. The larger percentage of the present-day SSA still has significant reliance on fossil fuels for power generation with most being off-grid.

Electricity among other types of energy has remained one of the primary markers of socio-economic development [9], [12], [13]. Since 2012, over 100 million people per year have gained access to electricity across the globe as compared to the rate of 60 million people per year in the previous decade [14]. Globally, about 1.3 billion people lack access to electricity with almost half of this population residing in the SSA [3]. Since the eighties, the electrification rate of SSA has remained low with no increase in the power supply per capita [15], [16]. Clean energy revolution has, however, being identified as the tool for winning the present energy poverty fight in the SSA [17]. Even though access to affordable, reliable, sustainable, and clean energy for all forms the seventh Sustainable Development Goal (SDG), it has been established that this seventh SDG is the solution to the first goal- "*ending poverty in all its forms everywhere*" [18]. These encompasses energy in entirety, involving the electricity and clean fuel technologies for all use [19]. The United Nations in the development of 2030 agenda for sustainable development has the set time around the corner. It is expedient to evaluate where in the region is in meeting the seventh SDG and potential roadmaps towards achieving this with renewable energy sources.

Renewable energy integration into the energy mix of sub-Saharan African countries is gradually gathering steam with a yearly increase in the percentage it occupies within the energy strata. In the SSA, report shows that wind, solar and biomass resource is gradually increasing in their integration into the national grid of the several countries [1]. For example, South Africa, as one of the countries in SSA has one of the most developed renewable energy programme in the region [20]. The programme has attracted about 195 billion rands to South Africa, 25 % of which is from foreign investors [21]. Also, renewable energy, including hydropower accounts for about 80 % of the electricity generated in Kenya [22]. Kenya has the first geothermal spring in Africa. Other sources for power generation include hydropower, oil thermal, wind power, and cogeneration from sugarcane bagasse [22]. This makes her electricity generation system one of the most sustainable globally. Strategic plans are being made by the country to meet the anticipated double increase in energy demand in 2020 and a six-time increase in 2030. Similarly, Ethiopia is endowed with abundance of renewable energy sources with a potential to generate over 60,000 megawatts from a nexus of wind, solar, hydropower, and geothermal. However, the country currently has 2,300 megawatts installed to serve her population of 95 million people, but with further plans to harness more of wind and geothermal resource in the country [23].

IOP Conf. Series: Earth and Environmental Science **331** (2019) 012039 doi:10.1088/1755-1315/331/1/012039

Empirical models are data driven models which unravel the hidden patterns and trends. They have been widely used in several studies with varying performance. Such models like Adaptive Neuro-Fuzzy Inference System (ANFIS), and artificial neural network (ANN) have found relevance in energy studies for short-term forecasting [24], [25] and long-term forecasting [26], [27]. These models learn patterns in data either using a supervised, unsupervised, semi-supervised, or deep learning approach [28]–[31]. These models can be autoregressive or input/output. The autoregressive models use the historical data at a specific delay to predict subsequent values within a trend. Many of these models fall into the category of self-organizing maps (SOMs). These SOMs models transforms the input of an arbitrary dimension into one or two-dimensional discrete map with neighbourhood constraint.

Energy is a subtle undertone of the industrial revolution with improved technologies for power generation prevalent in each revolution. It is saddening to realize that SSA countries were mere spectators in the first, second, and third industrial revolution. The fourth industrial revolution is closing-up on the sub-Saharan region at an alarming speed. To ensure that SSA countries are not spectators once again but key contributors, it is expedient to determine the percentage of renewable energy in the SSA energy mix in 2020 and a roadmap towards increasing this percentage as region prepares for the next decade. This study backcast and forecast the percentage of renewable energy sources exempting the hydropower in the energy mix of the SSA region using single layer perceptron non-linear autoregressive artificial neural network (NAR-ANN) with enhanced smoothing technique to achieve a percentage for 2020. The study also briefly discussed the roadmap to ensuring more of renewable energy is integrated into the energy mix in the SSA region.

2. Methodology

2.1. Data collection

Time series data of the percentage of renewable energy exempting the hydro resource was collected between 1971 and 2015 from the world bank database. The data forms a representation of the percentage of renewable energy nexus in the total energy mix of the 46 sub-Saharan African countries.

2.2. The Artificial Neural Network (ANN)

The ANN is a black-box modelling technique which imitates the biological nervous system in its mode of operation. The technique maps the input-output space by receiving inputs and processing them through a network of interconnected processing elements called neurons [32], [33]. Each neuron receives input signals with an associated weight from relative neurons or the input layer. A non-linear activation function which processes the signal in each neuron exist. ANN models are categorized by the number of layers as either a perceptron (with single layer) or a multi-layer perceptron (comprising multi-layers). They can also be categorized according to the input/output processing as either backpropagation, recurrent or gradient descent models. A typical perceptron consists of an input layer, a hidden layer, and an output layer. ANN has found its usefulness in intelligent non-linear function approximation, and time series forecast [34]. Basically, it has been commonly used as a forecasting technique due to its robustness and ability to learn different data patterns without a prior understanding of the system being modelled. Despite its several advantages, one of its setbacks is its black box nature. It does not explain explicitly its mode of operation unlike conventional models of data fitting.

2.3. Modelling technique

The non-linear autoregressive (NAR) neural network approach was used in this study for a short-term forecast of the percentage of renewable energy mix in the next decade. The NAR model is significantly useful in forecasting time series data from historical data. The autoregressive equation for a given series y_t can be represented by [35]:

$$y_t = a_1 y_{t-1} + a_2 y_{t-2} + \cdots + a_n y_{t-n} + \varepsilon_t \tag{1}$$

where $a_1, a_2, \dots a_p$ are called the autoregressive parameters and ε_t are normally distributed error term with a mean of zero and a finite variance σ^2.

From the least square method, the autoregressive model parameters were calculated using:

$$\varphi_0 = \bar{y}_0 - \sum_{i=1}^{p} \varphi_i \bar{y}_i, \tag{2}$$

$$(\varphi_1, \varphi_2, \dots, \varphi_p)^T = \left(L_{\eta\eta}\right)^{-1}_{p\times p} \left(L_\eta\right)_{p\times 1} \tag{3}$$

where the observation mean, $\bar{y}_i, i = 0,1, \dots, p$ can be calculated using:

$$\bar{y}_i = \frac{1}{n-p} \sum_{t=1+p}^{n} y_{t-p} \qquad i = 0,1, \dots p \tag{4}$$

and the $L_{\eta\eta} = (S_{ij})_{p\times p}$ is a matrix in the pth order, however, $L_\eta = (S_1, S_2, \dots, S_p)^T$ is a column vector in the pth order, whose constituting elements can be determined using:

$$S_{ij} = \sum_{t=p+1}^{n} (y_{t-i} - \bar{y}_i)(y_{t-j} - \bar{y}_j) \qquad i,j = 1,2, \dots, p \tag{5}$$

$$S_i = \sum_{t=p+1}^{n} (y_t - \bar{y}_0)(y_{t-i} - \bar{y}_i) \qquad i = 1,2, \dots, p \tag{6}$$

Estimating the coefficient of the autoregressive characteristic equation (3), $\hat{\varphi}_0, \hat{\varphi}_1, \hat{\varphi}_2, \dots, \hat{\varphi}_p$, the autoregressive prediction model equation then becomes:

$$\hat{y}_{n+l|n} = \hat{\varphi}_0 + \sum_{i=1}^{p} \hat{\varphi}_i y_{n+l-i|n} \tag{7}$$

such that $\hat{y}_{n+l|n}$ is the l −step-ahead prediction at a time $n+l$.

The ANN modelling proceeds by a single-layer perceptron. A log-sigmoid transfer function was used for the hidden layer. The training process was performed with 1000 epochs as maximum possible iterations. Multiple trainings were performed towards minimizing the mean square error and avoiding underfitting and overfitting. Multiple trainings were experimented and the different number of neurons in the hidden layer tried with different training functions. A delay of 10 and a total number of neurons ($N_h = 30$) was used for the hidden layer; Levenberg-Marquart backpropagation algorithm gave the best results. A total of 31 iterations was reached before one of the stopping criteria (maximum gradient

IOP Conf. Series: Earth and Environmental Science **331** (2019) 012039 doi:10.1088/1755-1315/331/1/012039

of 4.48×10^{-9}) was reached. It has been established in the literature that too many neurons in the hidden layer could cause overfitting or lack of generalization, thus resulting into large verification errors [36]. This was avoided in this study as much as possible. The script for the model was written on Matlab (2015a) and run on a computing device with a configuration of 4GB RAM, core i7 2.20GHz processor. A 10-year prediction was performed and the result compared with the observed values. Performance evaluation of the result using known metrics was carried out. Based on this, a decade forecast of the percentage of the renewable energy mix in the SSA was performed. Strategies towards achieving the increase in the renewable energy quota in the region was developed.

3. Results and Discussions

3.1. Historical renewable energy profile of the SSA

Shown in Figure 1 is a timeseries plot of the percentage of renewable energy integration for electricity generation in the SSA region for 45 years. This percentage includes all renewable sources except hydropower. It can be observed that the SSA region still has a low percentage of RE in sources for electricity generation. The data shows about three decades of RE for electricity generation in the region. The first decade in this data shows no significant change in the percentage. However, a significant increase is observed from four years after the first decade (from 1985). A decade from 1985 (up until 2005), there is a gradual increase in the percentage of RE in electricity generation. At 2005, the SSA achieved close to 1% renewable sources in electricity generation. However, a rapid increase in the use of renewable sources for electricity generation was observed in the third decade (from 2006 till 2015). This can be traced to a change in energy policy to accommodate more renewable energy sources in the energy mix at national levels. This was experienced in certain countries like South Africa, which inaugurated the Renewable Energy Independent Power Producer Procurement Programme (REI4P) in 2015 [37]. Similarly, renewable energy has been on the increase in other countries like Senegal, Ethiopia and so on with more plans to expand this at micro-grid level.

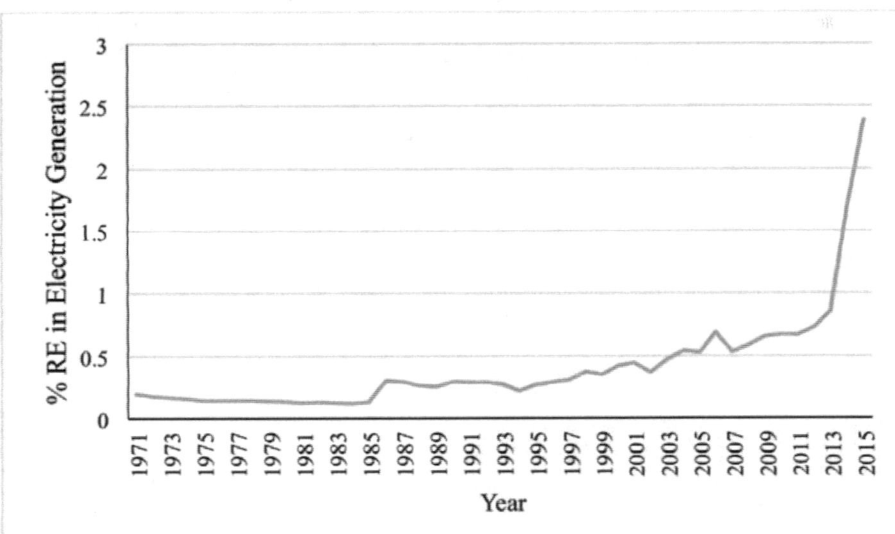

Figure 1. Percentage of Renewable energy in electricity generation for SSA between 1971 and 2015.

IOP Conf. Series: Earth and Environmental Science **331** (2019) 012039 doi:10.1088/1755-1315/331/1/012039

3.2. Model results

From the range of the data (2015), a backcast of 10 years and a forecast of 5 years was performed. The backcast was compared with the actual renewable energy percentage (without the hydropower) as shown in Figure 3. A 5-year ahead forecast was performed between 2016 to 2020. The result shows a gradual increase in the percentage of renewable energy to be expected in the energy mix of SSA region. Shown in Figure 3 is result of the ANN forecast compared to the observed values. Common statistical performance measures were used to compared the actual and predicted value for the 10 years data, like root mean square (RMSE), mean absolute deviation (MAD), and mean absolute percentage error (MAPE) and calculated as using equations (1) to (3). The result is as presented in Table 1.

Root Mean Square Error (RMSE)

$$RMSE = \sqrt{\frac{\sum_{1=1}^{N}[y_i - \hat{y}_i]^2}{N}} \tag{8}$$

Mean Absolute Deviation (MAD)

$$MAD = \frac{1}{N}\sum_{i=1}^{N}|y_i - \mu| \tag{9}$$

Mean Absolute Percentage Error (MAPE)

$$MAPE = \frac{1}{N}\sum_{i=1}^{N}\left|\frac{y_i - \hat{y}_i}{y_i}\right| \times 100\% \tag{10}$$

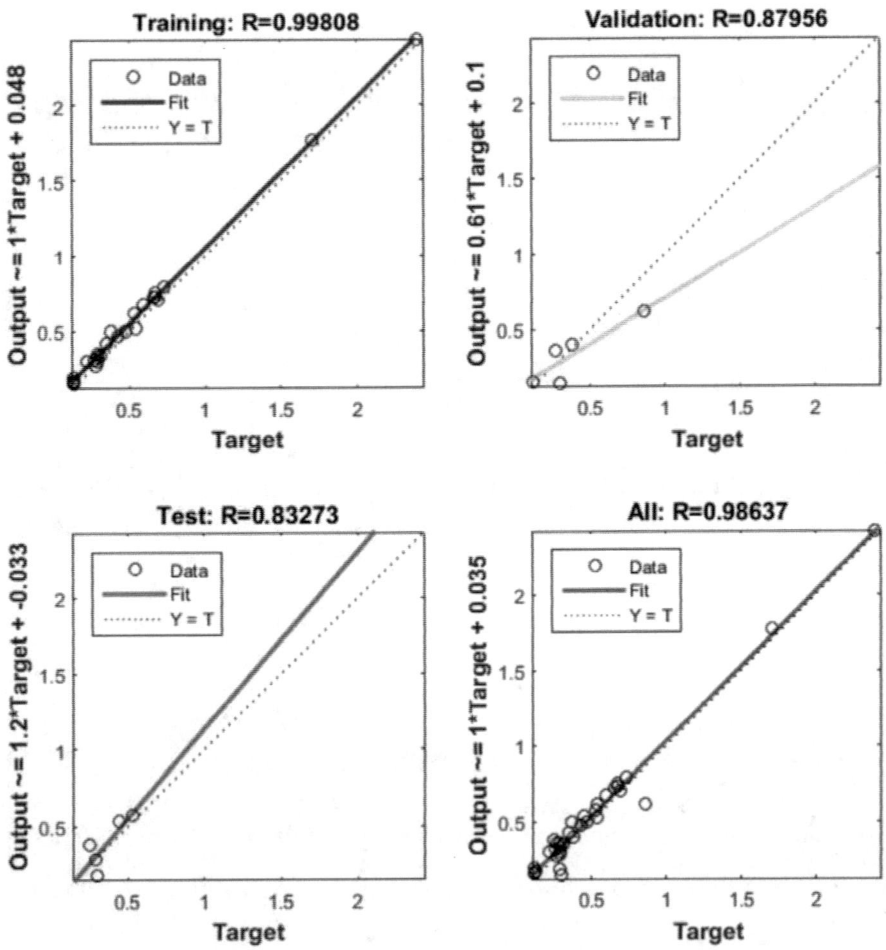

Figure 2. Regression plot for the SLP-ANN model (this comprises of the training, testing, validation, and model overall regression)

Table 1. Comparison results between actual and predicted percentage renewable energy.

Performance Metric	
RMSE	0.28993
MAD	0.18186
MAPE	14.61928

The regression value at model training, testing and validation obtained were 0.99808, 0.83273, and 0.87956 respectively. This values as close to unity signifies that the model proceeds effectively in training, testing, and validation. The overall regression for the model being 0.98637 also shows the accuracy of the model.

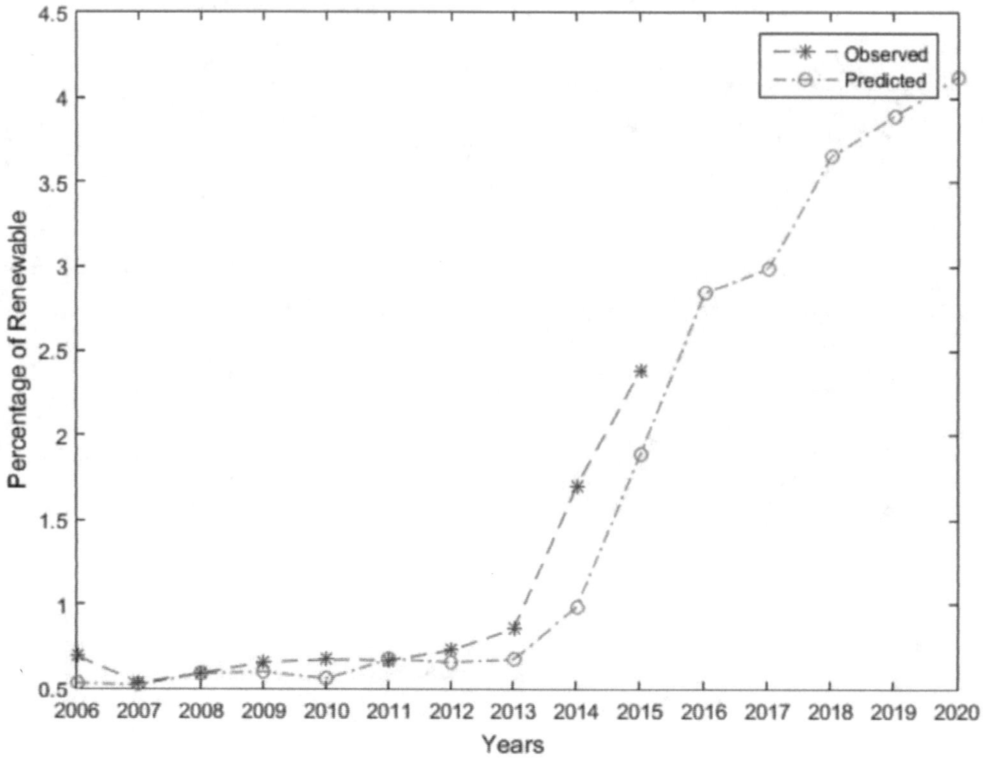

Figure 3. A forecast of percentage renewable energy between 2006 and 2020 in the SSA.

The error deviation from the mean was observed to be less than 0.5 and the MAPE value shows that the forecast is off by 14.62%. This shows reliability of the 5-year forecast. Presently, to the best of the authors knowledge, there have been no documentation on the percentage of renewable energy exempting the hydropower in the sub-Saharan region for 2020 (the last period in Figure 3) which necessitated this study. However, the result of this study was related to the REmap 2030 Africa analysis, a country-based analysis, which investigated the demand profile, resource endowment, and the potential for harvesting accounted for. The analysis showed that the African continent has explored less than 50% of her renewable energy potentials. The Remap 2030 analysis identified hydropower as occupying 14% of the African total electricity supply in 2013 (Figure 4) and it has been the predominant source of generation in the eastern and central regions [1]. However, other sources like photovoltaic (PV), concentrated solar power (CSP), and wind were envisaged to increase significantly within the African energy mix by 2030. They are envisaged to occupy close to half of the electricity generation with 20% for hydropower. Modern renewables are envisaged to share close to 22% of the African energy mix in 2030.

North Africa is envisaged to lead the trail in 2030 as show in Figure 4. The SSA region comprises largely of other African regions exempting the Northern Africa. Small hydro is expected to dominate the mix of the SSA in 2030, which is exempted in this study. Other renewable energy sources occupy a small quota except the solar PV in the Southern Africa region. However, a backcast from 2030 to the 2020, the focus of this study, we can say that the percentage of renewable energy sources exempting the hydropower in 2020 will be close to the predicted in this study. A marginal variation can exist which may not be statistically significant from the result obtained in this study.

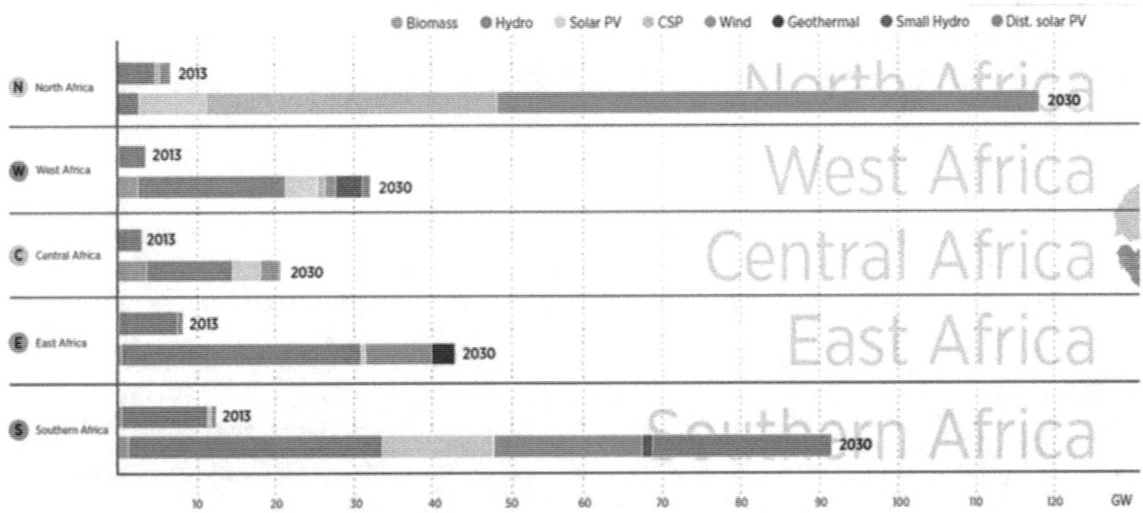

Figure 4. Renewable energy prospect of Africa in 2030. Source [1]

3.3. Achieving Increase in Renewable energy in SSA.

Renewable energy is highly profiled on a global scale. This is not only due to its low-carbon nature but also due to its abundance compared to the non-renewable sources. Increasing its quota in the energy mix of the SSA largely depends on government policies and reforms. It has been established that the region has a high resource availability index [1]. The sustainability of renewable energy largely depends on diversification of sources in the region. Almost 87% of the world population are living without electricity access, the rural areas of SSA and part of Asia significantly contribute to this value [19]. The encouragement of renewable energy micro-grids will significantly reduce this deficit, most especially in SSA where resource availability is not a problem. The industrial sector constitutes one of the largest energy consuming sectors in the SSA. Energy efficiency in this sector through the adoption of modern facilities will reduce energy consumption rate, which in turn reduce pressure on energy generated. From the 2018 World Bank report, there was observed for the first time a fall in the percentage of people without energy access with Eastern Africa being the driving region [19]. This percentage can be further reduced by encouraging modular off-grid systems which rests on effective strategic planning at national level.

4. Conclusions and Recommendations

This study was able to forecast the percentage of RE in the energy mix of the sub-Saharan region from 2016 till 2026. This study does not take into consideration the socio-technical aspect of RE and the dynamics of the renewable energy system, like change in policy, political instability, environmental conflicts in the RE space relative to existing policies, which could result in the closure of production plants. Historical data was only used for the intelligent forecast. It can be established from the result that RE percentage in the energy mix of the SSA will increase in the next decade. The actual percentage of RE in the energy nexus could be more, or less than the forecast in this study. This largely depends on the existing government policies, socio-economic factors, efficient siting of RE facilities for electricity generation, environmental concerns (especially in wind energy harvesting) and many other factors. Further studies with the use of systems dynamics which integrates several obvious and latent variables in the renewable energy system is recommended. This will give a holistic view of the renewable energy integration in the electricity generation.

Acknowledgements

The authors appreciate the University of Johannesburg for travel support for the conference. The authors will also like to appreciate Enel Foundation, a subsidiary of Enel Green Power for granting the first author a privilege of participation in the Open Africa Power 2019 programme, from where the idea of this study emanated.

References

[1] IRENA, "Africa 2030: Roadmap for a renewable energy future," Abu Dhabi, 2015.
[2] A. Aly, S. S. Jensen, and A. B. Pedersen, "Solar power potential of Tanzania: Identifying CSP and PV hot spots through a GIS multicriteria decision making analysis," *Renew. Energy*, vol. 113, pp. 159–175, 2017.
[3] P. Rolffs, D. Ockwell, and R. Byrne, "Beyond technology and finance: pay-as-you-go sustainable energy access and theories of social change," *Environ. Plan. A*, vol. 47, no. 12, pp. 2609–2627, 2015.
[4] P. P. Silva, P. A. Cerqueira, and W. Ogbe, "Determinants of Renewable Energy Growth in Sub-Saharan Africa :," *Energy*, vol. 156, 2018.
[5] S. Batchelor, E. Brown, J. Leary, N. Scott, A. Alsop, and M. Leach, "Solar electric cooking in Africa: Where will the transition happen first?," *Energy Res. Soc. Sci.*, vol. 40, no. January, pp. 257–272, 2018.
[6] C. G. Monyei, A. O. Adewumi, and K. E. H. Jenkins, "Energy (in)justice in off-grid rural electrification policy: South Africa in focus," *Energy Res. Soc. Sci.*, vol. 44, no. May, pp. 152–171, 2018.
[7] J. Corfee-Morlot, P. Parks, J. Ogunleye, and F. Ayeni, "Achieving Clean Energy Access in Sub-Saharan Africa," 2019.
[8] S. Karekezi and W. Kithyoma, "Renewable energy in Africa: Pospects and Limits," 2003.
[9] A. Brew-Hammond, "Energy access in Africa: Challenges ahead," *Energy Policy*, vol. 38, no. 5, pp. 2291–2301, 2010.
[10] I. Dunmade, N. Madushele, P. A. Adedeji, and E. T. Akinlabi, "A streamlined life cycle assessment of a coal-fired power plant: the South African case study," *Environ. Sci. Pollut. Res.*, 2019.
[11] IEA, "World Energy Outlook 2018," Paris, 2018.
[12] B. Pillot, M. Muselli, P. Poggi, and J. B. Dias, "Historical trends in global energy policy and renewable power system issues in Sub-Saharan Africa: The case of solar PV," *Energy Policy*, vol. 127, no. December 2018, pp. 113–124, 2019.
[13] A. G. Dagnachew, P. L. Lucas, A. F. Hof, D. E. Gernaat, H. S. de Boer, and D. P. van Vuuren, "The role of decentralized systems in providing universal electricity access in Sub-Saharan Africa – a model-based approach," *Energy*, vol. 139, pp. 184–195, 2017.
[14] IEA, "World Energy Outlook 2017," 2017.
[15] J. Klugman, "Human Development Report 2011 – Sustainability and Equity: A Better Future for All," 2011.
[16] F. Birol, "World Energy Outlook 2016," 2016.
[17] J. Corfee-Morlot, P. Parks, J. Ogunleye, and F. Ayeni, *Achieving clean energy access in sub-Saharan Africa A case study for the OECD, UN Environment, World Bank project:*

"Financing. 2018.
[18] R. V. Nalule, *Energy Poverty and Access Challenges in Sub-Saharan Africa: The role of regionalism*. Palgrave Macmillan, 2019.
[19] The World Bank, "Tracking SDG7: The Energy Progress Report 2018," 2018.
[20] D. Richard and A. Colin, "Renewable energy gathers steam in South Africa," *Renew. Sustain. Energy Rev.*, vol. 41, pp. 390–401, 2015.
[21] S. Louise, M. Khodani, K. Karin, M. Mbali, and F. Stephen, "Renewable Energy : Facts and Futures The energy future we want," 2017.
[22] J. K. Kiplagat, R. Z. Wang, and T. X. Li, "Renewable energy in Kenya : Resource potential and status of exploitation," *Renew. Sustain. Energy Rev.*, vol. 15, pp. 2960–2973, 2011.
[23] G. Lin and W. Butterfield, "Power Africa in Ethiopia," 2016.
[24] A. Buga *et al.*, "Short-term forecast of generation of electric energy in photovoltaic systems," vol. 81, no. November 2016, pp. 306–312, 2018.
[25] F. Ziel and R. Weron, "Day-ahead electricity price forecasting with high-dimensional structures: Univariate vs. multivariate modeling frameworks," *Energy Econ.*, vol. 70, pp. 396–420, 2018.
[26] B. Akdemir and N. Çetinkaya, "Long-term load forecasting based on adaptive neural fuzzy inference system using real energy data," *Energy Procedia*, vol. 14, pp. 794–799, 2012.
[27] X. Yuan, Q. Tan, X. Lei, Y. Yuan, and X. Wu, "Wind power prediction using hybrid autoregressive fractionally integrated moving average and least square support vector machine," *Energy*, vol. 129, pp. 122–137, 2017.
[28] K. P. Amber, R. Ahmad, M. W. Aslam, A. Kousar, M. Usman, and M. S. Khan, "Intelligent techniques for forecasting electricity consumption of buildings," *Energy*, vol. 157, pp. 886–893, 2018.
[29] K. Lee, D. Booth, and P. Alam, "A comparison of supervised and unsupervised neural networks in predicting bankruptcy of Korean firms," *Expert Syst. Appl.*, vol. 29, no. 1, pp. 1–16, 2005.
[30] S. Silva, L. Vanneschi, A. I. R. Cabral, and M. J. Vasconcelos, "A semi-supervised Genetic Programming method for dealing with noisy labels and hidden overfitting," *Swarm Evol. Comput.*, vol. 39, no. September, pp. 323–338, 2018.
[31] M. A. Waris, A. Iosifidis, and M. Gabbouj, "CNN-based edge filtering for object proposals," *Neurocomputing*, vol. 266, pp. 631–640, 2017.
[32] H. K. Ghritlahre and R. K. Prasad, "Application of ANN technique to predict the performance of solar collector systems - A review," *Renew. Sustain. Energy Rev.*, vol. 84, no. December 2017, pp. 75–88, 2018.
[33] T. A. Tutunji, "Parametric system identification using neural networks," *Appl. Soft Comput. J.*, vol. 47, pp. 251–261, 2016.
[34] A. Khashei-Siuki and M. Sarbazi, "Evaluation of ANFIS, ANN, and geostatistical models to spatial distribution of groundwater quality (case study: Mashhad plain in Iran)," *Arab. J. Geosci.*, vol. 8, no. 2, pp. 903–912, 2015.
[35] P. A. Adedeji, S. Akinlabi, O. Ajayi, and N. Madushele, "Non-linear autoregressive neural network (NARNET) with SSA filtering for a university enegy consumption forecast," in *16th Global Conference on Sustainable Manufacturing- Sustainable Manufacturing for Global Circular Economy*, 2019, pp. 176–183.
[36] S. A. Kalogirou, "Artificial neural networks in energy," *Int. J. Low Carbon Technol.*, no. March, pp. 201–216, 2015.
[37] Ministry of Lands Mines and Energy, "Renewable Energy and Energy Efficiency Policy and Action Plan," 2016.

A Mini Review on the Impact of Sewage Disposal on Environment and Ecosystem

G U Fayomi [1, 2*], S E Mini[2], O S I Fayomi[3, 4], T. Owodolu[3], A.A.Ayoola[5], O. Wusu[6]

[1]Strategic Business Services, Covenant University, Ota, Nigeria
[2]Department of Geography, University of South Africa, Johannesburg, South Africa
[3]Department of Mechanical Engineering, Covenant University, P.M.B 1023, Ota, Ogun State, Nigeria
[4]Department of Chemical, Metallurgical and Materials Engineering, Tshwane University of Technology, Pretoria, South Africa
[5]Department of Chemical Engineering, Covenant University, P.M.B 1023, Ota, Ogun State, Nigeria
[6]Department of Sociology and Centre for Environmental Studies and Sustainable Development, Lagos State University, Ojo, Nigeria

*Corresponding author: fayomiuche@gmail.com; ojo.fayomi@covenantuniversity.edu.ng +2348033752329

Abstract. Human activities contribute immensely to the production of wastewater which emanates from residences, industries and agricultural practices that pollutes the environment and water bodies. Up to 80-90% of the wastewater produced in developing countries is disposed of into surface and groundwater which is a major cause of environmental pollution that threatens human health. According to UN report, over 80% of the wastewater produced in the world and over 95% in some least developed countries is released without being treated into the environment. This paper reviews the methods of sewage disposal, the impact on the environment, aquatic contaminant and the human health. Emphasis is placed on the impact of disposal of various contaminants in aquifers and water bodies which could make water unsafe for drinking and to perform other domestic and recreational activities

Keywords: Eco system, Environment, housing, Sustainable

1. Introduction

The environmental impact of sewage disposal are numerous and constitutes major problems faced in our ecology resulting from human activities such as waste from households, industries and agriculture leading to pollution [1] Sewage is wastewater which consists of solid and liquid layer collected and conveyed through sewers, treated or untreated and disposed of into large bodies of water [2]

According to the United Nations (UN) report, it is estimated that there would be an increase of 1.1 billion in population in the next 15 years and over 6.3 billion people will live in urban centers by 2050 [3-5]. Urban settlements are expanding at a faster pace with a large number of people migrating from rural areas to cities which is as result of access to basic amenities and improved standard of living [6]. In most developing countries, it is estimated that 90% of sewage is disposed of into watercourses like rivers, lakes, streams and ponds without being treated thereby resulting in pollution of the environment [7]. Large amount of these wastewaters seeps into the soil, polluting aquifers and groundwater which is as a result of leakage in on-site sanitation facilities and lack of sewerage pipes for disposing wastewater in low and middle income countries [8].

Disposal of untreated waste into the environment can cause oxygen depletion as a result of microbial decomposition of organic matter, destructing the marine ecosystem, and also, affect the quality of water, making it unsafe for drinking and can lead to diseases such as dysentery, cholera and typhoid [9-10]. Wastewater in most countries especially in developing countries are not being treated before it is disposed of into large bodies of water and this can pose significant threat to life below water, life on land, poor access to clean water and sanitation [11-12]. However, sewage which is a wastewater could contaminate drinking water in different ways which could be from septic tanks by infiltrating into the ground and polluting groundwater, partially treated wastewater, sludge, and leakages in sewer lines [13]

Hence, this paper focus on the review concepts of Sewage Disposal, the impact of untreated sewage on human health, environment and the aquatic ecosystem and the measures to be taken to ensure a safe and hygienic environment.

2. Critical Concept of Sewage and Environment

Increase in the volume of sewage produced is as a result of growth in population and industrialization imparting pressure on land and water resources [18]. Sewage which is also referred to as wastewater containing water and solid waste suspension is disposed of from households, buildings and industries and conveyed in large pipes called sewers [19, 20]. This wastewater is mostly left untreated and it either infiltrates into the ground polluting aquifers or is disposed of into watercourses. Disposing of this wastewater is dependent on the region and availability of natural water in a particular area. [1]. One of the major challenges we are faced with in this twenty first century is the lack of quality water and improved sanitation [12]. Over one third of the freshwater available is used for agricultural, industrial and domestic purposes and these activities also result in the contamination of natural water. Pollution of natural water by chemical pollutants is a major concern today [16, 24]. Chemical pollutants from Emerging contaminants, pharmaceuticals, fertilizers including pathogens are washed down by storm water into lakes, rivers and oceans that are meant to be for recreation and drinking [21-23].

A study by [20] carried out in Blantyre city shows that poor design of on-site sanitation systems and inadequate means of collecting and disposing of solid wastes are major challenges of sewage disposal. Improper disposal and collection of solid waste in this particular area is as a result of insufficient vans by the Blantyre City assembly which could not cater for more than half of the population. Also, most pit latrines, septic tanks and sewer lines are designed poorly and are not properly maintained which has posed threat to water resources in this particular area [25, 27]. [11] affirmed that efficient refuse and solid waste management has been a major problem in Port-Harcourt, a city in Nigeria over the past years. Dumping of various wastes which are produced from the industries, household and commercial waste such as food wastes, textiles, plastics, and scrap metals is one of the causes of traffic congestion in the city and has made water unsafe for drinking due to infiltration. Moreover, increase in population would lead to an increase in the production of wastes if the situation is not properly attended to [14]

Malaysia is faced with the problem of Solid Waste Management. Rapid rise in her economic growth, attitude of the populace, had been highlighted as the bottleneck of solid waste management [15]. Increase in population in this area is as a result of migration from slums and rural areas, affluence, which are major factors that needs to be addressed in order to have an effective way of managing and disposing wastes. According to [1], chemical pollution has an impact on the quality of water. The need for water cannot be overemphasized in ensuring food security, preserving the environment and in alleviating poverty. However, when access to clean and safe drinking water is not guaranteed, there is tendency for outbreaks of waterborne and epidemic diseases such as diarrhea, skin rash, cholera [24]

Various methods of sewage disposal are adopted in different cities in Nigeria. These methods include the use of septic tanks, flow of sewage into storm water drainages, open ground, water ponds and streams [19]. In Katsina metropolis where decentralized system is practised, all of these methods are adopted but the use septic of tanks causes ailment on human health resulting from unpleasant smell from cleared drainages which pollutes the air people breathe in and make them uncomfortable in their jurisdiction; accumulation of stagnant water serving as home for mosquitoes to breed [10]

IOP Conf. Series: Earth and Environmental Science **331** (2019) 012040 doi:10.1088/1755-1315/331/1/012040

3. Wastewater and Contaminant

The use of antibiotics is an integral part of human lives. It comprises of natural products which has been vastly used as medicine for human and veterinary and helps in changing the model of the way we live, prominent in treating and preventing infectious diseases [9]. In many cases, when these drugs are not used and get expired, justice is not done when disposing them. They are disposed and flushed in toilets and found in landfills and wastewater plants (WWTPs) which when disposed into large bodies of water, contaminate the aquatic environment.

The presence of antibiotics in the environment affect natural microbial environment by promoting the development of antibiotics resistant pathogen posing severe threat on human and veterinary health [25]. The event of these antibiotics in the ecosystem distorts the structure and functioning of microbial communities in soil and water ecosystems. Pharmaceuticals are known to be emerging contaminants which has the ability to induce physiological effects in human at low doses [21]. They are disposed of into large bodies of water through municipal wastewater and decomposition of pharmaceuticals in wastewater forms metabolites rather than resulting in their full mineralization. The major concern of this emerging contaminant in wastewater effluent is the exotoxicological effects on human health due to the presence of these compounds in drinking water. These wastewater containing pharmaceuticals and their metabolites can seep into the aquifers making water unsafe for human [15].

In areas where there is no access to water for agriculture, wastewater is mostly used as replacement to irrigate farmlands [14]. This wastewater contains great amount of nutrients which helps in the growth of crops and also contain toxic pollutants which are capable of creating havoc during production of agricultural produce [10]. Due to the non-biodegradable nature of heavy metals, their presence in the soil and crop pose serious environmental problem. Abundance of these heavy metals in soil from wastewater for irrigation may impact the food chain [23]. Moreover, heavy metals like copper and Zinc are nutrients essential to the human body but can be toxic if consumed in high concentration [22] Also intake of heavy metals like Cd and Pb has been shown to have carcinogenic effects [17].

3.1 Sewage Disposal Methods and Its Implication

The sewage disposal method adopted in a residence or community determines their health status [8]. In a developing country like Nigeria, sewage produced in rural and urban areas are majorly discharged into pit latrines and into septic tanks, soak ways and cesspools in towns and municipalities. The table 1 shows the sewage disposal methods used in Nigeria and its negative implication.

Table 1: Method of Sewage Disposal and its Implication

Method of Sewage Disposal	**Negative Implication**
Septic tank	Pollution of soil and water environment [19]
Sewage flow into Stormwater	Air pollution, Risk of causing flooding [20]
Sewage into ponds	Stagnant water which is breeding area for vector diseases [21]

Direct dumping of untreated waste in rivers	Presence of toxic substances in the food chain, via the plants and animals that feed on it directly or indirectly [9]

In urban areas, the pattern of sewage disposal involves the use of septic tanks, pit latrines, sewer lines, soil pit and defecation in bushes. Figure 1 shows the pattern of sewage disposal in urban areas in Nigeria.

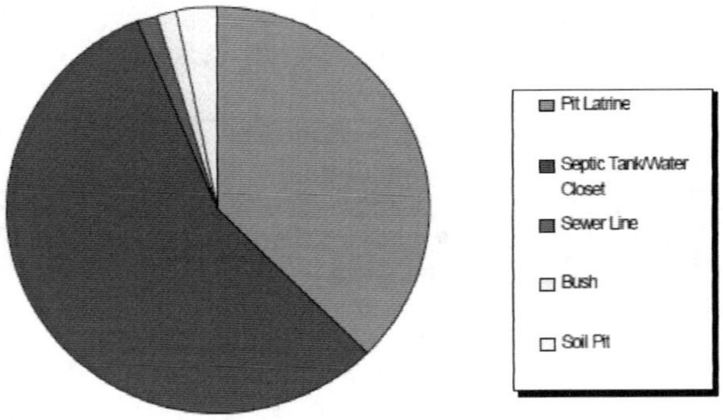

Figure 1: Sewage Disposal Pattern in Urban Areas of Nigeria [28]

The diagram in figure 1 shows the different methods of disposing waste in Nigeria having the septic tanks to be the most used. There is also abundance of pit latrines with sewer lines, though the primary concern of this study, are minimal. However, the traditional methods such as soil pit and defecating in the bush is still in use in some part of the country.

Different technologies are used in treating sewage which includes the Mechanical, Biological treatment technology. In the Biological treatment technology, microorganisms are made to feed on the organic material present in the sewage and decomposing them. This aim is to decrease the Biological Oxygen Demand (BOD) level [7, 19] The Mechanical treatment technology combines the biological, physical and chemical processes. Activated Sludge Process (ASP), Extended Aeration (EA), Moving Bed Biofilm Reactor (MBBR), Sequential Batch Reactor (SBR), Vermifiltration Technology are all technologies within the Mechanical Treatment technology for treating wastewater [27, 28]

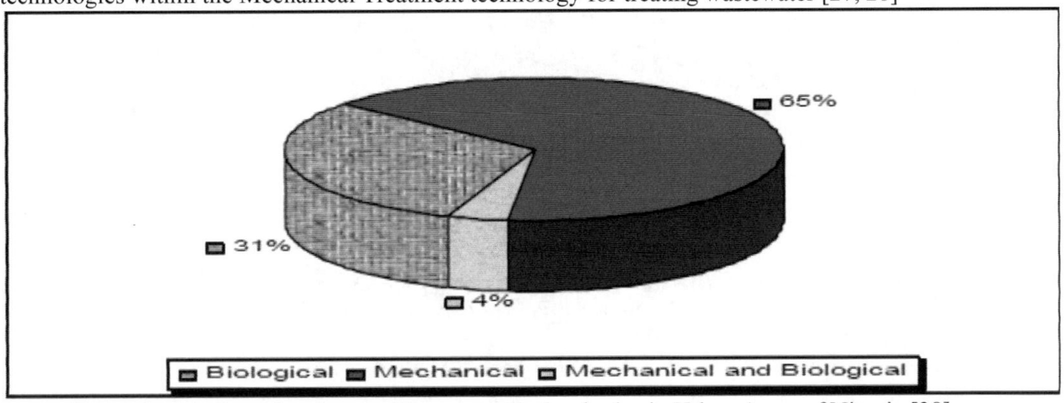

Figure 2: Relative Abundance of Sewerage Technologies in Urban Areas of Nigeria [28]

Figure 2 shows the various types of sewer system treatment used in Nigeria. It has been found that about 65% of most sewer system technologies involve the use of Mechanical technology for treatment, and are mainly found in industrial areas. Biological treatment method follows this (around

31 percent). The method of treatment using biological and mechanical methods in the country is the least (approximately 4%) means of sewerage treatment Nigeria [28]

Ensuring a safe, clean and hygiene environment is eminent to any sector in Nigeria. In order to achieve this, the following must be done.

- In order to prevent pollution of surface and ground water, it is important to treat sewage in order to alleviate the widespread of waterborne diseases such as cholera, dysentery caused by pathogenic organisms.
- It is everyone's priority to have a proper way of disposing waste so that flies, animals and human would not have any contact with them.
- Septic tanks should be properly installed and maintained so as to avoid leakages that cause offensive odour to the environment.
- Laws and regulations should be formulated and enforced on every citizen of the country on proper disposal of waste and also sensitize them on the effect of improper disposal on their health, the environment and the aquatic ecosystem.

4 CONCLUSION

The report on epidemic and waterborne disease has been a source of worry to every right thinking citizen of Nigeria. The review presented the major causes of sewage disposal and its implication on the ecosystem and mismanagement of sewage and sewage facilities can be said to be the reason for dilemma in our health sector. Adequate and proper attention should be given to the way we dispose our waste in various locations and critical review has been undertaken to access the way forward.

Acknowledgement

The author acknowledged the financial support by Covenant University CUCRED for the publication of the manuscript.

Reference

[1] Schwarzenbach, R. P., Egli, T., Hofstetter, T. B., von Gunten, U., & Wehrli, B. (2010). Global Water Pollution and Human Health. *Annual Review of Environment and Resources*, 35, 109–136.

[2] Zhang, Q. H., Yang, W. N., Ngo, H. H., Guo, W. S., Jin, P. K., Dzakpasu, M., Ao, D. (2016). Current status of urban wastewater treatment plants in China. *Environment International*, *92-93*, 11–22.

[3] Nations, U., & United Nations. (2013). *Realizing the right to development and a new development agenda*.

[4] Massoud, M. A., Al-Abady, A., Jurdi, M., & Nuwayhid, I. (2010). The challenges of sustainable access to safe drinking water in rural areas of developing countries: case of Zawtar El-Charkieh, Southern Lebanon. *Journal of Environmental Health*, *72*(10), 24–30.

[5] Lin, C., Eshel, G., Roehl, K. E., Negev, I., Greenwald, D., Shachar, Y., & Banin, A. (2006). Studies of P accumulation in soil/sediment profiles used for large-scale wastewater reclamation. *Soil Use and Management*, 22,. 143–150.

[6] Tabraiz, S., Nasreen, S., Qureshi, L. A., Zeeshan, M., Ahmad, I., Ahmed, S., & Hassan, Z. (2016). Sewage land disposal and unpaved drains: threat to groundwater quality. *Desalination and Water Treatment*, 57, 20464–20469.

[7] Ahmed, M. B., Zhou, J. L., Ngo, H. H., Guo, W., Thomaidis, N. S., & Xu, J. (2017). Progress in the biological and chemical treatment technologies for emerging contaminant removal from wastewater: A critical review. Journal of Hazardous Materials, 323, 274–298

[8] Ajibola, O. O. E. (2019). Groundwater Quality Assessment in a University Sub-Urban Community in Delta State, Nigeria: Part A-Physicochemical Analysis. *Journal of Applied Sciences and Environmental Management*, 23, 151-159.

[9] Ajibuah B. J, Terdoo F. (2013). Pattern and Disposal Methods of Municipal Waste Generation in Kaduna Metropolis of Kaduna State, Nigeria. *International Journal of Education and Research,* 1, 1-14

[10] Alghobar, M. A., & Suresha, S. (2015). Evaluation of Nutrients and Trace Metals and Their Enrichment Factors in Soil and Sugarcane Crop Irrigated with Wastewater. *Journal of Geoscience and Environment Protection*, 03, 46–56.

[11] Ayotamuno, J. M., & Gobo, A. E. (2004). Municipal solid waste management in Port Harcourt, Nigeria. *Management of Environmental Quality: An International Journal*, Vol. 15, pp. 389–398.

[12] Connor, & Richard. (2015). *The United Nations world water development report 2015: water for a sustainable world.* UNESCO Publishing.

[13] Gothwal, R., & Shashidhar, T. (2015). Antibiotic Pollution in the Environment: A Review. *CLEAN - Soil, Air, Water*, 43, 479–489.

[14] Hajjami, K., Ennaji, M. M., Fouad, S., & Cohen, N. (2013). Détection, numération des œufs de parasites et évaluation de leur viabilité dans les boues d'épuration de la station d'épuration de Settat. *European Journal of Water Quality*, 44, 55–66.

[15] Heberer, T., Massmann, G., Fanck, B., Taute, T., & Dünnbier, U. (2008). Behaviour and redox sensitivity of antimicrobial residues during bank filtration. *Chemosphere*, *73*(4), 451–460.

[16] Huntington, T. G. (2006). Evidence for intensification of the global water cycle: Review and synthesis. *Journal of Hydrology*, 319, 83–95.

[17] Jaishankar, M., Tseten, T., Anbalagan, N., Mathew, B. B., & Beeregowda, K. N. (2014). Toxicity, mechanism and health effects of some heavy metals. *Interdisciplinary Toxicology*, *7*(2), 60–72.

[18] Murtaza, G., Ghafoor, A., Qadir, M., Owens, G., Aziz, M. A., Zia, M. H., & SAIFULLAH. (2010). Disposal and Use of Sewage on Agricultural Lands in Pakistan: A Review. *Pedosphere*, 20, 23–34.

[19] Nowak, R., & Imperowicz, A. (2016). Liquid waste from septic tanks as a source of microbiological pollution of groundwater. *Inżynieria Ekologiczna*, 10, 60–67.

[20] Palamuleni, L. G. (2002). Effect of sanitation facilities, domestic solid waste disposal and hygiene practices on water quality in Malawi's urban poor areas: a case study of South Lunzu Township in the city of Blantyre. *Physics and Chemistry of the Earth, Parts A/B/C*, 27, 845–850.

[21] Petrie, B., Barden, R., & Kasprzyk-Hordern, B. (2015). A review on emerging contaminants in wastewaters and the environment: current knowledge, understudied areas and recommendations for future monitoring. *Water Research*, *72*, 3–27.

[22] Radanielson, A. M., Gaydon, D. S., Rahman Khan, M. M., Chaki, A. K., Rahman, M. A., Angeles, O.Ismail, A. (2018). Varietal improvement options for higher rice productivity in salt affected areas using crop modelling. *Field Crops Research*, *229*, 27–36.

[23] Radwan, M. A., & Salama, A. K. (2006). Market basket survey for some heavy metals in Egyptian fruits and vegetables. *Food and Chemical Toxicology*, 44, 1273–1278.

[24] Singh, K. P., Mohan, D., Sinha, S., & Dalwani, R. (2004). Impact assessment of treated/untreated wastewater toxicants discharged by sewage treatment plan ts on health, agricultural, and environmental quality in the wastewater disposal area. *Chemosphere*, *55*(2), 227–255.

[25] Sorensen, J. P. R., Lapworth, D. J., Nkhuwa, D. C. W., Stuart, M. E., Gooddy, D. C., Bell, R. A., … Pedley, S. (2015). Emerging contaminants in urban groundwater sources in Africa. *Water Research*, *72*, 51–63.

[26] Fayomi G.U, Wusu O, Mini S.E, Fayomi O.S.I, Kilanko O. 2018. Data analysis on the level of exposure to pollutions in industrial zone: A case study of Ewekoro and Ota Township Data in Brief, 19, 859-864

[27] Fayomi, O.S.I. Olukanni, D.O. Fayomi G.U. And Joseph, O.O. 2017. In situ assessment of degradable carbon effusion for industrial waste water treatment, Cogent Engineering 2017, 4: 1291151 http://dx.doi.org/10.1080/23311916.2017.1291151

[28] Adesogun S. 2013. Sewage Technology in Nigeria: A Pragmatic Approach Science. *Journal of Environmental Engineering Research* 1-10

The impact of basic forensic accounting skills on financial reporting credibility among listed firms in Nigeria.

N Chukwu[1], T O Asaolu[2], O R Uwuigbe[1], U Uwuigbe[1], O E Umukoro[1] ., L Nassar[2] and O Alabi[1]

[1]Department of Accounting, Covenant University, Ota, Ogun State, Nigeria.
[2]Department of Accounting, Obafemi Awolowo University, Ile-Ife, Nigeria.

Abstract: The paper aims at evaluating the impact of basic forensic accounting skills viz: communication skills, technical & analytical skills, accounting & auditing skills and psycho-social skills on the financial reporting credibility of listed organisations in Nigeria. The scope of the study is composed of firms quoted on the Nigerian Stock Exchange as at January, 2018 while the study population is the users and auditors of the financial reports produced by these entities. The users of financial reports are represented by investment advisers which a segment of capital market operators while the auditors are represented by the Big 4 Audit Firms operating in Nigeria as at January, 2018. The research design employed is the survey approach by which primary data was collected from the sample respondents described above while the multivariate analysis was conducted to highlight the impact of the selected independent variables on financial reporting. It was found that, with the exception of communication skills which exhibited a negative but significant impact, all the other basic forensic accounting skills exert positive and significant impact on the methods by which an organization reports her financials. It is prescribed that aspiring forensic accountants should be adequately trained to be able to practice the profession satisfactorily.

Keywords: Auditing skills, basic forensic, financial reporting credibility, multivariate psycho-social skills.

1. Introduction

It is essential for an organization to present its score card to its owners and other stakeholders at the end of a business period. Apart from the need to measure the ability, effectiveness and efficiency of the enterprise managers to execute the objectives entrusted to them, the information provided by the financial report is also required by other individuals and entities for critical decision-making purposes. It is important, therefore, for the financial report to be factual and represent the true financial health during a given period under consideration [9] [10].

The accuracy of the financial report is important not only for internal but also for external purposes [25], [26]. For internal purposes, there is need for corporate stewardship to identify good projects which will ultimately increase investment efficiency and share values [8]. For outer purposes, precise announcing will likewise help capital suppliers to survey the company's venture open doors just as screen administrative activities [7]. Additionally, accurate financial reporting also assists corporate lenders to assess the quality of the credit while capital providers will better compare the returns on their investment as a basis for either additional investment or divestment. Unfortunately, however, over the past few decades, the issue of accuracy of financial reporting has become problematic.

This is apparent from the public feature getting instances of financial summary misrepresentation which have pulled in the consideration of universal network and shocked the capital markets. These frauds have constituted a challenge and made a negative impact not only on investors' confidence but have also dealt a devastating effect on corporate reputation while also jeopardizing its going-concern profile [19]. For instance, [6] qualities the breakdown of Enron, WorldCom, Tyco, Adelphia, to corporate money related revealing extortion where over $460 billion was said to have been lost. In Nigeria, Cadbury Nig Plc whose books were creatively controlled by the board, Cadbury was credited to have lost N15 billion. On account of nine banks in Nigeria, around one trillion naira was additionally said to have been lost through various sorts of budgetary control and ill-advised detailing [13].

The level of confidence reposed on financial report is determined by the quality and reliability of such reports. However, there are two factors that underpin the credibility of financial reports. As observed by [12], the first is the quality of the report prepared by the management of the enterprise while the second is the care and skill displayed by the auditors and other gatekeepers in their statutory duties. The former is the obligation of management while the latter is the duty of the gatekeepers, especially the statutory auditor.

The whole essence of forensic accounting practice is to be able to improve on these two conditions. This is in line with the observation of [12], "the validity of financial summaries arranged by management of constrained risk organizations and reviewed by outer examiners remain the essential methods for illuminating the investors and different partners about the money related execution, advancement and position of a firm". Hence, financial statement credibility depends not only on management readiness to display transparency but also on the independence, objectivity, efficiency and effectiveness of the statutory auditors in applying their training and experience to discharge their statutory responsibilities when financial statements are submitted to them for audit. It stands to reason that the objective of accurate financial reporting will not be attained unless each of the two parties perform their duties satisfactorily [13]. In other words, while forensic accounting techniques can be deployed to prevent fraudulent practices in the preparation of the accounts, statutory auditors are also at liberty to avail themselves of forensic accounting skills to ensure a better quality audit which will ultimately improve the confidence level inherent in the attestation function. It is a complementary arrangement and the desired result will not be achieved unless each party plays its own role satisfactorily. This is essentially the rationale for the confidence-building. "genuine and reasonable" see frequently communicated by the outside reviewer which is intended to energize clients of budget reports who wish to settle on a speculation or different choices. In any case, by the developing spate of corporate money related revealing outrages referenced before, the significance of the believability building provision as contained in this 'genuine and reasonable view' sentiment is being called to address lately. The issue of credibility is, therefore, a fundamental challenge currently facing the accounting profession world-wide.

What is fraudulent financial reporting and why does it occur? According to [4], financial summary extortion is a cognizant twisting of the real monetary position of an endeavor. This can come through misquoting or discarding sums or neglecting to unveil what ought to be uncovered, with the goal to delude clients who usually depend on the substance of the financial report for motivations to reach a beneficial conclusion. [4] adduced reasons for this fraudulent financial reporting. One of the reasons is inadequate accounting controls. Another one is the fact that enterprise executives are often under intense pressure to meet apparently unrealistic performance targets and expectations, either by way of earnings or profits. This is the motivation behind why most monetary detailing fakes are focused to demonstrate a superior budgetary viewpoint for the organization than depicting the real situation. In addition, a significant number of enterprise managers also enter into stringent performance covenants with their bankers and sundry capital providers which they would rather not default. Operating executives have also been found to have 'perks', incentives and bonuses that are dependent on the financial performance of the enterprise [25]. In specific industries, commissions and bonuses are also paid to sales staff if their performance is adjudged to be satisfactory. These are the issues that drive the temptation for these personnel to submit inflated sales figures to boost such commissions.

It is due to the consideration of the above stated issues that this research intends to evaluate the impact basic forensic accounting skills has on financial reporting credibility of quoted companies in Nigeria. The specific skills that have been identified for this study are viz: communication skills, accounting and auditing skills, technical & analytical skills and psycho-social skills.

2. Literature Review

While the Oxford Advanced Learners Lexicon characterizes the word 'forensic' as having a place with, utilized in, or appropriate to a court of judicature or for open exchange or discussion, [15] contended that forensic accounting is the procedure of deciphering, condensing and introducing complex money related issues unmistakably, concisely and genuinely in a courtroom as a specialist witness should the need emerge. Meanwhile, according to [20], from organisation, through government offices, regulatory authorities, and the judicial system, evidence abound to the effect that a high level of expertise is necessary to undertake the analysis of complicated financial transactions and events. The consensus here, therefore, is that forensic accounting has been tossed into the bleeding edge of the campaign against money related trickery.

There are strong pointers to the fact that one in each three workers in corporate organizations gets involved in one type of fraud or another. This can be in form of book-keeping fraud, negative scheming, deception or even outright embezzlement which the auditors, whether internal or external, is not in a position to detect. Unfortunately, such occurrences are inimical to accurate financial reporting because they distort the true picture of the enterprise. Forensic accounting practice aims at forestalling this development not only by assisting the external auditor to detect them but also by enabling the internal auditor to prevent them [16].

While commenting on Financial Statement Misrepresentation, The Association of Certified Fraud Examiners stated this definition: "The conscious deception of the money related condition and consequences of activity of an undertaking achieved through the purposeful error or exclusion of sums or exposures in the budget reports to hoodwink financial summary clients". [22] have also identified the target of financial report extortion: "Expanding the market estimation of the business to fulfill the over the top desires of the board, putting forth money related expressions predictable with earlier spending plans and getting unjustifiable profit by anticipating dishonestly the estimation of the business".

The Webster's Dictionary, likewise characterizes the word 'Forensic' as: "Having a place with, utilized in or appropriate to court, of judicature or to open dialogs, banter and at last question goals; it is additionally characterized as a bookkeeping investigation that is reasonable to the court which will frame the reason for discourse, banter and eventually debate goals". According to [10], forensic accounting is the coordination of bookkeeping, evaluating and insightful aptitudes. Forensic accounting has likewise been characterized as the assurance and foundation of truth in help of legitimate case just as the utilization of legal systems to distinguish and research a wrongdoing and to uncover the personality of hoodlums and guilty parties [14].

Meanwhile, [5] likewise characterizes forensic activity as the action that comprises of get-together, handling, breaking down, confirming of and providing details regarding information to get realities or proof in a palatable setting in the region of legitimate budgetary debate. [23] also maintains that since the approach in a statutory audit engagement depends on a standards and rules issued by residential administrative bodies and organizations, the assignment is designed to be less controversial than that of a forensic investigation. In his own definition [14] reasoned that a scientific evaluator is part cop, part legal advisor, part bookkeeper, and part analyst while [7] noticed that a legal scientific inspector ought to have a very much created proficient distrust ('sniffer' frame of mind and insightful personality), expository, coherent and individual respectability, aptitude in interior controls, and intuition in talking systems.

The model below by [24] clearly demonstrates the interconnectedness between forensic accounting and other specialized fields such as criminology, accounting, investigative auditing and litigation services, all of which are relevant skill sets for the practice of forensic accounting. Figure 2.3.1 also establishes the fact that forensic accountants also need a broad-based working knowledge of other professions in the business world to enable him function satisfactorily. These various skills and professional disciplines also impact positively on the practice of forensic accounting.

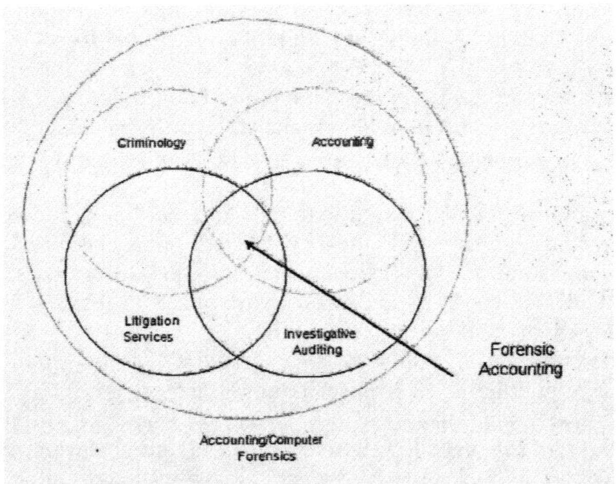

Figure 1. Relationship between Forensic Accounting and Other Disciplines
Source: [23].

In terms of specific tasks performed, there are distinct features between forensic accounting and traditional auditing function. Although they are inherently interrelated, Table 1 establishes the essential differences between forensic accounting and statutory audit activity. The table is seen on the next page and it highlights differences between both concepts.

Table 1. Comparison between forensic accounting and statutory auditing.

Items for analysis	Forensic Accounting	Auditing
Why, when and where the services take place	This proves fraud in the business in an apparent risk prone environment	Keeps on affirming the cutting edge of a business and consent to a proficient market hypothesis.
Extent of the activity	Reveal accounting related figures, ratios and financial information to support legal and administrative decisions	Opine on the accounting explanations of business substances considering all criteria utilized in the preparation.
Tasks performed	Itemized arranging of undertakings aimed at documenting deterministic and calculative analysis.	Sampled and/or probabilistic techniques to fill in as a base of finishing up the financial summary.
Periodicity	Whenever essential and especially as indicated by the stipulated by judge or customer	Covering the monetary year to substantiate the exercises of the accounting time frame.
Reporting	Analyzed or master reports	Fiscal reports, the board letters or auditors' report.

Historically, although [4] was the first to write about: "Forensic Accounting: Its Place in Today's Economy'', [4] have archived the way that legal bookkeepers have been around for about 200 years.
They contend that the most punctual reference to the subject was found in 1824 in a bookkeeper's promoting round in Glasgow, Scotland. These unique bookkeepers gave declaration in court and in assertion procedures. Nonetheless, enthusiasm for the field of measurable bookkeeping did not spread through the United States and England until from the get-go in the twentieth century.

One of the first users of the foundations to utilize the administrations of such insightful specialists was the Internal Revenue Service. This was the place the main prominent scientific bookkeeper, Frank J. Wilson initiated the battle to convict Alphonse "Scarface" Capone, the celebrated mobster who was gotten on a tax avoidance conspire in 1931[1]. In a different record by [4], Sherlock Holmes, the fake character made by Sir Arthur Conan Doyle is additionally viewed as the pioneer of scientific

IOP Conf. Series: Earth and Environmental Science **331** (2019) 012041 doi:10.1088/1755-1315/331/1/012041

bookkeeping. However, it is well documented that the FBI later decided to utilize them and went ahead to employ about 500 of them during World War II. Then, the scientific bookkeeping calling kept on developing amid the last 50% of the century, as GAAP and expense laws ended up across the board and compulsory. [3] likewise takes note of that the expert history of legal bookkeeping really initiated in 1992 when the American College of Forensic Examiners was set up while the American Board of Forensic Accountants began in 1997.

Inadequacies of traditional accounting birthed forensic accounting and this has helped to close the gap of dissuading, identifying and exploring cheats in the financial reporting function in all its ramifications [14]. This implies that, on consultancy basis, the forensic accountant can be engaged whether by management, users of financial services or even regulators, to clear whatever doubts exist concerning financial reporting integrity and collect evidence of wrong-doing accordingly. In words, their scope is not limited to the financial statement which is only the end product of financial processes; they will also assess the authenticity or otherwise of the transactions themselves, the people as well as the environment. All of these will help to establish the existence or otherwise of financial fraud, the people involved as well as the particular component of the business that is at risk.

In carrying out his function, a forensic accountant could be confronted with the situation in which he has to reconstruct incomplete financial in order to settle miscellaneous challenges such as insurance claims, over-valuation of inventories, money laundering and other sources of potential disputes [21]. Another difference between traditional auditing and forensic accounting has been alluded to by [4]. According to them, "the forensic accountant is a bloodhound (which the auditor is not) of accounting. These hunting dogs sniff out misrepresentation and criminal exchanges in bank, corporate substance or from some other association's monetary records. They dog for the indisputable confirmations. Outer reviewers discover the purposeful errors just however the scientific bookkeepers discover the mis-articulations intentionally. Outside evaluators take a gander at the numbers however the scientific inspectors look past the numbers"

2.1 Basic Forensic Accounting Skills

The thrust of this paper has to do with the requisite skills required by the prospective forensic accountant. As earlier mentioned, there are a myriad of miscellaneous skills required by the forensic accountant in order to distinguish himself from the pack of traditional accountants and auditors. According to [11], successful forensic accountants must have analytic abilities, strong written and verbal communication skills, a creative mindset, and business acumen. They must also be able to interview and obtain information from otherwise uncooperative people. He must also possess a large dose of internal professional watchfulness.

For [24], since the criminological bookkeepers are exceptionally put to probably reveal money related mis-directions, he/she needs an inside and out information of fiscal reports and the capacity to break down and censure them. He demands that scientific bookkeepers ought to most likely sufficiently comprehend the inward control and corporate administration frameworks of associations and have the capacity to survey their inalienable dangers and inadequacies. According to [24], the knowledge of sociology and psychology are also indispensable for the forensic accountant since he needs a working knowledge of impulsive criminal behaviour as well as the motivations for criminal and financial deception.

The contribution of [13] is in the area of possess research and information gathering skills, as well as knowledge of technological advancements, which he insists the forensic accountant also needs to enable him accumulate evidence from different sources. [13] does not discountenance the need for the forensic accountant to be up to speed with cybercrimes and a working knowledge of relevant standards, regulations and statutes for specific local jurisdictions. They also need to have a reasonable understanding of basic accounting and auditing skills to be able to appreciate the various ramifications of the environment in which data is processed and reported. [23] also happed extensively on the basic qualities already discussed above such as knowledge of financial statements, understanding of fraud schemes and internal control systems of corporations, including operational risks.

2.2 Hypotheses Development: Financial Reporting and Basic Forensic Accounting Skills

Although there are numerous skill sets that will enhance the ability of the forensic accountant, four of the skills are fundamental to the practice of the profession. There are basic fundamentals to the practice of forensic accounting which, by implication, also affect financial reporting credibility.

Writing on Contextualizing the South African Forensic Accountant [20] expressed that the aptitudes expected of a legal bookkeeper incorporate the accompanying: essential bookkeeping abilities, composing aptitudes, verbal relational abilities, logical aptitudes, identity attributes, PC measurable abilities, information mining abilities, talking with aptitudes, and a foundation in law requirement [16], [18]. [22] additionally stressed the way that, so as to end up a fruitful criminological bookkeeper, one ought to have investigative capacities, relational abilities, an imaginative outlook and business information. These skills can be organized and discussed under the following headings to enable us develop appropriate theoretical hypothesis viz: technical and analytical skills, effective communication skills, accounting & auditing skills and psycho-social skills.

2.2.1 Technical and Analytical Skills

The vast majority of forensic accounting techniques rely heavily on a working knowledge of information and communication technology. An aspiring forensic accounting practitioner therefore, needs to have technical and analytical skills. This is in line with the view expressed by the [9] on the desirable traits and characteristics of scientific and legal accountants. According to them, technical and analytical proficiency, detail-orientation, ethical mindset, responsiveness, insightfulness, intuitiveness, inquisitiveness, skepticism and persistence occupy the front burner on the niche grid with technical and analytical characteristics being the most critical of all the traits, an assertion that had earlier been made by [22]. Analytical skills are paramount as the forensic accountant often has to deal with unusual relationships that to be disentangled and explained. For instance, some forensic accounting experts opine that if that if the evaluators of HealthSouth had been versed in logical methods, they may have identified the over $1 billion misrepresentation that they neglected to discover for a long time [3]. Further empirical impetus was given to the pre-eminence of this factor by the studies conducted by [14]. According to them, explanatory aptitude remains the principal characteristic that scientific bookkeepers are relied upon to have.

H_0: There is no critical relationship between Technical & Analytical Skills and financial reporting credibility based on listed firms in Nigeria.

2.3 Effective Communication Skills

2.3.1 *Written Communication Skills*

There is doubt, whatsoever, that written communication is very vital as the need always arises for the forensic accountant to write down his work in a professional manner. A properly written findings can be a pivotal weapon in the course of prosecution and may help to impress the judge/jury [3].

2.3.2 Verbal Communication Skills

Apart from composed correspondence, the scientific bookkeeper should likewise have the capacity to clarify his discoveries verbally. The need may likewise emerge for him to go into the field and address organization work force who could possibly be included with the speculated misrepresentation [3]. Moreover, measurable bookkeepers are regularly called to be master observers and affirm amid prosecution which is a key part of the training. He ought to almost certainly present complex proof to the jury in a reasonable way [23].

H_0: There is no significant relationship between effective communication skills and financial reporting credibility among listed firms in Nigeria.

2.3.3 Accounting and Auditing Skills

[220 expressed the view that bookkeeping is generally viewed as the "language of companies" hence the need for the forensic accountant to have strong basic accounting and auditing skills, in all its ramifications. This argument has been corroborated in the study conducted by [17] which concluded that there is requirement for scientific bookkeepers to have strong information and aptitudes in the zone of fundamental bookkeeping and examining.

H_0: There is no critical relationship between basic accounting and auditing skills and financial reporting credibility among listed firms in Nigeria.

2.3.4 Psycho-Social Skills

Psycho-Social Skills has to do with a broad range of skills that are necessary for a successful practice of forensic accounting [15] personality traits, interviewing skills, psychology and social interaction, etc. These skills enable the forensic accountant to understand the mindset of the fraudster and be able to analyse and interpret his actions and inactions. It is psycho-social skills that enables the forensic accountant to understand the environmental factors around him. [1], argued that a forensic accountant must have interest and a solid distrust of the budgetary framework around him. He should most likely distinguish applicable actualities, assess judgment and translate plan; pursue a reasonable, moral and coherent way when playing out his work just as convey his discoveries and musings to companions and customers. As commented by [11],"An extortion bookkeeper (as forensic accountant are frequently called) ought to likewise watch and listen cautiously. By this, you build up an ability to recognize lies whether they include misrepresentation or not. Educationally, 'not all liars are fraudsters, however all fraudsters are liars'. Enlighteningly, in the expressions of [13], "systems, processes, technologies and institutions don't commit fraud, people do." It is absolutely necessary, therefore, to be able to fathom the psychology and sociology of human beings.

H_0: There is no crucial relationship between psycho-social skills and financial reporting credibility among firms listed in Nigeria.

3. Research Methods

The research design adopted is the questionnaire, through which the impact of the various basic forensic accounting skills on financial reporting credibility was assessed. The questionnaire containing fifteen questions each were presented to a sample of forty users of financial reports who are corporate investment advisers operating within the Lagos metropolis as at January, 2018, out of which thirty one responded. The responses were obtained through a five-point Likert-type scale which was subsequently converted into quantitative data and used to do the chi-square analysis model and test the strength of the relationship between the dependent and independent variables which are as follows:
Where:

Dependent Variable:

Financial Reporting Credibility

Independent Variables:

Basic Forensic Accounting Skills

Technical &Analytical Skills

Communication Skills

Accounting &Auditing Skills

Psycho-Social Skills

4. Presentation And Discussion Of Results

This section is devoted to the presentation of the results obtained from the field work as well as discussion of same. The objective here assesses the effect of forensic accounting skills on financial reporting credibility through the Cross Tabulation Analysis as shown in Table 4.

IOP Conf. Series: Earth and Environmental Science **331** (2019) 012041 doi:10.1088/1755-1315/331/1/012041

Table 4. Presentation of results

Basic Forensic Accounting Skills	**Financial Reporting Credibility: Impact Assessment**					
Variable **Analytical skills**	Strongly disagree Freq. (%)	Disagree Freq. (%)	Undecided Freq. (%)	Agree Freq. (%)	Strongly agree Freq. (%)	Total Freq. (%)
	0	0	0	0	0	0
	0	0	0	1(100)	0	1(100)
	0	0	1(25.0)	3(75.0)	0	4(100)
	0	0	0	71(81.6)	16(18.4)	87(100)
	0	0	1(1.0)	82(79.6)	20(19.4)	103(100)
	X^2 =24.382(0.000); Correlation coefficient (R) =0.333 (0.000)					
Communication Skills	Strongly disagree	Disagree	Undecided	Agree	Strongly agree	Total
	0	0	0	0	0	0
	0	0	0	0	0	0
	0	0	1(10.0)	5(50.0)	4(40.0)	10(100)
	0	0	1(0.9)	84(76.4)	25(22.7)	110(100)
	X^2 =17.970 (0.001); R= 0.290 (0.001)					
Accounting &Auditing Skills	Strongly disagree	Disagree	Undecided	Agree	Strongly agree	Total
	0	0	0	0	0	0
	0	0	0	0	0	0
	0	0	0	10(83.3)	2(16.7)	12(100)
	0	0	0	84(77.1)	25(22.9)	109(100)
	0	0	2(2.7)	63(85.1)	9(12.2)	74(100)
	X^2 = 3.081 (0.079); R= -0.133 (0.066)					
Psychosocial Skills	Strongly disagree	Disagree	Undecided	Agree	Strongly agree	Total
	0	0	0	0	0	0
	0	0	0	0	1(100)	1(100)
	0	0	0	6(100)	0	6(100)
	0	0	2(2.0)	84(85.7)	12(12.2)	98(100)
	0	0	0	67(74.4)	23(25.6)	90(100)
	X^2 = 13.022 (0.043); R= 0.250 (0.043)					

4.1. Interpretation & Discussion of Results

By way of summary, the result of Table 4.1shows that there is a critical positive relationship between basic forensic accounting skills and financial reporting credibility. The individual variables have also been subjected to analysis.

4.1.1. Technical & Analytical Skills

Further analysis of the result shows a significant positive correlation, *chi-squared (X^2) =24.382 (0.000); Pearson Correlation coefficient (R) =0.333; P-value<0.01)* between technical & analytical skills of the forensic accountant and financial reporting credibility. It is observed that as many as 82 (19.4%) of the respondents that strongly agree that a forensic accountant needs analytical skills to resolve both structured and unstructured issues also believed that financial credibility can be enhanced through analytical skills. However, while 71(81.6%) of the respondents supported both arguments, only one person (25.0%) of the respondents was undecided while 20 (19.4%) strongly agreed. 16(18.4%) of the respondents that agree on the need for analytical skill of the forensic accountant also strongly agree on the need for financial reporting credibility enhancement. It is observed that although one of the respondents strongly supported the need for acquisition of analytical skills by the forensic accountant he was undecided about its impact on financial reporting credibility of listed companies. Based on the overwhelming support as shown by this analysis, it stands to reason that the null hypothesis is rejected, paving the way for acceptance of the alternate. This verdict is in line with the view of [3] both of whom argued on the critical gap to be filled by analytical skills in the practice of forensic accounting and how it can improve financial reporting credibility.

4.1.2. Effective Communication Skills

In the case of communication skills of the forensic accountant, it was observed to be significantly associated with financial reporting credibility, *chi-squared* (X^2) *=17.970, p-value <0.01); Correlation coefficient (R) =0.290; P-value<0.01).* Further evidence from the result also shows that 84 respondents (77.1%) who agreed that forensic accountants need effective communication skill were also in support of the fact that it will improve the credibility of financial reporting by listed companies. They were also strongly supported by another 25 respondents (22.7% of the entire respondents). In addition, another 68respondents (90.7%) who strongly agreed on the need for good communication skills by forensic accountants are also on the same page concerning its potential for positive impact on financial reporting credibility while 7 respondents (9.3%) strongly agreed on both issues. Evidently, 25 respondents (22.7%) who agree on the need for forensic accountants to have good communication, both oral and written, were in strongly support its perceived effect on financial reporting. It is noteworthy that 5(50.0%) and 4(40.0%) of the respondents who were undecided about the need for communication skills by the forensic accountant, nonetheless, agree and strongly supported the argument on its impact on financial reporting credibility of listed companies respectively while one respondent (10.0% of the total respondents) was undecided about both issues. This conclusion paves the way for a logical rejection of the null hypothesis earlier formulated. While this conclusion is in line with the argument presented by [3] on the need for effective written communication ability, made a case for oral communication ability.

4.1.3. Accounting & Auditing Skills

As observed from the result on Table 4.1, majority of the respondents (84 or 77.1%) who agreed that forensic accountants need to be skilled in basic accounting and auditing also agree on the need to improve financial reporting credibility and were also strongly supported by another 25(22.9%) of the respondents. Further analysis of the result indicates that 63(85.1%) who were in strong support of the need for accounting and auditing skills critical in forensic accounting also supported the need for financial credibility improvement and 9 (12.2%) strongly agreed on the two issues though 2(2.7%) were undecided about this particular skill in forensic accounting practice. Furthermore, 10(83.3%) who were undecided about the importance of accounting and auditing skills and in forensic accounting were in support of the need to improve financial reporting credibility and were strongly supported by 2 (16.7%) of the respondents. Particularly, a significant and direct correlation could be found between accounting and auditing skills and credible financial reporting by listed companies, C*hi-squared* (X^2) *=3.081, P-value <0.10); Correlation coefficient (R) =0.133; P-value<0.10).* This empirical result leads us to reject the null hypothesis and accept the alternate which is in sync with the view of [17] on the importance of accounting and auditing skills to the forensic accountant

4.1.4. Psycho-Social Skills

The result of the relationship between psychological and social (psychosocial) skill in forensic accounting practice and financial reporting credibility *(*X^2*= 13.022; p-value <0.05; R= 0.250 (p-value<0.05)* suggest a significant level of positive association. Detailed analysis shows that most of the respondents(84 or 85.7%) agreed on the impact of psycho-social skills in relation to financial reporting credibility among listed companies, However, 12 (12.2%) were in strong support while 2(2.0%) were indeterminate about these issues. Also 23(25.6%) of those who showed strong support on the impact of psycho-social skills on financial reporting credibility were also supported by 67(74.4%) of the total respondents. However, 6(100%) who were undecided about psychosocial skill acquisition by forensic accountants displayed considerable support for the concern to improve financial reporting credibility. Here again, the null hypothesis is rejected and the alternate accepted and this is supported by several researchers on the subject [22]

5. Conclusion

The following findings are pertinent from this study:

1. There is a significant positive correlation, *chi-squared* (X^2) *=24.382 (0.000); Correlation coefficient (R) =0.333; p-value<0.01)* between technical & analytical skills of the forensic accountant and financial reporting credibility. An overwhelming number of respondents are united on the fact that possession of technical & analytical skills will improve financial reporting credibility.
2. In the case of communication skills of the forensic accountant, it was also observed to be significantly associated with financial reporting credibility, *chi-squared* (X^2) *=17.970, p-value <0.01); Correlation coefficient (R) =0.290; p-value<0.01)*. This is true for both verbal and written communication.
3. A significant and direct correlation could also be found between accounting and auditing skills and credible financial reporting by listed companies, C*hi-squared* (X^2) *=3.081, p-value <0.10); Correlation coefficient (R) =0.133; p-value<0.10)*.
4. In the case of psycho-social skills, the result of the relationship between psychological and social (psychosocial) skill in forensic accounting practice and financial reporting credibility (X^2= *13.022; p-value <0.05; R= 0.250 (p-value<0.05)* suggest a significant level of positive association.

In view of the strong and significant association between these various skills and financial reporting credibility, it is advised that deliberate efforts should be made to develop these skills not only by the professional bodies but also the regulatory agencies charged with responsibility for financial reporting. This can be done either as a prerequisite to practice forensic accounting profession or as part of continuing professional education.

Acknowledgement

Authors acknowledge Covenant University who has solely provided the platform for this research and has also fully sponsored publication of this research work.

References

[1] Bhasin, M.L. 2007. Forensic accounting and auditing – perspectives and prospects, accounting world magazine, http://www.iupindia.in/107/AW_Forensic_Accounting Auditing_40.html

[2] Crumbley, D. L. 2001. Forensic Accounting: Older than you think, JFA, **2** (2). 181-202

[3] Crumbley, D.L; Heitger, L.E; Smith, G.S (2005). Forensic and investigative accounting, CCHGroup, Chicago,IL.

[4] Enofe, A.O., Okpako, P.O., Atube, E.N. 2013. The impact of forensic accounting on fraud detection. *European Journal of Business and Management,* 5(16), 61-72

[5] Cabole 2009, Forensic Accounting: A paper presented in the Hilton Hotel Lagos. Certified public accountant (CPA) *Journal New York* accessed on-line File//user/document/accessed on 1/6/2013.

[6] Cotton, D.L. 2003. Fixing CPA ethics can be an inside job. Retrieved june7, 2007, from Http:/www.Washington post.com.

[7] Diamond, D., & Verrecchia, R. 1991. Disclosure, liquidity, and the cost of capital. *Journal of Finance*, **46**(4): 1325–1359.

[8] Chen, F., Hope, O.K., Li, Q., & Wang, X. 2011. Financial reporting quality and investment efficiency of private firms in emerging markets. *The Accounting Review,* **86**(4):1255

[9] FASB 2008.Financial accounting series, Statement of Financial Accounting Standards, No 1570-100: Exposure draft on an improved conceptual framework for financial reporting. Norwalk.

[10] IASB 2008. Exposure draft on an improved conceptual framework on financial reporting: The objective of financial reporting and qualitative characteristics of decision-useful financial reporting information. London.

[11] McNichols, M. F & Stubben, S. R. 2008. Does earnings management affect firms' Investment decisions? *The Accounting Review*, **83**(6): 1571–1603

[12] Luke (2013). Application of forensic accounting: a tool for confidence in auditors' reports. *Research Journal of Finance and Accounting.* **4**(17).105-110

[13] Dabor, E. L; and Adeyemi, S. B. 2009. Corporate and credibility of financial statements. *Nigeria Journal of Business System, Government and Ethics.* **4***(1), 13-14.*

[14] Degboro, D. and Olofinsola, J. 2007. Forensic accountants and the litigation support engagement. *Nigerian Accountant,* **40***(2): 49-52.*

[15] Howard, S. and Sheetz, M. 2006. Forensic accounting and fraud investigation for non- experts. New Jersey: John Wiley and Sons Inc.

[16] Marcus, E. 2010. Forensic investigation and evidence handling. *International Journal of Accounting. Australia.***32**, 45-49. Http//www.forensicinvestigation.net

[17] Ijeoma, N. B. 2015. Empirical analysis on the use of forensic accounting techniques in curbing creative accounting. *International Journal of Economics, Commerce and Management,* **3**(1).1-15

[18] Singh, P. 2012. "Forensic accounting concept in India" *International Journal of Trade and Commerce- IIARTC,* **1** (1)*:100-105*

[19] Warshavsky, M; Marcus, G. S; and Woodbury, L. 2012. Earnings Quality and the Beneish Model: Analysing earnings quality as a financial forensic tool (Accessed online20th May 2019).

[20] Rezaee, Z., Crumbley, D., and Elmore, R. 2004. Forensic accounting education: A survey of academicians and practitioners. *Advances in Accounting Education Teaching a Curriculum Development*, **6**: 193-232

[21] Owojori, A.A and Asaolu T. O. 2009.The Role of Forensic Accounting in Solving the Vexed Problem of Corporate World. *European Journal of Scientific Research* .29 (2)83-187*International Journal of Business and Social Science* **4** *(7)183-187*

[22] Ozkul, F.U., & Pamukc,A. 2012. Fraud detection and forensic accounting. *.Istanbul, Turkey.*

[23] Smith, G. and Crumbley, D. 2009. How divergent are pedagogical views toward the Fraud /forensic accounting curriculum? Global Perspectives on Accounting Education *Journal of Theoretical and Applied Information Technology.* **6**(1) 1-24

[24] Ramaswamy, V. 2005. Corporate governance and the forensic accountant. *Retrieved from::http://www.nysscpa.org/cpajournal/2005/305/essentials/p68.htm.*

[25] Uwuigbe, U., Eluyela, E.F., Uwuigbe, O.R., Teddy, O., Irene, F. (2018) Corporate governance and quality of financial statements: A study of listed Nigerian banks Banks and Bank Systems 13(3), 12-23

[26] Uwuigbe, U., Teddy, O., Uwuigbe, O.R., Emmanuel, O., Asiriuwa, O., Eyitomi, G.A., Taiwo, O.S. (2018). Sustainability reporting and firm performance: A bi-directional approach*) Academy of Strategic Management Journal, 17(3), 1-16*

Effect of Chemicals and Cations on the Activity of Partially Purified alpha amylase produced by *Aspergillus niger*

O M Atolagbe*, **A A Ajayi**. and **G I Olasehinde**.
Department of Biological Sciences, College of Science & Technology
Covenant University, Ota, Ogun State, Nigeria.
*Correspondence: bunmi.atolagbe@covenantuniversity.edu.ng

Abstract: Amylases are one of the most important enzymes used in the industry. They are responsible for the hydrolysis of starch and glycogen. This research aimed to investigate the effect of cations (Ca^{2+} and K^{+}) and chemicals (EDTA and $HgCl_2$) on alpha amylase activity. Alpha amylase was produced by *Aspergillus niger* under submerged fermentation conditions in a basal salt medium. The enzyme was partially purified with activated carbon which gave a 1.57 purification fold. The effects of ethylene diamine tetratacetic acid (EDTA), mercuric chloride ($HgCl_2$), Calcium ion (Ca^{2+}) and Potassium ion (K^{+}) on the partially purified enzymes were then investigated. The enzyme was inhibited to varying degrees by EDTA and $HgCl_2$ with inhibition of approximately 62% and 52% occurred with 8mM of EDTA and $HgCl_2$ respectively but stimulated by Ca^{2+} and Potassium ion K^{+}. Activated carbon partially purified alpha amylase produced by *A. niger* with a high yield. The cations and chemicals had varying effects on the enzyme.

1. Introduction

Amylases are an important class of enzymes which are well-known and are used for the hydrolysis of starch or glycogen[1]. They are employed in starch processing industries for the breakdown of starch into simple sugars[2]. They hydrolyse α 1-4 glycosidic bonds of starch, glycogen, amylopectin and other related compounds. They can be produced either by solid state fermentation or submerged fermentation[3,4].

The use of microbes for the production of enzymes is economical. This is because microorganisms can be easily manipulated to produce metabolites e.g. enzymes[5]. Fungi are however preferred over bacteria for enzyme production because of their filamentous nature, which helps in the penetration of the mycelia through solid substrates[6]. Recent advances in enzyme technology has brought about an increased interest in the use of fungi for microbial enzyme production.

The progress in the knowledge of enzymes, has brought about an acquired importance of fungi in several industries. This is because they may serve to improve various aspects of the final industrial products. There are approximately 200 species of *Aspergillus* which are known to produce enzymes[7]. The several and diverse reactions of *Aspergillus* niger has brought about a remarkable interest in its use as a species of great commercial interest with a highly promising future. *Aspergillus niger* is also already widely applied in modern biotechnology[7,8]. Moreover, the use of *A. niger* is not only because of its production of various enzymes, but it is one of the few species of the fungus kingdom classified as GRAS (Generally Recognized as Safe) by the Food and Drug Administration (FDA).

IOP Conf. Series: Earth and Environmental Science **331** (2019) 012042 doi:10.1088/1755-1315/331/1/012042

The aim of this study was to study the effect of the chemicals (EDTA and $HgCl_2$) and cations (Ca^{2+} and K^+) on the activity of the partially purified enzyme.

2. Materials and methods

The fungal isolate (*Aspergillus niger)* employed for this work was isolated by exposing potato dextrose agar (PDA) plates to the air in the Microbiology laboratory of the Department of Biological Sciences, Covenant University, Ota, Ogun State, Nigeria. The plates were incubated at room temperature (25 °C) for 3-5 days. Distinct colonies of the organism were sub-cultured till pure cultures were obtained. A 96-hr-old culture of the isolate was used.

Preparation and inoculation of Basal Salt medium

The basal salt medium contained the following reagents:

KH_2PO_4 (2.0g/L), $(NH_4)_2SO_4$ (1.4g /L), $MgSO_4.7H_2O$ (0.3g/L), $CaCl_2$ (0.3g/L), Urea (0.3g/L), Tween 80 (1ml/L), $FeSO_4.7H_2O$ (5mg/L), $MnSO_4$ (1.6mg/L), $ZnSO_4$(1.4mg/L), $CoCl_2$ (2.0mg/L). One milliliter of the aqueous spore suspension containing approximately $6x10^6$ spores per ml of the fungal isolate was inoculated into conical flasks containing 1000ml growth medium. The spores were counted using the Neubauer counting chamber. One gram (1g) of substrate per 100ml of basal salt medium was used according to the method [9].

Extraction of Enzyme from Culture Media

Enzyme solution was obtained from the basal salt media after 7 days of innoculation by sieving the culture content with muslin.

Purification of the Enzymes

The extracted enzyme was purified according to the method described by [10] whereby experiments were conducted in 10 ml centrifuge tubes. Samples (10 ml) of crude enzyme (α-amylase pH 6.0) was treated with various concentrations of activated charcoal (1–4 % w/v) and incubated at 30 °C for 30 min with occasional stirring. The mixture was then centrifuged at 2500 rpm in a bench centrifuge for 10 min.

Enzyme Assay

α-Amylase activity was determined using the method described by [11]. The reaction mixtures consisted of 2 ml of 0.05% (w/v) starch in 0.2 M citrate phosphate buffer (pH 6.0) as substrate and 0.5 ml of enzyme. Controls consisted of only 2 ml of the prepared substrate. The contents of both experimental and control tubes were incubated at 35 °C for 20 min. The reaction in each tube was terminated with 3ml of 1 N HCl. Enzyme (0.5 ml) was then added to the control tube. Two millilitre of the mixture from each of the sets of experimental and controls was transferred into new sets of clean test tubes. Three millilitre of 0.1 N HCl was added into the contents of each test tube after which 0.1 ml of iodine solution was added. Optical density readings were taken with a spectrophotometer at 670 nm. One unit of enzyme activity was defined as the amount of enzyme which produced 0.01 % reduction in the intensity of the blue colour of starch-iodine complex under conditions of the assay.

Effect of Chemicals and Cations

The effect of ethylene diamine tetraacetic acid (EDTA) and mercuric chloride ($HgCl_2$) at different molar concentrations on the activities of the enzyme was investigated. Concentrations of 0 - 8 mM, each of ethylene diamine tetraacetic acid (EDTA) and mercuric chloride ($HgCl_2$) were prepared in 0.05% (w/v) starch in citrate phosphate buffer pH 6.0. The substrate (starch) was incubated initially with each test chemical at 4 °C for 3h before being employed for the enzyme assays.

Calcium ion (Ca^{2+}) and Potassium ion (K^{+}) were also employed in this investigation. Concentrations of 0 - 40 mM were employed for the cations Ca^{2+} and K^{+}. The reaction mixtures for consisted of 2 ml substrate and 0.05 ml enzyme. These were incubated at 35 °C for 20 mins and then analysed for α-amylase activities as described above.

3. Results and Discussion

The partially purified fractions gave a purification fold of approximately 1.57 as shown in (table 1).

Table 1: Purification of partially purified amylase from *A. niger*

Fraction	Total Protein (mg)	Total Activity (Units)	Specific Activity (Units/mg)	Yield (%)	Purification fold
Crude	287	542	1.89	100	1
Activated Carbon (1%)	230	513	2.23	94.6	1.18
Activated Carbon (2%)	162	461	2.85	85.1	1.51
Activated Carbon (3%)	158	456	2.89	84.1	1.53
Activated Carbon (4%)	150	445	2.97	82.1	1.57

Effect of Chemicals and cations

Inhibition of approximately 62% and 52% occurred with 8mM of EDTA and $HgCl_2$ respectively (fig 1 and 2). On the other hand, both cations Ca^{2+} and K^{+} had a stimulation of approximately 98% and 28% respectively on α-amylase with 40mM of the cations (fig 3 and 4).

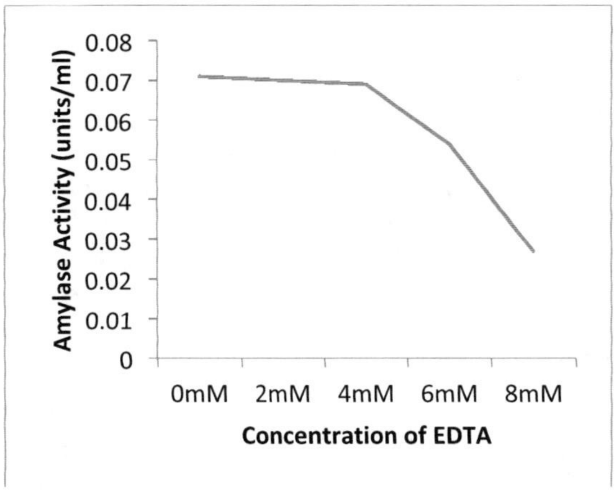

Fig 1: Effect of EDTA on partially purified α-amylase from *A. niger*

IOP Conf. Series: Earth and Environmental Science **331** (2019) 012042 doi:10.1088/1755-1315/331/1/012042

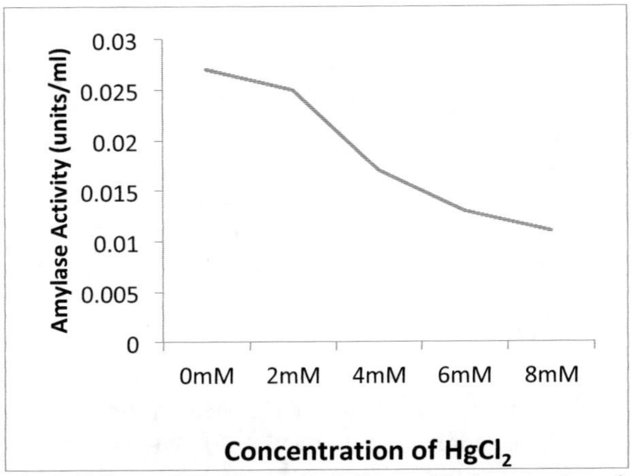

Fig 2: Effect of $HgCl_2$ on partially purified α-amylase from *A. niger*

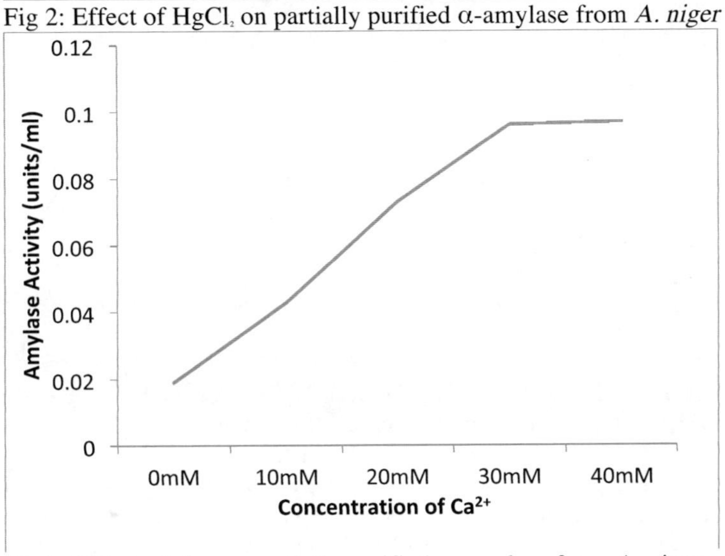

Fig 3: Effect of Ca^{2+} on partially purified α-amylase from *A. niger*

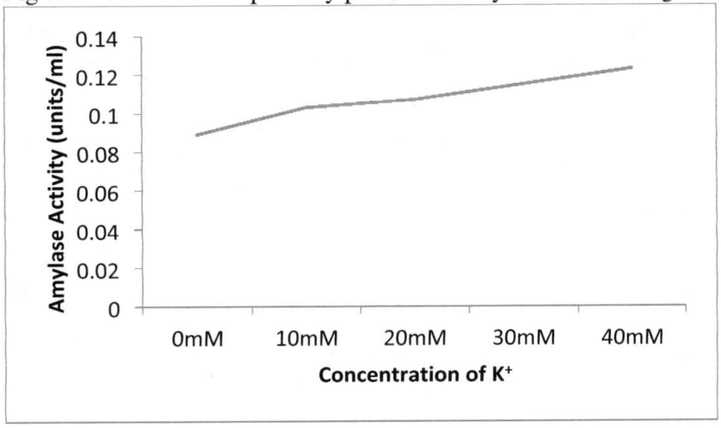

Fig 4: Effect of K^{2+} on partially purified α-amylase from *A. niger*

Production of alpha amylase under submerged fermentation conditions was employed in this study. This method has been widely employed for production of amylases and result obtained are in agreement with previous authors[12,13]. The cations employed in this work enhanced the activities alpha amylase under investigation. It was observed that Ca^{2+} and K^{+} had a stimulation of about 98% and 28% increase in activity respectively. This agrees with the work of previous authors[14-16]. The activation of enzymes by cations is usually through a variety of mechanisms. The metal ion may produce direct activation of enzymes in a reaction whereby the reaction is affected by a change of surface charge on the enzyme protein, particularly in the case of polyvalent cations, where they may produce very marked changes in the electro-kinetic potential of proteins[17]

The chemicals EDTA and Hg^{2+} were inhibitory to the enzymatic activities of the alpha amylase under investigation. Similar inhibition of enzymatic activities by metal ions have been reprted[18,19]. EDTA is a metal chelating agent that forms a complex with an inorganic group and it consequently reduces the affinity of the enzyme for its substrate[17].

4. Conclusion

The results suggest that Ca^{2+} and K^{+} serve as cofactors which will help to enhance the enzyme activity. This research will recommend that reactions set up in bioreactors with enzyme use should avoid the use of chemicals such as EDTA and Hg^{2+}. This is because they will interfere with the activity of the enzymes thereby slowing down the rate of reaction.

References

[1]. Krishna, P. R., Srivastava, A. K., Ramaswamy, N. K., Suprasanna, P. and S.F.D Souza (2011). Banana peel as substrate for alpha-amylase production using *Aspergillus niger* NCIM 616 and process optimization. *Indian J Biotechnol* 11: 314-319.

[2]. Alva, S., Anumpama, J., Savla, J., Chiu, Y. Y., Vyshali, P., Shruti, M., Yogeetha, B. S., Bhavya, D., Purvi, J., Ruchi, K., Kumudini, B. S. and Varalakshmi, K. N. (2007). Production and characterization of fungal amylase enzyme isolated from Aspergillus sp. JGI 12 in solid state culture. *Afr.* J. *Biotechnol.* **6**: 576-581.

[3]. Egas, M. C., Da Costa, M. S., Cowan, D. A. and E.M. Pires (1998). Extracellular alpha-amylase from *Thermus filiformis* Ork A2: Purification and biochemical characterization, *Extremophiles* 2: 23-32.

[4]. Khan, J. A. and Yadav, S. K. (2011). Production of alpha amylases by *Aspergillus niger* using cheaper substrates employing solid state fermentation. *Int. J. Pl.An and Env.Sci.* **1**(3): 100-108.

[5]. Aiyer, P. V. (2005). Amylases and their applications. *Afr. J. Biotechnol.* **4**(13): 1525- 1529.

[6]. Ramachandran, S., Patel, A. K., Nampoothiri, K. M., Francis, F. and Nagy, V. (2004). A potential raw material for the production of alpa-amylase. *Bioresour. Technol.* **93**: 169-174.

[7]. Soares I., Távora Z., Barcelos R.P. and Suzymeire B (2012). Microorganism Produced Enzymes in the Food Industry, Scientific, Health and Social Aspects of the Food Industry, Dr. Benjamin Valdez (Ed.), ISBN: 978-953-307-916-5, InTech, Available from: http://www.intechopen.com/books/scientific-health-and-social-aspects-of-the-food-industry/microorganismproduced-enzymes-in-the-food-industry

[8]. Andersen M.R., Nielsen M.L. and Nielsen J. (2008). Metabolic model integration of the bibliome, genome, metabolome and reactome of *Aspergillus niger. Mol. Sys. Biol.* **4(178)**: 1-13.

[9] Peter-Albert, C.F .,Ajayi, A.A ., and Awosika, F.A . (2015). Studies on Xylanase Production by *Aspergillus niger* on Tomato Pomace Medium. In: Mojekwu, J.N., Ogunsumi, L.O., Ojigi, L. M. Atepor, L., Thwala, D.W., Sackey, S. Awere E., and Bamfo-Agyei, E. (Eds) African Journal of Applied Research .(AJAR) Journal, Vol.1,No.1 ISSN 2408-7920 January 2015, Cape Coast, Ghana. 286-298.

[10]. Kareem, S.O., Akpan, I, Popoola, T.O.S. and Sanni L.O. (2011). Activated Charcoal—A Potential Material in Glucoamylase Recovery. *Enzyme Res.* http://dx.doi.org/10.4061/2011/483943

[11]. Adejuwon, A. O., Adejuwon, M.A., Ajayi, A.A., Bamkefa, B.A., Omololu-Aso, J., Alao, O.O., and Adesina, F.C. (2013). Effect of some nitrogen sources of growth medium on α-amylase production by *Penicillium solitum* and *Aspergillus rubrum* isolated from yam (*Dioscorea alata*). *Researcher* **5**(2):1-4.

[12]. Vidyalakshmi R., Paranthaman R. and Indhumathi J. (2009). Amylase Production on Submerged Fermentation by Bacillus. *Wor.J. Chem.* **4(1):** 89-91.

[13]. Mathew J. J., Vazhacharickal P.J., Sajeshkumar N.K. and Ashokan A (2016). Amylase production by *Aspergillus niger* through submerged fermentation using starchy food byproducts as substrate. *Int. J. Herb.***4(6):** 34-40.

[14]. Ahvazi B., Boeshans K. M., Idler W., Baxa U. and Steinert P.M. (2003). Roles of calcium ions in the activation and activity of the transglutaminase 3 enzyme. *J. Biol. Chem.* **278** (26): 23834 – 23841

[15]. Ajayi, A.A., Adejuwon, A.O., Awojobi, O.K. and Olutiola, P.O. (2007). Effect of Cations and Chemicals on the Activity of Partially Purified Cellulase from Tomato (Lycopersicon esculentum Mill) Fruits Deteriorated by Aspergillus flavus Linn. *Pak. J. Nutr.* **6** (2): 198-200.

[16]. Gohara D.W. and Di Cera E. (2016). Molecular Mechanisms of Enzyme Activation by Monovalent Cations. *J. Biol. Chem.* **291**: 20840–20848.

[17]. Dixon M. and Webb E.C (1971). Enzymes. William Clowes and Son. Great Britain pp 950.

[18]. Naika G.S. and Tiku P.K. (2011). Influence of ethylenediaminetetraacetic acid (EDTA) on the structural stability of endoglucanase from *Aspergillus aculeatus*. *J. Agric. Food Chem.* **59** (13): 7341-7345

[19]. Adejuwon A. O., Ajayi A.A. and Ladokun O.A. (2015). Alpha amylase from *Rhizopus arrhizus* Fisher. *J.f Pla. Sci..* **10**: 99-107.

Mapping of Uranium-238 Deposit and its Contribution to Indoor Radon Gas in Ota, Ogun State, Nigeria

Adewoyin, O. O., Omeje, M., Onumejor, C. A., Adagunodo, T. A., Ndubuisi, A., Nwaluka, V. O., Akinpelu, A., Joel, E. S.

Department of Physcis, Covenant University, PMB 1023, Ota, Ogun State, Nigeria

Corresponding author's e-mail: olusegun.adewoyin@covenantuniversity.edu.ng

Abstract. In the present study, the distribution of uranium-238 was mapped from the background radiation of the area of study in order to determine its contribution to the concentration of indoor radon. The background radiation was studied using the handheld gamma spectrometer (RS 125) while the indoor radon concentration was measured with a Durridge RAD7 instrument. The results obtained showed variation in the distribution of 238U, 232Th and 40K in the study area with values ranging between 17.9 ± 0.3 and 58.05 ± 0.3, 2848 ± 10541 ± 0.8 and 3.25 ± 0.2 and 11.48 Bq/kg respectively. Furthermore, the indoor radon concentration was determined for the office on the ground floor, the result obtained was noted to be between 20 and 160 Bqm^{-3} which is far higher than the result obtained for the office on the first floor which is BDL-19 Bqm^{-3}. In each cases, the results are lower than the international recommended safe limit of 200 Bqm^{-3}. Therefore, it can be concluded that the background radiation from the soil in the premises and under the building contributes to the high concentration of indoor radiation in the office.

Keywords: uranium-238, radon concentration, background radiation, RAD7, gamma spectrometer

1. INTRODUCTION

Studies have shown that the main source of human exposure to ionizing radiation is natural radioactivity due to its health implications [1, 2]. These radioactive substances consist mainly of ^{238}U, ^{232}Th and ^{40}K [3]. These substances can hardly be avoided since they are found in geological formations like soils, water, plants and so on. Among the radioactive materials, ^{238}U is known to disintegrate to ^{226}Ra, which in turn gives birth to ^{222}Rn, popularly known as Radon [4]. The health risk associated with radon gas has made it a subject of concern for researchers in the last few years. This is because a high concentration of this gas in workplaces and homes presents a high level of hazard to the occupants. According to the report by the International Agency for Research on Cancer (IARC), radon has been classified to be in the first place on human carcinogen and second place in the cause of lung cancer [5]. Radon gas contributes significantly to the indoor air contamination all around the world [6]. The variation in the levels of radon in homes and places of work is due largely to the uneven distribution of geological features [7, 8], the characteristics of the building and some other environmental indicators.

Zhou et al. (2019) observed that the radiological hazard common in uranium mines is the inhalation of radon and its progeny [1]. Studies on epidemiology conducted on humans exposed to radon at work and at homes revealed that the health risk associated with radon gas increases with increasing exposure [3]. The primary source of indoor radon is the ground on which the building is sited, from where radon penetrates with soil gas through cracks and faults in the foundation by pressure-driven convection [9]. Similarly, another source of radon could be domestic water from drilled wells or the escape of this gas from building materials. It is very important to control the level of radon exposure at home as a result of the high occupancy factor. It is worthy of reporting that all indoor environments, such as school, offices and so on, subject people to radon exposure. Moreover, the health risk associated with exposure to radon-222, especially by working class men and women is a subject of concern and it is on this basis that this study is conducted to investigate the contribution of uranium-238 in the environment to the concentration of radon in some of the offices at the administrative block of a pharmaceutical company.

IOP Conf. Series: Earth and Environmental Science **331** (2019) 012043 doi:10.1088/1755-1315/331/1/012043

2. GEOLOGY AND GEOGRAPHICAL LOCATION OF THE STUDY AREA

The study site is situated at Sango Ota, Ogun State in southwest Nigeria. It lies within the latitude 6° 38□N to 6° 41□N and longitude 03° 8□E and 3° 12□E. It is accessible through Agbara –Atan road, off Badagry–seme express road it is also accessible via Iyana Ipaja to if road. The study area for this project is within the eastern Dahomey basin which is in the sub-humid tropical region of southwest Nigeria. It extends along the mainland edge of the inlet of Guinea. It has a distinct wet and dry period, the dry season is from November-March while the wet portion of the year runs between April and October. There has been extensive discussion on the geology of this study area by various authors such as [10, 11]. They discussed the types of rocks in this area from the formation of early tertiary to late cretaceous in age. The basin stratigraphy has also been put in six lithostratigraphic formations which are Benin Formation, Ilaro, Oshosun Akinbo, Ewekoro and Abeokuta from the youngest to the oldest [12] discussed the formation of Abeokuta to consist of the Araromi, Ise and Afomo Formations. This formation is mainly made out of poorly arranged grit mud stone with shale clay layers and siltstone.

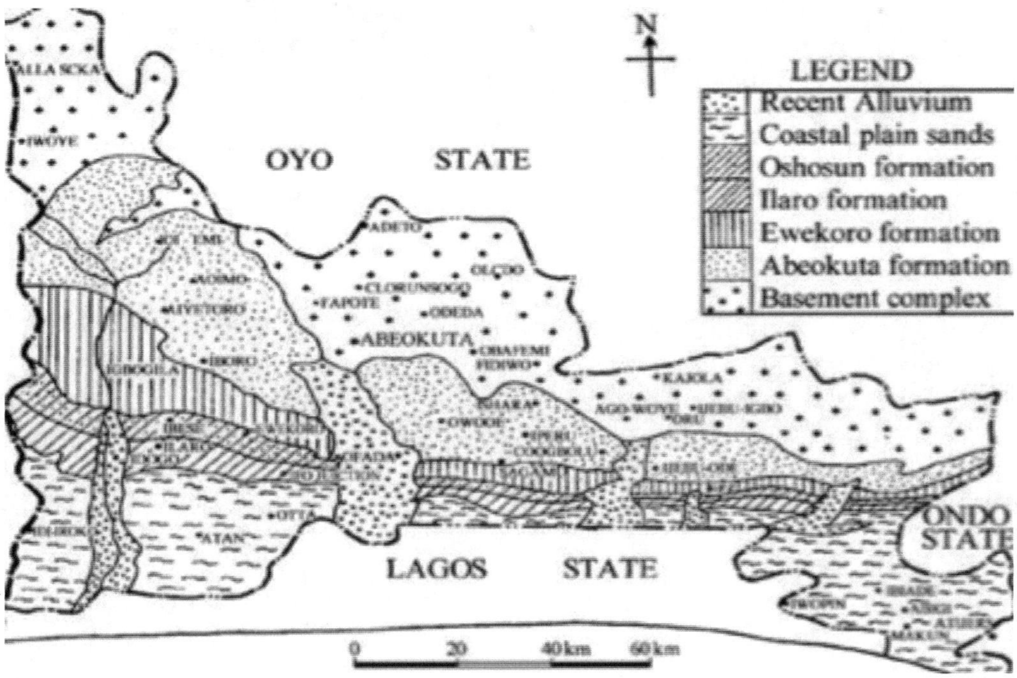

Figure 1: Geological map of the study area [12]

3. METHODOLOGY

The instrument used for this work are the handheld Gamma Spectrometer (RS-125) and the Durridge RAD7 electronic radon detector. The background radiation of the premises was surveyed in order to map the activity concentration of uranium-238 in the area of study. Similarly, indoor radon accumulation levels were measured continuously on hourly basis at the different sampling locations in the pharmaceutical company. It has a large detector for survey or search and also an integrated design, direct assay readout, a data storage and lastly a scan mode. Before the commencement of measurement, the gamma spectrometer is allowed for 15 to 30 minutes in order to re-condition itself with the background radiation of the area of study, this will ensure precise results. At all measurement points the radiation detector is kept 1 m above the ground level and measurements taken four (4) times at each station and their mean calculated in order to acquire very accurate data. At each measurement station the location coordinates were noted, an interstation spacing of 15 m was maintained all through the

IOP Conf. Series: Earth and Environmental Science **331** (2019) 012043 doi:10.1088/1755-1315/331/1/012043

survey. Readings were taken at an interval of 90 seconds at each station. 15 stations were surveyed around the administrative block. It takes 45 seconds for the RAD7 sniffing process to take place, during this process, air sample passes through the small drying tube and flows into the measurement chamber for radon concentration readout. The equipment is set up in the location of interest for 48 hours, during which it measures the concentration of radon at the investigated location. At the expiration of the 48 hours, a print-out comes from the printer of the RAD 7 that captures information such as time and date of the measurement, serial number of the machine, the average test values, the bar diagram of each reading and the cumulative spectrum of alpha energy.

4. RESULT AND DISCUSSION

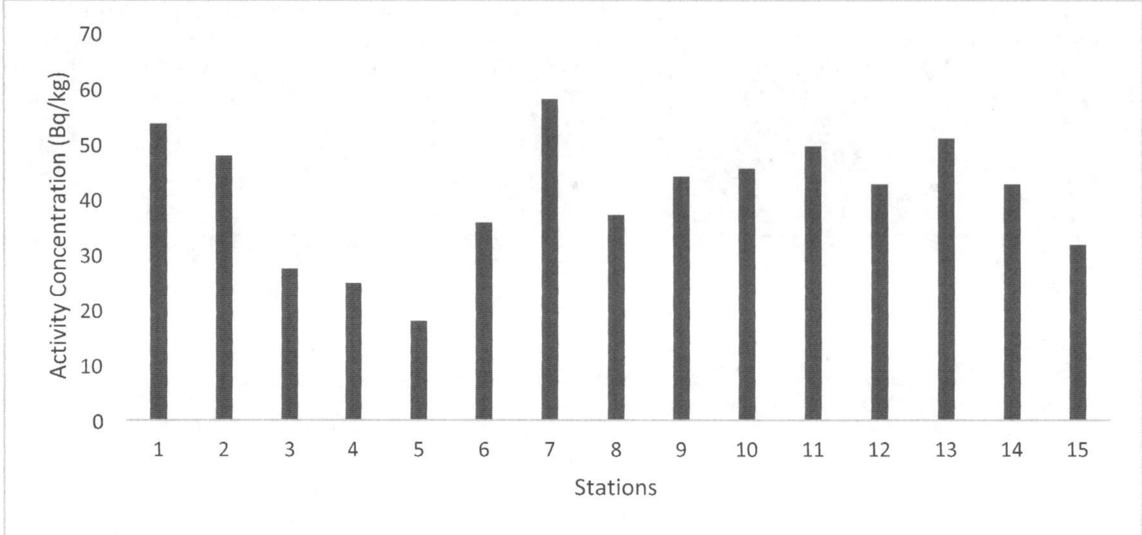

Figure 2: Distribution of 238U in the Area of Study

The distribution of uranium-238 in the study area is presented in Figure 2. It can be observed that the deposit of uranium-238 ranged between 17.9 ± 0.3 and 58.05 ± 0.3 Bqkg-1 as seen in stations 5 and 7 respectively. This variation in the distribution of 238U reported in this study could be due to anthropogenic sources or variation in the geological compositions of the study area [5].

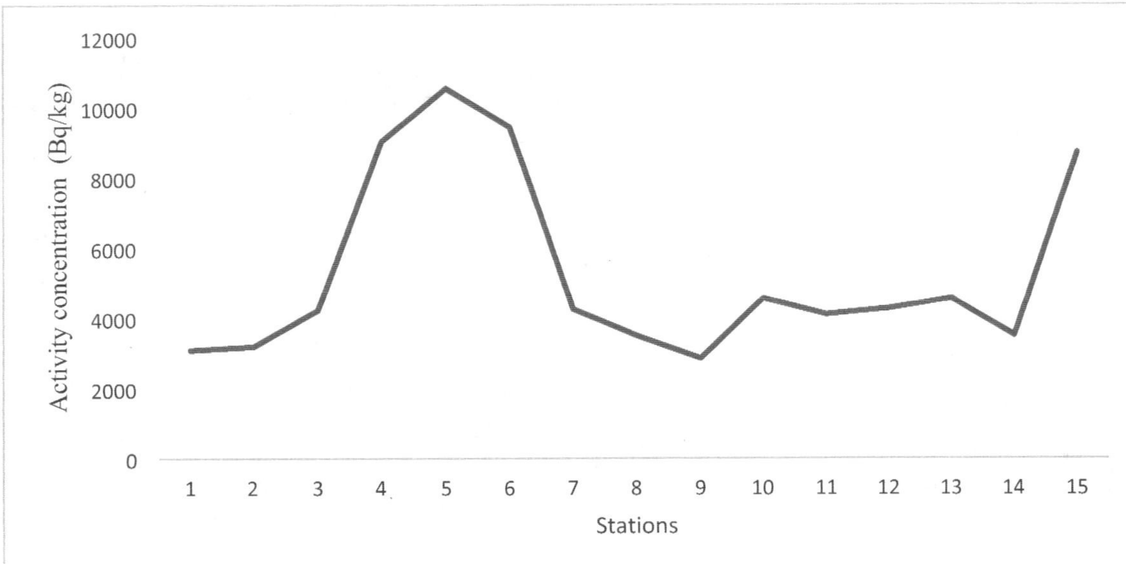

Figure 3: Distribution of 232Th in the area of study

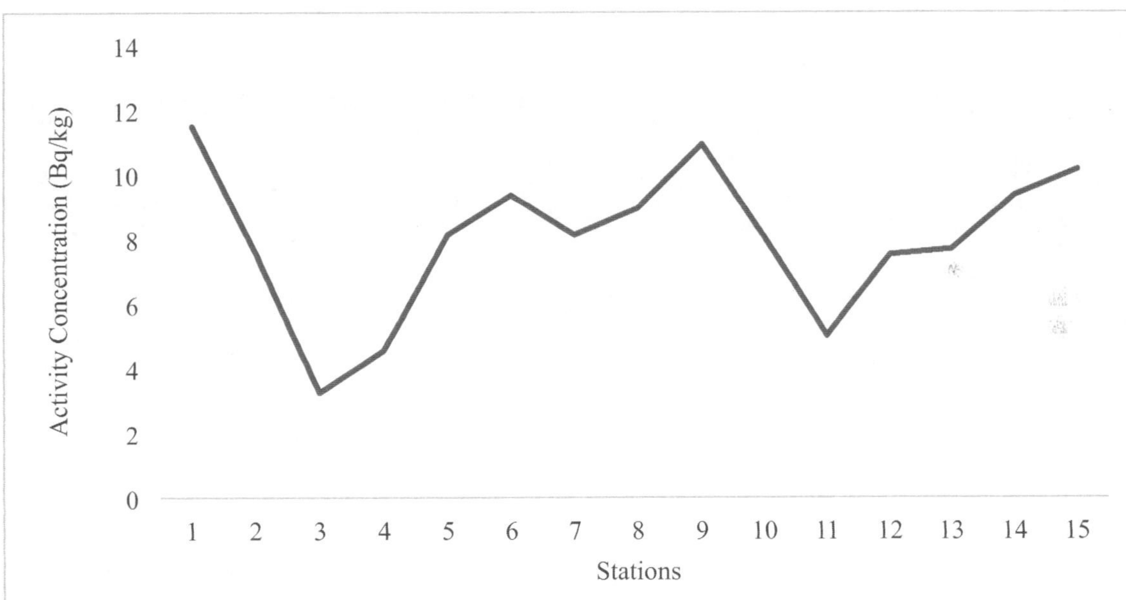

Figure 4: Distribution of 40K in the Area of study

Figures 3 and 4 present the activity concentrations of Thorium-232 and Potassium-40 in the study area. It could be seen that their activity concentrations ranged between 2848 ± 0.8 and 10541 ± 0.8 Bq/kg and between 3.25 ± 0.2 and 11.48 Bq/kg respectively. For the Thorium-40, the highest value was recorded at station 5 while the lowest was noted at station 9. Similarly for Potassium-40, the lowest and highest activity concentrations was observed at stations 3 and 1 respectively. The variations in the distribution of activity concentrations of these radionuclide materials could be as a result of uneven geological compositions or as a result of introduction of foreign geological materials during the process of construction of the buildings [9][4].

Radon Distribution in the Study Area

In order to investigate the influence of background radiation on the concentration of indoor radon in the area of study, the only office on the ground floor was examined. Another office on the first floor was also investigated which was used as a control for this study. Figure 5 presents the indoor radon concentration in the office on the ground floor. The concentration of radon was noted to range between 20 and 160 Bqm^{-3}. The highest value in this result is much closer to the international recommended safe limit of 200 Bqm^{-3}. The occupants of the office of this nature could be at risk of lung cancer and other related sicknesses if they are exposed over a long period of time. Similarly, the result of the indoor radon concentration measured on the first floor is presented in Figure 6, the result observed ranged from values below detectable limits (BDL) to 19 Bqm^{-3}. This result is far below the recommended safe limit of 200 Bqm^{-3}. These results revealed that background radiation due to underground soil can contribute to the indoor concentration of radon.

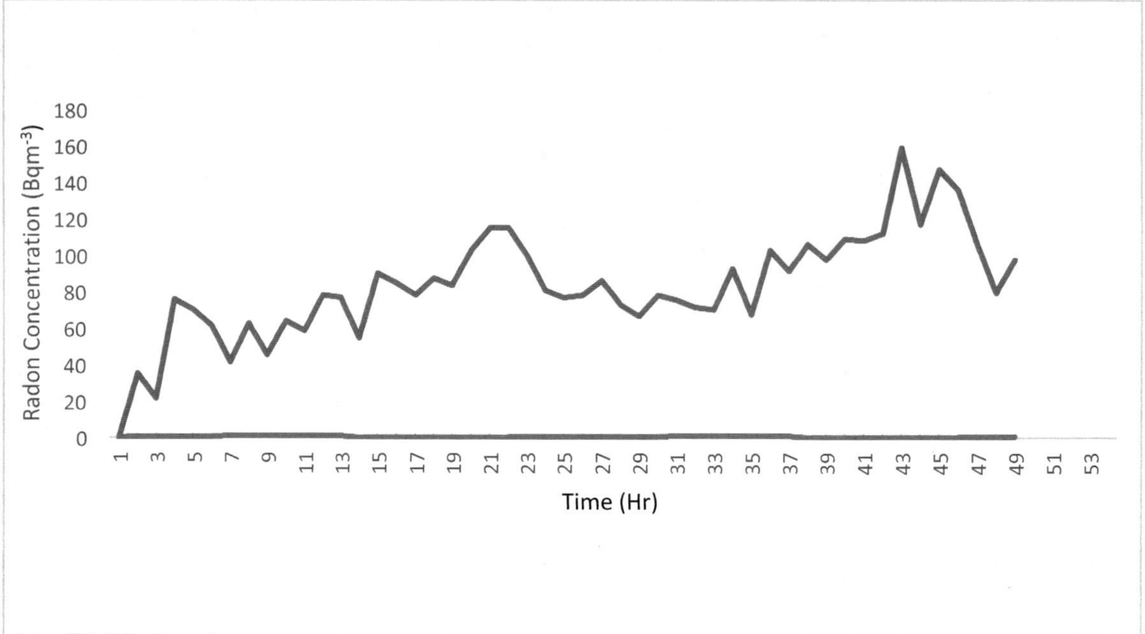

Figure 5: Indoor Radon Concentration on the Ground floor.

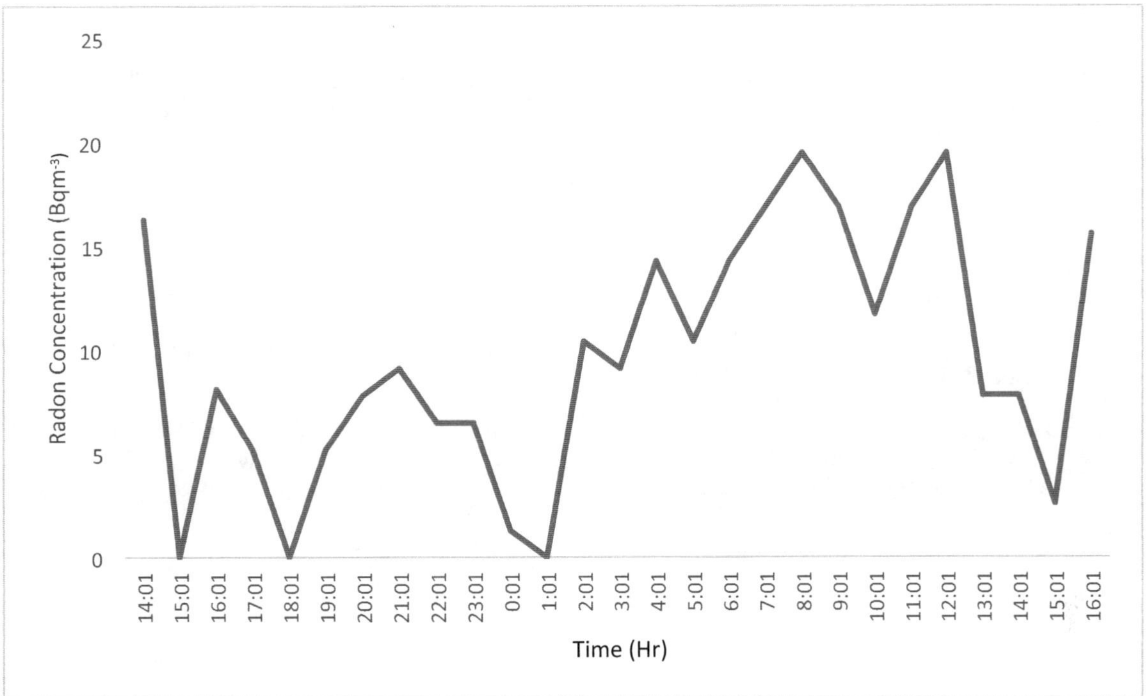

Figure 6: Indoor Radon Concentration on the First floor

5. CONCLUSION

The background radiation of the area of study was investigated in order to map the distribution of uranium-238 and observe its contribution to the concentration of indoor radon in the study area. The background radiation was studied using the handheld gamma spectrometer (RS 125) while the indoor radon concentration was measured with the aid of a Durridge RAD7. The results obtained showed variation in the distribution of 238U, 232Th and 40K in the study area with values ranging between 17.9 ± 0.3 and 58.05 ± 0.3, 2848 ± 10541 ± 0.8 and 3.25 ± 0.2 and 11.48 Bq/kg respectively. Furthermore, the indoor radon concentration was determined for the office on the ground floor, the result obtained was noted to be between 20 and 160 Bqm-3 which is far higher than the result obtained for the office on the first floor which is BDL-19 Bqm-3. Therefore, it can be concluded that the background radiation from the soil under the building contributes to the high concentration of indoor radiation in the office.

ACKNOWLEDGEMENTS

The authors wish to appreciate the management of Covenant University for supporting this study. In addition, appreciation goes to the management and staff of May and Baker Nigeria Plc for sponsoring this research.

References

[1] UNSCEAR (2000): Sources and Effects of Ionizing Radiation United Nations Scientific Committee on the Effects of Atomic Radiation

[2] WHO (2009): H. Zeeb, F. Shannoun (Eds.), WHO Handbook on Indoor Radon: a Public Health Perspective, World Health Organisation

[3] Alonso, H., Rubiano, J. G., Guerra, J. G., Arnedo, M. A, et al. (2019): Assessment of radon risk areas in the Eastern Canary Islands using soil radon gas concentration and gas permeability of soils. Science of the Total Environment, 664: 449-460.

[4] Zhou, Q., Liu, S., Xu, L., Hui, Z. et al. (2019): Estimation of radon release rate for an underground

uranium mine ventilation shaft in China and radon distribution characteristics. Journal of Environmental Radioactivity, 198: 18-26.

[5] Elío, J., Crowley, Q., Scanlon, R., Hodgson, J. and Zgaga, L. (2018): Estimation of residential radon exposure and definition of radon priority areas based on expected lung cancer incidence. Environ. Int., 114: 69-76

[6] Todea D., Cosma C., Dicu T., Roşca L., Cucoş (Dinu), A., Rişteiu M., Iancu D., Papuc I., Rădulescu D.(2013): Lung cancer risk induced by residential radon in Cluj and Alba Counties, Romania, Environ. Eng. and Manag. J. 12 (6): 1281–1285

[7] Ciotoli, G., Voltaggio, M., Tuccimei, P., Soligo, M., Pasculli, A., Beaubien, S. E. and Bigi, S. (2017): Geographically weighted regression and geostatistical techniques to construct the geogenic radon potential map of the Lazio region: a methodological proposal for the European Atlas of Natural Radiation. J. Environ. Radioact., 166: 355-375

[8] Joel, E.S. Omeje M., Adewoyin, O.O., Ehi-Eromosele, MA Saeed, M.A. (2018): Comparative analysis of natural radioactivity content in tiles made in Nigeria and imported tiles from China, Scientific Report (Springer Nature), Nature Publishing Group (Springer) 8(1): 1842

[9] Omeje M., Joel E., Adewoyin, O. et al. (2018): A study of natural radioactivity in some building materials in Nigeria. Radiation Protection Dosimetry (Oxford Press, ScholarOne), 02 August 2018, ncy121, https://doi.org/10.1093/rpd/ncy121 (Article in Press).

[10] Jones, H. A. and Hockey, R. D. (1964): The geology of part of southwestern Nigeria. Geol. Survey Nig. Bull. 31:101

[11] Omatsola, M. E. and Adegoke, O. S. (1981): Tectonic evolution and Cretaceous stratigraphy of the Dahomey Basin. Nig. J. Min. Geol. 18(01):130-137.

[12] Adewoyin, O. O., Joshua, E. O., Akinwumi, I. I. Omeje, M., Joel, E. S. (2017): Evaluation of geotechnical parameters using geophysical data. Journal of Engineering and Technological Science, 49(1): 95-113.

***** Formatting Issue - Best Available Paper/Graphic *****

Investigating the indoor Concentration of Radon-222 and its possible health Implications on the staff of a Pharmaceutical Company in Ota, Ogun State, Nigeria

Adewoyin, O. O., Omeje, M., Onumejor, C. A., Adagunodo, T. A., Ndubuisi, A., Nwaluka, V. O., Akinpelu, A., Joel, E. S.
Department of Physics, College of Science and Technology Covenant University
PMB 1023, Ota, Ogun state, Nigeria
Corresponding author's e-mail: olusegun.adewoyin@covenantuniversity.edu.ng

Abstract. The present study investigated the concentration of radon in some offices of a pharmaceutical company in Ota, Ogun State. A Durridge RAD 7 equipment was engaged for the indoor radon measurement. The measurement of radon concentration was carried out for 8 days in 4 different offices in the main administrative building of the company (one office on the ground floor and three on the first floor). The results obtained for the four locations ranged between 19 and 160 Bqm^{-3} which is still below the world set limit of 200 Bqm^{-3} as recommended by International Commission on Radiological Protection. However, the office on the ground floor reported the highest concentration of radon-222, which is 160 Bqm^{-3}. Therefore, it can be concluded that, since, the highest concentration of radon-222 observed in this study is below the international set limit, then, the occupants of the offices considered for this study are safe.

Keywords: Radiological protection, RAD 7, radon-222, Gamma Spectrometer

1. Introduction

Natural Occurring Radioactive Material (NORM) has become an international concern due to its impact on human health [1]. These radioactive substances; mainly ^{238}U, ^{232}Th and ^{40}K originate from the earth subsurface which forms a major part of man's existence [2]. They are found in geological formations like soils, water, plants and so on, since man depends on these materials for living, exposure to radiation can hardly be avoided. Among the radioactive materials, ^{238}U is known to disintegrate to ^{226}Ra, which in turn gives birth to ^{222}Rn, popularly known as Radon [3]. Radon has the symbol Rn and atomic number 86, and it is a chemical element. It is a noble gas that is radioactive, uncolored, does not have a smell and tasteless [4]. Radon is the product of radium's immediate decay. Its most stable isotope, ^{222}Rn, is only about 3.8 days old and has made radon one of the rarest elements, as it declines very fast [5]. Under normal conditions, radon is gaseous and easily inhaled. Radon gas is regarded as a risk to health. It often contributes the largest radiation dose to a person's background, but the radon gasses level differs from location to location due to local geological differences [6,7]. In spite of its short lifespan, natural radon gas, for example uranium mineral, could build on low areas like basements and crawl areas, especially due to its high density. In ground water radiation can also occur–in some spring and hot waters for example [8]. There has been a connection between high radon breathing level and the incidence of lung cancer in epidemiology studies. Radon is an indoor air quality contaminant worldwide. Radon is the second most common cause of lung cancer following cigarette smoking, resulting in 21,000 lung cancer death each year [9][4][10]. While radon is the second most common cause of lung cancer, according to USEPA estimates, it is the number one non-smoker cause. As the radon decay itself, it produces decay products called radon daughters (also known as radon descent), which are other radioactive elements [11]. In contrast to gas-based radon itself, radon daughters are solid and adhere to surfaces, such as airborne dust particles. These particles may also cause lung cancer if contaminated dust is inhaled. Radon was the fifth radioactive element to be found after uranium, thorium, radium and polonium [7] that radium compounds were emanating from a radioactive gas called Radium. [12] researched the improvement on indoor radon accumulation rate in CST laboratories at Covenant University Ogun state, Ota, Nigeria. In this study, the level of radon measured ranged from 0 to 57.3 Bqm^{-3} for all study locations which is well below the world set limit of 200 Bqm^{-3}. Their study further discovered that the indoor relative humidity has high influence on indoor radon. Studies have revealed that uranium-238 contributes to the concentration of radon-222 [2-4]. Since uranium-238 occurs naturally, there is need to measure its contribution to the indoor gas so as to mitigate its health effect on the people that are always in such buildings. It

IOP Conf. Series: Earth and Environmental Science **331** (2019) 012044 doi:10.1088/1755-1315/331/1/012044

is on this basis that the contribution of uranium-238 to the concentration of radon-222 in the premises of a pharmaceutical company was measured in order to investigate its health implications on the staff.

2. Geology And Geographical Location Of The Study Area

The study site is situated at Sango Ota, Ogun State in southwest Nigeria. It lies within the latitude 6° 38 N to 6° 41 N and longitude 03° 8 E and 3° 12 E. It is accessible through Agbara –Atan road, off Badagry–Seme express road it is also accessible via Iyana Ipaja to if road. The study area for this project is within the eastern Dahomey basin which is in the sub-humid tropical region of southwest Nigeria. It extends along the mainland edge of the inlet of Guinea. It has a distinct wet and dry period, the dry season is from November-March while the wet portion of the year runs between April and October. There has been extensive discussion on the geology of this study area by various authors such as [13, 14]. They discussed the types of rocks in this area from the formation of early tertiary to late cretaceous in age. The basin stratigraphy has also been put in six lithostratigraphic formations which are Benin Formation, Ilaro, Oshosun Akinbo, Ewekoro and Abeokuta from the youngest to the oldest [15] discussed the formation of Abeokuta to consist of the Araromi, Ise and Afomo Formations. This formation is mainly made out of poorly arranged grit mud stone with shale clay layers and siltstone.

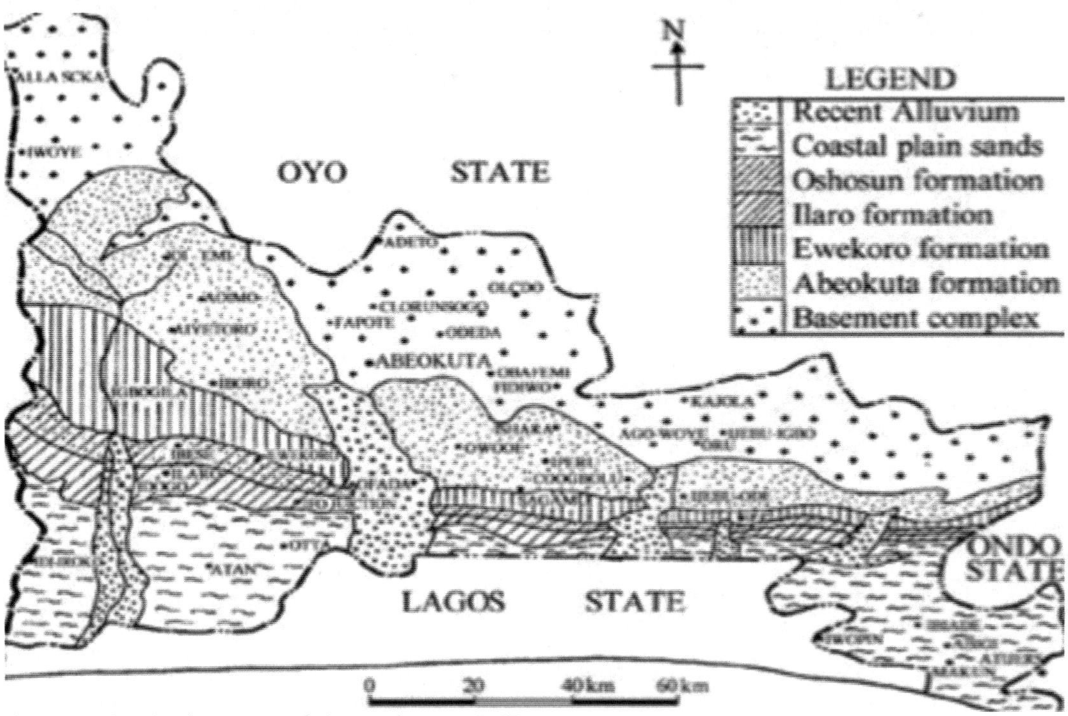

Figure 1: Geological map of the study area [15]

3. METHODOLOGY

The instrument used for this work is Durridge RAD7 electronic radon detector. Indoor radon accumulation levels were measured continuously on hourly basis at the different sampling locations in the pharmaceutical company between 23rd and 31st October 2018. It takes 45 seconds for the RAD7 sniffing process to take place, during this process, air sample passes through the small drying tube and flows into the measurement chamber for radon concentration readout. The equipment is set up in the

IOP Conf. Series: Earth and Environmental Science **331** (2019) 012044 doi:10.1088/1755-1315/331/1/012044

location of interest for 48 hours, during which it measures the concentration of radon at the investigated location. At the expiration of the 48 hours, a print-out comes from the printer of the RAD 7 that captures information such as time and date of the measurement, serial number of the machine, the average test values, the bar diagram of each reading and the cumulative spectrum of alpha energy.

4. Result And Discussion

As earlier declared, four (4) offices were considered for this study, one on the ground floor and the remaining three are located on the first floor of the main office complex. From the data acquired, the graphs of radon concentration were plotted against time in the four cases as presented in Figures 2 and 3.

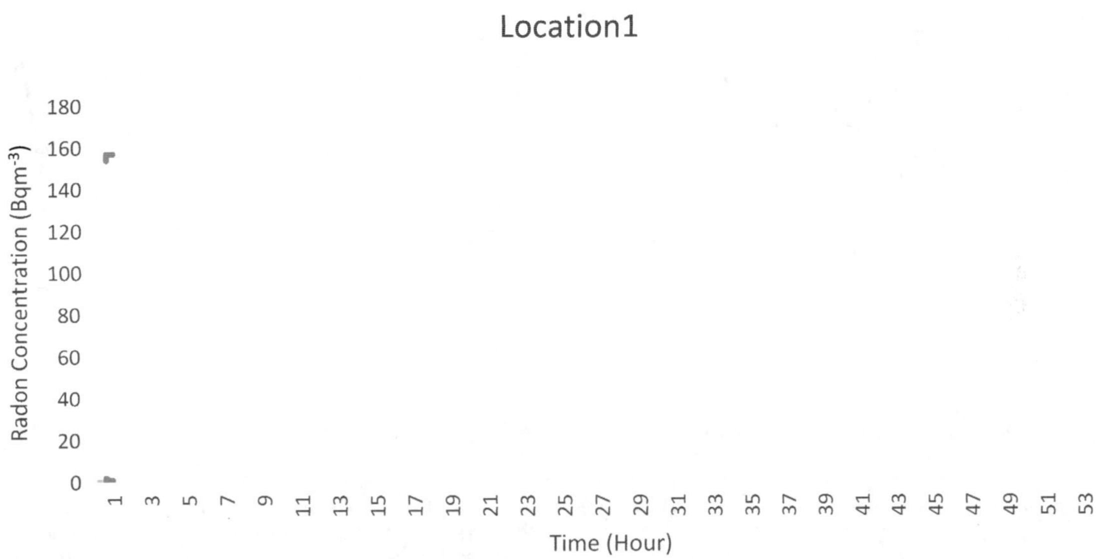

Figure 2: Graph of location 1

At location 1, the concentration of radon over the 48 hours period of data acquisition ranged between 20 and 160 Bqm^{-3}. The highest level of radon concentration, 160 Bqm^{-3}, was observed at this location. The measured result is below the world set limit of 200Bqm^{-3} as recommended by International Commission on Radiological Protection, and [16] so the office is considered safe for the occupants. The activity concentration of uranium-238 in the soil beneath the foundation of these building is suspected to have contributed mainly to the high level of radon concentration detected in the office [11]. Furthermore, indoor air is the major source of radon build-up in soil, concrete and rocks and building material [8]. The change in the concentration of radon indoors reflects the effect of ventilation in each office. The office considered as location 1 does not have any means of ventilation, besides the air conditioning cooling system that is used. Thus, the higher the concentration of radon a man is exposed to, the more the risk of having lung cancer and other related sicknesses increases [9,10].

*** Formatting Issue - Best Available Paper/Graphic ***

Figure 3: Graph of Location 2

In location 2 (Figure 3) the highest value of radon concentration was found to be 19Bqm-³ and the lowest level is below detectable limit (BDL) which makes the office safe for the occupants. As a result, they have lower risk of having lung cancer when compared to the occupants of the office in location 1. The reason for the lower radon concentration measured in this location could be largely, as a result of the fact that the office is located on the first floor of the building, where influence of background radiation from the environment is minimal. Furthermore, the level of ventilation in the office is more than what obtained in the other offices. Therefore, the workers that have their office in location are safe from the radiation poisoning that could lead to other serious diseases.

***** Formatting Issue - Best Available Paper/Graphic *****

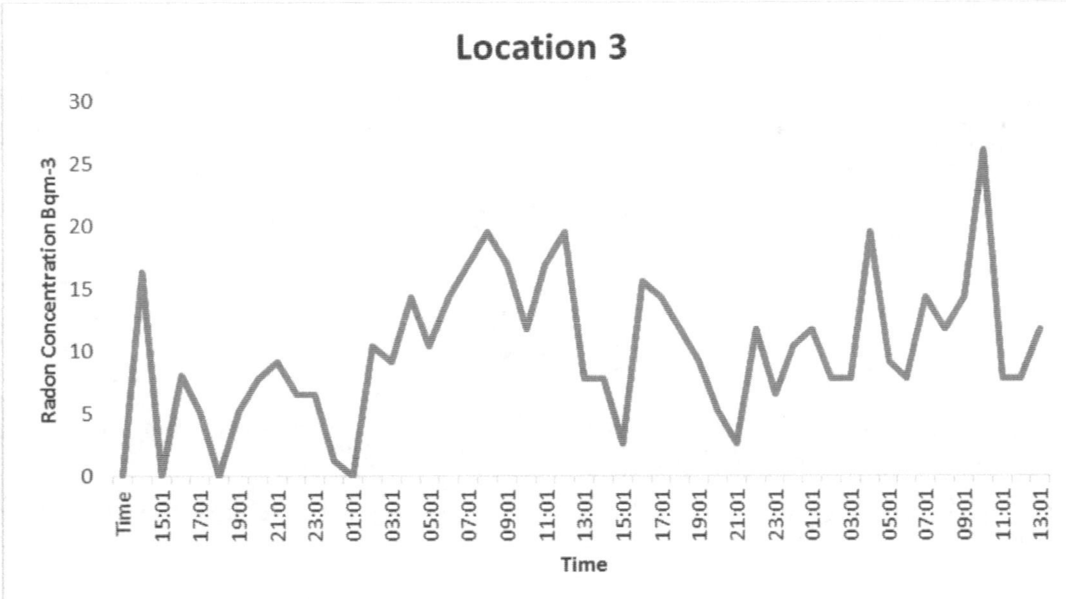

Figure 4: Graph of location 3

In locations 2, 3 and 4, the results obtained for radon concentration ranged between below detectable limit (BDL) and 19 Bqm-3, BDL and 27 Bqm-3, and BDL and 25 Bqm-3 respectively. A lower level of radon concentration was noted in locations 2, 3 and 4, when compared to location 1. This could be because locations 2, 3 and 4 are on the first floor of the office complex, thus reducing the influence of background radiation that emanates from the environment on the radon concentration [3]. Another reason for the low level of radon concentrations in these offices could be because of the level of a more reasonable level of ventilation provided in the offices. Unlike location 1 that has no room for ventilation, locations 2, 3 and 4 are more ventilated. Therefore, the workers that have their offices in locations 2, 3 and 4 are safer from the risk of radon exposure than those in location 1. Thus, they are less likely to have cancer of the lungs by reason of their stay in the offices except if they have exposure to radiation and radon gas from other sources besides the office.

5. Conclusion

The indoor radon concentration level were investigated at the offices of a pharmaceutical company in Ota, Ogun State, Nigeria. Radon concentration was measured on hourly basis using the RAD 7 equipment. The result obtained revealed a variation in radon concentration within the office complex. The values obtained ranged between 19 and 160 Bqm-3. The highest value of 160 Bqm-3 of radon concentration was reported at the office on the ground floor, this could be due largely to the contribution of uranium-238 from the background radiation or the poor ventilation of the office. On the other hand, the result obtained from the offices on the first floor took a different turn because the values reported were 19, 27 and 25 Bqm-3. The difference observed in this result could be attributed to the degree of ventilation available in the offices and the lower level of influence by background radiation. Finally, the results obtained in all the locations are far lower than the international safe limit of 200 Bqm-3. Thus, there may be no health implication by reason of exposure to radon gas of these concentrations.

Acknowledgements

The authors wish to appreciate the management of Covenant University for supporting this research. We will also want to like to express our gratitude to the management and staff of May and Baker Nigeria Plc, Ota, Ogun State for sponsoring this research endeavour. Our appreciation also goes to all our colleagues that provided the needed support for the successful completion of this project.

References

[1] UNSCEAR (2000): Sources and Effects of Ionizing Radiation United Nations Scientific Committee on the Effects of Atomic Radiation

[2] Alonso, H., Rubiano, J. G., Guerra, J. G., Arnedo, M. A, et al. (2019): Assessment of radon risk areas in the Eastern Canary Islands using soil radon gas concentration and gas permeability of soils. Science of the Total Environment, 664: 449-460.

[3] Zhou, Q., Liu, S., Xu, L., Hui, Z. et al. (2019): Estimation of radon release rate for an underground uranium mine ventilation shaft in China and radon distribution characteristics. Journal of Environmental Radioactivity, 198: 18-26.

[4] Zhang, W. Xu, L. C., Li, X. J., Hu, P. H. (2013): Investigation on radon concentration and precipitation rate in uranium mine exhaust Shafts and tailings reservoirs in China. The Proceedings of the 11th Conference Symposium on Radioactive Monitoring in Environment and Effluent. Beijing

[5] Rahman, S., and Ghauri, B. M. (2010). Comparison of seasonal and yearly average indoor radon levels using CR-39 detectors. Radiation Measurements, 45(2): 247-252.

[6] Xie, D., Wang, H. Q. and Kearfott, K. J. (2012): Modeling and experimental validation of the dispersion of 222Rn released from a uranium mine ventilation shaft. Atmos. Environ., 60: 453-459

[7] Bekteshi, S. Kabashi, S., Ahmetaj, S., Xhafa, B., Hodolli, G., Kadiri, S., Alijaj, F. Abdullahu, B. (2017): Radon concentrations and exposure levels in the Trepça underground mine: a comparative study. J. Clean. Prod., 155: 198-203

[8] Burghele, B., Tenter, A., Cucos, A., Dicu, T. et al. (2019): The first large-scale mapping of radon concentration in soil gas and water in Romania. Science of the Total Environmnent, 669: 887-892.

[9] IARC (1988): IARC monographs on the evaluation of the carcinogenic risk to humans Man-made Fibres and Radon, ISBN 43, 92 832 1243 6

[10] IARC (2012) IARC Monographs on the Evaluation of Carcinogenic Risks to Humans, vol. 100D Radiation, ISBN 978 92 832 1321 5

[11] Finne, I. V., Kolstad, T., Larsson, M., Olsen, B. et al. (2019): Significant reduction in indoor radon in newly built houses. Journal of Environmental Radioactivity, 196: 259-263.

[12] Usikalu, M. R., Onumejor, C. A., Akinpelu, A., and Ayara, W. A. (2018). Improvement on indoor radon accumulation rate in CST laboratories at Covenant University, Ota, Nigeria. Technology, 9(10): 135-148.

[13] Jones, H. A. and Hockey, R. D. (1964): The geology of part of southwestern Nigeria. Geol. Survey Nig. Bull. 31:101

[14] Omatsola, M. E. and Adegoke, O. S. (1981): Tectonic evolution and Cretaceous stratigraphy of the Dahomey Basin. Nig. J. Min. Geol. 18(01):130-137.

[15] Aizebeokhai, A. P. and Oyeyemi, K. D. (2014): Application of Geoelectrical Resistivity Imaging and VLF-EM for Subsurface Characterization in a Sedimentary Terrain, Southwestern Nigeria. Arabian J. Geosci.

[16] WHO (2009): H. Zeeb, F. Shannoun (Eds.), WHO Handbook on Indoor Radon: a Public Health Perspective, World Health Organisation

Solar Powered Soil Condition Activated Irrigation System with Automated Bird Repellant

B.Nnebedum, O.E. Oyewande, O. Adepitan, M. Omeje, Akinplu A.

E-mail: Maxwell.omje@covenantuniversity.edu.ng

Abstract. Irrigation plays a very important role in agriculture. Not all geographical regions utilize irrigation which is an important activity that is expensive. For the regions where irrigation is used, it commonly amounts to significant energy consumption. Traditional methods of farming are time-consuming and can only be efficient on a small scale. This project was implemented to make irrigation progressively inventive, easy to use, more efficient and cheaper than the current irrigation system and to solve traditional farming problems like an attack on crops by bird thereby increasing agricultural productivity. This system incorporates a soil moisture sensor to guarantee the application of water to soil adequately. It uses a PIR sensor and a buzzer to solve the problem of attack on crops by birds. This project is powered by a solar panel which is one of the cheapest form of renewable energy and environmentally friendly. The main objective of this project is to upturn the agricultural productivity of the nation by automating the irrigation process. Bird attacks on farmlands are also responsible for destroying a vast amount of crops. One of the traditional problems of agriculture is that farmers frequently neglect to water their plants; it can be as a result of inattentiveness or because of the fact that the farmers are far from home. The utilization of the soil moisture sensor limits the human presence. The microcontroller utilizes the real-time readings of the sensors and automates the irrigation and bird repellent process. From the results, , it was observed that the irrigation time was higher in totally dry clay compared to partially wet clay and wet clay respectively. For clay, the irrigation time is in order of magnitude dry clay > partial wet clay > wet clay of the irrigation other soil samples. For the loamy soil, the dry soil was higher. It follows the same trend as the clay in magnitude. It can be found in sandy soil that the wet sample was higher than the totally dry and partially wet sandy soil in terms irrigation time
Keywords: Microcontroller, Sensors, Irrigation, Buzzer, Solar

1. Introduction

Agriculture is a sure way to the prosperity of a nation. Every region of the world which has developed a contemporary economy has done so by setting up a compelling establishment in agriculture. Almost 70% of Africa's inhabitants are associated with agriculture. Therefore, agriculture remains Africa's surest wager for developing comprehensive economies and creating decent employment opportunities(AASR). The bulk ofAfrica's populace depends on subsistence farming (about 65%). In Africa, the average farmer is a woman with no fertilizer, no high yield seeds, no irrigation and no medication for the animals [6]. Agriculture is the major stream of income for the main part of the Nigerian family and it is an important division of the nation's economy.

The agricultural sector is very significant to Nigeria's economy, the importance of this sector cannot be exaggerated as it is a spur for the provision of food, a critical expansion to the gross domestic product, creating jobs, providing crude resources for agro-allied ventures, and creation of foreign wages. A sectoral examination of the genuine GDP (gross domestic product) conducted in 2006 showed that the agricultural segment of Nigeria has contributed about 41 percent of the GDP compared with the 41.2 percent it contributed in 2005. (CBN,2011) [1].

By the year 2042, the total populace is anticipated to increase to 9 billion individuals [2]. There will be an enormous challenge in providing abundant high quality and nutritious foods for such populace, particularly in the light of the pattern to utilize arable land for biofuel production. Also, with the traditional methods of agribusiness a few downsides like wastage of seed, ill-advised splashing of pesticide on yield so soil standard get corrupts likewise food end up unsafe to human body, if the major issue of water which is looking by every country, particularly India, Brazil, China, and South

Asian nations is just due to the inappropriate utilization of water and water management insufficiency [2].

Nigeria should be where food supply ought to be in abundance because of its vast Agricultural land. The atmosphere and soil in Nigeria support the development of numerous foods and cash crops. Regardless of all these, farming practices in Nigeria aren't comparable to what it used to be in time past, and the commitment of the Agricultural segment to the Nigerian economy is very low. This is because of a decent number of reasons. Some of which are:

1. poverty {insufficient funds and capital},
2. Bad road network{transportation},
3. poor power supply,
4. lack of modern farm machines{inadequate},
5. poor irrigation, fertilization, and drainage system
6. poor agricultural policies,
7. global warming. e. t. c

Despite Nigeria's rich agricultural resources, the agricultural sector has been advancing at an incredibly low rate. Less than 70 percent of the country's cultivable land is underdeveloped with rural and conventional agriculturists using simple cultivating techniques, with resultant low yields to build up or develop most of this land [3].

The primary objective of this project is to of addressing the drawbacks of the current irrigation system, the difficulties of power utilization and water management for increased productivity in the agricultural sector of the nation by designing a solar powered soil condition activated irrigation system with automated bird repellent, clean inexpensive power supply for huge scale cultivating, helping agriculturists and the country overall. Solar energy is easily accessible, non-polluting and an encouraging means of diminishing the use of non-renewable energy sources and sustainable advancement [7-10].

2. Materials and method

The solar-powered soil condition activated irrigation system with automated bird repellent is designed to specific requirements. These requirements are categorized as follows:

i. Hardware requirements
ii. Software requirements

this section we discuss how to format the title, authors and affiliations. Please follow these instructions as carefully as possible so all articles within a conference have the same style to the title page. This paragraph follows a section title so it should not be indented.

2.1. Hardware requirements

The hardware framework is separated into two sections namely; the central control unit and the sensory units. The central control unit houses the majority of the components and is the brain of the whole framework. The sensory units comprise of the soil moisture sensor utilized for getting information about the soil moisture content and the passive infrared(PIR) sensor for getting information on motion that is movement of birds in the field. These sensors feedback the information to the central control unit remotely. The central control unit and the sensory unit are designed to be powered by a solar panel. The block diagram of the system is shown in the Figure 1 below

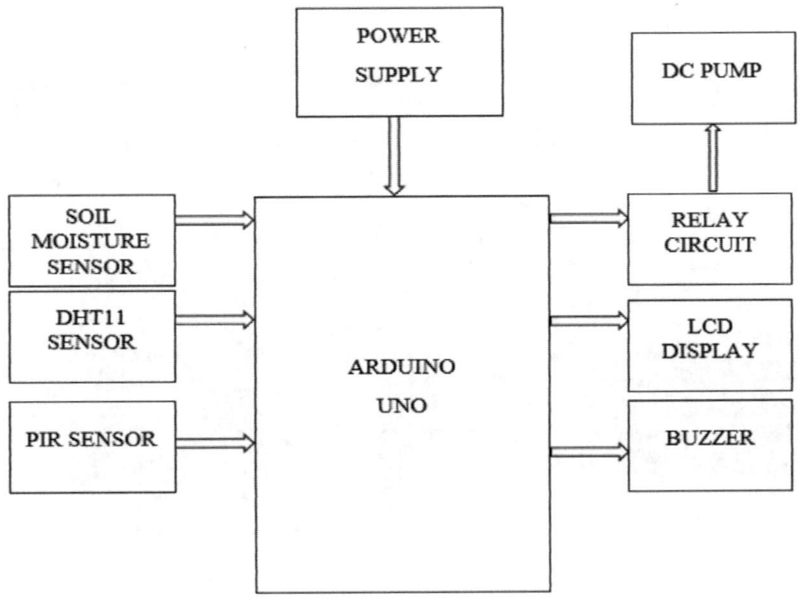

Figure 1: Block diagram of the system

Figure 2 below shows the circuit diagram of the system. It consists of all the components required to build the system and it shows how the systems circuit will be constructed.

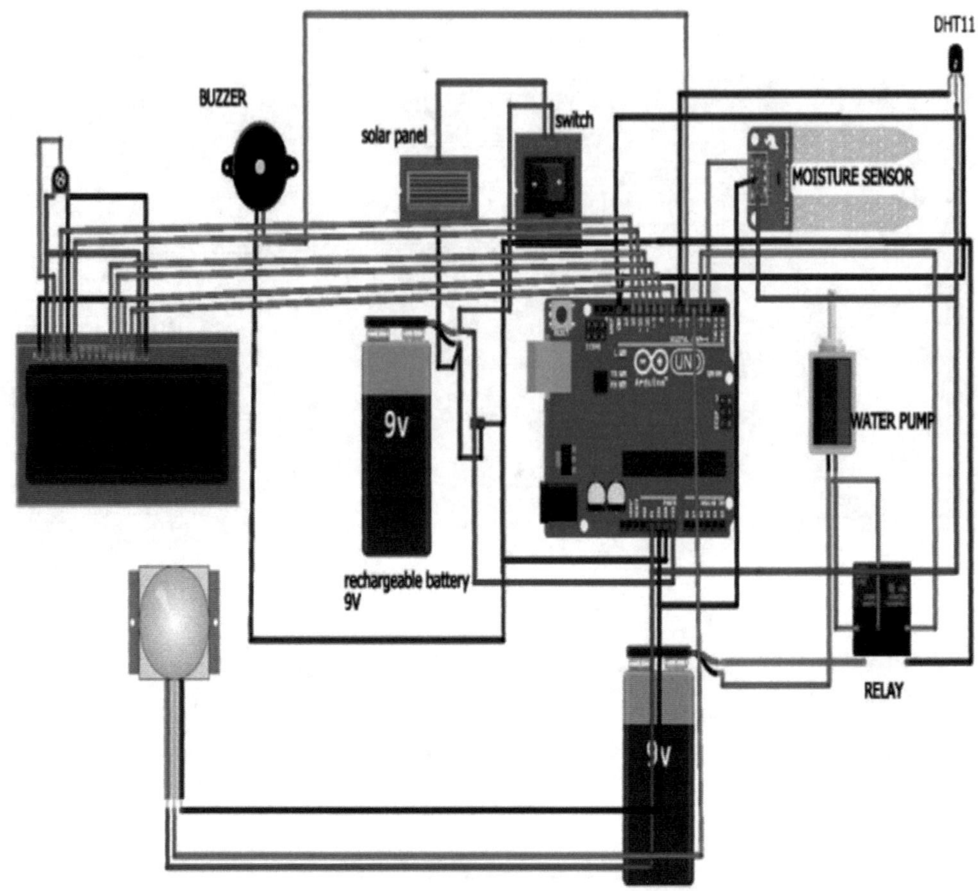

Figure 2: Circuit diagram of the system [2]

2.2 Software requirement

This consists of the microcontroller programming of the ATmega238 using Arduino IDE software and a personal computer. The Arduino Uno is the microcontroller utilized for the construction of this project. It comprises of an Arduino integrated development environment (IDE), and it is programmed using C programming language. The microcontroller chip programmed is the Atmega238 chip. Outlined below are the functions of the programs;

i. Obtain information from the analog moisture sensors
ii. Signal conversion from analog to digital
iii. Read the digital signal
iv. Send control signals based on the read signal.

The Arduino programming workflow using the Arduino IDE is described below in Figure 3 and Figure 4

```
#include <LiquidCrystal.h>


const int rs = 12, en = 11, d4 = 5, d5 = 4, d6 = 3, d7 = 2;
LiquidCrystal lcd(rs, en, d4, d5, d6, d7);

uint8_t PIR = 8;
uint8_t buz = 9;
uint8_t relay = 7;

void setup() {

    Serial.begin(9600);
  lcd.begin(16, 2);
  lcd.print("Smart Irrigation");

  pinMode(PIR , INPUT);
  pinMode(buz , OUTPUT);
  pinMode(relay , OUTPUT);
  digitalWrite(buz , LOW);
  digitalWrite(relay , HIGH);

}

void loop() {
  pir();

} ///
```

Figure 3: The Arduino coding

```
void pir() {

 unsigned int moist = digitalRead(A0);

  if (moist==0) {
 digitalWrite(relay , HIGH);
 Serial.println("no moist");
  }

  else
{
  digitalWrite(relay , LOW);
}

  if (digitalRead(PIR)) {
    lcd.clear();
    lcd.print("Bird alert!");
    digitalWrite(buz , HIGH);
    delay(200);
  }

  else {
    lcd.clear();
    lcd.print("Smart Irrigation");
    digitalWrite(buz , LOW);
    delay(200);
  }
```

Figure 4: The Arduino coding

3. Results of the construction of the project using soil samples

3.1 Construction

The construction of the power supply unit was carried out using a solar panel and a battery. The power supply unit consists of a 10watts solar panel and a 12volts battery. This unit supplies power to the soil condition activated solar powered irrigation system with automated bird repellent. Figure 5 shows the diagram of the power supply unit of the system.

IOP Conf. Series: Earth and Environmental Science **331** (2019) 012045 doi:10.1088/1755-1315/331/1/012045

Figure 5 Soldered components of the control and sensory circuit.

This is the connection of the components of the system to a vero board. After testing the various components with a multimeter to ensure that they supply the required voltage and current for the system to function properly, and testing them together on a bread board, the components are connected to the Arduino and soldered to the vero board permanently.

Figure 6: displays the soldered components of the system to a vero board

Figure 6 displays the system in the testing stage. When the components are connected to the Arduino and the Arduino is connected to the power source, the LCD screen comes on and displays a blank screen. When the soil moisture sensor is put inserted into a soil with low soil moisture the microcontroller sends an output to the DC pump which supplies water to the soil. When the soil moisture of the soil is normal the soil moisture sensor sends an output to the microcontroller which sends an output to the DC pump to cut off water supply. Also when motion is detected by the PIR (passive infrared) sensor the microcontroller triggers the buzzer which goes off. When motion is no longer detected by the PIR (passive infrared sensor) the buzzer goes off.

Figure 7: When motion is detected

Figure 7 displays the output of the system when motion is detected. When motion is detected by the PIR (passive infrared) sensor, the microcontroller sends an output to the buzzer which goes off and the LCD which displays "Bird alert".

Figure 8: When irrigation is required

Figure 8 displays the output of the system when water content is low. When the soil moisture is low the microcontroller send output to the DC pump which supplies water to the plants and the LCD screen which displays "Smart Irrigation".

Table 4.1 The reading of Soil Samples, Soil Type, Initial Soil State and Irrigation Time(Sec.)

CATEGORY	SOIL SAMPLE	SOIL TYPE	INITIAL SOIL STATE	IRRIGATION TIME (SECS)
A	1	SANDY	Totally dry	12
	2	SANDY	Partially Wet	10
	3	SANDY	Wet	6
B	1	LOAMY	Totally dry	20
	2	LOAMY	Partially Wet	13
	3	LOAMY	Wet	7
C	1	CLAY	Totally dry	30
	2	CLAY	Partially Wet	20
	3	CLAY	Wet	12

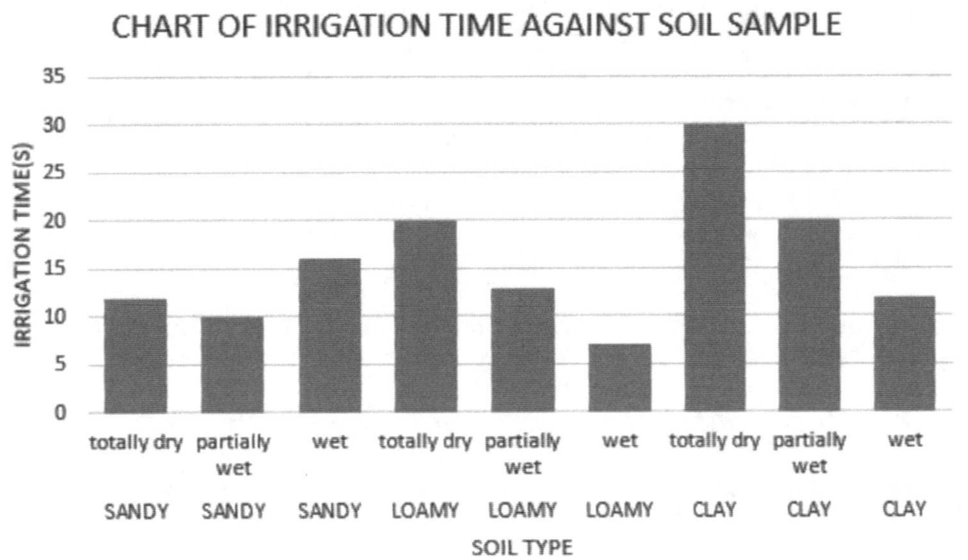

Figure 9: Chart of irrigation time against soil samples

In Figure 9, it was observed that the irrigation time was higher in totally dry clay compared to partially wet clay and wet clay respectively. For clay, the irrigation time is in order of magnitude dry clay > partial wet clay > wet clay of the irrigation other soil samples. For the loamy soil, the dry soil was higher. It follows the same trend as the clay in magnitude. It can be found in sandy soil that the wet sample was higher than the totally dry and partially wet sandy soil in terms irrigation time.

4. Conclusion

The soil condition activated solar irrigation system with automated bird repellent was developed with the aim of addressing the drawbacks of the current irrigation system, the difficulties of power utilization and water management for increased productivity in the agricultural sector of the nation. To accomplish the objectives of this project, a solar panel was used to power the system, this empowers the system to work efficiently without requiring an AC source. Management of water is accomplished by the use of the soil moisture sensor which decides when the soil needs to be irrigated. The effectiveness of this system us that power is preserved in two ways:

i. The project is direct current (DC) controlled, and the system is triggered only when the microcontroller receives data sent from the sensory unit data.
ii. The system is totally powered by a solar panel.

The completion of this project work has brought about the following achievement:

i. Construction of an irrigation framework.
ii. Successful design and implementation of a Solar Powered Soil Condition Activated Irrigation System with Automated bird repellent.
iii. Designed to defeat current drawbacks of the irrigation system.

Acknowledgement

Authors would like to thank Covenant University Centre for Research, Innovation and Discovery for their financial support of this present study.

References

[1] **Eregha, P.B., Babatolu, J.S., Akinnubi, R.T. (2014).** Climate Change and Crop Production in Nigeria: An Error Correction Modelling Approach. *International Journal of Energy Economics and Policy*, 4(4):pp.297-311 ISSN: 2146-4553

[2] **Katariya, S.S, Gundal, S.S., Kanawade, M.T. and Khan, M. (2015).** Automation in agriculture. *International Journal of Recent Scientific Research*, 6 (6): pp.4453-4456.

[3] **Ayinde, A. E., Luke, O. O., Idris, K. O., Oluwole, A. A. (2012).** Effect of irrigation and drought on agricultural, Oyo. *International Journal of the Bioflux Society.*

[4] **Kelvin, A.U. (2017).** Nigeria agriculture and sustainability: problems and solutions.

[5] **Onyeneke, R.U. and Madukwe, D. K.(2010).** Adaptation measures by crop Farmers in the southeast rainforest zone of Nigeria to climate change. *Science World Journal*, 5(1): ISSN 1597-6343.

[6] **Burt, C., Clemmens, Albert, Solomon, K., Howell, Terry andStrelkoff, S.T. (1999).** Closure to "irrigation performance measures: efficiency and uniformity". *Journal of Irrigation and Drainage Engineering-ASCE*, 98(2):1999;0733-9437.

[7] **Brugiati, E., Prof. Jinyue, Y., Dr. Hailong L.** Groundwater balance and level variation induced by PV and wind pumping system," School of Sustainable Development of Society and Technology (HST).

[8] **Nasiakou, A.,Manolis, V., Dimitris, Z. (2016).** Smart Energy for Smart Irrigation, Computers and Electronics in Agriculture.

Sustainable Smart City and its Promising Urban Value: An Overview

G U Fayomi [1, 2*], S E Mini[2], O S I Fayomi[3, 4], C C Mbonu[3], N E Udoye[3]

[1]Strategic Business Services, Covenant University, Ota, Nigeria
[2]Department of Geography, University of South Africa, Johannesburg, South Africa
[3]Department of Mechanical Engineering, Covenant University, P.M.B 1023, Ota, Ogun State, Nigeria
[4]Department of Chemical, Metallurgical and Materials Engineering, Tshwane University of Technology, Pretoria, South Africa
*Corresponding author: fayomiuche@gmail.com; +2348033752329

Abstract. Development from conventional urban city to the smart city requires a level of sustainable development. Smart city involves multiple fields of human capacity in knowledge of artificial intelligence, information and communication technology and management in form of environment sensing, innovation, education, culture, deep and machine learning, networking, information processing and digital technology, and city governance in order to enhance performance in resource management, improve well-being and economic development in the aspect of individuals standard of living which impacts on transportation, water supply, waste disposal, healthcare, etc. This paper highlights actions that can combine sustainability and innovation to result to a sustainable smart city and measures put into play in order to achieve a smart city which claims to be sustainable.

Keywords: Urban, City, Wastes, Technology, Sustainable

1. Introduction

In search of a better standard of living, rural to urban migration has increase in the years passed and will keep increasing. This has a great impact on pollution, energy demand, employment, etc. The increasing rate of urbanization from a global aspect sets up new challenges for the urban planners and policy makers [1]. Studies by [2] highlighted that climate change, ecosystem destruction, increased rate of natural disasters and many more are environmental and socioeconomic crises faced by most part of the world due to increased urbanization. This brings into play more methods to combat the negative effects of increased urbanization. Development from conventional urban city to the smart city to address the negative impact of urbanization requires a level of technological development and innovation which involves ICT, this provides new ways to solve urban problems [3]. Studies by [4] reveals that the quality of living among members of a city, economy, easy to access a means to transport individual and commodity, traffic management, the environmental state, and interaction with government, can be step up as information and communication technology plays a vital role in achieving that. Understanding what smart cities are were derived from [5] as a result of multi-dimensional approach of human capacity in knowledge of artificial intelligence, information and communication technology and management in form of environment sensing, innovation, education, culture, deep and machine learning, networking, information processing and digital technology, and city governance in other to enhance

IOP Conf. Series: Earth and Environmental Science **331** (2019) 012046 doi:10.1088/1755-1315/331/1/012046

performance in resource management, improve well-being and economic development in the aspect of individuals standard of living which impacts on transportation, water supply, waste disposal, healthcare, etc. In the study of smart cities and its makeup, [6] highlighted that smart cities integrates Internet of Things which includes sensors as one of the means to generate big data where the big data are analysed through artificial intelligence. These functions work together and are linked to each other as shown in the figure1. This sensor is used to collect data. Data from numerous neighbourhoods can now be obtained in order to better understand the urban fabric, this digital concept which are linked to artificial intelligence (AI) and machine learning which is an element of AI, making collection of near real-time data possible. This allows planners in the urban region and policy makers also, to make better decisions as the digital concept gives a deeper understanding of the city's evolution, adaptation and response to various conditions.

Research on sustainability has taken turns in providing different understanding on what sustainability is and that it means for a city to be referred to as a sustainable city. This study acknowledges the understanding of sustainable cities by [7] , these are cities characterized by its conformance in accomplishing UN's 11th sustainability development goal (sustainable cities and community) which envelops some other sustainability development goals on reduction of poverty and hunger, improved health and well-being, provision quality education which is inclusive, solving issues of gender inequality, provision of clean water and environment, provision of energy which is affordable and clean, increased production with respect to consumption, reduction of action that affect the climate condition negatively, provision of peace and justice, and partnership for the goals. Possibility to develop a smart city that doesn't having the functions of sustainability is certain. Sustainable smart city is the application of technology to a sustainable city as explained by [8] although outlined that smart city is a factor of transition from sustainable city. This paper outlines that a sustainable smart city can be due to management and innovation to sustain sustainability in a smart city. This leads to the creation of a sustainable smart city. This paper looks forward to studies the makeup of a smart city which consist of the concept and features in a city to achieve smartness, and the measures put into play in order to ensure sustainability of a smart city, although [9] portrayed points to show the difference between sustainable cities and smart cities. This paper highlights actions that can combine sustainability and smartness to result to a sustainable smart city building on the discoursed theory and concept of various authors.

2. Concept of a Smart City and Enabler

The smart city also referred to as a digital city: [9]. It is a city with increased application of digital technology in addressing human needs and economic growth, information city:[10] clams that information and communication technology(ICT) is the bases of a smart city and affect the other sectors of urban development to smart city, intelligent city: [11] believes smart city is a function of intelligence involving artificial intelligence and sensors, etc., innovative city, knowledge-based city:[3] sees human capacity development as the basic function for smart city development and sustenance referring a smart city as a knowledge-based ecosystem, ubiquitous city: [1] says smart city as a global phenomenon, or wired city involves multiple fields according to [8].

From the brake down of authors point of view of what a smart city is, better understanding of a smart city has been derived. Over the years' researchers have worked on different sectors of a smart city. [5] enumerated how universities manages the knowledge management governance issues in 20 smart city projects to conclude that university is responsible for knowledge based ecosystems such as the smart city project. Smart city technology as outlined

IOP Conf. Series: Earth and Environmental Science **331** (2019) 012046 doi:10.1088/1755-1315/331/1/012046

by [20] consists of internet of things, sensors in city infrastructures, building and other service units such as the metro bus stop, etc., resource management technologies such as renewable energy devices etc., means to store big data received from the environment means provided by internet of things (IOT), visualization technology for individuals, etc. These technologies are put into consideration while planning a smart city.

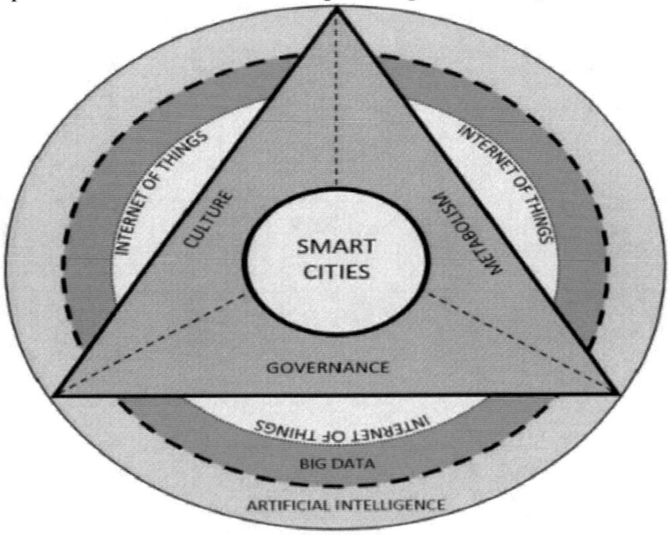

Figure 1: AI and Big data integration in smart cities to ensure liveability [6]

City planning is experiencing major changes now when new expectation and demand are made on sustainability in cities and resource efficiency. City planning therefore no longer only involves designing areas, building buildings, infrastructure and services, but also takes into consideration new perspectives, such as digitalisation, integration, quality of life, needs of citizens, new technologies, and equality. New ways for city planning are needed to plan and build a smart city district that integrates ICT in infrastructures and utilizes it in novel ways to control and manage the smart city functions [12].[3] highlighted six smart city development enablers which includes

- Smart city governance.
- Urban pro-activeness.
- Service innovation.
- Urban openness.
- Partnership development.
- Integration of infrastructure.

Studies by [8] explains that a city can be sustainable if the conditions of productivity does not deteriorate over time hence striking a balance between environmental protection and development of urban area with the mind set of equality in income, employment, shelter, basic service, social infrastructure and transportation in the urban area. [8] in figure 2 outlined the difference between sustainable framework and smart city framework illustrated in the graph below.

IOP Conf. Series: Earth and Environmental Science **331** (2019) 012046 doi:10.1088/1755-1315/331/1/012046

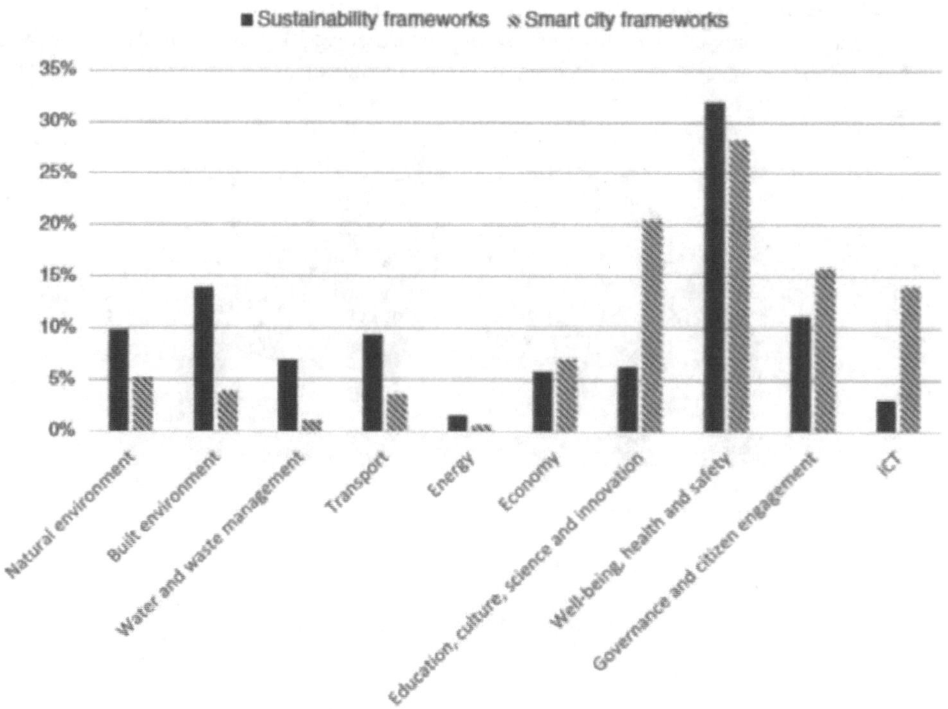

Figure 2: Difference between sustainable cites and smart cities under different sectors [8]

This study by [8] shows the difference between sustainable cites and smart cities, moreover the sustainable smart city brings together the attributes of both the smart and the sustainable city i.e., having sustainable development with the help of technology. It was also attested that sustainable development occurs when the cities production conditions are properly maintained with good standing and growth. This development comes with the protection of the environment, improves equality, and creates employment, provision of social infrastructure and basic services. It is of great importance according to [4] for a smart city to be able to sustain the above implication of development as discussed by [8] with the help of technology. [3] studied that the physical infrastructure, innovation ecosystem and quality of life of a smart city are determined six factors, also affirmed by [1] as Giffinger's six implementation domain as illustrated in figure 3 for a sustainable smart city which includes: smart environment, smart mobility, smart people, smart economy, smart governance, and smart living .

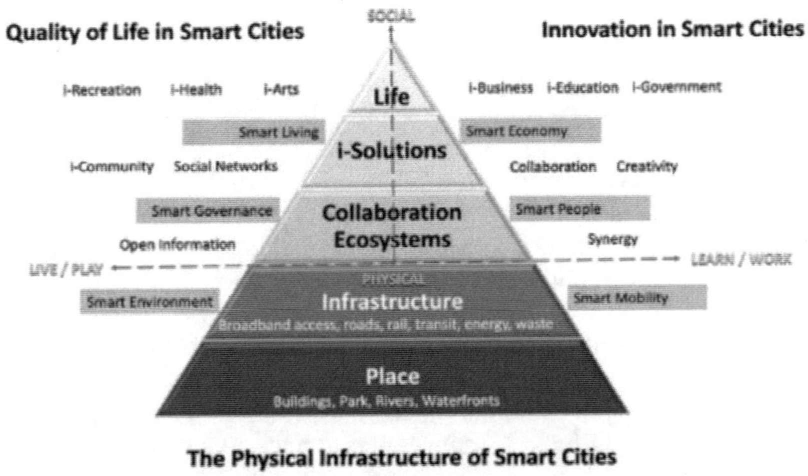

Figure 3: Giffinger's smart city [3]

3. Smart keys and progression

3.1. Smart environment

This involves the use of technology to impact positively on waste disposal [13], food growth (i.e. in terms of plant growth monitoring, pest control and soil condition, bio-science and microbiology practitioners can also develop special treatment for reduction of the use of toxic substance such as toxic pesticides and toxic fertilizers.), pollution control (this involves the use of sensors to detect and prevent wildfires, improve prediction, and simulation of pollution in the city). To measure the coherence of the zero carbon objectives projected for 2050, [14] proposed a zero means of reversing model greenhouse gas emissions, housing quality (Increased security via connected cameras and smart refrigerators can learn how home dwellers behaved and increased comfort at home to reduce food waste). Facility management and smart electricity grid (This includes intelligent energy supply in which the architecture enables the deployment of renewable energy efficiency system).

3.2. Smart mobility

This was made to make mobility more favourable therefore managing traffic, tracking vehicle, smart metros etc. It involves the deploying sensors in specific infrastructures such as roads, rail, airports, seaports, bridges and subways, etc. These sensors offer valuable information on fluidity, reducing accidents, improving public transportation and facilitating parking [3].

3.3. Smart people

The skills and competences of the individual or group can be defined as human capital. Universities and other higher education present in a smart city develops human capacity, this leads to citizens adaptation of technology and security [15].

3.4. Smart economy

Innovative business for the smart city based on smart environment, mobility and people. Investigations on how metropolitan cities set up patterns of its business model to become a smart city [16].

3.5. Smart Governance

IOP Conf. Series: Earth and Environmental Science **331** (2019) 012046 doi:10.1088/1755-1315/331/1/012046

Smart governance requires innovative ICT infrastructure, planning process, sustainable governance, and collaboration utilization to achieve these goals [17]. Investigations was carried out carried out

3.6. Smart living

In a smart city the smart living is a makeup of Smart Environment, smart mobility, smart people, smart economy, and smart governance. Smart living must include improved education, health, security, empower citizen participation and social services. Smart structures, water management, mobility, security and energy grid are also physical makeup of smart living [18, 21]. The aspect of health is one of the basic necessity in a smart city due to this, medical practitioners, health researchers has come in collaboration with urban city planners and urban city policy makers, with the objective to identify and decode parameters which has impact in public health and urban liveability [19]. At the point when a city can be called smart, it must have fulfilled certain criteria and attained a point of technical development. The quest for sustainable smart city is a reality by combing the futures of a sustainable city with that of a smart city. This involves the use of modern technology and ICT to sustain equity between environmental protection and productivity. The studies from literature presented by [1] draw this paper to a conclusion that sustainable smart cities are attainable and must not be separated. The table below highlights the area of sustainability as set up by UN which is handed by the uses of technology put in to use in a city.

Table1: Giffinger's implementation domain and the SDGs covered

Giffinger's six implementation domain studies by [1]	Addressed sustainable development goals set up by UN [16]
Smart environment	Provision of clean water and environment (SDG6, 14, and 15), provision of energy which is affordable and clean(SDG7), sustainable cities and communities(SDG11).
Smart mobility	Industry, innovation and infrastructure(SDG9)
Smart people	Provision quality education which is inclusive (SDG4), solving issues of gender inequality (SDG5)
Smart economy	reduction of poverty and hunger (SDG1&2), decent work and economic growth(SDG8)
Smart government	Increased production with respect to consumption (SDG12), reduction of action that affect the climate condition negatively (SDG13), peace, justice and strong institutions (SDG16), partnerships for the goals (SDG17).
Smart living	Provision for good health standing and which includes well-being of citizens(SDG3),

The six implementation domain by Giffinger can bridge the caps between sustainable and smart cities to address UN's sustainable goals as high by [20].

Conclusion

This paper have identifies actions that combine sustainability and innovation on sustainable smart city and measures to achieve a smart city which claims to be sustainable. Concept factor to foster sustenance of smart city from management and innovation was clearly identified. This falls in the hands of knowledge through human capacity development to improve the body for smart city governance. Innovation begins from improved productivity, maintenance of the smart city to its driven mechanism.

Acknowledgement

The author acknowledged the financial support by Covenant University CUCRED for the publication of the manuscript.

Reference

[1] Dameri, R. P., Benevolo, C., Veglianti, E., & Li, Y. (2019). Understanding smart cities as a glocal strategy: A comparison between Italy and China. *Technological Forecasting and Social Change*. https://doi.org/10.1016

[2] Yigitcanlar, T., Kamruzzaman, M., Foth, M., Sabatini-Marques, J., da Costa, E., & Ioppolo, G. (2019). Can cities become smart without being sustainable? A systematic review of the literature. *Sustainable Cities and Society*, 45, 348–365. https://doi.org/10.1016/j.scs.2018.11.033

[3] Appio, F. P., Lima, M., & Paroutis, S. (2019). Understanding Smart Cities: Innovation ecosystems, technological advancements, and societal challenges. *Technological Forecasting and Social Change*. https://doi.org/10.1016/j.techfore.2018.12.018

[4] Ismagilova, E., Hughes, L., Dwivedi, Y. K., & Ravi Raman, K. (2019). Smart cities: Advances in research—An information systems perspective. *International Journal of Information Management*. https://doi.org/10.1016/j.ijinfomgt.2019.01.004

[5] Ardito, L., Ferraris, A., Petruzzelli, A. M., Bresciani, S., & Del Giudice, M. (2019). The role of universities in the knowledge management of smart city projects. *Technological Forecasting and Social Change*. https://doi.org/10.1016/j.techfore.2018.07.030

[6] Allam, Z., & Dhunny, Z. A. (2019). On big data, artificial intelligence and smart cities. *Cities*. https://doi.org/10.1016/j.cities.2019.01.032

[7] Martin, Chris J., James Evans, and Andrew Karvonen. 2018. "Smart and Sustainable? Five Tensions in the Visions and Practices of the Smart-Sustainable City in Europe and North America." *Technological Forecasting and Social Change*. https://doi.org/10.1016/j.techfore.2018.01.005.

[8] Ahvenniemi, H., Huovila, A., Pinto-Seppä, I., & Airaksinen, M. (2017). What are the differences between sustainable and smart cities? *Cities*. https://doi.org/10.1016/j.cities.2016.09.009

[9] Hussain, A., Wenbi, R., da Silva, A. L., Nadher, M., & Mudhish, M. (2015). Health and emergency-care platform for the elderly and disabled people in the Smart City. *Journal of Systems and Software*. https://doi.org/10.1016/j.jss.2015.08.041

[10] Caragliu, A., Del Bo, C., & Nijkamp, P. (2009). *Smart Cities in Europe*. Proceedings of the 3rd Central European Conference in Regional Science—CERS 2009, Kosice, 10, 49-59

[11] El-Haddadeh, R., Weerakkody, V., Osmani, M., Thakker, D., & Kapoor, K. K. (2019). Examining citizens' perceived value of internet of things technologies in facilitating public sector services engagement. *Government Information Quarterly*. https://doi.org/10.1016/j.giq.2018.09.009

IOP Conf. Series: Earth and Environmental Science **331** (2019) 012046 doi:10.1088/1755-1315/331/1/012046

[12] Axelsson, K., & Granath, M. (2018). Stakeholders' stake and relation to smartness in smart city development: Insights from a Swedish city planning project. *Government Information Quarterly*. https://doi.org/10.1016/j.giq.2018.09.001

[13] Rybnytska, O., Burstein, F., Rybin, A. V., & Zaslavsky, A. (2018). Decision support for optimizing waste management. *Journal of Decision Systems*, 27, 68–78.

[14] Contreras, G, and Federico P. 2019. "Economic and Policy Uncertainty in Climate Change Mitigation: The London Smart City Case Scenario." *Technological Forecasting and Social Change*. https://doi.org/10.1016/j.techfore.2018.07.018

[15] Yeh, H. (2017). The effects of successful ICT-based smart city services: From citizens' perspectives. *Government Information Quarterly*, 34, 556–565.

[16] Schiavone, F. Francesco, P. and Daniela M. 2019. "Business Model Innovation for Urban Smartization." *Technological Forecasting and Social Change*. https://doi.org/10.1016/

[17] Truong, H., & Dustdar, S. (2012). A survey on cloud-based sustainability governance systems. *International Journal of Web Information Systems*, 8, 278–295.

[18] Peña, M., Biscarri, F., Guerrero, J. I., Monedero, I., & León, C. (2016). Rule-based system to detect energy efficiency anomalies in smart buildings, a data mining approach. *Expert Systems with Applications*, 56, 242–255.

[19] Dritsa D, and Biloria. A. 2018. "Towards a Multi-Scalar Framework for Smart Healthcare." *Smart and Sustainable Built Environment*. https://doi.org/10.1108/sasbe-10-2017-0057

[20] Parks D, and Harald R. 2019. "From Sustainable to Smart: Re-Branding or Re-Assembling Urban Energy Infrastructure?" *Geoforum*. https://doi.org/10.1016/j.geoforum.2019.02.012.

[21] Thibaud, M., Chi, H., Zhou, W., & Piramuthu, S. (2018). Internet of Things (IoT) in high-risk Environment, Health and Safety (EHS) industries: A comprehensive review. *Decision Support Systems*, 108, 79–95.

Caragliu, A., Del Bo, C. and Nijkamp, P. (2009) Smart Cities in Europe. Proceedings of the 3rd Central European Conference in Regional Science—CERS 2009, Kosice, 10, 49-59

Design and Construction of Density Based Traffic Control System

M. R. Usikalu, A. Okere, O. Ayanbisi, T. A. Adagunodo and I. O. Babarimisa
Department of Physics, Covenant University, P.M.B 1023, Ota, Ogun State, Nigeria

Correspondence Email: moji.usikalu@covenantuniversity.edu.ng

Abstract. Congestion is a serious issue due to vechicular traffic. One of the known causes of traffic congestion is the amount of time spend waiting for the red light to change to green. The changing of traffic light is hard-coded and it is not reliant on traffic volume. There is therefore need to simulate and optimize traffic control to better accommodate density based traffic rather than time based. This system attempts to lessen possibilities of traffic jams brought about by traffic lights to a reasonable degree. This project, a density based traffic control system is been implemented to solve this problem. The system entails programming an Arduino using Arduino integrated development environment (IDE) to enable traffic lights give the right of access to the road by selecting the lane with the high number of cars. The traffic lights are modified to chip away at an auspicious premise until there is a signal identified by the infrared sensors. The sensor identifies an object (i.e. a vehicle, a motorcycle etc) and signals the Arduino to control the traffic lights for its individual path. Once there is no sign identified by any of the four sensors the traffic lights keep on dealing with an auspicious premise. The mean response time of the sensor was found to be 0.39 seconds. Further research is recommended to produce the device on a large scale to be deployed to all roads in the country.

Keywords: Traffic, Density, Control, IR sensors.

1. Introduction

One of such over-stretched infrastructure is the road, a circumstance which has resulted to increase in traffic. In spite of the fact that traffic lights have dependably been utilized for controlling the movement of traffic (pedestrian or automobile), traffic management in many real urban areas around the globe has kept on being a subject of concern [1]. A traffic light is a signalling device which controls a traffic flow at road intersections. It consists of three basic lights which include red, yellow and green. Red signal is used to halt traffic from proceeding; yellow signal alerts vehicles for brief stop, while green signal alerts vehicles for procession in the indicated directions [2]. Traffic congestion is a situation that happens on road networks in which vehicles travel slower than usual due to the increased physical use of vehicles (traffic) on the road at that moment. Also known as traffic jam, traffic congestions may result from roads being blocked, bad roads, accidents on the road, lack of proper traffic light system to control vehicles, inappropriate driving by road users etc. This would make the journey longer due to the slow movement of the traffic and increased queuing of vehicles [3]. Urban communities started to make traffic tenets to minimize crashes, while traffic flags and cops were utilized to coordinate option to proceed at major urban convergences. Traffic control development in urban streets has aided easy movement and use of automobile in mega cities [4]. Most of the major roads have effective traffic control system [5], which has facilitated easy flow of vehicles in urban regions. Railroads were invented serving as a temporary solution to the developing issue of street traffic control. However, this creates congestion at terminals inside urban communities.

Therefore, the aim of this study is to design a density based traffic control using Arduino in order to solve the problem of traffic congestion.

2. Methodology and System Design

In the design and construction of this system all components work simultaneously. The system works just as a typical traffic light system. The uniqueness of this is in the event that the thickness of vehicles in a specific path of the road is high. At that point, the sensor in that specific path turns out to be low else it is read as a high signal. The signal from the IR (Infrared) is used by the system to control the traffic jamming of the lane. In the event that we get a low signal from any of these sensors, at that point the green LED shines to that specific way and gives a red to every other way. The Arduino IDE is programmed using C language. The block diagram of the density based traffic control system is shown in Figure.1. It is divided into four different sections with each section representing a lane.

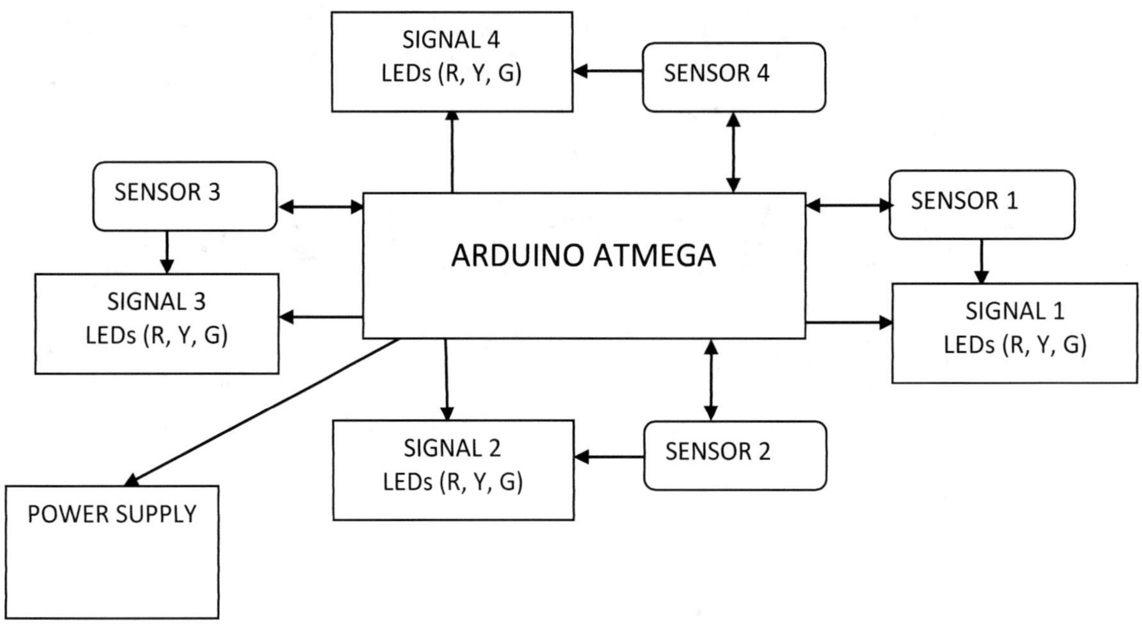

Figure 1: Block diagram of a density based traffic control system.

2.1 Design of the Project

The construction of the system is divided into three stages as stated: power supply stage, sensor switching stage, and Arduino connection stage. The power supply for this system is 9 V DC battery as shown in Figure 2. It has a nominal voltage of 9 V, a discharge resistance of 620 Ω s and a cut off voltage of 5.4 V. According to the power needed for the components of the thickness based traffic light control framework, supply of +5 V regarding GND is created. The total hardware worked with TTL (Transistor-Transistor Logic) rationale dimension of 0 V to 5 V. Power applied to the V_{in} pin is stepped down to 5 V by the on-board regulator on the Mega. The minimum voltage is about 6.2 V because of the regulator dropout (that is, the regulator needs at least 1.2 V above its 5 V output in order to operate) 9 -12 V is recommended. Applying a higher voltage to the system will not provide any more power to the Mega and its peripherals/shields. Instead, the excess power is dissipated in the regulator as heat.

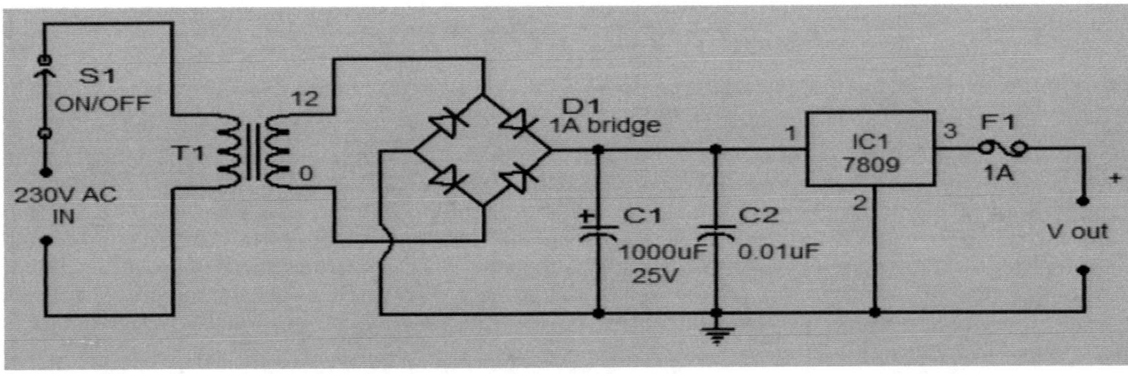

Figure 2: 9V power supply for the Arduino Atmega (retrieved from [11])

The infrared sensors (IR) are the most important components of this project. The sensors act like a switch as it controls the switching of the LEDs. The IR sensors have been applied to several traffic systems [6-9]. Other uses of electronic device for environmental factors have been listed in [10]. The IR system is designed in such a way that its receiver and transmitter are mounted on either side of the road such that it gets activated whenever automobile passes between the two sensors. The sensors enable this system to be automated unless it is a regular traffic control system which has been rendered ineffective in densely populated areas. The infrared sensors have a detection range of 2 cm – 30 cm. They determine if there is a heavy traffic on one lane and allow the flow of traffic in preference to other less dense lanes. Figure 3 shows the connection of the sensor and Arduino.

The sensor detects a minimum of 3.3 V and a maximum of 5 V to the Arduino board. The board then reads the signal in zeros and ones. When the board processes the signal from the sensor as high, it performs the function that it is programmed to do. After the sensor senses signal at one lane all other lanes will be at a stop, but if it is still sensing at one lane and another sensor is triggered it would not change until the traffic at the lane that was called first is done and sends a low signal to the Arduino board for it to switch to the lane of the second command. This process continues until no signal is detected at the sensors anymore and returns to its timely controlled state.

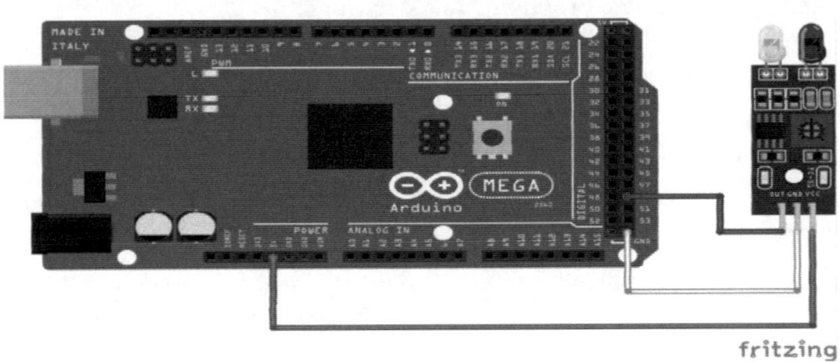

Figure 3 Infrared Sensor interface with Arduino mega

Brief description of the board

Microcontroller	Atmega2560
Operating Voltage	5 V
Input Voltage (recommended)	7 V-12 V
Input Voltage (limit)	6 V-20 V

IOP Conf. Series: Earth and Environmental Science **331** (2019) 012047 doi:10.1088/1755-1315/331/1/012047

Digital I/O Pins	54 (of which 15 provide PWM output)
Analog Input Pins	16

The Arduino mega board can be powered via USB connection or with an external power supply. The power supply gives 9 V used to power the board. This power can be by an adapter or battery of required voltage. The board can operate on an external supply of 7 V – 12 V. If there is a voltage supply less than 7 V, the board will become unstable and will not be able to power the components connected to it. Voltage above 12 V may damage the board due to overheating. There are 54 digital input output pins which out of this 16 are used in this project. The LEDs are connected to 12 of these pins i.e. 3 LEDs for each lane and 4 are for the sensors on each of the lanes. The Arduino is programmed using the Arduino IDE.

```
#include<TimerOne.h>

int signal1[] = {23, 25, 27};
int signal2[] = {48, 50, 34};
int signal3[] = {10, 9, 8};
int signal4[] = {6, 5, 4};

bool Sensor1 = true;
bool Sensor2 = false;
bool Sensor3 = false;
bool Sensor4 = false;

bool sen1 = true;
bool sen2 = true;
bool sen3 = true;
bool sen4 = true;

bool found = false;

uint8_t current;


const int triggerpin1 = 27;
const int ProxSensor1 = 29;
const int triggerpin2 = 34;
const int ProxSensor2 = 52;
const int triggerpin3 = 8;
const int ProxSensor3 = 11;
const int triggerpin4 = 4;
```

This part of the code declares the variables to a name that is used throughout the program.The variables declared correspond to the components used in the physical project design. By declaring them as variables, they can be called any time in the program when needed.

```
void setup() {

  Serial.begin(9600);
  for (int i = 0 ; i < 3; ++i) {
    pinMode(signal1[i], OUTPUT);
    pinMode(signal2[i], OUTPUT);
    pinMode(signal3[i], OUTPUT);
    pinMode(signal4[i], OUTPUT);
  }




  // Declaring Proximity sensor pins as output
  pinMode(triggerpin1, OUTPUT);
  pinMode(ProxSensor1, INPUT);
  pinMode(triggerpin2, OUTPUT);
  pinMode(ProxSensor2, INPUT);
  pinMode(triggerpin3, OUTPUT);
  pinMode(ProxSensor3, INPUT);
  pinMode(triggerpin4, OUTPUT);
  pinMode(ProxSensor4, INPUT);

```

This part is the initial process of the circuit, this states that the LEDs are declared as output and the sensors are declared as the input to control the LEDs. The "pinMode" identifies the connections on the digital input output parts of the Arduino

IOP Conf. Series: Earth and Environmental Science **331** (2019) 012047 doi:10.1088/1755-1315/331/1/012047

```
if (Sensor1) {

  current = 1;

  digitalWrite(signal2[0] , HIGH);
  digitalWrite(signal3[0] , HIGH);
  digitalWrite(signal4[0] , HIGH);

  while (countt != 4) {

    digitalWrite(signal1[a] , HIGH);
    digitalWrite(signal2[b] , HIGH);
    if ( (millis() - tim) >= 3000) {
      digitalWrite(signal1[a] , LOW);
      digitalWrite(signal2[b] , LOW);
      tim = millis();
      --a;
      ++b;
      ++countt;

    }
    Sensor1 = false;
    Sensor2 = true;
    rst(countt , a , b , tim);
    if (found) {
      found = false;
      break;
    }
```

This part of the code controls the sensor. Once the sensor has detected an obstacle within its observation radius it will send a signal to Arduino to command the traffic to flow for the lane to which it corresponds. Not until the obstacle has left the radius on the sensor before it continues sequentially on a timely basis. However, the sensors at any other lane are being triggered to control traffic on the other lanes. The code can be briefly explained using the flow chart in Figure 4.

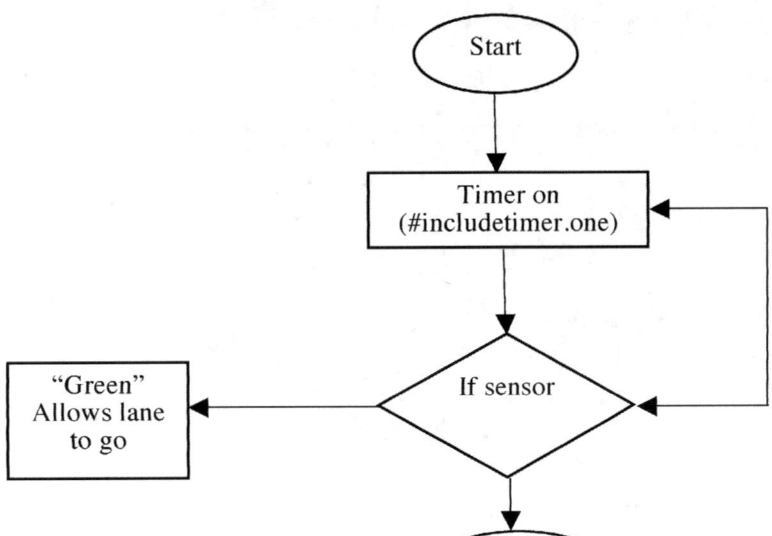

Figure 4: Flowchart of coding process

3. Results and Discussion

The construction of the project was done firstly on the breadboard before being transferred to the vero-board. The LEDs which are red (5 mm), yellow (5 mm) and green (5 mm) are connected in series with resistors of 220 Ω each connected to the negative legs of the LEDs. Figure 5 shows the connections of the LEDs and resistors. The connections of the LEDs and the resistors are created to represent traffic lights for each lane i.e. there will be four of these for each of the lanes. The Arduino is placed at the

IOP Conf. Series: Earth and Environmental Science **331** (2019) 012047 doi:10.1088/1755-1315/331/1/012047

middle as the controlling system that will send the information for this operation to be effective. Figure 6 shows the arrangement of the LEDs on the breadboard for each lane. The legs of each of the components are then wired to the digital input and output pins of the Arduino board. The LEDs at the lanes will be connected to the pins on the board between pins 3 and 53. The legs of the IR sensors are 3 in number. The V_{cc} leg is connected to the 5 V pin on the arduino board, the GND leg is connected to the GND pin on the board and the legs of the resistors simultaneously. The OUT leg is connected to one of the digital input and output pins corresponding to the traffic light it is to control. Figure 7 shows the connections of the components to the arduino board. The power is connected using an Arduino power cord and a 9 V battery. Although the alternative power supply used is the USB cable to be able to send the codes and power to the board simultaneously. The Arduino is then programmed to enable the traffic flow on a timely basis. Figure 8 shows a lane going and the rest are stopped. So also when the sensor at lane 2 is being signalled, this will turn the traffic lights to green i.e. a GO on that lane and making the other lanes to stop for lane 2 to have the right of way. The sensor at lane 3 also detecting an obstacle turns the traffic light on that lane to green for go and stops all other lanes. When two sensors are detected at the same time as shown in Figure 9 the sensor that detected the signal first will be given the right of way and it will switch to the next sensor once the first one stops detecting the signal.

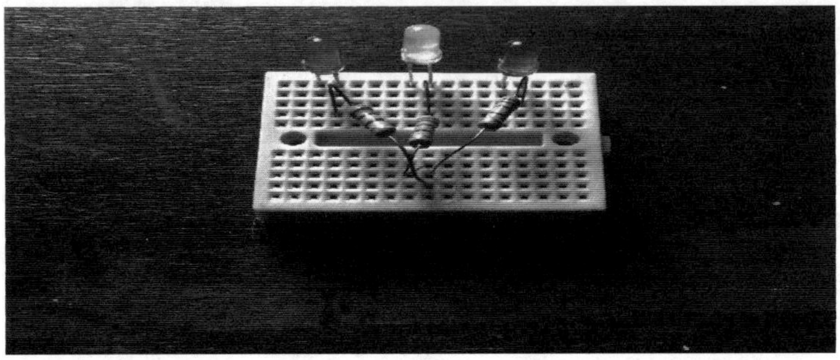

Figure 5: LEDs and resistor connection on the breadboard

Figure 6: Arrangement of LEDs for each lane

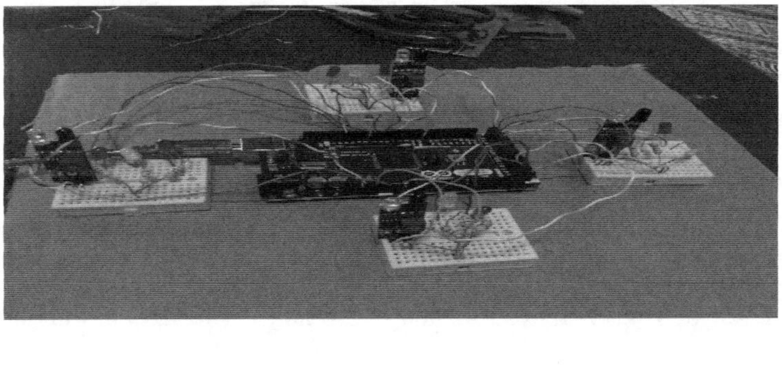

IOP Conf. Series: Earth and Environmental Science **331** (2019) 012047 doi:10.1088/1755-1315/331/1/012047

Figure 7: Connection of components to the Arduino board

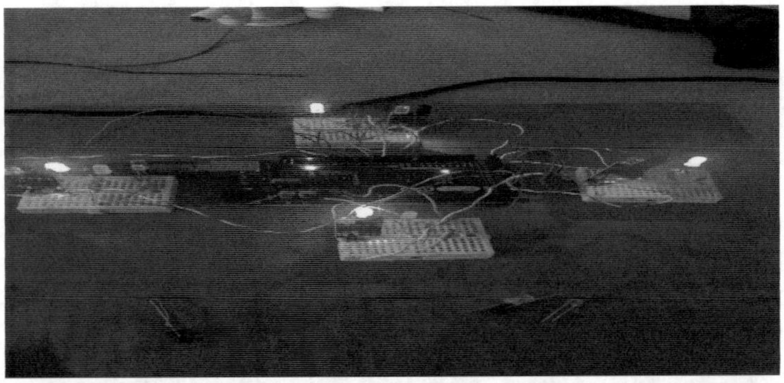

Figure 8: Traffic flow at signal 1

Figure 9: Signalling sensors and response

The response time of the sensor is the time taken for the sensor to produce an output. This is the time it takes once it detects an obstacle and sends a signal to the Arduino and changes the traffic light colour for execution. The result obtained here is in agreement with the report of [9]. Table 1 shows the average response time and Figure 10 is the pictorial chart of the time.

Table 1: Response time of the device

Number of Attempts	Time (s)
1	0.32
2	0.41
3	0.44
4	0.31
5	0.31
6	0.32
7	0.54
8	0.54
9	0.29
10	0.39
Mean	0.387

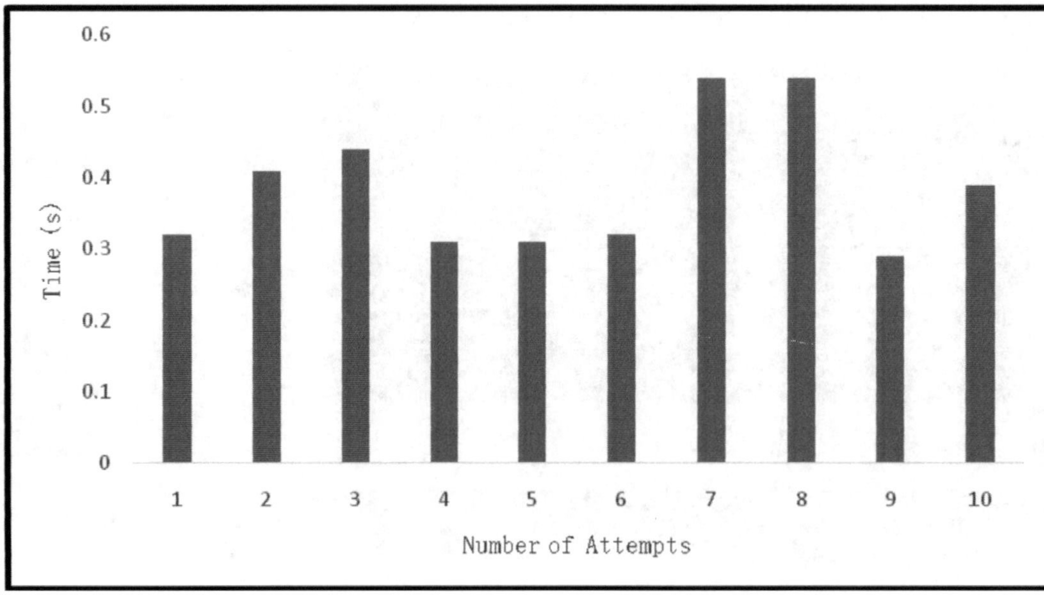

Figure 10: Response time of the device

4. Conclusion

The density based traffic control system has been designed, constructed and tested to ensure validation of its function and operations. In this research, we have succeeded in minimizing the traffic congestions created by the fixed time based traffic light system. The system is effective and the cost of production is very low. Future work is recommended in order to produce the device on a large scale and deploy to all roads in order to reduce traffic congestion in places like Lagos where traffic congestion has become a big issue.

Acknowledgments

Authors acknowledged the publication support received from Covenant University.

References

[1] Udoakah1 Y. N. and Okure I. G. (2017) Design and implementation of a density-based traffic light control with surveillance system, *Nigerian Journal of Technology (NIJOTECH)*, 36(4): 1239 – 1248
[2] Kham N and New (2014) Implementation of modern traffic light control system", *International Journal of Scientific and Research Publications*, 4(6): 1-6
[3] Traffic Control Systems Handbook (2005). Prepared for federal highway administration by Dunn Engineering Associates in association with Siemens Intelligent Transportation Systems
[4] Kell J. H. and Fullerton I. J (1998) Manual of Traffic Signal Design Institute of Transportation Engineers, Prentice-Hall, Inc., 138pp
[5] Sinhmar P (2012), "Intelligent traffic light and density control using IR sensors and microcontroller", *International Journal of Advanced Technology and Engineering Research (IJATER)*, 2(2): 30-35.
[6] Nwoye C. D., Usikalu M. R., Babarimisa I. O, Achuka J. A and Ayara W. A. (2017) Construction of An Automatic Power Switch using Infrared Motion Sensor, *Journal of Informatics and Mathematical Sciences*, 9(2): 331–337
[7] Ayara W. A, Omotosho T. V, Usikalu M. R, Singh M. S and Suparta W. (2017) Development of a solar charged laboratory bench power supply, *Journal of Physics: Conference Series*, 852(1): 012044
[8] Usikalu M. R, Shittu A. H and Obafemi L. N (2018) Construction of an intelligent and efficient light control system, *International Journal of Mechanical and Production Engineering Research and Development (IJMPERD)*, 8(4): 1057-1066

[9] A. Dakhole and M. Moon (2013) Design of intelligent traffic control system based on ARM, *International Journal of Advance Research in Computer Science and Management Studies*, **1(6)**: 76-80.

[10] Adagunodo T.A., Ajigbotosho J.J., Obafemi L.N., Usikalu M.R., Akinwumi S.A., Ayara W.A. (2018) Construction of an in-situ Smart Device that Measures some Basic Environmental Factors for Agricultural Monitoring. IOP Conference Series: Earth and Environmental Science, 173: 012023. https://doi.org/10.1088/1755-1315/173/1/012023.

[11] Retrieved from www.circuitdiagram.org

IOP Conf. Series: Earth and Environmental Science **331** (2019) 012048 doi:10.1088/1755-1315/331/1/012048

Board Expertise and Sustainability Reporting in Listed Banks in Nigeria

O E Umukoro[1], O R Uwuigbe[1], U Uwuigbe[1], Adegboye A[1], O Ajetunmobi[1], and C Nwaze[1]

[1]Department of Accounting, Covenant University, Ota, Ogun State, Nigeria.

Abstract: Despite the growing evidence on the determinants of sustainability reporting, there exist limited and inconclusive studies on the impact of board expertise on sustainability reporting. This study investigates the influence of environmentally sensitive, certified or educated board members on the disclosure of sustainability report. Based on the static panel data regression estimators for 10 Nigerian Deposit Money Banks over the period of 2014-2016, the study revealed that highly educated directors have an altogether constructive influence on the sustainability report disclosure while controlling for corporate administration and firm-level qualities. In addition, we find that the executive and non-executive directors have low experience in environmental issues resulting in an insignificant effect on the disclosure of sustainability reporting. This paper suggests that firms should allow more directors with environmental background, who have a lower motivation to boost transient returns since they are likely to influence environmental performance.

1. Introduction

The sole aim of every profit-making organisation is to continue to make a profit and meet the needs of all her stakeholders, it, therefore, has become very pertinent for management to engage in practices and corporate strategies that would aid the fulfilment of her going concern objective. In the light of recent global happenings, various international organisations and companies seeking to belong to that global space are gradually conforming to the needs of stakeholders and global organizations by carrying out their operations in such a way that transparency and sustainability are evident from their reports and activities.

According to [15], "A sustainability report is a tale disseminated by a firm about the environmental, economic and societal effects brought about by its ordinary exercises. A sustainability report also shows the firm's qualities and management model and exhibits the connection between its procedure and its duty to an economical world view". This simply means a sustainability report must be in alignment with the vision and strategic plans of a given organisation [16].

The results from previous researches on the subject matter remain inconclusive as [3] are of the opinion that sustainability reporting instigate no critical influence on the arrival of benefit of Listed firms in Nigeria while [11] uncovered that there exit critical adverse connection between Environmental Accounting and ROCE and EPS while a critical progressive connection between Environmental Accounting and Net Profit Margin and Dividend per Share.

The inconclusive nature of the results obtained by prior studies [4; 11 and 15] has given rise to this study. Sustainability reporting has become a necessity in every country and sector, this is major because researches such as [15] have empirically averred there exists a progressive relationship between sustainability reporting and revenue generation in listed banks In Nigeria. The CBN in the year 2012 released nine sustainability banking principles to guide listed banks in maintaining sustainable practice, the principles cover a holistic approach to ensuring that all humans are treated fairly, rural areas and women are not left behind in banking operations and services, they are also to ensure that banks do not negatively affect the environment through their operations, they are also to encourage collaborative partnership and ensure their financial reports encapsulate their sustainability practice.

Although some prior researches [11; 14; 12] statistically proves that Sustainability reporting is a necessary tool in revenue generation, return on assets and company performance, the studies did not empirically ascertain if there's any relationship or effect between board magnitude, board independence, board members knowledge of sustainability reporting, qualifications of the board, directors level of education board magnitude on sustainability reporting. It is therefore important to examine other factors that would aid the strict adherence to the CBN sustainability guidelines.

Corporate administration has been characterized by various people as ways and methods of how a given organisation is directed and controlled. The board members of a given organisation have been entrusted with the sole aim of ensuring a given organisation is ethically acting in the shareholders' interest, they also monitor the progress of the organisation hence it is therefore important to examine if the expertise of board members have any huge impact on sustainability reporting in listed banks in Nigeria.

This paper is composed in the accompanying sections. Section 2 provides a review of prior and hypothesis development. Section 3 deals with the research methodology adopted and data gathering procedures. Section 4 describes the data and discusses the empirical findings while section 5 reports the final conclusions and suggestions.

2. Review of Prior Studies

[2] utilized a bi-directional methodology and found a progressive and bi-directional connection between supportability procedures and firm execution. Their sample size was limited to developed countries, their scores were centred on 4 Indices Diversity, Environment, Ethics and Community proxied for sustainability reporting and Return on assets for performance of firm for time- frame 2006-2010. Similarly, [14] in his investigation inspected the effect of sustainability reporting about firm execution in Nigeria. The researcher's sample size was limited to 64 corporations stated on the NSE for the time frame 2002-2012. He statistically obtained an ever-increasing relationship between sustainability reporting and Profit for Assets, ROE, Earnings per offer and net overall revenue. [1] measured company performance using proxies such as MBV, size of the firm and Return on Capital Employed (ROCE). The sample size utilised for the study was 500 UK firms and the study statistically concluded that an impactful ever increasing relationship exists between giving back to the society, MBV and ROCE.

[3] result reveal that performance of economic disclosures do not citically influence the return on assets of Nigerian listed corporations and this contradicts previous results stated above. Similarly, [5] examined if directors from related industries (DRIs) on a company's board help sustainability reporting and bridge information gap. The study concluded that DRIs have a significant impact on firm value. In the same vein, [8] studied the role of board ability in ecological issues (estimated by the nearness of non-official chiefs with past involvement in natural issues, EEDs) and executive systems on GHG outflows. They utilized the utilization of emanation information of FTSE 350, they measurably presumed that the nearness of EEDs on the board decreases GHG discharges and board individuals with better-arranged executives have better ecological execution.

[12] analysed the effect of diversity of board resource on firm reputation. board members were separated into groups such as business specialists, bolster pros, political directors and other network

persuasive, with an end goal to explore whether business, specialized ability or political ties in the meeting room influence partners' conclusion and, in this way, firm notoriety. This investigation affirms that not every outside executive is similarly compelling in enhancing firm notoriety and that particular sorts of independent directors, particularly business specialists, help increment it.

2.1 Hypothesis Development

Following the study background, we stipulate the study alternative hypotheses below.

H_1: There exists significant association between executive directors' expertise and sustainability reporting.

H_2: There exists significant relationship between non-executive directors' expertise and sustainability reporting.

H_3: There exists significant association between directors' education and sustainability reporting.

3. Data and Method

3.1 Methodology

To test our postulated hypotheses, we adopt a convenient sampling technique to draw the appropriate sample size for this study. Convenient sampling technique is useful to select appropriate samples following the availability level of the desired data [7]. We draw 10 banks from 15 banks listed on the Nigerian Stock Exchange after the filtering criteria was adopted. We then drop banks with insufficient variables desired for this study. In addition, the desired variables are hand-collected from financial statement and the corporate governance section of the firms webpage. We consider 3 years from 2014-2016 because the banks have fully implemented and disclosed all reports on Sustainability Reporting with the directives of the Nigerian Stock Exchange Commission.

We identify the disclosure of sustainability report as the dependent variable. The proxy for sustainability report disclosure was adapted from prior literature such as [10]; [13]. The measure is patterned according to the Global Reporting Initiative framework subjected to three categories such as i) Economic Indicators ii) Environmental Indicators and iii) Social Indicators. We use content analysis to derive the disclosure score by the selected banks to construct the disclosure index. The overall disclosure score (TDS) for sustainability indices is generated as follows:

$$TSD_i = \sum_{s=1}^{N} d_s$$ While the Sustainability Reporting Disclosure *(SRD)* = TSD_i/M

Where;

d_s = 2 for full disclosure, 1 for partial disclosure and 0 for no disclosure against economic, environmental and social indicators respectively.

N=9 for the economic indicator i

N=30 for the environmental indicator i

N=45 for the social indicator i

M=Maximum possible score of 168

We then identify three basic independent variables representing the level of the directorship expertise relating to environmental issues. Following the work [6], we explore the profile of individual directors on the board to indicate the level of directors' experience in environmental sustainability. However [6] only explore the experience of non-executive directors in environmental issue. We then explore both the experience of executive and non-executive directors in sustainability issues in a Nigerian bank. In addition, we assume that higher educated directors will be more knowledgeable in

IOP Conf. Series: Earth and Environmental Science **331** (2019) 012048 doi:10.1088/1755-1315/331/1/012048

the areas of environmental issues. Therefore, we include the number of higher educated directors in the study model.

Consequentially, we further include other variables that comprise the corporate governance mechanisms and firm characteristics to control for perplexing features. We include board magnitude, board independence and board composition to the difference in individual banks corporate governance structure. We control for firm size as it could affect the environmental engagement of the banks. In addition, we control for the TobinQ and the return on equity (*ROE*) for market valuation modelling.

3.2 Model Specification

We use OLS regression with company and year controls to test the effect of board expertise on the sustainability reporting. Pooled regression model always discards the panel data structure by basically pooling collectively data on the units (Kunst, 2010). Thus, the unit difference that arises due to the resultant of the different coefficient is usually ignored. The coefficient time and unit subscript are assumed to be common, which eliminate any possibility of individual heterogeneity.

The model is expressed in explicit form as follow

$$SRD_{it} = \beta_0 + \beta_1 Ede_{it} + \beta_2 Nede_{it} + \beta_3 EDU_{it} + \beta_4 \sum others_{it}$$

All variables included in the model are clearly explained in Table 1.

Table 1

Definition of Variables

Variables	Definition	Measurement
Dependent Variable		
SRD	Unweighted Sustainability Reporting Index	As explained earlier
Independent Variable		
Ede	Executive Directors' Expertise	The Total number of Executive Members with Experience in Environmental Issues
Nede	Non-executive Directors' Expertise	The Total number of Non-executive Members with Experience in Environmental Issues
EDU	Directors' Education Level	Number of Directors with Higher Education Certification such as M. Sc, M.Phil. or Ph. D
Other Variables		
BM	Board Magnitude	The total number of Board Members
BIN	Board Independence	The ratio of Independent Directors to Total Board Size
BC	Board Composition	The ratio of Non-executive Directors to Total Board Size
TobinQ	TobinQ	The book value of total assets minus the book values of equity plus the market value of equity all divided by the book value of total assets.
ROE	Return on Equity	Total profit for the year/ Total Equity
Size	Firm Size	Natural logarithm of Total Asset

IOP Conf. Series: Earth and Environmental Science **331** (2019) 012048 doi:10.1088/1755-1315/331/1/012048

4. Empirical Results

4.1 Descriptive Statistics

Table 2 reports the descriptive statistics of the study variables. From the Table, the mean of the sustainability report disclosure is 0.344 with a maximum of 0.62 and a minimum of 0.196. Therefore, the mean level of the Disclosure Index verifies that the majority of the banks possesses reasonable sustainability reporting.

In the case of the board environmental expertise, the minimum level of the executive directors with environmental issues is none revealing that some banks do not have executive directors with sustainability experience while the maximum number of the executive with environmental experience is 2 directors. On the other hand, there is a minimum of one person with environmental experience sitting as a non-executive director while the maximum number of such as directors in the Nigerian banks is 5 directors. In addition, it is indicated that between 3 to 9 directors possess Higher Education Certification such as M. Sc, M.Phil. or Ph. D.

Table 2: Descriptive Statistics

VARIABLES	Obs.	Mean	S. D	Min	Max
SRD	30	0.344	0.110	0.196	0.620
Ede	30	1.133	0.571	0	2
Nede	30	2.533	1.106	1	5
Educate	30	5.633	1.326	3	9
BM	30	15.07	2.947	7	19
BIN	30	0.128	0.0683	0	0.250
BC	30	0.650	0.114	0.500	0.917
TobinQ	30	0.970	0.0931	0.840	1.210
ROE	30	0.136	0.0734	0.0154	0.262
Lnasset	30	21.38	0.611	20.45	22.28

Table 3 reports the Pearson correlation matrix for the dependent variable and independent variables adopted in the analysis. The Table indicates low correlation among the variables, especially among the independent variable. Hence, there is no indication of serious multicollinearity in the models.

Table 3: Correlation Matrix

	1	2	3	4	5	6	7	8	9	10
1. SRD	1									
2. Ede	0.0121	1								
3. Nede	-0.235	0.702***	1							
4. Educate	0.0507	-0.0239	-0.144	1						
5. BM	-0.200	-0.149	-0.244	0.209	1					
6. BIN	0.638***	0.132	-0.0671	-0.210	-0.275	1				
7. BC	0.263	-0.0834	-0.0724	-0.0351	-0.103	0.00400	1			
8. TobinQ	-0.167	0.367*	0.489**	-0.240	-0.556**	0.151	-0.159	1		
9. ROE	-0.0576	0.312	0.315	-0.231	-0.157	0.0642	-0.178	0.685***	1	
10. Size	0.452*	-0.117	-0.271	0.214	-0.0180	0.499**	0.376*	-0.0342	0.185	1

$^{*}p < 0.05$, $^{**}p < 0.01$, $^{***}p < 0.001$

IOP Conf. Series: Earth and Environmental Science **331** (2019) 012048 doi:10.1088/1755-1315/331/1/012048

4.2 Regression Results

Table 5 reports the Multiple Regression results of the consequence of members expertise on sustainability reporting for the selected banks. The R-squared value identifies the percentage of dependent variability explained by the independent variables in the regression model. The study R-squared basically for Model 4 is 0.658. Thus, the result reports 65.8% of the variability in Sustainability Reporting Disclosure is explained by the independent variables. In addition, F-test is 4.272 with the p-value of 0.003 which indicates a significant linear association between the explanatory variables that are the board environmental expertise and explained variable that is sustainability reporting disclosure.

Table 4: Regression Results using OLS

VARIABLES	1	2	3	4
Ede	-0.00962			0.00761
	(0.0291)			(0.0388)
Nede		-0.0116		-0.0142
		(0.0155)		(0.0209)
Ede * Nede			-0.00390	
			(0.00639)	
Educate	0.0259*	0.0257*	0.0254*	0.0249*
	(0.0141)	(0.0134)	(0.0135)	(0.0143)
BM	-0.0116*	-0.0115*	-0.0114*	-0.0116*
	(0.00658)	(0.00650)	(0.00654)	(0.00667)
BIN	1.333***	1.317***	1.327***	1.293***
	(0.324)	(0.303)	(0.307)	(0.334)
BC	0.307*	0.317*	0.303*	0.313*
	(0.163)	(0.160)	(0.159)	(0.166)
TobinQ	-0.788**	-0.739**	-0.739**	-0.733**
	(0.285)	(0.291)	(0.297)	(0.300)
ROE	0.738**	0.737**	0.710**	0.719*
	(0.351)	(0.337)	(0.336)	(0.357)
Lnasset	-0.0494	-0.0534	-0.0506	-0.0514
	(0.0426)	(0.0411)	(0.0407)	(0.0432)
Constant	1.732*	1.786*	1.720*	1.750*
	(0.888)	(0.861)	(0.852)	(0.900)
Observations	30	30	30	30
R-squared	0.650	0.657	0.654	0.658
Rmse	0.0762	0.0754	0.0758	0.0772
F-test	4.872	5.032	4.966	4.272
Prob > F	0.00168	0.00138	0.00150	0.00329

Standard errors in parentheses *** p<0.01, ** p<0.05, * p<0.1

Under the hypothesis one, it is postulated to test whether there exists significant affiliation between Executive Expertise Directorship and Sustainability Reporting. Thus, the finding reveals that from Model 1 & 4 in Table 4 reports no significant association between executive directors' expertise and sustainability reporting at 5% level of significance and this is in tandem with [3]. This indicates that

the listed banks do not have sufficient executive directors with sustainability inclined. Hence, it is suggested that a sizeable number of executive directors with environmental experience are required.

IOP Conf. Series: Earth and Environmental Science **331** (2019) 012048 doi:10.1088/1755-1315/331/1/012048

Under hypothesis two, it is assumed to test whether there exists significant affiliation between non-executive directors' expertise and sustainability reporting. Thus, the finding reveals that from Model 2 & 4 in Table 4 reports that there is no significant adverse relationship between non-executive directors' expertise and sustainability reporting at 5% level of significance and this result is at variance with [8]; [12]. This suggests that there exists a deficient number of non-executive directors with environment experience that could enhance sustainability reporting. Thus, there is a necessity for more non-executive directors with environmental experience in the Nigerian banks to augment sustainability reporting.

Under hypothesis three, it is postulated to test whether there exists significant and positive affiliation between directors' education and sustainability reporting. Thus, the finding reveals that from Model 1-4 in Table 4 report that there exists significant positive relationship between directors' education and sustainability reporting at 5% level of significance. This designates that possession of higher education certifications tends to enhance the awareness of the board regarding sustainable banking practice. Thus, this study then advocates that listed banks should encourage their directors to possess higher education degree.

Table 5: Regression Results using Logistics Analysis

VARIABLES	1	2	3	4
Ede	-4.106			-3.609
	(2.576)			(2.571)
Nede		-1.237		-0.689
		(1.058)		(0.864)
Ede * Nede			-0.820*	
			(0.492)	
Educate	2.223*	1.157	1.760	2.236*
	(1.283)	(0.903)	(1.192)	(1.243)
BM	0.262	0.0843	0.312	0.353
	(0.554)	(0.408)	(0.532)	(0.554)
BIN	73.23**	41.77*	63.60*	75.91**
	(36.77)	(24.25)	(35.12)	(36.53)
BC	29.58	18.76	28.01	32.36*
	(18.55)	(14.76)	(20.16)	(18.89)
TobinQ	-30.18	-7.318	-19.28	-28.52
	(22.37)	(15.09)	(20.70)	(22.32)
ROE	68.95*	36.51	57.49	71.58*
	(38.98)	(25.38)	(38.29)	(38.72)
Lnasset	-5.115	-2.880	-4.500	-5.530*
	(3.359)	(2.664)	(3.357)	(3.325)
Constant	86.82	39.99	66.72	91.19
	(66.43)	(45.66)	(60.05)	(64.36)
Observations	30	30	30	30
Pseudo R-squared	0.500	0.453	0.502	0.521
chi-squared	19.11	17.30	19.17	19.89
Prob > chi2	0.0143	0.0271	0.0140	0.0186

Standard errors in parentheses *** p<0.01, ** p<0.05, * p<0.1

4.3 Robustness Check

IOP Conf. Series: Earth and Environmental Science **331** (2019) 012048 doi:10.1088/1755-1315/331/1/012048

To find the robustness of the study, we further extend our regression model by classifying the sustainability reporting to high and low quality. We then allocate a dummy variable that is 1 stand for any value greater than 0.5 signifying high-quality sustainability reporting and 0 stands for any value less than 0.5 representing low-quality sustainability reporting. Since the dependent variable is represented by the dummy variable, we then employ logistic analysis for the robustness check. From Table 6, these results confirm the robustness of the main test as the desired variables retain their results.

5. Discussion and Conclusion

This paper explores the effect of board expertise and level of director education on sustainability performance. Additionally, following the prior studies, we further include other variables that comprise the corporate governance mechanisms and firm characteristics to control for perplexing features. We proxy for sustainability performance using content analysis to derive the disclosure score by the selected banks to construct the disclosure index (i.e. 2 for full disclosure, 1 for partial disclosure and 0 for no disclosure against economic, environmental and social indicators respectively).

We use availability sampling technique to derive the study sample size of 10 out of 15 Nigerian banks for the period 2014-2016. We use OLS regression with company and year controls to test the effect of Board Expertise on the Sustainability Reporting. We find that the board expertise has no significant influence on sustainability reporting. However, we find that the educated directors possess the experience to influence sustainability practices of the banks.

Based on the results of this study, we then recommend that Nigerian Banks should adopt a structure that could enhance sustainability practices for the banks. Stressing on the level of directors' education could enhance effectiveness in term of information flow and it is expected to reflect the disparity within the society. Our results contribute to the increasing literature stressing the importance of board environmental and education practices in view of sustainability activities for the banks.

References

[1] Adeneye, Y., A & Ahmed, M. (2015). Corporate social responsibility and company performance. *Journal of Business Studies Quarterly,* 7 (1), 151-166.

[2] Ameer, R. & Othman, R. (2012). Sustainability practices and corporate financial performance: A study based on the top global corporations. *Journal of Business Ethics*, 108(1), 61-79.

[3] Asuquo, I., A, Dada, E., T & Onyeogaziri, U., R. (2018). International Journal of Business & Law Research 6(3):1-10.

[4] Central Bank of Nigeria (2012). Nigerian Sustainability Banking Principles.

[5] Dass, N., Kini, O., Nanda, V., Onal, B., and Wang, J. (2014). Board expertise: Do directors from related industries help bridge the information gap? *Review of Financial Studies,* 27, pp. 1533-1592.

[6] Green, C. P., & Homroy, S. (2018). Female directors, board committees and firm performance. *European Economic Review, 102*, 19–38.

[7] Hasimi, F., Darina, N., & Amran, A. (2015). Corporate Governance and Sustainability Practices in Islamic Financial Institutions : The Role of Country of Origin, *31*(15), 36–43. https://doi.org/10.1016/S2212-5671(15)01129-6

[8] Homroy, S., & Slechten S. (2016). Board expertise, networked boards and environmental performance. Department of Economics, Lancaster University, LA1 4YX.

[9] Kunst, R. M. (2010). Econometric Methods for Panel Data.

[10] Laskar, N. (2016). Impact of Corporate Sustainability Reporting on Firm Performance: An

Empirical Examination in Asia. *Journal of Asia Business Studies*, *11*. https://doi.org/https://doi.org/10.1108/JABS-11-2016-0157

[11] Makori, D., M & Jagongo, A. (2013). Environmental accounting and firm profitability: an empirical analysis of selected firms listed in Bombay stock exchange, India. *International Journal of Humanities and Social Science*, 3(18), 248-256.

[12] Meca, E., G & Palacio, J., C (2018). Board composition and firm reputation: The role of business experts, support specialists and community influential. *Business Research Quarterly*. www.elsevier.es/brq

[13] Munshia, D., & Duttab, S. (2016). Sustainability Reporting Quality of Indian and American Manufacturing Firms, *11*(2), 245–260. https://doi.org/10.5937/sjm11-9593

[14] Kwaghfan, A. (2015). Impact of sustainability reporting on corporate performance of selected quoted companies in Nigeria. University of Nigeria. Nsukka, Nigeria.

[15] Uwuigbe, U., Obarakpo, T., Uwuigbe, R., Ozordi, E., Osariemen, A., Akpevwenoghene, G., & Oluwagbemi. T (2018). Sustainability reporting and firm performance: a bi-directional approach. *Academy of Strategic Management Journal* 17 (3), 1-16.

[16] Uwuigbe, O.R., Uwuigbe, U., Jafaru, J., Igbinoba, E.E., Oladipo, O.A. 2016. Value relevance of financial statements & share price: A study of listed banks in Nigeria, Banks and Bank Systems, 11(4), pp. 135-143

IOP Conf. Series: Earth and Environmental Science **331** (2019) 012049 doi:10.1088/1755-1315/331/1/012049

Human waste biogas production: Processes and Utilization

Emetere M.E.[1,2], and Pindar H..[1]
[1]Department of Physics, Covenant University Canaan land, P.M.B 1023, Ota, Nigeria.
[2]Department of Mechanical Engineering and Science, University of Johannesburg, APK, South Africa

E-mail: emetere@yahoo.com

Abstract. Human waste biogas production is not fashionable at the moment due to its quality compared to other animal waste. However, human waste is more abundant than animal waste. More so, the possibility of having scarcity of human waste in the coming years is almost not possible. This research seeks ways of optimizing the quality of biogas from human waste. The results show that the processed human wastes have improved biogas quality. It is recommended that standalone energy users in urban and rural centre to adopt this technique to utilize biogas usage from human waste.

Keywords: energy, biogas, human waste, renewable energy, biofuel

1. Introduction

Biogas is a gas that has a synthesis of around 50-70% Methane (CH_4), 30-half Carbon dioxide (CO_2) with the rest of the gases being: H_2, O_2, H_2S, N_2 and water vapour, produced from the anaerobic processing of natural waste. Biogas comprises mostly of CH_4 and CO_2, with low dimensions of H_2S and different gases. The biogas yield in animal waste and biomass is presented in Figure 1. It is a known fact that biogas from human waste is quite low [1-2]. However, it is abundant and may not go extinct as the waste or biomass presented in Figure 1.

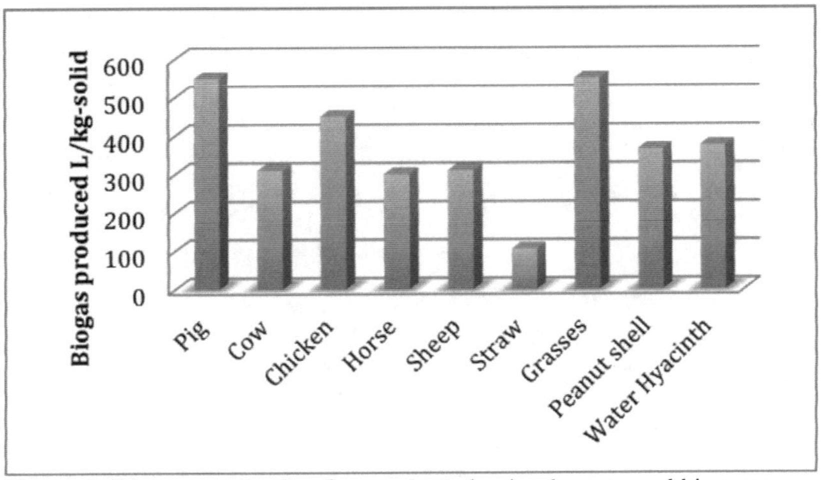

Figure 1: Biogas production from selected animal waste and biomass

IOP Conf. Series: Earth and Environmental Science **331** (2019) 012049 doi:10.1088/1755-1315/331/1/012049

The results in Figure 1 are due to the rate of anaerobic biodegradation in the waste or biomass. Anaerobic creation of biogas does not deliver any hostile smell, causing reduction of pollution, hence making it environmentally friendly, cost efficient, cheap and readily available, and reduces greenhouse effect [3-5]. Biogas resulting from biomass or waste is targeted to replace fossil fuel usage to mitigate air pollution [6-7].
The richest forms of biogas or biofuel materials are cow waste (i.e. cow dung) due to the already digested state from the multiple stomachs of the host [8-9]. Cow dung is easily gotten in large quantity from rural areas. There is the possibility that the continuous usage of animal waste may go extinct due to its other application as manure for rural farmers. In other words, the only other alternative is human waste, and this is richly available in the sewage sludge. Unlike the cow dung sewage sludge is not processed and still has to be refined for it to be a good source of biogas. This therefore poses a problem for future biogas users.

2. Experimental Design, Materials and Methods

Pipes are connected to open sewage slurry, the fumes from the slurry is collected into a cylinder. The biogas naturally travels through the pipe by diffusion. This process takes close to an hour. However, if automatic pumps are used to suck the biogas in the sewage slurry, the process is faster i.e. 5 minutes. The component of the biogas in the cylinder was tested for methane (CH_4), hydrogen sulphide (H_2S), carbon (II) oxide (CO_2), and moisture before passing it through a connecting pipe to the purification cylinder containing silica gel. The silica gel is a drying agent that would suck-up the moisture content of the human waste biogas. After about thirty minutes, the biogas is harvest and tested the second time.

3. Results and Discussion

According to Figure 2, the pre-processed methane levels started at a steady concentration of 167 ppm and maintained this levels up until 500 seconds (8 minutes) into the readings from the first cylinder, where it fell to 150 ppm. Methane readings eventually after 1000 seconds recorded 180 ppm. This reveals that the methane production in the sewage slurry is dynamic with time. According to figure 3, the CO_2 contained in the sewage fumes was constantly fluctuating. It started off at 124ppm and constantly fluctuated till it dropped drastically after 500 seconds of reading. It peaked again at 600 seconds and dropped at 800 seconds, after which it was on a constant increase with just minor fluctuations.

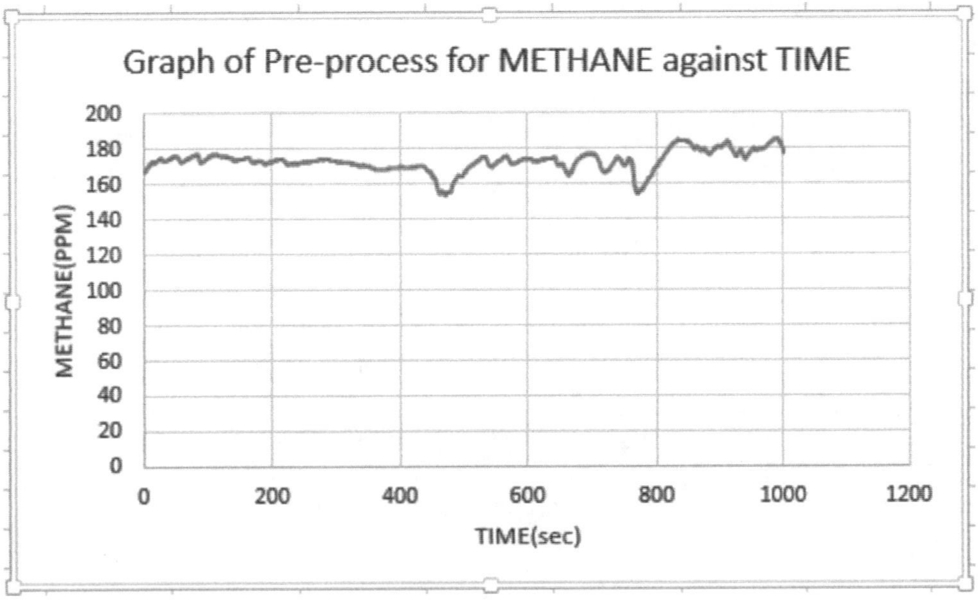

Figure 2:The graph of pre-process for Methane.

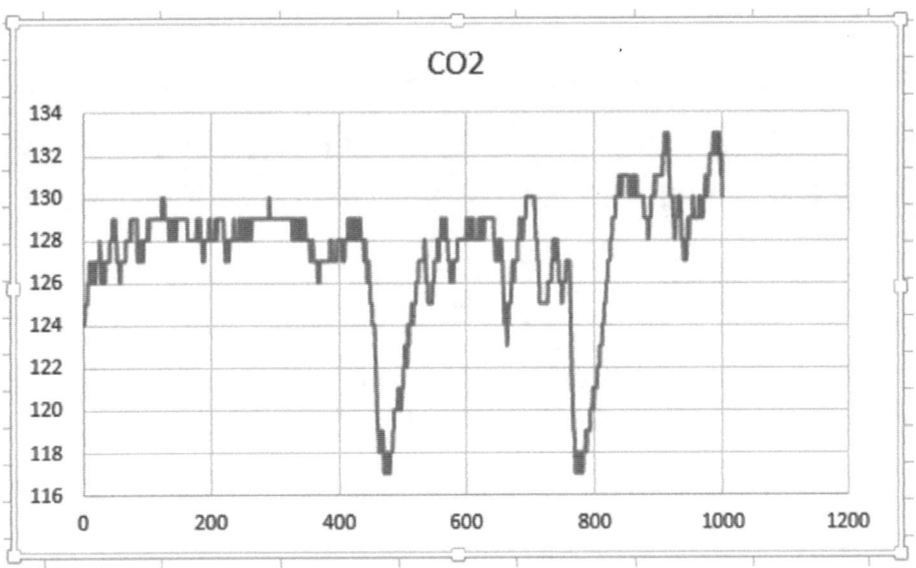

Figure 3: The graph of pre-process for CO2

The significance of this is that the CO_2 content in the pre-processed fumes was unstable and this could have been as a result of a number of reasons that may not be captured in this report because the history of the sewage sludge is not known by the authors. According to figure 4, the undehydrated hydrogen-sulphide reading recorded steady levels throughout the 1000 seconds of the experiment except for some minor fluctuations. This implies that the H_2S levels contained in the pre-processed fumes were steady and in rich supply. According to figure 5, looking at the range of value between which the concentration of humidity, it can be inferred that the humidity maybe almost uniform in larger part of the experiment.

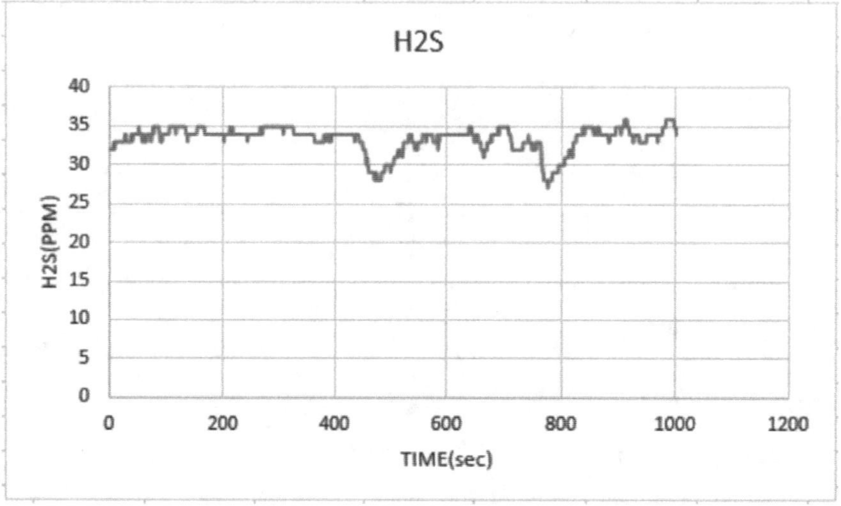

Figure 4: The graph of pre-process for H_2S.

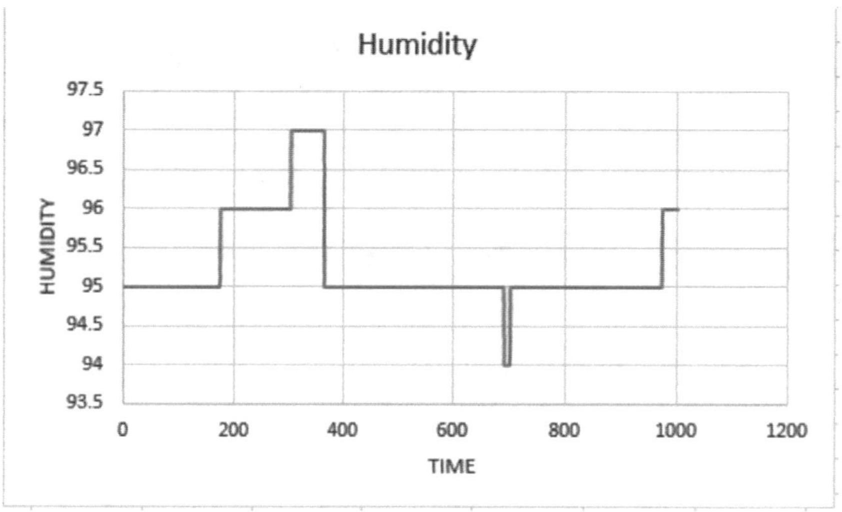

Figure 5: The graph of pre-process for Humidity.

When the captured biogas was processed. The results is presented in Figures 6-9.

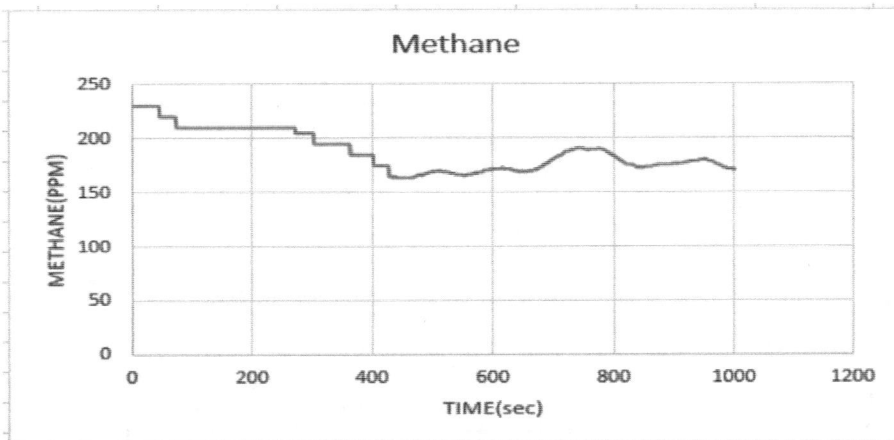

Figure 6: The graph of post-process for methane.

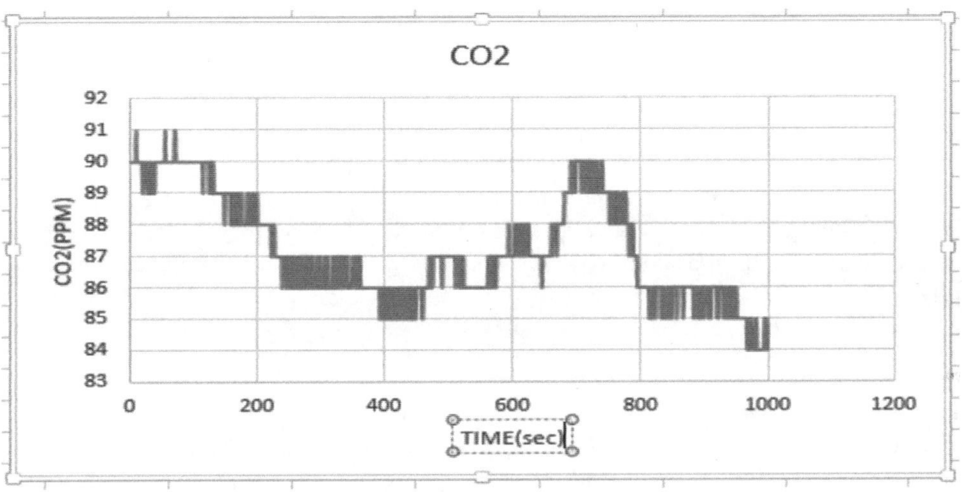

Figure 7: The graph of post-process for CO2.

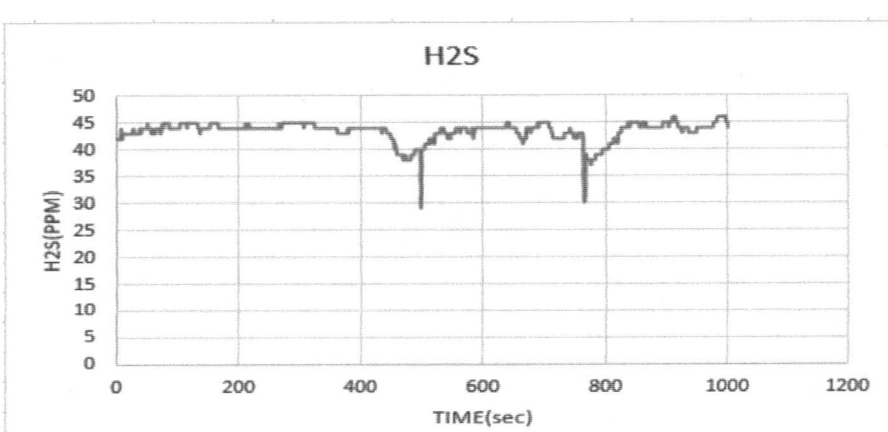

Figure 8: The graph of post-process for H2S.

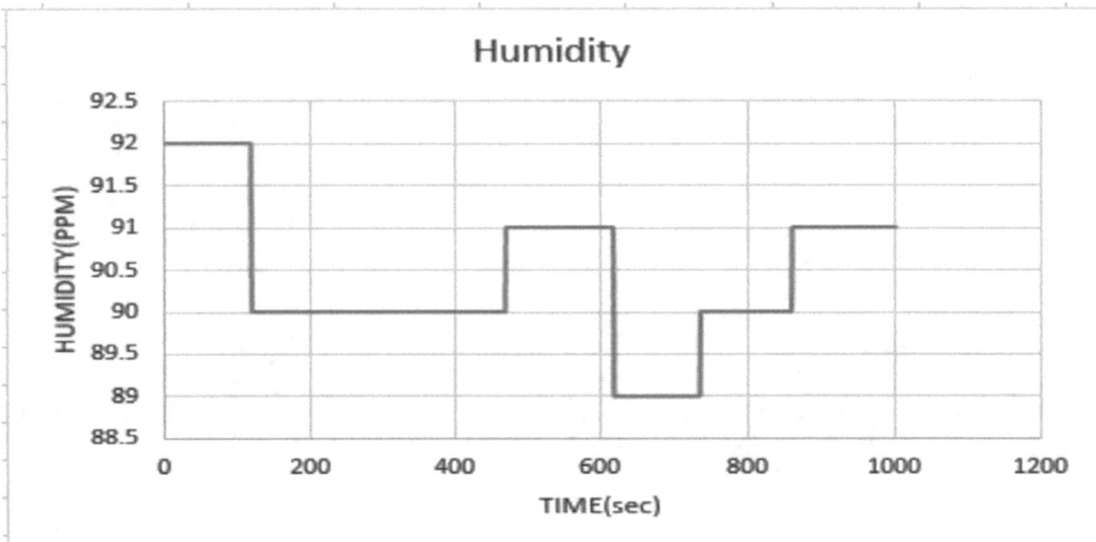

Figure 9:The graph of post-process for humidity.

According to the post-process graphs, it shows methane has improved in its level of concentration, from 180ppm to 230ppm maximum value in the pre-process and post-process respectfully. CO_2 levels, drastically reduced in the post process from 124ppm to 91ppm maximum value in the pre-process and post-process respectfully. Hydrogen-sulphide levels also improved a little but maintained its recorded reading without much fluctuation from 36ppm to 46ppm maximum value in the pre-process and post-process respectfully.

4. Conclusion

Biogas from human waste can be treated and utilized as shown in this research. Hence, the quality of biogas from human waste can be further optimized for better quality. This research considers only the biogas quality and not its quantity. However, it is recommended that the quantity of the biogas from human waste can be improved if the fluid in toilet system is changed from water to pressurized air.

Acknowledgement

The authors appreciate Covenant University for partial sponsorship. The authors acknowledge NASA for primary dataset.

5. References

[1]. D. Andriani, A. Wresta, A. Saepudin, B. Prawara, A review of recycling of human excreta to energy through biogas generation: indonesia case, Energy Procedia, 68 (2015), pp. 219-225
[2]. Owamah, H.I.; Dahunsi, S.O.; Oranusi, U.S.; Alfa, M.I. Fertilizer and sanitary quality of digestate biofertilizer from the co-digestion of food waste and human excreta. Waste Manag.2014,34, 747–75
[3]. Abouelenien F, Fujiwara W, Namba Y, etal. (2010) Improved methane fer-mentation of chicken manure via ammonia removal by biogas recycle. Bioresource Technology 101: 6368–6373.

[4]. Bashir MJK, Aziz HA, Yusoff MS, etal. (2010) Application of response surface methodology (RSM) for optimization of ammonia nitrogen removal from semi-aerobic landfill leachate using ion exchange resin. Desalination 254: 154–161.
[5]. Bilgili MS, Demir A and Ozkaya B (2007) Influence of leachate recircula-tion on aerobic and anaerobic decomposition of solid wastes. Journal of Hazardous Materials 143: 177–183.
[6]. M. E. Emetere & M.L. Akinyemi (2017) Atmospheric Dynamics of Air Pollution Dispersion and Sustainable Environment in Nigeria, Journal of Urban and Environmental Engineering, 11 (1): 51-57
[7]. Emetere Moses E., Akinyemi M.L. & Akinwumi S.A. (2016) Aerosols Loading Trends And Its Environmental Threats Over Cotonou-Benin, Nature Environment and Pollution Technology, 15 (3), 1111
[8]. P. Abdeshanian, M.G. Dashti, M.S. Kalil, W.M.W. Yusoff, Production of biofuel using biomass as a sustainable biological resource, Biotechnology, 9 (2010), pp. 274-282
[9]. Momoh, O.L. Yusuf; Nwaogazie, L. Ify, Effect of Waste Paper on Biogas Production from Co-digestion of Cow Dung and Water Hyacinthin Batch Reactors, J. Appl. Sci. Environ. Manage., 12 (4) (2008), pp. 95 – 98

IOP Conf. Series: Earth and Environmental Science **331** (2019) 012050 doi:10.1088/1755-1315/331/1/012050

Processing human biogas using drying agent

Emetere M.E.[1,2], **and Pindar H..**[4]
[1]Department of Physics, Covenant University Canaan land, P.M.B 1023, Ota, Nigeria.
[2]Department of Mechanical Engineering and Science, University of Johannesburg, APK, South Africa

E-mail: emetere@yahoo.com

Abstract. Biogas production in cow dung supersedes other bio-waste product. Human waste is one of the lowest biogas sources. More so, the nature of the human excretal deposition matters. There are higher biogas productions in pit latrines than in water closet (WC) system. However, it is known that WC users are the majority in modern building design. The main disadvantage of the WC system ingenerating biogas is that the solid excretal is broken into pieces by water dissolution and collision with the sewage pipe. The chemistry of the event is that the little biogas produced has high percentage of moist. In this research we deployed drying agents to process the biogas from WC system.

Keywords: energy, biogas, human waste, renewable energy, biofuel

1. Introduction

Feasible power sources in renewable energy options begins from ordinary resources, for instance, light, wind, storm, tides, and geothermal warmth, which are endless. Roughly 16% of energy users begin with renewables, 10% starting with standard biomass, mainly used for warming, and 3.4% for hydroelectricity. New renewable energy (wind, solar, geothermal and biofuels) accounts for an additional 3% and is growing at an all-time high [1]. In Europe, Asia and the US, energy generation via wind has attain 238 GW at end of 2011at a rate of 30 per cent, and is used extensively. Photovoltaic (PV) was at a global limit of 67 GW at the completion of 2011, and photovoltaic control stations are excellent in Germany and Italy [2]. Although a number of sustainable energy sources are large, inexhaustible progress has been made in rural and remote regions where life is important for human improvement on a regular basis [3-4]. Most global energy users in developing countries adopt the off-grid or standalone energy consumption due to inability for most users to tap into national grid system [5]. The use of fossil fuel generator is the major but there are emphases to limit its use due to air pollution. The renewable energy options that are widely used include wind power, hydropower, solar energy, biomass, biofuels, geothermal energy etc. Figure 1 shows the distribution of the energy source usage.

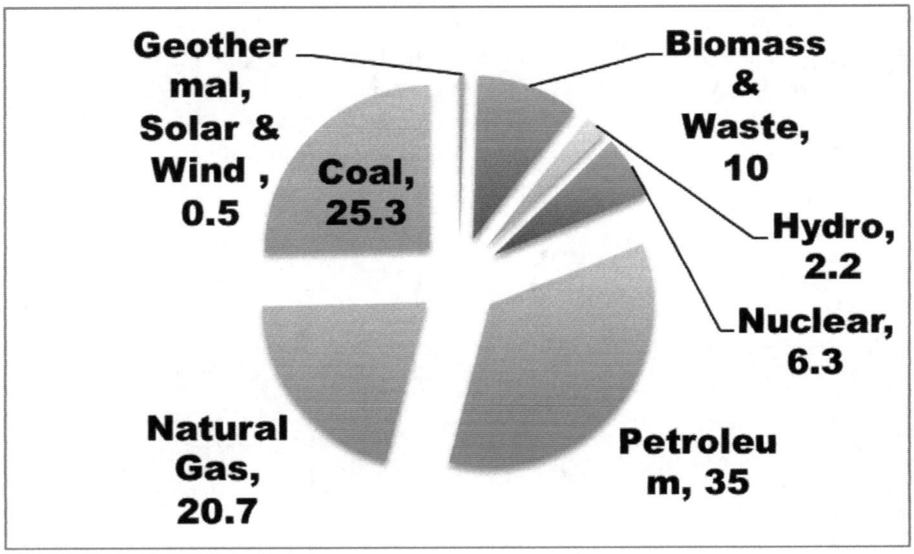

Figure 1: Energy usage distribution

The sources of renewable energies are in abundant in specified locations of the world. For example the tropical climatic zones have access to sun – preferring solar energy. The ocean belt location prefers the wind energy etc. However, what is common to all zones of the world is waste from human or animal (bio-waste). Biogas production in cow dung supersedes other bio-waste product [6]. Human waste is one of the lowest biogas sources [7]. From observation, the nature of the human excretal deposition is key towards ensuring maximum utilization of human waste biogas. There are higher biogas productions in pit latrines than in water closet (WC) system because solid excretal is broken into pieces by water dissolution and collision with the sewage pipe. This means that the little biogas than can be produce would have excess water moisture content. Since, there are above 600 million homes globally that patronize the water closet method. The successes of purifying the generated biogas are a viable project that will alleviate poverty and anthropogenic emissions to a large extent. Therefore, in this research, we examined ways of actualizing biogas production from water closet systems.

2. Experimental Design, Materials and Methods

Pipes are connected to open sewage slurry, the fumes from the slurry is collected into a cylinder. The biogas naturally travels through the pipe by diffusion. This process takes close to an hour. However, if automatic pumps are used to suck the biogas in the sewage slurry, the process is faster i.e. 5 minutes. The component of the biogas in the cylinder was tested for methane (CH_4), hydrogen sulphide (H_2S), carbon (II) oxide (CO_2), and moisture before passing it through a connecting pipe to the purification cylinder containing silica gel. The silica gel is a drying agent that would suck-up the moisture content of the human waste biogas. After about thirty minutes, the biogas is harvest and tested the second time.

3. Results and Discussion

When the biogas was harvested from the open sewage slurry, the component that was tested is presented in Figures 2-5, which shows that the locations initial methane levels were at 471ppm; the concentration of CO_2 started off at 354ppm and then dropped throughout the experiment; initial concentration of H2S was 440ppm and abruptly increased to 447ppm after which it constantly dropped throughout the remaining time of the experiment; the level of humidity started off at a steady 95 g/m3 and after 47 seconds there was a slight increase in humidity to 96 g/m3 which cold have been as a result of exposure of the cylinder to moist air through the sensor openings. This is an indication that the location has appreciable biogas constituents. After a minute the concentration from the sludge began to drop, and this can be due to the fumes escaping into the atmosphere.

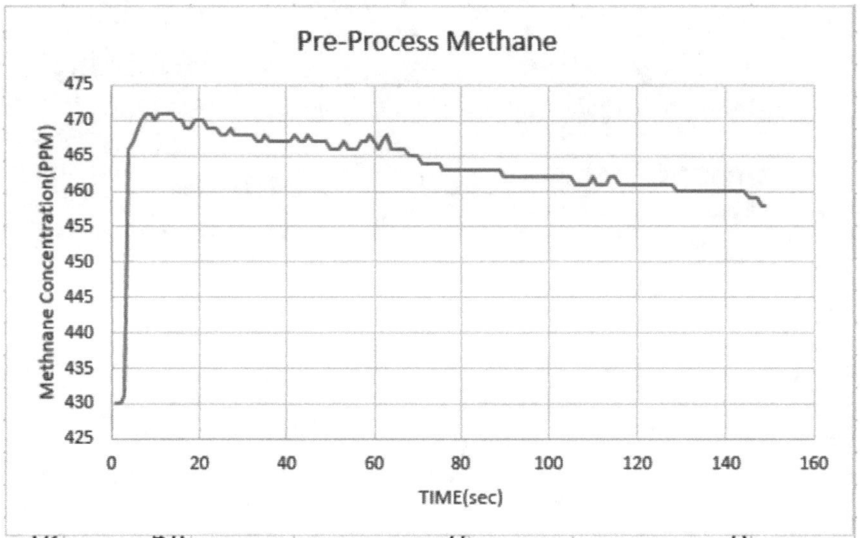

Figure 2: Graph of Pre-Process Methane.

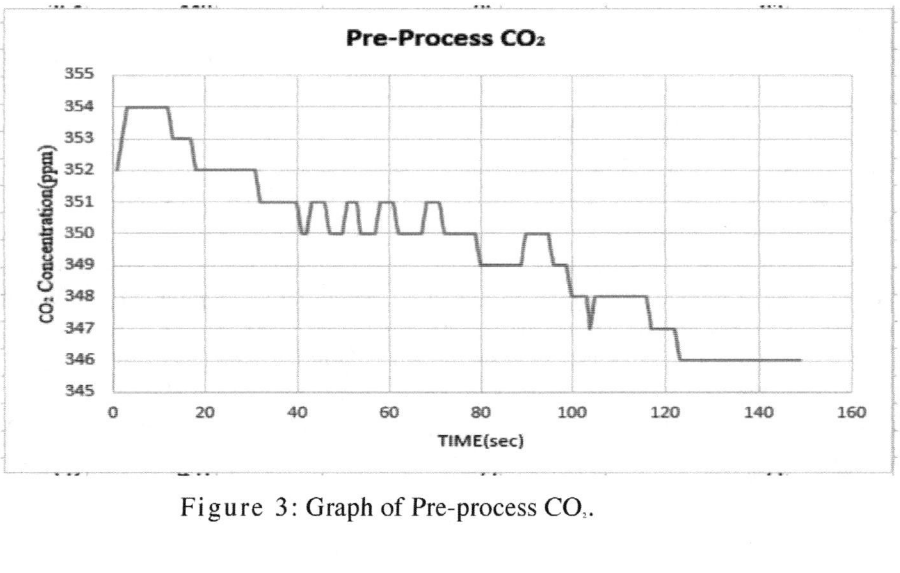

Figure 3: Graph of Pre-process CO_2.

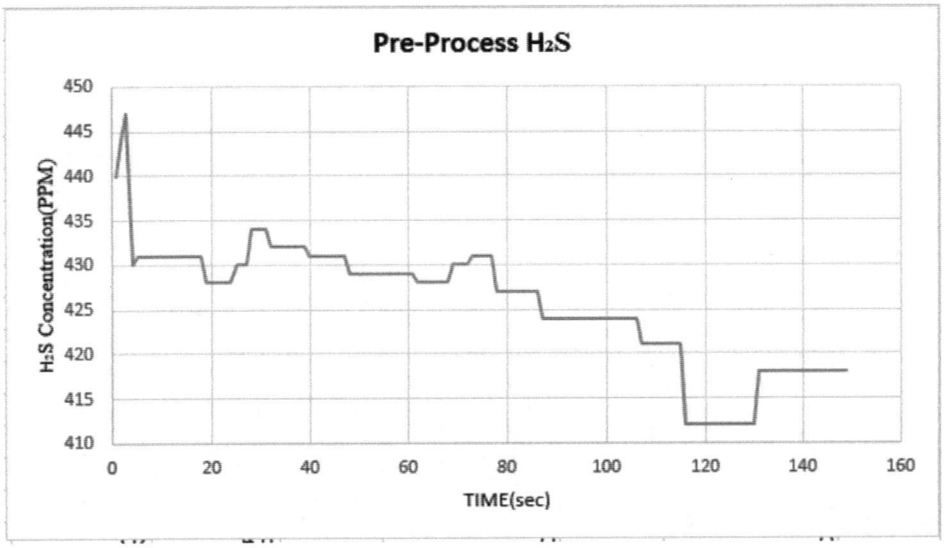

Figure 4: Graph of Pre-Process H_2S.

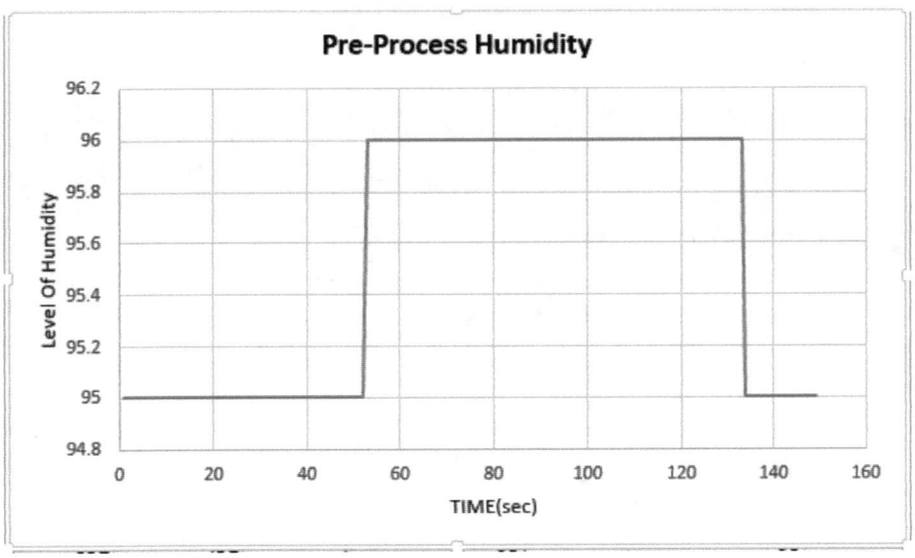

Figure 5 : Graph of Pre-process Humidity.

According to the post-process graphs (Figures 6-9), it shows methane has improved in its level of concentration, from 471ppm to 539ppm maximum value in the pre-process and post-process respectfully. CO_2 levels, increased in the post process from 354ppm to 454ppm maximum value in the pre-process and post-process respectfully. H_2S levels also improved a little but maintained its recorded reading without much fluctuation from 447ppm to 451ppm maximum value in the pre-process and post-process respectfully. Also, Humidity levels were drastically reduced in this experiment from 96units to 90 units in the pre-process and post-process respectfully.

IOP Conf. Series: Earth and Environmental Science **331** (2019) 012050 doi:10.1088/1755-1315/331/1/012050

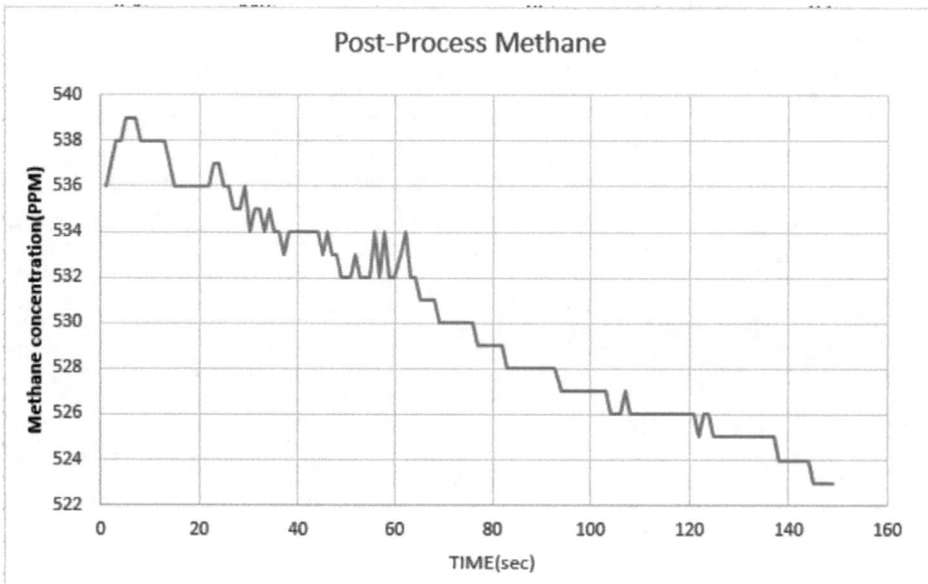

Figure 6: Graph of post-process for Methane.

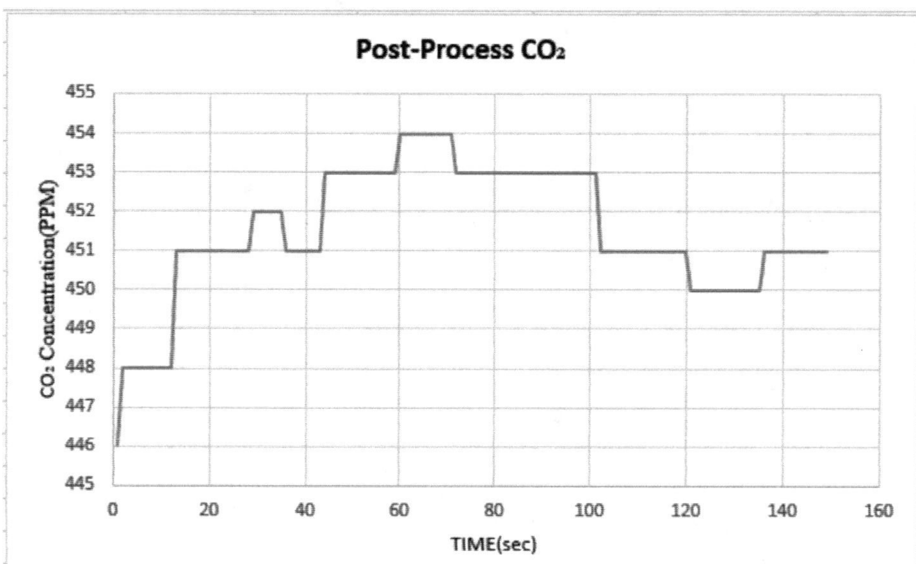

Figure 7: Graph of post-process for CO_2.

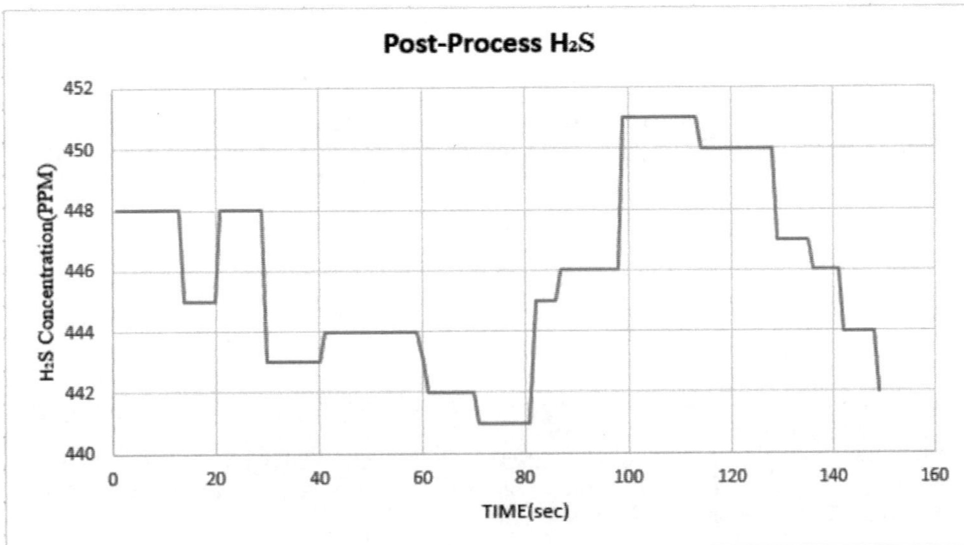

Figure 8: Graph of post-process for H_2S.

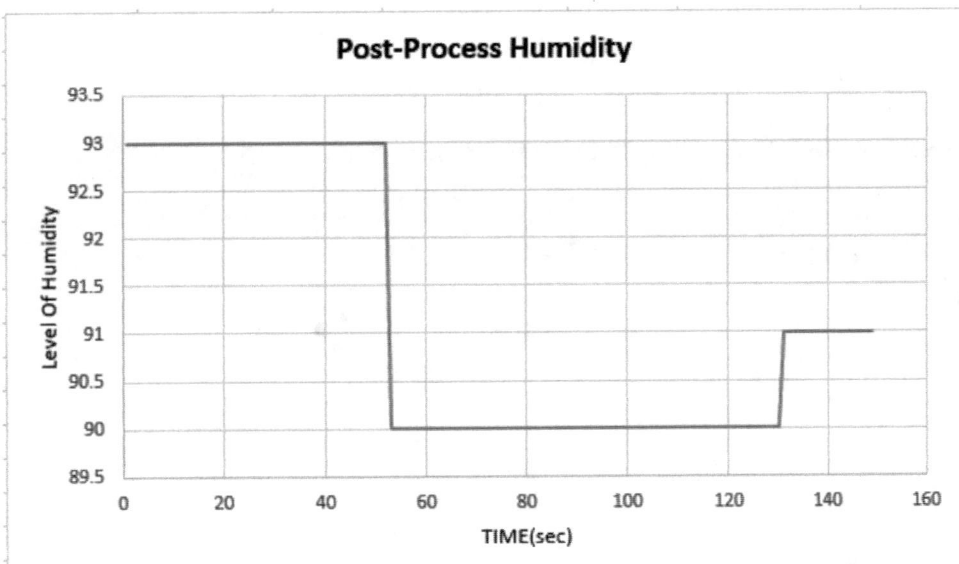

Figure 9: Graph of post-process for Humidity.

4. Conclusion

The use of drying agents to purify human waste biogas was found to very successful via the quality of biogas obtained. Hence, research affirms a process that takes fumes from sewage slurry filled with human waste (that are originally not rich in biogas) and purifies it for numerous uses. Adopting several techniques that includes sewage slurry design and concentrated silica gel can optimize this process. For future or further research, this project can be improved by increasing the storage capacity of the tank to accommodate more gas so that the stability of the biogas can be monitored for onward usage.

Acknowledgement
The authors appreciate Covenant University for partial sponsorship. The authors acknowledge NASA for primary dataset.

5. References

[1]. Khan, M. A. A. (2011). National Economic & Environmental Development Study (NEEDS), ENVORK: A Research and Development Organization.
[2]. Colville F (2014) Solar power portal. UK solar PV industry reaches 5GW installed capacity. Availableat: http://www.solarpowerportal.co.uk/guest_blog/uk_solar_pv_industry_reaches_5gw_installed_capacity_3467 (accessed 4/06/ 2019).
[3]. Moses E. Emetere, Marvel L. Akinyemi, and Etimbuk B. Edeghe, (2016). A Simple Technique for Sustaining Solar Energy Production in Active Convective Coastal Regions, International Journal of Photoenergy 2016, 3567502, 1-11, http://dx.doi.org/10.1155/2016/3567502
[4]. Emetere M.E., & Akinyemi M.L., (2015) Prospects of Solar Energy in the Coastal Areas of Nigeria, 2015 PIAMSEE: AIP Conference Proceedings, 1705 (1), 020035
[5]. Emetere M.E., Okoro U., Etete B., and Okunbor G., (2016), Free energy option and its relevance to improve domestic energy demands in southern Nigeria, Energy Reports, 2, 229–236.
[6]. Momoh, O.L. Yusuf; Nwaogazie, L. Ify, Effect of Waste Paper on Biogas Production from Co-digestion of Cow Dung and Water Hyacinthin Batch Reactors, J. Appl. Sci. Environ. Manage., 12 (4) (2008), pp. 95 – 98
[7]. D. Andriani, A. Wresta, A. Saepudin, B. Prawara, A review of recycling of human excreta to energy through biogas generation: indonesia case, Energy Procedia, 68 (2015), pp. 219-225

Short review on the prospects of human biogas utilization in Nigeria

Emetere M.E.[1,2]**, and Adesina T. A.**[1]
[1]Department of Physics, Covenant University Canaan land, P.M.B 1023, Ota, Nigeria.
[2]Department of Mechanical Engineering and Science, University of Johannesburg, APK, South Africa

E-mail: emetere@yahoo.com

Abstract. Energy challenge in Nigeria is really disturbing due to many factors that borders on the inability of the national grid to meet-up with the huge energy demands in the most populated country in Africa. About 85% industries and domestic users patronize fossil fuel generators. This development had increased the air pollution load over the country. In this research, fourteen years dataset from the Nigeria Metrological Agency (NIMET) was used to simulate the energy accruable from solar energy in Nigeria. The study is poised to enhance higher patronage of solar technology in the research site.

Keywords: energy, biogas, human waste, renewable energy

1. Introduction

In today's energy-intensive society, it is a must to probe and profit from new renewable and eco-friendly energy sources [1]. Various cellulosic biomass (cattle dung, agricultural residues, etc.) are available in rural areas of developing countries in abundance that have very good potential to meet the demand for energy, particularly in the domestic sector. In India alone, there is an estimated over 250 million bovine animals in India alone, and if 33% of the waste created yearly is accessible for biogas generation, in excess of 12 million biogas plants can be introduced [2]. Biogas technology provides a very attractive way to use certain biomass categories to meet domestic energy needs in urban and rural areas.

The high demand for fuelwood which is a big business in places like Nigeria, has monumentally grown with the rise in the population of the world, leading to the loss of trees and forests, through a process called deforestation. In any case, individuals appear to overlook that the wood radiates from the trees in the backwoods, and notwithstanding certain guidelines concerning the utilization of woods, the exercises of unlawful lumberjacks can't be totally regulated or diminished. The deforestation level is higher than the forestation endeavours, and this has brought about ecological corruption. The best obstruction to the recognition of these guidelines is the nonappearance of elective wellspring of fuel, for example, the lamp oil and other oil local fuel [3].

Biogas is a gaseous fuel containing 60% methane, 40% carbon dioxide and small amount of hydrogen sulphide, nitrogen and hydrogen. It is obtained from biomass-plant and animal materials, by the process of anaerobic (absence of oxygen) digestion or fermentation, of which the in-feed to the biogas plant includes: urban waste (garbage), urban refuse, agricultural waste, cow dung-case study. Biogas is used as biofuels that originates from biogenic material, can be used for cooking, heating, generating electricity and running a vehicle [4].

IOP Conf. Series: Earth and Environmental Science **331** (2019) 012051 doi:10.1088/1755-1315/331/1/012051

The richest forms of biogas or biofuel materials are cow waste (i.e. cow dung) due to the already digested state from the multiple stomachs of the host [5]. Cow dung is easily gotten in large quantity from rural areas. But in an area like urban settlements where cow dung is not easily gotten, the only other alternative is human waste, and this is richly available in the sewage sludge. More so, the continuous use of cow dung (or its alternative i.e. bird droppings) for biogas production would be a huge problem to local farmers who adopts the natural manure from animal waste.
This review is examines the prospects of biogas generation from human waste in Nigeria. According to Worldometers [6], the population of Nigeria is approximately two hundred and one million (200 million) people (Figure 1). The distribution of the human population is displayed in Figure 2. This means that Nigeria has a huge biogas potential from human waste if it is properly explored. The conversion of human waste to biogas has taken the center stage in research with amazing results in literatures [7-10].

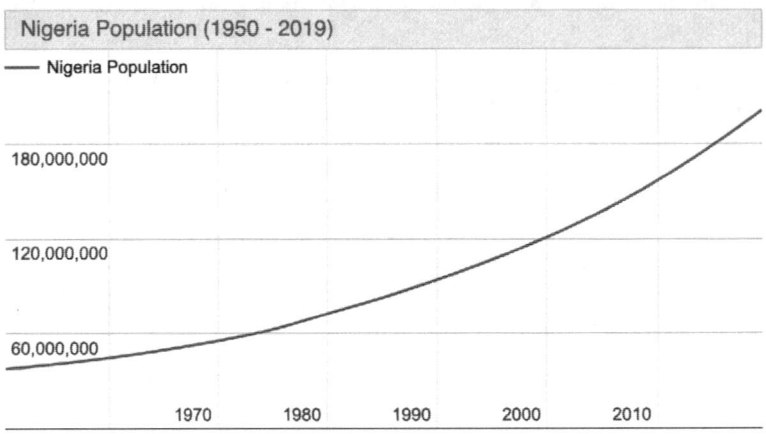

Figure 1: Progression of Nigeria population [6]

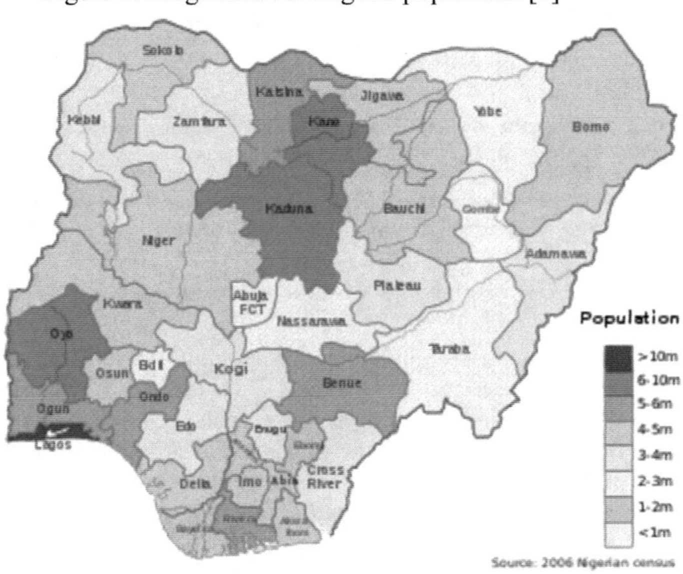

Figure 2: Distribution of Nigeria population

2. Sources of Biogas

Biogas refers to a gas created by the natural degradation/crumbling of natural issue without oxygen. Natural waste, for example, decomposing plant and creature tissue, creature fertilizer, and kitchen waste can be changed over into a vaporous fuel known as the 'Biogas'. Biogas starts from organic material (i.e. resulting from biological activities) and is a type of biofuel, which is commonly used to depict auxiliary sustainable powers that are gotten by bottle concoction handling or the bioconversion of biomass particularly carbonaceous waste materials [11].

Biogas is a vaporous fuel acquired from biomass by the procedure of anaerobic assimilation or maturation, when microscopic organisms corrupt natural material without oxygen. The in-feed or raw material to the biogas plant or bio-purification chamber for the production of biogas includes:

a. Manure (for example cow dung)
b. Urban waste (sewage sludge)
c. Green waste
d. Garbage
e. Agricultural waste.

Methane (CH2) and CO2 carbon dioxide constitute biogas and can also include hydrogen sulphide (H2S), dampness and siloxane traces. Oxygen may be used to burn the gasses methane, hydrogen and carbon monoxide (CO). This release of vitality allows biogas to be used as a fuel. For any warming reasons, for example, cooking, biogas may be used as fuel in any nation. In anaerobic digesters it can also be used where in a gas motor it is commonly used to convert power and warmth into the vitality of the gas. Like gas, biogas can also be compacted and used for control of motor vehicles. In the UK, for instance, biogas is evaluated to have the capability of supplanting around 17% of vehicle fuel. Biogas is an inexhaustible fuel, so it meets all prerequisites for practical power source assignments in specific bits of the world. Biogas can in like manner be tidied and climbed to vaporous oil models when it advances toward getting to be bio-methane [4,5]. The biogas sources and the corresponding yield is presented in Figure 3 below.

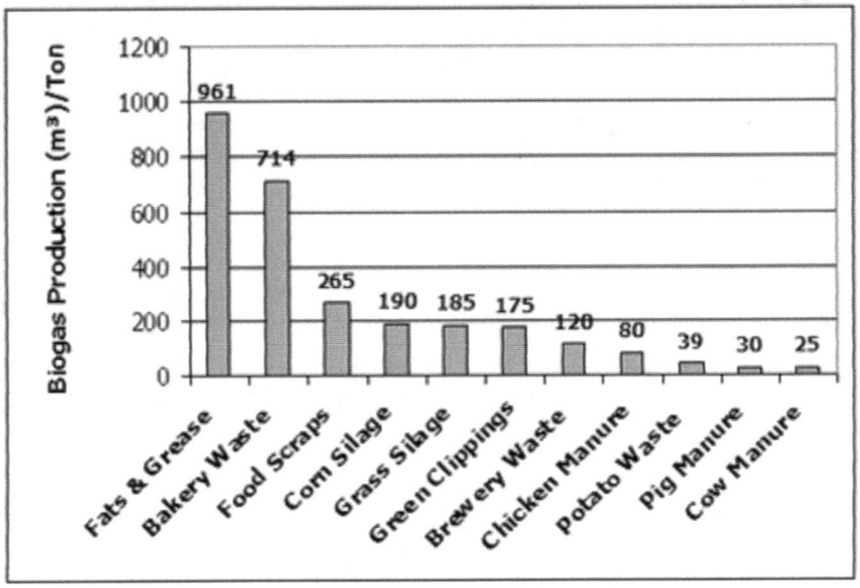

Figure 3: Biogas sources and yields [12]

In order to haven an in-depth of biogas, it is very vital to have a good understanding of organic waste; what they are, their examples, and their environmental impacts. A natural waste in this way is whatever originates from plants or creatures that is biodegradable, which are substances that will rot generally rapidly because of the activity of microscopic organisms and separate into components, for example, carbon that are reused normally. The types of biowaste includes: agro-food Industries waste, green waste; waste-water treatment sludge, grease etc. Organic waste has both negative and positive impacts to the environment and its entire component (biotic and abiotic). A great part of the land utilized for waste transfer can't be reused later on due to defilement. This happens when litter in landfills is packed and the air is crushed out. The garbage separates anaerobically (without oxygen), which implies that acids are delivered. The acids influence other refuse things, for example, plastic, to make a poisonous blend known as leachate. Leachate gathers at the base of landfills where it at that point saturates the ground water and from that point into the conduits [13-14].
The use of waste transforms waste into a kind of useful imperative. One ton of waste (2,000 lbs.) has as much warmth as 500 lbs of charcoal. Not all of junk is biomass; maybe half of its compelling content comes from plastics supplied through oil and petroleum gas. Power plants which wastes are essential and referred to as waste plants. On the other hand, real burnable garbage and not coal is the fuel used to fire their boilers, as do coal-ended factories. Waste power costs more than coal and other sources of vitality. The main ideal position for solid waste expenditure is that it reduces 60% to 1990% of waste dumped on sites, thus lowering sites' exchange costs. It likewise utilizes the vitality in the rubbish, as opposed to covering it in a landfill, where it stays unused [10-11].

Another source of biogas is the landfill gas. Landfill gas is an astounding blend of different gases made by the action of microorganisms inside a landfill, which is a carefully structured misery in the ground (or based over the ground, resembling a football field) into which wastes are put. The point of having this landfill is to maintain a strategic distance from any pressure driven (or water-related) association between the squanders and nature, most particularly the groundwater Bacteria and growths are not fussy eaters. They are saprophytic. An organism on a spoiling log is changing over cellulose to sugars to encourage itself. Although this procedure is obstructed in a site, the methane gas substance is still created as waste rots [16].
For safety and ecological reasons, new guidelines expect methane deposits to collect. Methane gas is boring and fragrant but is not harmless. If it saturates nearby homes and is touched off, the gas can cause flames or blasts. Methane gas can be collected, sanitized and used as fuel by waste dumps. Methane is a decent source of vitality, which is the main fixation in gas petrol. Most gas heaters and stoves use methane from services. The East Kentucky Power Cooperative began to recover methane from three waste dumps in 2003. The gas company is currently using gas at five sites to generate 16 MW power–enough for 7,500 or 8,000 homes to be controlled [16,17].

3. Human waste: future trends

The human waste is vital for biogas production. The composition of the solid human waste is made up several components as presented in Table 1.

Table 1: Component of solid human waste

Component	Percentage
Nitrogenous matter or Protein	2-25%
Carbohydrate	25%
FAT	2-15%
Organic humus	25-54%

Hence, the human waste has the following advantages: biogas as a fuel option in contrast to wood, oil, LPG and power; slime from the bio-gas reactor could be utilized as manure. Natural nitrogen from waste will be changed into smelling salts nitrogen, a type of nitrogen which plants can take-up effectively. The robustness of sewage sludge design determines the maximum biogas production. The water closet system have disadvantage of having excess water over the solid human excreta (Figure 4). More so, the excretal is broken down into smaller beats due to collision with the walls of the connecting pipes and sewage tank. This reduces the biogas production and quality. The other option is the latrine system that allows the excretal to be in its original form as presented in Figure 5. The advantage is that the biogas yield and quality is high compared to the water closet system.

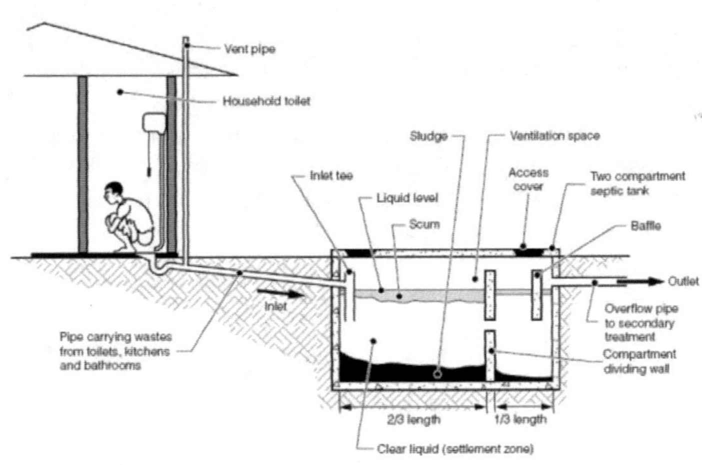

Figure 4: Water closet system [18]

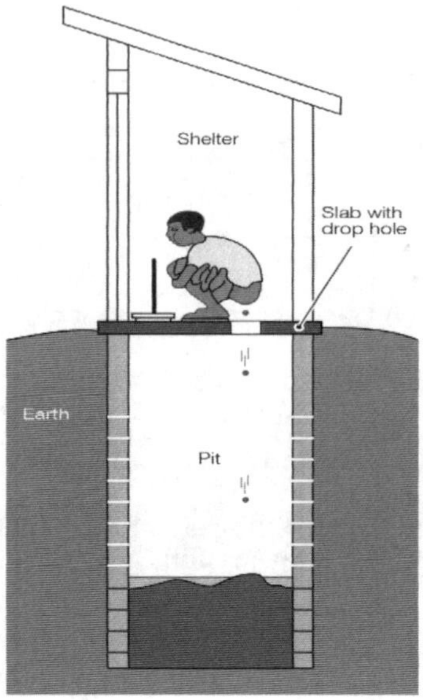

Figure 5: Latrine toilet [18]

More than 80 million Nigerians live in the rural areas and 30 million of the rural dwellers adopt the latrine system. Hence, the biogas generation have huge potentials if adequate or robust sewage design systems are used.

4. Conclusion

Biogas generation from human have been discussed as a viable alternative biogas source due to availability. Domestic energy needs for cooking and mini-power generating plant is a possibility. This will greatly reduce anthropogenic emissions due to biomass burning. In other words, biogas production from human waste may likely gain potential because it's a lot cheaper than solar energy generation – both in the short and long run. Based on the human population and the current toilet system adopted by over 40 million of the populace, the kick-off cost for individuals and small businesses will be appreciably low. It is recommended that the human waste biogas can be enhanced if homes can adopt robust sewage sludge system.

Acknowledgement

The authors appreciate Covenant University for partial sponsorship. The authors acknowledge NASA for primary dataset.

5. References

[1]. Emetere M.E., Okoro U., Etete B., and Okunbor G., (2016), Free energy option and its relevance to improve domestic energy demands in southern Nigeria, Energy Reports, 2, 229–236

[2]. S. Mittal, E.O. Ahlgren, P.R. Shukla, Barriers to biogas dissemination in India: a review, Energy Pol, 112 (2018), pp. 361-370

[3]. N.L. Lam, K.R. Smith, et al. Kerosene: a review of household uses and their hazards in low- and middle-income countries, J. Toxicol. Environ. Health B: Crit. Rev., 15 (6) (2012), pp. 396-432

[4]. Achinivu Ezinne Chidiebere, Isolation and Recovery of Lignin from Lignocellulosic Biomass Using Recyclable Protic Ionic Liquids (PILs) for a Cost-Effective Biomass Processing Technique, PhD thesis submitted to North Carolina State University

[5]. Momoh, O.L. Yusuf; Nwaogazie, L. Ify, Effect of Waste Paper on Biogas Production from Co-digestion of Cow Dung and Water Hyacinthin Batch Reactors, J. Appl. Sci. Environ. Manage., 12 (4) (2008), pp. 95 – 98
[6]. Worldometers, https://www.worldometers.info/world-population/nigeria-population/ (Access 2/06/2019)
[7]. D. Andriani, A. Wresta, A. Saepudin, B. Prawara, A review of recycling of human excreta to energy through biogas generation: indonesia case, Energy Procedia, 68 (2015), pp. 219-225
[8]. Biogas From Human Waste, Workshop held in Delhi. (August22-23.1986).https://www.ircwash.org/sites/default/files/352.1-4961.pdf
[9]. Owamah, H.I.; Dahunsi, S.O.; Oranusi, U.S.; Alfa, M.I. Fertilizer and sanitary quality of digestate biofertilizerfrom the co-digestion of food waste and human excreta.Waste Manag.2014,34, 747–75
[10]. Mudasar, R.; Kim, M.H. Experimental study of power generation utilizing human excreta.Energy Convers. Manag.2017,147, 86–99
[11]. P. Abdeshanian, M.G. Dashti, M.S. Kalil, W.M.W. Yusoff, Production of biofuel using biomass as a sustainable biological resource, Biotechnology, 9 (2010), pp. 274-282
[12]. Integrated Energy Industries, Biogas process, https://www.integratedenergyindustries.com/biogas-process.html (accessed 4/06/2019)
[13]. Z. Salem, K. Hamouri, R. Djemaa, K. Allia, Desalination, 220 (2008), pp. 108-114
[14]. S. Renou, J.G. Givaudan, S. Poulain, F. Dirassouyan, P. Moulin, J. Hazard. Mater., 150 (2008), pp. 468-493
[15]. Rose C, Parker A, Jefferson B, Cartmell E ., The Characterization of Feces and Urine: A Review of the Literature to Inform Advanced Treatment Technology, Critical Reviews in Environmental Science and Technology. 45 (17) (2015), pp. 1827–79.
[16]. O.N. Agdag, D.T. Sponza, Co-digestion of mixed industrial sludge with municipal solid wastes in anaerobic simulated landfilling bioreactors, J. Hazard. Mat., 140 (2007), pp. 75-85
[17]. N.J. Themelis, P.A. Ulloa, Methane generation in landfills, Renew. Energy, 32 (2007), pp. 1243-1257
[18]. WEDC Water, Engineering and Development Centre The John Pickford Building School of Civil and Building Engineering Loughborough University. p. 22. ISBN 978 1 84380 143 6. (accessed 4/06/2018)

IOP Conf. Series: Earth and Environmental Science **331** (2019) 012052 doi:10.1088/1755-1315/331/1/012052

Solar energy estimation and generation in Ibadan Nigeria

Emetere M.E.[1,2], Emetere J.M.[3], and Diana E.[4]
[1]Department of Physics, Covenant University Canaan land, P.M.B 1023, Ota, Nigeria.
[2]Department of Mechanical Engineering and Science, University of Johannesburg, APK, South Africa
[3]Department of Mathematics, Federal University of Technology, Minna
[4] Department of Biochemistry, Covenant University Canaan land, Ota, Nigeria.

E-mail: emetere@yahoo.com

Abstract. Energy challenge in Nigeria is really disturbing due to many factors that borders on the inability of the national grid to meet-up with the huge energy demands in the most populated country in Africa. About 85% industries and domestic users patronize fossil fuel generators. This development had increased the air pollution load over the country. In this research, fourteen years dataset from the Nigeria Metrological Agency (NIMET) was used to simulate the energy accruable from solar energy in Nigeria. The study is poised to enhance higher patronage of solar technology in the research site.

Keywords: solar, energy, solar radiation, meteorology, solar energy

1. Introduction

In 1929, the first electricity supply company in Nigeria was established but electricity generation began in Nigeria around 1896. It was later in 1972 that all isolated electricity-generating corporations in Nigeria were amalgamated with the Niger Dams under the unified operations i.e. National Electric Power Authority (NEPA). The monopoly of NEPA was broken Electric Power Sector Reform Act (EPSRA) in 2005. In 2010, Power Holding Corporation of Nigeria (PHCN) emerged and was made up of eleven distribution companies, six-generation companies and one transmission company [1]. An annual report from PHCN shows that Nigeria Electricity consumption per capital within 2010 and 2014 is $149kWh$ [2].

The energy crisis in Nigeria has been a major issue for both the Nigerian government and the citizens for the past four decades [2]. This energy issue has caused businesses to run into loss while homes depend solely on the off-grid supply (i.e. using diesel/gas/petrol-powered electric generators), thereby accumulating huge costs and contributing to greenhouse gas pollution amongst other dangerous environmental problems [3]. The Nigerian government has a plan of increasing electricity production to 40,000MW by the year 2020 [2].

Analysts have projected that Nigeria could generate 600,000MW by deploying Solar PV panels from just 1% of Nigeria's land mass [4]. Despite the affordability of solar device set-up, there is the technical challenge that may hinder the reality of this projection [5]. The challenges includes: shadows- which may be natural or man made; dust, dirt, frost and snow on modules- which depends on location and also on rainfall and module inclination; reflectance of module surface; PV conversion efficiency (which depends on temperature and irradiance); and sunlight spectrum. The known types of PV systems are: grid-connected PV systems and off-grid PV systems. In the study area, the off-grid PV system is widely used. In regards to a consumer solar installation, there are two main photovoltaic technologies used in solar panels across the variety of manufacturers. Specifically, mono-crystalline silicon and poly-crystalline silicon. In this study, the adaptation both types of PV panels in different scenarios were simulated to assist solar device companies and government to invest in solar generation in the study area.

2. Experimental Design, Materials and Methods

The dataset that was used for this research was obtained from the Nigerian Metrological Agency (NIMET). Fourteen years dataset (1981-1994) was used for the research. The sunshine hour and solar irradiation dataset was extracted from the bulk dataset.

The efficiency of the solar cell is mainly dependent on the temperature of solar cell they acquire during sunshine. Hellman [6] gave the approximated cell temperature without considering the cooling via airflow under the PV panel as:

$$T_c = T_a + (T_{NOCT} - 20^0C)\frac{G}{800W/m^2} \quad (1)$$

where Ta is the ambient temperature, T_{NOCT} is the normal operating cell temperature (47 °C), and G the global irradiance. The efficiency of the solar module can then be estimated by different methods and one of them is:

$$\eta = \eta_{STC}(1 - \beta_P(T_c - T_{STC}) + \gamma \log G) \quad (2)$$

where η_{STC} is the efficiency in STC, γ=0 [7]. Hence, the power in this regard is given as:

$$P = \frac{G}{1000W/m^2} P_{max}(1 - \beta_P(T_c - T_{STC})) \quad (3)$$

βP is the power temperature coefficient (typically around 0.45–0.48 %/°C), T_{STC} is the temperature at standard test conditions (25 °C), P_{max} is the nominal power of the module in STC. Hence the open circuit voltage (v_{OC}) at this point is given as [8]:

$$V_{oc} = V_{oc}^{*}(1 - \beta_P(T_c - T_{STC})) + V_t n_p n_s ln\left(\frac{G}{G_o}\right) \quad (4)$$

G is irradiance, Go = monthly average daily extraterrestial radiation for a PV panel composed of np cells in parallel and ns in series.

The monthly average daily extraterrestial radiation is calculated as:

$$G_o = \frac{24}{\pi} I_{sc}\left[1 + 0.33cos\left(\frac{360D_n}{365}\right)\right] \times \left[cosLcos\delta sin\omega_s + \frac{2\pi\omega_s}{360} sinLsin\delta\right] \quad (5)$$

ω_s = sunset hour angle, I_{sc} = solar constant, L = latitude of location, δ = declination angle, D_n = day of the year starting from January

$$\omega_s = cos^{-1}(-tan\, L\, tan\, \delta) \quad (6)$$

$$\delta = 23.45 \sin\left[\frac{360(284-N_d)}{365}\right] \quad (7)$$

Unlike, the conditions in equation (3), power in the PV panel can be calculated as:

$$P = \frac{G}{\frac{1000W}{m^2}} P_{max} \quad (8)$$

where Pmax is the peak power, in kWp. From literatures, monocrystalline solar cells has peak power (Wp) of 140W ± 5% and voltage at maximum power (Vmp) of 17.5V[9] and for poly crystalline solar cells, the peak power (Wp) is 250W ± 5% and voltage at maximum power (Vmp) of 30.5V[10]
The ratio between actual power and the theoretical power is called Performance Ratio (PR). This terminology may also be referred to as module PR (PRm) or system PR (PRs).

3. Results and Discussion

The solar irradiance over the research was analysed between 1981 and 2004. It was observed that 1984 had the highest solar irradiance (Figure 1). The solar irradiance between 1988 and 1996 was almost uniform i.e. showing that the research area does not have significant solar irradiance perturbations.

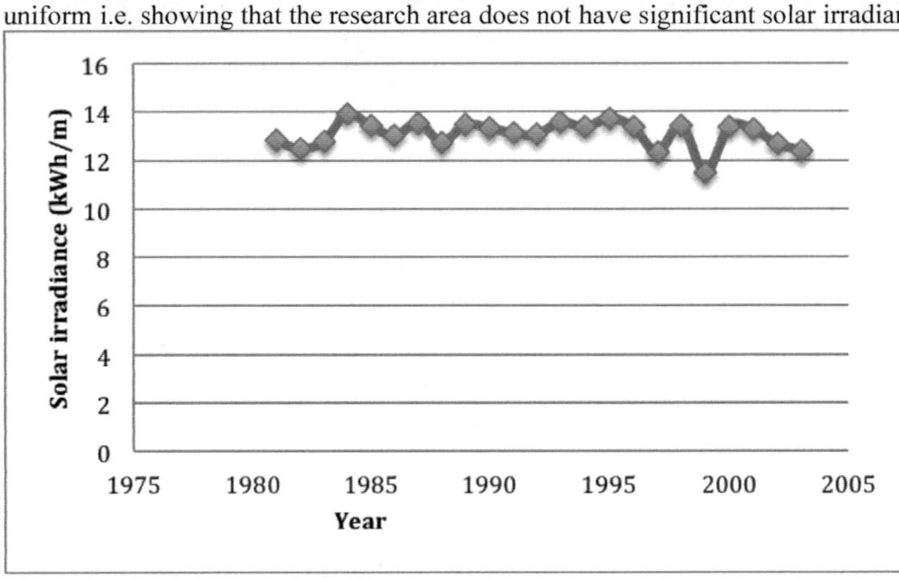

Figure 1: solar irradiance in Ibadan

This is particularly interesting because it means that the results can be projected (in form of forecast) with minimal errors. The ambient and solar cell temperatures are presented in Figure 2. This result suggests that the solar cell (due to its inherent properties) retains temperature. The thermal gradients between the ambient and solar cell temperatures are good reasons to make the PV panel work at lower efficiency. The solar module efficiency is presented in Figure 3. The highest and lowest efficiency was observed in 1984 and 1998 respectively. The solar module efficiency for PV panels (for the years considered in the dataset) showed that the outcome from the PV system is not proportional to the solar irradiance. This shows that solar cell anomalies can be adduced to the material type. Hence, the optimization of the PV panel is essential for perfect conversion of solar irradiance to electricity.

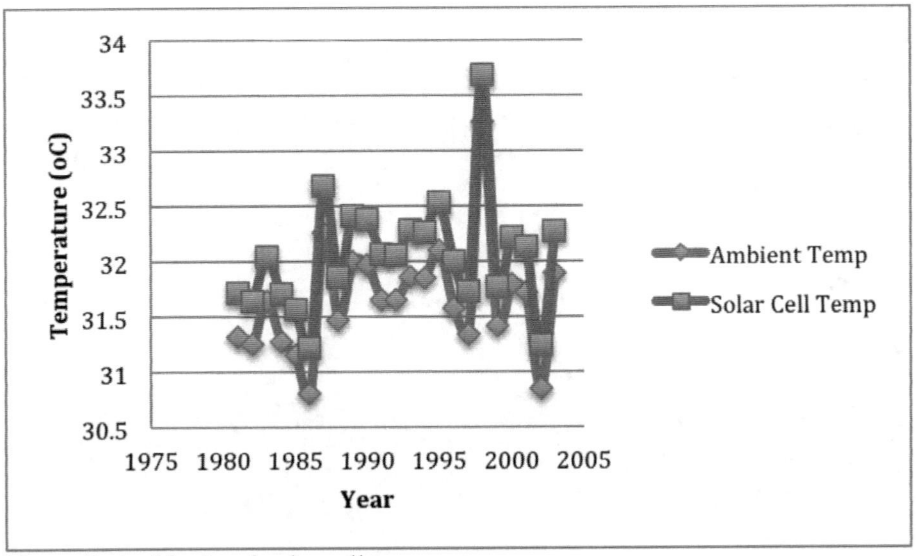

Figure 2: Ambient and solar cell temperatures

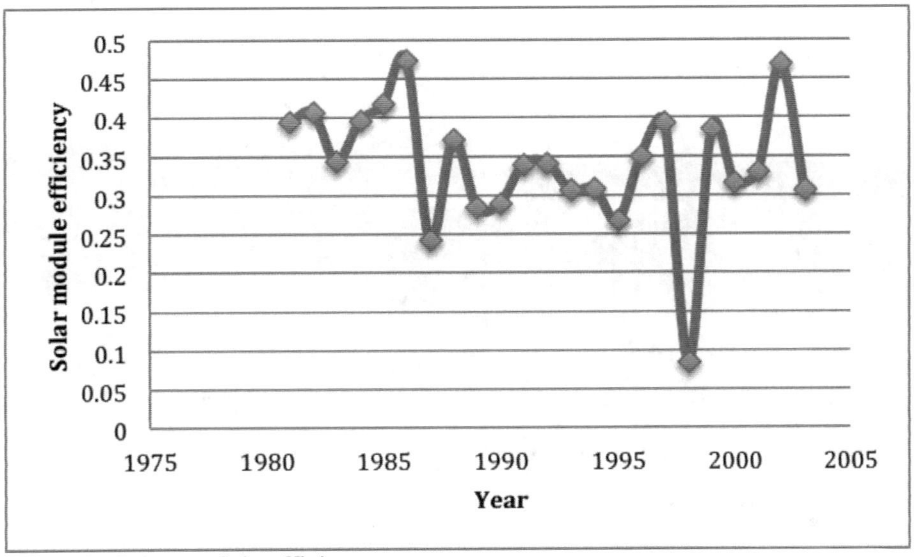

Figure 3: Solar module efficiency

The meteorological dataset was used to calculate the power outcome in normal (equation 8) and optimized (equation 3) state as shown in Figure 4. It is observed that optimized state is more stable to accommodate solar irradiation perturbation. Same scenario occurred in the polycrystalline PV panel displayed in Figure 5. This scenario is essential in prolonging the lifespan of the PV system.

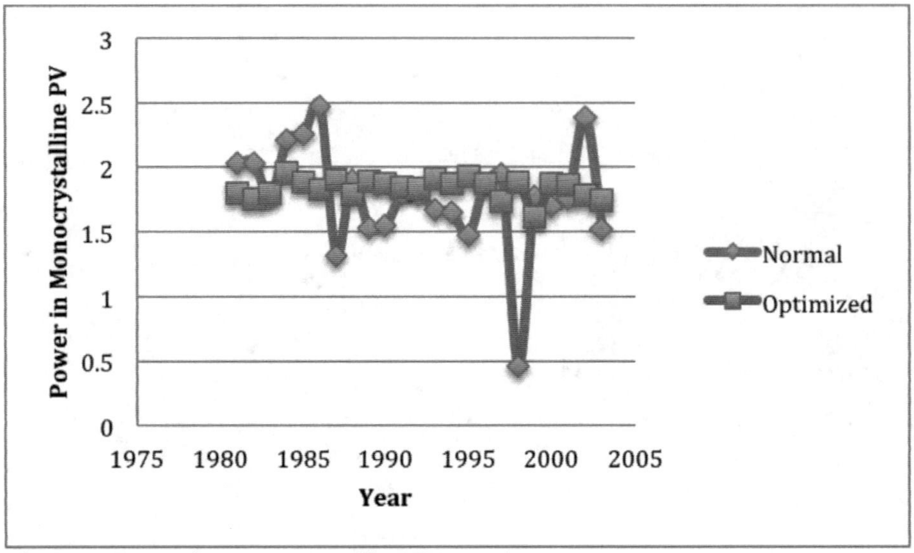

Figure 4: Power in mono-crystalline PV panel

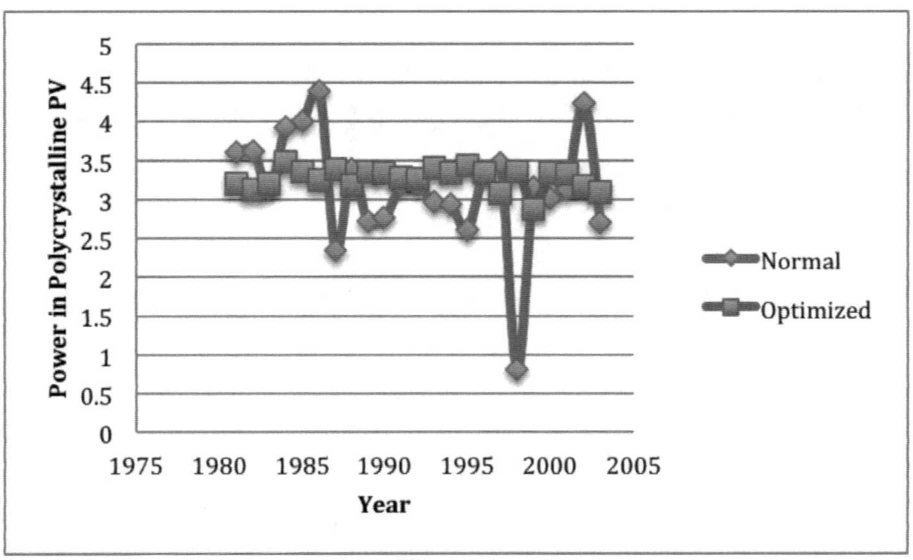

Figure 5: Power in poly-crystalline PV panel

However, the out puts presented in Figures 4 and 5 do not reveal a perfect scenario of optimal working solar cells.

4. Conclusion

The study shows the theoretical analysis of PV systems in Ibadan Nigeria using meteorological dataset. The direct perturbations noticed in the dataset were seen to reflect in the outcome of the PV system i.e. solar cell temperature, solar module efficiency, and power output. The thermal gradients between the ambient and solar cell temperatures are good reasons to make the PV panel work at lower efficiency. Also, it was reported that the solar cell anomalies observed in PV systems could be adduced to the material type of the solar cells. More so, it was observed that solar irradiance fluctuation have significant effect on the power output of the PV system. Hence, it is recommended that PV systems brought to the research site should incorporate spectra filtering layer to stabilize the solar irradiance fluctuations.

Acknowledgement
The authors appreciate Covenant University for partial sponsorship. The authors acknowledge NASA for primary dataset.

5. References

[1]. NERC (2019), Power Generation in Nigeria, https://www.nercng.org/index.php/home/nesi/403-generation (Accessed 3/06/2019).
[2]. NESISTATS (2019). Nigeria Power Baseline Report – NESISTATS, https://mypower.ng/wp-content/uploads/2018/01/Baseline-Report.pdf (Accessed 3/06/2019).
[3]. Moses Emetere, Marvel L Akinyemi, ME Ojewumi, (2018). Systemic Study of Aerosol Loading and Retention over Praia-Cape Verde: Satellite and Ground Observation Analysis, Pol. J. Environ. Stud., 27 (1): 1-8
[4]. Power Nigeria (2019). https://www.power-nigeria.com/en/industry-news/nigeria-scaling-up-solar-energy-deployment-nigerian-govt-must-lead.html
[5]. D.C. Jordan, S.R. Kurtz, (2011). Photovoltaic Degradation Rates-an Analytical Review Progress in Photovoltaics: Research and Applications, doi:10.1002/pip.1182
[6]. Hellman Hannu-Pekka (2011). Photovoltaic power generation modelling, sgemfinalreport.fi/files/WP611_photovoltaics%20HP%20Hellman.pdf (Accessed 3/06/2019).
[7]. Mattei M. & Notton G. & Cristofari C. & Muselli M. & Poggi P. Cal- culation of the polycrystalline PV module temperature using a simple method of energy balance, Renewable Energy, Volume 31, Issue 4, April 2006, Pages 553-567, ISSN 0960-1481, DOI: 10.1016/j.renene.2005.03.010. Available from: http://www.sciencedirect.com/science/article/pii/S096014810500073X
[8]. IEC60891, Photovoltaic Devices. Procedures for Temperature and Irradiance Corrections to Measured IV Characteristics." IEC60891, 2010.
[9]. Energymatters (2019). Trina Solar Honey 250Watt Polycrystalline Solar Panel, https://www.energymatters.com.au/trina-solar-honey-250watt-polycrystalline-solar-panel-p-2733.html (Accessed 3/06/2019).
[10].AKT (2019). AKT-140-M Solar Panel Kit Review, http://www.alternative-energy-tutorials.com/solar-panel-review/140w-akt-solar-panel.html (Accessed 3/06/2019).

IOP Conf. Series: Earth and Environmental Science **331** (2019) 012053 doi:10.1088/1755-1315/331/1/012053

Modified Angstrom Solar Radiation Model for Abeokuta Nigeria

Emetere M.E.[1,2], Emetere J.M.[3], and Dania O.E.[4]
[1]Department of Physics, Covenant University Canaan land, P.M.B 1023, Ota, Nigeria.
[2]Department of Mechanical Engineering and Science, University of Johannesburg, APK, South Africa
[3]Department of Mathematics, Federal University of Technology, Minna, Nigeria
[4]Department of Biochemistry, Covenant University Canaan land, Ota, Nigeria

E-mail: emetere@yahoo.com

Abstract. Sustainable power supply still remains the challenge in developing countries. Unfortunately the growing energy need is directly proportional to the population growth. Hence, there is the need to step-up on the alternative energy use. The adoption of solar technology in the research site is encouraging-partly because of the low cost of purchasing solar devices. However, there is the issue of short lifespan of the photovoltaic (PV) panel that operates in the area. This research used the modified Angstrom solar radiation model to determine that the main cause of short PV lifespan is solar spectrum signature over the research area. It is recommended that solar companies should start incorporating spectra filtering layer to protect the PV panel.

Keywords: energy, solar, solar radiation, meteorology, solar energy, renewable energy

1. Introduction

The scorching sun in sub-Saharan Africa has been regarded by many for decades as having little or no economic value and a possible hindrance to development. This perception has fundamentally changed though due to the emergence of technologies in solar energy utilization [1]. One of the main drivers of development is electricity. In Nigeria, less than 40% of the country is connected to the National Grid yet, less than 60% of the energy needs of this group is generated and distributed [2]. Erratic and insufficient power supply has been a major problem bedevilling Nigeria resulting in the use of fossil fuel generators both in residences and industries with the attendant emissions that increase the rate of air pollution accompanied with possible adverse health effects on the populace [3,4]. The discovery of the use of the sun's energy in the production of photovoltaic effect was made in 1831 [5]. According to Nasir [6], the utilization of solar energy is dependent on its availability and appropriate technology. The amount of energy radiated by the sun in a year is greater than that used by man since the beginning of time. Okafor and Joe-Uzuegbu [7] reported that the 6.25 hours average daily sunshine and 5.25kW/m^2 per day average daily solar radiation in Nigeria translates to about 4.851 × 10^{12}kWh of energy per day or 1.804 × 10^{15}kWh per annum. Standalone photovoltaic (PV) system is a collection of interconnected electrical components used in the generation of electricity from sunlight. One of its components, the solar panel is responsible for trapping the solar energy [8]. Most citizens welcomed the advent of standalone solar PV installations into the country though; the cost of purchase and installation is still quite high for the average citizen to afford. It offers advantages such as environmental friendliness, quiet operations, high reduction in cabling for external lighting, does not require transformer, high or low tension wiring making its installation on individual establishment and premises easier [2, 9]. The challenges facing the utilization of solar energy system in Nigeria include high initial investment cost, variability and intermittency of radiation, operation and maintenance cost, ineffective quality of products, insecurity of solar plant infrastructure etc. [10]. The research site Abeokuta, the capital of Ogun State, Nigeria (Figure 1), also known as the Gateway State can serve as a model for other states in the south-western region of the country. In this research, the focus is to use the modified Angstrom solar radiation model to determine the causes of short PV panel lifespan.

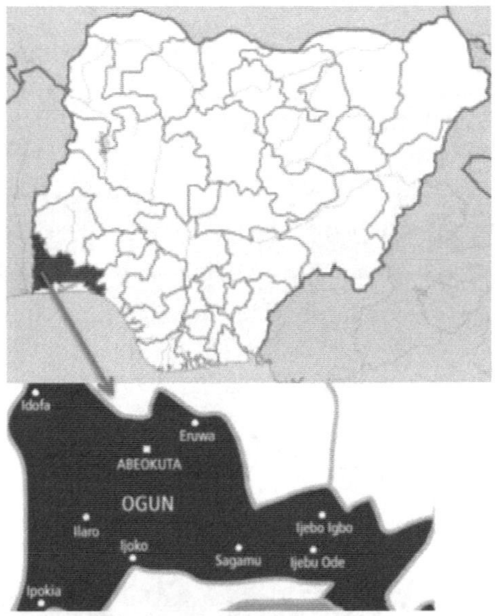

Figure 1: Map of Abeokuta

2. Methodology

The dataset that was used for this research was obtained from the Nigerian Metrological Agency (NIMET). Fourteen years dataset (1981-1994) was used for the research. The sunshine hour and solar irradiation dataset was extracted from the bulk dataset. The dataset where used to obtain the required parameter for the solar radiation model. The average monthly global radiation model as derived by Angstrom[11] and modified by Page[12] and Prescott [13]. Is given as:

$$\frac{H}{H_o} = a + b\left(\frac{S}{S_o}\right) \quad (1)$$

H =monthly average global radiation, Ho = monthly average daily extraterrestial radiation, S = monthly average daily bright sunshine hour, So = maximum possible monthly average daily hour, a=b= constants.

The monthly average daily extraterrestial radiation is calculated as:

$$H_o = \frac{24}{\pi} I_{sc}\left[1 + 0.33cos\left(\frac{360D_n}{365}\right)\right] \times \left[cosLcos\delta sin\omega_s + \frac{2\pi\omega_s}{360} sinLsin\delta\right] \quad (2)$$

ω_s = sunset hour angle, I_{sc} = solar constant, L = latitude of location, δ = declination angle, D_n = day of the year starting from January

$$\omega_s = cos^{-1}(-tan\,L\,tan\,\delta) \quad (3)$$

$$\delta = 23.45sin\left[\frac{360(284-D_n)}{365}\right] \quad (4)$$

$$S_o = \frac{2}{15}cos^{-1}(-\,tan\,L\,tan\,\delta) \quad (5)$$

The regression model was used to obtain the value of 'a' and 'b'

3. Results and Discussion

The monthly variation was used to calculate the constants ('a' and 'b') for Lagos. For January, the modified Angstrom model (MAM) is given as:

$$\frac{H}{H_o} = 8.211 - 0.068\left(\frac{S}{S_o}\right) \quad (6)$$

The linear regression for this result is 27.64%. Though the result is low, however, it should be noted that the relationship depends on the measured data provided by NIMET. The results for February is displayed as:

$$\frac{H}{H_o} = 10.03 - 0.071\left(\frac{S}{S_o}\right) \quad (6)$$

The linear regression for this result is 42.76%. The results for March is displayed as:

$$\frac{H}{H_o} = 10.04 + 0.039\left(\frac{S}{S_o}\right) \quad (7)$$

The linear regression for March result is 6.1%. The results for April is displayed as:

$$\frac{H}{H_o} = 7.08 + 0.109\left(\frac{S}{S_o}\right) \quad (8)$$

The linear regression for April result is 18.55%. The results for May is displayed as:

$$\frac{H}{H_o} = 6.048 + 0.104\left(\frac{S}{S_o}\right) \quad (9)$$

The linear regression for May result is 28.5%. The results for June is displayed as:

$$\frac{H}{H_o} = 6.94 - 0.022\left(\frac{S}{S_o}\right) \quad (10)$$

The linear regression for June result is 1.65%. The results for July is displayed as:

$$\frac{H}{H_o} = 4.88 + 0.063\left(\frac{S}{S_o}\right) \quad (11)$$

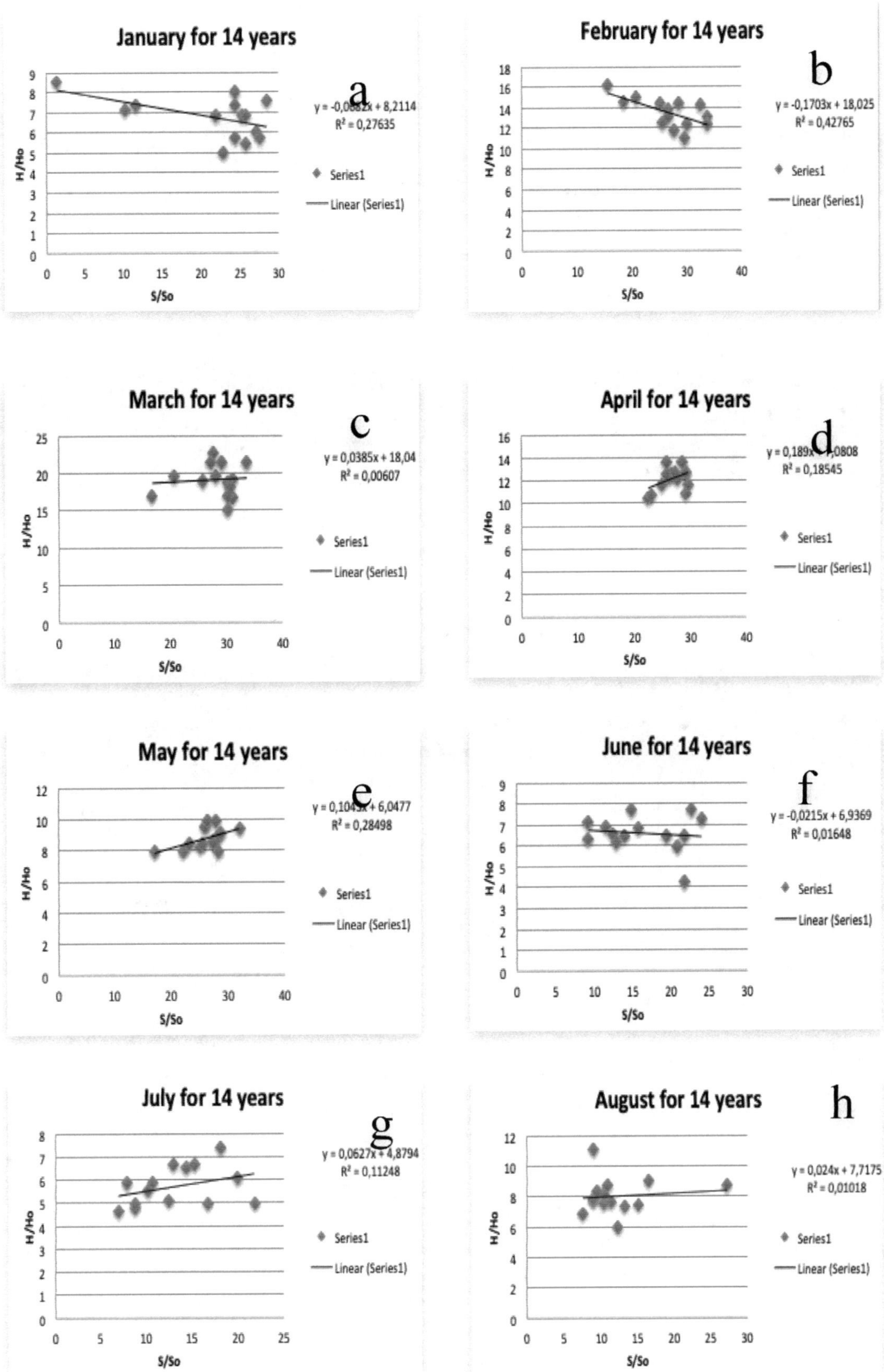
January for 14 years
a
R² = 0,27635
Series1
Linear (Series1)
H/Ho
S/So
February for 14 years
b
y = -0,1703x + 18,025
R² = 0,42765
Series1
Linear (Series1)
H/Ho
S/So
March for 14 years
c
y = 0,0385x + 18,04
R² = 0,00607
Series1
Linear (Series1)
H/Ho
S/So
April for 14 years
d
R² = 0,18545
Series1
Linear (Series1)
H/Ho
S/So
May for 14 years
e
R² = 0,28498
Series1
Linear (Series1)
H/Ho
S/So
June for 14 years
f
y = -0,0215x + 6,9369
R² = 0,01648
Series1
Linear (Series1)
H/Ho
S/So
July for 14 years
g
R² = 0,11248
Series1
Linear (Series1)
H/Ho
S/So
August for 14 years
h
y = 0,024x + 7,7175
R² = 0,01018
Series1
Linear (Series1)
H/Ho
S/So

IOP Conf. Series: Earth and Environmental Science **331** (2019) 012053 doi:10.1088/1755-1315/331/1/012053

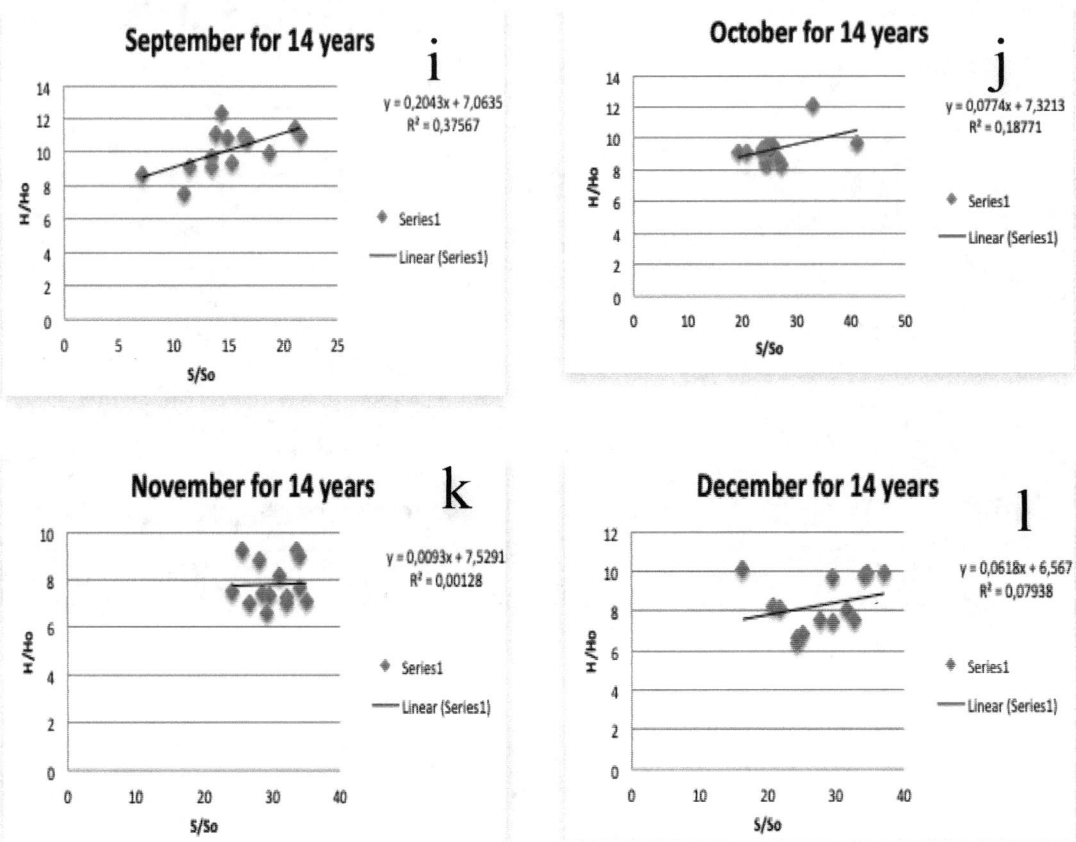

Figure 2: Extraction of constants from the modified Angstrom model

The linear regression for July result is 11.25%. The results for August is displayed as:

$$\frac{H}{H_o} = 7.72 + 0.024\left(\frac{S}{S_o}\right) \quad (12)$$

The linear regression for August result is 1.02%. The results for September is displayed as:

$$\frac{H}{H_o} = 7.064 + 0.204\left(\frac{S}{S_o}\right) \quad (13)$$

The linear regression for September result is 37.57%. The results for October is displayed as:

$$\frac{H}{H_o} = 7.32 + 0.077\left(\frac{S}{S_o}\right) \quad (14)$$

The linear regression for October result is 18.77%. The results for November is displayed as:

$$\frac{H}{H_o} = 7.53 + 0.009\left(\frac{S}{S_o}\right) \quad (15)$$

The linear regression for November result is 0.12%. The results for December is displayed as:

$$\frac{H}{H_o} = 6.57 + 0.062\left(\frac{S}{S_o}\right) \quad (16)$$

The linear regression for December result is 7.93%. The average of the twelve months is given as:

$$\frac{H}{H_o} = 8.78 + 0.069\left(\frac{S}{S_o}\right) \quad (17)$$

It is observed that February had the highest linear regression. Also, constants ‘a’ and ‘b’ has high variability on a monthly basis. This means that the solar spectrum over this region is the main cause of low lifespan of PV panel. This result is far from the known models discussed in literature [14-16].

4. Conclusion

The modified Angstrom solar radiation model has been proven to be very reliable. Hence, the results in this research are reliable to give scientist and energy professional clue into the solar energy sustainability over the research site. The results show that the solar activity of Nigeria is dynamic with high monthly variability (via the values of 'a' and 'b'). This means that the solar spectrum over this region is the main cause of low lifespan of PV panel. It is recommended that solar device companies should seek ways of incorporating spectrum filter layer over the surfaces of PV systems.

Acknowledgement

The authors appreciate Covenant University for partial sponsorship. The authors acknowledge NASA for primary dataset.

5. References

[1]. Joseph A-A (2014). Solar energy in sub-Saharan Africa: the challenges and opportunities of technological leapfrogging. Thunderbird International Business Review.
[2]. Akinboro, F., & Adejumobi, L. A. (2012). Solar Energy Installation in Nigeria: Observations, Prospect, Problems, and Solution. Transnational Journal of Science and Technology, 2(4), 73-84
[3]. Moses E. Emetere, Marvel L. Akinyemi, and Etimbuk B. Edeghe, (2016). A Simple Technique for Sustaining Solar Energy Production in Active Convective Coastal Regions, International Journal of Photoenergy 2016, 3567502, 1-11, http://dx.doi.org/10.1155/2016/3567502
[4]. Emetere M.E., & Akinyemi M.L., (2015) Prospects of Solar Energy in the Coastal Areas of Nigeria, 2015 PIAMSEE: AIP Conference Proceedings, 1705 (1), 020035
[5]. Webb, J. 1995. By the light of the sun. New Sci. J. 121, 40
[6]. Nasir, A. (2001). A Technology for Helping to Alleviate the Energy Problems; Solar Energy for Cooking and Power generation. In 3rd Annual Engineering Conference of Federal University of Technology, Minna, Nigeria.
[7]. Okafor E.N.C and Joe-Uzuegbu C.K.A. (2010). Challenges to development of renewable energy for electric power sector in Nigeria. International Journal of Academic Research, Vol.2. No.2 March 2010
[8]. Abdullahi B. S., Timothy O.O and Adekunle J.O (2017). Solar photovoltaic system: a case study of Akure, Nigeria. World Scientific News 83(2017):15-28
[9]. Dare-Abel O.A (2010). Solar powered buildings in Nigeria: hallenges and opportunities for the future. Sustainable Development and Environmental Protection-Strategies for developing nationa. 2010:152-153
[10]. Ohunakin, O. S., Adaramola, M. S., Oyewola, O. M., and Fagbenle, R. O. (2014). Solar energy applications and development in Nigeria: drivers and barriers. Renewable and Sustainable Energy Reviews, 32, 294-301
[11]. Angstrom, "Solar and terrestrial radiation," Quarterly Journal of the Royal Meteorological Society, vol. 50, no. 210, pp. 121–126, 1924.
[12]. J. K. Page, "■e estimation of monthly mean values of daily total short wave radiation on-vertical and inclined surfaces from sun shine records for latitudes 400 N–400 S," in Proceedings of the United Nations Conference on New Sources of Energy, vol. 98, no. 4, pp. 378–390, 1961.
[13]. J. A. Prescott, "Evaporation from water surface in relation to solar radiation," Transactions of the Royal Society of South Australia, vol. 64, pp. 114–118, 1940.
[14]. R. B. Benson, M. V. Paris, J. E. Sherry, and C. G. Justus, "Estimation of daily and monthly direct, diffuse and global solar radiation from sunshine duration measurements," Solar Energy, vol. 32, no. 4, pp. 523–535, 1984.

[15]. H.P.GargandS.N.Garg,"Correlationofmonthly-averagedaily global, diffuse and beam radiation with bright sunshine hours," Energy Conversion and Management, vol. 25, no. 4, pp. 409–417, 1985.
[16]. V. Bahel, R. Srinivasan, and H. Bakhsh, "Solar radiation for Dhahran, Saudi Arabia," Energy, vol. 11, no. 10, pp. 985–989, 1986.

Modified Angstrom Solar Radiation Model for Lagos Nigeria

Emetere M.E.[1,2], Emetere J.M.[3], and Dania O.E.[4]
[1]Department of Physics, Covenant University Canaan land, P.M.B 1023, Ota, Nigeria.
[2]Department of Mechanical Engineering and Science, University of Johannesburg, APK, South Africa
[3]Department of Mathematics, Federal University of Technology, Minna, Nigeria
[4]Department of Biochemistry, Covenant University Canaan land, Ota, Nigeria

E-mail: emetere@yahoo.com

Abstract. Solar energy generation in tropical climate system is said to be a viable project because of the availability of solar radiation. The use of solar photovoltaic (PV) panel in the research area is encouraging at the moment but the issue has been that the lifespan of the PV panel is somewhat very low. This research aims to determine the solar radiation model for Lagos to enhance the optimization of the production of solar PV panels for Lagos Nigeria. This would enhance higher patronage in the commercial hub of Nigeria.

Keywords: energy, solar, solar radiation, meteorology, solar energy, renewable energy

1. Introduction

In the tropical climate system, solar energy generation is adjudged to be at its peak due to abundant sunshine. However, this can be argued because different solid-state devices are currently emerging which requires low solar radiation to attain maximum solar energy generation. Solar devices are devices that make use of the solar irradiance to generate heat or electricity. Solar devices can be divided into either passive or active solar devices. In recent time, the adoption of solar device (solar photovoltaic (PV) panel) is gaining higher patronage in the research site. At the moment, only standalone users patronize PV panels because of low energy supply via the national grid system [1].
The research site (Figure 1) is the commercial hub of Nigeria. Hence, the adoption of source of energy generation for standalone users is lucrative and laudable. At the moment, fossil fuel generators are still majorly in use in the research site. With the high state of air pollution in the region [2], the use of fossil fuel should be mitigated to allow for the drastic reduction of anthropogenic emissions. The renewable energy source awareness is encouraging, as most researchers have demonstrated that the populace have shown great enthusiasm for the adoption of solar devices for energy generation [3]. However, early users of solar panels have complained that the lifespan of the PV panel is low at the moment (average of 4 to 7 years). Hence, the huge cost of setting up a solar PV panel and its maintenance would not be economical if the system packs-up within a short time of useage. Based on the above, the patronage of solar devices have reduced.
There are many factors affecting the lifespan of PV panels in tropical region [4]. However, the main is the solar spectra filtering on PV panels. The solar cells in the PV panels react to this spectra and shut-down with time. In this research, we will derive the solar radiation model over the research site. It is believed that this initiative would guide PV panel companies to better understand why patronage is low at the moment in the research site.

IOP Conf. Series: Earth and Environmental Science **331** (2019) 012054 doi:10.1088/1755-1315/331/1/012054

Figure 1: Map of the research site (Lagos).

2. Methodology

The dataset that was used for this research was obtained from the Nigerian Metrological Agency (NIMET). Fourteen years dataset (1981-1994) was used for the research. The sunshine hour and solar irradiation dataset was extracted from the bulk dataset. The dataset where used to obtain the required parameter for the solar radiation model. The average monthly global radiation model as derived by Angstrom[5] and modified by Page[6] and Prescott [7] is given as:

$$\frac{H}{H_o} = a + b\left(\frac{S}{S_o}\right) \quad (1)$$

H =monthly average global radiation, Ho = monthly average daily extraterrestial radiation, S = monthly average daily bright sunshine hour, So = maximum possible monthly average daily hour, a=b= constants.

The monthly average daily extraterrestial radiation is calculated as:

$$H_o = \frac{24}{\pi} I_{sc}\left[1 + 0.33\cos\left(\frac{360D_n}{365}\right)\right] \times \left[\cos L\cos\delta\sin\omega_s + \frac{2\pi\omega_s}{360}\sin L\sin\delta\right] \quad (2)$$

ω_s = sunset hour angle, I_{sc} = solar constant, L = latitude of location, δ = declination angle, D_n = day of the year starting from January

$$\omega_s = \cos^{-1}(-\tan L \tan \delta) \quad (3)$$

$$\delta = 23.45 \sin\left[\frac{360(284-D_n)}{365}\right] \quad (4)$$

$$S_o = \frac{2}{15}\cos^{-1}(-\tan L \tan \delta) \quad (5)$$

The regression model was used to obtain the value of 'a' and 'b'

3. Results and Discussion

The monthly variation was used to calculate the constants ('a' and 'b') for Lagos. For January, the modified Angstrom model (MAM) is given as:

$$\frac{H}{H_o} = 6.457 + 0.142\left(\frac{S}{S_o}\right) \quad (6)$$

The linear regression for this result is 14.8%. Though the result is low, however, it should be noted that the relationship depends on the measured data provided by NIMET.
The result for February is displayed as:

$$\frac{H}{H_o} = 9.043 + 0.038\left(\frac{S}{S_o}\right) \quad (6)$$

The linear regression for this result is 8.16%.
The result for March is displayed as:

$$\frac{H}{H_o} = 8.37 + 0.075\left(\frac{S}{S_o}\right) \quad (7)$$

The linear regression for March result is 8.16%. The result for April is displayed as:

$$\frac{H}{H_o} = 7.68 + 0.098\left(\frac{S}{S_o}\right) \quad (8)$$

The linear regression for April result is 10.8%. The result for May is displayed as:

$$\frac{H}{H_o} = 7.57 + 0.065\left(\frac{S}{S_o}\right) \quad (9)$$

The linear regression for May result is 13.4%. The result for June is displayed as:

$$\frac{H}{H_o} = 5.882 + 0.107\left(\frac{S}{S_o}\right) \quad (10)$$

The linear regression for June result is 21.52%. The result for July is displayed as:

$$\frac{H}{H_o} = 4.925 + 0.119\left(\frac{S}{S_o}\right) \quad (11)$$

IOP Conf. Series: Earth and Environmental Science **331** (2019) 012054 doi:10.1088/1755-1315/331/1/012054

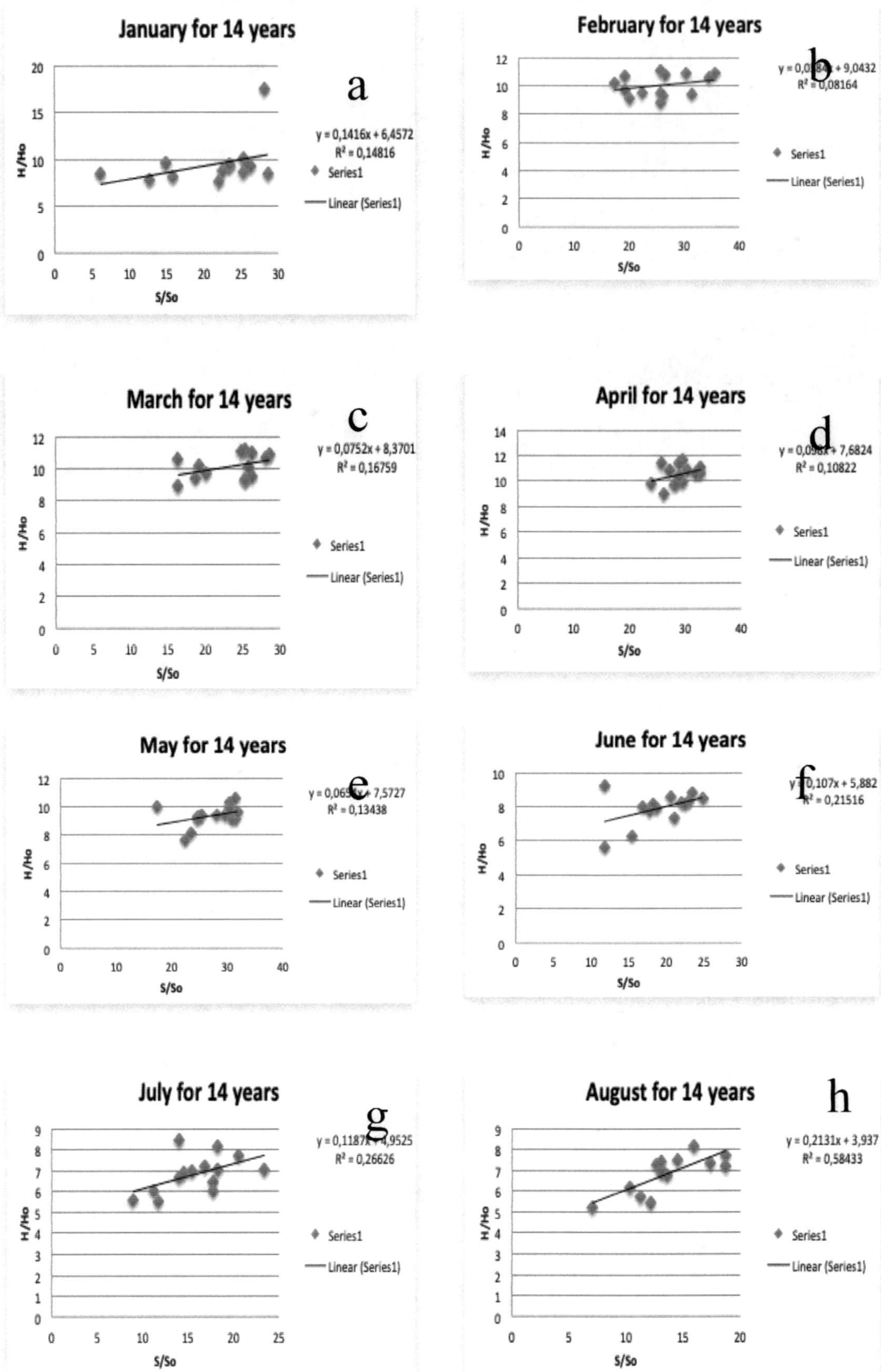

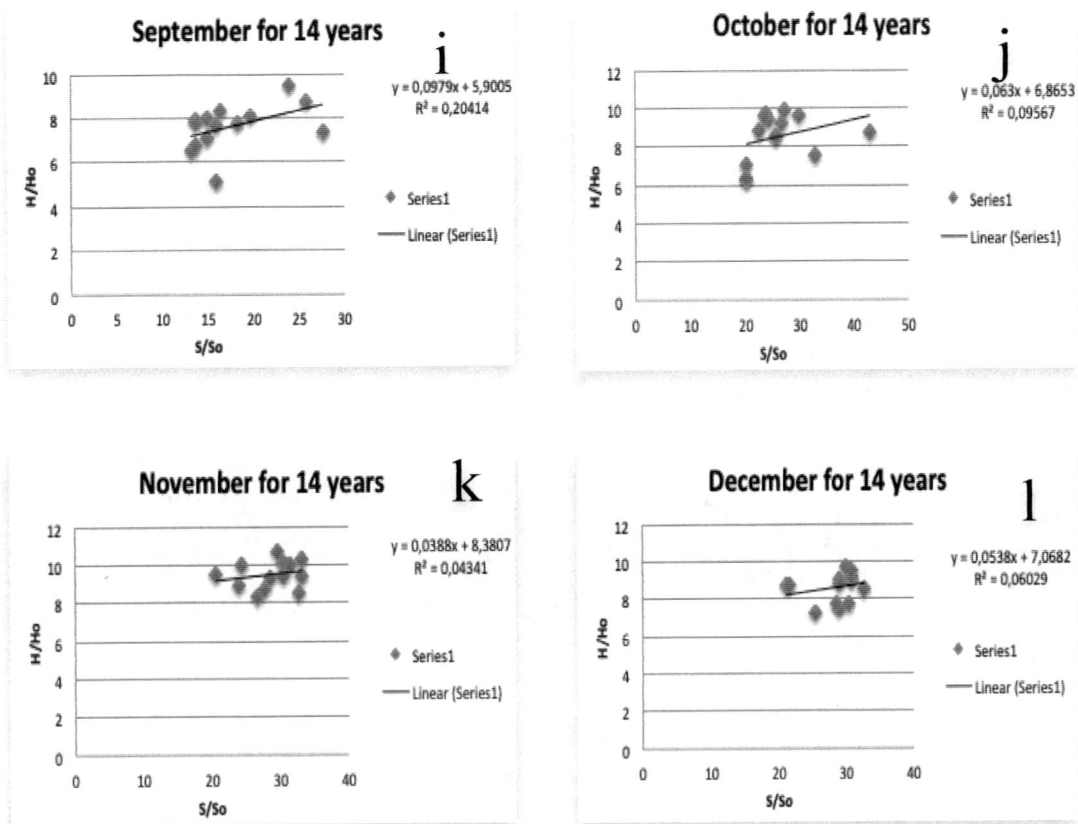

Figure 2: Extraction of constants from the modified Angstrom model

The linear regression for July result is 26.63%. The result for August is displayed as:

$$\frac{H}{H_o} = 3.937 + 0.213\left(\frac{S}{S_o}\right) \tag{12}$$

The linear regression for August result is 58.43%. The result for September is displayed as:

$$\frac{H}{H_o} = 5.901 + 0.098\left(\frac{S}{S_o}\right) \tag{13}$$

The linear regression for September result is 20.41%. The result for October is displayed as:

$$\frac{H}{H_o} = 6.865 + 0.063\left(\frac{S}{S_o}\right) \tag{14}$$

The linear regression for October result is 9.58%. The result for November is displayed as:

$$\frac{H}{H_o} = 8.381 + 0.039\left(\frac{S}{S_o}\right) \tag{15}$$

The linear regression for November result is 4.34%. The result for December is displayed as:

$$\frac{H}{H_o} = 7.068 + 0.054\left(\frac{S}{S_o}\right) \tag{16}$$

The linear regression for December result is 6.03%. The average of the twelve months is given as:

$$\frac{H}{H_o} = 6.84 + 0.087\left(\frac{S}{S_o}\right) \tag{17}$$

It is observed that August had the highest linear regression. Also, constants 'a' and 'b' has high variability on a monthly basis. This result is far from the known models discussed in literature [8-12].

4. Conclusion

It was shown that the variability of solar radiation over Lagos is high via the monthly output of constants ‘a’ and ‘b’. This means that the solar PV panel recommended in this location should be have a rigid solar cell that can be flexible to operate in dynamic weather condition as shown in the results of this research. Now solar device companies have a working directory to solve low PV panel lifespan in Lagos Nigeria.

Acknowledgement

The authors appreciate Covenant University for partial sponsorship. The authors acknowledge NASA for primary dataset.

5. References

[1]. Moses E. Emetere, Marvel L. Akinyemi, and Etimbuk B. Edeghe, (2016). A Simple Technique for Sustaining Solar Energy Production in Active Convective Coastal Regions, International Journal of Photoenergy 2016, 3567502, 1-11, http://dx.doi.org/10.1155/2016/3567502
[2]. Emetere M.E., & Akinyemi M.L., (2015) Prospects of Solar Energy in the Coastal Areas of Nigeria, 2015 PIAMSEE: AIP Conference Proceedings, 1705 (1), 020035
[3]. Finelib, (2019). Lagos Alternative Energy Companies, https://www.finelib.com/cities/lagos/business/energy/alternative-energy (Accessed 1/06/2019)
[4]. D.C. Jordan, S.R. Kurtz, (2011). Photovoltaic Degradation Rates-an Analytical Review Progress in Photovoltaics: Research and Applications, doi:10.1002/pip.1182
[5]. Angstrom, “Solar and terrestrial radiation,” Quarterly Journal of the Royal Meteorological Society, vol. 50, no. 210, pp. 121–126, 1924.
[6]. J. K. Page, “The estimation of monthly mean values of daily total short wave radiation on-vertical and inclined surfaces from sun shine records for latitudes 400 N–400 S,” in Proceedings of the United Nations Conference on New Sources of Energy, vol. 98, no. 4, pp. 378–390, 1961.
[7]. J. A. Prescott, “Evaporation from water surface in relation to solar radiation,” Transactions of the Royal Society of South Australia, vol. 64, pp. 114–118, 1940.
[8]. P. C. Jain, “Global irradiation estimation for Italian locations,” Solar and Wind Technology, vol. 3, no. 4, pp. 323–328, 1986.
[9]. M. A. Alsaad, “Characteristic distribution of global solar radia- tion for Amman, Jordan,” Solar and Wind Technology, vol. 7, no. 2-3, pp. 261–266, 1990.
[10]. S. Jain and P. C. Jain, “A comparison of the Angstrom-type correlations and the estimation of monthly average daily global irradiation,” Solar Energy, vol. 40, no. 2, pp. 93–98, 1988.
[11]. P. V. C. Luhanga and J. Andringa, “Characteristics of solar radiation at Sebele, Gaborone, Botswana,” Solar Energy, vol. 44, no. 2, pp. 77–81, 1990.
[12]. I. A. Raja and J. W. Twidell, “Distribution of global insolation over Pakistan,” Solar Energy, vol. 44, no. 2, pp. 63–71, 1990.

Corporate Governance and Financial Statement Fraud among Listed Firms in Nigeria

O R Uwuigbe[1], O Olorunshe[1], U Uwuigbe[1], E Ozordi[1], O Asiriuwa[1], T Asaolu[2], O Erin[1]

[1]Department of Accounting, Covenant University, Ota, Ogun State, Nigeria.
[2]Department of Management and Accounting, Obafemi Awolowo University, Osun State, Nigeria

Abstract: The study looked into the association which exists amid financial statement fraud and governance among business organizations in Nigeria. A population of 122 non-financial companies registered on Nigeria stock exchange was limited to 20 firms employing the rule of thumb based on stratified and simple random technique for a period of 2012-2016. The method of data analysis is panel regression. The dependent variable, fraud in the financial statement was measured using the Beneish M-score model while the independent variable was measured using audit committee independence, board structure. Findings show that an insignificant association exist amid audit committee independence, the composition of the board and financial statement fraud. This research design suggests regarding the reduction of the occurrence of financial statement fraud, less emphasis should be placed on audit committee independence, board composition and independent non-executive directors' effectiveness.

Keywords: Corporate governance, Financial statement fraud, Audit committee, Board Independent non-executive directors

1. Introduction

Businesses are always under the risk of fraud from different sources as they relate to the firm. Although most fraud perpetrated by external sources are usually massive and serious. However, this is mostly done in collaboration with insiders. Majority of the fraud cases recorded in organizations indicate that fraud performed at the management level have a stringent and breath-taking effect on the organizational goals. This can be empirically attested to by constant collapse cases of reputable corporations such as WorldCom to Enron and Cadbury, Oceanic Bank, Intercontinental Bank, Access bank and much more, exploring that the successful penetration of such malicious act was inspired by directly or indirectly by top management.[5] asserted that corporate fraud is the number one threat to business organizations which is also a reflection of ability (or the lack) of its managers and/or deficiencies in the corporate mechanism. Whenever this malicious act arises by organizations' management it clearly depicts a failure in the existing corporate governance structure guiding its affairs, whereas these structures are in place to help moderate top management excesses.

The collapse of Enron could be said to be the beginning of the serious financial scandals however, the series of corporate failures and distresses continue to grow daily. Other notable companies affected are Cadbury, Cresta bank, Intercontinental bank, Oceanic bank, Parmalat, Northern Rock bank, Lehman Brothers, Nebraska bank, Etisalat, Skye Bank and Diamond Bank of Nigeria. Some of these organizations, for instance, Enron and WorldCom have obliterated billions of dollars in investor esteem while moving towards liquidation, [3]. There are several kinds of frauds however among all the kinds, the highest is financial statement fraud. This kind of fraud has the most critical money related effect on organizations compared to the several other kinds of fraud, [28]. Asides the monetary impact, such fraud has an effect on the reputation of the firm. Consequently, Farrell and Franco (1999) opined that fraud related cost might be hard to appraise in light of the fact that not all misrepresentation is found and not all extortion found is accounted for; in spite of this, various endeavors have been made to estimate fraud, [12] estimated that six percent of companies in the United States lost revenue in 2002 due to financial statement fraud; 491 companies in Australia and New Zealand responded to the KPMG survey and it was discovered that half had encountered fraud costing $457 million, (KPMG, 2004)

[20] discovered that fraud relating to financial statement would prompt a decline in the market value of the firm and less of revenue for the company, investors begin to lose trust in the companies and the companies would find it difficult to obtain the financial resources needed.

According to [39], corporate governance became a global phenomenon following the consistent collapses of eminent companies. [30] contends that poor corporate governance has prompted contradictions amongst board and management offering ascend to board quarrels; unsuccessful board oversight capacities; fake and self-centered rehearses among individuals from the board, administration and staff; the domineering impact of director or MD/CEO, particularly in organizations that have family-control. Poor corporate governance has additionally prompted frail inward controls; rebelliousness with set down inside controls and working systems. Different endeavors have been made towards handling this worldwide challenge with some of these endeavors being the institution of the Sarbanes-Oxley Act, extreme legal penalty e.g. death penalty in some countries like China.

The International Auditing and Assurance Standards Board (IASB) in [15] clearly indicated in the International Standard on Auditing (ISA 240) that the most extreme duty regarding the recognition and aversion of money related articulation misrepresentation dwells with those accused of the administration of the organization ([16]; [40]. The weakness of corporate governance has provided incentives for financial statement frauds; the company is not a victim of financial statement fraud but an instrument used to perpetuate the fraud.

Regardless of the different governance structures and frameworks, misappropriation of funds and alteration of financial reports in the interest of the management has remained prevalent in most developing economies including Nigeria [41]. Thereby making the subject matter attract extensive attention and reaction throughout the years because of the inexhaustible collateral damage draining the long-term success of companies; Financial statements are supposed to be reliable tools used by investors when making investment decisions and also used by the stakeholders of the company when appraising the financial performance of the firm [6].

When financial statement fraud occurs, the company is not the victim but rather the instrument of fraud; the perpetrators are within the firm, holding a sufficiently senior position to be able to browbeat other employees into participating in the fraud. Since financial statement fraud occurs within the firm, the best controls to counter it have to be within the firm with one of those counters being the establishment of corporate governance of the company.

Based on the above problem, this study scrutinized the influence of Corporate Governance on the prevention and recognition of Financial Statement Fraud in listed firms in Nigeria.

2. Literature Review

2.1 Theoretical Framework

Numerous theories articulating and driving the aforementioned. However, this study would be evaluating the agency theory.

2.1.1 Agency theory

[39] explained that agency theory is a model employed to scrutinize the association amid management (agent) and shareholders (principal) in which the principal employs the agent to apply necessary measures to maximize the wealth of shareholders. According to [26] the agency connection exist were a firm owned by a party (the principal) gives the daily operations of its function to another party (An agent) to help direct in line with his or her expectations. This claim was further buttressed that such relationship is contractual and the managers (agent) must make decisions, carry out functions that corroborates owners' expectations. The perspective of this theory is solely pinned on the deliberate need of an agent (director) to safeguard and always have the shareholders interest at heart. The theory

clearly takes its stand point on the premises that a relationship exists amid two parties in an ordinary phenomenon and it is expected that the agent (directors) who is charged with the responsibility of carry out daily activities or services execute them on behalf of the principal (owner) and not that alone but also to satisfy of the owner's expectations [26].

In furtherance, [33] argues that the decisions made by the management are sometimes contrary to the interests of the shareholders.

In achieving the objectives of this research, agency theory would be espoused in scrutinizing the role of corporate governance in the detection and prevention of fraudulent financial reporting. Furthermore, the theory helps explain the contractual association amid the managers (who are the preparers and manipulators of the financial statements) and the shareholders (who majorly rely on those information). However, the assumption of the stewardship theory is unrealistic as it assumes the interests of the director (agent) would match the expectations of capital owners. The fraud triangle theory focuses only on the reasons for committing fraud. Therefore, the agency theory has been seen to best set the foundation for this study.

2.2 Summary of Empirical Review

[10] explored the link amid fraud in the financial statement and corporate governance of registered firms in Italy, spanning through the period of 2001-2011 with a sample size of 26 companies. The result shows that having an audit committee who confides their operations within the corporate governance codes stipulated in Italy have a higher proximity to reduce risk and the occurrence of fraud and the more audit meetings the more likely the reduction of risk occurrences.

[43] scrutinized the association amid the structure of corporate governance and the probability of fraudulent financial reporting across 227 listed public companies in Malaysia for the period of 2010-2011. This was done using an integration of Beneish M-score model and Altman's Z-score model. The results of the analysis indicated that an effective corporate governance structure would reduce the possibility of fraud.

[4] examined the impact of monitoring characteristics have on the financial reporting quality of oil marketing firms listed on Nigeria Stock Exchange for the period 2000-2011. The research discovered that separation of power, directors' independence, shareholdings of managers and audit committee independence significantly affect the quality of financial report of firms in the oil marketing in Nigeria.

[37] carried out a research to analyze the influence of independence, knowledge and experience of members of the audit committee on financial reporting quality. Based on the research, it was discovered that the ineffectiveness of the independence, expertise and experience of the audit committee is the cause of ineffective performance of oversight functions.

Furthermore, [16] investigated on how corporate governance affects the reduction of fraud in organisations. The research explored how corporate governance affect the level of fraud. Three salient roles were explored: internal audit (IA), internal control (IC) and external audit (EA). They were investigated based on three dimensions: existence, implementation and effectiveness. The result of this study shows that IA is the most significant tool for reduction of fraud among the three components used and also indicates effectiveness as the most significant dimension for the reduction of fraud.

[7] explored the connection amid the combination of the board and financial statement fraud in 75 companies in North Carolina. The mechanisms for measuring board composition were number of board members, audit committee and outside directors. The result of the analysis showed that audit committee had an insignificant association with fraud related to financial statement, though a larger board and an upsurge in the tenure of external directors would minimize financial statement fraud.

[19] examined the association amid board characteristics, audit committee and earnings management. An adverse association was found between audit committee and abnormal accruals, an adverse

relationship was also found amid firm characteristic and abnormal accruals. The result depicts that board independence and a rise in the proportion of outside directors in the audit committee would reduce the manipulation of earnings.

[27] investigated the degree to which directors and ownership features relate to earnings management. 570 non-financial companies registered on Vietnam stock exchange spanning from 2010-2014. Result of the analysis showed an insignificant association amid the fraction of non-executive directors and manipulation of earnings.

[25] explored the consequence of the features of audit committee on the quality of financial disclosure. The selected sample was 101 firms listed on the Nigerian stock exchange between 2010-2014. Multivariate regression was introduced to confirm the relationship. The result showed that the independence of the audit committee, financial knowledge and share ownership significantly influence the quality of financial reporting.

2.3 Gap in Literature

Based on the empirical review, numerous scholars have explored the association amid corporate governance mechanisms and the prevention of financial statement fraud. However, few studies have been carried out in Nigeria regarding the role of corporate governance in the prevention and detection of fraud; the few studies conducted in Nigeria have not explored non-financial firms. In lieu of the aforementioned lacuna this study would explore the non-financial companies registered on the Nigerian Stock Market.

2.4 Development of Hypotheses

Based on the inclusive outcomes from extant studies and theoretical pronouncement, the following hypotheses were stated in their null form:

Hypothesis one

$H_{0:}$ Audit committee has no impact on financial statement fraud mitigation of listed firms in Nigeria.

Hypothesis Two

$H_{0:}$ Board composition has no influence on the occurrence of financial statement fraud of listed firms in Nigeria.

3. METHODOLOGY

This study engaged the panel data as the study requires collation of several data from several companies over a period of time. It is a quantitative technique of research used to test the feasibility of any association amid constructs. To examine the direction of association amid two or more constructs correlation matrix was employed to clearly explore the relationship amid the firm's governance structure (independent variables) and financial statement fraud of listed consumer non-financial firms in Nigeria (dependent variable). Panel regression is then used to envisage the degree of variation amid the dependent and independent construct. Financial statement fraud of the non-financial firms was measured using the Beinish M-score model. This study therefore utilizes both descriptive statistics (such as means and standard deviation to describe and summarize the data) and econometric analysis using ordinary least square to perform the linear regression as often used in recent studies.

The Beneish M-score model was used to discover the occurrence or tendency of firms to manipulate their financial statements. The Beneish model is presented mathematically as:

M= -4.84+0.920(DSRI)+ 0.528(GMI) + 0.404(AQI) + 0.892(SGI) + 0.115(DEPI) – 0.172(SGAI) + 4.679 (TATA) – 0.327(LVGI)
If the M score is greater than -2.22, it means the financial statement of the company has been manipulated.

3.1 Population and Sample Size
The population for this research work comprises of the 122 companies listed on Nigerian Stock Exchange between the time periods of 2012-2016, this period was chosen due to availability of data, as well as the implementation of IFRS (International Financial Reporting Standard), which brought about huge leap for financial reporting in Nigeria. The sampling technique utilized was stratified sampling which is appropriate when selecting a particular sector. The sample size of 12.2 was determined based on a sample size rule of thumb by John Curry (10% of population). However, this was rounded up to 20 samples.

3.1.1 Model Specification and Variable Selection
In an attempt to provide empirical evidences on the aforementioned objectives, (corporate governance and financial statement fraud), a linear regression model equation was used. The regression equation can be computed as :

$Y = \beta_0 + \beta X_1 + \mu_{it}$ ....(1)

Where,

Y= Financial statement fraud (Dependent variables)

X= Corporate Governance (Independent variable)

β= Coefficient

μ_{it}= Error term

Explicitly, equation 1 can be well-defined as:
Financial statement fraud = f (Corporate Governance) + c ....(2)
To enhance a better predictability and analysis of the relationship existing between the two constructs, equation 2 can be expanded further by the introduction of the constructs of corporate governance, hence formulating Equation 3. Therefore, the equation becomes:
Financial Statement Fraud = f (Audit committee; Board composition; Independent non-executive directors)..(3)

The above can be deduced to:

$FSF_1 = ACOMM_1 + BCOMP_1$..(4)

Therefore, the OLS linear Regression Equation is:

$FSF_1 = \beta_0 + \beta_1 ACOMM_1 + \beta_2 BCOMP_1 + \mu_{it}$..(5)

Where,

FSF= Financial statement fraud

ACOMM= Audit committee i.e. audit committee effectiveness.

BCOMP= Board composition, the proportion of foreign directors to the entire directorship base.

β= Coefficient of parameter

μ= Error term

it = Time coefficient

Table 1: MEASUREMENT OF VARIABLES

S/N	Variables	Meaning	Measurement
1	FSF	Financial statement fraud	Beneish model
2	ACOMMI	Audit committee independence	Number of independent non-executive directors in the audit committee.
3	BCOMP	Board composition	Percentage foreign members to total number of board members.

Table 2: Descriptive statistics

	MSCORE	ACOMMI	BCOMP
Mean	-4.458068	11.37481	0.034541
Median	-4.846456	0.000000	0.002222
Maximum	5.195931	66.66667	3.000000
Minimum	-12.80684	0.000000	0.000000
Std. Dev.	2.040825	17.32094	0.310851
Skewness	0.712358	1.605078	9.486943
Kurtosis	9.513047	4.921469	91.00508
Jarque-Bera	172.2422	54.23895	31406.50
Probability	0.000000	0.000000	0.000000
Sum	-414.6003	1057.857	3.212272
Sum Sq. Dev.	383.1768	27601.38	8.889819
Observations	93	93	93

Source:-views 7 Output (2019)

Table 2 shows a brief descriptive statistic of both constructs of the selected listed firms stating the average of the variables, the standard deviation, the minimum and maximum of variables used.

The mean value of financial statement fraud of sampled firms stands as 4.458068 with a standard deviation of 2.040825 (based on the Beneish model). This shows an average of the sample firms manipulate their financial statements indicating that the deviation between the companies is quite large. The Audit committee independence (ACOMMI) is ascertained by the amount of Independent non-executive directors in the audit committee. The above table depicts a mean of 11.37481, with maximum and minimum values of 66.66667 and 0.000000 respectively with a standard deviation of 17.32094. This indicates that approximately 10% of the sample firms have an independent committee.

Board Composition (BCOMP) was measured by the proportion of foreign directors on the board to the total number of the board. The mean value was 0.034541, the maximum value was 3, minimum value 0 and the standard deviation was 0.310851. This indicates that 3% of the board of the sample size included foreign directors.

Table 3: Random Effect Model (REM)

Variable	Coefficient	Std. Error	t-Statistic	Prob.
ACOMMI	-0.027123	0.037125	-0.730587	0.4671
BCOMP	1.211494	0.694696	1.743918	0.0848
FSIZE	-0.032950	0.325153	-0.101338	0.9195
LEVERAGE	0.001950	0.006993	0.278917	0.7810
C	-4.489306	2.427487	-1.849364	0.0679
	Effects Specification			
			S.D.	Rho
Cross-section random			0.762518	0.1443
Idiosyncratic random			1.856534	0.8557
	Weighted Statistics			
R-squared	0.081972	Mean dependent var		-3.359948
Adjusted R-squared	0.027328	S.D. dependent var		1.905049
S.E. of regression	1.868653	Sum squared resid		293.3167
F-statistic	1.500104	Durbin-Watson stat		2.404332
Prob(F-statistic)	0.198530			
	Unweighted Statistics			
R-squared	0.096704	Mean dependent var		-4.470385
Sum squared resid	345.4199	Durbin-Watson stat		2.041663

Source: E-views Output (2018)

From table 3 above, it can be observed that the R^2 which shows an approximate of 10% variations in the explainable construct are expressed by the explained construct, leaving the remaining 90% unaccounted for.

Audit committee independence (ACOMMI) exhibited an adverse and insignificant association with financial statement fraud with a value of -0.730587, Board composition (BCOMP) had an optimistic and insignificant association with financial statement fraud with a value of 1.743918. FSIZE and LEVERAGE had t-values of -0.101338 and 0.278917 respectively, indicating an adverse and insignificant association for FSIZE and an optimist and insignificant relationship for LEVERAGE. Regarding the overall significance, the combined variables showed an insignificant association with the dependent variable with a probability (F-static) of 0.198530.

3.1.2 Hypotheses Testing

The verdict rule for accepting the hypotheses of this study is set at 5% level of significance and a t-value of 1.96 therefore, the null hypotheses would be rejected if the value of probability (P-value) is less than 0.05 and the t-value less than 1.96, and the alternative hypotheses would then be selected.

Hypothesis 1

There is no significant association amid audit committee independence and financial statement fraud of listed firms in Nigeria.

The outcome above reveal that audit committee independence had a t-stats of (-0.730587) and a p-value of (0.4671) which is below the verdict rule benchmark. Hence, the study accepts the null hypothesis implying the independence of the audit committee has no statistical influence on financial statement fraud of the sampled companies.

Hypothesis 2

There is no significant relationship between board composition and financial statement fraud of listed firms in Nigeria.

The outcome above reveal that board composition holds a t-stats of (1.743918) and a p-value of (0.0848) which is below the verdict rule benchmark. Hence, the study accepts the null hypothesis is embraced connoting the absence of significant association amid board composition and financial statement fraud of registered corporations in Nigeria.

3.2 Discussion of findings

Based on the results, an adverse and insignificant association derived amid independence of audit committee and financial statement fraud, showing a report stat of (--0.730587) and a p-value of (0.4671). The finding aligns with null hypothesis and this outcome is inconsistent with the result of [37] and [4] who found an optimistic and significant association amid audit committee independence and the quality of financial reporting. This corroborates with [44] who demonstrate that manipulation of the earnings and fraction of outside directors has an insignificantly association. Therefore, the fraction of independent directors does not influence financial statement fraud because their function contributes little in monitoring the committee activities due to threat on independence, time constraints, and adequate information. The result of the analysis is consistent with that of [7] who found an insignificant relationship between audit committee and the possibility of financial statement fraud occurrences.

The outcome suggests, an optimistic but insignificant association amid board composition and financial statement fraud of listed firms reflecting a t-stats of (1.743918) and a p-value of (0.0848). Hence the study **rejects the alternative hypothesis H_2** which connotes a significant association exist amid construct. This is consistent with the result of [8] who discovered a positive and insignificant association exist amid board composition and financial statement fraud.

4. Conclusion and Recommendations

This study examined the association amid corporate governance and financial statement fraud of listed firms in Nigeria. The model was analysed using some companies registered on the Nigerian Stock Exchange. The empirical analysis concludes that the corporate governance mechanisms which included audit committee independence and board composition not significantly related to financial statement fraud. Although board composition has a positive relationship with financial statement fraud, but the result concludes that the relationship is insignificant.

The study therefore recommend that the board should focus more on other characteristics of the audit committee such as financial expertise as this may significantly reduce the occurrence of financial statement fraud. In addition, the study recommends that managers should focus more on the experience and other demographic characteristics of the board members to improve the quality of financial reporting.

ACKNOWLEDGEMENT: Authors acknowledge Covenant University who has solely provided the platform for this research and has also fully sponsored publication of this research work.

References

[1] Abdullahi, R. & Mansor, N. (2015). Concomitant Debacle of Fraud Incidences in the Nigeria Public Sector: Understanding the Power of Fraud Triangle Theory. *International Journal of Academic Research in Business and Social Sciences.* 5(5), 312-326.

[2] Abdullahi, R. & Mansor, N. (2015). Fraud triangle theory and fraud diamond theory. Understanding the convergent and divergent for future research. *International Journal of Academic Research in Accounting, Finance and Management Sciences.* 5(4). 38-45.

[3] Akintoye, I.R., Asaolu, T. (2008) The effect of misinterpretation on trading volume in an informational efficient market. European Journal of Social Sciences. 5(4), 253-260

[4] Aliyu, S.K. & Ishaq, A.S. (2015). Board characteristics, independent audit committee and financial reporting quality of oil marketing firms: Evidence from Nigeria. *Journal of Finance, Accounting and Management.* 6(2), 34-50.

[5] Ashamu SO. (2014): Fraud management in the Nigerian banking industry: Evidence from Nigeria. Journal of Technology, Entrepreneurship and Rural Development. ;5(1):125-137.

[6] Asiriuwa, O., Aronmwan, E.J., Uwuigbe, U., Uwuigbe, O.R. (2018) Audit committee attributes and audit quality: A benchmark analysis. *Business: Theory and Practice* 19, 37-48

[7] Beasley, M. S. (1996). An empirical analysis of the relation between the board of director composition and financial statement fraud. *The accounting review*, 71(4): 443-465.

[8] Bushgan .T. (2005) Corporate governance, earnings management and the information content of accounting earnings: Theoretical model and empirical tests. (Doctoral Thesis, Bond university Queensland. Australia).

[9] Cressey, D. R. (1953). *Other People's Money*. Montclair, NJ: Patterson Smith, pp.1-300.

[10] D'onza, G. &Lamboglia, R. (2014). The Relation between the corporate governance characteristics and financial statement frauds: An empirical analysis of Italian listed companies. Retrieved from *www.iacmaster.it/iacgconference2012/wp-content/uploads/2012/04/D-Onza.pdf*

[11] Farrell. B.R & Franco J.R.(1999). ''The Role of the Auditor in the Prevention and Detection of Business Fraud: SAS No 82'', *Western Criminology Review;*.2 (1) *Available: http://www.westerncriminology.org/documents/WCR/v02n1/farrell/farrell.html*

[12] Holtfreter, K. (2004). Fraud in US organizations: an examination of control mechanisms. *Journal of Financial Crime,* 12, (1), 88-95.

[13] Hooper, M. J,. & Pornelli, C. M. (2010). Deterring and detecting financial fraud: A platform for action.http://www.thecaq.org/docs/reports-and-publications/deterring-and-detecting Retrieved on 28th January, 2018.

[14] Humera, K. (2011). A literature review on corporate governance. 2011 International conference on E-business, Management and Economics, *IPEDR*, 25. IACSIT Press, Singapore.

[15] International Federation of Accountants (IFAC) (2009).: International Auditing and Assurance Standards. Board (IAASB) *The Auditor's Responsibilities Relating to Fraud in an Audit of Financial Statements*, International Standard on Auditing (ISA) 240. New York, NY: IFAC.

[16] In'airat, M. (2015). The role of corporate governance in fraud reduction- a perception study In the South Arabian business environment. *Journal of Accounting and Finance.* 15(2):119.

[17]Kent, P. & Stewart, J. (2008). Corporate governance and disclosures on the transition to international financial reporting standards. School of Business.Retrieved from http:// epublications.bond.au/business_pubs/130 on 1st February 2018.

[18] Kenyon, W. & Tilton, P. D. (2006*). Potential red flags and fraud detection techniques: A Guide to Forensic Accounting Investigation.* New Jersey: First Edition, John Wiley & Sons, Inc.

[19] Klein, A. (2002). Audit committee, board of director characteristics and earnings management. *Journal of Accounting and Economics.*33, 375-400

[20] Lev B. (2003) Corporate earnings: facts and fiction, Journal *of Economic Perspectives*, 17 (2), 27-50.

[21] Lister, L. M. (2007). A Practical Approach to Fraud Risk: Internal Auditors.

[22] Madan, L.B. (2013). Corporate accounting fraud: A case study of Satyam computers limited. *Open Journal of Accounting.* 2, 26-38.

[23] Marshall, J.B. (2015). Corporate practices: An overview of the evolution of corporate governance codes in Nigeria. *International Journal of Business & Law Research.3(3).*

[24] Muhammad, U.K. Ayoib & C. Noor, A.A. (2016) Audit committee independence, financial expertise, share ownership and financial reporting quality: Further evidence from Nigeria. *International Journal of Economics and Financial Issues* 6(S7) 125-131.

[25]Murdock, H. (2008). The Three Dimensions of Fraud: Internal Auditors. Retrieved from www.emerald.comon the 28th January, 2018.

[26] Musa, M. A . (2012) The influence of institutional shareholder activism as a corporate governance monitoring mechanism in Malaysia. *International Journal of social, management, economics and business engineering.* 6(11), 584-588.

[27] Nguyen, H. T. (2016). The impact of board of directors and ownership characteristics on earnings management of publicly listed firms in Vietnam. *Master Thesis Business Administration.*

[28] Oghojafor B.E, Olayemi O.O, Okonji P.S & Okolie I.U (2010). Poor corporate governance and its consequences on the Nigerian banking sector. *Serbian journal of management.* 5(2), 243-250

[29] Ogoun, S & Obara, C. (2013). Curbing occupational and financial reporting fraud: An alternative paradigm. *International Journal of Business and Social Science.* 4(9), 123-132.

[30] Ojeme, S. (2010). Forensic Accounting Reduce Fraud in Financial Industry. Retrieved from http://www.gaflbcstocwrLcom

[31] Okpara, J. (2011). Corporate governance in a developing economy: barriers, issues, and implications for firms. *Corporate governance,* 11, (2), 184-199.

[32] Organisation for Economic Co-operation and Development (OECD). (2004). *OECD Principles of corporate governance.* Paris, France: OECD Publishing.

[33] Patton, A. & Baker (1987). Why won't directors rock the boat? . *Harvard Business Review*, 65: 10-18.

[34] Rae, K., & Subramaniam, N. (2008), Quality of internal control procedures: Antecedents and moderating effect on organisational justice and employee fraud. *Managerial Auditing Journal*, 23(2), 104-124.

[35]Reed, S. (2014). Retrieved from http://www.cgteam.com/blog/six-strategies-for-fraud-prevention-in-your-business on 30th January, 2018.

[36] Sarbanes-Oxley Act of 2002, Public Law 107-204, 107th Cong., 2nd sess. 24 July.

[37] Shab, H. (2013). Independence, expertise and experience of audit committees: Some aspects of Indian corporate sector. *American International Journal of Social Science* 2(5). 102-113

[38] SEC (2008) *Revised code of corporate governance in Nigeria (The Securities and Exchange Commission).*

[39] Uwuigbe, O.R (2011). Corporate governance and financial performance of banks: A study of listed banks in Nigeria. (Doctoral Thesis, Covenant University. Nigeria).

[40] Uwuigbe, O.R., Uwuigbe, U., Jafaru, J., Igbinoba, E.E., Oladipo, O.A. (2016) Value relevance of financial statements & share price: A study of listed banks in Nigeria. *Banks and Bank Systems,* 11(4), pp. 135-143

[41] Uwalomwa Uwuigbe, DF Eluyela, Olubukola Ranti Uwuigbe, T Obarakpo, I Falola (2018) Corporate governance and quality of financial statements: a study of listed Nigerian banks. *Banks and Bank Systems*, 13(3), pp. 12-23

[42] Vona, I. W. (2008). *Fraud risk assessment: Building a fraud audit program*: Hoboken, New Jersey: John Wiley and Sons.

[43] Wan, A. A Wan M. R. & Roshayani, A. P. (2014). Disclosure of corporate governance structure and the likelihood of fraudulent financial reporting. *Social and Behavioral Sciences* 145: 243 – 253.

[44] Xie, B., Davidson, W. N., & Dadalt, P. J. (2003). Earnings management and corporate governance: The roles of the board and the audit committee. *Journal of Corporate Finance*, 9(3):295-316.

APPENDIX

S/N	YEARS	COMPANIES	M-SCORE	ACOMMI	BCOMP
1	Livestock	2012	-5.012111	0	0
		2013	-0.8013553	0	0
		2014	-5.4625893	0	0
		2015	-2.3697291	0	0
		2016	-5.5676578	0	0
2	Guiness	2012	-5.9392472	0	0.004167
		2013	-4.9317578	0	0.004167
		2014	-4.1679453	0	0.003333
		2015	-5.8475065	0	0.004167
		2016	-6.3949612	0	0.001667
3	BERGER	2012	-4.5721704	0	0.002857
		2013	-4.8734897	0	0.002222

		2014	-5.134857	16.66667	0.001538
		2015	-6.6952967	16.66667	0.002
		2016	-6.1177236	33.33333	0.002
4	Dangote cement	2012	-2.0088213	0	0.001111
		2013	-2.8898037	0	0.001111
		2014	-3.3903176	0	0.001667
		2015	5.1959309	28.57143	0.0025
		2016	-2.29052	50	0.002308
5	Dangote sugar	2012	-3.7450687	0	0
		2013	-5.5249995	33.33333	0.001111
		2014	-5.498109	50	0.001
		2015	-3.1211864	66.66667	0.001111
		2016	-0.6750711	66.66667	3
6	Unilever	2012	-4.4597989	16.66667	0.002857
		2013	-3.9982307	33.33333	0.004286
		2014	-5.1258762	33.33333	0.001429
		2015	-3.2638602	16.66667	0.00375
		2016	-2.1260876	16.66667	0.002222
7	Nestle Plc	2012	-4.1136747	0	0.005
		2013	-3.9684279	16.66667	0
		2014	-5.6644788	16.66667	0.00625
		2015	-4.7167199	16.66667	0.005
		2016	-4.6425849	16.66667	0.005
8	FMN	2012	-4.8847436	0	0.003571
		2013	-4.7084118	0	0.002857
		2014	-5.4404339	0	0.002857
		2015	-4.6273586	0	0.003571
		2016	-3.65718	0	0.003571
9	PZ Plc	2012	-6.2247514	66.66667	0.0025
		2013	-5.7198887	0	#DIV/0!
		2014	-6.1423799	50	0.001818
		2015	-2.33187	0	#DIV/0!
		2016	-3.5807988	33.33333	0.001818
10	Mobil	2012	-3.7409932		0
		2013	-4.6002573	0	0.003333
		2014	-4.1681998	0	0.003333
		2015	-3.9387583	0	0.003333
		2016	-1.31611	0	#DIV/0!
11	Julius Berger	2012	-1.7985079	16.66667	0.004167
		2013	-3.2614708	16.66667	0.004167
		2014	-3.1478225	16.66667	0.004286
		2015	-3.2461519	14.28571	0.004545
		2016	-3.4168819	16.66667	0.004545
12	CUTIX Plc	2012	81.802471	N/A	0

		2013	-3.0264458	N/A	0
		2014	-5.6062271	25	0
		2015	-5.8812243	50	0
		2016	-3.8092384	0	0
13	CHELLARAM Plc	2012	-6.0229695	0	0.005714
		2013	-1.4743607	0	0.005714
		2014	-3.4209127	N/A	0
		2015	-5.5026762	0	0.005
		2016	-2.301273	0	0.003333
14	CAP Plc	2012	-4.9961494	0	0
		2013	-3.0670475	0	0
		2014	-4.1298401	0	0
		2015	-4.846456	16.66667	0
		2016	-0.6079323	16.66667	0
15	LEARN Plc	2012	-6.6691247	0	0
		2013	-5.9212949	0	0
		2014	-6.5488742	16.66667	0
		2015	-1.6759891	16.66667	0
		2016	-12.806841	16.66667	0
16	Nascon Plc	2012	-5.1272036	0	0.001111
		2013	-6.9339308	0	0.001111
		2014	-6.1281851	0	0.002222
		2015	-0.7565547	40	0.002
		2016	-4.5319091	33.33333	0.002
17	Honeywell	2012	-5.7218524	0	0.002222
		2013	-5.6089255	0	0.002222
		2014	-3.9909487	0	0.002
		2015	-5.0611675	0	0.001429
		2016	-5.0417033	0	0.002857
18	NB	2012	-4.5562407	16.66667	0.004615
		2013	-5.6059513	0	0.004615
		2014	-5.4365375	0	0.004
		2015	-6.2700223	0	0.00375
		2016	-2.90116	33.33333	0.00375
19	BOC GASES	2012	-5.48081	0	0.003333
		2013	-6.46173	0	0.003333
		2014	-6.70263	0	0.003333
		2015	-6.37811	0	0.005
		2016	-5.65519	0	0.0025
20	FIRST ALLUMINIUM	2012	-5.10748	0	0
		2013	-5.20993	0	0
		2014	-5.93121	0	0.003333

		2015	-1.63224	0	0.001667
		2016	-2.957	N/A	N/A

Ownerships Structures and Accounting Conservatism among Nigeria Listed Firms.

O Asiriuwa[1], R T Akperi[1], O R Uwuigbe[1], U Uwuigbe[1], L Nassar[2], S Ilogho[1], S Eriabe[1]

[1]Department of Accounting, Covenant University, Ota, Ogun State, Nigeria.
[2]Department of Management and Accounting, Obafemi Awolowo University, Osun State, Nigeria

Abstract: The article aims to examine the effect of ownership structure on accounting conservatism among Nigeria listed firms. Correlational research design was employed in this study and the sample size comprises of all 15 firms from the banking sector listed on the Nigerian Stock Exchange (NSE) for the years 2013 – 2017. The study employed the panel regression technique to estimate the coefficients of the variables in the model specified.
The study finds that there is a positive and insignificant relationship between managerial ownership and accounting conservatism; a positive but an insignificant relationship was observed between institutional ownership and accounting conservatism. However, foreign ownership revealed a negative but not significant relationship with accounting conservatism. We recommend that financial statements to include more detailed analysis of the structure of their ownership and shareholding to provide more information for decision making for users of the accounting information such as researchers and potential investors. Also, Regulatory bodies should ensure all companies comply with the concept of conservatism by imposing strict penalty on erring companies.

Keywords: Accounting Conservatism, Ownership Structure, Nigeria

1. INTRODUCTION

The transparency and timeliness of accounting information has become of greater importance in financial reporting owing to the rise of globalization and internationalization in the capital markets. One of the distinct areas that affects these accounting principles is the idea of conservatism in accounting [27].

Accounting conservatism can be seen as a method used by firms in reducing the risk in accounting reporting that arises from events such as district contracts and implementation of poor decisions [19].

The anticipated relationship between ownership structure and accounting conservatism arises from the existence of previous literature enumerating a relationship between the knowledge of directors in anticipating benefit, delivery time and profitability of projects, and their exposure to such information spurring excessively risky investment decisions [3], [35].

The goal of accounting conservatism, by means of early loss recognition in financial reporting, is to increase the investment profitability of a business [22]. [17]. This postulation is further supported by the thoughts of [31] and [3] suggesting that Accounting conservatism revealed a positive effects on the success of the decisions made based on accounting information. Studies have linked the winding up of large companies to accounting practices directly opposing the principle of conservatism in the years preceding their liquidation [32].

Active monitoring theory postulates that institutional shareholders have significant tendencies to involve themselves in overseeing management of firms and are often longer term, and so this monitoring leads to an enhancement in firm value and quality of accounting data to favor the other shareholders and public stakeholders [5].

A wide range of literature examined the connection between conservatism in accounting and other variables with the use of different models, theorem and research models, and most of such studies have been centred on selected variables such as government ownership, agency, agency conflict and equity market demand influence.

Their findings have indicated a correlation between accounting conservatism and chosen variables in developed countries as in the studies of [26] in Malaysian companies, [2] in Jordan, [1] in the United States, [8] also in the United states, [6] in the United kingdom, [5] in Korea amongst many others all generating mixed results. However, little empirical data exists that investigates the existence and

extent of a relationship between apportionment of equity holding in quoted companies and the amount of conservatism applied in their financial reporting practices in developing countries specifically in Nigeria, and very few have used institutional ownership, managerial ownership and foreign ownership as the set of independent variables potentially influencing accounting conservatism. Based on this, the study examine the impact of ownership structure on account conservatism in Nigerian quoted firms'.
The remaining part of this paper is structured as follows: Section 1 is an introduction as above. Section 2 contains the literature review and hypothesis development. Section 3 outlines details of the research design adopted. Section 4 presents the results from the estimates and analyses the results. Section 5 concludes the study and provides recommendations.

2. Literature Review and Hypotheses Development

2.1 Concept of Accounting Conservatism

According to [36] in a general sense, conservatism is a likelihood that accountants, when met with uncertain economic conditions in financial transactions, opt to anticipate smaller amounts for the worth of assets and revenues, but larger projections for the amounts of liabilities and expenses, the aim being to prevent the overstatement of incomes in financial reporting and the understatement of costs and losses.

Conservatism can easily be associated with principle of realisation, in that it encourages that an income should not receive recognition prior to its actual realization. This principle as seen by contemporary academicians and accounting standard setters, is a concept under which accountants apply a sensible amount of prudence in the recognition of transactions that are affected by economic uncertainty.

Modern accounting researchers agree that conservatism has the impact of speeding up the recognition of losses and declaration of profit recognition [36]. [36] further argues that this does not imply a weakness to the prudency rationale behind conservatism because it appears to simply postpone the recognition of incomes, but that it does so only in order to verify the materialization of such incomes before they are reported.

The rationale behind encouraging accounts conservatism is to further confirm the truthfulness of the incomes reported in financial statements and eliminate the threat of ghost incomes in the statements which may have been earned but were never received and are therefore of no financial benefit to the business in question in terms of their liquidity. [20], [29] in their work, stated that the desire for accounts conservatism is encouraged by agency conflicts, citing an example in the idea that varying management ownership structures amount to varying agency costs to businesses, and therefore varying demands for conservatism. The effect of the agency principle on conservatism rallies around the idea that the managers, being the agents of the shareholders for the running of the business have little or no personal incentives to be conservative in their reporting ad their investments are not involved in the company and to have reported higher profits would imply that they have done a better job, hence to apply conservatism in delaying the recognition of the profits will be to their disadvantage. [24] further espouse that the differences in the degree of conservatism applied in reporting within the same regions imply that the level of conservatism applied is not regional but is dependent of the internal goals of financial reporting held valuable by each individual organization

2.2 Hypotheses Development

2.2.1 Managerial Ownership and Accounting Conservatism

The empirical evidence on the influence of managerial ownership on accounting conservatism is inconclusive with opposing opinions.

The crux of the agency problem rests in the fact that for the sake of their own interests which are separate from those of the stakeholders, managers may make both accounting and real decisions to deceive the ownership of the company into making choices to favour them by concealing the true state of the firm's economic performance [23], [34].

Depending on the extent to which managers are constrained in accruals manipulation, they may likely resort to making such fraudulent adjustments in the area of real earnings management which may be even more advantageous to them [14]. In addition to concealing the real performance of the company, earnings management hides the true patterns in the company's actual performance in terms of revenue and earnings growth which will in turn create an impression that shareholders and the public can anticipate improvements in the business growth and demand of its products.

According to [23] however, recent literature presents evidence that conservatism can serve as a preventive measure on the harm caused by earnings management practices. In support of these claims, [16] say that applying conditional conservatism by way of timely recognition of losses and delaying economic gains recognition reduces the chances that earnings management will successfully be carried out. [33] reported a significant relationship between managerial ownership and accounting conservatism. Also, [2] found a positive association between managerial ownership and accounting conservatism.

H_1: There is no significant relationship between managerial ownership and accounting conservatism in Nigeria listed firms.

2.2.2 Foreign Ownership and Accounting conservatism

Foreign investors tend to gravitate towards firms with high investments and abundant information which they often associate with low levels of information asymmetry [11]; [21]; [20]. Additionally they have stronger motives, experience and skill sets to independently analyse companies, and this increased monitoring abilities are in turn poised to influence their demand for the accuracy of accounting information, and cause them to turn to enforcing more conservatism as a means to achieving that end.

[5] suggested that foreign ownership could either boost or reduce financial reporting quality. Their results indicated that foreign ownership is positively associated with financial reporting conservatism at all three models. Studies such as [2]; [4] reveal a positive relationship between foreign ownership and accounting conservatism.

H_2: There is no significant relationship between foreign ownership and accounting conservatism in Nigeria listed firms.

2.2.3 Institutional ownership and conservatism.

Institutional shareholders are major professional shareholders and are owners of another part of the company's shares, who in spite of the first group, can access to internal valuable information on company's future outlook, business strategies through direct relation with managers [28]. [7] concluded that using abilities of institutional investors to monitor corporate management and performance follows amount of their investment.

The higher level of institutional ownership, the better management, which it is a direct relationship. Theoretically, institutional investors may have incentives for active monitoring on management and consequently, increasing shareholders' value in comparison with the shareholders with lower investments in a company block holders have a larger financial stake in the wellbeing of a company, they also pull enough weight through their rights to vote to remove and rearrange the structure of management within the company [7]. Large shareholders on the outside therefore possess a lot of power to control the behavior of the management and prevent themselves from being victims of the agency conflict [37].

Institutional investors may have incentives for monitoring financial reporting. Financial statements, especially profit and loss statements, are important source about company's information. Empirical evidences show that presence of institutional investors has a positive relationship with market reaction to dividend distribution [18].

[5] posited that institutional investors are best armed with the skills, knowledge and all round resources to efficiently and effectively monitor and control the activities of the management in a company wherein they have invested. Therefore, the lagers shareholders present an opposition for the managers and can prevent them from being able to manipulate profits to levels that best suit their interest. In spite of all this, the heavy involvement of institutional investors in the affairs of a company is expensive and will only show its fruits in the long term in the aspect of conservatism, therefore not all institutional investors involve themselves in the corporate governance of a company [25].

Several studies have failed to show how investors that are not part of the board of directors will affect the amount of accounting conservatism of the firm. [10] opined that the policy of the firm can be affected by the institutional shareholders. They conclude that the policy of the firm can be influence through their voting right Conversely, [2], [5] and [4] in their study found a positive relationship between institutional ownership and accounting conservatism.

H_3: There is no significant relationship between institutional ownership and accounting conservatism in Nigeria listed firms.

2.3 Underpinning Theory

IOP Conf. Series: Earth and Environmental Science **331** (2019) 012056 doi:10.1088/1755-1315/331/1/012056

2.3.1 Stakeholder theory

This concept posits that the managers of a company, of ethical necessity, are required to be conscious of the effects of their actions on all the parties with a stake in the company. This applies even more specifically to the shareholders who are one of the most directly affected by the overall wellbeing of the company. [13] defines stakeholders "as any person or group that may or may influenced by the accomplishment of the purpose of a corporation". The term embodies various resource providers specifically unions, shareholders, employees, suppliers, creditors, consumers, and regulatory agencies. The stakeholder theory was criticized to drastically overemphasize the major role of managers to be accountable to stakeholders without stating the way to keep their interests functioning in line with each other, but remain relevant as equity demands for conservatism determine the amount of conservatism that the managers of companies take due care to present in their financial statements.

2.3.2 Transient and Active Monitoring Investors theory.

According to the findings of [9] larger shareholders are more often than not institutional and therefore institutions with significant shareholding have a say in the way organizations are run. The effect of this can however function in both the positive and negative directions because institutional investors come in two different forms, the transient investors with highly diversified portfolios and little interest in the management of the firms and short term incentives for investment, heavily dependent on accounting information to know when to withdraw their investments from the firms: they are more likely to pose a danger to the conservatism in financial reporting because they provide the equity demand incentive for managers to manipulate data in order to retain their investment or attract more of it. The Active monitoring investors have long term investment intentions with less diversified portfolios and a stronger interest in the activities of management. They are more prone to heavy monitoring on the management and strict scrutiny on how financial reports are prepared [5], [25]

This study will be centered on the Stakeholders theory on the basis that it encompasses the debt owed by management to its stakeholders the most primary of which being the shareholders) the due caution of their interests being considered in the actions taken in the running of the firms in which they have entrusted their investment.

3. METHODOLOGY

3.1 Research design

This research was carried out using correlational research design. This involves the collection of previously prepared quantitative data from valid secondary sources about the population and performing mathematical operations on them to test the hypothesis. The research was carried out within Nigeria.

3.2 Population and Sample size

The research population, being the total group of parties to be involved in this study includes all the banks in Nigeria quoted in the Nigerian Stock Exchange. There were 15 quoted banking firms in the NSE as at 31st December 2017.

The Judgmental sampling technique was used for this study. Owing to the fact that the population being analyzed is very small, the entire population of 15 banks was analyzed in this survey, using data from their financial statements from the years 2013 to 2017.

3.3 Model Specification

This study draw inspiration from the study [2]. We expect a functional relationship between ownership structure (managerial ownership, Institutional ownership, foreign ownership) and accounting conservatism.

$$\textit{Accounting Conservatism} = f(\textit{managerial ownership, institutional ownership, foreign ownership}) \quad (1)$$

In econometric form and considering the panel structure of the data, equation one is transformed as:

$$AC_{it} = \alpha_0 + \alpha_1 ForeiO_{it} + \alpha_2 InstO_{it} + \alpha_3 ManO_{it} + \alpha_4 sizit + \alpha_5 profit + \varepsilon_{it} \quad (2)$$

Where,
AC = Accounting conservatism;
ForeiO = Foreign Ownership
InstO = Institutional ownership =
ManO = Managerial ownership
Fsize = Firm Size
Profit = Profitability
ε = Error term;
i = corporations; t = the financial year.

Table 3.1: Measurement of Variables

S/N	Variables	Measurement	Apriori Sign
	Dependent Variable		
1	Accounting Conservatism	Measured using Accrual Model Formula which is express as: $\frac{Income+Depreciation\ Expenses-Net\ Operating\ Cash\ flow}{Total\ Asset}$	
	Independent Variable		
2	Foreign Ownership	Measured as the proportion of shares owned by foreigners' to the total number of shares in issue	+
3	Managerial Ownership	Measured as the proportion of shares owned by board of directors to the total number of shares in issue	+
4	Institutional Ownership	Measured as the proportion of shares owned by institutional investors to the total number of shares in issue	+
	Control Variables		
5	Firm Size	Measured as the natural logarithm of total asset	+
6	Profitability	Measured as profit after tax	

Source: Authors Computation (2019)

3.4 Method of Data Analysis

The analysis of the relevant figures extracted from the financial statements will be carried out first with the use of a mathematical formula for the measurement of conservatism, and a panel regression model will be adopted for the study to test the hypothesis with the aid of the STATA 13.0 software package to establish relationship or absence thereof between accounting conservatism and percentage of ownership structure as per the data in the financial statements, reason being the reliability.

4. EMPIRICAL RESULTS

4.1 Descriptive Statistics

Table 1: Descriptive Statistics

Variable	Obs	Mean	Std. Dev.	Min	Max
CONSERVATISM	72	4.21e+07	4.20e+07	-1.97e+07	1.78e+08
MANOWN	71	.1052354	.1549379	0	.71858
FOROWN	71	.176166	.2460458	0	.8589
INSTOWN	71	.4376099	.2641433	0	1
PROFIT	72	3.20e+07	3.60e+07	-2.26e+07	1.67e+08
FIRMSIZE	73	12.11552	.3373502	11.06684	12.67576

Source: STATA 13.0 output (2019)

The descriptive statistics of the variables used in the analysis presented in Table 1 explains the, minimum, maximum, standard deviation and mean values of the data distribution. The mean value of organization's conservatism in financial reporting is 4.21e+07. It had a standard deviation of 4.20e+07, a maximum absolute value of 1.78e+08 and a minimum absolute of value of -1.97e+07 revealing a large range and by so doing presenting a significant extent of variability in the conservatism of the financial statements of the banks.

The managerial ownership shows a mean of 0.1052354, with maximum and minimum values of .71858 and 0 respectively while the standard deviation is .1549379. This means an average of 10 percent of the shares of quoted banks in Nigeria is held by the management, however there exists a significant range in the variance of the percentage in different banks. The foreign ownership showed an average of.1776166, with maximum and minimum values of .8589 and 0 respectively while the standard deviation is .2460458. This implies that the average quoted bank in Nigeria is 18% owned by foreign investors although the range of differences to extent of this ownership is very wide. The institutional ownership has a mean value of .4376009, indicating that nearly half of the quoted banking industry is owned by institutions which is in agreement with the postulation that institutions possess the large funds required to make large share purchases in companies, or at least, more so than individuals. It has a standard deviation of .2641433, with minimum and maximum values of 0 and 1 respectively.

The Firm Size, a control variable in the analysis has a mean of 12.11552 and a standard deviation of .3373502 with a maximum value of and a minimum value of 12.67576 and 11.06684 respectively. Profit as computed in the financial statements is the second control variable with a mean of 3.20e+07 and a standard deviation of 3.60e+07, as well as a maximum and minimum values of 1.67e+08 and -2.26e+07 respectively.

Robustness test

Test for Normality (Histogram)

Normality typically depicts a symmetrical, bell-shaped curve, having its greatest frequency of scores at the center with reducing frequencies toward the left and right sides [15].

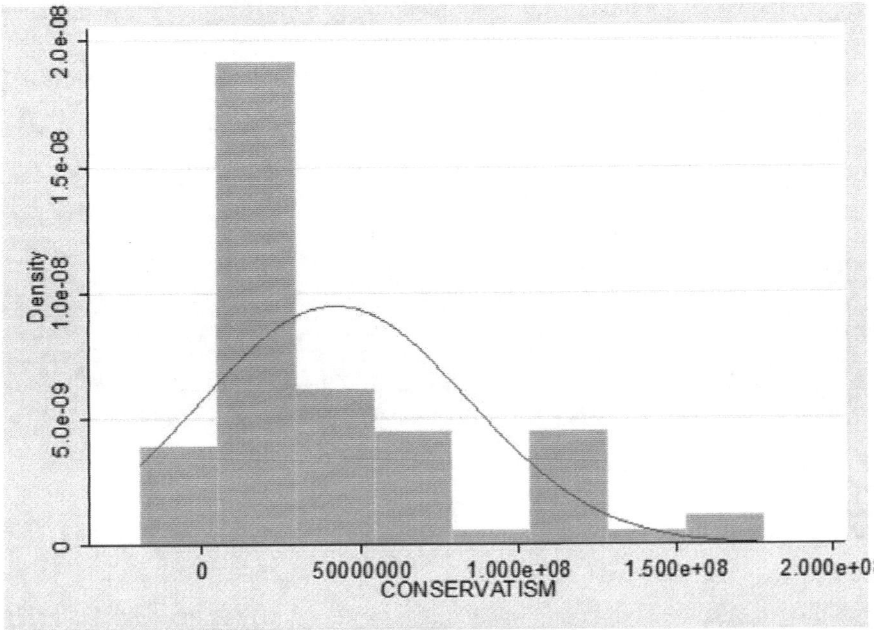

Figure 1: Test for Normality (Histogram)

Computed output (STATA 2019)

The table above shows the frequency distribution of the data analyzed presented as a histogram. The series range is separated into a number of equal length intervals and a count of the number of observations in the sample that lie within the intervals is presented by the histogram. Normality is said to exist when the outcomes of the observations are normally distributed, that is they take the expected bell shape, larger in the middle and tapering down toward the left and right. In this study however, the peak of the curve leans toward the left indicating some abnormality in the distribution, hence the assumption of normality can be said to have been significantly violated. However the extent of violation does not completely negate the premises that the score from the analysis are related. In order to further verify the normalcy of the data distribution, normal Q-Q plot tests will be carried out.

IOP Conf. Series: Earth and Environmental Science **331** (2019) 012056 doi:10.1088/1755-1315/331/1/012056

Normal Q-Q plot.

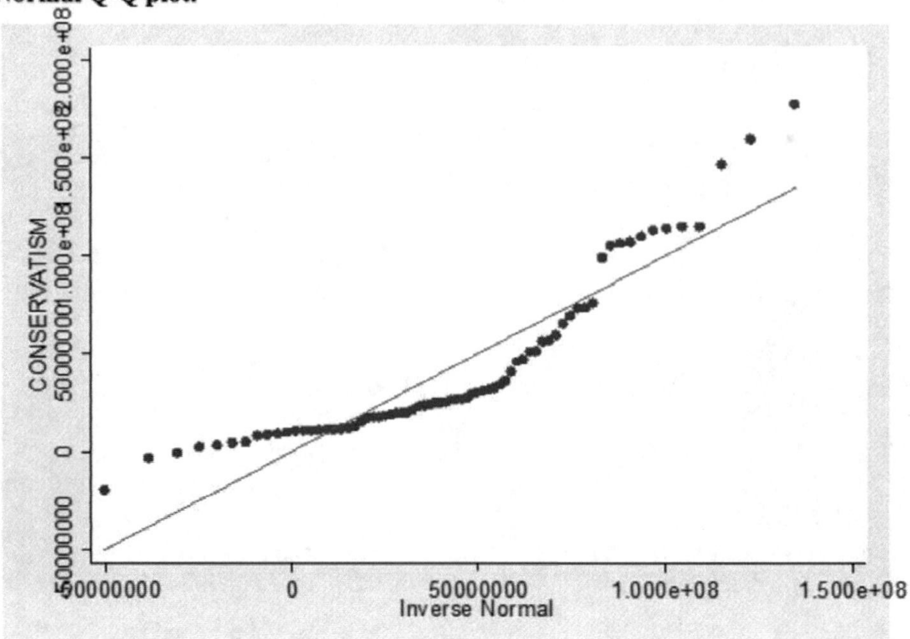

Figure 2. Normal Q-Q plot.

Computed Output (STATA, 2019)

Normal Q-Q plot affords a graphical way to ascertain the extent of normality. The dotted line indicates the tenets that data ought to ideally align with if the distribution is normal. As shown, most of the sample does not align with the line, and so the distribution is not normal.

4.2 Correlation Matrix

Table 2. Pearson Correlation Matrix

	CONSER~M	MANOWN	FOROWN	INSTOWN	PROFIT	FIRMSIZE
CONSERVATISM	1.0000					
MANOWN	-0.2848	1.0000				
FOROWN	-0.2634	0.0174	1.0000			
INSTOWN	-0.3308	-0.1464	0.3821	1.0000		
PROFIT	0.9605	-0.2804	-0.2753	-0.3426	1.0000	
FIRMSIZE	0.7668	-0.3568	-0.1829	-0.2845	0.7197	1.0000

Computed output (STATA, 2018)

The correlation between account conservatism and managerial ownership revealed a negative relationship indicated by -0.2848 shown in the table 2. This implies that an increase in managerial ownership will lead to a reduction in conservatism.

The association between accounts conservatism and managerial ownership is a negative relationship indicated by -0.2634 shown in the table 2. This implied that an increase in foreign ownership, will lead to a reduction in account, but for the purpose of hypothesis testing, the regression analysis will be used.

The association between accounts conservatism and firm size is a strongly positive one with the coefficient of correlation being 0.9605 as shown in the table 2, which by implication indicates that as profit increases, so does conservatism. But for the purpose of testing the hypothesis, regression analysis will be later applied.

The association between accounts conservatism and institutional ownership is a negative relationship indicated by -0.3308 shown in the table 2. This means that as the institutional ownership increases, the conservatism reduces, but for the purpose of hypothesis testing, the regression analysis will be used.

The association between account conservatism and firm size according to the results of the correlation analysis are strongly positive with a correlation coefficient of 0.7668, implying that there is a positive relationship between firm size and accounts conservatism.

4.3 The Hausman tests.

This was carried out to establish whether or not a fixed or random effects model will be used to carry out the final regression analysis upon which the hypotheses testing will be based. The decision rule from the test is that if the difference in coefficients is systematic, then a random effects model is to be used, but if the difference in coefficients is systematic then a fixed effects model should be used for the data analysis.

Table 3. Hausman Tests

	Coefficients			
	(b) fixed	(B) random	(b-B) Difference	sqrt(diag(V_b-V_B)) S.E.
MANOWN	1.33e+07	3933066	9370992	2.81e+07
FOROWN	-3649988	-655225.8	-2994762	2.28e+07
INSTOWN	-2.09e+07	1556272	-2.24e+07	1.82e+07
PROFIT	.9997299	.9906662	.0090637	.0717279
FIRMSIZE	5.08e+07	2.17e+07	2.90e+07	2.19e+07

```
          b = consistent under Ho and Ha; obtained from xtreg
B = inconsistent under Ha, efficient under Ho; obtained from xtreg

Test:  Ho:  difference in coefficients not systematic

          chi2(4) = (b-B)'[(V_b-V_B)^(-1)](b-B)
                  =       2.48
        Prob>chi2 =     0.6480
```

Computed output (STATA, 2019)

Based on the decision rule for Hausman data testing, the observations will be analyzed using a random effects model for multiple regression, the results of which will be the premise for the final hypothesis testing. The result shows that the correlated Hausman random-effect likelihood value (0.6480) exceeds 5 percent (0.05). The random effect model is therefore accepted as the grounds for discussion of the results.

4.4 Regression Analysis

Table 4. Panel Regression Analysis (Random Effect)

```
Random-effects GLS regression                   Number of obs      =        70
Group variable: SN                              Number of groups   =        15

R-sq:  within  = 0.7436                         Obs per group: min =         1
       between = 0.9909                                        avg =       4.7
       overall = 0.9346                                        max =         5

                                                Wald chi2(5)       =    914.77
corr(u_i, X)   = 0 (assumed)                    Prob > chi2        =    0.0000
```

CONSERVATISM	Coef.	Std. Err.	z	P>|z|	[95% Conf.	Interval]
MANOWN	3933066	1.11e+07	0.35	0.723	-1.78e+07	2.57e+07
FOROWN	-655225.8	6043204	-0.11	0.914	-1.25e+07	1.12e+07
INSTOWN	1556272	6009666	0.26	0.796	-1.02e+07	1.33e+07
PROFIT	.9906662	.0557123	17.78	0.000	.881472	1.09986
FIRMSIZE	2.17e+07	6431718	3.38	0.001	9126799	3.43e+07
_cons	-2.54e+08	7.76e+07	-3.28	0.001	-4.06e+08	-1.02e+08
sigma_u	0					
sigma_e	11739601					
rho	0	(fraction of variance due to u_i)				

Computed output (STATA, 2019)

4.5 Test of Hypothesis and Discussion

From the table above, the P>|z| (Prob) value for managerial ownership is 0.723 which is far above the significance level of 5% which is being used in this work. Although the slope of the coefficient is a positive value of 1.11e+07, the basis of analysis rests more importantly on the probability value, hence the null hypothesis is accepted over the alternative hypothesis. On the basis of the probability therefore, managerial ownership has no significant relationship with accounting conservatism. This is consistent with the ideas of [12] who advocate that managers use financial reporting discretion to manage earnings in both directions in order to satisfy their own goals. Including management in the shareholding of a company can therefore swing both ways in either improving or reducing accounts conservatism, an indices of earnings quality.

The result from the above table also revealed, the P>|z| (Prob) value for foreign ownership is 0.914 which exceeds the significance level of 5% being applied for this analysis. The slope of the coefficient value is –655225.8, which shows a significant negativity, the decision is however being taken on the probability which clearly exceeds 0.05, and is therefore the null hypothesis will be accepted and the alternative hypothesis rejected. By implication, this result implies that variations in foreign ownership do not affect the extent of accounting conservatism applied in the preparation of financial reports in the banking sector. This result is supported by the postulations of [30] attributing variations in incentives for accounts conservatism to ownership concentration and not nature of ownership. By implication, it is the unity of incentive of the largest shareholder group to influence management decisions that matters and not the size of the groups. The studies of [6] acknowledge that there are 2 groups of investors, the transient and the active monitoring investors, the transient ones being passive and readily willing to sell off their investments in a sudden turn of market conditions and the active monitoring ones with long run incentives to maintain their shareholding in the company. Hence, subject to the nature of foreign investors, the impact on conservatism could be both positive and negative, suggesting a likely reason for the absence of a significant relationship.

The P>|z| (Prob) value for institutional ownership is 0.796, which is clearly above the significance level of 5% being used in this research. The slope of the coefficient was a strong positive 1556272, but the decision rule rests on the probability hence the reason for accepting the null hypothesis, meaning in real terms that variations in the proportions of institutional ownership in listed banks in Nigeria has no effect on the level of accounting conservatism. This is in consistent with the theory of Transient and the Active Monitoring investors [25]. The transient investors due to their short run incentives and investment volatility, rather than monitor managers, sell their investments as soon as the market turns unfavourable, hence the incentive for the management to manipulate data increases in order to keep the transient investors happy. However, with the active monitoring investors, keen attention is paid to the management of the company because they have long term intentions to maintain their investment in such companies, hence the demand more conservatism in their financial reporting.

5. Conclusion and Recommendations

The study was driven by the premise that the structure of the organizations ownership has an influence on the reliability of its financial statements, in this case being measured by accounting conservatism. The aim is to provide additional data to back up the reliability of financial statements by way of the ownership structure informing the viewer of annual reports on how reliable/ conservative the financial statements of the company will likely be, and for individual investors with small financial capability and hence little power to influence management, it can be helpful in deciding where is safest to invest.

Based on the above analysis, data collection, reviews of literature and interpretation of data, among the quoted banks in the Nigerian Stock Exchange, there is no significant relationship between ownership structure of firms (specifically managerial, foreign and institutional ownership) and accounting conservatism. Although many literatures support the presence of relationships between these variables and accounting conservatism, for probable reasons, some of which have been stated above, a relationship between those variables is not traceable in the observations collected from the financial statements in the industry.

Consequently, based on the findings, the study recommends that in assessing the tendency of ownership structure and accounting conservatism, there is need for:

Financial statements to include more detailed analysis of the structure of their ownership and shareholding to provide more information for decision making for users of the accounting information such as researchers and potential investors.

Also, more attention should be placed on accounting conservatism by regulatory bodies as a means of improving the quality of earnings reported by companies.

ACKNOWLEDGEMENT

Authors acknowledge Covenant University who has solely provided the platform for this research and has also fully sponsored publication of this research work.

References

[1] Ahmed, A. S., & Duellman, S. (2007). Accounting conservatism and board of director characteristics: An empirical analysis. *Journal of accounting and economics*, *43*(2-3), 411-437.

[2] Alkurdi, A., Al-Nimer, M., & Dabaghia, M. (2017). Accounting conservatism and ownership structure effect: Evidence from industrial and financial Jordanian listed companies. *International Journal Of Economics and Financial Issues*, 608-619.

[3] Aminu, L., & Hassan, U.S. (2017). Accounting conservatism and financial performance of Nigerian deposit money banks: An analysis of recent economic recession. *Scholedge International Journal of Business Policy & Governance, 4*(12), 124-135.

[4] Amos, B., Ibrahim, G., Nasidi, M., & Ibrahim, K. Y. (2016). The Impact of Institutional Ownership Structure on Earnings Quality of Listed Food/Beverages and Tobacco Firms in Nigeria. *Journal of Arts, Science & Commerce, 7*(1), 20- 30.

[5] An, Y., & Naughton, T. (2015). Does foreign ownership increase financial reporting quality?. *Asian Academy of Management Journal*, *20*(2), 1-24.

[6] Ball, R., & Shivakumar, L. (2006). The role of accruals in asymmetrically timely gain and loss recognition. *Journal of accounting research*, *44*(2), 207-242.

[7] Bataghva, H., & Garkaz, M. (2015). The relationship between acccounting conservatism, institutional investors and profit misprice in pharmamceutical companies listed in the Terhan stock exchange. *International Journal of Basic Sciences and Natural Research*, 390-395.

[8] Caskey, J., & Laux, V. (2016). Corporate governance, accounting conservatism, and manipulation. *Management Science*, *63*(2), 424-437.

[9] Chen, S., Chen, X., Cheng, Q., & Hutton, A. (2009). *Accounting conservatism and large shareholders: an empirical analysis*. Tech. rep., Working Paper, University of Texas at Austin.

[10] Cronqvist, H., & Fahlenbrach, R. (2008). Large shareholders and corporate policies. *The Review of Financial Studies*, *22*(10), 3941-3976.

[11] Fan, J., & Wong, T.J. (2002). Corporate ownership structure and the in formativeness of accounting earnings in East Asia. *Journal of Accounting and Economics, 33*, 401-425.

[12] Fischer, P. E., & Louis, H. (2008). Financial reporting and conflicting managerial incentives: The case of management buyouts. *Management Science*, *54*(10), 1700-1714.

[13] Freeman, R. E., Wicks, A. C., & Parmar, B. (2004). Stakeholder theory and "the corporate objective revisited". *Organization science*, *15*(3), 364-369.

[14] Graham, J.R., Harvey, C.R., & Rajgopal, S. (2005). The economic implications of corporate financial reporting. *Journal of Accounting and Economics, 40*, 3-73.

[15] Gravetter, F. J., & Wallnau, L. B. (2000). Statistics for the behavioral sciences Stamford. *CT: Wadsworth*.

[16] Guay, W., & Verrecchia, R. (2006). Discussion of an economic framework for conservative accounting and Bushman and Piotroski. *Journal of Accounting and Economics, 42*, 149-165.

[17] Jinadu, O., Uwuigbe, U., Uwuigbe, O. R., Asiriuwa, O., Eriabie, S., Opeyemi, A., & Osiregbemhe, I. S. (2018). Ownership Structure and Corporate Performance of Multinational Banks: Evidence from Nigeria. *Academy of Strategic Management Journal*, 5,

[18] Jung, K., & Kwon, S. Y. (2002). Ownership structure and earnings informativeness: Evidence from Korea. *The International Journal of Accounting*, *37*(3), 301-325.

[19] Kwon, Y. K. (2005). Accounting conservatism and managerial incentives. *Management Science, 51*, 1626-1632.

[20] Lafond, R., & Roychowdhury, S. (2008). Managerial ownership and accounting conservatism. *Journal of Accounting Research*, *46*(1), 101-135.

[21] LaFond, R., & Watts, R. L. (2008). The information role of conservatism. *The Accounting Review*, *83*(2), 447-22478.

[22] Lalbar, A., Ghaemmaghami, M., Pourmansoor, M. H., & Karamali, M. (2012). A study of the relationship between accounting conservatism and investment efficiency in Tehran Stock Exchange. *Journal of Basic and Applied Scientific Research*, *2*(6), 5719-5725.

[23] Lara, J. M. G., Osma, B. G., & Penalva, F. (2012). Accounting conservatism and the limits to earnings management. 1- 55.

[24] Li, W., He, T., Tang, G., & Marshall, A. (2017). Revisiting conditional accounting conservatism in state-controlled firms. *Asian Review of Accounting, 25*(4), 486-501.

[25] Lin, F., Wu, C. M., Fang, T. Y., & Wun, J. C. (2014). The relations among accounting conservatism, institutional investors and earnings manipulation. *Economic Modelling*, *37*, 164-174.

[26] Nawang, D., & Selahudin, N. (2015). Accounting conservatism in governmetn linked companies: A conservatism estimating model. *Journal of Education and Social Sciences*, 84-87.

[27] Neag, R., & Maşca, E. (2015). Identifying accounting conservatism–A literature review. *Procedia Economics and Finance*, *32*, 1114-1121.

[28] Nowravesh, I., & A. Ebrahimi, K. (2005). The investigation and determination of the relationships between the structure of the stockholders with information and the usefulness of accounting performance measures. *Accounting and Auditing Review, 42*, 97-124

[29] Shuto, A., & Takada, T. (2010). Managerial ownership and accounting conservatism in Japan: A test of management entrenchment effect. *Journal of Business Finance & Accounting*, *37*(7□ 8), 815-840.

[30] Song, F. (2015). Ownership structure and accounting conservatism: A literature review. *Modern Economy*, *6*(04), 478.

[31] Suijs, J. (2008). On the value relevance of asymmetric financial reporting policies. *Journal of Accounting Research*, *46*(5), 1297-1321.

[32] Suleiman, S. (2014). Corporate governance mechanisms and accounting conservatism. *Journal of Management Policies and Practices*, *2*(2), 113-127.

[33] Utomo, D., Pamungkas, I. D., & Machmuddah, Z. (2018). The moderating effects of managerial ownership on accounting conservatism and quality of earnings. *Academy of Accounting and Financial Studies Journal*, *22*(6), 1-11.

[34] Uwuigbe, U., Eluyela, E. F., Uwuigbe, O. R., Teddy, O., Irene, F. (2018). Corporate governance and quality of financial statements: A study of listed Nigerian banks. *Banks and Bank Systems, 13*(3), 12-23.

[35] Uwuigbe, U., Teddy, O., Uwuigbe, O. R., Emmanuel, O., Asiriuwa, O., Eyitomi, G. A., & Taiwo, O. S. (2018). Sustainability reporting and firm performance: A bi-directional approach. *Academy of Strategic Management Journal*, *17*(3), 1-16.

[36] Wang, L. Y., & Liu, J. X. (2003). Implementation deviation of the accounting information disclosure rule inside and outside listing corporation: From AH shares of company evidence which exist dual reporting difference. *Economic Research*, *11*, 71-78.

[37] Yu, K. (2013). Block ownership and accounting conservatism. *Corporate Ownership & Control, 10,* 272-293.

Review On Probiotics Potentials, Nutritional Composition Of Bambara *Nut (Vigna Subterranea (L.) -An Underutilized LegumE.*

Nwinyi, Obinna C*[1]. Umane, Phoebe O[1]
[1]Department of Biological Sciences (Microbiology unit), Covenant University, Km 10 Idiroko Road, Canaan Land, PMB 1023 Ota, Ogun State, Nigeria.

E-mail: nwinyiobinna@gmail.com, obinna.nwinyi@covenantuniversity.edu.ng

Abstract. Bambara nut is one of the underutilized legumes. Conversely this underutilized legume could be utilized to solve the ravaging food security issues in sub-Saharan Africa owing to its balanced nutritive composition. Bambara nut limited use, is because of the ant-nutritional factors, hard to cook phenomenon and difficulty in dehulling. Because of high global demand on food security, there is an urgent need to review possible methods that could reduce the limiting factors to the widespread use of this crop and enunciate other benefits derivable from this legume. In this review, we appraised the nutritional composition of Bambara nut from different geographic locations, commonly associated bacteria species that predominate mostly during Bambara fermentation and their potential uses as probiotics. Expedient areas for further research were highlighted in this paper

1. Introduction

Bambara groundnuts (Vigna subterranea) varieties are of two types; V. subterranea var. spontanea (the wild type) and V. subterranea var. subterranean (the domesticated type). Bambara groundnut is an under-utilized legume mostly grown by the female gender in tropical and sub-tropical areas where it serves as a means of livelihood[1,2]. This legume is also cultivated in some Asian countries such as India, Malaysia, Philippines and Thailand [3]. It belongs to Fabaceae family under the genus Vigna. The crop takes its name from the Bambara tribe in Mali and in the Sahelian provinces of West Africa [4-6]. It has been rated as the third most significant pea in West Africa, in terms of its versatility [7-8]. The husk may be crumpled and it is 1.5cm in length enclosing 1 to 2 seeds [9]. However, the pea itself is to a great extent oval or round. The pods are yellowish-white when unripe, and yellowish-brown or purple when ripe [10]. The seeds come in different sizes and colours such as dark-brown, black, white, red, creamy or mixture of the colors. It may be spotted with or without hilum coloration [11]. The nuts appear smooth, round and tough.

Bambara groundnut has various vernacular names based on different tribes/areas. These include but not limited to: indhlubu, Nzama, Njama (Malawi), jugo beans, Ndlowu, Nduhu nwa tzidzimba (South Africa), Ntoyo cibemba, Juga bean (Republic of Zambia), Njogo bean(Central Africa). In Nigeria the vernacular names include: Gurjiya or Kwaruru, Okpa and Epa- Roro. In Zimbabwe, it bears Nyimo beans while is called Aboboi akyii in Ghana. [12-14].

2. Nutritional composition

Bambara nut has nutritional benefits which could improve the level of malnutrition and boost food security. The Bambara nut seed is comprised of (%) 63 carbohydrate, 19 protein and 6.5 fat [14, 16-18]. Other works on the nutritional benefits of Bambara nuts obtained from various geographical location revealed similar trends. [19] reported Bambara nut from the South East Nigeria to comprise of (%) 2.86 moisture, 32 protein, 7.4 fat, 5.78 ash, 2.68 crude fibre and 53 carbohydrates . In Northern Nigeria, [20] revealed that Bambara nut from Kano contained approximately (%) moisture content 13, ash 3.52, protein 19, fat 7; fibre 6 and carbohydrate 63. In the North central (Benue State), [21] reported of Bambara nut seeds containing approximately (%) 6 moisture content, 4 ash content; 6 % crude fibre, 20% protein content, 6% lipid composition and 56% carbohydrate. Also Commercial Bambara nut collected from South- West Nigeria by [22] revealed to possess moisture content of (%) 7 Protein (%) 18 Fat (%) 6, Ash (%) 4 , Crude fiber (%) 4 and Carbohydrate 60%. Bambara seeds are highly proteinous, with malted seeds consisting of ten dissimilar proteins and dry seeds consisting of twelve different protein [4].

Boiled bambara nuts has antidiarrheic properties, it's mostly applied by the Lio tribe of Kenya. Bambara leaves serves in healing infected wounds. Its sap is used on the eye for epilepsy treatment [23] . Amongst others, the roots when eaten has an aphrodisiac effect, Igbos in Nigeria make use of its plant for venereal ailments. In Senegal, cataract is managed using a mixture of minced bambara nut and water. The raw seed suppress nausea when eaten by pregnant women, and also suppresses vomiting [14,16]. In Botswana, black land races treat impotence with Bambara plant [24]. Bambara nut possesses other nutrients, these include: Vitamin A, thiamine, riboflavin, niacin and carotene. The nut also contains trace amounts of vitamin C [25-26]. The seeds also contain micronutrients such as zinc, calcium, iron and potassium [27], but have trace quantities of phosphorus and magnesium. Seed with red coloration have about two-fold measure of iron as cream seeds; thus, they are fundamental for the remedy of iron deficiency [26]. Germinating Bambara nuts hold nutrients and also take out anti-nutrients efficiently at the same time [28] Studies have shown high levels of flavonoids, saponins and alkaloids present in the seed could assist in the general well - being of human [29]. In their study, they showed the beneficial effects of consuming Bambara nut which was inferred to due to flavonoids, saponins and alkaloids that occur at high levels in the seed. Anthocyanins, phenols and carotenoids could occur at low concentrations; of which their presence could be favorable since they act probably in synergy with other secondary metabolites.

2.1. Health benefits

- Reduction in dangers of stomach cancer

- Fatty acids present in nuts provide antioxidant properties capable of suppressing development of carcinogenic constituents in the stomach. Thus, the nuts could prevent stomach cancer [30].

- Antimicrobial action

- Bambara milk can serve as a probiotic. Probiotics minimize proliferation and activity of pathogenic organisms [26, 30].

- Rich radix of lysine

- Lysine Production

Bambara nut produce lysine naturally which humans are unable to produce [30]. As stated in [31] Bambara nut contains high levels of Lysine content. Lysine required for body enhancement and critical roles in making carnitine that functions in creation of energy and brings about decreased cholesterol level in blood via breaking down fatty acids. Furthermore, lysine aids in bone development by infusing calcium and assist in collagen formation that is essential in creation of connective tissue (e,g, tendon, skin. and cartilage) and bones[30]. Fights Kwashiorkor

Bambara groundnut contains high levels of protein that aid in the prevention of protein deficiency typically seen in undernourished children often known as Kwashiorkor [30]. Wholly cereal-based meals can be supplemented with Bambara as a protein source. Thus, malnutrition due protein deficiencies could be prevented.

2.2. Limiting factors to use of Bambara nut

Anti-nutritional factors: There is a connection between colours of the seeds and levels of anti-nutrients found in them. For example, cream Bambara nuts have low tannin content while brown and red nuts have high tannin content [13-14]. Other anti-nutrients present include: trypsin inhibitor, tannic acid, tannin, phytin oxalate, haemagglutinin and phytic acid. These lead to the decrease in nutritive significance of this legume [8, 28]. **Figure 1** shows anti-nutrients that could be present in different forms of bambara nut during its preparation for other uses. Hard to cook phenomenon (HTC): HTC factor is a principal reason behind the underutilization of this legume. The alterations in the cotyledons and seed coat result in the HTC factor. These factors could advance under storage conditions such as high temperature (30-40 °C) and high relative humidity (RH) (> 75%) in tropical and subtropical areas [32-36]. Boiling time could be increased to about 3-4 hours due to this HTC factor and this is a major problem for those residing in rural areas that use fire wood for its preparation. Due to long duration of cooking, it demands more use firewood/increase in costs of the woods that are rarely available in most rural areas [37]. Aside from the upsurge in the cost of wood, deforestation of trees will be on the increase [38].

Bambara nuts ought to be preserved at low temperature and humidity to avoid the HTC factor. Some latest methods to deal with this HTC phenomenon include chemical methods (cooking aids), physical (grinding, broiling and canning) and biological methods (germination and fermentation) [26,39].

2.3. Cultural disposition

Among the Mzimba culture, Bambara nut consumption is tied to some beliefs. It is believed that if the right persons does not handle it, it could result to evil. The Malawi culture views Bambara nut as difficult to ingest, and with terrible scent and in some cases taste-less. Their convictions influence them to stay away from the Bambara nut. In some instance, the bullet shape of Bambara nut portends danger that one could be shot via strayed bullets. In some cultures they believe bambara nut plays a role in witchcraft [40].

2.4. Ignorance of appropriate agronomic practices and prospects as a cash crop

Bambara nut are generally developed by rustic ranchers and the greater part of them have no training on the great benefits of this crop (Ibrahim and Ogunwusi, 2016). Consequently they need to be trained on recent agronomic methods to improve production of this legume. Most farmers do not consider the legume to be a cash crop because typically, it is grown around the farmers' households (William et al., 2016).

2.5. Availability and access to markets

According to [8]; Bambara nut could be limited in supply due to several reasons. These include but not limited to:

Inaccessibility of nuts lasting through the year because of little conservation and preparations employed to store the nuts when harvest time is over

Little scale agriculturists have a tendency to cultivate some other crops that they believe would be in demand by customers than Bambara nut

Effect of Hoarding: The bambara nut can withstand environmental dryness when there is scarcity of water supply. Different crops cannot survive such conditions thus farmers will rather save the legume for personal use than distribute it for sales.

Export rejection and poor marketing networks for local and especially international markets limit the widespread use of bambara nut. Government support and sponsorship is also at low level as governments do not seem to value Bambara nut as a cash crop. The government can assist in creating awareness about the benefits of bambara nut as a cash crop rather than the lack of awareness of the benefits the crop is having. Urbanisation is deemed an additional factor as it appears to confine the crop to mostly rural, subsistence-agriculture life [8].

2.6. Bambara nut a good means of achieving Food Security

Farming Bambara crop is not capital intensive; hence it could provide new vistas of opportunities for agriculturists and farmers. It is a remarkable wholesome quality crop. Since is mostly cultivated by female genders, it provides a functional way to economically empower women, thus improve the living conditions of communities. The seeds are drought resistant, pest- and disease-resistant. Bambara nut is perfect for intercropping and crop rotation; they do not require huge land space but do enhance nitrogen fixation and consequently soil fertility [27].

2.7. Fermented Bambara nut

[41] stated that fermentation modifies leguminous seeds physically, helps lower anti-nutrients, improves nutrients, digestibility of seeds and also adjusts flavor. Fermentation has been able to cause a decrease in under-utilization of several legumes [42]. A traditional condiment such as those produced from African locust bean(Parkia biglobosa) daddawa/Iru can be produced from Bambara groundnuts [41-48] proposed that bambara flour could made from using Rhizopus oligosporus and Rhizopus nigricans as starter cultures during bambara nut fermentation. The product of fermentation could act as good supplement with enhanced nutritional composition. The researchers inferred that the product could provide avert obesity, cardiovascular diseases, heart attack and other related ailments that are predominant in developing nations. [49] reported that when bambara nut is fermented into Bambara-Ogi, It can adequately serves as food for infants since the pattern of amino acids created by this blend compares favourably to those recommended for infants.

In all fermentation procedures, there is need to properly comprehend the microbial ecology of legume fermentations. This usually requires one to know fermentation substrates i.e. seeds of the different plants and the end products [43]. For example, Bacillus subtilis, B. licheniformis (Antai and Ibrahim, 1986), Staphylococcus saprophyticus, and Leuconostoc spp. are predominant organisms during African locust bean fermentation into Dawadawa/Iru [50-51]. Dadawa/Soyiru contains Bacillus spp., Staphylococcus spp. [52], Soy-daddawa contains B. pumilis, B. licheniformis and Bacillus subtilis (Omafuvbe et al., 2002) and Bambara-daddawa contains Staphylococcus spp., Streptococcus, Enterococcus, Bacillus subtilis, Bacillus pumilus and B. lichenformis [46]. Traditionally, fermented legumes have differences in quality and stability of product due to lack of starter cultures (Achi, 2005). Conversely with Hazard Analysis Critical Control Points (HACCP) generated for improvement of Microbial Safety Quality during the fermentation processes, the variations can be averted [53].

2.8. Common organisms associated with Fermentation of the Bambara nut

Information on succession studies of the fermentation protocols for Bambara groundnut solid state fermentation or liquid state fermentation has been scarcely reported. However few studies on Bambara nut fermentation revealed that the predominant organisms are mostly Bacillus species. The notable bacteria species include: Bacillus subtilis, Bacillus licheniformis, Bacillus pumilus, Staphylococcus spp. and Streptococcus Enterococcus [47].

2.9, Common bacteria species associated with fermented bambara nut with probiotic potential

Globally, statistics indicate that about 1.8 million people die annually as a result of food borne diseases caused by contamination. Gastrointestinal disorders are major health challenge when microbes, chemical, or physical agents are ingested.

Bambara groundnut an underutilized crop could have probiotic potential with high nutritional composition [4, 47]. According to Cutting, (2011); for over 50 years Bacillus species have been known to provide probiotics potentials. Some Bacillus species with probiotic function from naturally fermented bambara groundnut have been identified and shown to prevent gastrointestinal disorders See Table 1, [54-55] See Table 1.0). These bacteria species could assume a critical part in the gut as a result of their high metabolic actions and ability to survive in the gastro-intestinal tract due to presence of certain properties. For example, the spore structure has an outermost layer called the exosporium which is made up of proteins, lipids and carbohydrates that is responsible for hydrophobic properties. The higher hydrophobicity and the lower energy barrier of spores results in significantly higher adhesion forces, compared to vegetative cells [56-59]. Various workers [60-62] reported of the probiotic potential of some organisms they worked with, although no detailed scientific proof to support these claims. Recently, scientific evidence of probiotic bacteria ability to prevent and treat variety of diseases have emerged[63]. Generally, probiotics have been known for their safety, but in rare cases bacteria-host contact and reactions may arise [64-65].

To ensure of the safety of potential probiotic organism the following tests could be carried out:

- Choice of antibiotic resistance schemes
- Evaluation of definite metabolic activities (e.g., D-lactate production, bile salt deconjugation)
- Assessment of reactions amid human analysis
- Epidemiological reconnaissance of antagonistic episodes in customers
- Toxin generation
- Hemolytic potential could be assessment

Table 1.0 Common bacteria species associated with fermented bambara nut with probiotic potential

	Predominant microorganisms	Product source	source
Naturally found during fermentation process	*Bacillus subtilis, Bacillus licheniformis Bacillus pumulus Staphylococcus spp. Streptococcus, Enterococcus*	Iru-like condiment from Bambara nut, Bambara daddawa	Fadahunsi and Olubunmi, 2010
Artificially induced (starter culture)	*Lactobacillus bulgaricus, streptococcus thermophiles L. plantarum*	Bambara Yoghurt	Murevantemas and Jideani 2013 Pahare et al., 2017 Olanipekun et al., 2012
	Rhizopus (R. oligosporus, R oryzae, R.nigricans	Bambara flour	

3. Bacillus spp as a Probable Probiotic

Bacillus species have been found to have probiotic abilities and can survive the low pH of the gastric environment, making them preferable to non-spore formers as a probiotic (Cutting, 2011).
The ability of spores to withstand heat makes spores more advantageous than vegetative cells. For example, Lactobacillus spp. and Bacillus strains have improved digestion, immune modulation and stimulates growth in many animals [66-67]. [68] reported Bacillus subtilis makes use of 4-5% its genome for antibiotic production. To function admirably as a probiotic, the organism must possess basic requirements for survival in the gastrointestinal tract, notwithstanding great adherence and bio remedial characteristics. B.subtilis has been reported to acclimatize to prevailing conditions in the

human gastrointestinal tract. They can cope with antimicrobials, create biofilms, sporulate anaerobically and could be viewed as gut commensals [69-70]. Bacillus spores can tolerate serious gastric acids, bile salts and other antagonistic conditions of the gastrointestinal tract [71-72]. Furthermore, Bacillus species have been reported to support human well-being by aversion and cure of Heliocbacter pylori contamination, looseness of the bowels, support of gut homeostasis and prevent gingivitis [73-75].

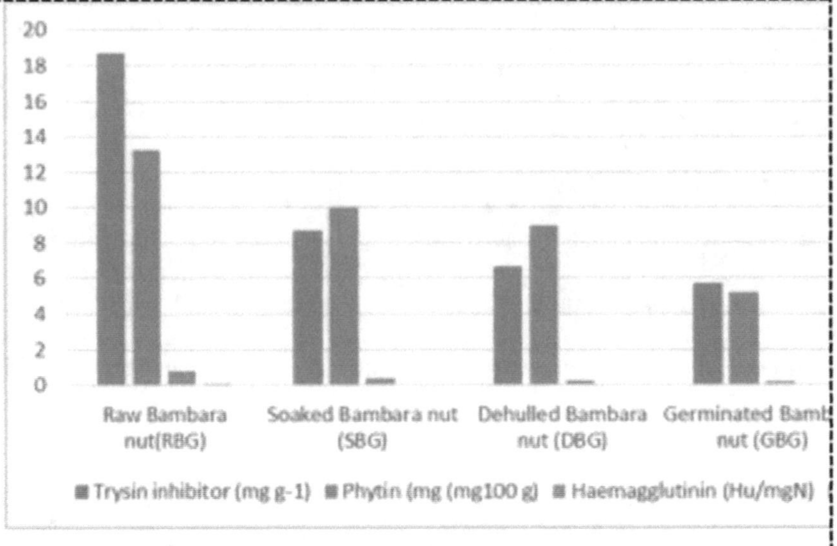

Figure 1. Figure 1 shows anti-nutrients that could be present in different forms of bambara nut during its preparation for other uses.

Source: (Omoikhoje et al., 2006)

3.1. Possible mechanisms of actions of Bacillus species:

Direct antagonism

The creation of antimicrobials by probiotics is thought to be one of the main mechanisms thwarting the development of pathogenic organisms in the gastrointestinal tract (Hong et al., 2005). Bacillus spp. convey different antimicrobial substances, for example, coagulin, bacteriocins, subtilin and anti-infection agents [45]. B. subtilis antimicrobial products are dynamic against an extensive variety of irresistible microorganisms, and unwanted side effects or the event of pathogen opposition are decreased because of the similarity of probiotic microbes to that in the human antimicrobial resistance framework. Bacillus subtilis are strain specific and can produce lytic catalysts. [62-71] revealed that B. subtilis had a noteworthy deterrent result on the growth of Candida spp. proposing it as a substitution treatment for oral candidiasis. In poultry farming, controlled trials have shown that oral utilization of Bacillus subtilis spores decreases diseases, for example, diseases caused by Salmonella enterica, Salmonella enteritidis, Clostridium perfringens and Escherichia coli (La Ragione et al., 2001; La Ragione and Woodward, 2003).

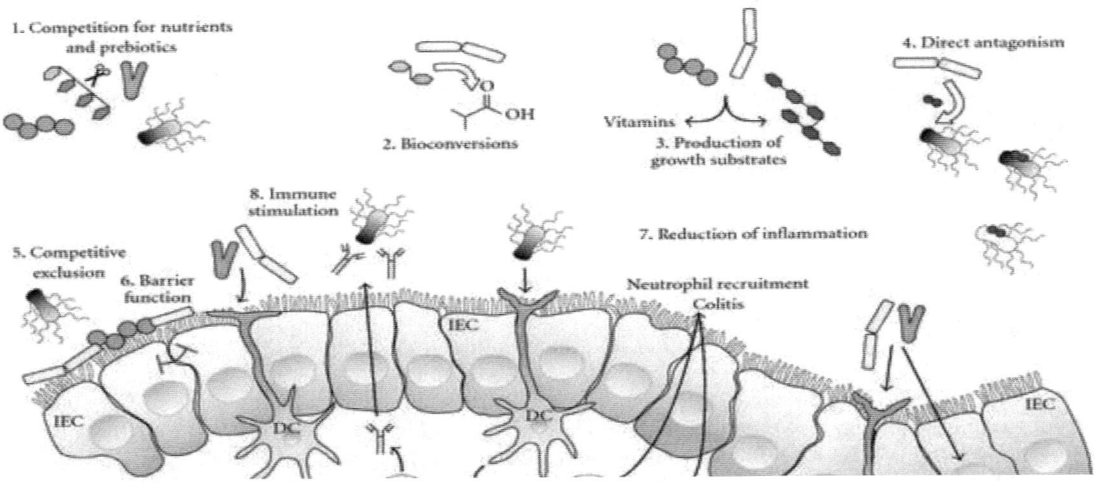

Figure 2: Mechanisms of Action for Bacillus Probiotics
Source: Custom Probiotics, 2018

3.2. Immunomodulatory Effect
Immune stimulation can be activated during treatment of oral infection using B. subtilis spores because the spores cause increased expression of activation markers on the lymphocytes. Spores also have the ability to cause humoral immunological responses (Sorokulova, 2013, Suva et al., 2016). Various studies report enhanced immune responses in the gastrointestinal tract of mice brought about by growing spores, furthermore, this immune incentive could be the hidden driver of the probiotic impact of spores (Cutting, 2011, Sorokulova, 2013).

3.3. Competitive Exclusion of Pathogens
Non-pathogenic probiotic organisms when ingested into the gastrointestinal tract, creates favourable conditions that support the growth of other beneficial microbes but hinder the growth of pathogenic organisms. Probiotic Bacillus species hinder the growth of pathogenic organisms by taking most of the nutrients and space, leaving little or none left to sustain growth of the pathogen. They also outcompete pathogenic microbes with adherence and help in hastening the development of natural flora in the intestine (Hong et al., 2005; Suva et al., 2016; Bernardeau et al., 2017). It should be known that Bacillus can also make use of extra-cellular enzymes and vitamin products to initiate a beneficial effect on growth and make up of commensal and useful species in the gut (Elshaghabee et al., 2017).
Some other mechanisms include detoxification, inhibition of intestinal inflammation (Hong et al., 2005; Bernardeau et al., 2017; Elshaghabee et al., 2017) and antidiarrheal effect (Suva et al., 2016)

3.4. How safe is Bacillus?
Bacillus species cause disease or deterioration of food. B. anthracis and B. cereus are both pathogenic strains of the genus Bacillus (Cutting, 2011, Elshaghabee et al., 2017). Certain Bacillus strains, such as B. subtilis, B. indicus, and B. licheniformis have been considered non-toxic and safe for consumption if in food, according to European standards (Sorokulova et al., 2008; Hong, 2008; Sorokulova, 2013; and Olmos and Paniagua-Michel, 2014).

4. Conclusion
Bambara nut is a legume categorized as underutilized however it holds considerable potential in solving the issue of food security in Nigeria and other sub-Saharan African countries. What is most important is to explore the benefits of this legume, assess its probiotic potential in model animals. In addition, elucidate on the proper phylogenetic studies on predominant bacteria species associated with the fermented Bambara nut. This could be a veritable way of establishing potential bacteria species that could serve as probiotic organisms. We anticipate that when these investigations are carefully established, widespread use of this underutilized crop will be adopted in most homes, more so, new food products developed from this crop.

Acknowledgements
The authors would like to acknowledge Covenant University, Ota, Nigeria for defraying the cost of this publication.

REFERENCES

[1]. Achi, O. K. (2005). Traditional fermented protein condiments in Nigeria. African Journal of Biotechnology, 4(13).
[2]. Adeleke, O. R., Adiamo, O. Q. and Fawale, O. S. (2018). Nutritional, physicochemical, andfunctional properties of protein concentrate and isolate of newly developed Bambara groundnut (Vigna subterrenea L.) cultivars. Food Science and Nutrition, 6(1), 229-242.
[3]. Adewumi, G. A., Oguntoyinbo, F. A., Keisam, S., and Jeyaram, K. (2013). Combination of culture-independent and culture-dependent molecular methods for the determination of bacterial community of iru, a fermented Parkia biglobosa seeds. Frontiers in Microbiology, 3, 436.
[4]. Adewumi, G. A., Oguntoyinbo, F. A., Romi, W., Singh, T. A., and Jeyaram, K. (2014). Genome subtyping of autochthonous Bacillus species isolated from Iru, a fermented Parkia biglobosa seed. Food biotechnology, 28(3), 250-268.
[5]. Adeyeye, E. I., Olaleye, A. A. and Adesina, A. J. (2013). Food Properties of Hull, Dehulled and Whole Seed Samples of Bambara Groundnut (Vigna Subterranea L. Verdc). Global Journal of Science Frontier Research Chemistry, 13(2), 9-28.
[6]. Alhassan, A. J., Dangambo, M. A. and Abdulmumin, T. M. (2015). Evaluation of the Proximate Contents of Bambara Groundnut Vigna subterranean (L.) Verdc Grown in MadobiLGA, Kano State, Nigeria. British Journal of Applied Science and Technology, 8(4), 361-365.
[7]. Alkaya, B., Laleman, I., Keceli, S., Ozcelik, O., Cenk Haytac, M. and Teughels, W. (2017). Clinical effects of probiotics containing Bacillus species on gingivitis: a pilot randomized controlled trial. Journal of Periodontal Research, 52(3), 497-504.
[8]. Anhwange, B. A., and Atoo, G. H. (2015). Proximate Composition of Indigenous Bambara nuts (Vigna subterranean (L.) Verdc). Journal of Pure and Applied Sciences, 2, 11-16.
[9]. Ani, D. P., Umeh, J. C. and Ekwe, K. C. (2013). Bambara groundnut as panacea for food security: profitability and production efficiency in Benue state, Nigeria. Acta Horticulturae.
[10]. Antai, S. P., and Ibrahim, M. H. (1986). Micro organisms associated with African locust bean (Parkia filicoidea Welw) fermentation for 'dawadawa'production. Journal of Applied Microbiology, 61(2), 145-148.
[11]. Azam-Ali, S. N., Sesay, A., Karikari, S. K., Massawe, F. J., Aguilar-Manjarrez, J., Brennan, M. and Hampson, K. J. (2001). Assessing the potential of an underutilised crop-a case study using bambara groundnut. Experimental Agriculture, 37, 433-472.
[12]. Bamshaiye, O. M., Adegbola, J. A., and Bamishaiye, E. I. (2011). Bambara groundnut: An under-utilized nut in Africa. Advances in Agricultural Biotechnology, 1, 60-72.
[13]. Baoling, H., Chengqun, L., Bo, W., and Liqin, F. (2007). A rhizobia strain isolated from root nodule of gymnosperm Podocarpus macrophyllus. Science in China Series C-Life Science, 50, 1-6.
[14]. Barimalaa, I. S., Achinewhu, S. C., Yibatama, I., and Amadi, E. N. (1994). Studies on the solid substrate fermentation of bambara groundnut (Vigna subterranea (L) Verde). Journal of the Science of Food and Agriculture, 66(4), 443-446.
[15]. Bernardeau, M., Lehtinen, M. J., Forssten, S. D. and Nurminen, P. (2017). Importance of the gastrointestinal life cycle of Bacillus for probiotic functionality. Journal of Food Science and Technology, 54(8), 2570-2584.
[16]. Chazovachii, B., Chitongo, L. and Ndava, J. (2013). Reducing urban poverty through fuel wood business in Masvingo city, Zimbabwe: A Myth or Reality. Bangladesh e-Journal of Sociology, 10(1), 59-70.
[17]. Chude, C., Amadi, E. R. and Iro, N. (2018). Fermentation of Bambara groundnut (Vigna subterranean (L) Verdc.) to complement infant foods. Food Science and Quality Management, 73, 33-37.

[18]. Cotthem, W. V. National Tree Planting Day – 3rd December 2011. In Environment Africa. 2011, Environment-Africa.
[19]. Custom Probiotics: Mechanisms of Action. (2018). Retrieved from https://www.customprobiotics.com/mechanisms-of-action
[20]. Cutting, S. M. (2011). Bacillus probiotics. Food Microbiology, 28(2), 214-220.
[21]. Didari, T., Solki, S., Mozaffari, S., Nikfar, S. and Abdollahi, M. (2014). A systematic review of the safety of probiotics. Expert Opinion on Drug Safety, 13(2), 227-239.
[22]. Doron, S. and Snydman, D. R. (2015). Risk and safety of probiotics. Clinical Infectious Diseases, 60, 129-134.
[23]. Durchschein, F., Petritsch, W. and Hammer, H. F. (2016). Diet therapy for inflammatory bowel diseases: The established and the new. World Journal of Gastroenterology, 22(7), 2179–2194.
[24]. Elshaghabee, F. M., Rokana, N., Gulhane, R. D., Sharma, C. and Panwar, H. (2017). Bacillus as Potential Probiotics: Status, Concerns, and Future Perspectives. Frontiers in Microbiology, 8, 1490.
[25]. Enwere, N. J. and Hung, Y. C. (1996). Some chemical and physical properties of Bambara groundnut (Voandzeia subterranea Thouars) seed and products. International Journal of Food Sciences and Nutrition, 47(6), 469-475.
[26]. Fadahunsi, I. F., and Olubunmi, P. D. (2010). Microbiological and enzymatic studies during the development of an 'Iru'(a local Nigerian indigenous fermented condiment) like condiment from Bambara Nut [Voandzeia subterranea (L) Thours]. Malaysian Journal of Microbiology, 2, 123-126.
[27]. Forsythe, L., Nyamanda, M., Mbachi Mwangwela, A. and Bennett, B. (2015). Beliefs, taboos and minor crop value chains: the case of Bambara Groundnut in Malawi. Food, Culture and Society, 18(3), 501-517.
[28]. Harrigan, W. F., and McCance, M. E. (1990). Laboratory methods in food and dairy microbiology. Academic Press Inc.(London) Ltd..
[29]. Hillocks, R. J., Bennett, C. and Mponda, O. M. (2012). Bambara nut, A review of utilisation, market potential and crop improvement. African Crop Science Journal, 20(1), 1-16.
[30]. Hoa, T. T., Isticato, R., Baccigalupi, L., Ricca, E., Van, P. H. and Cutting, S. M. (2001). Fate and dissemination of Bacillus subtilis spores in a murine model. Applied and Environmental Microbiology, 67(9), 3819-3823.
[31]. Hong, H. A., Duc, L. H. and Cutting, S. M. (2005). The use of bacterial spore formers as probiotics. FEMS Microbiology Reviews, 29(4), 813-835.
[32]. Hong, H. A., Huang, J. M., Khaneja, R., Hiep, L. V., Urdaci, M. C. and Cutting, S. M. (2008). The safety of Bacillus subtilis and Bacillus indicus as food probiotics. Journal of Applied Microbiology, 105(2), 510-520.
[33]. Hong, H. A., Khaneja, R., Tam, N. M., Cazzato, A., Tan, S., Urdaci, M., Brisson A., Gasbarrini, A., Ian Barnes, I. and Cutting, S. M. (2009). Bacillus subtilis isolated from the human gastrointestinal tract. Research in Microbiology, 160(2), 134-143.
[34]. Ibrahin, H. D. and Ogunwusi, A. A. (2016). Industrial potentials of Bambara nut. Journal of Poverty, Investment and Development, 22, 12-16.
[35]. Jideani, V. A. and Diedericks, C. F. (2014). Nutritional, therapeutic, and prophylactic properties of Vigna subterranea and Moringa oleifera. In O. Oguntibeju (Ed.), Antioxidant-Antidiabetic Agents and Human Health (pp 188-207). Croatia, Intech Publishers.
[36]. Kaptso, K. G., Njintang, Y. N., Nguemtchouin, M. M. G., Scher, J., Hounhouigan, J. and Mbofung, C. M. (2015). Physicochemical and micro-structural properties of flours, starch and proteins from two varieties of legumes, Bambara groundnut (Vigna subterranea). Journal of Food Science and Technology, 52(8), 4915-4924.
[37]. Khodadad, A., Farahmand, F., Najafi, M. and Shoaran, M. (2013). Probiotics for the treatment of pediatric Helicobacter pylori infection, a randomized double blind clinical trial. Iranian Journal of Pediatrics, 23(1), 79–84.
[38]. La Ragione, R. M. and Woodward, M. J. (2003). Competitive exclusion by Bacillus subtilis spores of Salmonella enterica serotype Enteritidis and Clostridium perfringens in young chickens. Veterinary Microbiology, 94(3), 245-256.
[39]. La Ragione, R. M., Casula, G., Cutting, S. M. and Woodward, M. J. (2001). Bacillus subtilis spores competitively exclude Escherichia coli O78, K80 in poultry. Veterinary Microbiology, 79(2), 133-142.

[40]. Lefevre, M., Racedo, S. M., Ripert, G., Housez, B., Cazaubiel, M., Maudet, C., Jüsten, P., Marteau, P. and Urdaci, M. C. (2015). Probiotic strain Bacillus subtilis CU1 stimulates immune system of elderly during common infectious disease period, a randomized, double-blind placebo-controlled study. Immunity and Ageing, 12(1), 24.

[41]. Lopetuso, L. R., Scaldaferri, F., Franceschi, F. and Gasbarrini, A. (2016). Bacillus clausii and gut homeostasis, state of the art and future perspectives. Expert Review of Gastroenterology and Hepatology, 10(8), 943-948.

[42]. Mabhaudhi, T., Modi, A. T. and Beletse, Y. G. (2013). Growth, phenological and yield responses of a Bambara groundnut (Vigna subterranea L. Verdc) landrace to imposed water stress, Under field conditions. South African Journal of Plant and Soil, 30(2), 69-79.

[43]. Marcel, A. and Bienvenu, M. J. (2014). Chemical and phytochemical compositions of Voandzeia subterranea seeds. Pakistan Journal of Biological Sciences, 17(9), 1083-1088.

[44]. Mazahib, A. M., Nuha, M. O., Salawa, I. S. and Babiker, E. E. (2013). Some nutritional attributes of Bambara groundnut as influenced by domestic processing. International Food Research Journal, 20(3), 1165–1171.

[45]. Mubaiwa, J., Fogliano, V., Chidewe, C. and Linnemann, A. R. (2017). Hard-to-cook phenomenon in Bambara groundnut (Vigna subterranea (L.) Verdc.) processing, Options to improve its role in providing food security. Food Reviews International, 33(2), 167-194. DOI, 10.1080/87559129.2016.1149864

[46]. Murevanhema, Y. Y. and Jideani, V. A. (2013). Potential of Bambara groundnut (Vigna subterranea (L.) Verdc) milk as a probiotic beverage—A review. Critical Reviews in Food Science and Nutrition, 53(9), 954-967.

[47]. Mwangwela, A., Bennet, B., and Chitio, F. (2012). Annual Report for Bambara Groundnut Promotion in Malawi, Mozambique and Tanzania.

[48]. Nti, C. A. (2009). Effects of Bambara groundnut (Vigna subterranea) variety and processing on the quality and consumer appeal for its products. International Journal of Food Science and Technology, 44(11), 2234-2242.

[49]. Nwodo, S. N. and Nwinyi, C. O. (2012). Proximate analysis of Sphenostylis stenocarpa and Voadzeia subterranean consumed in South-Eastern Nigeria. Journal of Agricultural Biotechnology and Sustainable Development, 4(1), 1.

[50]. Odunfa, S. A. (1981). Microorganisms associated with fermentation of African locust bean (Parkia filicoidea) during iru preparation. Journal of Plant Foods, 3(4), 245-250.

[51]. Ogbadu, L., and Okagbue, R. N. (1988). Bacterial fermentation of soya bean for 'daddawa'production. Journal of Applied Microbiology, 65(5), 353-356.

[52]. Ogbonna, D. N., Sokari, T. G., and Achinewhu, S. C. (2001). Development of an owoh-type product from African yam beans (Sphenostylis stenocarpa)(Hoechst (ex. A. Rich.) Harms.) seeds by solid substrate fermentation. Plant Foods for Human Nutrition, 56(2), 183-194.

[53]. Oguntoyinbo, F. A. (2012). Development of hazard analysis critical control points (HACCP) and enhancement of microbial safety quality during production of fermented legume based condiments in Nigeria. Nigerian Food Journal, 30(1), 59-66.

[54]. Olanipekun, B. F., Otunola, E. T. and Oyelade, O. J. (2015). Effect of Fermentation on Antinutritional Factors and in Vitro Protein Digestibility of Bambara Nut (Voandzeia subterranean L.). Food Science and Quality Management, 39, 98-110.

[55]. Olanipekun, B. F., Otunola, E. T., Adejuyitan, J. A. and Adeyanju, J. A. (2012). Proximate and fatty acid composition of Bambara groundnut (Voandzeia subterranean L. Thouars) as influenced by fermentation with a combination of Rhizopus oligosporus and R. nigricans. Transactions Journal of Science and Technology, 2, 77-87.

[56]. Olmos, J. and Paniagua-Michel, J. (2014). Bacillus subtilis a potential probiotic bacterium to formulate functional feeds for aquaculture. Journal of Microbial and Biochemical Technology, 6(7), 361-365.

[57]. Omafuvbe, B. O., Abiose, S. H., and Shonukan, O. O. (2002). Fermentation of soybean (Glycine max) for soy-daddawa production by starter cultures of Bacillus. Food Microbiology, 19(6), 561-566.

[58]. Omoikhoje, S. O., Bamgbose, A. M., and Aruna, M. B. (2006). Determination of the Nutrient and Anti-Nutrient Components of Raw, Soaked, Dehulled and Germinated Bambaragrounnut Seeds. Journal of Animal and Veterinary Advances.

[59]. Pahane, M. M., Tatsadjieu, L. N., Bernard, C. and Njintang, N. Y. (2017). Production, nutritional and biological value of Bambara groundnut (Vigna subterranea) milk and yoghurt. Journal of Food Measurement and Characterization, 11(4), 1613-1622.

[60]. Popoola, T. O. S., and Akueshi, C. O. (1985). Microorganisms associated with the fermentation of soybean for the production of daddawa (a condiment). Nigeria Food Journal, 2, 194-196.
[61]. Rijkers, G. T., De Vos, W. M., Brummer, R. J., Morelli, L., Corthier, G., and Marteau, P. (2011). Health benefits and health claims of probiotics: bridging science and marketing. British Journal of Nutrition, 106(9), 1291-1296.
[62]. Singh, J. (2014). Bambara Groundnut - health benefits. Retrieved May 17, 2017, from http://www.eherb.com/article/1295/bambara-groundnut--health-benefits/
[63]. Singhi, S. C. and Kumar, S. (2016). Probiotics in critically ill children. F1000Research, 5.
[64]. Slashinski, M. J., McCurdy, S. A., Achenbaum, L. S., Whitney, S. N. and McGuire, A. L. (2012). "Snake-oil,""quack medicine," and "industrially cultured organisms," biovalue and the commercialization of human microbiome research. Medical Ethics, 13(1), 28.
[65]. Sorokulova, I. (2013). Modern status and perspectives of Bacillus bacteria as probiotics. Journal of Probiotics and Health, 1(4), 1-5.
[66]. Sorokulova, I. B., Pinchuk, I. V., Denayrolles, M., Osipova, I. G., Huang, J. M., Cutting, S. M. and Urdaci, M. C. (2008). The safety of two Bacillus probiotic strains for human use. Digestive Diseases and Sciences, 53(4), 954-963.
[67]. Stein, T. (2005). Bacillus subtilis antibiotics, structures, syntheses and specific functions. Molecular Microbiology, 56(4), 845-857.
[68]. Sumi, C. D., Yang, B. W., Yeo, I. C. and Hahm, Y. T. (2014). Antimicrobial peptides of the genus Bacillus, a new era for antibiotics. Canadian Journal of Microbiology, 61(2), 93-103.
[69]. Suva, M. A., Sureja, V. P. and Kheni, D. B. (2016). Novel insight on probiotic Bacillus subtilis, Mechanism of action and clinical applications. Journal of Current Research in Scientific Medicine, 2(2), 65-71.
[70]. Thakur, N., Rokana, N. and Panwar, H. (2016). Probiotics, Selection criteria, safety and role in health and. Journal of Innovative Biology, 3(1), 259-270.
[71]. Vermeiden, S. J., Campbell, B. M., and Matzke, G. E. (1996). The consumption of wood by rural households in Gokwe Communal Area, Zimbabwe. Human Ecology, 24(4), 479-491.
[72]. Villena, J. and Kitazawa, H. (2017). Probiotic Microorganisms: A Closer Look.
[73]. William, A., George, N., O'Reilly Patrick, J., Sean, M., Aryo, F. and Halimi, A. (2016). Adoption of Bambara groundnut production and its effects on farmers welfare in Northern Ghana. African Journal of Agricultural Research, 11(7), 583-594.
[74]. Yao, D. N., Kouassi, K. N., Erba, D., Scazzina, F., Pellegrini, N. and Casiraghi, M. C. (2015). Nutritive evaluation of the Bambara groundnut Ci12 landrace [Vigna subterranea (L.) Verdc.(Fabaceae)] produced in Côte d'Ivoire. International Journal of Molecular Sciences, 16(9), 21428-21441.
[75]. Zhao, C., Lv, X., Fu, J., He, C., Hua, H. and Yan, Z. (2016). In vitro inhibitory activity of probiotic products against oral Candida species. Journal of Applied Microbiology, 121(1), 254-262.

IOP Conf. Series: Earth and Environmental Science **331** (2019) 012058 doi:10.1088/1755-1315/331/1/012058

Remediation of soil polluted with spent oil using cow dung

Nwinyi, Obinna C* Akinmulewo, B. A ,

Department of Biological Sciences, College of Science and Technology, Covenant University, KM 10 Idiroko Road, Canaan Land, PMB 1023 Ota, Ogun State, Nigeria

E-mail: nwinyiobinna@gmail.com, obinna.nwinyi@covenantuniversity.edu.ng

Abstract. Spent oil is any petroleum-based or synthetic oil that contain impurities or loss of major properties thus affecting its unique purpose. In this study, we selected and contaminated a site in Covenant University with spent oil. This site was amended with cow dung and studied for 56 days. Spent oil contaminated soil without stimulation with cow dung served as the control. This study was done during the dry season period and the sites left to natural edaphic factors. Physicochemical parameters such as pH, moisture content were monitored to determine the influence of cow dung as biostimulatory agent when compared with control (untreated) site. Within the period of this investigation, we recorded decrease in the pH value on the amended site, the value ranged between 8.60-7.77 while the control increased from 8.30 -8.42. The moisture content (%) were 95.0- 82.1 and 90.8-103.0 for the amended and control sites respectively. Following these obtained dynamics, we isolated organisms using enrichment technique from the contaminated sites and the control. We characterized the isolates using phenotypic characteristics and comparison with standard reference organisms. The bacterial isolates obtained include: *Arthrobacter Mycobacteria Pseudomonas* and *Corynebacterium* The axenic cultures ability to utilize spent oil was monitored via indirect estimation using pH and Optical density dynamics for 240hrs. All the organisms exhibited growth in the MS medium supplemented with spent oil. The pH and optical density (OD) from bacterial species obtained from the control sample ranged between 6.91 ± 0.20- 6.56± 0.29 and the OD 0.278±0.150-0.826 ± 0.33. For the amended sample, the bacterial species showed decline in pH that ranged between 7.13±0.30 - 6.33±0.10 while the recorded OD values ranged between 0.190±0.03 - 0.621±0.50. Comparing the results obtained for the control and the amended soil. It was obvious that organisms from the amended soil (either in-situ or in-vivo) showed more metabolic activities on the spent oil. This study suggests that use of cow dung in appropriate concentrations could be very useful in bioremediation of soil contaminated with spent oil.

1. Introduction

Spent oil or waste lubricating oil is oil obtained during servicing of machines and automobiles [1]. Spent oil is similar to unused oil, except that impurities such as trace metals have been added to the oil due to high temperature and pressure of the operating engines, thus contributing significantly to chronic hazards of the spent oil [2,3]. Most companies that operate in Nigeria use machines that requires lubricating oil, thus the release of the spent oil into the environment is one major source of environmental pollution. This challenge is more worrisome where the regulatory framework/ law for citing of companies and mechanic workshop are not obeyed. Most mechanic workshops are manned by ignorant artisans that may not be aware of the environmental implications of indiscriminate discharge of spent oil. When spent oil are released into the environment it could affect many species of flora and fauna in the environment [4,5]. Spent oil could be characterized in appearance as black to brown liquid mixture. It could contain low to high molecular weight (C15-C18) aliphatic and aromatic hydrocarbon, polychlorinated biphenyls, lubricative additives and decomposition products.
Hydrocarbons could be removed from the soil environment via several ways. These methods include: biosparging, bioventing, composting bioaugmentation, biostimulation, mycoremediation and phytoremediation. Furthermore, other conventional technique used for remediation include excavating

IOP Conf. Series: Earth and Environmental Science **331** (2019) 012058 doi:10.1088/1755-1315/331/1/012058

up contaminated soil and moving it to landfill sites or capping the contaminated areas of soil. Among these, bioremediation is utmost in terms of hydrocarbon removal [6]. It is effective, economic and eco-friendly and could lead to the complete mineralization of hydrocarbon [7]. One promising method is the application of animal manure, chemical fertilizer that will stimulate soil microbes. Chemical fertilizers have the disadvantage in that it lowers the soil fertility. An alternative is to use organic fertilizers (poultry/ goat manure), to biostimulate the polluted areas for a clean-up or remediation process[8].

In the reports of [9-10] organic nutrients such as chicken droppings, periwinkle shells have been used for the bioremediation of petroleum polluted soils in Nigeria.

Cow dung is composed of reservoir of nutrients and energy that could support microbial growth and subsequently remediation of various organic pollutants [11]. Cow dung can improve the soil structure and fertility by contributing diverse species of microorganisms which include but not limited to *Acinetobacter* , *Bacillus* and *Pseudomonas* that are essential for natural biogeochemical processes [12,13]. Numerous bacteria genera have exhibited great potentials in using hydrocarbon substrates [14-18]. These bacterial species are common to pristine and oil polluted environments [19-21]. Based on the need to evolve a cost effective method of reclamation of sites polluted with spent oil, this study was set out to use cow dungs in stimulating soil polluted with spent oil for reclamation purposes. In addition, we herein report that bacterial species associated with the cow dung were competent in removing the spent oil via slurry experimentation.

2. Materials And Methods

Chemicals, Reagents and Instrumentation

The nutrient agar, nutrient broth were sourced from Micro master, India. The reagents $(NH_4)_2SO_4$, $MgSO_4.7H_2O$, $Ca(NO_3)_2.4H_2O$, K_2HPO_4, KH_2PO_4, and NaCl are of analytical grade and obtained from Merck, Germany. The Voges Proskauer medium, urease base agar, starch agar, methyl red, peptone water were generously provided by the Microbiology laboratory of Covenant University, Ota. The spent oil was obtained from a mechanic workshop in Ota, Ogun State, Nigeria. Mechanical/orbital shaker (Model HZQ-X300), Centrifuge (Model 80-2), pH meter, UV spectrophotometer (Genesys 10 UVS spec model) were used in this investigation.

Media Preparation

Chloride-free minimal salts (MS) medium as described by (Nwinyi *et al.*, 2014; 2016) was used for the e degradation experiments . The medium consists of 0.5g $(NH_4)_2SO_4$, 0.1 g $MgSO_4{\bullet}7H_2O$, 0.076 g $Ca(NO_3)_2.4H_2O$ and 40mM phosphate buffer (pH 7.25). Solid MS medium was made by the addition of 1.8% Agar (Difco Laboratories, Detriot, MI USA). MS medium was supplemented with the spent oil for the enrichment and degradation assay.

Study site

The selected site for this study was soil surrounding the College of Science and Technology. The site was condoned off and marked out in two transect. The two sections of the soil were contaminated with spent oil collected from a motor mechanic shop in Ota. A section of the marked study site was inoculated with cow dung while the other section without the cow dung served as a control. Prior to the contamination with the spent oil, the soil samples were collected using a hand trowel and placed in a black nylon bag for physicochemical evaluation. The samples were then transported to the laboratory for further analysis.

The pH fluxes and moisture content analysis of soil samples

The collected soil samples were sieved, removing stones and sent for analysis on some physicochemical parameters such as conductivity, pH and moisture content.

Moisture determination

The crucible were washed and dried. The weight of each crucible was measured and referred to as (W1). 5g of each soil sample were weighted into two sterile crucible dish (W2) and placed in the oven

IOP Conf. Series: Earth and Environmental Science **331** (2019) 012058 doi:10.1088/1755-1315/331/1/012058

at 105 ° C for 3hours. This was allowed to cool before it was reweighed as (W3). The $W3_1$, were placed in the oven again for 30 minutes and allowed to cool and reweighed as ($W3_2$). This was repeated until similar weight value is obtained (W3*). The difference in weight between the initial weight and constant weight gained represents the moisture content. The mathematical calculation of moisture content is given below.

$$Moisture\ Content\ (\%) = \frac{W3 - W1}{W2 - W1} \times 100$$

Where W1 = initial weight of empty crucible,
W2 = weight of crucible + soil before drying,
W3= final weight of crucible + soil after drying

pH determination

Five (5g) of soil samples were dissolved in distilled water. However since it was not deionised water, the pH of the distilled water was measured following the calibration of the pH meter. Then soil samples were homogenised using glass rod. This was allowed for about 5 minutes before taking readings.

Isolation and characterization of spent oil degrading bacterial species

Following the physiological changes obtained from the physicochemical analysis, liquid culture spent oil degradation was conducted. For this, 20g each of the soil samples collected (soil biostimulated with cow dung) were sieved and inoculated into 250ml conical flask containing 100ml MS media. This was homogenized by gently stirring to allow soil to evenly spread and mix with the medium. Aliquots of spent oil were added to supplement as carbon and energy source. The bioreactor was incubated in an orbital shaker (Model HZQ-X300) for 2weeks. Following the end of the incubation period, 0.1ml of the enrichment broth was pipette and streaked over minimal salt agar (MS-Agar) with spent oil sprayed on the surface. This was incubated in the dark for 5days at ambient temperature. Colonies that appear were further selected and periodically transferred to obtain a pure culture. The pure isolates were maintained on MS agar sprayed with spent oil to keep the isolates stressed in other to continue to express the catabolic plasmids required for spent oil degradation. Using morphological examinations, notable bacterial strains from distinct colonies were selected for identification and degradative potentials. They selected bacterial strains were maintained on agar slant and placed in the refrigerator at 4°C. The pure cultures were tentatively named as C1, C2, C3, S1, S2, S3 classified phenotypically using standard methods and comparison with standard reference organisms. The following tests were carried out Gram stain, morphology, catalase, oxidase, colony motility, methyl red, Voges Proskauer, indole, nitrate reduction, gelatin hydrolysis, spore test, starch hydrolysis, Urease activity, citrate, sugar utilization , growth in 5%NaCl and growth at pH 6.0

Inoculum preparation for degradation studies

The obtained pure cultures were incubated on a freshly prepared sterile nutrient broth in cotton wool stoppered Balch tubes and incubated at 30°C for 120h. This was done to produce critical mass of cells required for the degradation studies. The cells were harvested by centrifugation at 35 x 100 rpm for 50 minutes, washed two times in phosphate buffer saline at pH of 7.25, transferred into a sterile eppendorf tubes and re-suspended in MS medium. The inocula were used for the growth and degradation assay.

Biodegradation studies

The biodegradation studies were performed in sterile 160mL serum bottles tubes containing MS medium (20 mL) supplemented with the spent oil. These were seeded with pure cultures at 10^5 cells/ml. Tubes were labelled, stoppered with sterile cotton and incubated in a rotary shaker (Model H2Q-X 300) at 65 rpm at 30.4°C for 12 days. For the abiotic control, MS medium and spent oil devoid of organism was incubated at the similar conditions as samples. Measurements of optical density (OD) and pH were carried out at 0, 72, 192, 240 hrs. Optical density was measured at 600 nm using (Genesys 10 UVS spec model).

Statistical analysis
Statistical tests (mean and standard deviation) were performed using the EXCEL spreadsheet

3. Results And Discussion

The pH fluxes of the amended soil (cow dung) was 8.6-7.8; control 8.2-8.4 for the 56 days of monitoring of the soil samples (See Figures 1and 2). The figures 3a and b, show the biodegradation potential of the obtained isolates. The pH and optical density fluxes shows the behavior of the organisms in spent oil when incubated at 240hrs.The moisture content within this period of investigation ranged between 95-82.1 for the amended soil while the control had 90.8-103.
Six bacterial species (C1, C2, C3, S1, S2 and S3) were isolated, characterized and selected for further biodegradation studies on spent oil. The isolates C1, C2 and C3 are non-motile, S2 and S3 are motile. S3 and C1 does not have pigmentation C3 are gram negative while others are gram positive. The data of Table 1a and 1b showed that isolates were similar to members of the genus *Corynebacterium spp, Pseudomonas, Arthrobacter* and *Mycobacteria* species.

Table 1a. Morphological characteristic of bacterial isolates

Distinguishing features	C1	C2	C3	S1	S2	S3	
Colony Shape	Round	Irregular	Irregular	Round	Irregular	Irregular	
Colony egde/margin	Entire	Undulate		Lobate	Entire	Filiform	Filiform
Pigmentation No	Green	Green	Yellowish green		No	No	
Opacity of Colony	Translucent		Opaque	Opaque	Opaque	Opaque	Opaque
Surface of Colony	Smooth	Smooth	Smooth	Smooth	Rough	Rough	
Consistency or Texture	Buttery	Brittle	Mucoid	Mucoid	Brittle	Brittle	

Table 1b. Cultural and Biochemical Features of Bacterial isolates capable of Degrading Spent oil

Distinguishing features	C1	C2	C3	S1	S2	S3
Gram's Reaction & Morphology	+ Short rods	+ Short rods	- Long rods	+ Short rods	+ Short rods	+ Short rods
Motility	Non-motile	Non-motile	Non-motile	Non-motile	Motile	Motile
Citrate	+	+	+	+	+	+
Catalase	+	+	+	+	+	+
Indole	-	-	+	+	-	-
H_2S Production	-	-	-	-	+	+
Methyl Red	-	-	-	-	-	-
Starch Hydrolysis	+	+	+	+	+	+
Urease	-	-	+	-	-	-
Oxidase	+	+	+	+	+	+
Glucose	-	+	+	+	+	+
Gas Production	-	-	+	-	+	+
Lactose	-	-	-	-	-	+
Gas Production	-	-	-	-	-	-

IOP Conf. Series: Earth and Environmental Science **331** (2019) 012058 doi:10.1088/1755-1315/331/1/012058

Sucrose	-	-	-	-	+	+
Gas Production	-	-	-	-	+	+
Growth at pH 6.0	+	+	+	+	+	+
Growth at 35 °C	+	+	+	+	+	+
Growth in 5% NaCl	+	+	+	+	+	+
Acid Fast Stain	-	-	-	-	-	-
Spore Stain	-	-	-	-	-	-

+: Positive Reaction
-: Positive Reaction
C1> Control 1
C2> Control 2
C3> Control 3
S1> Stimulated soil 1
S2> Stimulated soil 2
S3> Stimulated soil 3

From the results of the biochemical tests and colonial morphology features, the most probable microorganisms include:
C1> *Corynebacterium spp*
C2> *Corynebacterium spp*
C3> *Pseudomonas spp*
S1> *Arthrobacter spp*
S2> *Mycobacterium spp*
S3> *Mycobacterium spp*

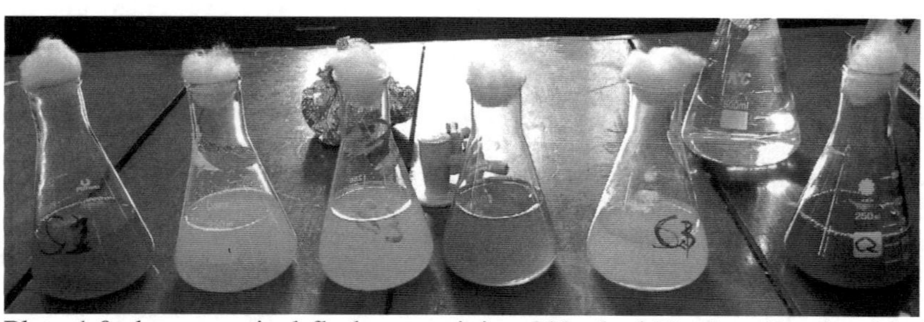

Plate 1.0 shows conical flasks containing 200ml of Nutrient Broth with the bacterial isolates. The turbidity and colour change depicts microbial growth.

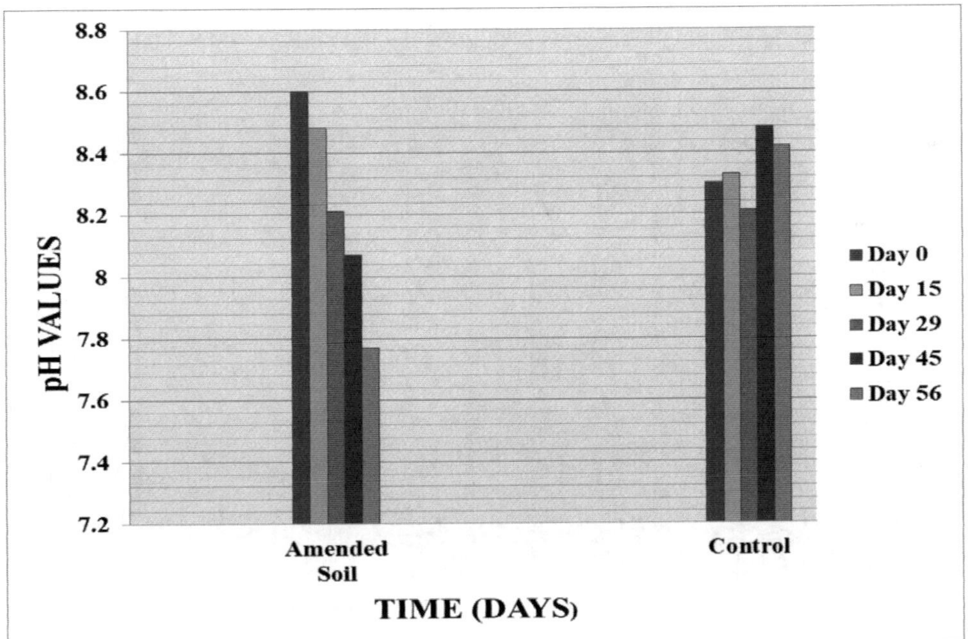

Fig. 1.0 shows pH values of soil contaminated with spent oil against different day intervals

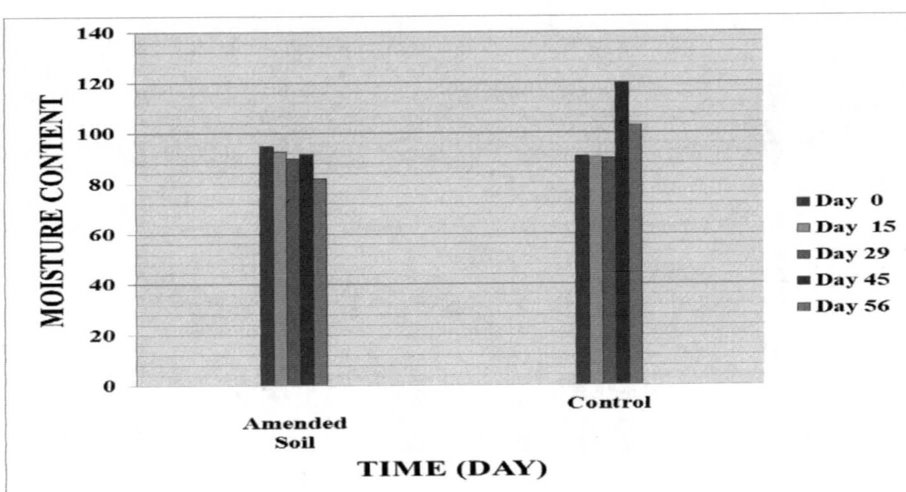

Fig. 2.0 shows the moisture content of contaminated soil samples against time (day)

3.

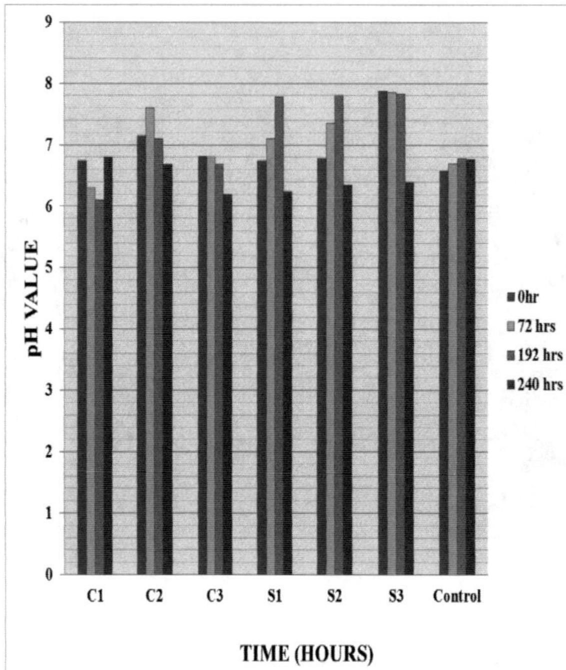

Fig. 3a. pH values for the bacterial isolates against time during biodegradation of spent oil

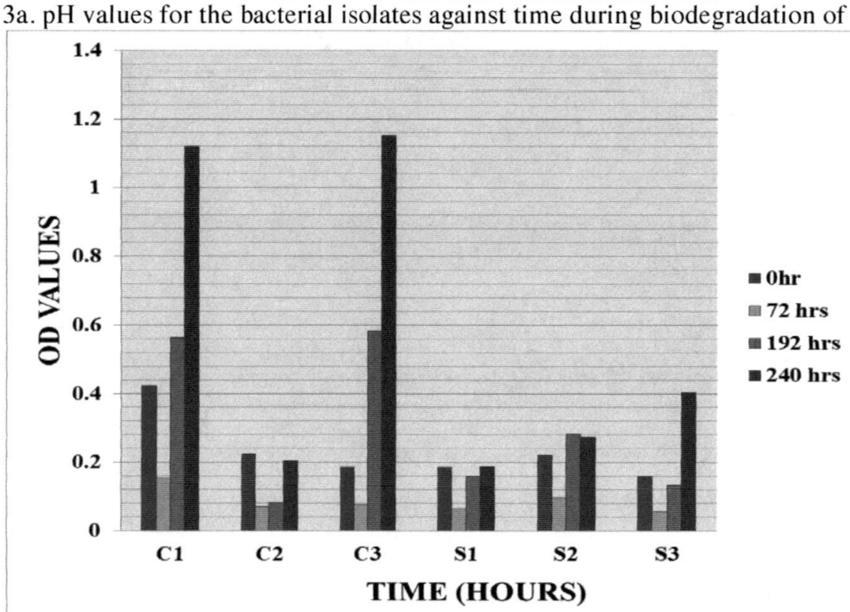

Fig. 3b. Optical densities of bacterial isolates against time during biodegradation of spent oil

Culture-dependent methods using the organisms' morphological and physiological behaviours have been used in characterizing these bacterial species, which include: Gram-negative bacteria, *Pseudomonas* and other gram positive *bacteria* genera *Corynebacterium*, *Arthrobacter* and *Mycobacterium* genera. These organisms are among some notable bacteria species competent in degrading hydrocarbon. Some specific organisms that are capable of degrading the hydrocarbons include *Bacillus subtilis, Bacillus* sp., *Alcaligenes* sp., *Flavobacterium sp.,Micrococcus roseus, Corynebacterium* sp., *Citrobacter freundii, Klebsiella, Arthrobacter, Acromobacter, Alcaligenes, Nocardia* and actinomycetes [22-24]. This study reports the effects of cow dung on the microbial utilisation of spent oil on polluted soil in-situ. The samples (control and stimulated soil) which were collected at different dimensions of collection site were analysed weekly for various pH and moisture content. pH refers to the hydrogen ion concentration in soil and moisture content refers to the maximum water content that does not restrict oxygen diffusion in the soil. From the results shown in Fig 1.0, there is a decrease in moisture content from 95 at day 0 to 82.1 at day 56; while in Fig 2, fluctuations where observed in moisture content from 90.8 at day 0 to 89.9 at day 29 and increases back to 103 at day 56. The decrease in pH shown in Fig.1 of the stimulated soil signifies the presence of organic acids produced as a result of microbial utilisation of hydrocarbons [25] while the fluctuations in pH shown in Fig.2, may have resulted from high metabolic activities possibly resulting in the production of intermediate metabolites by indigenous microbes [26].

In this study, enrichment methods were used to isolate the bacterial species where the spent oil served as carbon and energy source [27]. From the enrichment culture, certain bacterial strains were isolated successfully. They were tentatively named *Arthrobacter*, *Corynebacterium spp*, *Mycobacterium spp* and *Pseudomonas spp* (See Table1a,b). The pH and optical density (OD) were taken at different time intervals; 0hr, 72hrs, 192hrs and 240hrs. From results obtained there were fluctuations in pH and OD, with pH showing significant decrease from 6.75 to 6.1 and a gradual increase to 6.8 in C1 tentatively identified as *Corynebacterium spp* and OD reads 0.422, 0.153, 0.565, 1.120 from 0hr – 240hrs.The increase could be from high metabolic activities possibly resulting in the production of intermediate metabolites produced by organisms [26]. The decreases in pH observed in Fig. 4.1 were attributed to the degradation of spent oil which could lead to a release of organic acids and final products that probably lowered pH of the mixture [26]. The result shown in Fig. 4.3 and Fig. 4.4 showed an increase in OD which indicates an increase in the number of degradative cells signifying the isolates' ability to utilize spent oil as a sole carbon source. This ability may be possibly due to production of biosurfactants by the organisms making the hydrocarbons more available for microbial utilisation [28-29]. The dominance of gram positive bacteria in this study is due to their stronger cell wall envelope than gram negative bacteria and a higher tolerance to hydrocarbon hence their ability to thrive in highly variable environment contaminated with spent oil [30].

[31] reported that cow dung contain a vast reservoir of nutrients and energy capable of supporting microbial growth, thereby enhancing microbial degradation of various pollutants These isolates *Arthrobacter*, *Pseudomonas*, *Corynebacterium* and *Mycobacterium* species with proper nutrient supplementation such as the stimulation of organic manure could be used for enhancement of bioremediation of oil polluted sites.

The results of this study have shown that there is potential in using cow dung in the bioremediation of contaminated sites.

4. Conclusion

This study has shown that the spent oil spilled sites could be remediated using cow dung, overall, the soil fertility and structure could be improved using this eco-friendly means of bioremediation.

Acknowledgements

The authors would like to acknowledge Covenant University, Ota, Nigeria for defraying the cost of this publication.

References

[1]. Abdulsalam, S., Adefila, S.S., Bugaje, I.M. and Ibrahim, S. (2012). Bioremediation of soil contaminated with used motor oil in a closed system. *Journal of Bioremediation and Biodegradation* **3(12):** 172

[2]. Adebusoye, S. A., Ilori, M. O., Amund, O. O., Teniola, O. D. and Olatope, S. O. (2007). Microbial degradation of petroleum hydrocarbons in a polluted tropical stream. *World Journal of Microbiology and Biotechnoly* **23**: 1149-1159.

[3]. Adesodun, J. K. and Mbagwu, J. S. C. (2008). Distribution of heavy metals and hydrocarbon contents in an alfisol contaminated with waste-lubricating oil amended with organic wastes. *Bioresource Technology* **99:** 3195-3204.

[4]. Agarry, C. N., Owabor, R. and Yusuf, O. (2010). Bioremediation of soil artificially contaminated with petroleum hydrocarbon oil mixtures: Evaluation of the Use of animal manure and chemical fertilizer. *Bioremediation Journal* **14(4):**189-191.

[5]. Akinde, S. B. and Obire, O. (2008) Aerobic heterotrophic bacteria and petroleum utilizing-bacteria from cow dung and poultry manure. *World Journal of Microbiology and Biotechnology* **24:** 1999-2002.

[6]. Amund, O. O. (2000). The oil-eating microbes: a remedy to the menace of oil pollution. An inaugural lecture delivered at the University of Lagos, Nigeria.

[7]. Berwick, P.G. (1984). Physical and Chemical Conditions for Microbial Oil Degradation. *Journal of Advanced Biotechnology and Bioengineering* **26:** 1294-1305.

[8]. Blodgette, W. C. (2001). Water-soluble mutagen production during the bioremediation of oil contaminated soil. *Florida Scientists* **60(1):** 28-36.

[9]. Bogan, B. W., Lahner, L. M., SulliVan, B. W. R. and Paterek, J. R. (2003). Degradation ostraight chain aliphatic and high – molecular weight polycyclic aromatic hydrocarbons by a strain of *Mycobacterium austroafricanum*. *Journal of Applied Microbiology* **94:** 230-239.

[10]. Bragg, J. R., Prince, R. C., Harner, E. J., and Atlas, R. M. (1994). Effectiveness of Bioremediation for the Exxon Valdez Oil Spill. *Nature* ***368*:** 413-418.

[11]. Burke, G., Ben, S. R., and Louis, T. (2000). *Handbook of Environmental Management and Technology*. John Wiley & Sons. p 597.

[12]. Butler, C. S. and Mason, J. R. (1997). Structure-function analysis of the bacterial aromatic ring–
hydroxylating dioxygenases. *Advances in Microbial Physiology* **38:** 47-84.

[13]. Chikere, C. B., Okpokwasili, G. C., Chikere, B. O. (2009). Bacterial diversity in a tropical crude oil-polluted soil undergoing bioremediation. *African Journal of Biotechnology* **8(11)**: 2535-2540

[14]. Foght, J. M. and Westlake, D. W. S. (1982). Effect of the Dispersant Corexit 9727 on the Microbial Degradation of Prudhoe Bay Oil. *Canadian Journal of Microbiology* **28:** 117-122.

[15]. Geerdink, M. J., Kleijntjens, R. H., Van Loosdercht, M. C. M., and Luyben, K. C. A. M. (1996) Soil in Slurry Process. *Journal of Environmental Engineering* ***122*:** 975-982.

[16]. Gopamma D. and Srinvivas N. (2011). Effects of Soil Treatments amended with Organic Manure on Lubricant Oil Dgradation **15(4):** 18-21.

[17]. Hamamura, N., Olson, S. H., Ward, D. M., and Inskeep, W. P. (2006). Microbial population dynamics associated with crude oil biodegradation in diverse soils. *Applied Environmental Microbiology* **72:** 6316-6324.

[18]. Ijah, U.J. and Antai, S. P. (2003) Removal of Nigerian light crude oil in soil over a 12 month period. International Biodeterioration and Biodegradation **51:** 93–99.

[19]. Jirasripongpun, K. (2002). The characterization of oil-degrading capabilities of bacteria isolated from soil, W.A.J. Biology and Applied Chemistry **38:** 9-15.

[20]. Moller, J., Gaarn, H., Steckel, T., Wedebye, E. B., and Westermann, P. (1995). Inhibitory Effects on Degradation of Diesel Oil in Soil - Microcosms by a Commercial Bioaugmentation Products. *Bulletin of Environmental Contamination and Toxicology* **54:** 913-918.

[21]. Nwinyi, O. C., (2011). Enrichment and identification of askarelo State, Nigeria. (PCB) blend degrading bacteria enriched from landfill sites in Edo State, Nigeria. ***Agriculture and Biology Journal** of **North** America* **2:** 89 – 100.

[22]. Nwinyi, O. C., Flynn, W. P., Thuy, T.and Olukayode, O. A. (2013). Aerobic Degradation of Naphthalene, Fluoranthene, Pyrene and Chrysene using indigenous strains of bacteria isolated from a former industrial site *Canadian Journal of Pure and Applied Sciences* **7:** 2303-2314.

[23]. Nwinyi, O. C., Ajaja, O. and Nwinyi, C. (2014). Growth Dynamics of Bacteria Isolated from Spent Engine Oil Contaminated Tropical Soil. *Research Journal of Environmental and Earth Sciences* **6:** 430-436.

[24]. Nwinyi,O.C, Ajayi O, O. Amund,O.O (2016) Degradation of Polynuclear aromatic hydrocarbons by two strains of *Pseudomonas* *Brazilian Journal of Microbiology* **47** (**3**):551-562.

[25]. Nwinyi, O. C., Olawore Y. A (2017). Biostimulation of spent engine oil contaminated soil using *Ananas comosus* and *Solanum tuberosum* peels. *Environmental Technology and Innovation* **8**: 373-388.

[26]. Obire, O., Anyanwu, E. E. and Okigbo, R. N. (2008). Saprophytic and crude oil degrading fungi from cow dung and poultry droppings as bioremediating agents. *Journal of Agricultural Technology* **4 (92):** 81 -89.

[27]. Plohl, K., Leskovšek, H. and Bricelj, M. (2002). Biological degradation of motor oil in water. *Acta Chimica Slovenica* **49:** 279-289

[28]. Rittmann, B. E. and Johnson, N. M. (1989). Rapid Biological Clean-Up of Soils Contaminated with Lubricating Oil. *Water Science and Technology* **21:** 209-219

[29]. Ryan, J. R., Loehr, R. C. and Rucker, E. (1991). Bioremediation of Organic Contaminated Soils. *Journal of Hazardous Materials* **28:**159-169.

[30]. U.S.EPA, (1996). Managing Used Oil Advice for Small Businesses. Retrieved from http://www.epa.gov/osw/conserve/materials/usedoil/usedoil.htm. Accessed March 10, 2015

[31]. Wackett, L. P., Hershberger, L. C. D. (2001). Biocatalysis and biodegradation: Microbial transformation of organic compounds. Washington: ASM Press.

[32]. Zhuang WQ, Tay JH, Maszenan AM, Tay STL (2002). Bacillus naphthovorans sp. Nov. from oil-contaminated tropical arine sediments and its role in naphthalene biodegradation. *Applied Microbiology and Biotechnology* **58:** 547-553.

Effect of Saw-dust on soils contaminated with waste lubricating oil

Nwinyi, Obinna C[a]* Ikhine, P. J[a],

[a]Department of Biological Sciences, College of Science and Technology, Covenant University, KM 10 Idiroko Road, Canaan Land, PMB 1023 Ota, Ogun State, Nigeria
E-Mail: nwinyiobinna@gmail.com, obinna.nwinyi@covenantuniversity.edu.ng

Abstract. In this study, we evaluated the influence of sawdust as biostimulatory agent on waste-lubricating oil site. To achieve this, two sites were contaminated with waste lubricating oil, with one of the site amended with sawdust and the other left to serve as control. This study was conducted during the dry season. Decreases in pH and percentage moisture content were used to monitor the metabolic activities between the amended and control sites. Both sites (amended and control) showed decrease in the pH values however the control site showed just slight decrease. The mean pH value obtained ranged between 7.68± 0.15- 8.08± 0.57 for the amended and control while the mean moisture content (%) 38.3 – 48.6 respectively for the amended and control sites. We screened for fungal species that showed growth on sawdust amended site using waste lubricating oil as source of carbon/energy. This was done through conventional enrichment culture methods. The isolated fungal strains tentatively named as P1, P2, P3, P4 and P5 were identified by visual observation and micro-morphological technique. From the cultural, and morphological characterization and comparison with respect to the standard reference of fungi, the fungal species were identified as members *Aspergillus niger*, *Rhizopus* sp., *Penicillium* sp., *Aspergillus flavus* and *Mucor* sp. Pure cultures of these fungal species were tested for their ability to utilize waste lubricating oil as carbon and energy source. The ability of these fungal species to use the waste lubricating oil was done by monitoring their physiological responses via Optical Density (OD) and pH gradient readings. The mean pH obtained range from 5.90-7.33 and the (OD) 0.715-1.978. The fluctuations in OD readings as well as pH values for the different microfungi may be due to variation in growth patterns of the different fungal species.

1. Introduction

The quality of one's environment affects the individuals' life on earth. The releases and accumulation of persistent and toxic chemicals could have a harmful impact on human health and the environment [1]. Waste-lubricating oil is a common and toxic environmental contaminant not obviously found in the environment [2]. Waste-lubricating oil or mineral based crankcase oil occur, as brown-to-black liquid produced when lubricating oil is subjected to high temperature and pressure [3].

Waste-lubricating oil is a mixture of several different chemicals which include low and high molecular weight aliphatic hydrocarbons, aromatic hydrocarbons, lubricative additives heavy metals and decomposition products [3,4]. The waste-lubricating oil, is obtained following servicing and subsequent draining from automobile and generator engines. These are often released into the environment by motor and generator mechanics through unchecked releases into gutters, water drains, open vacant plots and farmlands [5-8]. When waste-lubricating oil is released into the environment it creates difficult condition within the soil matrix owing to poor aeration low pH and immobilization of soil nutrients[9-10]. In the environment, waste-lubricating oil spillages could spread horizontally into the ground waters [11], cause immense destruction to the mangrove and rainforest crops, farmland and aquaculture such as fish and periwinkle [12-13]. The toxicity of waste-lubricating oil or the refined products could be difficult to assess due to limited information on the effects of the additives, their synergistic or antagonistic effects of the mixtures [14-15].

Land is becoming scarce, polluted soil with waste-lubricating oil could be amended by adding materials to the soil that could improve its physical and chemical properties (Anon., 2012). Soil contamination with petroleum products have been treated with technologies such as thermal treatment, dig and dump method and stabilization/solidification technology [16]. Nonetheless, these technologies have their drawbacks [17]. These limitations include: the high cost, its operability at full scale and the destruction of soil texture and characteristics [18]. Bioremediation-mycoremediation presents as alternative technology to the pervious technologies currently in use.

IOP Conf. Series: Earth and Environmental Science **331** (2019) 012059 doi:10.1088/1755-1315/331/1/012059

Several soil improvement materials exist. One of such is sawdust. The economic disposal of saw dust is a serious challenge in the wood industries [19]. Enormous quantities of sawdust are produced yearly by sawmills. This could be diverted to soil conditioning/amendments efforts. Sawdust contain about 40% of lignin, 60% cellulose along with various waxes, resins and oils. [20]. The high lignin content makes sawdust potentially a good source of humus and thus good for soil amendments. Recently, fungi have received considerable attention for their bioremediation potential due to the range of enzymes they produce. These enzymes have capacity to degrade lignin and a range of recalcitrant pollutants. Due to the hyphae possessed by fungi, it has advantage over bacteria in the decontamination of soil polluted with hydrocarbons. Due to the overwhelming challenges that emanate from spent oil when released to the environment, authors aimed at using the saw-dust to biostimulate the activities of soil fungi in the remediation of spent oil contaminated site.

2. Materials And Methods

Chemicals and Reagents

All chemicals and reagents were of analytical grade. Media used include: Potato Dextrose agar (manufactured by Biomark laboratories India), Chlorine-free MS medium (as described by [21], Agar-agar (manufactured by Biomark Laboratories India) to support the growth of fungi and Ethanol (manufactured by Sigma-Aldrich, Germany). Then $(NH_4)_2SO_4$, $MgSO_4.7H_2O$, $Ca(NO3)_2.4H_2O$, K_2HPO_4, KH_2PO_4, Na_2HPO_4 of analytical grades were obtained from LOBA Chemie lab, Mumbai,India, Qualikems lab, India, and Guangdong Guanghua Sci-tech Co.,Ltd., China. Waste-lubricating oil, were obtained from a mechanic workshop in Ota, Ogun State Nigeria.

Instrumentation

Some of the instruments used in this study include: Genesys 10 UVS spectrophotometer, Incubator-shaker (Guangzhou healthy ling, Model HZQ-X 300), Centrifuge (Table centrifuge machine, model 800D), and pH meter (Jenway, 3505).

Stock solutions and Media preparation

Chloride free minimal salts (MS) medium as described by Nwinyi et al., (2014) with modification was used for enrichment and degradation experiments. The medium consists of 34.84g KH_2PO_4, 27.22g K_2HPO_4, 1.0g $MgSO_4.7H_2O$, 5.0g $(NH_4)_2SO_4$, 2.0g Na_2HPO_4 ,0.76g $Ca(NO3)_2.4H_2O$ and 4.8g of agar-agar dissolved in 1000 mL of deionized water. The slurry MS medium was supplemented with the spent oil.

Study Location

The study was carried out as field study experiment in an open field within College of Science and Technology, Covenant University, Ota, Ogun State, Nigeria. A 1×1.5sqm of the test field in CST building, Covenant University was chosen. This test site was divided into two pieces. To both pieces, were added about 500ml of waste lubricating oil each and thoroughly mixed. One parts was amended with sawdust and the other without sawdust to serve as control. The experiment was left to stand for 40 days under the influences of natural factors. Moisture and pH were determined every 7days.

Determination of moisture content

The crucible were washed and dried. The weight of each crucible was measured and referred to as (W1). 5g of each soil sample were weighted into two sterile crucible dish (W2) and placed in the oven at 105°C for 3hours. This was allowed to cool before it was reweighed as (W31). The W31, were placed in the oven again for 30 minutes and allowed to cool and reweighed as (W3). This was repeated until similar weight value is obtained (W3*). The difference in weight between the initial weight and constant weight gained represents the moisture content. The mathematical calculation of moisture content is given below.

$$\text{Moisture Content (\%)} = \frac{W3 - W1}{W2 - W1} \times 100$$

Where W1 = initial weight of empty crucible,
W2 = weight of crucible + soil before drying,
W3= final weight of crucible + soil after drying

pH determination

Two (2g) each of the amended soil sample and control were dissolved in 10ml of deionized water and the pH value measured.

Enrichment of Fungal Isolates

IOP Conf. Series: Earth and Environmental Science **331** (2019) 012059 doi:10.1088/1755-1315/331/1/012059

Soil samples were collected from the study site and the enrichment slurry prepared as described by Nwinyi et al., (2014). The fungi diversity was obtained by conventional enrichment culture methods. For this, 10g each of soil samples were mixed with 50ml freshly prepared MS broth into two 250 mL conical flasks, 5ml of spent oil was added. The resulting mixture was stirred, stoppered with sterile cotton wool and incubated at 25 ± 2°C for 14 days. Thereafter, a loop full of the solution was streaked over freshly prepared sterile potato dextrose agar (PDA) and sprayed with waste lubricating oil and incubated at 25 ± 2°C for 5 days.

Preliminary screening of hydrocarbon degrading fungal isolates

For the estimation of hydrocarbon utilizers. A minimal salt broth (MSB) 34.84g KH_2PO_4, 27.22g K_2HPO_4, 1.0g $MgSO_4.7H_2O$, 5.0g $(NH_4)_2SO_4$, 2.0g Na_2HPO_4 , 0.76g $Ca(NO3)_2.4H_2O$ **a**nd 5.0g of agar – agar dissolved in 1000ml of deionized water was prepared. Twenty 20ml of the minimal salt broth was measured into 6 test bottles (bioreactors) which was supplemented with 2ml of spent engine. The isolated fungal strains from the enrichment setup were seeded into 5 test tubes, while the 6th test tube with no fungi served as control. Each of the bioreactors were stoppered with sterile cotton wool to allow aeration and prevention of cross contamination. All the test tubes were then incubated at room temperature (25 °C - 27°C) for 12 days. The test tubes were slightly shaken periodically to facilitate oil-cell phase contract.

Isolation and mass production of Hydrocarbonoclastic Fungi in Minimal Salt Slurry

Pure cultures from the waste lubricating oil–enriched media were isolated by plating out 2.0 ml of the enriched cultures onto prepared PDA plates, sprayed with waste-lubricating oil on the surface. This was incubated at room temperature (25 + 2°C) for 5 days. Colonies were periodically transferred to fresh PDA agar as they appeared to obtain pure culture. Fungi isolates were identified based on their cultural properties, colonial characteristics, and the microscopic features of their sporulating structures. The fungi isolates were compared with standard reference organisms (Bannet and Hunter, 1972).

Harvesting of fungal isolates

Pure cultures of fungal isolates were inoculated on freshly prepared sterile mineral salt slurry (30ml) in balch tubes stoppered with sterile cotton wool to enable aeration and incubated at room temperature for 5 days. The different fungal isolates were harvested by centrifugation at 400rpm for 60 min, fungal inocula were washed twice in phosphate buffer saline at pH of 7.25 and transferred into a sterile balch tubes and kept briefly in a refrigerator at 6oC for use during the growth / degradation studies (Verdin et al., 2004).

Characterization of Hydrocarbonoclastic Fungi Species and determination of their degradation potentials

Most fungi have been reported to be identified by visual observation and micro-morphological techniques (Thenmozhi et al., 2013). Based on this, the different fungal colonies were observed on the plated potato dextrose agar after incubation for 5 days at room temperature. The cultural and microscopic examination of the fungal isolates were carried out. These include: morphology of the sporulating structure and lactophenol cotton blue stain

Pure cultures of tentative fungal species namely: S1, S2, S3, S4 and S5 were tested for their ability to utilize spent oil as carbon and energy source using MS Medium slurry supplemented with spent engine oil (4% v/v) as carbon source for the determination of hydrocarbon degradation. The enrichment slurry was prepared as described by Nwinyi et al., (2014). For this, the slurry media composed of (per liter of distilled water) 34.84g KH_2PO_4, 27.22g K_2HPO_4, 1.0g $MgSO_4.7H_2O$, 5.0g $(NH_4)_2SO_4$, 2.0g Na_2HPO_4 ,0.76g $Ca(NO3)_2.4H_2O$ and 5g of agar- agar.. About 1mL of the different fungal harvested inocula were transferred into 30ml of the different MS slurry tubes. In each of the aseptically prepared MS slurry tubes, 2mL of waste-lubricating oil were used to amend it. Incubation was carried out at room temperature for 12 days; however the tubes were intermittently shaken at 95 rpm using shaker (Model H2Q-X 300) to facilitate oil phase contact. Abiotic controls were set up with the biotic controls (MS slurry medium and spent oil devoid of organism) and incubated at the same conditions as the test samples. Determination of the extent of growth rate of the organism in MS slurry was measured by the turbidity (Optical density) and pH fluxes at 0, 3, 6, 9 and 12 days intervals. Optical density was measured at 600 nm using Genesys 10 UVS Spectrophotometer.

3. Results And Discussions

The Table 1.0 shows the cultural and microscopic characteristics of the obtained fungal isolates S1-S5. The most probable organisms include: *Aspergillus spp, Rhizopus, Penicillum* and *Mucor* species (See plate 1.0-4.0).

Table 1 Cultural and Microscopic characterization of the Soil Fungi

Isolates	Cultural characteristics	Microscopic characteristics	Most probable organisms
S1	Fluffy colonies with elevated mycelia that appeared black in coloration. Observing from underneath the plates showed a pale yellow color	Conidia in chains borne externally at the end of the vesicle. Long conidiophore present arising from septate hyphae	*Aspergillus niger*
S2	White cottony colonies	Sporangiospores within spherical shaped sporangium	*Rhizopus sp.*
S3	Mature colonies were observed, it showed velvety texture, leaf green in coloration. Reverse color is cream	Conidiosphores are formed on a brush –like conidia head. Microscopic appearance in blue green	*Penicillum spp.*
S4	Greenish and dusty colonies observed, reverse color appear pale yellow	Septate hyphae, non- septate conidiosphore with columnar conidial head bearing spherical spore	*Aspergillus flavus*
S5	White entire growth. Cottony with orange colored sporangium	Globular sporangium, spores are oval collumella present,non-septate hyphae and absence of rhizoids	Mucor sp.

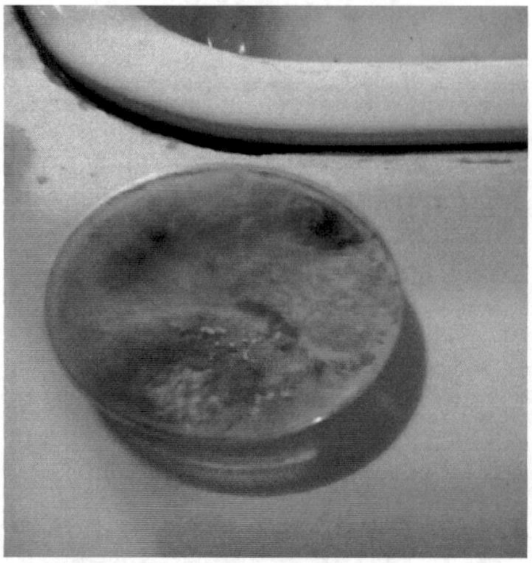

Plate 1.0: Growth of *Aspergillus flavus* on PDA sprayed with spent engine oil.

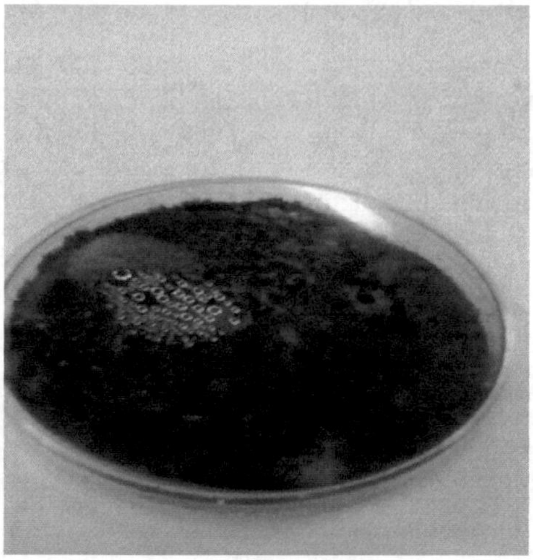

Plate 2.0 : Growth of *Mucor spp*. on PDA sprayed with spent engine oil.

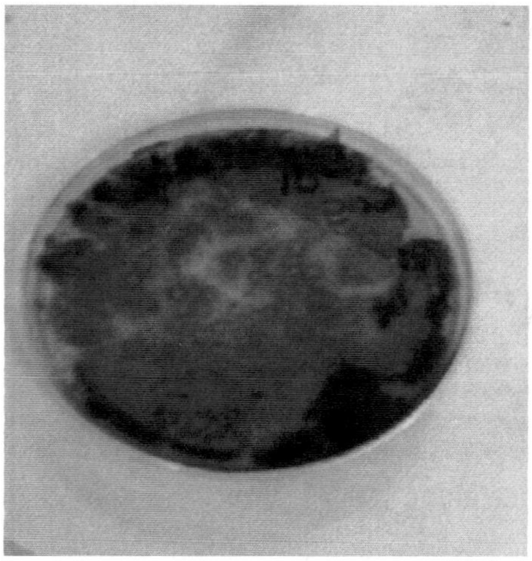

Plate 3.0: Growth of *Rhizopus spp*. on PDA sprayed with spent engine oil.

Plate 4.0: Growth of *Aspergillus niger* on PDA sprayed with spent engine oil.

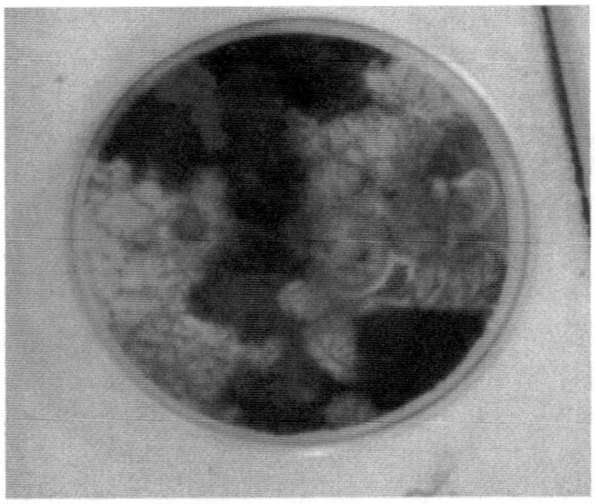

Plate 5.0: Growth of *Penicillium spp.* on PDA sprayed with spent

The Figures 1-5 shows the fluctuations in Optical densities and pH values of the different isolated fungal species growth on spent engine oil with an increasing OD and decreasing pH values.

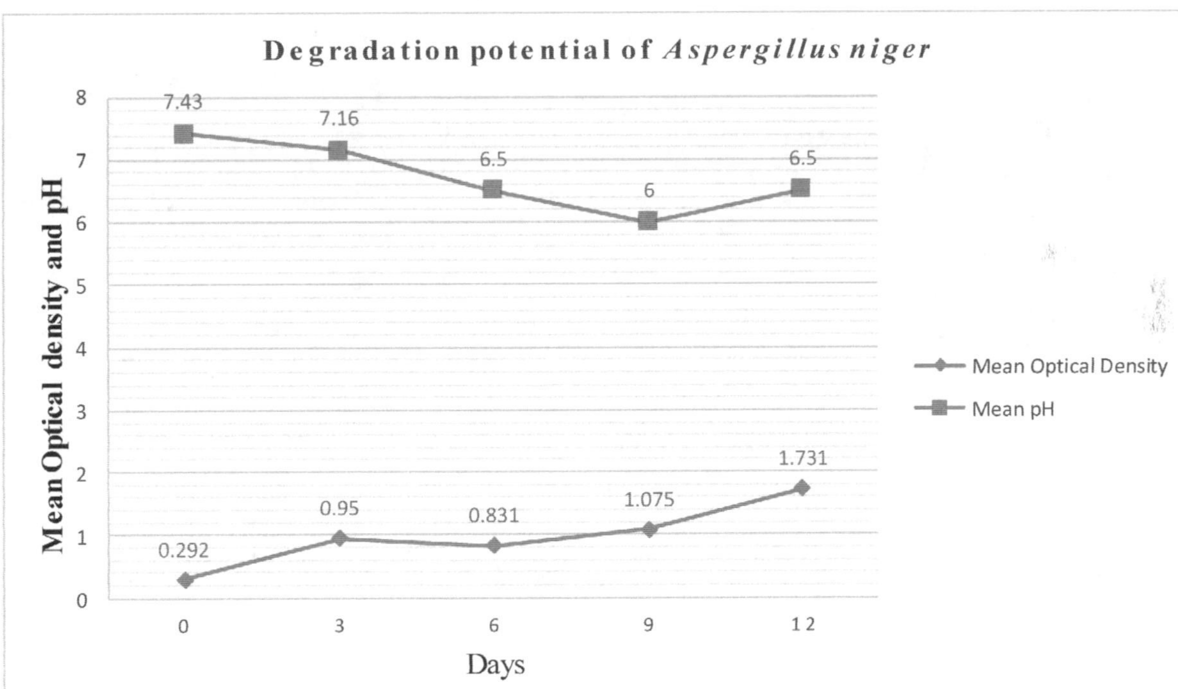

Figures 1.0: Data represent the mean value of pH and Optical Density (OD) against different time intervals with the same concentrate of spent engine oil (2ml) for 12 days by *Aspergillus niger*

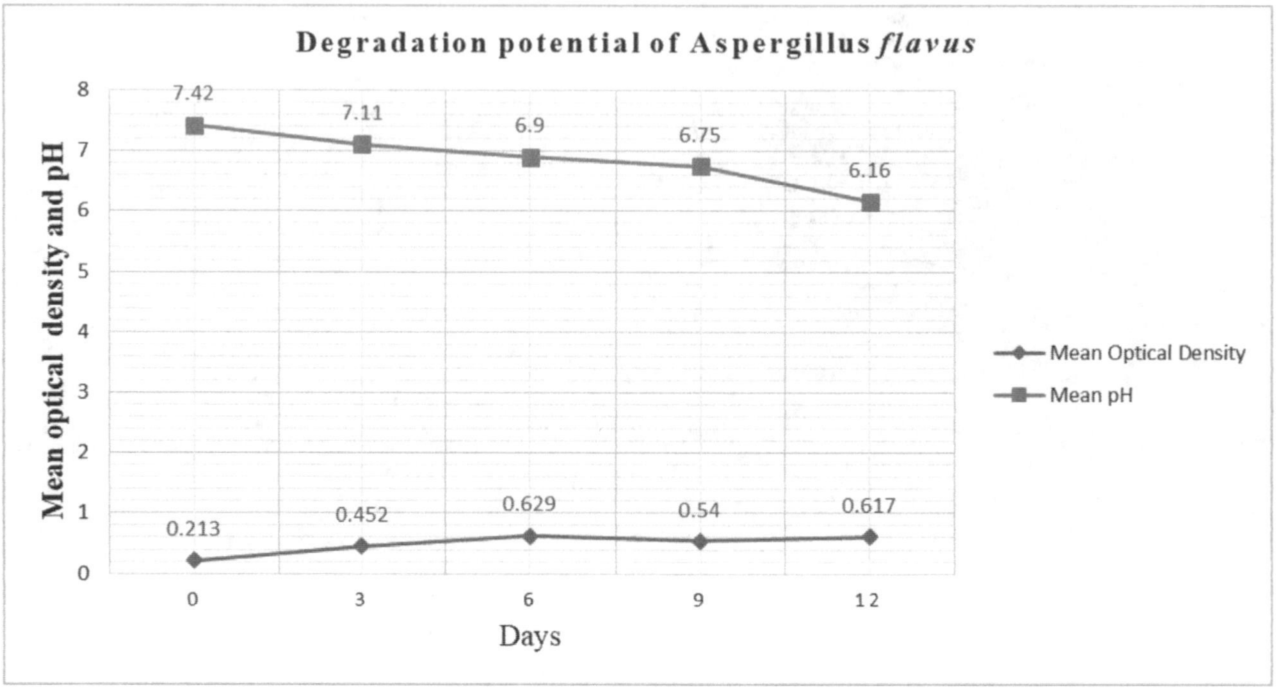

Figure 2.0: Data represent the Mean value of pH and Optical Density (OD) against different time

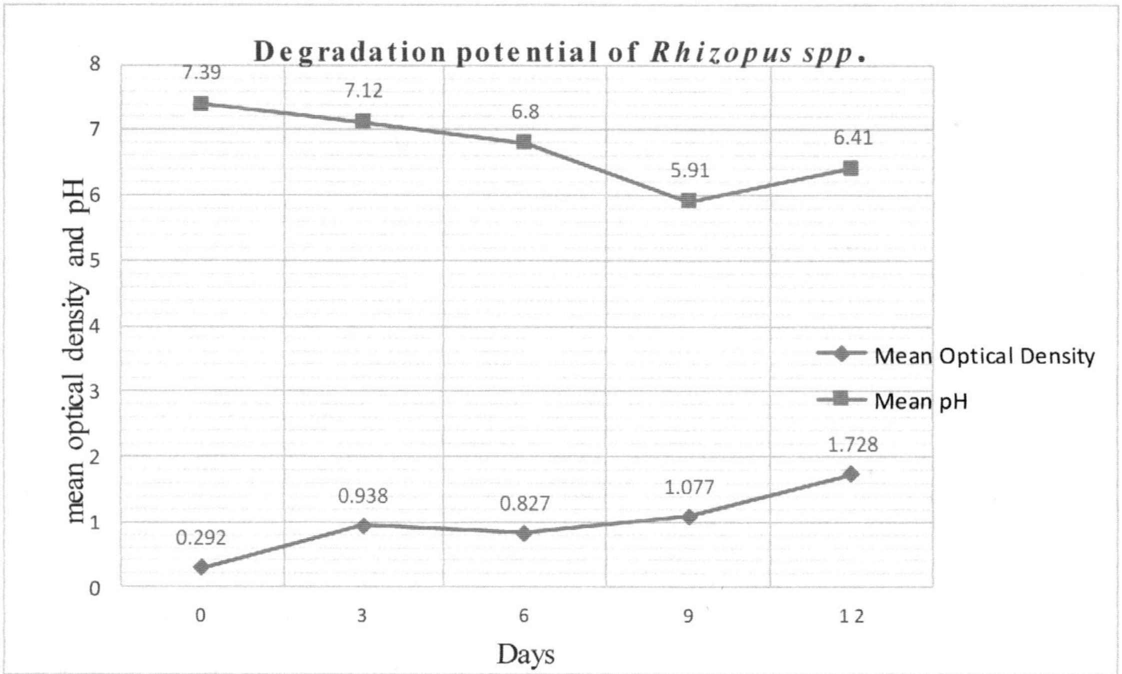

intervals with the same concentration of spent engine oil (2ml) for 12 days by *Aspergillus flavus*
Figure 3.0: Data represent the mean value of pH and Optical Density (OD) against different time intervals with the same concentration of spent engine oil (2ml) for 12 days by *Rhizopus spp*.

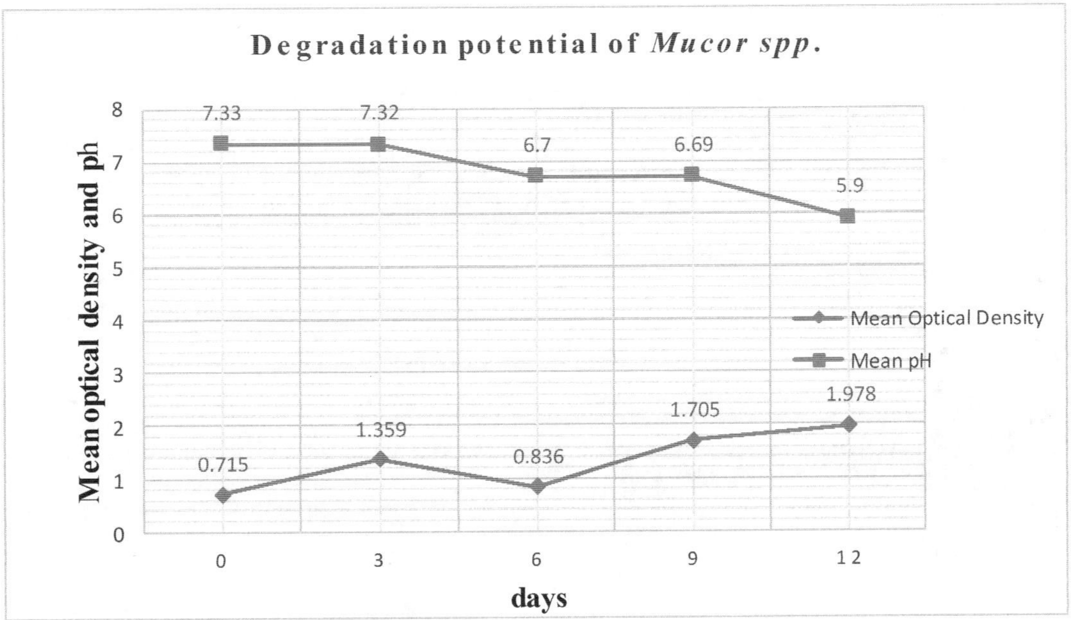

Fig 4.0: Data represent mean value of pH and Optical Density (OD) against different time intervals

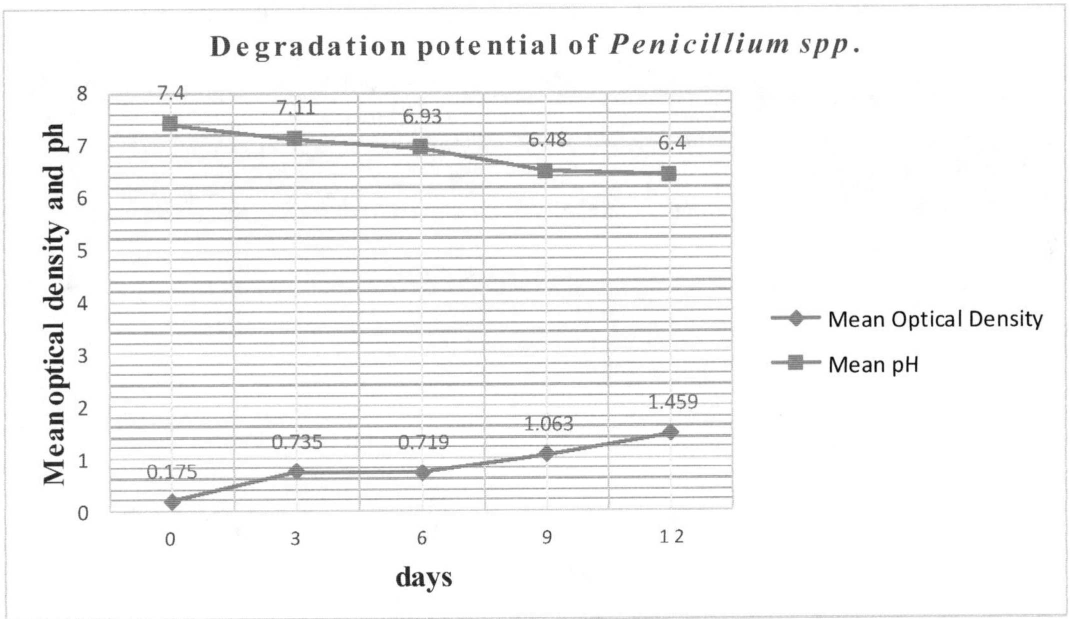

with the same concentration of spent engine oil (2ml) for days by *Mucor spp*.

Figure 5.0: Data represent the mean value of pH and Optical Density (OD) against different time intervals with the same concentration of spent engine oil (2ml) for 12 days by *Penicillium spp*.

IOP Conf. Series: Earth and Environmental Science **331** (2019) 012059 doi:10.1088/1755-1315/331/1/012059

Bar chat and Area chat showing the differences in moisture content and pH values respectively of the amended soil and the control.

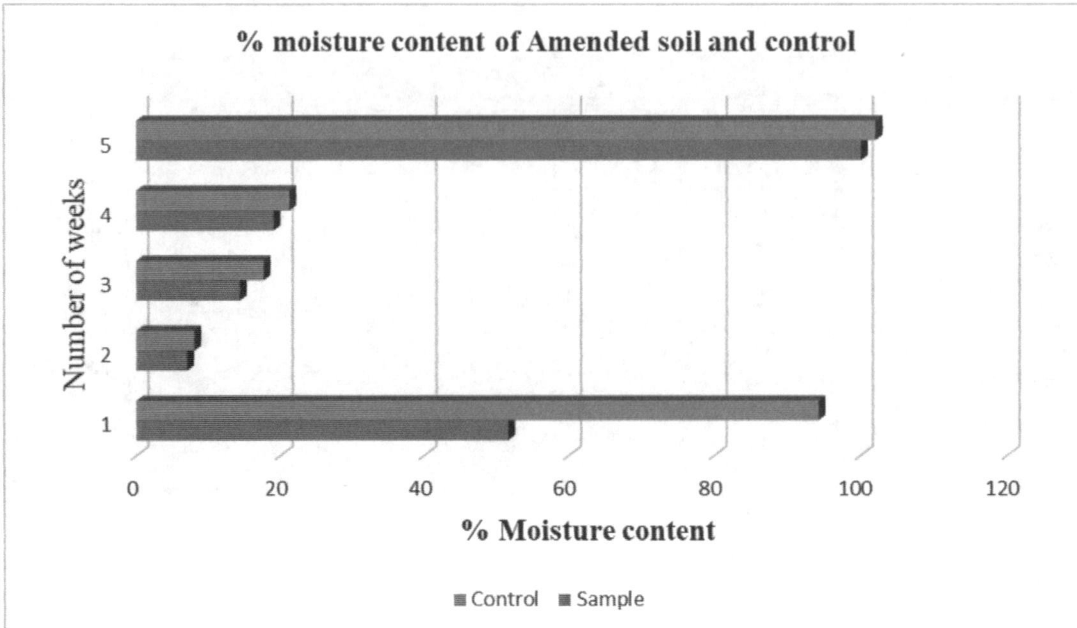

Figures 6.1: Bar chart showing variation in % moisture content of amended soil and contaminated soil.

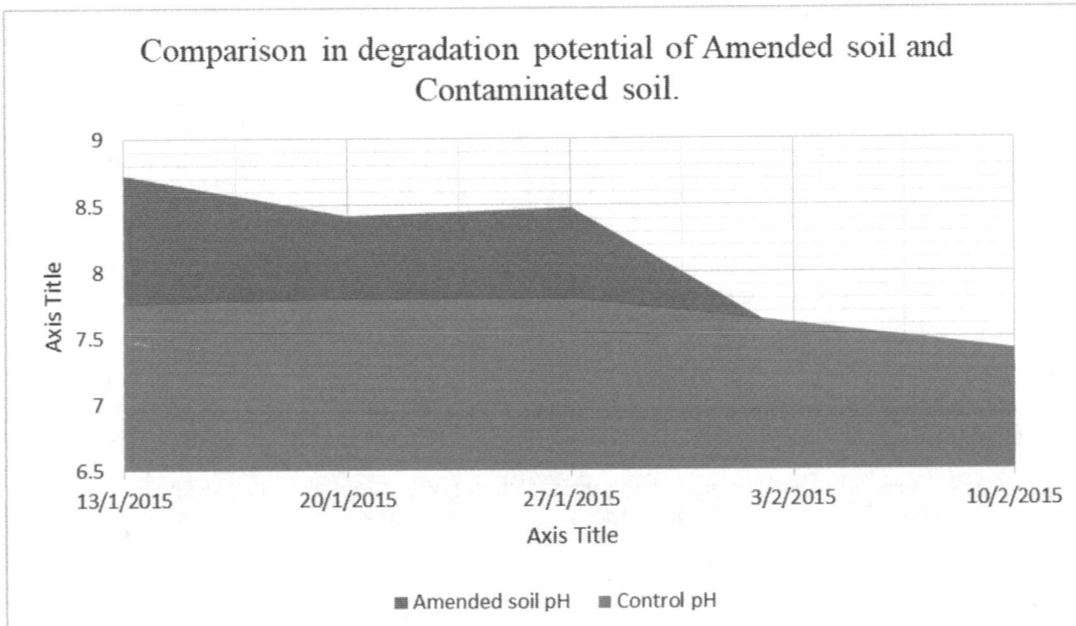

Fig 6.2: Bar chart showing the disparity in the degradation potential of amended soil and contaminated soil taking decrease in pH as an indicator of Degradation.

From the field study, it was observed from the moisture content and pH fluxes, that the amended soil showed evidence of the disappearance of the hydrocarbons. The decrease in pH observed for the contaminated soil could be due to the presence of microbial flora capable of utilizing waste-lubricating oil (Fig 6.1 and 6.2). This observation is in line with the reports of Ref [22] that opined that when hydrocarbon occur in the ecosystem, with the right micro-organisms capable of utilizing the hydrocarbons and the metabolites produced, there will be disappearance of the hydrocarbons. The moisture content recorded as shown in Fig 6.1 did not give clear evidence of degradation probably due to variations in the edaphic factors and the limitations on the indirect method of assessment. In this study, it was obvious that the fungal species *Penicillum, Mucor, Rhizopus, Aspergillus* species showed potential to utilize spent oil when the fluxes of pH and optical readings were measured (See plates 1-5). Okerentugba and Ezeronye (2003), reported that *Penicillium spp., Aspergillus spp., and Rhizopus spp.* isolated were capable of degrading hydrocarbons especially when single cultures were used. Furthermore, Batelle (2000) showed that fungi were better degraders than the traditional bioremediation techniques including bacteria. Different species of fungi respond differently to hydrocarbon sources as sole sources of carbon. The fungal species identified were *Aspergillus flavus, Aspergillus niger, Muscor sp., Rhizopus sp.* and *Penillicium sp.*
The (Figs 1-5), demonstrate the mean changes in the pH and optical density of spent engine oil degradation study by the fungi species. From the assessment of the fungal growth (increase in biomass) using the optical density at 600nm, it showed that the fungi species had different physiological dynamics. The results obtained from this work showed that the fungal species isolated from Sawdust (mostly considered as waste of the wood mill industries) were capable of degrading spent engine oil. *Apergillus flavus* showed a mean pH value of (7.42-6.16) and the mean values of OD (0.213-0.617) for the 12day incubation period. *Penicillium sp.* showed decreasing mean pH values of (7.40-6.40) and an increasing mean OD values of (0.175-1.457) within the 12 days incubation period. This result agrees with previous studies on *Aspergillus species* in utilizing hydrocarbons with the

resultant releases of CO_2, water and energy required to create cellular biomass (Thenmozhi et al., 2013). This explains the reason for the decrease in pH values as CO_2 and H_2O forms weak acid.
From the onsite treatment, there was as slight reduction in soil pH which was insignificant compared to the control. Similar observations were made in the reports (Atuanya, 1987; Osuji and Nwoye, 2007). Petroleum hydrocarbon mediated decrease in soil pH has been attributed to the production of organic acids by microbial metabolism (Osuji and Nwoye, 2007).The strain identified as *Mucor sp.* showed a decreasing pH of 7.33-5.90 and increasing OD of 0.715- 1.978 (Fig.4.0). In (Fig.1.0), *Aspergillus niger* showed a pH of 7.43 – 6.50 and OD value of 0.292- 1.731. The *Rhizopus sp.* exhibited a decreasing pH values of 7.39- 6.41 and the OD 0.292- 1.728 at the same incubation periods (Fig. 3.0). The differences obtained in the readings were probably due to the difference in growth rates of each fungus and the continuous increase in the optical density means exponential growth of the fungal isolates and their ability to utilize spent as a sole carbon source.

4. Conclusion

This study have shown that *Penicillum sp.*, *Aspergillus niger*, *Mucor sp.*, and *Rhizopus* species have the potential of utilizing waste lubricating oil as carbon and energy sources. The observed significant increases in the turbidity may likely be explained by increase in cell growth (cell biomass) of the fungal organisms when compared with the controls. This showed that the fungal species have the potentials to access energy from the hydrocarbons. Also, the decrease in pH may be due to acidic products of the degradation of spent engine oil.

Acknowledgements
The authors would like to acknowledge Covenant University, Ota, Nigeria for defraying the cost of this publication.

References

[1]. Adekunle, O. and Adeduntan, S. (2003). Challenges Environmental Sustainability in democratic government proceeding of the 11th annual conference of environmental and behaviour association of Nigeria, pp. 279-288.

[2]. Adenipekun, O., Oyetungi, O.S. And Kassim, C.Q. (2008). Effect of spent engine oil on the growth parameters and chlorophyll content of *Corchorus olitonus linn. Environmentalist* 28:446-450.

[3]. Akowuah, G.A., Zhari, I. and Norhayati, I. (2004). Sinensetin, eupatorin, 3′-hydroxy-5, 6, 7, 4′-tetramethoxyflavone and rosmarinic acid contents and antioxidative effect of *Orthosiphon stamineus* from Malaysia. Food Chem 87: 559– 566.

[4]. Anjali, V., Mamta, K., Namita, D. and Nandkishor M. (2014). Phytoremediation Potential of *Phragmites Karka* for Arsenic Contaminated Soil and Water. *IOSR Journal of Environmental Science, Toxicology and Food Technology*, **8**(1):84-92.

[5]. Anoliefo, G.O. and Edegbai B.O. (2000). Effect of spent engine oil as a oil contaminant on the growth of two eggplant species; *Solanum melongena L.* and *S. incanum*. *J. Agric. Forestry Fisheries,* **1**: 21-25.

[6]. Anoliefo, G.O. and Vwioko, D. E. (1995). Effects spent lubricating oil on the growth of *Capsicum annum L.* and *Lycopersicon esculentum* Miller. *Environmental Pollution*, **88**: 361-364.

[7]. Anoliefo, G. O. and Vwioko, D.E. (2001). Tolerance of *Chromolaena odorata* grown in soil contamination with spent lubrication oil. *Journal of Tropical Biosciences*, **1**:20-24.

[8]. Anon. (2012). Choosing a soil amendment. Retrieved on October 30, 2013. Available on www.extcolostate.edu/pubs/garden.

[9]. Anon. (2013). Definition and meaning of crude oil. Retrieved June 20, 2013. Available in www.businessdictionary.com/definition.

[10]. Atlas, R.M. and Bartha, R. (1992). Hydrocarbon biodegradation and oil spill bioremediation. *Adv. Microb. Ecol.* **12**:287-338.

[11]. Atlas, R.M. (1985). Effects of Hydrocarbons on microorganisms and bioremediation in Arctic ecosystems, in Petroleum Effects in the Arctic Environment, F.R. Engelhardt's Ed., pp. 63-99, Elsevier, London, UK.

[12]. ATSDR (Agency for Toxic Substances and Disease Registry), 1997. Toxicology. Profile for used mineral base crankcase oil. Department of Health and Human Services, Public Health Service Press, Atlanta, GA, USA.

[13]. Atuanya, E.J. (1987). Effect of oil pollution on physical and chemical properties of soil: a case study of waste oil contaminated delta soil in Bendel State, Nigeria. *Journal of Applied Sciences*, **55**:155-176.

[14]. Bannet, H.C. and Hunter, B.B. (1972). Illustrated genera of Imperfect Fungi, 3rd edition. Minneapolis Burgress Publishing Company Minneapolis.

[15]. Battelle, C.D. (2000). Mushrooms: Higher Macrofungi to clean up the environment. *Batelle Environment Issues http://www.clu-in.org/ products/tins/tinsone.cfm.accessed on February 2014.*

[16]. Dominguez-Rosado, R.E. and Pichtel, J. (2004). Phytoremediation of soil contaminated with used motor oil; enhanced microbial activities from laboratory and growth chamber studies. *Environ. Eng. Sci.*, **2**:157-168.

[17]. Fouépé, T. A., Kengni, T. L., Gurunadha Rao, V. V. S. and Ndam, N. J. R. (2009). Transfer of moisture through the unsaturated zone in the tropical forest using the neutron probe. *International Journal of Environmental Science and Technology*, **6**(3):379-388.

[18]. Molla, B., Salah, W., Alemayehu, D. and Mohammed, A. (2004). Antimicrobial resistance pattern of Samonella serovars isolated from apparently healthy slaughtered

camels (*Camelus dromedrius*) in eastern Ethiopia. *Berliner und Munchener Tierarztliche Wochenschrift.*117:39-45.

[19]. Nilanjana, D. and Preethy, C. (2010). Review Article Microbial Degradation of petroleum Hydrocarbon contaminants: An Overview. *Biotechnology Research International* 2011:1-13.

[20]. Nwinyi, O.C., Okonkwo, C., Nwinyi, C.E., Ajanaku, K. O., Siyanbola, T. O., Ogunniran, K. O., Ehi-Eromosele, C. O., Akinsiku, A. A. and Ayano, T. (2014). Evaluating the diesel biodegradation potential of wild microfungi isolated from decaying wood in Nigeria. *South Asian Journal of Experimental Biology,* **4** (3):110-117.

[21]. Nwinyi, O. C., Olawore Y. A (2017). Biostimulation of spent engine oil contaminated soil using Ananas comosus and Solanum tuberosum peels. *Environmental Technology and Innovation* **8**: 373-388.

[22]. Nwoko, C.O., Okeke, P.N., Agwu, O.O. and Akpan, I.E. (2007). Performance of *Phaseolus vulgaris L.* in a soil contaminated with spent engine oil. *African Journal of Biotechnology,* **6** (16): 1922-1925.

[23]. Nwuche, C. O. and Ugoji, E. O. (2008). Effects of heavy metal pollution on the soil microbial activity. *Int. J. Environ. Sci. Tech.*, **5** (3):409-414.

Optimization of solar photovoltaic efficiency using angular inclination: Effect on voltage output

Emetere M.E.[1,2], and Elumole O..[1]
[1]Department of Physics, Covenant University Canaan land, P.M.B 1023, Ota, Nigeria.
[2]Department of Mechanical Engineering and Science, University of Johannesburg, APK, South Africa

E-mail: emetere@yahoo.com

Abstract. The physical optimization processes of maximizing the solar photovoltaic (PV) panel via the angular adjustment of PV inclination to the sun were examined. The efficiency may vary within geographic locations. In this study, the measured dataset of a location in the tropical region of Nigeria was examined with the inclination ranging from 0° to 45°. It was observed that at 30° in the south –east direction yielded a more stable result that is essential basically for the stability of the solid state content of the PV panel. Though there were spikes of higher voltage in the 30°, it is recommended that PV users in the region to adopt 30° to enhance efficiency and longevity of the PV panel.

Keywords: energy, solar, voltage output, solar energy, renewable energy

1. Introduction

Over the years, the need for the generation of power has given rise to various means of harnessing energy. The implementation of windmills, gas turbines, hydroelectricity and so much more have become answers to the provision of the highly sought power. The adoption of renewable energy in low-income countries is encouraging because energy source never gets depleted. A good example is solar energy -energy that never runs out, readily available as soon as the sun rises. The photovoltaic cells are a solar-based device that converts solar energy into electricity [1-2].

The use of solar power has spread so wide that it is now used as an alternative source of power in workplaces, schools and so on. It is used to power large areas via solar farms (an area also known as a power station where a large-scale photovoltaic system is designed to supply merchant power into the electricity grid). These solar farms need to be put under observation so that there are no fluctuations in the output power generated. Manually testing each panel to determine its efficiency can be very tedious, and this is where the essence of the data logger comes in. The data logger is designed to automatically detect the current and voltage a panel is generating under hot, cold or humid conditions and determines when the efficiency of a panel reduces by analysing its output current and voltage [3-4].

The photovoltaic effect is a process which, when exposed to sunlight, produces voltage or current in a photovoltaic cell. It is composed of solar cells or different semiconductors, the p - type and the n - type semiconductorare combined to create a p - n junction. By joining these semiconductors, in the junction region where electrons move to the p-side and holes move to the n-side, an electric field is formed. The field causes particles positively charged to move in one direction and particles negatively charged to move in the other. Light comprises of photons that are electromagnetic radiation quantities [5]. A solar cell can absorb these photons. In the p-n junction, energy from the photon is transferred to an atom of the semiconducting material when light of an appropriate wavelength is incident on these cells. This induces the electrons to jump to a higher state of energy called the conduction band. This leaves a "hole" behind in the valence band from which the electron jumped. Electrons hold the semiconducting material together when unexcited by forming bonds with surrounding atoms that disable their movement. However, these electrons are free to move through the material in their excited state in the conduction band. Electrons and holes move in the opposite direction because of the electrical field that exists as a result of the p-n junction. The freed electron tends to move to the n-side rather than being attracted to the p-side [6]. This electron motion creates in the cell an electrical current. There's a "hole" left after the electron moves. This hole can move as well, but it moves to the p-side instead of moving to the n-side. It is this process that creates a cell current.

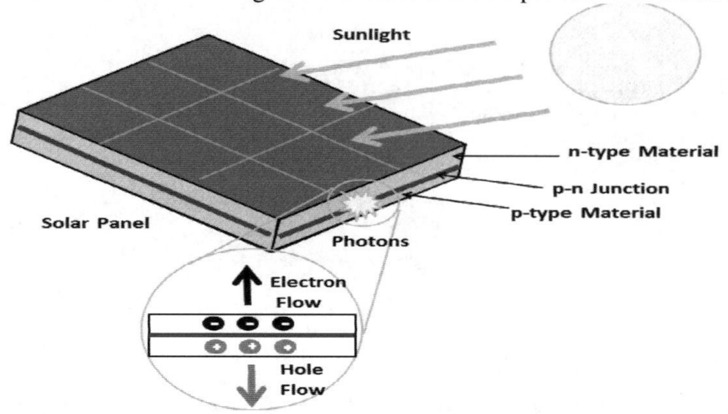

Figure 1: visual representation of the photovoltaic effect [7]

The focus of this study is to determine the best direction for optimal solar energy generation. From literature, we understand that the best direction is the south-east direction [8]. Hence, the angular inclination of the solar cell was investigated with the aim of determining the most stable voltage generation direction to elongate the lifespan of the PV panel.

2. Experimental Design, Materials and Methods

A manual logger was designed to measure the voltage and current simultaneously [3]. A polycrystalline 10V solar panel was used for the experiment. The PV panels were connected to the logger facing the south-east direction. The direction was determined using a GPS hand-held device. The angle of inclination was 0°, 30° and 45° as presented in figures 2-4 below. The dataset was processed using the MATLAB software.

Figure 2: Condition 1 (0°)

Figure 3: Condition 2 (30°)

Figure 4: Condition 3 (45°)

3. Results and Discussion

The current versus time plot for 0° is presented in Figure 5. It is observed that the current generation was optimal at great part of the day. However, the current drop suggests that the solar radiation dropped.

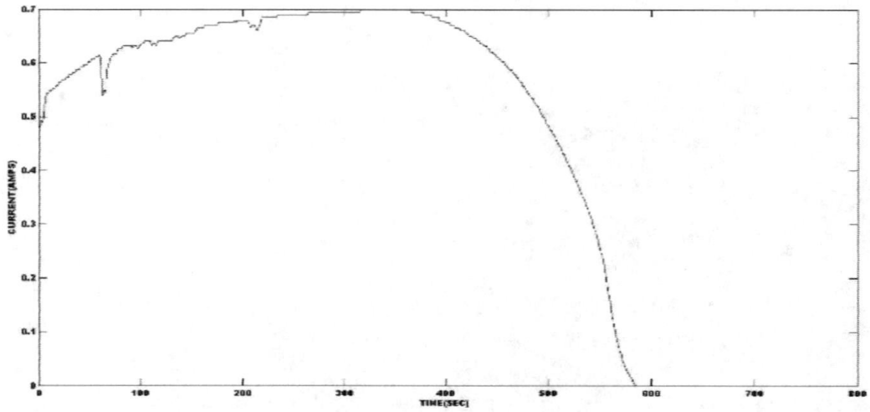

Figure 5: graph of current against time for 0◦

The parabolic nature of the plot shows that current depreciates at higher time due to solar irradiance of the geographical region and the time of the day. The voltage generated is presented in Figure 6.

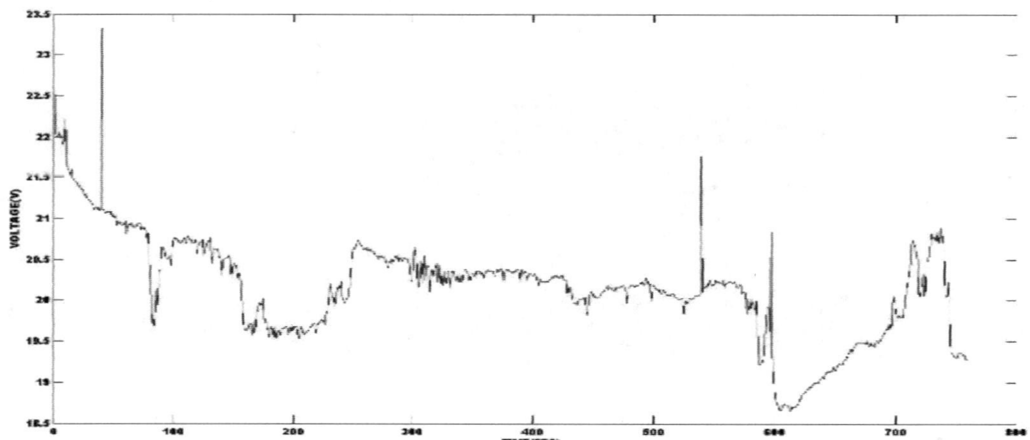

Figure 6: graph of voltage against time for 0◦

The graph in Figure 6 shows the plot of voltage against time, the spectral nature of the graph simply shows the rise and fall in the voltage of the panel as affected by the rise and fall of solar irradiance. This scenario affects the solid-state arrangement of the PV panel [8].

The current versus time plot for 30◦ is presented in Figure 7 below. The graph in Figure 7 above shows the graph of time against current that exhibits a step sinusoidal form. The form taken by this graph indicates that at certain points in time, current was constant.

IOP Conf. Series: Earth and Environmental Science **331** (2019) 012060 doi:10.1088/1755-1315/331/1/012060

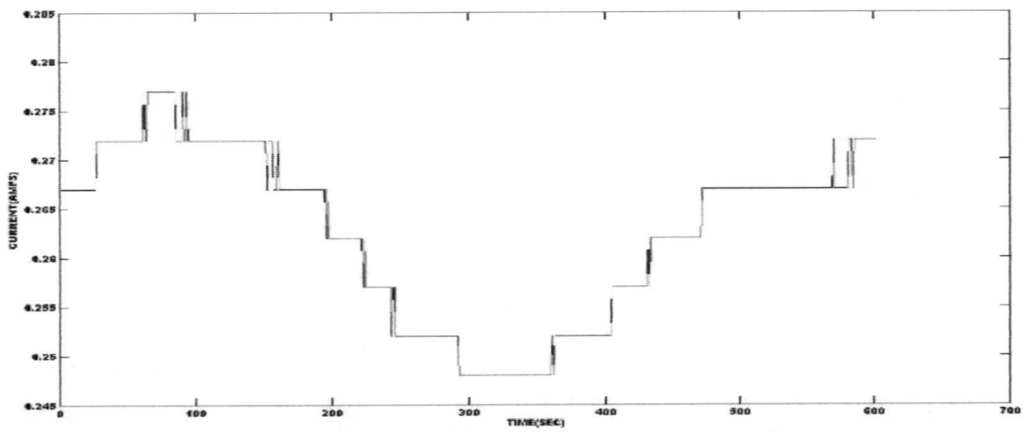

Figure 7: graph of current against time for 30°

The voltage generated is presented in Figure 8. This plot indicates that the voltage rises and drops considering factors such as solar irradiance, temperature of the panel and time of day. The voltage production was almost uniform for tewo-third of the experimentation time. This result is particularly interesting because the solid-state operation in the PV panel would not be over-burdened by abnormal excitations.

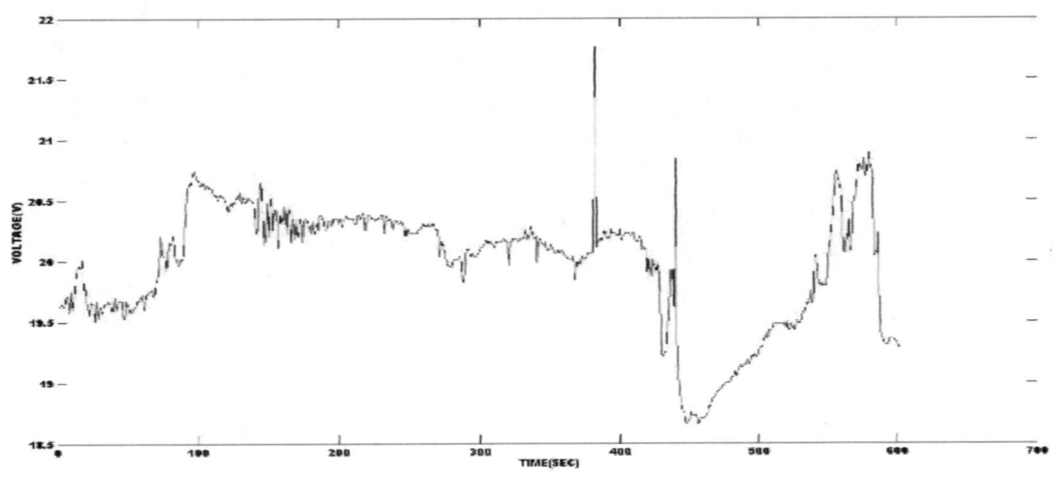

Figure 8: graph of voltage against time for 30°

The current versus time plot for 45° is presented in Figure 9. It shows the spectral form that explains that current rises and drops at certain points in time due to variations in solar irradiance.

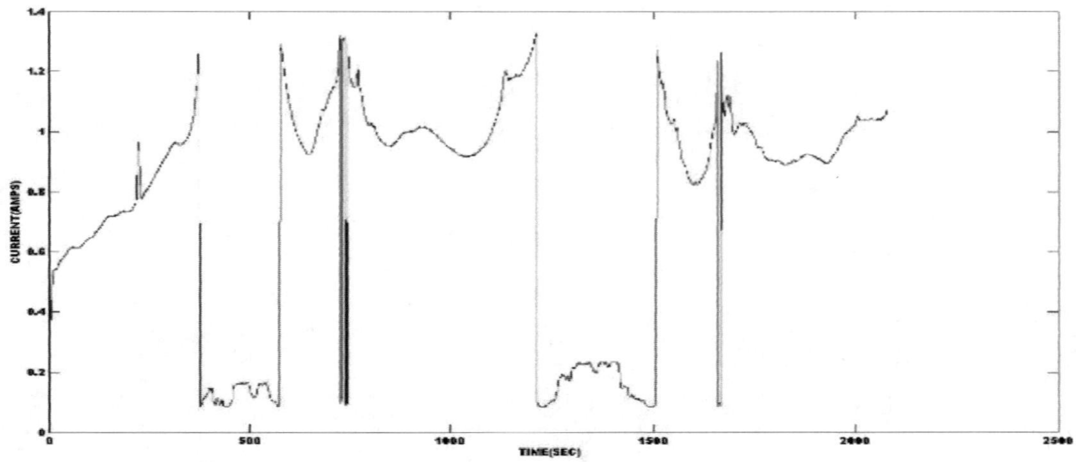

Figure 9: graph of current against time for 45◦

This emerging current production, leads to the voltage production as presented in Figure 10. Aside that the spectral form of this graph indicates the rise and drop of the panel voltage as it reacts to solar irradiance, it is interesting to not that the abnormal voltage spikes is dangerous for the PV panel.

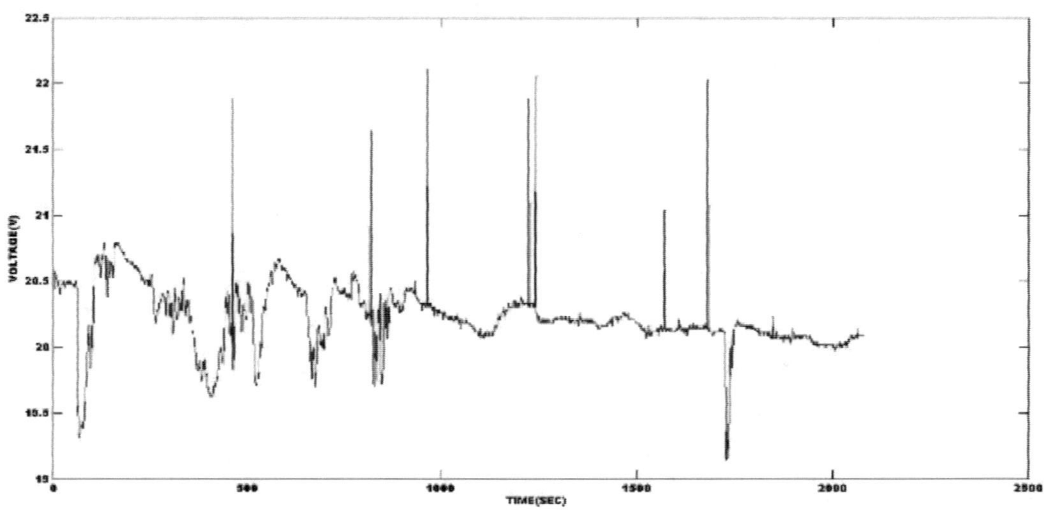

Figure 10: graph of voltage against time for 45◦

4. Conclusion

From the experimentation, it was proven that the maximum voltage in a solar PV panel is not as important to the longevity of the panel. The efficiency of a solar panel should be measured with respect to the voltage stability and the duration of voltage stability. It was observed that at 30◦ in the south –east direction yielded a more stable result that is essential basically for the stability of the solid state content of the PV panel. Though there were spikes of higher voltage in the 30 ◦, it is recommended that PV users in the region to adopt 30° to enhance efficiency and longevity of the PV panel.

Acknowledgement
The authors appreciate Covenant University for partial sponsorship. The authors acknowledge NASA for primary dataset.

5. References

[1]. Moses E. Emetere, Marvel L. Akinyemi, and Etimbuk B. Edeghe, (2016). A Simple Technique for Sustaining Solar Energy Production in Active Convective Coastal Regions, International Journal of Photoenergy 2016, 3567502, 1-11, http://dx.doi.org/10.1155/2016/3567502
[2]. Emetere M.E., & Akinyemi M.L., (2015) Prospects of Solar Energy in the Coastal Areas of Nigeria, 2015 PIAMSEE: AIP Conference Proceedings, 1705 (1), 020035
[3]. K. N. Akposionu, and A. O. C Nwokoye (2012), Design and Fabrication of A Low-Cost Data Logger for Solar Energy Parameters, Journal of Energy Technologies and Policy, Vol.2, No.6, ISSN 2224-3232
[4]. Pratik Painjan, Rahul Nangare, Anupsing Pardeshu, "Design of Solar Insolation Level Detector and Data Logger", International Journal of Current Engineering and Technology E-ISSN 2277-4106, P-ISSN 2347-5161.
[5]. Cousera, https://www.coursera.org/lecture/solar-cells/semiconductors-udcqR (accessed 4/06/2019).
[6]. A.A.F. Husain, W.Z.W. Hasan, S. Shafie, M.N. Hamidon, S.S. Pandey, A review of transparent solar photovoltaic technologies, Renew. Sustain. Energy Rev., 94 (2018), pp. 779-791, 10.1016/j.rser.2018.06.031
[7]. Solar education (2012). https://energyeducation.ca/encyclopedia/Main_Page (accessed 4/06/2019).
[8]. M. Guo, H. Zang, S. Gao, T. Chen, J. Xiao, L. Cheng, Z. Wei, G. Sun, Optimal tilt angle and orientation of photovoltaic modules using HS algorithm in different climates of China, Appl. Sci., 7 (2017), p. 1028

Optimization of solar photovoltaic efficiency using angular inclination: Effect of dust deposition on PV Module

Emetere M.E.[1,2], **and Elumole O.**[1]
[1]Department of Physics, Covenant University Canaan land, P.M.B 1023, Ota, Nigeria.
[2]Department of Mechanical Engineering and Science, University of Johannesburg, APK, South Africa
E-mail: emetere@yahoo.com

Abstract. The West African belt is known for Sahara dust influx that is a great concern as it increases the aerosol loading over the region. In this study, we examine the effect of dust deposition on solar photovoltaic (PV) panels. A local logger was constructed to determine the dust deposition and its corresponding current generation in a PV panel. The PV panel were placed at different inclination i.e. ranging from 0° to 45°. It was observed that at 45°, the dust deposition was higher than other PV angle inclination because the PV surface was exposed to both horizontal and vertical dust deposition.

Keywords: energy, solar, solar radiation, meteorology, solar energy, renewable energy

1. Introduction

Solar radiation can be captured and turned into useful forms of energy, such as heat and electricity, using a variety of technologies. Photovoltaic (PV) modules are used to capture the energy from the sun and convert it to useful electricity. Solar power could be considered as clean energy for the future as it possesses such interesting operation features such as: being generated from a clean and renewable source of energy; reduced or no harmful emissions released into the atmosphere; and its ability to generate electricity without needing fuel, it has no variable cost therefore it is inexpensive [1-2].

A solar cell is an electronic device that converts sunlight directly into electricity. Light on the solar cell produces a current as well as a voltage to generate electricity. This process requires, first, a material in which light absorption raises an electron to a higher energy level and Secondly, the moving of this higher - energy electron from the solar cell to the outer circuit. In the external circuit, the electron then dissipates its energy and returns to the solar cell [3-4].

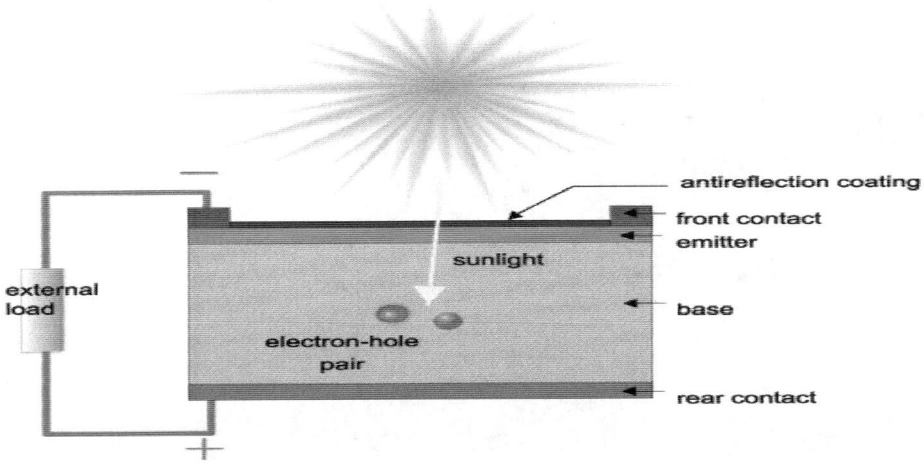

Figure 1: mode of operation of a photovoltaic cell [5]

Researchers in the past have shown that dust deposition on PV panels lower its efficiency [6-8]. The research area is under the influence of Sahara dust influx and other anthropogenic dust emissions [9-10]. The study is to determine if the angular inclination of the PV panel would either increase or decrease dust deposition.

2. Methodology

A manual logger was designed to measure the dust deposition and current simultaneously [11]. A polycrystalline 10V solar panel was used for the experiment. The PV panels were connected to the logger facing the south-east direction. The direction was determined using a GPS hand-held device. The angle of inclination was 0° , 30° and 45° as presented in figures 2-4 below. The dataset was processed using the MATLAB software.

Figure 2: Condition 1 (0°)

Figure 3: Condition 2 (30°)

Figure 4: Condition 3 (45°)

3. Results and Discussion

The current versus time plot for 0° is presented in Figure 5. It is observed that the current generation was optimal at great part of the day. However, the current drop suggests that the solar radiation dropped. It is observed that when the dust deposition was high (Figure 6), the current production drops. This is against the original understanding that lower solar irradiation is directly proportional to current production [1]. The current versus time plot for 30° is presented in Figure 7 below. The graph in Figure 7 above shows the graph of time against current that exhibits a step sinusoidal form. The form taken by this graph indicates that at certain points in time, current was constant.

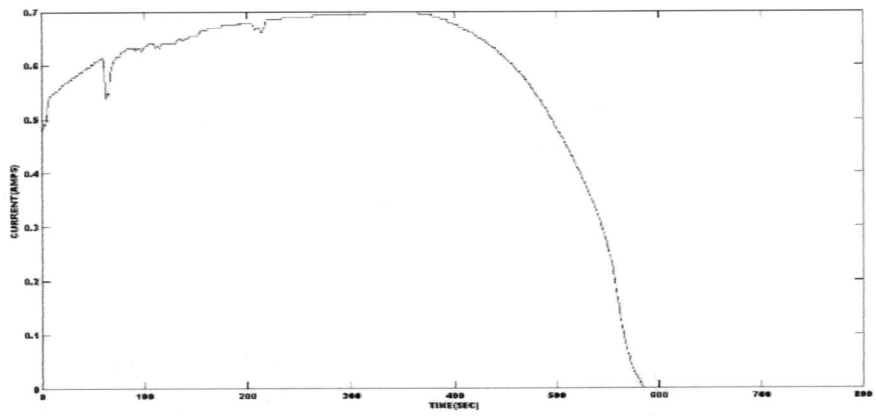

Figure 5: graph of current against time for 0°

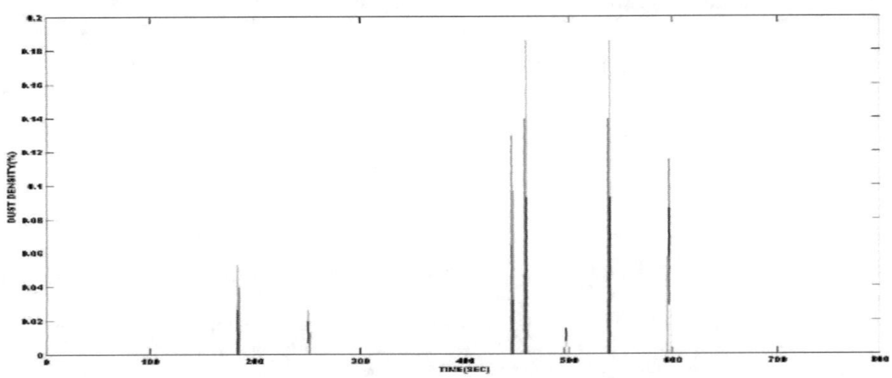

Figure 6: graph of dust density against time for 0°

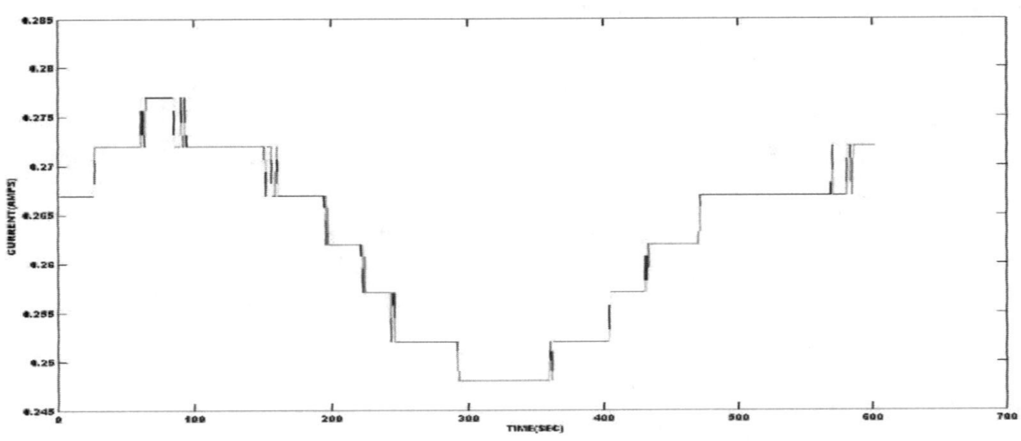

Figure 7: graph of current against time for 30°

As already discussed, the low dust deposition had higher current production. However, the dust deposition noticed towards the end of the experiment did not affect the current production. It can be inferred that in this peculiar case, the solar irradiance rise was higher than the early period of the experiment.

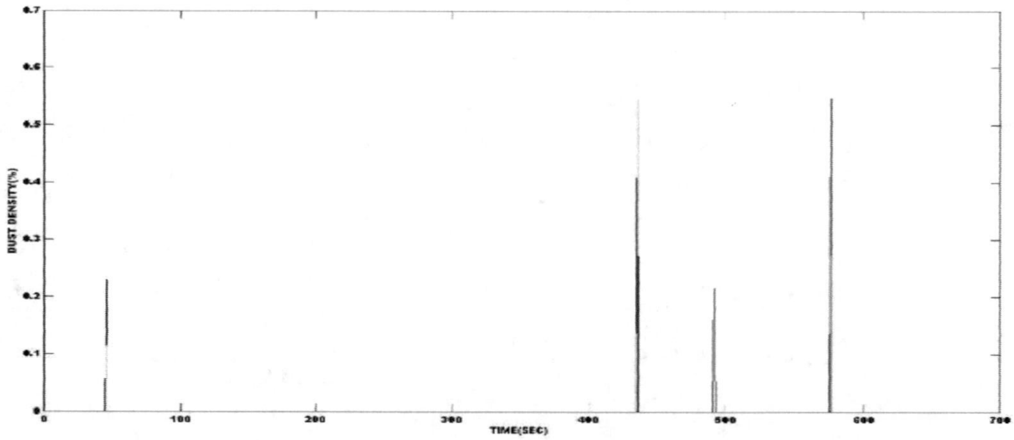

Figure 8: graph of dust density against time for 30°

The current versus time plot for 45° is presented in Figure 9. It shows the spectral form that explains that current rises and drops at certain points in time due to variations in solar irradiance. The performance of the current production can be traced to the dust deposition presented in Figure 10.

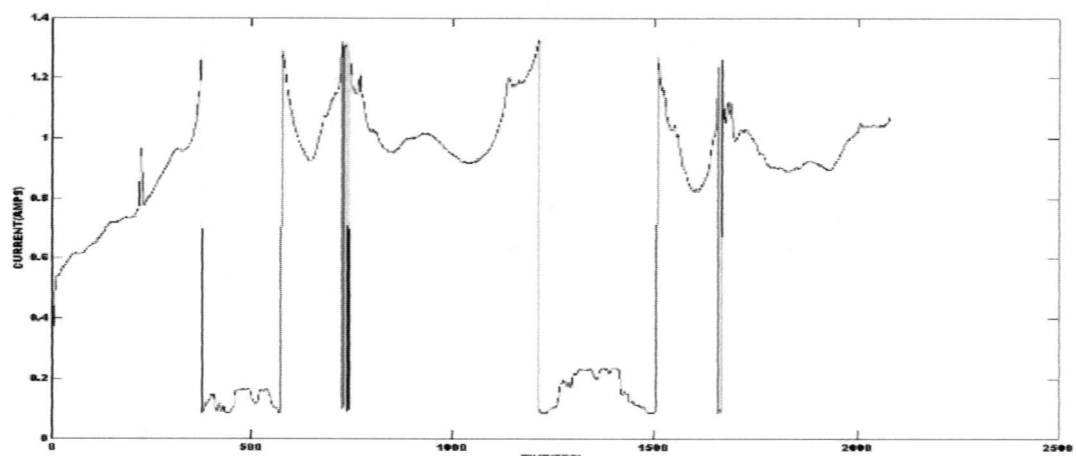

Figure 9: graph of current against time for 45°

Particularly, it is observed that when the PV panel is inclined at 45°, the dust deposition was higher than other PV inclination. In this case, the dust deposition may not be vertical only, other anthropogenic emissions from the horizontal direction were deposition on the PV panel.

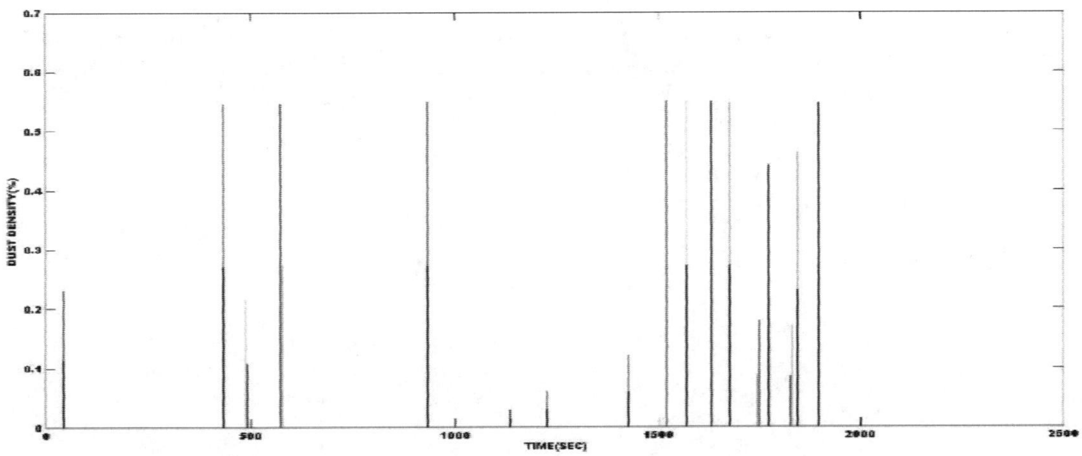

Figure 10: graph of dust deposition against time for 45°

4. Conclusion

From the experimentation, it was proven that dust deposition had significant influence on the current production of the solar PV panel. This is known fact already reported by scientist. In this case, we examined the best inclination that would have lower dust deposition. It was observed that at 45°, the dust deposition was higher than other PV angle inclination because the PV surface was exposed to both horizontal and vertical dust deposition..

Acknowledgement

The authors appreciate Covenant University for partial sponsorship. The authors acknowledge NASA for primary dataset.

5. References

[1]. Moses E. Emetere, Marvel L. Akinyemi, and Etimbuk B. Edeghe, (2016). A Simple Technique for Sustaining Solar Energy Production in Active Convective Coastal Regions, International Journal of Photoenergy 2016, 3567502, 1-11, http://dx.doi.org/10.1155/2016/3567502

[2]. Emetere M.E., & Akinyemi M.L., (2015) Prospects of Solar Energy in the Coastal Areas of Nigeria, 2015 PIAMSEE: AIP Conference Proceedings, 1705 (1), 020035

[3]. Cousera, https://www.coursera.org/lecture/solar-cells/semiconductors-udcqR (accessed 4/06/2019).

[4]. A.A.F. Husain, W.Z.W. Hasan, S. Shafie, M.N. Hamidon, S.S. Pandey, A review of transparent solar photovoltaic technologies, Renew. Sustain. Energy Rev., 94 (2018), pp. 779-791, 10.1016/j.rser.2018.06.031

[5]. Pveducation (2019). https://www.pveducation.org (accessed 4/06/2019).

[6]. M. El-Shobokshy, F. Hussein, Degradation of photovoltaic cell performance due to dust deposition on its surface, Renew. Energy, 3 (1993), pp. 585-590

[7]. D. Goossens, Z.Y. Offer, A. Zangvil, Wind Tunnel experiments and field investigations of aeolian dust deposition on photovoltaic solar collectors, Sol. Energy, 50 (1993), pp. 438-486

[8]. H. Jiang, L. Liu, K. Sun, Experimental investigation of the impact of airborne dust deposition on the performance of solar photovoltaic (PV) modules, Atmos. Environ., 45 (2011), pp. 4299-4304

[9]. M. E. Emetere & M.L. Akinyemi (2017) Atmospheric Dynamics of Air Pollution Dispersion and Sustainable Environment in Nigeria, Journal of Urban and Environmental Engineering, 11 (1): 51-57

[10]. Emetere Moses E., Akinyemi M.L. & Akinwumi S.A. (2016) Aerosols Loading Trends And Its Environmental Threats Over Cotonou-Benin, Nature Environment and Pollution Technology, 15 (3), 1111

[11]. K. N. Akposionu, and A. O. C Nwokoye (2012), Design and Fabrication of A Low-Cost Data Logger for Solar Energy Parameters, Journal of Energy Technologies and Policy, Vol.2, No.6, ISSN 2224-3232

On the need for the development of low wind speed turbine generator system

[1,2]**Oluseyi O. Ajayi**, [2]**Olorunfemi Ojo**, [3]**Ahmed Vasel**
[1]Mechanical Engineering Department, Covenant University, P.M.B. 1023, Ota, Ogun State, Nigeria
[2]Electrical and Computer Engineering Department, Tennessee Technological University, Cookeville, Tennessee, USA
[3]Mechanical Engineering Department, Tennessee Technological University, Cookeville, Tennessee, USA
Correspondence: oluseyi.ajayi@covenantuniversity.edu.ng; Jojo@tntech.edu

Abstract.The study focused on the need for the development of a low cut-in wind speed turbine system with a view to challenge for concerted efforts towards improvement in design in such a way that favours operation at wind speeds as low as 1.5 to 2.0 m/s. It reviewed some existing reports on rotor designs with major focus on improved blade design for enhanced turbine performance and structural integrity. It found that despite the advances in turbine blade designs, there is the need to further the design to include those that can drive generators not just for small wind turbine applications but also for operation under lower cut-in speeds (less than 2.0 m/s). It further demonstrated the need to have improvement in wind turbine generator design that eliminates/reduce lubrication parts and possibly employ electro-magneto-mechanical principles for operation.

Key words: Wind Power, Rotor Design, Wind Turbine, Turbine Blade Design, Generator

1. Introduction

Wind power technology has over the years gone through several developmental changes. From the use of wind energy as windmills over 2000 years ago, it has grown to become pivotal in the power sector and engaged for the production of gigawatts of electricity across the globe. Today, wind power technology is one of the growing electricity resource technology in China, the United States of America and Europe. Global cumulative installed wind power capacity has constantly increased over the last two decades. Fig. 1 displays the trend of increase. It shows that there is a polynomial increase in the additions of new turbine plants to wind energy markets per year. Further analyses show that the annual growth in capacity averages about 22% within the last two decades while it is likely that the cumulative increase will exceed one terawatts (i.e. 1.0 TW) in the next three to five years. Based on this, some countries have contributed tremendously to this achievement. For instance, in the whole of Europe, five countries accounted for 68% of cumulative installed wind power capacity as at the year 2018. Germany, Spain and the United

Kingdom (UK) are the first three nations with 59,311 MW, 23,494 MW and 20,970 MW respectively, while France (with 15,309 MW) and Italy (with 9,958 MW) are the fourth and fifth in the rank [1]. Of this lot, Germany and Spain's percentage contribution to Europe's wind power capacity as at 2018 accounted for 31% and 12% respectively. The total for Europe was 189,229 MW in 2018. However, of this total capacity, onshore wind installations dominate across Europe. The percentage contribution of onshore wind turbine capacity for electricity generation is 99.1% [1]. This shows that adoption of offshore installations still has a long way ahead to further its development.

China, the United States of America (USA), India, Brazil and Canada are other five leading nations, outside the top five of Europe, that are making good contributions to global wind power capacity. For example, China's total installed wind power capacity as at the year 2017 stood at 188,232 MW. This figure competes closely with that of the whole of Europe. USA on the other hand, has total installed capacity that is far above that of the best of Europe. Its value as at 2017 was 89,077 MW. Fig. 2 shows the contribution of the top five countries outside Europe and reveal that the values for the rest of the world excluding the top ten countries enumerated above was 83,008 MW as at 2017. Other nations are equally making developmental progress towards the adoption of wind energy utilities for power generation.

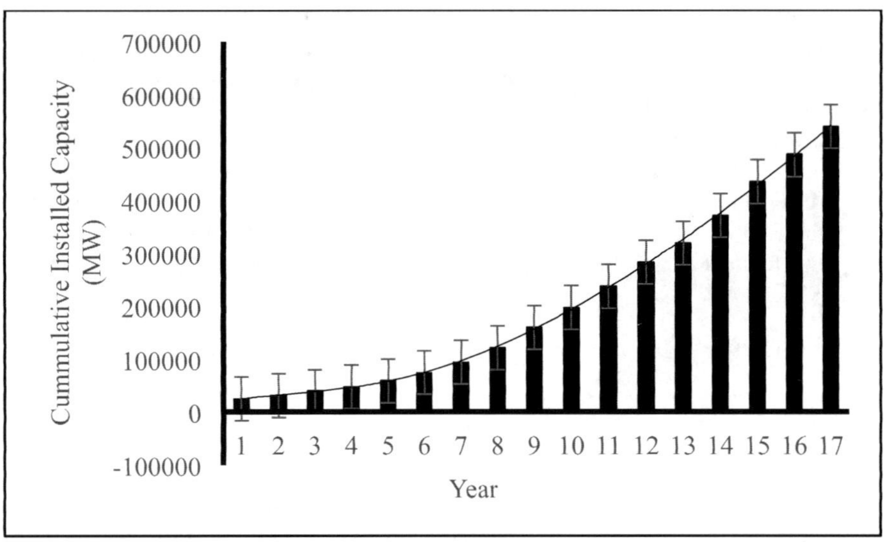

Fig. 1: Global cumulative installed wind power capacity [2]

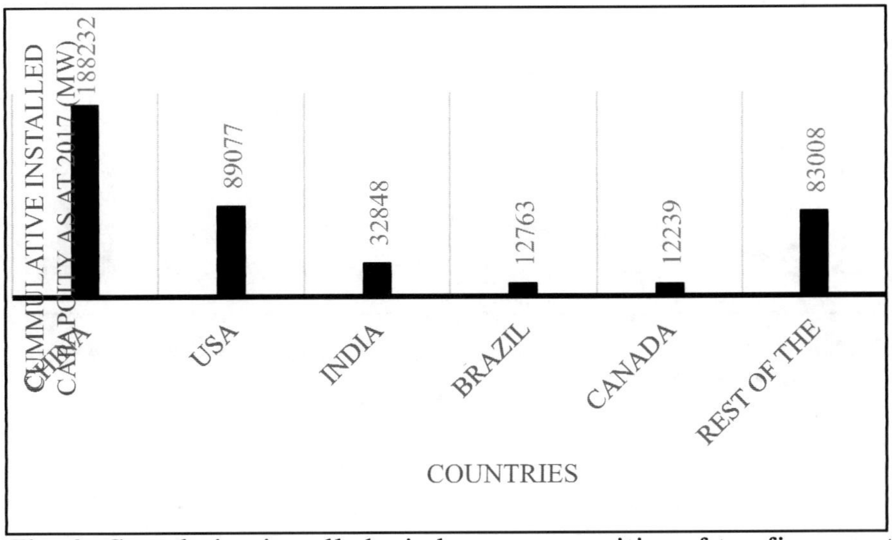

Fig. 2: Cumulative installed wind power capacities of top five countries outside Europe [2]

2. Advances in wind turbine design and development

Despite the progress in wind power industry, there are limiting factors, which have slowed/hindered a heightened growth of the wind energy market. These factors include high initial costs of hardware and installation. A few years ago, precisely between 2001 and 2012, a 1 MW wind turbine cost a little over US $ 1 million and today, the cost has not fared significantly better. This high cost of installed wind turbine may be a major hindrance to poorer nations adopting the technology of wind energy for electricity production. Based on this, research efforts focusing on hardware improvement are ongoing to bring down prices and reduce the cost of energy around the world. One major route adopted was to develop small wind turbine machines capable of producing few kilowatts (kW) of wind power either as standalone, embedded generation or for domestic use. Other arguments that have favoured development of small scale wind turbines is the fact that large wind farm operation may lead to climate changes which include increased average temperature and precipitation [3-4]. Hence, Tummala *et al.* [5] suggested that large-scale wind turbine farms may not be a sustainable viable option at the long run. Consequently, several researches have focused on small-scale wind turbine development. Part of these includes the various initiatives that range from using natural materials, like timber [6-7] as turbine rotor material to aero-structural design of the rotor/blade [8]. Thus, promoting the technology of small wind turbine system. The part that the shape and design of turbine rotors play in the magnitude of its performance, energy yield and aeroacoustics characteristics cannot be overemphasized. The design usually focus among other things on the interplay that occurs between the two most important forces that act on the airfoil. These forces are the lift and drag, and the ratio of their coefficients is a very important parameter in the design process. Fig. 3 show the directions of the lift and drag forces along the line of airflow.

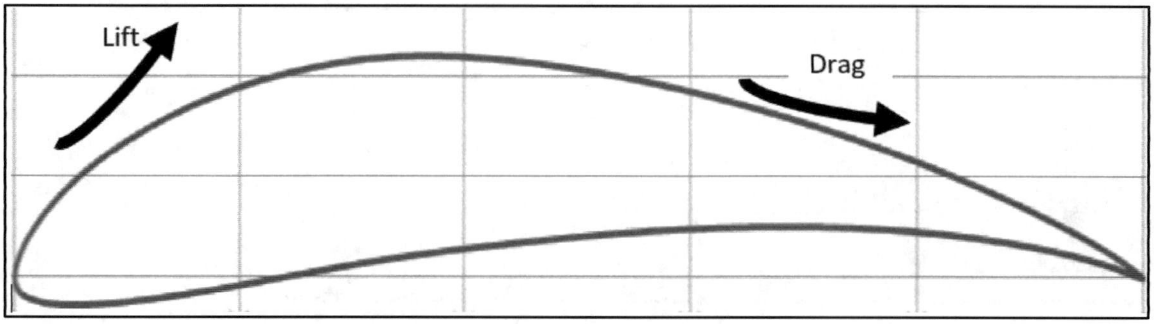

Fig. 3: Schematics of airfoil showing the lift and drag forces in their direction of impact along the airflow direction

Based on this, several studies have looked into the impact of rotor designs on the performance of wind turbines. For instance, [9] developed shrouded-twin-rotor turbine design to enhance the performance and power generation from a wind turbine. Elsakka *et al.* [10] studied the influence of the angle of attack of a vertical axis wind turbine's rotor on power generation and the aerodynamic forces using computational fluid dynamics (CFD) analysis technique. It demonstrated that the sinusoidal variable pitch configuration has effect on the performance of the turbine. Hosseini and Goudarzi [11] developed and investigated the performance of a hybrid vertical axis wind turbine (VAWT) in order to obtain an extended operational range and enhance self-starting ability. A combined 2-bladed modified Savonius batch-type and 3-bladed Darrieus rotors were employed for the hybrid system's design and CFD analysis. The results demonstrated that the design gave an improved efficiency with an operable tip speed ratio of between 2.5 and 4.5, and maximum coefficient of performance of 41.4%. Rezaei *et al.* [12] studied the effect of aeroelasticity and its corresponding dynamics as it relates to wind turbine rotor's geometric nonlinearities, which originates from the blade deformation under cyclic loading. It developed a multi-flexible-body model for the aeroelastic analysis on the 3-bladed turbine assembly. The results were compared with those of a single-blade model, and found that the single-blade model gave the behavioural understanding of a wind turbine. However, the study asserted that employing the entire system model for analysis is necessary for complete dynamic performance or stability analysis. Lositano *et al.* [13] demonstrated the effect of cambered tubercle leading edge (TLE) blades on the performance, under steady conditions, of a 5 kW 3-bladed H-rotor Darrieus VAWT. It showed that the LTE had a negative impact on the performance of the rotors.

In the same vein, Wang and Zou [14] demonstrated a method to analyse the impact of geometric variations on the aerodynamic performance of a wind turbine rotor. Genç *et al.* [15] studied the impact of the pre-stall flow control on the lift and drag coefficients of a wind turbine airfoil. It found that, introduction of a roughness element improves the aerodynamic performance of the blades. Menon and Ponta [16] investigated how pitch control contributes to the mitigation of fatigue damage of turbine rotors. It used rapid pitch control to regulate short-term fluctuations in load and wind conditions. Other studies include Hua *et al.* [17], Manganhar *et al.* [18], Sengupta *et al.* [19] and Yang *et al.* [20]. Each of the studies focused on improved blade design for enhanced turbine performance and structural integrity.

However, despite the advances in turbine blade designs, there is the need to further the design to include those that can drive generators not just for small wind turbine applications but also for operation under very low cut-in speeds (less than 2.0 m/s). This may require that the rotor is designed in such a way that the desire is to have an optimised chord and twist distribution with airfoils having glide ratio that is high at a range of angles of attack. When this is achieved and operated successfully, several low wind speed regions across the globe may be able to take advantage and adopt wind energy technology for power generation.

3. Cut-in wind speed as a limiting factor to popular turbine models

In addition to the limiting factors of large wind turbine models, another issue of importance is the fact that majority of the commercial megawatt (MW) turbine models have their cut-in wind speeds from 3.0 m/s (see Table 1 for instance). Table 1 displays that none of the popular wind turbine models has the cut-in wind speed below 3.0 m/s. This is in addition to the height requirement, which has made the large turbines unsuitable for application in residential and commercial areas.

Table 1. Speed requirement for some popular wind turbine models [21-25]

S/N	Turbine Model	Cut-in speed (m/s)	Cut-out speed (m/s)	Rated speed (m/s)	Power rating (kW)	Hub height (m)
1	PGE 20/25	3.5	25	9.0	25	24/30/36
2	Enercon	3.0	25	12.0	3000	120/135
3	GE 1.5 sle	3.5	25	14.0	1500	65/80
4	G.E 1.5 xle	3.5	20	11.5	1500	80
5	AV 927	3.0	25	13.1	3300	60-80
6	AV 928	3.0	25	11.6	2500	80
7	V 90	4.0	25	15.0	3000	80
8	SWT-3.6-107	3.0	25	13.0	3600	80

This speed requirement is a major shortcoming when consideration is given to the fact that several potential sites, especially across the sub-Saharan Africa (SSA) region, have average surface wind speed below this value.

Moreover, the cut-in wind speed is a very important technical parameter. This is because it is related to the magnitude of the initial torque that the turbine rotor shaft is required to overcome to produce an output. The relationship between the torque and the wind speed is given as equation (1):

$$\tau = \left(\frac{1}{2}r\rho A\right)C_T v^2 \quad (1)$$

where: r = rotor radius, A = swept area, ρ = air density, v = wind speed, C_T = torque coefficient

Based on equation (1), the torque is proportional to the square of the wind speed. Thus, suggesting that whatever can be done to reduce the initial torque will also have an impact on the cut-in wind speed. Another important parameter is the tip speed ratio (TSR) (λ). This is equally related to the wind speed (v) and rotor speed (u) as:

$$\lambda = \square/_{v} = \frac{\pi DN}{60v} \qquad (2)$$

where: N is the number of rotation of the blade per minute, *D* is the rotor diameter

However, it has been shown empirically that the optimum TSR for maximum power output, depending on the number of blades (n), occurs at [26]:

$$\lambda_{max.\ power} = \frac{4\pi}{n} \qquad (3)$$

Relating equation (2) to (3) gives that:

$$\frac{nDN}{240} = v \qquad (4)$$

This also suggests that the wind speed required at optimum TSR for maximum power output is related to the number of revolution per minute (N). Hence, maximum power can be achieved with few revolutions, if the cut-in wind speed can be reduced.

4. Wind turbine generator technologies

Moreover, aside the reduction of cut-in wind speed, another possible means of achieving output at reduced torque is to eliminate/reduce the lubricating parts. This is because the lubricating parts introduces mechanical losses into the system. Thus, efforts tailored towards achieving these sets of targets may require appropriate generator design. However, there are different types of wind turbine generators that have been the focus of previous studies. The aim of such studies is on criteria that include power quality, size, cost, weight and speed range [27]. Another factor of design is the determination of the operating principle of the generator, whether synchronous or induction (asynchronous) and may be multi-phase or single phase, direct current (DC) or alternating current (AC). The general differences in wind turbine generator models are summarised with Fig. 4. It shows that whatever the AC generator type, it either operates by squirrel-cage, permanent magnet, wound rotor or high temperature superconductor (HTS). According to Cao *et al.* [28], small-scale wind turbines generators use either DC machines, squirrel-cage induction machines or synchronous machines, while the medium to large-scale turbines employs doubly-fed induction generator (DFIG), permanent magnet, switched reluctance and HTS.

4.1. Synchronous Generator

This kind of electrical machines function to convert mechanical energy from prime mover to electrical energy at fixed frequency or voltage. Its name was derived from the fact that the machine moves with constant speed termed synchronous speed. It operates by electromagnetic induction as a result of relative motion between the magnetic field flux and the conductor, thereby generating induced electromotive force in the conductor. It is an efficient generator system. The make-up consist of two parts namely the stator, which is made up of the (stationary) armature coils (or conductor windings), and the rotor that consist of the (permanent) magnetic field windings. Fig. 5 is a schematic representation of a synchronous generator, while Fig. 6 shows the schematics of the utilisation of the synchronous generator (SG) in a wind turbine.

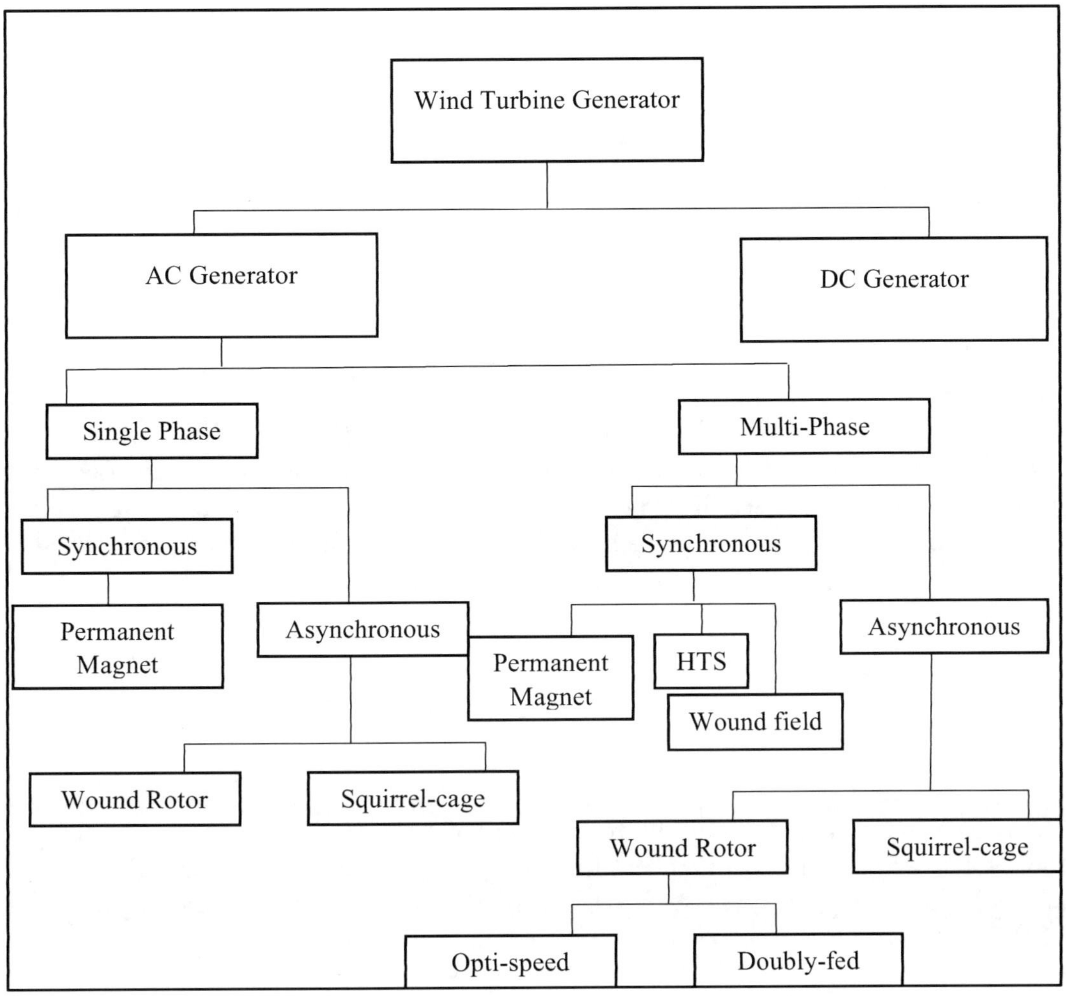

Fig. 4: Division wind turbine generator system

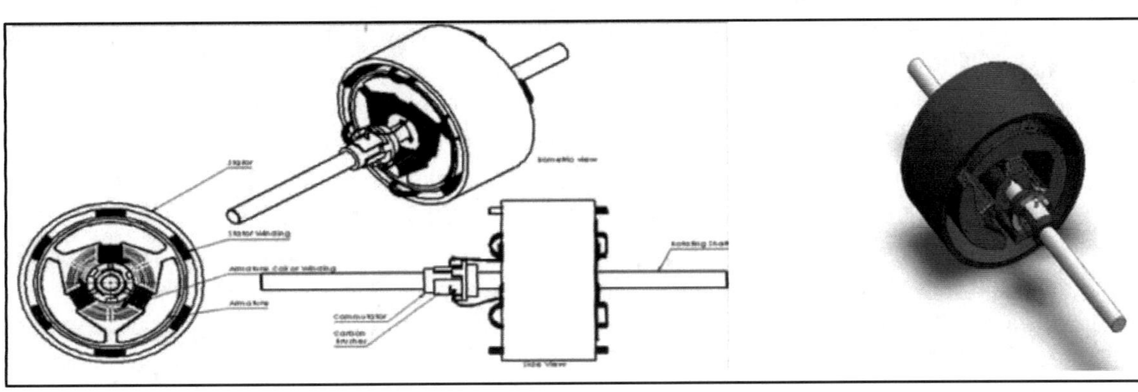

Fig. 5: Schematic representation of a synchronous generator

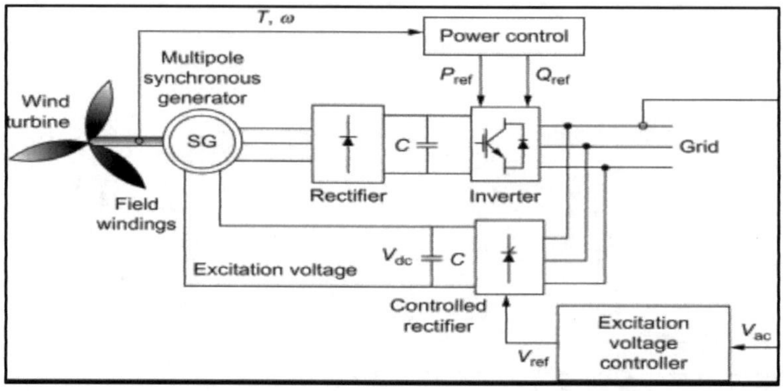

Fig. 6: Schematic of the synchronous generator (SG) in a wind turbine [28]

Synchronous generators are commonly employed in gas, steam, hydro and reciprocating turbines as well as in wind turbines. Its use in variable speed wind turbines is due to the turbines' low rotational speed [29]. Worth noting is the fact that the frequency (f) of the induced voltage in the armature conductor is proportional to the rotation speed (N) of the rotor. This is given as:

$$f \propto N \quad (5)$$

$$f = kN = \frac{p}{120}N \quad (6)$$

where: k is the constant of proportionality given as $\frac{p}{120}$, p is the number of magnetic poles.

Hence, going by equation (6), for a constant frequency of 50 or 60 Hz, the revolution per minutes will be inversely proportional to the number of poles. A 2-poles and 50 Hz system will rotate at 3000 rpm while 8-poles, 50 Hz will rotate at 750 rpm. However, power in the prime mover (turbine blades) is related to the torque (τ) and rotation (or synchronous) speed (N) as:

$$P = \tau \times N \quad (7)$$

Worth noting therefore is the fact that the cut-in wind speed (related from equations (1) to (7)) is a very important design criterion that affects the generator performance. Hence, the need to consider the design of wind turbine generators in such a way that the power requirement to overcome the initial torque will be achieved possibly at lower speeds that 3.0 m/s. When this is achieved, sites with low wind speed profiles will benefit from the adoption of wind turbine for electricity generation. This may require the use of magneto-electromechanical system as speed booster. Further to this, a major disadvantage of the SG is that maximum power output occurs at the optimum tip speed ratio. Thus, at the TSR, the rotational speed is low and this will require a high number of poles, which could make the system bulky, expensive and may require speed limiter or continuous variable transmission in order to regulate the speed when the turbine is a direct drive machine. This shows that there is still the need to further the synchronous generator design for small scale wind turbine power generation, in order to have a system that is compact and capable of operating at very low wind speed.

IOP Conf. Series: Earth and Environmental Science **331** (2019) 012062 doi:10.1088/1755-1315/331/1/012062

4.2. Asynchronous Generator

This type of generators use a principle similar to that of the induction motor in which a 3-phase set of voltage, applied to a stator winding, produces magnetic field in the opposite (anticlockwise) direction. In this type of generator however, the rotor moves faster than the synchronous speed. In this case, producing negative torque. Moreover, just like the synchronous type generators, it has a fixed stator winding in which rotating magnetic field produces induced voltage. The construction is usually based on the squirrel-cage induction machine. One major advantage of this type of generator is the fact that it can produce electricity at the frequency and voltage of the grid system, thus, requires no inverter or rectifier. Further to this, when it is connected to a 3-phase mains supply, it can produce electricity without a voltage controller or exciter, which could make it require additional circuitry. However, operating it, as a standalone facility, usually requires the addition of capacitors connected to its windings for self-excitation. This is its major disadvantage. The capacitors function to supply reactive power (which would have been supplied by the grid, if connected to it) to the generator. Fig. 7 is a schematic representation of an asynchronous generator.

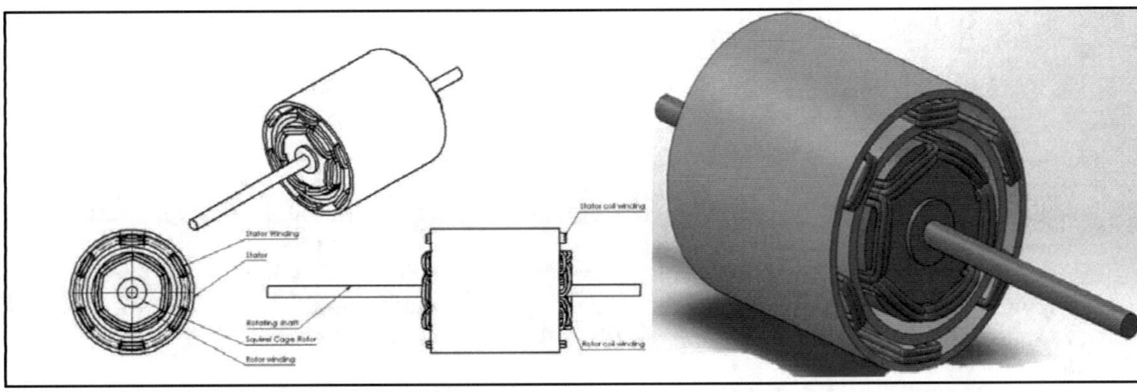

Fig. 7: Schematic representation of an asynchronous generator

Hence, based on the aforementioned, for both synchronous and asynchronous generators, the magnitude of the torque plays an important role in the power generation, hence the need for designs that is capable of generation at low wind speeds.

Conclusion

The study explored the various efforts at the development of wind turbines systems. It reviewed some of the existing reports on turbine rotor designs and also considered associated generators that are employed with wind turbine system for power generation. The outcome showed that, despite the progress in the design of wind turbine rotors and generators, there is still the need to further the designs towards the development of lower cut-in wind speed turbine system that is capable of generating power below 2.0 m/s. If this is achieved, several sites with low wind speed regimes across the globe will be able to employ wind turbine models for power generation. To achieve this will require:

- Improved rotor design with a focus on lower cut-in wind speed below 2.0 m/s

- Concerted research efforts to have the rotor designed in such a way that the chord and twist distribution will be optimised and the airfoil will have high glide ratio at a range of angles of attack instead of at single peak angle or few angles.
- Improvement in wind turbine generator design that eliminates or reduce lubrication parts and possibly engage electro-magneto-mechanical principle of operation.

Acknowledgement

The authors wish to thank the Covenant University Management and the United States' Fulbright African Research Scholar Programme for the support granted to see the fruition of this research.

References

[1] WindEurope (2019). Wind energy in Europe in 2018: Trends and Statistics, available online [https://windeurope.org/wp-content/uploads/files/about-wind/statistics/WindEurope-Annual-Statistics-2018.pdf, accessed March 26, 2019]

[2] Statista (2019). Global cumulative installed wind power capacity from 2001 to 2017 (in megawatts). Available online [https://www.statista.com/statistics/268363/installed-wind-power-capacity-worldwide/accessed April 27, 2019]

[3] Wang, C, Prinn, R.G. (2010). Potential climatic impacts and reliability of very large-scale wind farms. *Atmospheric Chemistry and Physics* 10(4), 2053–61

[4] Fiedler, B.H. Bukovsky, M.S. (2011). The effect of a giant wind farm on precipitation in a regional climate model, *Environmental Research Letters* 6(4), 045101.

[5] Tummala, A. Velamati, R.K. Sinha, D.K. Indraja, V. Krishna, V.H. (2016). A review on small scale wind turbines, *Renewable and Sustainable Energy Reviews* 56, 1351-1371

[6] Mishnaevsky, L. Jr., Freere, P. Sinha, R. Acharya, P. Shrestha, R. Manandhar, P. (2011). Small wind turbines with timber blades for developing countries: Materials choice, development, installation and experiences, *Renewable Energy* 36, 2128-2138

[7] Astle, C. Burge, I. Chen, M. Herrler, T. Kwan, L. Zibin, N. Wood, D. (2013). Timber for small wind turbine blades *Energy for Sustainable Development*, 17, 671-676

[8] Pourrajabian, A. Afshar, P.A.N. Ahmadizadeh, M. Wood, D. (2016). Aero-structural design and optimization of a small wind turbine blade, *Renewable Energy* 87, 837-848

[9] Kumar, V. Saha, S. (2019). Theoretical performance estimation of shrouded-twin-rotor wind turbines using the actuator disk theory, *Renewable Energy* **134**, 961 – 969

[10] Elsakkaa, M. M. Inghama, D. B. Maa, L. Pourkashanian, M. (2019). CFD analysis of the angle of attack for a vertical axis wind turbine blade, *Energy Conversion and Management* **182**, 154–165

[11] Hosseinia, A. Goudarzi, N. (2019). Design and CFD study of a hybrid vertical-axis wind turbine by employing a combined Bach-type and H-Darrieus rotor systems, *Energy Conversion and Management* **189**, 49–59

[12] Rezaei, M.M. Zohoor H. Haddadpour, H. (2018). Aeroelastic modeling and dynamic analysis of a wind turbine rotor by considering geometric nonlinearities, *Journal of Sound and Vibration* **432**, 653 – 679

[13] Lositano, I. C. M. Danao, L. A. M. (2019). Steady wind performance of a 5 kW three-bladed H-rotor Darrieus Vertical Axis Wind Turbine (VAWT) with cambered tubercle leading edge (TLE) blades, *Energy* **175**, 278 – 291

[14] Wang, X. Zou, Z. (2019). Uncertainty analysis of impact of geometric variations on turbine blade performance, *Energy* **176**, 67 – 80

[15] Genç, M. S. KOCA, K. Açikel, H. H. (2019). Investigation of pre-stall flow control on wind turbine blade airfoil using roughness element, *Energy* **176**, 320 – 334

[16] Menon, M. Ponta, F. L. (2017). Dynamic aeroelastic behavior of wind turbine rotors in rapid pitch control actions, *Renewable Energy* **107**, 327 – 339

[17] Hua, X. Zhang, C Wei, J. Hub, X. Wei, H. (2019). Wind turbine bionic blade design and performance analysis, *J. Vis. Commun. Image R* **60**, 258–265

[18] Manganhar, A. L. Rajpar, A. H. Luhur, M. R. Samo, S. R. Manganhar, M. (2019). Performance analysis of a savonius vertical axis wind turbine integrated with wind accelerating and guiding rotor house, *Renewable Energy* **136**, 512 – 520

[19] Sengupta, A. R. Biswas, A. Gupta, R. (2019). Comparison of low wind speed aerodynamics of unsymmetrical blade H-Darrieus rotors-blade camber and curvature signatures for performance improvement, *Renewable Energy* **139**, 1412 – 1427

[20] Yang, Y. Zeng, P. lei, L. (2019). Aero-structural investigation of sesquiplane wind turbine blade, *Energy Procedia* **158**, 779 - 784

[21] Ajayi, O.O. Fagbenle, R.O. Katende, J. (2011a). Assessment of wind power potential and wind electricity generation using WECS of two sites in South West, Nigeria, *International Journal of Energy Science*, 1 (2), 78-92

[22] Ajayi, O.O. Fagbenle, R.O. Katende, J. (2011b). Wind profile characteristics and econometric analysis of wind power generation of a site in Sokoto State, Nigeria, *Energy Science and Technology*, 1 (2), 54-66

[23] Ajayi, O.O. Fagbenle, R.O. Katende, J. Okeniyi, J.O. (2011). Availability of wind energy resource potential for power generation of Jos, Nigeria, *Frontiers in Energy*, (4), 376 – 385

[24] Ajayi, O.O, Fagbenle, R.O. Katende, J. Aasa, S.A. Okeniyi, J.O. (2013). Wind profile characteristics and turbine performance analysis in Kano, North-Western Nigeria,

International Journal of Energy and Environmental Engineering, Springer, 4 (27), doi:10.1186/2251-6832-4-27

[25] Ajayi, O.O. Ohijeagbon, O.D. (2015). Feasibility and techno-economic assessment of standalone and hybrid RE for rural electrification in selected sites of south eastern Nigeria, *International Journal of Ambient Energy*, 38 (1), 55 – 68. DOI: 10.1080/01430750.2015.1035799

[26] Ragheb, M. 2009. Optimal rotor tip speed ratio, available on the web [https://netfiles.uiuc.ed u/mragheb/www/NPRE%20475%20Wind%20Power%20Systems/Optimal%20Rotor%20T ip%20Speed%20Ratio.pdf, accessed November 15, 2010]

[27] Goudarzi, N, Zhu, W.D. (2013). A review on the development of wind turbine generators across the world, *International Journal of Dynamics and Control* 1, 192 – 202

[28] Cao, W. Xie, Y. Tan, Z. (2012). Wind turbine generator technologies. In: Carriveau, R. (ed.). Advances in wind power, Intech, 177 – 202.

[29] Khaligh, A. Onar, O.C. (2018). Energy Sources. In: Rashid, M.H. (ed.), Power electronics handbook, 4th ed., Elsevier Inc., 725 – 765.

IOP Conf. Series: Earth and Environmental Science **331** (2019) 012063 doi:10.1088/1755-1315/331/1/012063

Corporate Social Responsibility & Brand Loyalty in The Nigerian Telecommunication Industry

J O Akhigbe[1] and F O Olokoyo[1]

[1]Department of Banking and Finance, Covenant University, Ogun State, Ota, Nigeria

Abstract. The existing literature on Corporate Social Responsibility with regards to consumer responsiveness towards the organization is currently split and filled with inconsistencies. This reality makes theory building quite difficult. Therefore, further studies will be necessary to build theory consensus and update the existing literature. This study attempts to evaluate the effects of corporate social responsibility on brand loyalty in a developing nation such as Nigeria, with special focus on its telecommunication industry. Data gathering was carried out through the distribution of 386 questionnaires to the customers of the four largest mobile telecommunication firms in Nigeria. Simple linear regression was used in testing the formulated hypothesis. Findings from the hypothesis testing reveals there is no significant relationship between corporate social responsibility and brand loyalty in the Nigerian telecommunication sector. However, favorable price point, good quality service and good customer care were some of the attributes the customers emphasized on as regards to what determines their brand loyalty. Based on the findings from this study, firms are recommended to integrate CSR policies into their core operations. CSR should be focused on providing solutions to Nigeria's social economic challenges. A concise policy should be initiated to ensure clear information and reporting channels are duly released to every stakeholder group. Without diligent reporting of CSR activities, customers can easily assume the firms do not engage in CSR practices.

Keywords: Corporate Social Responsibility, Brand Loyalty, Telecommunication, Consumer Behavior.

1. Introduction

The biggest segment of the Information and Communication sector in Nigeria is the telecommunication industry. The Nigerian telecommunication market is commonly referred to as one of the biggest in Africa. Overtime, the industry slowly morphed into an oligopolistic form of market structure (a relatively few number of companies decide the market prices and exert control over majority of the market). The Nigerian telecom industry is filled with multinational firms. These multinational firms include; MTN, this is a South African firm, Airtel, this is a strong multinational firm from India, Globacom, a Nigerian based multinational telecom firm, and 9mobile, this firm was formerly known as Etisalat.

As Africa's biggest mobile telecommunication market, Nigeria currently has about 168 million active subscribers (NCC, 2019). The economic recession that swept Nigeria from 2015 to 2017 had a major role to play in the declining figure of internet users recorded in 2017, this fall was as a result of consumers reacting to the harsh economic realities. Consumers later jumped onto other OTT channels to meet their communication needs, another reason for the fall in figure of internet users is the regulatory policy which required all mobile operators to disconnect every single unregistered SIM cards on its platform, MTN NG was later fined heavily for not adhering strictly to this policy.

Table 1: Overview of the subscriber base of Nigerian GSM sector

S/N	Telecom Companies	Number of Subscribers
1	Globacom Nigeria	43,300,000
2	MTN Nigeria	66,900,000
3	9Mobile	15,400,000
4	Airtel Nigeria	43,100,000
	Total	168,700,000

Source: Nigerian Communications Commission, 2019

The 2019 Nigeria Communications Commission's report indicated that MTN is in peak position and controls the mobile market with 39.94% market share and 66.9 million active subscribers. Closely following MTN NG in second place is Globacom NG with 43.3 million active subscribers and 25.23% market share. However, Airtel now controls 25.24% of the mobile market with 43.1 million active subscribers. While 9mobile currently controls 9.6% of the market share with 15.4 million active subscribers.

Considering the growth trajectory trend of the Nigerian mobile industry and also the high performance rate of the mobile industry operators, there is an expected influx of new operators into the mobile industry space in the future and with the existing operators frenetically looking for ways to maintain and also enhance their mobile market shares, this reality has made customer retention hugely significant for the firms' viability. Another reason why customer retention is increasingly very important in this highly competitive industry is the fact that gaining new customers is quite a difficult and expensive endeavor for any firm to undertake.

A robust insight into consumers' behavior is a tool every firm needs to aim itself with, especially in this highly competitive Nigerian mobile industry. This insight or robust understanding of consumers' behavior will aid the operators to remain connected with every development in the mobile market, thus, helping them to deliver quality services to its customers and thereby ensuring the firms survival chances.

The available literature on consumer behavior is ever expanding, in today's market environment, holding onto acquired market share has become a challenge and a task firms must weather to maintain relevance and viability. Craig argued that customers play a significant role in keeping an enterprise a going-concern [1]. His argument was hinged on the notion that customers are the revenue generating elements of the firm, by patronizing, and using word of mouth to influence others to patronize the concerned firm as well. Therefore, firms are forever eager to understand the psychology behind consumers' buying decisions. Therefore, study of consumer buying behavior is gate way to success in the market [2].

Firms all over the world are now being tasked by its customers to be more socially responsible. This social responsibility awareness on the part of the consumers has lead firms to think beyond their economic interests. Some of the Corporate Social Responsibility (CSR) initiatives that firms undertake include; manufacturing aerosol products by utilizing technologies that are environmentally-friendly, being vigilant to employees' welfare, satisfaction and empowerment, establishing recycling and pollution prevention policies, and pushing for the prosperity aim of the community organizations (e.g. working closely with groups) [3].

The social identity theory and the consumer-company identification study reveal that there exist tremendous advantages for firms seen to be socially responsible by its stakeholders. With regards to community engagement, firms that make meaningful impact through its CSR activities have a higher chance of gaining favorable reputation and positive corporate image perception amongst the community members where the firms carry out their operations. Several studies have linked active and immersive CSR participation to increased favorable consumer perception, higher propensity to pay at premium prices. Therefore, it is believed to be in the firms' self-interest to actively participate in CSR activities, as these CSR participations can possibly strengthen the corporate image of the firms and subsequently bring to reality the enviable achievement of 'triple bottom-line'

The Triple Bottom Line (TBL) concept is when firms seek to affect the society positively, this includes the environment and the business community. The main objective of the triple bottom line is to engender sustainability through the careful joggling of the economic, environmental and societal objectives.

With the Nigerian telecommunication industry growing at a fast rate, firms are always looking for new strategies to attract and retain customers, such strategies include corporate social responsibility, CSR affords firms the opportunity to gain and maintain their grounds while simultaneously developing the host communities and by extension the country at large. [4] affirm that firms that actively practice corporate social responsibility are getting positive commendation, which was not the case in previous years, therefore, this has prompted analysts to attribute higher value to CSR strategies. They went further to add that researchers such as [5] are proponents of the stakeholder's management theories which they opined that CSR strategies could lead to higher performance through the protection and the strengthening of corporate reputation.

1.1 Hypothesis Development

Overtime, a significant amount of studies was undertaken to determine if corporate social responsibility has any influence on consumers' buying behavior. However, these studies revealed split reports on the impact of CSR on consumers' buying behavior. A significant amount of studies that were undertaken in this subject tend to take the firms' perspective, however, fewer studies have taken the perspective of the consumers. This study intends to deepen existing literature on the influence of corporate social responsibility and its effect on brand loyalty.

The controversial nature of the existing literature has made theory and general consensus building difficult. Hence, the need to undertake more studies in different conditions and settings to further deepen the literature. [6] Argued that given the competitive nature of modern business environments and how difficult it is to attract and retain customers, which has pushed many firms to actively engage in corporate social responsibility as a means to strengthen their customer base, theoretically as well as practically speaking, it becomes important to find if such CSR engagements has actually helped achieve the stated objective.

Studies have revealed that there is still a lot to be learnt as regards the link between corporate social responsibility and its effect on consumers' behavior. Researchers have opined that due to varying economic, business and organizational systems and with various cultures necessitating different customers' expectation of business behaviors, therefore, the results from the western countries regarding the impact of corporate social responsibility on brand loyalty will likely not be suitable to developing nations such as Nigeria.

Another justification for this study is that it examines corporate social responsibility from the perspective of developing nations such as Nigeria, thereby helping to validate theories and deepen the existing literature in the subject.

IOP Conf. Series: Earth and Environmental Science **331** (2019) 012063 doi:10.1088/1755-1315/331/1/012063

2. LITERATURE REVIEW

2.1 The Concept of Corporate Social Responsibility

Corporate social responsibility is viewed by (Wood, 1991), as a firm's principles of social responsibility, its social responsiveness, and policies with regards to the firm's social relationships. [7] view corporate social responsibility as holding firms accountable for its practices and the attendant effects on people, and the environment. So the general view is that corporate social responsibility is the connection between a firm and the society it operates in.

2.2 History of Corporate Social Responsibility

The evolution of CSR began as early as the 1930s [8] stating that corporate social responsibility is usually viewed from varying perspective depending on the literature, some earlier views of CSR were concentrated on strictly the benefits of CSR to the community, others on the accountability perspective of CSR; where corporate social responsibility is explored on the basis of its accountability to stakeholders.

There was a rapid increase in corporate social responsibility awareness level in the 1990s, this was due to privatization, less government interference, mergers and acquisitions in the 1980s, as capital flows became more entrenched due to globalization and better communications technology. This lead to informed stakeholders demanding for more accountability from the firms, and thereby actively rewarding firms that were more socially and environmentally conscious and subsequently punishing firms that do not practice or integrate corporate social responsibility initiatives into the firm's core operations.

Over the last two decades, the influence of corporate social responsibility momentum is quite evident through the current business language, such as; corporate citizenship, sustainability; triple bottom line etc. The language is ever evolving, fine-tuning and expanding.

2.3 Nigeria's Telecommunication Sector: Current Corporate Social Responsibility Practices

A study by [9] revealed that a lot of firms generally focus their corporate social responsibility endeavours in areas such as healthcare, infrastructural development, communities' empowerment.

In line with the above study, [10] showed that a lot of firms in Nigeria still practice philanthropic model of CSR, which encompasses CSR activities that consider social issues generally, rather than CSR, which is in line with business strategy. This is generally due to the present state of decadence in several sectors of the Nigerian economy; the despicable road and rail system, the outdated educational system. Majority of CSR activities engaged in Nigeria are thus primarily targeted at filling institutional gaps

.

2.4 Theoretical Framework

The study adopted both the Stakeholder's Theory and Institutional Theory.

2.4.1. The Stakeholder's Theory.

[11] argue that, the stakeholder theory is hinged on the proposition that a firm ultimately strive to provide a balance among the interests of its numerous stakeholders with the aim of ensuring some degree of satisfaction amongst them.

A primary proposition of the stakeholder's theory is that firms should voluntarily agree to be socially and environmentally responsible, thereby avoiding stakeholders' pressures, and ensuring that their responsibility is not forced upon them by the law or the union contracts [12]. The stakeholder's theory maintains that the firm's obligation extends beyond the traditional responsibilities to the shareholders, and that the responsibilities of the firms should extend beyond the shareholders to other stakeholders such as customers, employees [13].

The most common stakeholders highlighted in the stakeholder's theory and their respective needs are as follows;

i Customers: Satisfy customer needs

ii Employees: empower the employees and treat them respectfully

iii Government: meet the stipulated legal requirements

iv Communities: empower and develop the communities

v Shareholders: realize profits for the investors

vi Environment: Avoid environment degradation

2.4.2. Institutional Theory

The concept of CSR de-emphasizes its dominance found in the west, as regards the concept and its operation when applied in societies that are characterised by institutional and deficiencies, as observed in developing countries such as Nigeria. [14] argued that CSR is context based, such that CSR engagements in developing countries are deployed with the ideology of filling institutional gaps.

This assertion is supported by this institutional theory, which states that the structural development of a firm can strongly be influenced by its institutional environment [15].

2.5 Empirical Framework

Majority of the studies carried out agree that firms should carry out corporate social responsibility practices. The studies also suggested these CSR practices ought to be heavily advertised to the general public because of the positive effect it has on consumer behaviour. One of the researchers in support of this notion is [16] who states that active engagement in corporate social responsibility practices increases consumer retention. Another empirical evidence to this narrative is [17] suggesting that firms should make its philanthropic practices more known to the general public, in order to illustrate its commitment to the community and thereby gain competitive advantage in marketplace.

IOP Conf. Series: Earth and Environmental Science **331** (2019) 012063 doi:10.1088/1755-1315/331/1/012063

Table 2: Selected Empirical Studies on CSR and Consumer Behaviour

Author(s)	**Key Results**
Abdeen (1991)	Concluded that the advertisement of CSR practices of firms enhance customer retention.
Mohr et al. (2001).	The study concluded that CSR has a significant influence on consumers' buying behaviour.
Mohr et al. (2001).	The study concluded that CSR has a significant influence on consumers' buying behaviour.
Alexandris et al. (2002)	The study revealed that CSR has a positive influence on customers' retention.
Girod and Michael (2003)	This study revealed CSR helps maintain brand loyalty.
Yu et al. (2005)	This study revealed positive link between customer satisfaction and customer loyalty.
Imran Ali et al., (2010)	Another study revealed that CSR has no significant influence on consumers' purchase intentions.

Author(s)	**Key Results**

In contrast, Imran Ali et al., (2010)	Found no linkage between CSR practices and customer purchase intentions.
Ali (2011)	Ali revealed that a significant positive relation exists between CSR and consumer buying intentions.
Kolkailah et al. (2012)	The study revealed a positive relationship between corporate social responsibility and corporate image.
(Nevine, Noha, 2017)	Another study carried out in Egypt finds that CSR has no significant effect on consumers' buying behaviour

3. Methodology

Explanatory stratagem was adopted to examine and analyse the effect corporate social responsibility has on brand loyalty in the Nigerian telecommunication industry. To obtain and measure the consumers' responses, the four leading telecommunication firms in Nigeria were selected. The reason behind the selection of the cellular industry is partly due to the systemic role it plays in the entire country. Another reason behind the selection of the telecom industry is due to its high penetrative rate in the country, leading to huge customer base. That huge customer base is now being used to test brand loyalty with CSR serving as the predictor.

The Nigerian cellular industry has been growing at a rapid rate since its inception in the year 2001. (Nigerian Communications Commissions – NCC). The subscriber base of the industry is currently estimated to be 168 million subscribers. The study has used the information from the firms' CSR to determine their effect on brand loyalty.

3.1 Sample

Taro Yamane formula was applied in determining the sample size of the study. Based on the findings from the calculations, 400 questionnaires were distributed to customers of the four major telecom firms in the country across multiple cities. These firms are MTN, GLO, AIRTEL & 9MOBILE.

Out of the 400 distributed questionnaires, 386 were retrieved back, amounting to 96.5% response rate.

3.2 Measurement

A 5-point Likert rating scale was applied in measuring the key variables of the study. The 5-point scale was applied in the measurement of customers' perception towards the firms' various corporate social responsibility practices. The customers were exposed to the various CSR actions of the firms; this was done to really determine if such CSR engagements can significantly affect the consumers' brand loyalty.

IOP Conf. Series: Earth and Environmental Science **331** (2019) 012063 doi:10.1088/1755-1315/331/1/012063

3.3 Method of Data Collection

The survey instrument was subjected to face and content validity. The primary data sources for this research were the responses from customers extracted from the administered questionnaires to the customers of Mtn, Glo, 9Mobile and Airtel. The questionnaire for the customers was adopted to gain data to evaluate the research questions.

The questionnaire is divided into two sections: A&B. Section A contains background questions, Section B consist of questions that are related to the study, that is, the effect of corporate social responsibility on brand loyalty, as it relates to information available to the customers.

3.4 Procedure for Data Analysis

Descriptive statistic was used to collect and analysed data. frequency count, and simple percentage were used to interpret demographic data and other univariate variables while simple regression and Pearson Product Moment Correlation Coefficient was used to test the hypothesis at 0.05 level of significance using the Statistical Package for Social Sciences (SPSS).

4. Results and Discussion

H_0 Corporate social responsibility has no significant influence on brand loyalty.

Table 3: Variables Entered/Removed[a]

Model	Variables Entered	Variables Removed	Method
1	CSR[b]	.	Enter

a. Dependent Variable: brand loyalty
b. All requested variables entered.
Source: SPSS Output version 23, (2019)

4.1 Reporting

This table 3 shows the variable used to predict another variable. As such it can be deduced that the corporate social responsibility is used to predict the brand loyalty.

Table 4: Model Summary

Model	R	R Square	Adjusted R Square	Std. Error of the Estimate	Change Statistics				
					R Square Change	F Change	df1	df2	Sig. F Change
1	.052[a]	.003	-.002	.45725	.003	.525	1	386	.470

a. Predictors: (Constant), CSR

4.2 Interpretation

From the model summary in the table above, the coefficient of determination (R^2) which measures the proportion of the dependent variable variance and how best it fits to the regression line. The r^2 of 0.003 simply connotes that 0.3 of the variability can be explained while 99.7% are explained by external

IOP Conf. Series: Earth and Environmental Science **331** (2019) 012063 doi:10.1088/1755-1315/331/1/012063

variables. Thus, there is no strong goodness of fit as 0.3% of the values falls along the regression line. This simply connotes that there are many outliers along the regression line.

Table 5: **ANOVA**[a]

Model		Sum of Squares	df	Mean Square	F	Sig.
1	Regression	.110	1	.110	.525	.470[b]
	Residual	80.465	385	.209		
	Total	80.575	386			

a. Dependent. Brand loyalty
b. Predictors: (Constant), CSR.

4.3 Interpretation

From the ANOVA table, the probability value of 0.470 is greater than 0.05 level of significance which from the rule of thumb the study rejects the alternate hypothesis and accept the null hypothesis.

Findings from the study revealed that the Nigeria's telecom firms' participation in corporate social responsibility activities does not have a significant influence on customers' brand loyalty *(r^2=0.003, 0.470>0.05)*. This implies firms' corporate social responsibility engagement in Nigerian telecom industry is not a key determinant in brand loyalty decision by consumers.

Cultivating brand loyalty is an important aspect of a business strategy, brand loyalty can aid in keeping an enterprise a going-concern. This enables a firm to plan for the future, it significantly helps during production by knowing or having an estimate on the required production volume.

Corporate social responsibility is one way of ideally ensuring brand loyalty, helping to fill the institutional gap by erecting infrastructure for example can go a long away in convincing the community of the good intentions of the firm, which should ideally lead to more patronage and eventual loyalty to that particular firm.

Empirical findings from the study revealed that majority of the customers are willing to move to a different telecom provider with better prices or rates and service quality at any given opportunity. With Nigeria being a developing nation and a large portion of its citizens falling within the low income bracket, the consumers do not have the luxury of being loyal to a particular firm. The consumers need to take advantage of any favourable price point in the market.

This finding is supported by [18] revealing that Egyptians consumers do not consider CSR in their buying decisions. There exists an insignificant relationship between consumer buying behaviour and CSR. The study revealed that Egyptians do not consider CSR in their buying decisions because of the low level of income they earn.

5. Conclusion and Recommendation

The study has analyzed the effect of corporate social responsibility practices on brand loyalty in the Nigerian telecommunication sector. The study revealed that customers from developing country such as Nigeria with relatively low income households are usually not big on brand loyalty. The study showed some customers do indeed believe that firms that engage in meaningful corporate social responsibility actions could influence their brand loyalty, but a large number of the respondents categorically affirmed that CSR do not influence their brand loyalty. This culture of not being loyal to any brands in this part of the world is revealed to be due to the customers' constant search for substitute goods with more favourable price points. The economic reality on ground in Nigeria dictates that prices and quality of service trumps other factors when it comes to making buying decisions. CSR in developing countries such as Nigeria essentially means filling the institutional gaps left by the government. This is carried out by the firms' involvement in infrastructural projects.

The results from this study are important to the cellular operators in staying competitive by designing products with competitive price points to meet the needs of Nigerian consumers. The study helped to deepen the existing literature on the effects of CSR on brand loyalty in an emerging economy.

Based on the findings, the study recommends that firms should have corporate social responsibility cultures and values, which will help firms create a framework to achieve their CSR objectives.

Another recommendation is that, firms should have their CSR embedded into the firms' core operations, this will help the personnel easily know their place in the firm,

A concise policy should be initiated to ensure clear information and reporting channels are duly released to every stakeholder group. Without diligent reporting of CSR activities, customers can easily assume it does not exist.

The corporate social responsibility practices of the telecom firms in Nigeria should be concentrated on providing solutions to Nigeria's socio-economic development challenges. Such as healthcare provision, infrastructure development, education etc.

There is therefore a need for the government to take a leading role in triggering CSR behaviour among telecom companies by creating fiscal incentives and setting regulations that enforce environmental protection and the efficient reporting of sustainable practices.

The study has a limitation, this research attempted to view corporate social responsibility as a predictor of brand loyalty in Nigerian telecommunication firms, also noting the role of CSR in filling institutional gaps in developing countries; however, the use of one country poses a limitation to this research, as Nigeria has its peculiar challenges that make it insufficient to generalize the results of this study to other developing countries.

Acknowledgements

We would like to express our profound gratitude to Covenant University for its support and sponsorship throughout this research work.

Appendix

1. Bolton R N Drew J H 1991 *A Multistage Model of Consumer Feelings and Purchase Likelihood. Psychol. Market*, 9:347–363.
2. Ali S M 2004 Corporate Social Responsibility. *Daily Times*. October 05, 2004.
3. Davis K 1973 The Case for and Against Business Assumption of Social Responsibilities. *Acad. Manage. J,* 16:312-322.

4. Deetz S 2003 Corporate Governance, Communication and Getting Social Values into the Decisional Chain. *Manage. Comm. Quart,* 16:606–611.
5. Dick A S Basu K 1994 Customer Loyalty: Towards an Integrated Conceptual Framework. *J. Acad. Market. Sci*, 22:99–113.
6. Beyer R 1972 The Bottom Line is No Longer Where it's at: An Accounting Executive Lists Five Considerations for Corporate Social Responsibility. *New York Times Magazine*, September 24, 1972.
7. Courneya K S and Hellsten L A 2001 Cancer Prevention as a Source of Exercise Motivation: An Experimental Test Using Protection Motivation Theory. *Psychol. Health Med.*, 6:59-64.
8. Grepott T Rams W Schindler A 2001 Customer Retention, Loyalty and Satisfaction in the German Mobile Cellular Telecommunications Market. *Telecommunication Pol.* 25 (4):249-269.
9. Gulliver P Horwath C 2001 Women's Confidence to Consume Adequate Servings of Milk Products: Validation of a Summated Rating Scale. *Hum. Nutr. Dietetics*, 14:287-29
10. Guo J Sun L Li X 2009 Corporate Social Responsibility Assessment of Chinese Corporation. *Int. J. Bus. Manage*, 4(4):54-57.
11. Gauntam V 2015 Service Quality Perceptions of Customers about Mobile Telecommunication Services: A Case of India. *Journal of Global Marketing*, 28(1):19-31.
12. Cronin J J and Taylor S A 1992 Measuring Service Quality: A Reexamination and Extension. *J. Market,* 56(3):55–68.
13. Daugherty E L 2001 *Public relations and social responsibility. In R.L. Heath (Ed.), Handbook of Public Relations*, 389-401. Thousand Oaks, California: Sage.
14. Drucker P F 1974 *Management: Tasks, Responsibilities, Practices*. New York: Harper & Row.
15. Das K 2009 Relationship Marketing Research. (1994-2006) an academic literature review and classification. *Marketing intelligence & planning*, 27(3):326-363.
16. Davis K 1973 The Case for and Against Business Assumption of Social Responsibilities. *Acad. Manage. J,* 16:312-322.
17. Deetz S 2003 Corporate Governance, Communication and Getting Social Values into the Decisional Chain. *Manage. Comm. Quart,* 16:606–611.
18. Dick A S Basu K 1994 Customer Loyalty: Towards an Integrated Conceptual Framework. *J. Acad. Market. Sci.,* 22:99–113.
19. Kolkailah S K AbouAish E and El-Bassiouny N 2012 The Impact of Corporate Social Responsibility Initiatives on Consumers' Behavioural Intentions in the Egyptian Market. *International Journal of Consumer Studies*, 36(4):369-384.
20. Liu Y and Zhou X 2009 Corporate Social Responsibility and Customer Loyalty: A Conceptual Framework. *Conference Proceedings* of the 6th International Conference on Service Systems and Service Management, 8-10 June.

21. Lam SY Shankar V Erramilli MK and Murthy B 2004 Customer Value, Satisfaction, Loyalty, and Switching Costs: An Illustration from a Business-to-Business Service Context. *Journal of the Academy of Marketing Science*, 32(3):293-311.
22. Maignan I 2001 Consumers' Perceptions of Corporate Social Responsibilities: A Cross-Cultural Comparison. *Journal of Business Ethics*, 30(1):57–72.
23. Min C M Ai YJ Choo ACP Wah WP and Yang YC 2012 A Study of the Effect of Corporate Social Responsibility Towards Consumer Buying Behaviour. Conference Proceeding of the International Conference of Management, Economics and Finance, 15-16 October.
24. Mohr LA Webb DJ and Harris KE 2001 Do Consumers Expect Companies to be Socially Responsible? The Impact of Corporate Social Responsibility on Buying Behaviour. *Journal of Consumer Affairs*, 35(1):45–72.
25. Schiffman LG Hansen H and Kanuk LL 2008 Consumer Behaviour: A European Outlook. *Pearson Education.*
26. Tichy NM McGill AR St. Clair L 1997 Corporate Global Citizenship: Doing Business in the Public Eye. San Francisco: *The New Lexington Press.*
27. Van Riel ACR Semijn J Pauwel 2004 Online Travel Service Quality: The Role of Pre-Transaction Services. *Total Quality Manage*, 15(4):475–493.
28. Waddock SA Samuel BG 1997 The Corporation Social Performance – Financial Performance Link. *Strat. Manage. J.* 18(4):303-319.
29. Waheed A 2005 Evaluation of the State of Corporate Social Responsibility in Pakistan and a Strategy for Implementation. *A Report Written for the Securities & Exchange Commission of Pakistan & the United Nations Development Program*
30. Yusof JM Musa R and Rahman S 2011 Functional Store Image and Corporate Social Responsibility Image: A Congruity Analysis on Store Loyalty. *International Journal of Social, Behavioral, Economics, Business and Industrial Engineering*, 5(5):486 – 493.
31. Bolton R N Drew J H 1991 *A Multistage Model of Customers' Assessments of Service Quality and Value*, J. Consum Res. 17:375–384
32. Bowen H R 1953 *Social Responsibilities of the Businessman*, New York: Harper & Row.
33. Biran M Wilson G T 1981 *Treatment of Phobic Disorders Using Cognitive Clinical Psychology*, 49:886-899.
34. Chaudhary R 2017 Corporate Social Responsibility and Employee Engagement: Can CSR Help in Redressing the Engagement Gap? *Social Responsibility Journal*, 13(2).
35. Aras G Aybars A and Kutlu O 2010 Managing Corporate Performance: Investigating The Relationship Between Corporate Social Responsibility and Financial Performance in Emerging

Markets. *International Journal of productivity and Performance management,* 59(3):229-254. Carson D 1977 Point of View: Companies as Heroes? Bah! Humbug!! *New York Times Magazine,* December 25, 1977.

36. Chang H H 2006 *Technical and management perceptions of enterprise information systems importance, implementation and benefits. Inf. Systems J.* 16(3):236-292.
37. Chin D and Kameoka, V.A. 2002 Psychological and Contextual Predictors of Educational and Occupational Self-Efficacy Among Hispanic Inner-City Adolescents. *Hispanic J. Behavioral Sci.* 24:448-464.
38. Courneya K S and Hellsten, L A 2001 Cancer Prevention as a Source of Exercise Motivation: An Experimental Test Using Protection Motivation Theory. *Psychol. Health Med*, 6:59-64.
39. Cronin J J and Taylor SA 1992 Measuring Service Quality: A Re-Examination and Extension. *J. Market.* 56(3):55–68.
40. Friedman M 1970 The social responsibility of business is to increase its profits. New York Times Magazine, 13: 3-33, 122, 124, 126.
41. Hart S L 1995 A Natural-Resource-Based View of the Firm. *Acad. Manage. Rev*, 20:986-1014.
42. Hart S L Mark B Milstein C 2003 Creating Sustainable Value. *Acad. Manage. Executive*, 17 (2):56-69.
43. Harrison P Shaw R 2004 Consumer Satisfaction and Post-Purchase Intentions: An Exploratory Study of Museum Visitors. *Int. J. Arts Manage*, 6(2):23-32
44. Holmes J H Kilbane CJ 1993 Cause-Related Marketing: Selected Effects of Price and Charitable Donations. *J. Non-profit Public Sector Market*, 1: 67-83.
45. Jayakumar A and Geetha K 2014 An Analysis of Consumer Perception Towards Corporate Social Responsibility Practices in Salem City. Proceedings of the Second International Conference on Global Business, Economic, Finance and Social Sciences.
46. Judge Jr William Q Thomas J Douglas F 1998 Performance Implications of Incorporating Natural Environmental Issues into the Strategic Planning Process: An Empirical Assessment. *J. Manage. Stud,* 35:241-260.
47. Kashyap R Mir R Iyer E 2006 Toward a Responsive Pedagogy: Linking Social Responsibility to Firm Performance Issues in the Classroom. *Acad. Manage. Learning Educ.,* 5(3): 366-376.
48. Kanji G K 1998 Measurement of Business Excellence. Total Quality Manage. *Bus. Excellence*, 9:633-643.
49. Kotler P T and Keller K L 2016 *Marketing Management*: 15th Edition. Pearson Education. Upper Saddle Valley, New Jersey, USA.

50. Nochai R and Nochai T 2014 The Effect of Dimensions of Corporate Social Responsibility on Consumers' Buying Behaviour in Thailand: A Case Study in Bangkok. Conference Proceeding of the International Conference on Economics, Social Science and Languages, May 14-15.
51. Nsikan JE Umoh, VA and Bariate M 2015 Corporate Social Responsibility and Mobile Telecommunication Competitiveness in Nigeria: The Case of MTN Nigeria. *American Journal of Industrial and Business Management*, 5(8):527-537.
52. Onlaor W and Rotchanakitumnuai S 2010 Enhancing Customer Loyalty Towards Corporate Social Responsibility of Thai Mobile Service Providers. *World Academy of Science, Engineering and Technology*, 4(6):1341-1345.
53. Rahin RA Jalaludin FW and Tajuddin K 2011 Consumer Behaviour Towards Corporate Social Responsibility in Malaysia. *Asian Academy of Management Journal*, 16(1):119-139.
54. Roscoe JT 1975 *Fundamental Research Statistics for the Behavioural Sciences*, 2nd edition, New York: Holt Rinehart & Winston.
55. Safi A and Ramay MI 2013 Corporate Social Responsibility and Consumer Behaviour: A Study from Pakistan. *Information Management and Business Review*, 5(4):194-202.
56. Schyns B 2004 The Influence of Occupational Self-Efficacy on the Relationship of Leadership Behaviour and Preparedness for Occupational Change. *J. Career Dev.*, 30:247-261.
57. Schyns B Von Collani G 2002 A New Occupational Self-Efficacy Scale and its Relation to Personality Constructs and Organizational Variables. *Eur. J. Worn Org. Psychol.*, 11:219-241.
58. Simon H 1953 Causal Ordering and Identifiability in Hood, W.C. Koopmans, T.C, *Studies in Econometric Method,* 49–74
59. Szymanski DM Henard DH 2001 Customer Satisfaction: A Meta-Analysis of the Empirical Evidence*, J. Acad. Market. Sci.,* 29(1):16– 35.
60. Drucker P F 1974 *Management: Tasks, Responsibilities, Practices.* New York: Harper & Row.
61. Drumwright M E 1996 Company Advertising with a Social Dimension: The Role of Non-Academic Criteria. *J. Market,* 60(4):71-86.
62. Dolandson, 1983 Constructing a Social Contract for Business. In T. Donaldson & P. Werhane (Eds.), *Ethical issues in business*: 153-165.
63. Freeman RE 1984 Strategic management: A stakeholder approach. Boston: Pitman.
64. Alexandris K, Dimitriadis N and Markata, D 2002 Can Perceptions of Service Quality Predict Behavioural Intentions? An Exploratory Study in the Hotel Sector in Greece. *Managing Service Quality*. 12(4):224- 231

References

[1] Craig E 2012 Does Media Attention Drive Corporate Social Responsibility? Journal of Business Research 65(11):1622-1627

[2] Crane A Matten D L J Spence 2013 Corporate Social Responsibility: In Global Context. *Social Responsibility Journal.*

[3] Mcwilliams A and Siegel D 2001 Corporate Social Responsibility: A Theory of the Firm's Perspective. *The Academy of Management Review* 26(1):117-127

[4] Ionnis I and Serafeim G 2010 The Impact of Corporate Social Responsibility on Investment Recommendations. *Academy of Management Proceedings*

[5] Freeman R E Harrison J S and Wicks A S 2007 Managing for Stakeholders: Survival, Reputation and Success. *The Academy of Management Annals* 4(1)

[6] Churchill Jr and Surprenant, C 1982 An Investigation into the Determinants of Customer Satisfaction, *J.Market. Res*. 19:491–504.

[7] Bolton R N Drew J H 1991 *A Multistage Model of Customers' Assessments of Service Quality and Value*, J. Consum Res. 17:375–384

[8] Carroll A B 1979 *A three-dimensional conceptual model of corporate performance. Acad. Manage. Rev*, 4:497-505.

[9] Chang H H Wang Y and Yang W 2009 The Impact of E-Service Quality Customer Satisfaction and Loyalty on E-Marketing: Moderating Effects of Perceived Value. *Total Quality Manage*, 20 (4):423-443.

[10] Agyemang O S and Ansong A (2017) Corporate Social Responsibility and Firm Performance of Ghanaian SMES: Mediating Role of Access to Capital and Firm Reputation. *Journal of Global Responsibility*, 8(1).

[11] Amin-Chaudhry A 2016 Corporate Social Responsibility: From a Mere Concept to an Expected Business Practice. *Social Responsibility Journal*, 12(1):190-207.

[12] Berger I E Cunningham P H and Kozinets, R V 1999 *Consumer Persuasion Through Cause-Related Advertising. Adv. Consum. Res*, 26:491-497.

[13] Famiyeh S 2017 Corporate Social Responsibility and Firm's Performance: Empirical Evidence. *Social Responsibility Journal*, 13(2).

[14] Elkington J 1994 Towards the Sustainable Corporation: Win-Win-Win Business Strategies for Sustainable Development. *California Manage. Rev.* 36 (2):90-100.

[15] Cooil B Keiningham TL Aksoy and L Hsu M 2007 A Longitudinal Analysis of Customer Satisfaction and Share of Wallet: Investigating the Moderating Effects of Customer Characteristics, *J. Market.* 71(1) 67-83.

[16] Abdeen A M 1991 *Social Responsibility Disclosure in Annual Reports. Bus. Forum.* (16):23-26.

[17] Daugherty E L 2001 Public Relations and Social Responsibility. *In R.L. Heath (Ed.), Handbook of Public Relations*, 389-401. Thousand Oaks, California: Sage.

[18] Arshad H 2006 Corporate Social Responsibility: The Rising Trend?

IOP Conf. Series: Earth and Environmental Science **331** (2019) 012064 doi:10.1088/1755-1315/331/1/012064

Comparative Analysis of Biogas Produced from Cow Dung and Poultry Droppings

Evangeline A Diagi[1], Marvel L Akinyemi [1], Moses E Emetere [1], Iyanuoluwa E Ogunrinola[1] and Amanda O Ndubuisi[1]

[1]Department of Physics, College of Science and Technology, Covenant University, Ota, Nigeria.

E-mail: marvel.akinyemi@covenantuniversity.edu.ng

Abstract. This study involves a comparative analysis of two waste substrates; cow dung and poultry droppings. The purpose of this study was to ascertain which of the waste substrates produces a greater yield of biogas, their individual retention period as well as the percentage of the methane content in the gas generated. The study was divided into two experiments which comprised of a water displacement set up in a laboratory and one 30 L fixed dome digester for each waste substrate. In both experiments waste substrates were mixed with water in ratio 1:1 and operated at a mesophilic temperature condition. In the laboratory, the daily gas yield of the individual substrates in ml was obtained, it was observed that cow dung produced an average of 29.9 ml of biogas per day and the poultry waste produced 60.7 ml per day. A gas analysis was performed to obtain the gas profile of gas produced from the two substrates. As deduced from the analysis, cow dung had a methane and CO_2 weight percentage of about 92 wt% and 6.68 wt% respectively, while poultry droppings had about 90 wt% methane and 6.56 wt% CO_2.With the aid of a constructed mini gas stove, a flame test was performed, which gave off a blue flame. Hence, the two substrates are efficient for biogas production.

1. Introduction

Energy is a necessary requirement for economic development, it clearly improves living conditions and standards in various parts of the world. It is an indispensable part of modern society. Despite the advances in technology, many individuals especially in the rural area of developing countries seek to meet their energy requirements for cooking through traditional means of direct combustion of biomass resources such as firewood, straws, and crop residues [1]. This has resulted in evident environmental, economic and public health issues. Therefore, utilization of a 'green' energy source, which is eco-friendly, one that would improve living conditions, health standards and the environment is necessary in meeting the energy demand, hence the research on biogas.

Biogas is a renewable form of energy that provides a cheap, clean and easily controlled source of energy for diverse purposes [2,3,4]. It gradually replaces fossil fuels and direct combustion of biomass as energy sources, thereby lessening its harmful impact on the environment and human health. The composition of biogas is dependent on the substrate fed into the digester. Generally, biogas comprises of 50-75% CH_4, 25-50% CO_2 along with traces of some other components such as water vapour H_2O, and hydrogen sulphide H_2S. Methane is the major constituent of biogas and though the contribution of methane molecules (CH4) to the greenhouse effect is 21 times greater than that of the carbon dioxide molecule, consuming methane in form of renewable energy, reduces its impact on the environment [5]. It has been said that an ideal substrate, is one which is available in large quantities and all year round [6]. Hence, the waste materials to be used in this study are ideal in nature. This research focuses on the design and implementation of a low-cost biogas technology for comparative analysis of cow dung and poultry droppings as sources of biogas.

IOP Conf. Series: Earth and Environmental Science **331** (2019) 012064 doi:10.1088/1755-1315/331/1/012064

2. Materials and Method

2.1. Substrate Waste Materials

The cow dung used in the research was obtained from Eliozu abattoir in Port Harcourt and chicken droppings from Omagwua poultry farm at Rokpokwu. For the purpose of the research, two major experiments were adopted; the main digester experiment and the laboratory water-displacement experiment in which a total solid content (%) test was performed for both substrates. A gas chromatography analysis was also performed on the gas generated from the individual waste substrates to obtain their gas compositions.

2.2. Total Solid Content (TS %).

An analysis to obtain the TS% of the individual waste substrate was carried out using a conventional oven drying method. The samples consisted of 20 kg of slurry for each of the waste substrates. The mass of the empty crucible was termed weight A. The slurry was placed in the empty crucible with the total mass constituting weight B. The crucible loaded with the slurry was then heated in an oven set at 105 °C for two to three hours after which it was placed in a desiccator to cool at room temperature. The crucible with sample was then weighed again and recorded as weight C. The percentage of TS in the samples were obtained using the formula:

$$\%T.S = \frac{weightC - weightA}{weightB - weightA} \times 100$$

where: Weight A- empty crucible
Weight B- crucible + sample quantity
Weight C- crucible + sample quantity after heating and cooling.

2.3. Digester Design

The digester design adopted for this study is a simple on-ground digester type (Figure 1). It is designed to operate at a mesophilic temperature range. The digester system is formed with a plastic gallon with a capacity of 30 kg and a tyre tube as its storage chamber. The digester design is structured to have an influent inlet, gas flow line and a storage chamber. It is connected in a manner such that during anaerobic digestion process gas formed in the digester is released through the outlet valve and then flows along the gas flow line in to the gas storage chamber or can be connected directly to the mini gas stove for use. The design is such that the effluent also referred to as the digestate formed after digestion process has ended, has an outlet from which it is removed and can then be used as organic fertilizer for crop growth.

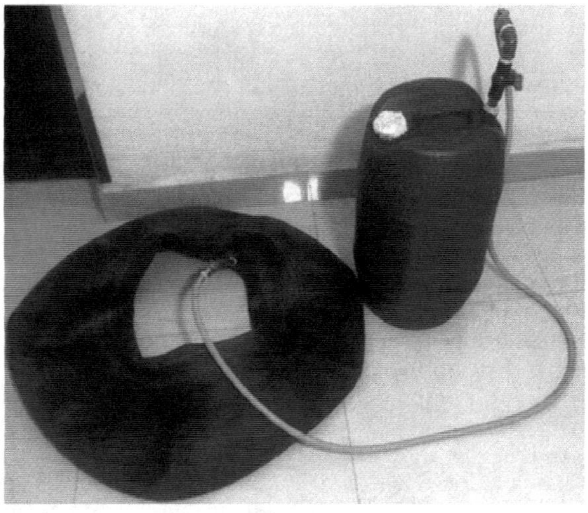

Figure 1. Fixed digester design.

2.4. Preparation of Influent: Slurry Mixture and Digester Loading

IOP Conf. Series: Earth and Environmental Science **331** (2019) 012064 doi:10.1088/1755-1315/331/1/012064

2.4.1. Substrate 1 (Fresh Cow Dung). About 10 kg of fresh cow dung was measured out for the purpose of the experiment. It was mixed with water to reduce the percentage of total solid content (TS%) in the ratio 1:1.

10 kg cow dung + 10 kg of water = 20 kg of slurry

The slurry was transferred into the digester with the aid of a funnel while a head space was made in the digester where the gas formed would reside temporarily. The inlet was tightly sealed and gas outlet valve was closed to ensure absence of oxygen in the digester and to avoid leakage of gas formed.

2.4.2. Substrate 2 (Poultry droppings). The poultry droppings contained saw dust (lignin) which indicates a higher TS%. It was then mixed with water in the ratio 1:1 and transferred into a digester.

10 kg poultry droppings + 10 kg water = 20 kg slurry

2.5. Laboratory Experiment (Water Displacement)

This experiment was carried out to obtain the gas yield and gas volume measurement of the waste substrates daily. A 500ml flask was used as the digester while a 400ml flask of served as the water tank and a measuring cylinder was used as the water collector. With the aid of a rubber hose, the digester was connected to the side of the 400ml flask (water tank), while a longer rubber hose was inserted through the top of the 400ml flask to the bottom giving a little allowance. The other end of the was directed into the 50ml measuring cylinder which serves as the water collector.

This experiment was designed such that the volume of water which was displaced, represents or equals the volume of biogas formed within the period of time. The set-up is shown in Figure 2. The waste substrates were mixed with water as expressed below:
225 g of cow dung + 225 ml of water.
225 g of poultry droppings + 225 ml of water.

The individual sample measured slurry were fed into the 500 ml flask, allowing anaerobic digestion (AD) process to commence while monitoring and recording the gas yield for residence period of 7 days. The set up was placed in an area easily accessible to sunlight to ensure that temperature of the mesophilic range is reached.

Figure 2. Water displacement setup: Slurry in flask, water in 400ml flask and measuring cylinder.

2.6. Gas Analysis

The biogas was analyzed by gas chromatography (GC) technique for molecular composition using GPA 2286 (Figure 3) as the standard test method at Laser laboratory. Gas chromatography (GC) technique involves the use of a gaseous mobile phase to transport sample components through either packed columns or hollow capillary columns containing a polymeric liquid stationary phase. With the aid of the GC machine, the percentage composition and gas profile of individual waste substrates was obtained.

Figure 3. Gas chromatography machine GPA 2268.

3. Results and Discussion

3.1. Total Solid (%) Result

Using the formulas indicated below, the total solid content as well as the moisture content for the individual substrates were obtained and presented in Table 1:

$$\%T.S = \frac{weightC - weightA}{weightB - weightA} \times 100$$

The moisture content was also calculated using the equation below and presented in Table 2:

$$\%\text{ Moisture content} = 100 - \left(\frac{weightC - weightA}{weightB - weightA} \times 100\right)$$

Table 1. Values to obtain TS(%).

Weight (grams)	Cow dung (g)	Poultry (g)
Weight A	31	31
Weight B	51	51
Weight C	40.5	44.5

Table 2. Composition result of individual substrate.

Composition	Cow dung (g)	Poultry (g)
Total solid (%)	47.5	67.5
Moisture content (%)	52.5	32.5

3.2. Water Displacement Experiment

The daily gas volume for 7 days was obtained using the water displacement experiment, where the volume of gas was equivalent to the volume of water displaced through the rubber hose into the measuring cylinder (water collector). The quantity of cow dung and poultry dropping slurry, and volume of containing vessels used are as follows:

Volume of digester	500 ml
Volume of water tank	400 ml
Volume of slurry	450 ml
Head space in digester	50 ml

3.2.1. Substrate I- Cow Dung. As seen in Figure 4, the daily gas production rapidly increases and then peaks on the fourth day, after which the volume of gas produced steadily declines. It is observed that the total volume of biogas produced from cow slurry after 7 days is about 188.5 ml. While the average quantity of biogas produced is 29.9 ml/day.

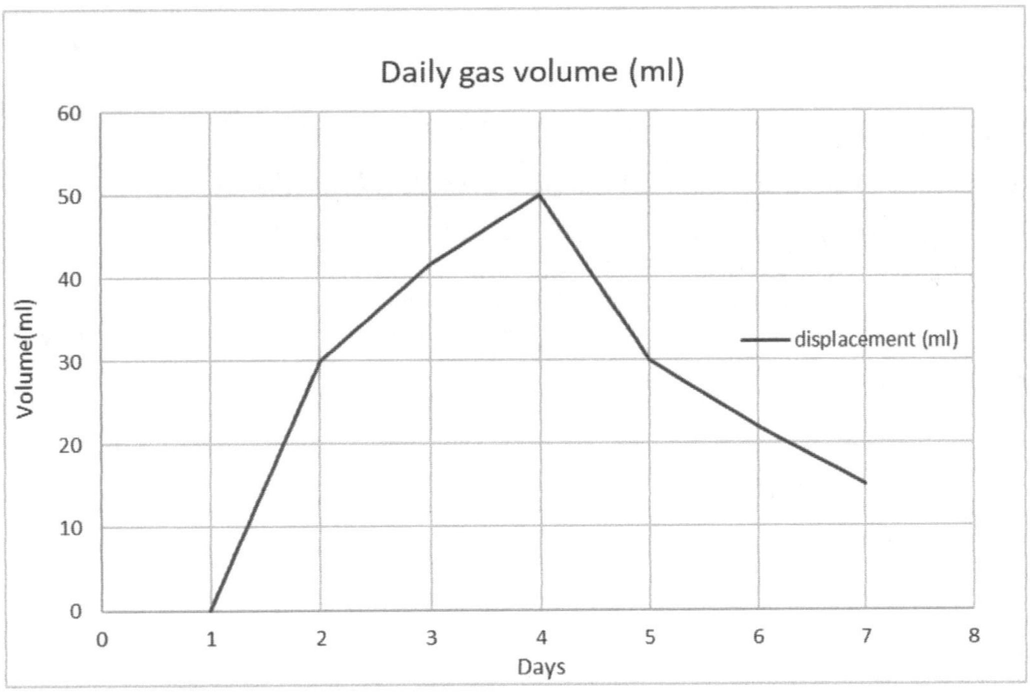

Figure 4. Daily gas yield of cow dung for 7 days (Substrate I).

3.2.2. Substrate II- Poultry Droppings. The total volume of biogas produced from poultry waste slurry is about 425 ml with an average biogas production of about 60.7 ml/day. It is observed in Figure 5 that the gas yield pattern for poultry droppings differs from that of the cow dung. Here, gas formation occurred so rapidly in the first two days, giving off all the gas generated from AD process. It then dropped, allowing for further digestion process to take place which subsequently resulted in a rapid gas yield once more, after which gas yield gradually declined. This suggests the different retention and residence time possessed by both substrates.

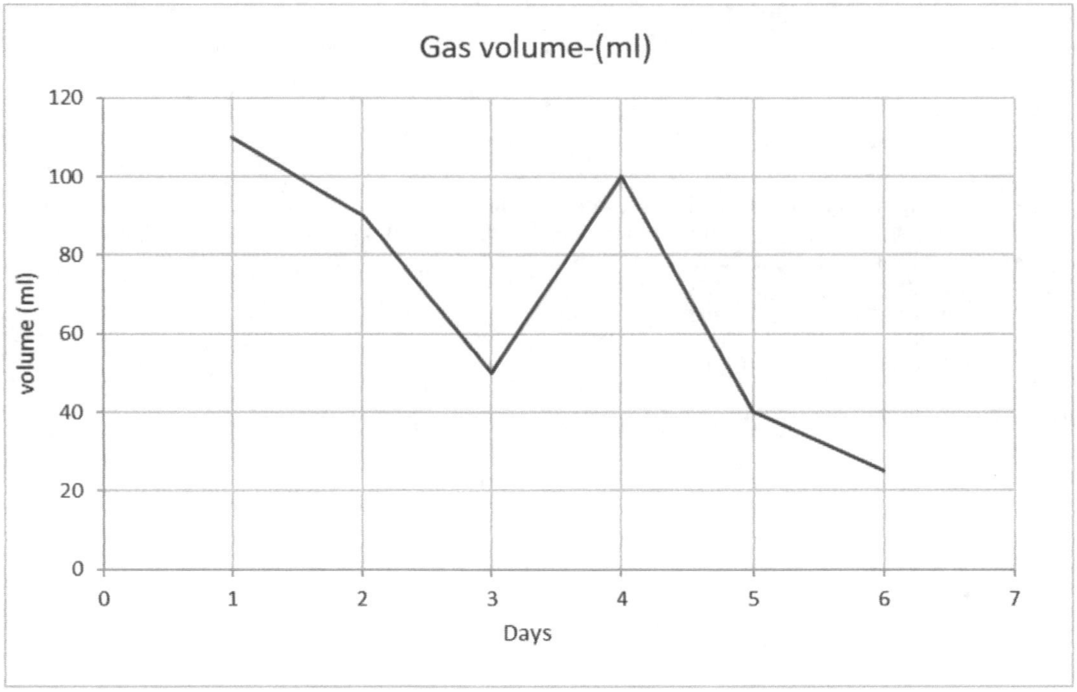

Figure 5. Daily gas yield of poultry droppings for 7 days (Substrate II).

3.3. Gas Analysis Result

The analysis result was obtained with the aid of a gas chromatography machine. It shows the constituents of the gas produced from the individual waste substrates. The gas profiles of each substrate are presented in the following tables and figures (Tables 3 and 4, and Figures 6 and 7). Tables 3 and 4 also show the mole and weight percentage of the various constituents of biogas generated from the two substrates.

Table 3. Composition of biogas sample (cow dung).

Component	Mole%	Wt%
CH_4	96.63	91.97
N_2	0.81	1.35
CO_2	2.56	6.68
Total	100	100
Sample Properties	-	Wt%
Molecular Weight (g/mole)	-	16.85
Gravity [air=1]	-	0.582

IOP Conf. Series: Earth and Environmental Science **331** (2019) 012064 doi:10.1088/1755-1315/331/1/012064

Table 4. Composition of biogas sample (poultry droppings).

Component	Mole%	Wt%
CH_4	96.10	91.24
N_2	1.33	2.20
CO_2	2.52	6.56
Total	100	100
Sample Properties	-	Wt%
Molecular Weight (g/mole)	-	16.91
Gravity [air=1]	-	0.584

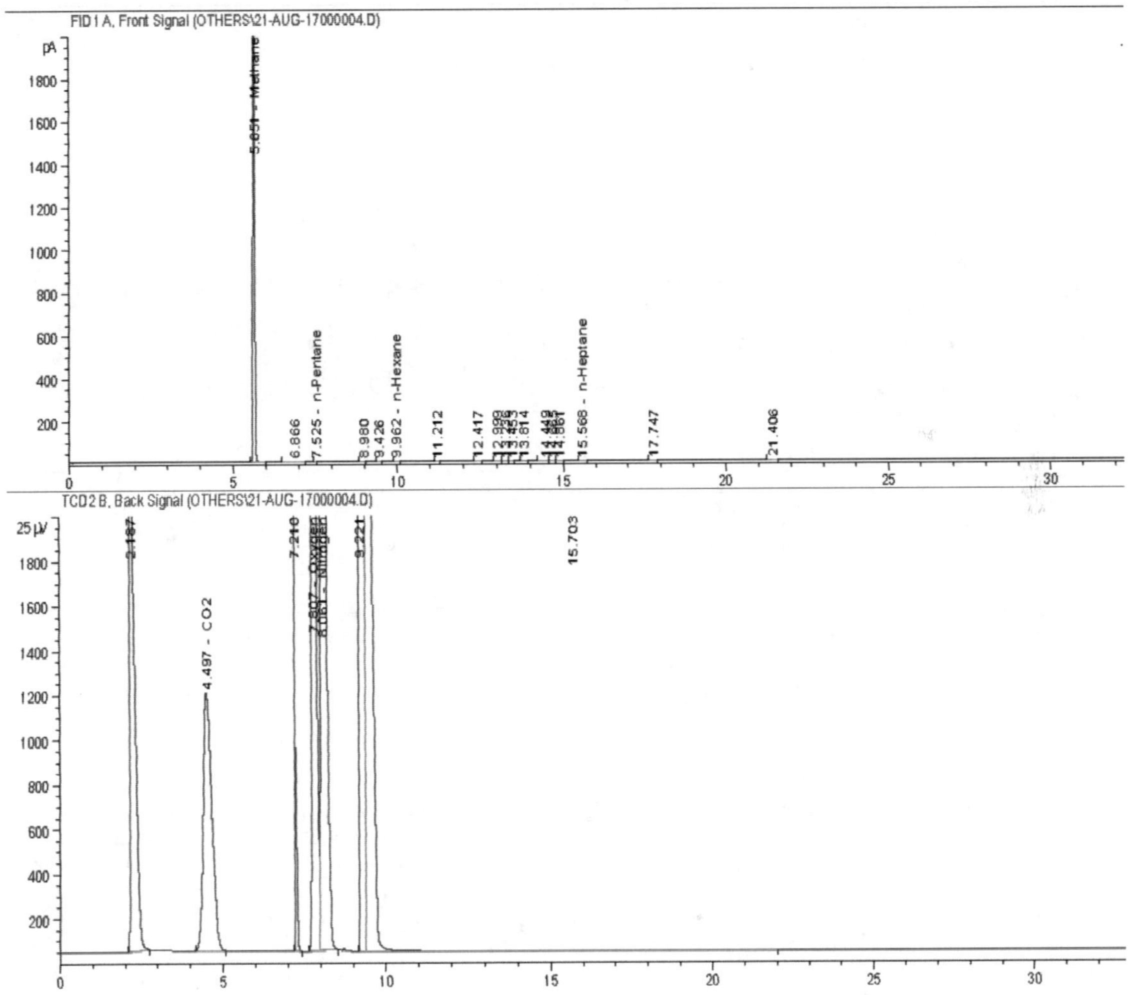

Figure 6. Gas profile of substrate I (Cow dung).

IOP Conf. Series: Earth and Environmental Science **331** (2019) 012064 doi:10.1088/1755-1315/331/1/012064

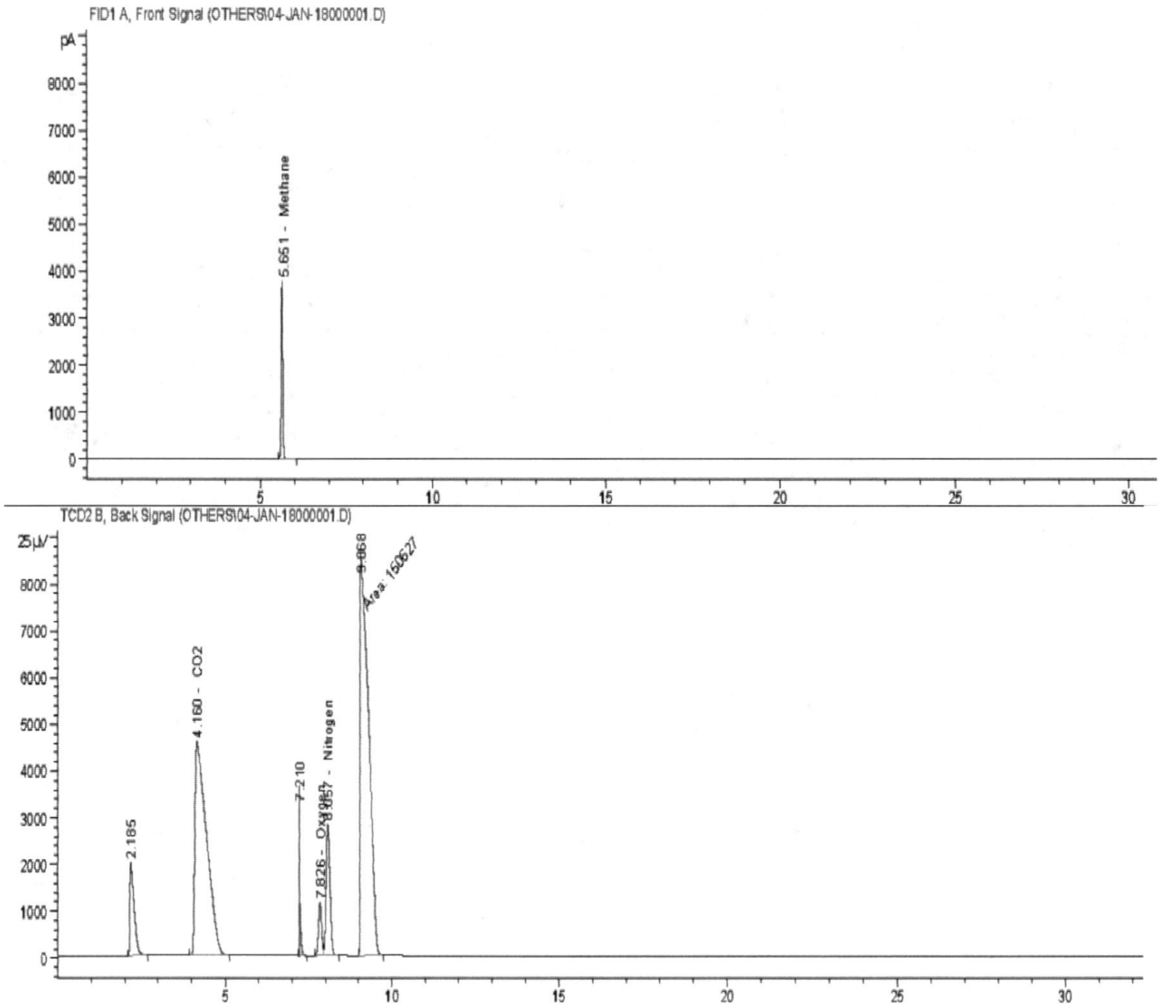

Figure 7. Gas profile of substrate II (poultry droppings).

Figures 6 and 7 show the peak levels of the various constituents of biogas generated from cow dung and poultry waste respectively. Methane being the major constituent is present and of a higher peak in cow dung than in poultry waste, as estimated from the graphs above.

4. Conclusion

This research has been useful in revealing vital information on the gas obtained from the anaerobic digestion process, the gas yield pattern, the gas volume per day and gas profile of generated gas. The research shows that both substrates are good for production of biogas although they differ in terms of their gas profile and yield rate patterns. It was shown from the laboratory experiment that cow dung takes a longer retention time of about 7 days after being placed in the digester before it begins to produce biogas unlike the poultry droppings which has a retention time of just about 3 days. The results obtained from the gas chromatography analysis, shows that cow dung has a higher percentage of methane than poultry droppings, although it yields gas slower and in a short period after the start time of the anaerobic digestion process than the poultry droppings.

5. Acknowledgement

IOP Conf. Series: Earth and Environmental Science **331** (2019) 012064 doi:10.1088/1755-1315/331/1/012064

The authors appreciate the Covenant University Management through the Covenant University Center for Research, Innovation and Discovery (CUCRID) for financial support.

6. References

[1] Surendra K C, Takara D, Hashimoto A G and Khanal S K 2014 Biogas as a sustainable energy source for developing countries: Opportunities and challenges *Ren. Sustainable Rev.* **31** 846-859

[2] Adekoya O, Akan W, Ndubuisi A, Adebayo T and Boyo H 2018 Evaluation of biogas production and pressure from composite of poultry droppings and lemon grass using strain gage rosette. *IOP Conf. Series: Earth Env. Sci.* **173** 012048

[3] Ben-Iwo J, Manovic V and Longhurst P 2016 Biomass resources and biofuels potential for the production of transportation fuels in Nigeria *Ren. Sustainable Rev.* **63** 172-192

[4] Manyi-Loh C, Mamphweli S N, Meyer E L, Okoh A I, Makaka G and Simon M 2013 Microbial Anaerobic Digestion (Bio-Digesters) as an approach to the decontamination of animal wastes in pollution control and the generation of renewable energy *Int. J. Env. Res. Pub. Health* **10** 4390-3317

[5] Dahunsi S O, Oranusi S, Efeovbokhan V E, Zahedia S, Ojediran J O, Olayanju A, Oluyori A P, Adekanye T A, Izebereb J O and Enyinnaya M 2018 Biochemical conversion of fruit rind of *Telfairia occidentalis* (fluted pumpkin) and poultry manure *Energy Sources, Part A: Recovery, Utilization, and Env. Effects* **40** 2799-2811

[6] Vijayaraghavan P, Vijayan A, Arun A, Jenisha J K and Vincent S G P 2012 Cow dung: a potential biomass substrate for the production of detergent-stable dehairing protease by alkaliphilic *Bacillus subtilis* strain VV. *SpringerPlus* **1** 76

IOP Conf. Series: Earth and Environmental Science **331** (2019) 012065 doi:10.1088/1755-1315/331/1/012065

Geoelectrical investigation of foundation failure in Akowonjo, Ogbomoso, Nigeria

Oyelowo Gabriel Bayowa [1], Theophilus Aanuoluwa Adagunodo [2], Olusegun Oladotun Adewoyin [2]

[1] Department of Earth Sciences, Ladoke Akintola University of Technology, Ogbomoso, Oyo State, Nigeria

[2] Department of Physics, Covenant University, Ota, Ogun State, Nigeria

obayowa@lautech.edu.ng; theophilus.adagunodo@covenantuniversity.edu.ng

Abstract. An electrical resistivity investigation was carried out in Akowonjo, Ogbomoso, Nigeria with the aim to obtain various lithological units and subsurface geological features within the study area. Fifty-four sounding points adopting Schlumberger array, with 100 m as the maximum value for half-current electrode spacing were occupied in the study area. Partial curve matching technique and automated platform (known as WinResist) were used to interpret the obtained data qualitatively and quantitatively. The resulting final layers parameters involving resistivity and thickness of each layer were used to map the two-dimensional geoelectric variations beneath the study area. The curve types obtained from iterations of WinResist, which is based on 1-D forward modelling approach include H, KH, HA and QH. Four subsurface layers were identified. These layers are topsoil, laterite, weathered/partly weathered layer and fractured/fresh bedrock. The topsoil had resistivity and thickness values ranging from 76 – 1858 Ωm and 0.4 – 4 m respectively. The lateritic layer lies under the topsoil with resistivity and depth values ranging from 649 – 2021 Ωm and 0.7 – 1.9 m, followed by a weathered/partly weathered layer with resistivity and depth values ranging from 17 – 880 Ωm and 1.9 – 25.2 m, respectively. The fourth layer is the fractured/fresh bedrock which depict varying resistivity from 260 – 33385 Ωm. It is revealed that thick clayey materials dominate the first and second subsurface layers within the study area. This is known to be the major cause of cracks experienced in the buildings in Akowonjo community, Ogbomoso, Nigeria. It is recommended that further foundations for civil engineering works in the study should be extended beyond the zone of water content fluctuation.

Keywords: Foundation failure, Geoelectrical investigation, Competent subsoil, Geoelectric section, Building collapse

1. Introduction

The structure's foundation is a medium that transmit the structure's load to the supporting ground without generating unnecessary vibrations, which could lead to differential settlement in the structure after construction. Most of these structures are constructed on pads (or isolated footing), combined footings, strips, rafts (or mats) for shallow foundations. When deep foundation is required to support massive structures, the foundations are either pile or caissons (or drilled shaft). Structural collapses have been on increase in the recent time. Most of these failures have occurred as a result of variations in subsurface geological sequence [1-2]. Geophysical techniques are capable of producing three-dimensional subsurface imaging without causing physical disturbances to the subsoil, while geotechnical assessment can only be used to access information about the point tested for in the subsurface [3-4].

Foundation failures resulting to building collapses have been one of the alarming issues in Nigeria for over four decades to date. Faulting in rocks and other geological formations has been reported to be one of the causes of these collapses [5]. Other geologic features that could have contributed to this differential settlement and total failure include faults, buried cavity as a result of decayed roots beneath a foundation, joints and subsurface inhomogeneity [6-18]

It is however imperative to carry out near surface geophysical investigations prior to construction of civil engineering structures (pre-foundation study) in order to avert the losses that usually accompany their failure [19]. The information gathered during the pre-foundation geophysical campaign is capable to provide aerial and subsurface information that could assist building engineers, town planners, civil engineers and other construction officials in order to know the appropriate design, exact location and type of foundation that suites individual structure [2]. Geophysical technique is also a proven tool in investigation into the cause of post-foundation deficiencies if such incidence occurs [7, 12, 20-21]. Some of the geophysical techniques that have been found useful independently or integratedly with other geophysical techniques for civil engineering purposes include electrical resistivity, electromagnetic, seismic refraction and magnetic. These techniques have been used in Engineering Geophysics to determine the overburden thickness, map subsurface structures and/or evaluate near surface competency for civil engineering constructions [10-18, 22-24].

Akowonjo is a developing community in Ogbomoso, facing rigorous challenge of cracks on the buildings around the community. The members of the community complained about rigorous cracks running through their buildings during the dry season and it roughly closes up during the wet season. This has necessitated the use of geophysical approach to investigate the cause of this problem. In this present study, geoelectrical investigation into the subsurface geology for civil engineering worthiness in Akowonjo, Ogbomoso, Nigeria was carried out with the aim to determine the cause(s) of foundation failure within the community. Some of the previous works from Ogbomoso (the proximity of the study area) can be found in Refs. [25-38]. All these studies have suggested presence of clayey unit within the weathered layer overlying the water table.

2. Study area location and geology

The study area is bounded by latitude 08° 8.998333' to 08° 9.018333' and longitude 04° 16.440000' to 04° 16.386667', which falls within the southwestern (SW) zone of Nigeria. Roads and footpaths networks serve as means of accessibility in the study area. Akowonjo community in Ogbomoso is within the tropical climate which is characterized by rainy and dry seasons. The rainy season varies from April to October, with a little break in August. The dry season varies from November to March. The annual mean estimation of temperature in the study area is about 27°C, with varying relative humidity from 60 to 80%.

The study area is embedded within the SW Basement Complex of Nigeria (Figure 1). This rock is an integral part of the Proterozoic Schist belt in western half of Nigeria [39]. The geologic sequences noticed in the study location include banded gneiss, quartzite and granite gneiss. The minerals of gneisses are arranged in bands, which are of mafic minerals that have been transformed under the actions of temperature and pressure. The quartzites has light colour and are mainly composed of quartz, which may be categorized as part of the migmatite-gneiss quartzite complex [40].

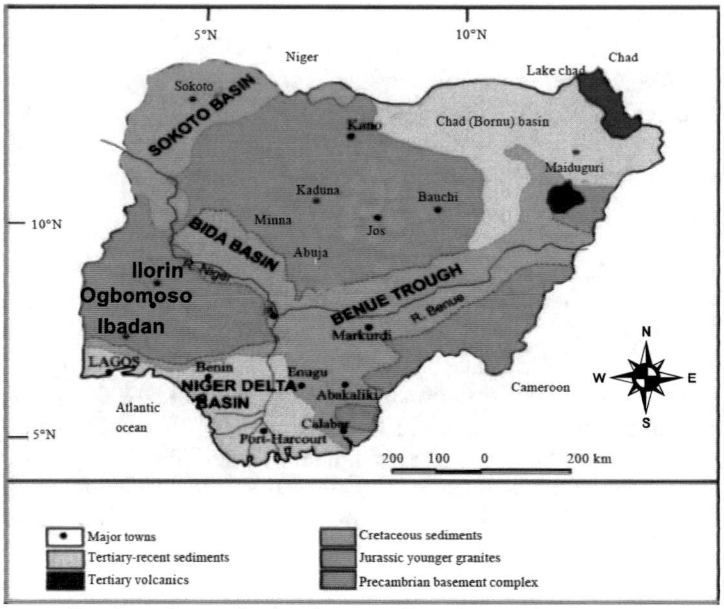

Figure 1: Generalized geology of Nigeria revealing the study area (Adapted from [41])

3. Materials and Method

The resistivity meter with its accessories was used to execute the current study. Fifty-four (54) Vertical Electrical Sounding (VES) points using Schlumberger configuration were occupied randomly in the study area (Figure 2), with a maximum of half electrode spacing of 100 m. Direct current was injected into the near surface via the current electrodes, while the potential difference across its terminals were received via the potential electrodes. This analogy is based on Ohm's law, which was adopted in order to acquire information about the electrical properties of the subsurface. VES is known for its simplicity, conveniency both on field and during interpretation, and its non-destructiveness in determination of depth to rock head (overburden) for formation purpose and investigation into the degree of saturation within the subsurface [42].

The VES data were plotted on a bilogarithm graph with the apparent resistivity (ρ_a) values on the ordinate and the electrode separation (AB/2) along the abscissa. Qualitative field interpretation known as partial curve matching was first carried out on the data set [43] in order to get the first layers parameters. An automated package known as WinResist was used to generate the final layers parameters, which were used for the subsurface interpretation in the study area.

4. Results and Discussion

The VES results in the present study are summarized in Table 1. Four different curve types present in the study area are H, KH, QH and HA. H curve-type takes 52% of the 54 VES sounding points. Other curve types are KH, QH and HA, with the percentage distribution of 22, 7 and 19% respectively. The maximum subsurface geoelectric layers delineated in this current research varied from 3 to 4 units. These layers are topsoil, laterite, weathered/partly weathered layer, and fractured/fresh bedrock.

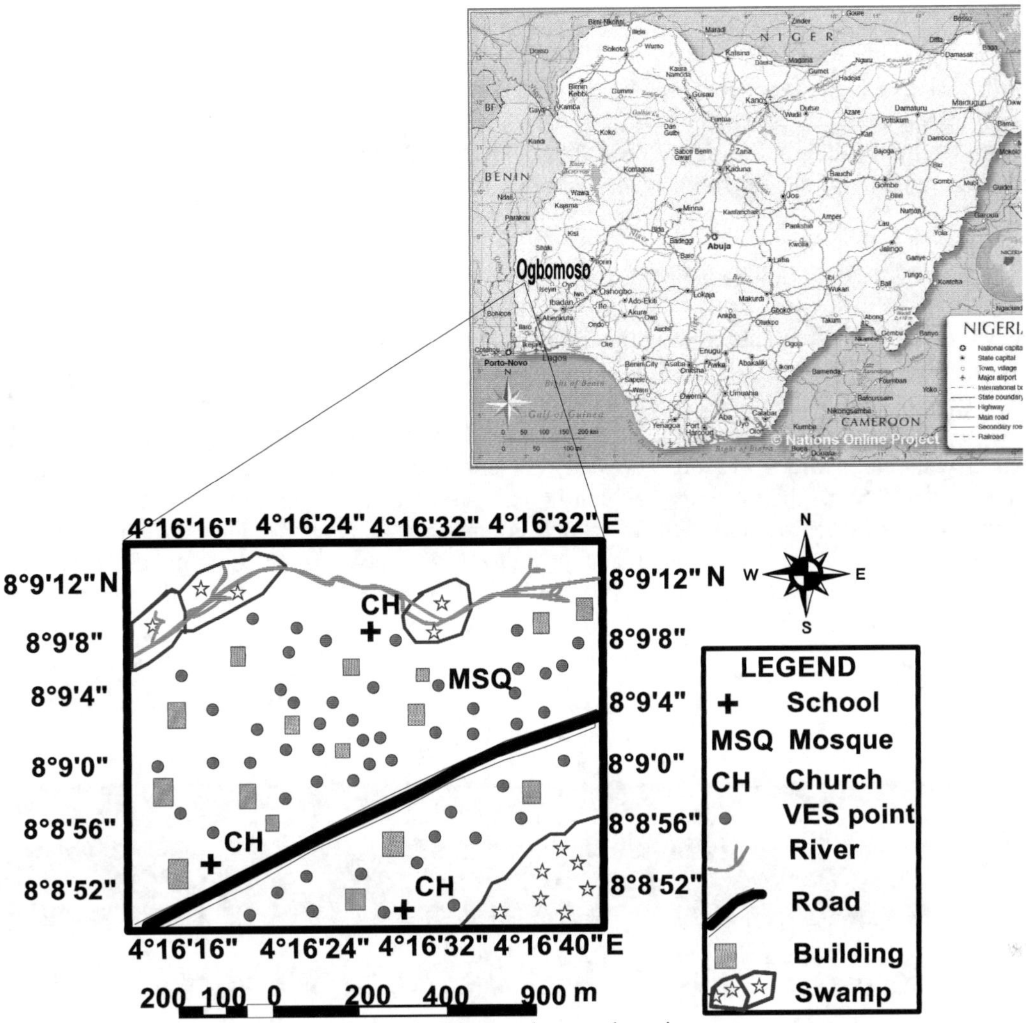

Figure 2: The study area location

Table 1: Classification of Resistivity Sounding Curves

Type of Curves	Resistivity Model	Number of Stations	Curve Type Percentage (%)
H	$\rho_1>\rho_2<\rho_3$	28	52
KH	$\rho_1<\rho_2>\rho_3<\rho_4$	12	22
HA	$\rho_1>\rho_2<\rho_3<\rho_4$	10	19
QH	$\rho_1>\rho_2>\rho_3<\rho_4$	4	7
Total		**54**	**100**

The depth sounding interpretation results are presented as geoelectric sections in order to have insight about the geologic sequence and structural disposition beneath the study area. Three geoelectric sections mapped in the study area trend in SW–NE, W-E and NW-SE orientation, respectively.

4.1 Geoelectric section along SW-NE direction

Figure 3 shows the geoelectric section which relates VES 24, 43, 17, 42, 6, 41, 8, 40 and 13 along SW-NE trending profile. The geoelectric section along this trend reveals four lithologic units. Topsoil,

IOP Conf. Series: Earth and Environmental Science **331** (2019) 012065 doi:10.1088/1755-1315/331/1/012065

which is made up of sandy clay/clay with varying layer resistivity from 76 to 340 Ωm formed the first layer. The depth of topsoil varies between 0.4 and 1.3 m. Laterites formed the second layer depicting resistivity values which vary between 649 and 732 Ωm. This layer's depth ranges between 1.0 and 1.2 m. The third layer which is weathered layer shows resistivity value ranging between 23 to 276 Ωm. The depth of the third layer varies between 2.0 to 17 m which is made of clay/sandy clay. The fourth layer is presumed to be the fractured/fresh bedrock. The resistivity values vary between 439 and 6540 Ωm. The only geologic structure observed in this section is the basement depression beneath VES 8.

4.2 Geoelectric section along W-E direction

Figure 4 shows the geolectric section which relates VES 27, 28, 45, 19, 54, 2, 35, 36 and 9 along W–E trending profile. The geoelectric section along this trend reveals four lithologic units. The first layer constitutes the topsoil and is made up of sandy clay with layer resistivity ranging between 147 and 750 Ωm. The depth beneath this unit varies between 0.6 and 1.4 m. The second layer constitutes the laterite and has a resistivity values ranging between 880 and 1151 Ωm. The second layer's depth ranges between 1.0 and 7.5 m. The third layer which is weathered/partly weathered layer is having resistivity values that vary between 49 and 249 Ωm which comprise of clay/sandy clay. The depth of the third lithologic unit ranges between 3.1 and 20.9 m. The fourth layer is presumably the fractured/fresh basement which has varying resistivity value between 479 and 3490 Ωm. The only geologic structure observed in this section is the basement depression beneath VES 45, 19, 2, and 36.

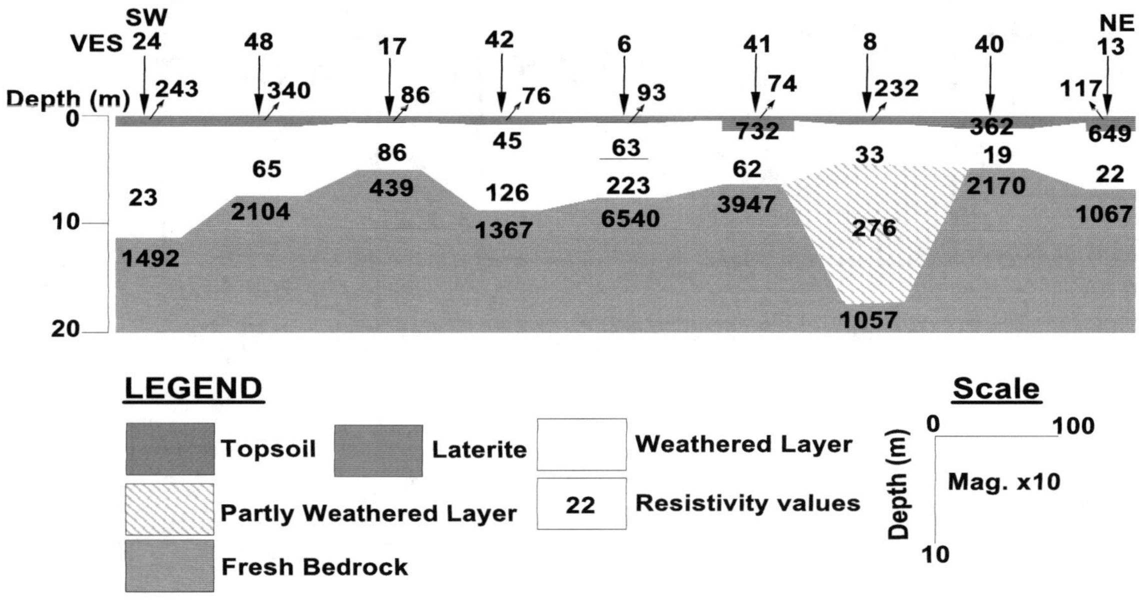

Figure 3: Geoelectric Section along SW-NE direction.

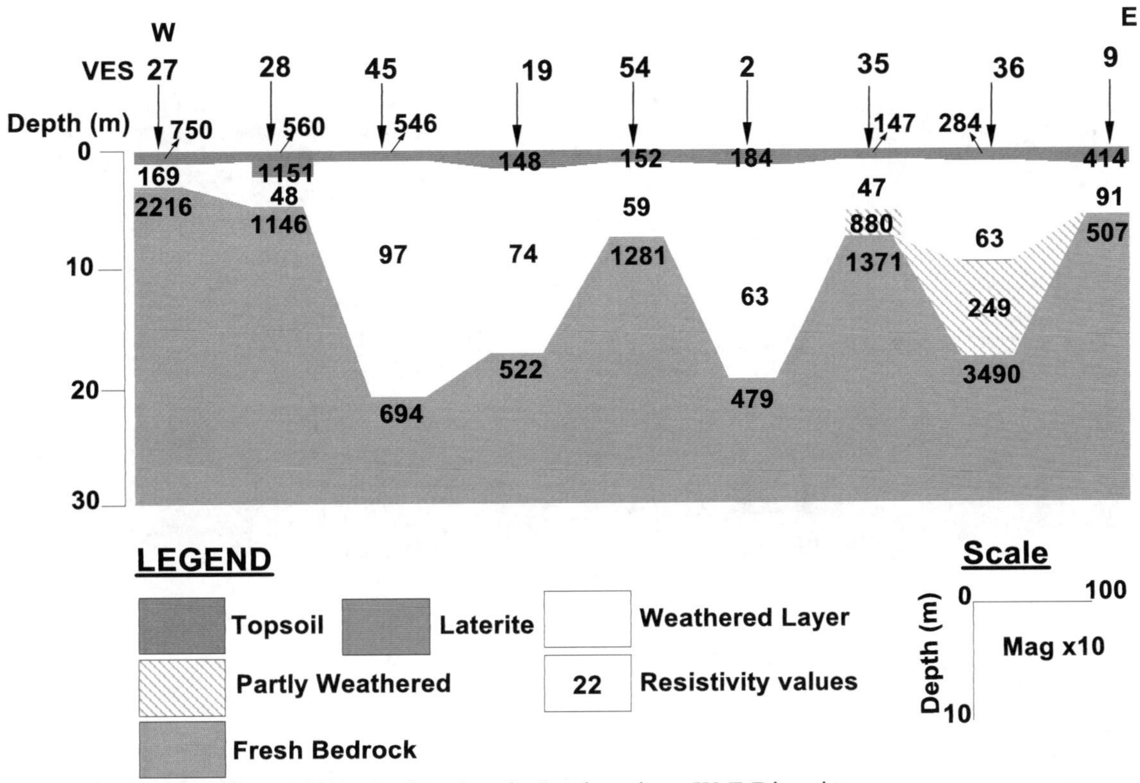

Figure 4: Geoelectric Section along W-E Direction.

4.3 Geoelectric section through NW-SE trend

Figure 5 shows the geolectric section which relates VES 5, 4, 3, 52, 2, 1, 6, 7 and 48 along NW–SE trending profile. The geoelectric section along this trend reveals four lithologic units. The first layer constitutes the topsoil and is made up of sandy clay/clay, with layer resistivity ranging between 93 and 1858 Ωm. The layer thickness ranges between 0.4 and 1.4 m. The second layer constitutes the lateritic layer with a resistivity value ranging between 681 and 1629 Ωm. The depth of the second layer ranges between 1.6 and 1.3 m. The third layer which constitutes the weathered layer has resistivity values which vary between 32 and 760 Ωm. The depth of the third layer ranges between 2.1 and 19.4 m, which is made up of clay/clayey sand. A layer which is interpreted as fractured/fresh bedrock is mapped as the fourth layer, with resistivity value ranging between 425 and 6202 Ω m. The only geologic structure observed in this section is the basement depression beneath VES 4, 3, 52, 2, and 6.

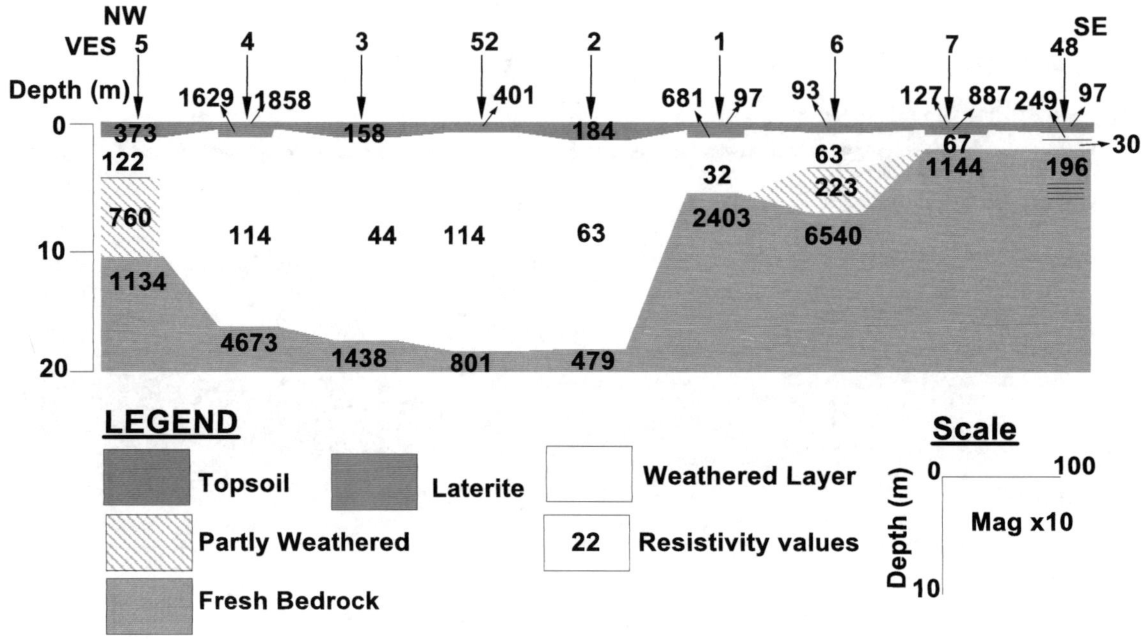

Figure 5: Geoelectric Section along NW-SE.

5. Conclusion

Geoelectrical investigation was done in Akowonjo area, Ogbomoso, SW Nigeria, where cracks were noticed on the columns of buildings, which could affect the stability of the structures in the study area. In doing this, a total of 54 VES points were carried out randomly with the utilization of Schlumberger array in the study area. Four geoelectric sequences were mapped. The topsoil has resistivity and thickness that varies from 74 – 1858 Ωm and 0.4 – 4 m. The second layer which is presumably clay/sandy clay (and in some zones which is revealed as laterite) has resistivity values varying between 649 and 2021 Ωm, and thicknesses from 0.7 to 1.9 m. The third layer is the weathered/partly weathered layer having resistivity value varying between 17 and 880 Ωm, with depth ranging from 1.9 to 25.2 m. The fourth layer which is presumably fractured/fresh bedrock with layer resistivity values between 260 and 33385 Ωm.

Pronounced low resistivity values were noticed across the topsoil and weathered layer, which is an indication that the upper lithologic sequences in the study area are incompetent to support mega structures. Low resistivity and high overburden of the study area indicates the incompetency of the subsoil for erection of engineering structures. From the interpretations of the data collected within the study area, it is affirmed that the cracks on the buildings' column is as a result of presence of clay and inhomogeneity within each stratum. Clay is known with high porosity and low permeability, which results to swelling and shrinking of weathered layers during the rainy and dry season respectively.

It is advisable to extend structures' foundation in the study area beyond the zone of water fluctuation. This will provide required skin friction adhesion beneath the dried zone and help to avoid possible damage that could be done through expansive soils to the structures. It will also inhibit upward transit of water during the rainy season, when the weathered layers are wet and swelling. Based on the types of structures to be constructed in the study area, in-situ compaction and dewatering ground treatments could be adopted before the reinforced concretes would be used for construction of shallow foundation, while piles could be considered for deep foundation in case of mega structures.

Acknowledgment

The partial support from Covenant University is acknowledged.

References

[1] Blyth F.G.H. and de Freitas M.D. (1988). A Geology for Engineers. Butler and Tannar Ltd, Frome and London. Pp. 292-293.

[2] Omoyoloye N.A., Oladapo M.I. and Adeoye O.O. (2008). Engineering Geophysical Study of Adagbakuja Newtown Development, Southwestern Nigeria. Medwell Online Journal of Earth Science, 2(2): 55-63.

[3] Adewoyin O.O., Joshua E.O., Akinwumi I.I., Omeje M., Joel E.S. (2017). Evaluation of Geotechnical parameters using geophysical data. J. Eng. Technol. Sci., 49(1): 95-113.

[4] Adewoyin O.O., Joshua E.O., Akinyemi M.L., Omeje M., Akinwumi S. (2019). Predicting dynamic geotechnical parameters in near-surface coastal environment. Cogent Engineering, 6: 1588081. https://doi.org/10.1080/23311916.2019.1588081.

[5] Adagunodo T.A., Sunmonu L.A., Oladejo O.P. and Olafisoye E.R. (2013). Groundmagnetic Investigation into the Cause of the Subsidence in the Abandoned Local Government Secretariat, Ogbomoso, Nigeria. *ARPN Journal of Earth Sciences*, 2(3): 101–109.

[6] Adenika C.I., Ariyibi E.A., Awoyemi M.O., Adebayo A.S., Dasho O.A., Olagunju E.O. (2018). Application of geotechnical approach to highway pavement failure: a case study from basement complex terrain southwestern Nigeria. International Journal of Geo-Engineering, 9:8. https://doi.org/10.1186/s40703-018-0076-0.

[7] Aroyehun M.T., Akintorinwa O.J. (2018). Application of geophysics in post foundation study: a case study of the Faculty of Social Science Building, Phase I, Federal University, Oye-Ekiti, Southwestern Nigeria. Journal of Geology and Geophysics, 7:1.doi.10.4172/2381-8719.1000324.

[8] Ako B.D. and Olorunfemi M.O. (1989). Geoelectric Survey for Groundwater in the Newer Basalts of Vom, Plateau State. Journal of Mining and Geology, 25 (1 & 2): 247- 250.

[9] Adelusi A.O., Akinlalu A.A. and Daramola B.W. (2014). Integrated Geophysical methods for Post-Construction Studies: Case Study of Omuo Comprehensive High School, Omuo Ekiti, Southwestern Nigeria. Global Journal of Science Frontier Research: Environmental and Earth Sciences, 14: 49.

[10] Adagunodo Aanuoluwa and Sunmonu Ayobami (2013). The Study of Basement Pattern of an Industrial Estate. ISBN: 978-3-659-35930-9. Published by LAP Lambert Academic Publishing GmbH & Co. KG, Heinrich-Bocking. Str. 6-8, 66121 Saarbrucken, Deutschland, Germany.

[11] Adagunodo T.A. and Sunmonu L.A. (2012). Groundmagnetic Survey to Investigate on the Fault Pattern of Industrial Estate Ogbomoso, Southwestern Nigeria. *Advances in Applied Science Research*, 3(5), 3142–3149.

[12] Adagunodo T.A., Sunmonu L.A. and Oladejo O.P. (2014). Effect of Constructing High-Rise Buildings without a Geophysical Survey. *Nigerian Journal of Physics*. Special Edition September 2014, 91–100.

[13] Adagunodo T.A., Sunmonu L.A. and Adeniji A.A. (2015). Effect of Dynamic Pattern of the Saprolitic Zone and its Basement on Building Stability: a Case Study of a High-Rise Building in Ogbomoso. *Journal of Applied Physical Science International*, 3(3), 106–115.

[14] Adagunodo T.A., Adeniji A.A., Erinle A.V., Akinwumi S.A., Adewoyin O.O., Joel E.S., Kayode O.T. (2017). Geophysical Investigation into the Integrity of a Reclaimed Open Dumpsite for Civil Engineering Purpose. Interciencia Journal, 42(11): 324 – 339.

[15] Adagunodo T.A., Sunmonu L.A., Erinle A.V., Adabanija M.A., Oyeyemi K.D., Kayode O.T. (2018). Investigation into the types of Fractures and Viable depth to Substratum of a

Housing Estate using Geophysical Techniques. IOP Conference Series: Earth and Environmental Science, 173: 012030. https://doi.org/10.1088/1755-1315/173/1/012030.

[16] Adagunodo T.A., Sunmonu L.A., Oladejo O.P., Hammed O.S., Oyeyemi K.D., Kayode O.T. (2018). Site Characterization of Ayetoro Housing Scheme, Oyo, Nigeria. IOP Conference Series: Earth and Environmental Science, 173: 012031. https://doi.org/10.1088/1755-1315/173/1/012031.

[17] Oyeyemi K.D., Aizebeokhai A.P., Adagunodo T.A., Olofinnade O.M., Sanuade O.A., Olaojo A.A. (2017). Subsoil Characterization using Geoelectrical and Geotechnical Investigations: Implications for foundation Studies. International Journal of Civil Engineering and Technology, 8(10): 302 – 314

[18] Adagunodo T.A., Sunmonu L.A., Oladejo O.P., Olanrewaju A.M. (2019). Characterization of Soil Stability to withstand Erection of High-Rise Structure using Electrical Resistivity Tomography. In: Kallel A. et al. (eds.). Recent Advances in Geo-Environmental Engineering, Geomechanics and Geotechnics, and Geohazards. Advances in Science, Technology and Innovation (IEREK Interdisciplinary Series for Sustainable Development). https://doi.org/10.1007/978-3-030-01665-4_38 © Springer Nature Switzerland AG 2019. Print ISBN 978-3-030-01664-7, Online ISBN 978-3-030-01665-4.

[19] Hammed O.S., Adagunodo T.A., Aroyehun M., Badmus G.O., Fatoba J.O., Igboama W.M. and Salami A.J. (2017). Geoelectric Survey of Foundation Beds of the Proposed Faculty of Engineering Building, OSUSTECH Permanent Site, Okitipupa, Nigeria. FUOYE Journal of Pure and Applied Sciences, 2(1): 126 – 137.

[20] Fajana A.O., Olaseeni O.G., Bamidele O.E., Olabode O.P. (2016). Geophysical and geotechnical investigation for post foundation studies, Faculty of Social Sciences and Humanities, Federal University Oye Ekiti. FUOYE Journal of Engineering and Technology, 1(1): 62-66.

[21] Bayode S., Omosuyi G.O., Abdullai H.I. (2012). Post-foundation Engineering Geophysical Investigation in part of the Federal University of Technology, Akure, southwestern Nigeria. Journal of Emerging Trends in Engineering and Applied Sciences, 3(1): 203-210.

[22] Adagunodo T.A. and Sunmonu L.A. (2012). Estimation of Overburden Thickness of Industrial Estate Ogbomoso, Southwestern Nigeria. *Advances in Applied Sciences Research*, 3(5), 3129–3141.

[23] Burland J.B. and Burbidge M.C. (1981). Settlement of Foundations on Sand and Gravel. Proceedings of the Institution of Civil Engineers, 78(1): 1325- 1381.

[24] Sunmonu L.A., Adagunodo T.A., Bayowa O.G. and Erinle A.V. (2016). Geophysical Mapping of the Proposed Osun State Housing Estate, Olupona for Subsurface Competence and Groundwater Potential. *Journal of Basic and Applied Research*, 2(2): 27–47.

[25] Akinola M.A., Akinlalu A.A., Adelusis O.A. (2017). Integrated geophysical investigation for pavement failure along a dual carriage way, southwestern Nigeria: a case study. Kuwait J. Sci., 44(4): 135-149.

[26] Sunmonu L.A. (2018). The good and the bad of faults: A geophysical perspective. 25th Inaugural Lecture Series, Ladoke Akintola University of Technology, Ogbomoso, Nigeria. ISBN: 978-2902-82-9.

[27] Sangodiji E.E., Olorunfemi M.O. (2013). Geophysical investigation of a suspected foundation failure at Ogbomoso, southwestern Nigeria. The pacific Journal of Science and Technology, 14(2): 522-536.

[28] Bayowa O.G., Fashola D.K., Adegoke A.B., Agesin A.A., Oyeniyi S.A. (2018). Geophysical investigation for groundwater potential around Ladoke Akintola University of Technology Campus, Ogbomoso, Southwestern Nigeria. Journal of Earth Science and Climatic Change, 9:8. Doi: 10.4172/2157-7617.1000485.

[29] Adewoye A.O., Adegbola A.A., Bolaji A.A., Opebiyi D.F. (2004). Engineering properties of foundational materials of Oyo-Ogbomoso road in southwestern Nigeria. Science Focus, 9: 42-47.

IOP Conf. Series: Earth and Environmental Science **331** (2019) 012065 doi:10.1088/1755-1315/331/1/012065

[30] Adabanija M.A., Oladunjoye M., Adeboye O. (2013). Groundwater prospecting and exploration in a low potential hard rock aquifer: case study from Ogbomoso North, southwestern Nigeria. Journal of Environmental and Earth Sciences, 3(14): 84-102.

[31] Adegbola A.A. and Adewoye A.O. (2012). On investigating pollution of groundwater from Atenda abattoir wastes in Ogbomoso, Nigeria. International Journal of Engineering and Technology, 2(9): 1570-1585.

[32] Adabanija M.A., Afolabi A.O., Olatunbosun A.T., Kolawole L.L. (2014). Integrated approach to investigation of occurrence and quality of groundwater in Ogbomoso North, southwestern Nigeria. Environ. Earth Sci., doi:10.1007/s/2665-014-3401-8.

[33] Adabanija M.A., Adetona E.A., Akinyemi A.O. (2016). Integrated approach for pavement deterioration assessments in a low latitude crystalline basement of south-western Nigeria. Environ. Earth Sci., 75:289. Doi:10.1007/s12665-015-5030-2.

[34] Sunmonu L.A., Adagunodo T.A. Olafisoye E.R. and Oladejo O.P. (2012). The Groundwater Potential Evaluation at Industrial Estate Ogbomoso Southwestern Nigeria. *RMZ-Materials and Geoenvironment*, 59(4), 363–390.

[35] Olafisoye E.R., Sunmonu L.A., Adagunodo T.A. and Oladejo O.P. (2013). Groundwater Contaminant's Investigation at Aarada Waste Disposal Site Using Geophysical and Hydro-Physicochemical Approach. *IOSR Journal of Environmental Science, Toxicology and Food Technology*, 2(4), 01–10.

[36] Oladejo O.P., Sunmonu L.A., Ojoawo A., Adagunodo T.A. and Olafisoye E.R. (2013). Geophysical Investigation for Groundwater Development at Oyo State Housing Estate Ogbomosho, Southwestern Nigeria. *Research Journal of Applied Sciences, Engineering and Technology*, 5(5), 1811–1815.

[37] Adagunodo T.A., Sunmonu L.A., Ojoawo A., Oladejo O.P. and Olafisoye E.R. (2013). The Hydro Geophysical Investigation of Oyo State Industrial Estate Ogbomosho, Southwestern Nigeria Using Vertical Electrical Soundings. *Research Journal of Applied Sciences, Engineering and Technology*, 5(5), 1816–1829.

[38] Adagunodo T.A., Sunmonu L.A. and Oladejo O.P. (2014). Electromagnetic Investigation into the cause(s) of Road Failure along Takie-Ikoyi Road, Ogbomoso. *International Journal of Business and Applied Sciences*, 1(1), 78–85.

[39] Olafisoye E.R., Sunmonu L.A., Ojoawo A., Adagunodo T.A. and Oladejo O.P. (2012). Application of Very Low Frequency Electromagnetic and Hydro-physicochemical Methods in the Investigation of Groundwater Contamination at Aarada Waste Disposal Site, Ogbomoso, Southwestern Nigeria. *Australian Journal of Basic and Applied Sciences*, 6(8): 401–409.

[40] Afolabi O.A., Abimbola A.F., and Kolawole L.L., (2011) Premilinary study of the geology and structural trends of Ogbomoso Township, Southwestern Nigeria. Science Focus: an International Journal of Biological and Physical Sciences, 2 (2): 21.

[41] Omeje M., Adagunodo T.A., Akinwumi S.A., Adewoyin O.O., Joel E.S., Husin W. and Mohd S.H. (2019). Investigation of Driller's Exposure to Natural Radioactivity and its Radiological Risks in Low Latitude Region using Neutron Activation Analysis. International Journal of Mechanical Engineering and Technology, 10(1): 1897 – 1920.

[42] Keary P. and Brooks M. (2002). An Introduction to Geophysical Exploration, KLBS Blackwell Sci. Publ., Oxford. Pp. 296.

[43] Keller G.V. and Frischknecht F.C. (1966). Electrical methods in geophysical prospecting. Pergamon Press, Oxford.

IOP Conf. Series: Earth and Environmental Science **331** (2019) 012066 doi:10.1088/1755-1315/331/1/012066

Wine production from *Hibiscus sabdariffa* calyxes using probiotics starter cultures

*** U. Omole[1] S. Oranusi[1]**

Department of Biological Sciences, Covenant University, Canaan land, Ota, Ogun State, Nigeria
*Corresponding Author: ufuoma.omole@stu.cu.edu.ng +2348130021559

ABSTRACT

Aqueous calyxes extract of Roselle, popularly known as Zobo drink in Nigeria is a non-alcoholic local beverage widely consumed by millions of people across different socio-economic classes. However, large-scale production of zobo is hindered by the rapid deterioration of the drink if not refrigerated. Thus the aim of this research was to extend the viability of the drink by fermentation into red wine using probiotic organism *Lactobacillus fermentum* and *Saccharomyces cerevisiae*. Both organisms produced wines of different physicochemical properties within 24hours fermentation and with pH 3.47 – 2.76, titratable acidity; 0.236%– 0.252%; Vitamin C; 5.22 mg/100 – 2.44 mg/100 and total dissolvable solids; 11.95 Mg/L – 9.97 Mg/L for *Lactobacillus fermentum* and pH 3.48 – 2.8, titratable acidity; 0.2% - 0.225%, Vitamin C; 8.56mg/100 – 6.33mg/100 and total dissolvable solids; 13.32 Mg/L – 9.71 Mg/L for *Saccharomyces cerevisiae*. There was also a gradual alcohol increase during fermentation process and at the end, both wines showed alcohol content of 5.71% and 5.61% for *Lactobacillus fermentum* and *Saccharomyces cerevisiae* respectively. Except for the alcohol content, the results shows that the formulation with the brewers yeast produced a better wine at the 12 hour fermentation time. The sensory evaluation was also conducted using a 5 judge panel on a 7 point hedonic scale and the result shows that the wine produced from fermentation by *Saccharomyces cerevisiae* was moderately accepted (value 3) by the panel and the mean heterotrophic microbial load of $3\pm0.2 \times 10^3$ cfu/ml indicated that formulated red wine is safe for human consumption. The outcome of this study will boost the economy of this country if adopted for large scale production as it will generate revenue that would have been lost on importation of wine.

Keyword: *Hibiscus sabdariffa*, Calyxes, Fermentation, Roselle, Zobo, Wine, Probiotics

1. Introduction

Roselle (*Hibiscus sabdariffa* L.) is a commonly grown plant and is extensively grown in Nigeria, particularly in the country's North-Eastern and middle belt areas [5]. It is an herb of economic importance as it can serve as a source of essential minerals and vitamins such as riboflavin, niacin, calcium and iron [18].

Zobo drink is a traditional refreshment normally created by either boiling or steeping the calyx of sorrel in a consumable, clean water and improved by sugar, which afterwards refrigerated before serving chilled to consumers [20]. The red and additionally purple succulent calyces are bubbled, sieved, improved with sugar, and flavoured with pineapple juice, ginger and lemon to deliver the drink. This drink has demonstrated its immense potential to be great source of carbohydrate, vitamin C and protein which is the significant reasons for drinking juice [19].

However, Zobo drink has not been successfully produced at commercial scale. This is associated with its short shelf life attributable to microbial activity. This short shelf-life (approx. twenty-four hours following production if not refrigerated) of the drink is associated with degradation of the nutrient component which erodes its health benefits and antioxidant property. This has generated an interest in the possible fermentation of this drink by probiotic organisms to produce red wine that can impart additional nutrient and extend viability. Also in 2016, Nigeria spent about N9 billion on the importation of champagne (sparkling wine) alone [7]. However, high duty on imported wines has stimulated interest in the promotion of our local drinks (kunu and zobo) in the Presidential villa [8] and there was decline in the importation of wine by 24% [9]. Possible wine production from Zobo drink will provide health benefits to its consumers due to abundance of phenolic compounds coupled with low alcohol content. Thus, the objective of this work is to produce red wine from *Hibiscus sabdariffa* calyxes by fermentative activities of probiotics organisms *Saccharomyces cerevisiae* and *Lactobacillus fermentum* as well as determination of microbial, physico-chemical and sensory properties of the wine.

2. Methodology

2.1 Sample collection

The samples used for the isolation of the fermenting organisms and the production of wine were obtained from suppliers opposite Canaan land, Ota and Oja Ota in Ogun state. Samples acquired were corn gruel, yogurt and queen pineapple for isolation of the fermentation organism *Lactobacillus fermentum* and *Saccharomyces cerevisiae.* During the fermentation process, hibiscus calyxes were used to produce wine and sugar serve as a sweetener and carbon source

2.2 Isolation and identification of probiotic organisms

De Man Rogosa and Sharpe Agar (MRS) and Potato Dextrose Agar (PDA) made according to manufacturer's instructions were used for the isolation and characterization of the probiotic microorganisms. A volume of 0.1ml of the dilution factors 10^{-3}, 10^{-5} and 10^{-9} were plated aseptically on MRS agar (for isolation of *Lactobacillus*) and PDA (for isolation of *Saccharomyces*). MRS agar plates were incubated in 5% CO_2 using an anaerobic jar while PDA plates were incubated at about room temperature of 25 ° C for 3-5 days. The colonies on the agar plates were observed for morphology, pigmentation, shape and size upon incubation, and subsequently sub-cultured and incubated until pure colonies were obtained after which tests were performed to tentatively identify the organisms. Isolated organisms were identified based on their morphological, biochemical, cultural and physiological features. The tests carried out were Gram reaction, test for urease, oxidase, catalase, citrate utilization, indole, fermentation of different sugars test, endospore tests and milk coagulation test.

2.3 Molecular identification

Cultures of pure colonies grown on liquid medium were centrifuged at 4600 x g for 5 min. The resultant sediments were resuspended in 520 µl of TE buffer (10 mM Tris-HCl, 1mM EDTA, pH 8.0). 15 microliters of 20% SDS and 3µl of Proteinase K (20 mg/ml) were included. The resulting mixture was placed in an incubator for 60 minutes at around 37 °C, then 100 µl of 5 M NaCl and 80 µL of a 10% CTAB solution in 0.7 M NaCl were introduce and thoroughly mixed. The suspension was positioned in an incubator for 10 min at 65 °C and maintained on ice for 15 min. An exact volume of chloroform: isoamyl alcohol (24:1) was introduced, followed by ice incubation for 5 min and centrifugation at 7200rpm for duration of 20 min. The aqueous phase was decanted to a new tube, isopropanol in the ratio 1: 0.6 was included and DNA was precipitated out at –20 °C for 16 h. DNA was obtained by centrifugation at 7200rpm for 10 min, rinsed with 500 µl of 70% ethanol, dried in the incubator at 37°C for 30 minutes and finally dissolved in 50 µl of sterile distilled water as stock DNA. The stock DNA was checked on a 1.5% Agarose gel ran on a voltage of 120V for about 40min. This later viewed under UV light to confirm the presence of the bacteria DNA. The stock DNA was also quantified on Nanodrop spectrophotometer 2000 to determine the quantity and also purity of the samples before PCR. Working DNA solution was diluted at 1:50 for subsequent PCR assay.

IOP Conf. Series: Earth and Environmental Science **331** (2019) 012066 doi:10.1088/1755-1315/331/1/012066

2.4 Production of wine from hibiscus calyces

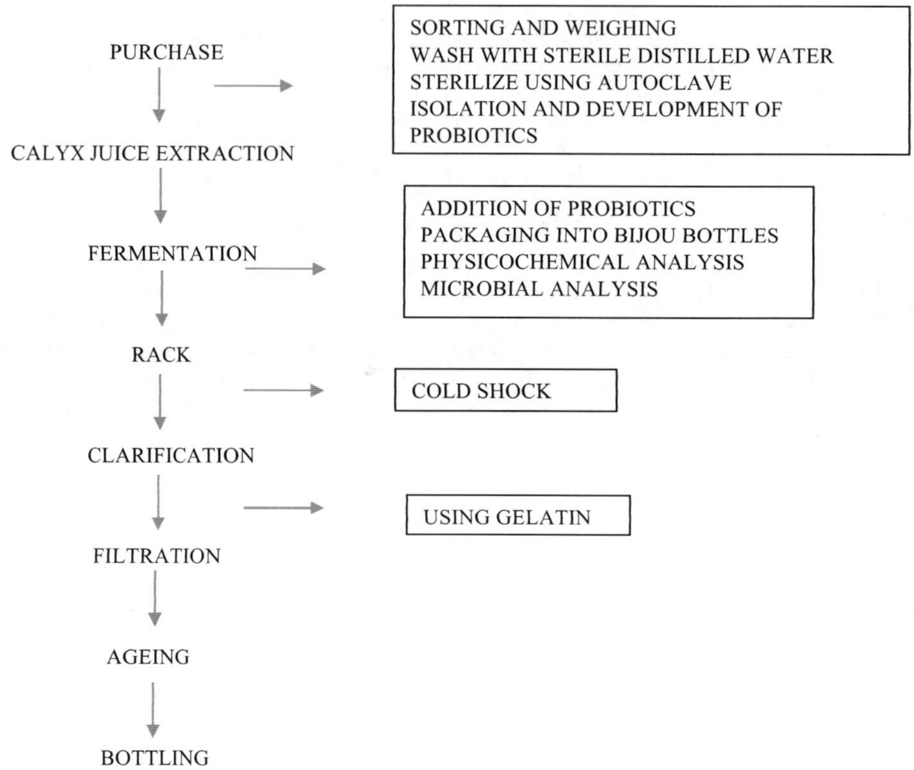

Fig 1: Flow diagram showing production line of red wine

2.5 Inoculum development

The isolates were developed using their respective medium for growth. De Man Rogosa and Sharpe broth for cultivation of *Lactobacillus fermentum* and Potato Dextrose broth for growing *Saccharomyces cerevisiae*. The yeast was introduced into the PD broth and later incubated at 27°C for 48 – 72 hours. After 72 hours, the broth was scaled up by 500 ml till a volume of 1.5 L was attained. After scaling up was complete, the inoculum was centrifuged at 4000 rpm for 15minutes and the sediment was used for fermentation. *L. fermentum* was inoculated into MRS broth and incubated at 30°C to 35°C for 3 – 5 days. Scaling up was also done till a volume of 1.5 L was acquired and then centrifuged at 4000 rpm for 15 minutes and the sediment was used for fermentation.

2.6 Preparation for calyx extract fermentation

The Hibiscus calyxes were sorted to remove dirt, after which 100 g were weighed in three places and wrapped in foil paper and sterilized. 100g of the sterile calyxes were introduced to 1000 ml of sterile deionized water in triplicates and allowed to stand for 30mins to aid extraction. The suspension were filtered using the sterile muslin cloth and 100g of sugar was added as carbon source to each of the filtrate to aid fermentation. Three formulations A, B and C were adopted. Formulation A was sealed without inoculation while for Formulation B and C, 1 ml of the organisms were inoculated into their respective flasks and corked.

A: Calyx extract + sugar
B: Calyx extract + sugar + *Saccharomyces cerevisiae*
C: Calyx extract + sugar + *Lactobacillus fermentum*

IOP Conf. Series: Earth and Environmental Science **331** (2019) 012066 doi:10.1088/1755-1315/331/1/012066

The three formulations were incubated anaerobically at 25^{0}C for 24 hours with 12 hour analysis of physicochemical properties from the time of preparation. The red wine obtained from the fermentation was clarified using gelatin

2.7 Physicochemical analysis

The formulations were shared into different sterile bijou bottles which were used to store the formulations for the different time zone (0 hours, 12 hours and 24 hours). The physicochemical parameters determined were the pH, total titratable acidity, total dissolved solids, alcohol content and vitamin C content. The pH was assessed by aid of pH meter in sample. Meter reading was recorded in triplicates and an average was calculated. For the estimation of total titratable acidity, 15 ml of the sample was measured and 100 milliliters of distilled water was added followed by 3 drops of phenolphthalein indicator and then titrated against 0.1N NaOH solution, until colour changed to pink. For total dissolved solids, a total of 50 cm^3 of the sample was measured and filtered, then poured into cleaned, dried and pre-weighed evaporating dish and put in the oven at 105°C until the sample dried completely, cooled inside desiccator and weighed until the constant weight is recorded. Using the specific gravity method, the formula for calculating alcohol content was calculated thus as described by Biri [10]. Alcohol content by volume (%) = (Original Gravity-Final Gravity) × 131.25.

2.8 Total viable cell count

At intervals of 0 hours, 12 hours and 24 hours, microbiological culture was done on Potato Dextrose Agar and De-man Sharpe Rogosa agar to check for the variation in the growth of the organism used to achieve fermentation.

2.9 Organoleptic evaluation

According to a procedure reported by Maragatham & Panneerselvam (2011), organoleptic assessment was performed for all the formulations [11]. Wines were evaluated by a panel of 5 judges who were very familiar with wine. The judges were made up of 3 males and 2 females, ages between 23 to 30 years Panelists were selected based on interest and availability and familiarity with wine (Drinks wine at least once in a week) lower values indicate greater acceptance on a 7-hedonic scale.

3. Results

The biochemical characteristics of organisms isolated from respective food source revealed the presence of *L. fermentum* and *S. cerevisiae* in table I. The identity is further confirmed with DNA sequencing and PCR amplification of the 16srRNA and the phylogenetic neighbor-joining tree is constructed based on the sequence (Fig. 2). During the experimental stage of this work, physicochemical analysis such as pH determination, temperature, titratable acidity, vitamin C content, total dissolved solids, specific gravity and alcohol content were analysed during the 0-24hours of fermentation (Table II) and after clarification (Table III). The sensory evaluation as judged by 5 panelist are elucidated in Table IV

TABLE I: Biochemical characteristics of Isolates

	L. fermentum	***S. cerevisiae***
Colony morphology	Rough, round	Raised, Oval
Cell shape	Rod	oval
Gram staining	+	Nil
Catalase test	-	+
Oxidase	-	-
Urease test	-	-
Citrate Utilization	-	-

Sucrose Utilization	+	+
Lactose Fermentation	+	+
Glucose Fermentation	+	+
Gas production	+	+
Spore staining	-	-
Indole test	-	-

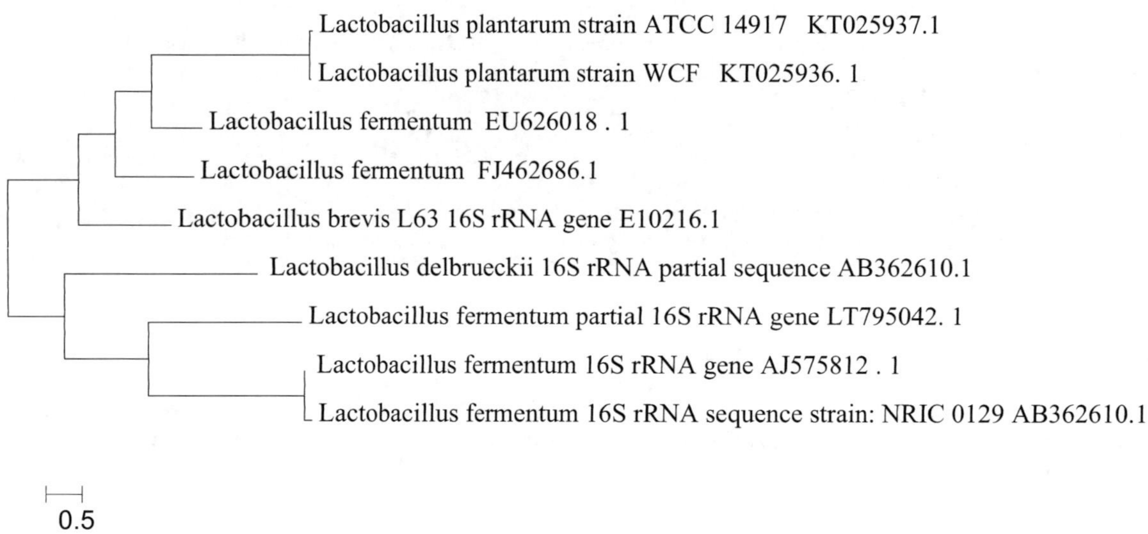

Fig 2: The Neighbor-Joining method was used to deduce the evolutionary history [17].

TABLE II: Physicochemical properties of the three formulations of the drink at different time regime

PARAMETERS / FORMULATIONS	FERMENTATION TIME(HR)	pH	ACIDITY (%)	VITAMIN C (Mg/100)	TOTAL DISSOLVABLE SOLIDS (Mg/L)	ALCOHOL CONTENT (%)	TOTAL VIABLE COUNT (cfu/ml)
A	0	3.50	0.195	10.54	14.31	0	0
	12	3.39	0.210	6.54	11.31	0	0
	24	2.79	0.225	4.20	5.16	0	0
B	0	3.48	0.2	10.5	13.31	0	2
	12	3.37	0.215	6.34	12.31	2.7	3

	24	2.80	0.225	6.33	9.71	5.65	5
C	0	3.47	0.236	10.53	11.95	0	3
	12	3.30	0.239	2.60	10.52	2.9	5
	24	2.76	0.252	2.44	9.97	5.71	7

Table III: Physicochemical Analysis of Wine after Clarification

	pH	**ACIDITY (%)**	**VITAMIN C (Mg/100)**	**TOTAL DISSOLVABLE SOLIDS (Mg/L)**	**ALCOHOL CONTENT (%)**
A	2.5	0.27	4.0	5.1	0
B	2.6	0.273	6.2	9.71	5.65
C	2.42	0.29	2.38	9.97	5.71

Table IV: Sensory evaluation of wine

Product	Formulation A	Formulation B	Formulation C
		Sensory attribute	
Aroma	3.2	2.80	3.00
Sweetness	3.0	2.60	2.60
Clarity	3.4	3.60	3.60
Alcohol content	3.6	3.00	3.20
Overall acceptability	3.30	3.00	3.10

IOP Conf. Series: Earth and Environmental Science **331** (2019) 012066 doi:10.1088/1755-1315/331/1/012066

4. Discussion

There is a dearth of information on the fermentation of aqueous extract of *H. sabdariffa* calyxes using *L. fermentum* and *S. cerevisiae* to produce red wine. The choice of these organisms were deliberate as both organisms have been implicated in the fermentation of grape for production of wine [12]. In several African cereal-based fermented food, *Lactobacillus fermentum* has been recognized as the predominant lactic acid bacteria (LAB) species that possess intrinsic functional features and can add to food quality and safety by providing one or more organoleptic, dietary or health benefits-The various biochemical characteristics were determined to identify both *Sacharomyces cerevisiae* and *Lactobacillus fermentum.* Sequencing of the isolates further confirm our organisms and the phylogenetic was constructed based on the obtained sequence which is elucidated in figure 2.

There were changes in the quality attributes and physicochemical properties of 'zobo drink' (formulation A) and formulation B and C following fermentation with *Sacharomyces cerevisae* and *Lactobacillus fermentun* respectively (table II). The pH reduced from 3.50 to 2.79 over 24 hours. Titrable acidity which measures the amount of acid in a particular product [13] increased across the 24hours (0.195-0.225) in formulation A. This indicate and confirm the high acid content usually observed in Roselle drinks. Roselle is a naturally acidic fruit endowed in organic substances with low pH: Oxalate, tartarate, malate and succinic acid [14]. Formulation B and C showed gradual increment in their acidic content and a corresponding reduction in their respective pH value. These observations are similar to an account from Alobo and Offonry [6] on roselle wine in which the titratable acidity increased and pH reduced at the end of aging The increase in titrable acidity and decrease in pH in formulation B and C can be attributed to the effect of the yeast and lactic acid bacteria in fermenting the sugars to produce organic acids [15]. Acidity is crucial in wine fermentation as it determines the quality of the wine and creates an environment that inhibit potential spoilage microorganisms but support the growth of desirable organisms [1]. Alcohol content at the onset of the fermentation was zero for the three different formulations (without any pre-inoculated starter culture-A; with *Sacharomyces cerevisiae* –B; and with *Lactobacillus fermentum* -C), but as time progresses, alcohol content gradually increased to 5.65 and 5.71 for formulation B and C respectively (Table II) while formulation A had no change due to the absence of microorganisms. These observations agrees with the findings of Archibong [1]. During fermentation, microorganism carry out metabolic activities which result in the release of metabolites such as organic acids. This may account for the increase in alcohol content recorded across the different formulations. Low alcohol content reported in this work is in tandem-with Yokotsuka *et al.*(1997) [16] who reported that Hibiscus *sabdariffa* calyces contained small amounts of acetaldehyde, pyruvic acid, and alpha- keto glutaric acid, which normally react with SO_2 to form bisulphate complexes in fermentation, leading to the low amount of alcohol produced at the end of fermentation [16]. Total soluble solids measures the sugars present in the fermented must and wine. Total dissolvable solids (Table II) decreased across the different formulation with the formulation A having a range of 14.31-12.00, formulation B (13.31-9.71) and formulation C (11.95-9.97). The decrease can be linked to the activity of organisms that utilizes these sugars to produce acids. The total dissolvable solid for formulation B and C were more reduced compared to formulation A. This could be attributed to the pre-inoculated starter culture in both B and C formulatios. This agrees with the report of Efiuvwevwere and Eka [4] and also Nwafor and Ikemebomah [2]. There was a marked decrease in vitamin C content in all formulations. The loss of vitamin C is known to optimize with heat, light and oxygen exposure. Oxidation happens in fruit juices during storage and is extremely dependent on the presence of oxygen in the head room or dissolve in the samples could also be ascribed to the observed reduction. The sensory panelists comprised 60% males and 40% females between the ages of 23 - 30 years, though it is generally known that women have more developed olfactory and sensory system in wines than men [3]. Among the test wines, formulation B was the most acceptable wine to the panelists in terms of the sensory attributes analyzed among the test wines and also the most preferred in terms of the product preference.

5. Conclusion

In conclusion, the result shows *Saccharomyces cerevisiae* and *Lactobacillus fermentum* were able to ferment extract of red calyces of *H. sabdariffa* to produce red wine that is consumable with the absence of microorganisms due to sterilization methods ensured. This research work elucidates that zobo drink can be fermented *by Saccharomyces cerevisae* and *Lactobacillus fermentum* to produce red wine. The composition of the low alcohol wines produced from *H. sabdariffa* calyces showed acceptable physicochemical properties with reduction in pH, increase in total titratable acidity, reduction in the carbohydrate value and decrease in the total soluble solid. This research shows that the most preferred probiotics to be used would be *Saccharomyces cerevisiae* as it has proved to be highly efficient in wine production from aqueous extract of *Hibiscus sabdariffa.* The outcome of this study, if adopted for large scale production, will boost the economy of this country as it will generate revenue that would have been lost on importation of wine. It will also create job opportunity for the teeming Nigeria youth thereby reducing the rate of unemployment in the country.

REFERENCES

[1]. Archibong, E. ., Ezemba, C. C., Chukwujama, I. C., & Archibong, U. E. (2015). Production of wine from mixed fruits : pineapple (*Ananas comosus*) and orange (*citrus sinensis*) using yeast. *World Journal of Pharmacy and Pharmacutical Sciences*, *4*(8), 126–136.

[2]. Nwafor, O. E., and Ikenebomeh, M. J. (2009). Effects of different packaging materials on microbiological, physicochemical and organoleptic quality of Zobo drink during storage of room temperature. *African Journal of Biotechnology*., 8: 2848-2852

[3]. Wilf (2011). Wilf's Wine press. *http://wwpress.blogspot.com/2011/08/wine-womenand-cardinal-rule-of-tasting.html;* accessed March 3, 2019

[4]. Efiuvwevwere, B. J.O. and Eka, A. E. (1991). Shelf and microbiological stability of acidified non inoculated (*Altenaria tenuis*) in tomato juice. *Journal of Food Processing Preservation* 15: 159-165

[5]. Oboh, G., Elusiyan, C. A. (2004). Nutrient Composition and Antimicrobial Activity of Sorrel Drinks (soborodo). *Journal of Medicinal Food*. **7**(3):340-342

[6]. Alobo, A.P. and Offonry S.U. (2009). Characteristics of Coloured Wine Produced from Roselle (*Hibiscus sabdariffa*) Calyx Extract. *Journal of Institute of Brewing* **115**(2): 91–94.

[7]. Murphy, S. (2016) Nigeria Spends Nine Billion Naira Per Annum on The Importation Of Champagne. Retrieved from https://www.drinks.ng/Nigeria-spends-nine-billion-naira-per-annum-on-the-importation-of-champagne. Accessed 15-05-2019, 20:40 hours

[8]. Oti, A. (2016). Zobo and Kunu Vs Champagne and Wine. Thisday Newspaper. Retrieved from https://allafrica.com/stories/201603280660.html

[9]. Ani, M. and Adams, S. (2019). Nigeria's Champagne Imports Down 24% since 2014, Mirrors Economic Slump. *Business Day*. **15**(297): 134

[10]. Biri, H. B., Pan, O. G., Yahaya, M. M., & Ezeribe, A. (2015). Wine from water melon juice using palm wine yeast isolate. *International Journal of Research in Engineering and Science*, *3*(1), 35–40.

[11]. Maragatham C., and Panneerselvam A. (2011). Standardization technology of papaya wine making and quality changes in papaya wine as influenced by different sources of inoculums and pectolytic enzyme. *Advances in Applied Science Research*. **2** (3): 37-46.

[12]. Walker, G.M. and Stewart, G.G. (2016) *Saccharomyces cerevisiae* in the Production of Fermented Beverages. *Beverages* **2**:30

[13]. Adeleke, R.O., and Abiodun, O.A. (2010). Physico-chemical Properties of Commercial Local Beverages in Osun State, Nigeria. *Pakistan Journal of Nutrition*. **9**(9): 853-855.

[14]. Wong, P.K., S. Yusof, H.M. Ghazali and Y.B. Che Man, 2002. Physico-chemical characteristics of roselle (*Hibiscus sabdariffa* L.). *Journal of Nutrition and Food Science*. 32: 68-73.

[15]. Ojokoh, A.O., Adetuye, F.A., Akiuyosoye E., and Oyetayo, V.O., (2002). Fermentation studies on roselle (Hibiscus sabdariffa) calyces neutralized with trona, in proceeding of 16th annual conference of Biotechnology society of Nigeria, pp. 90-92.

[16]. Yokotsuka, K, Yajima, M., Seki, T., and Mattews M., (1997). Changes in colour parameters during fermentation and storage of red wines using Thai Roselle under different pHs and SO_2 concentrations. *Food Science and Technology International*. Tokyo, **3**(2): 105-109.

[17]. Saitou, N. and Nei, M. (1987). The neighbor-joining method: a new method for reconstructing phylogenetic trees. *Molecular Biology and Evolution*. 4(4):406-25

[18]. Muhammed, F.S. and Umar, B. M. (2007). Production and organoleptic assessment of a sweetened Sorrel drink. Namoda Tech-Scope- A. *Journal of Applied Science and Technology*. 7: 7-13.

[19]. Braide, W., Oranusi, S and Peter-Ikechukwu, A. I. (2012). Perspectives in the hurdle techniques in the preservation of a non-alcoholic beverage, Zobo. *African Journal of Food Science and Technology*. 3:46-52.

[20]. Mohammed, S. F. and Ismail, B. B. (2014). Comparison on Two Methods of Preparation of Zobo Drink on the Survival of Bacillus species. *American Journal of Food Technology* 9(4): 200-208

IOP Conf. Series: Earth and Environmental Science **331** (2019) 012067 doi:10.1088/1755-1315/331/1/012067

Corporate ethical standard and the quality of sustainability reporting: empirical evidence from commercial banks in nigeria

Sharon O Oluseyi-sowunmi[1], Akintola A Owolabi[2], Francis O Iyoha[1] and Olubukola R Uwuigbe[1]
[1]Department of Accounting, Covenant University, Km 10 Idiroko road, Ota Nigeria.
[2]Pan–Atlantic University, Lagos Business School, Lagos, Nigeria.

Abstract. Compliance with legal requirements is mandatory for corporate entities in Nigeria, but decision making in a situation that is not legally binding relies on the ethical standard of the company. Sustainability reporting in Nigeria is voluntary, therefore the quality of disclosure is at the discretion of company leadership. This study evaluated the ethical behaviour of Nigerian commercial banks and how it affects their sustainability disclosure quality. The Focus was on the proportion of each bank's corporate annual reports that contain environmental disclosure, social responsibility disclosure and governance disclosure. Information on the banks' websites that relate to sustainability policies or activities were also considered. This work includes an extensive review of relevant literature, hinging the study on legitimacy theory. The Cross-sectional research design was utilized in undertaking the study. A sample of fourteen (14) commercial banks was selected from the companies listed on the Nigerian stock exchange and analysed for a period of 2008-2017 financial years. Pearson Correlation and Multivariate Linear model analysis were employed to test the hypotheses. Findings revealed a positive relationship between corporate ethical standard and sustainability disclosure of Nigerian commercial banks. The level of corporate ethical standard in Nigerian banks causes significant positive change in environmental reporting quality, social responsibility reporting quality and governance reporting quality. It is hereby, recommended that company leadership should build strong corporate ethical culture since it directly affects their sustainability. While quality sustainability reporting practice is beneficial to the reporting entity, stakeholders and environment.

1. Introduction

The ethical standard of a company evaluates its actions in terms of what is morally right or acceptable, beyond legal requirements. It is good to comply with legal requirements, as expected, but when a decision is taken in a situation that is not legally binding, the appropriate moral decision is commendable and has a positive effect on reputation. A reputable company is expected to take responsible decisions and ensure their corporate representations uphold the law. Company policies and protocols should reflect the required laws binding on them and the standard of their ethical behavior [1]. Laws differ from one country to another and more stringent in some. Companies in developing countries tend to take advantage of laxities in their regulation [2]. The International Financial Reporting Standards (IFRS) state the standard requirements for a financial report. IFRS is the required reporting standard in Nigeria, issued by the International Accounting Standard Board (IASB) and enforced by the Financial Reporting Council of Nigeria (FRC). Also, in Nigeria The Companies and Allied Matters Act (CAMA), Securities and Exchange Commission (SEC) have financial reporting requirements for companies and the Central Bank of Nigeria has financial reporting requirements for banks in Nigeria.

The role of corporate reports in making investment decisions has necessitated the keen interest of investors in the reporting process. Past corporate failures have errored the value placed on the firm by stakeholders. Specifically, the Nigerian financial sector suffered a crisis between 2008 and 2009, which was attributed to low or weak accountability [3]. Shareholders now have a greater demand for more transparency and accountability. More stakeholders have become interested in a broader scope of analysis and reports for their decision making. A broader scope that includes legally unrequired contents such as environmental, social and governance reports [4]. Corporate reports state the results of operations in the stated reporting period, also sustainability reports are expected to present the environmental preservation activities, social responsibility projects and corporate governance compliance of the reporting entity. This practice attracts additional costs on the business. The additional cost leads to the question of 'reason' in a system of voluntary sustainability disclosure practice. Why would a company engage in sustainability practice and reporting? Disclosing this legally unrequired (in Nigeria) content is an improvement on transparency and accountability of the reporting entity to its stakeholders [5]. This study questions the impact of corporate ethical standard on the quality of their sustainability reporting.

1.1. Research Objectives

The aim of this work is to empirically examine the relationship between corporate ethical standard and sustainability disclosures of commercial banks in Nigeria. The specific objectives are to:

1. determine the relationship that exists between corporate ethical standard and the quality of sustainability disclosures published by commercial banks in Nigeria.
2. investigate whether corporate ethical standard impacts the quality of sustainability disclosures published by commercial banks in Nigeria.

2. Literature Review

Corporate ethics is the application of moral principles to the conduct of business and in the governance of personal behaviour at work. The purpose of ethics is to provide the insight for addressing situations of moral complexity in business. The ethical implications of business decisions must be considered before choosing a course of action [6]. A company that is seen as ethical usually exhibits significant level of transparency, has integrity, values and respects people [7]. Company policies and protocols should reflect the required laws binding on them [8]. Laws differ from one country to another and more stringent in some. Companies in developing countries tend to take advantage of laxities in their regulation [9]. Regulations on sustainability reporting in Nigeria are mild. Sustainability reporting guidelines that pertain to ethics and integrity are contained in the fourth edition (G4) of the Global Reporting Initiative (GRI) guidelines. Sections 56, 57 and 58 of G4 state standard codes of conduct and codes of ethics. [10]. For this study, stated code of conduct, occurrence or frequency of corporate fines and dues arising from violations, tax compliance, litigations against the company or its staff on duty would be considered in assessing corporate ethical standard.

Activities of firms have been seen to have some adverse effect on the environment and society. The attention of managers has been called to this and the way they respond to the situation has to be reported to the stakeholders. Corporate reporting advancement occurs as society evolves and the expectations of stakeholders rise. Sustainability reporting is the practice of measuring, disclosing, and being accountable to all stakeholders for the performance of the business; towards achieving the goal of sustainable development [11].

IOP Conf. Series: Earth and Environmental Science **331** (2019) 012067 doi:10.1088/1755-1315/331/1/012067

There are Nigerian empirical studies on environmental and social responsibility reporting, most of which focused on Oil and gas, manufacturing or financial sectors [11-14]. Researchers justified their choice of these sectors. Studies that focused on oil and gas or manufacturing sectors explained that these sectors have more adverse effects on the environment than other sectors. Other studies that gathered data and based their findings on the financial sector noted that the Central Bank of Nigeria (CBN) has more strict regulations as regards corporate governance for Nigerian banks.

The Nigerian banking sector is thoroughly regulated due to its sensitive to the economy and finances of the country [15]. The Central Bank of Nigeria states mandatory policies for corporate governance required of banks in Nigeria, which includes some policies on social and environmental reporting. Even with this, there are aspects of sustainability not included in the policy, which remain voluntary [16]. Companies in Nigeria that publish comprehensive sustainability reports are mostly multinationals that have international market standards to uphold and companies in the oil and gas industry that are considered to have major impacts on the environment [17]. Many of the Nigerian quoted companies only have short social responsibility reports or sustainability activities mentioned in the directors' report. This study considers the environmental, social and governance disclosures of the sampled data, to appropriately capture the level of sustainability reporting in each commercial bank.

Some benefits of sustainability reporting have been identified to include financial performance, gaining legitimacy and recently there have been discussions about building trust and earning reputation through sustainability reporting practice [18]. Public relations experts agree that a way to override criticisms is by broadcasting success stories, through the media or through published corporate reports. Publishing stories in the media could be in favour of or against the reputation of the company depending on who is reporting. On the other hand, it is easier for companies who practice voluntary sustainability disclosure to report when their sustainability performance is high and not when the performance is low. The relevance of corporate reporting can only be justified when information derived from the report influences the economic decisions of stakeholders and users of the report [19].

Legitimacy theory is an extension of social contracting theory. Legitimacy theory proposes that companies respond to demands of divergent interest groups by legitimizing their actions. Deephouse and Carter [20] opined that legitimacy is assessed based on expectations of social systems, values and regulations. Studies have confirmed that there is relationship between litigation or prosecutions arising from environmental hazards and different measures of social responsibility reporting; which are proxies for legitimacy threats [21]. This study is in line with the principle of legitimacy theory since the research focus is on voluntary sustainability disclosure and ethical threats [22].

3. Research Methodology

Secondary data were gathered to measure corporate ethical standard (ETS), data was gathered on clearly stated policies, fulfilment of Tax obligation, compliance with required reporting standards, any contingent liabilities arising from legal default or litigation against the firm, malpractice and scandal, scoring -1 where the company defaults or +1 where the company has not defaulted. Secondary data were also gathered to measure corporate Environmental, Social and Governance (ESG) disclosures by content analysis of integrated corporate annual reports and stand-alone sustainability reports published on the websites of the selected commercial banks. Disclosure quality was measured by classifying texts into categories on pre-selected criteria of qualitative disclosure as applied by Toms et.al. (2005). Quality scores are awarded on each report using a 0–5 qualitative scale as follows: no disclosure: 0, general rhetoric: 1, specific endeavour, policy only: 2, specific endeavour, policy specified: 3, implementation and monitoring but qualified result not published: 4, implementation and monitoring and qualified result published: 5. Data on disclosure quantity was gathered by sentence count to determine disclosure volume. Then to estimate the percentage of the report devoted to sustainability disclosure, the total number of sustainability sentences was divided by the total number of sentences in the corporate report.

The control variables are the power of shareholders, company size and Financial performance. Power of shareholders was measured by taking share capital as a percentage of net worth. Company size was measured by gross income or interest income, this captures the size of the bank's operations. Financial performance was captured by computing the Return on equity (ROE).

Sixteen (16) commercial banks are registered with the central bank of Nigeria, fourteen (14) of these are quoted on the Nigerian Stock Exchange. Convenience sampling was employed, for the availability of data, the 14 quoted commercial banks were sampled for this study. Data were gathered for ten (10) years (2008 – 2017) to capture long term effect as suggested by Toms et. al. (2005). 2008 being post-consolidated period and 2017 being the most recent year ended of the audited and published full annual report. The gathered data resulted in 140 observations.

3.1 Validity and Reliability of Instruments
The major sources of data, corporate annual reports of companies were carefully selected to include only audited annual reports, signed by reputable auditing firms. All the annual reports included in this study were qualified by the external auditor as representing a true and fair view of the affairs of the company. The researcher ensured that the annual reports included in the analysis were only those signed by the chairman of the board of directors of the reporting company. The D'Agostino-Pearson omnibus test for normality was carried out on the data.

3.2 Model Specification
The model applied by Toms, et. al. was adopted in the following implicit function and regression equation (2005):

$$SUSQ = f(ETS, POWS, GI, ROE) \quad (1)$$

$$SUSQ\ (ERQ, SRQ, GRQ) = \beta_0 + \beta_1 ETS + \beta_2 POWS + \beta_3 GI + \beta_4 ROE + \varepsilon \quad (2)$$

Where:

β_0: Intercept

$\beta_1, \ldots, \beta_4$: Coefficient of Slope Parameters

ε: Error term

SUSQ is Sustainability Reporting Quality measured by ERQ, SRQ and GRQ.

ERQ is environmental reporting quality

SRQ is social responsibility reporting quality

GRQ is governance reporting quality

ETS is ethical standard

POWS is power of shareholders

GI is Gross income as a control for company size

ROE is Return on Equity representing financial performance

3.3. Research Hypotheses

To achieve the objectives of this study, research hypotheses have been developed. Each hypothesis is stated in null and the alternate forms as follows:

1. H_0: Corporate ethical standard does not have a significant relationship with the quality of sustainability disclosures published by commercial banks in Nigeria.
   H_1: Corporate ethical standard has a significant relationship with the quality of sustainability disclosures published by commercial banks in Nigeria.
2. H_0: Corporate ethical standard has no significant impact on the quality of sustainability disclosures published by commercial banks in Nigeria.
   H_1: Corporate ethical standard has a significant impact on the quality of sustainability disclosures published by commercial banks in Nigeria.

4. Results

	N	Minimum	Maximum	Mean		Std. Deviation	Skewness		Kurtosis	
	Statistic				Std. Error			Std. Error		Std. Error
ERQ		1	5	2.35	.103	1.223	.546	.205	-.571	.407
SRQ		1	5	4.33	.099	1.166	-1.857	.205	2.49	.407
GRQ		2	5	3.43	.067	.797	.367	.205	-.304	.407
ETS		-1	1	-.03	.085	1.003	.058	.205	-2.026	.407
ROE		-.225	.997	.245	.029	.347	-4.044	.205	2.516	.407
POWS		.001	.960	.21	.024	.28	1.497	.205	.934	.407
GI		.03	512.4	145.267	10.019	118.552	.883	.205	.274	.407
N	140									

Table 1. Descriptive Statistics

The maximum score obtainable for ESG reporting quality (5) was achieved in the data set, in each of the three categories. Some of the analyzed reports scored the lowest mark (1) for the reporting quality of environmental and social impacts while, corporate governance reporting quality was at least the score '2', which means that all the reports analyzed in the sample had disclosure on corporate governance. The Gross Income (GI) was captured in billions of naira. The smallest bank with the minimum GI had ₦30,000,000 ($83,330) in their lowest period. The biggest bank with the maximum GI ₦512,400,000,000 ($1,423,330,000) in their peak period. For all the variables, the skewness revolves around 0 which means the data set is symmetrical and the kurtosis is of less than 3 shows the data set has a normal peak.

4.1. Normality Test

The histograms in figures 1, 2 and 3 show the normal distribution of the data set for the dependent variables of this study, that is, environmental reporting quality, social reporting quality and governance reporting quality.

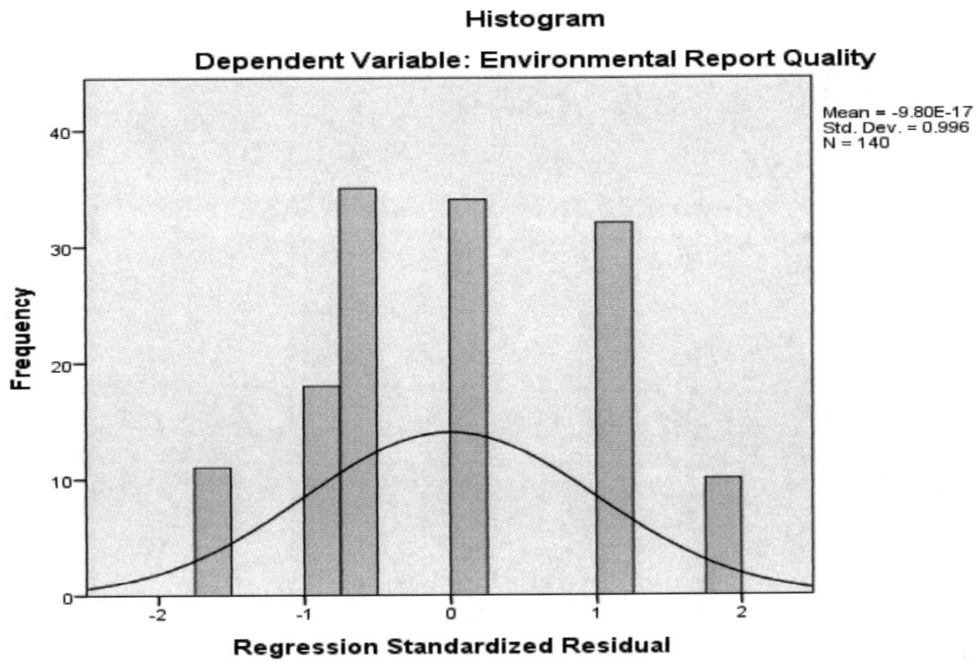

Figure 1: Dependent variable: Environmental report quality

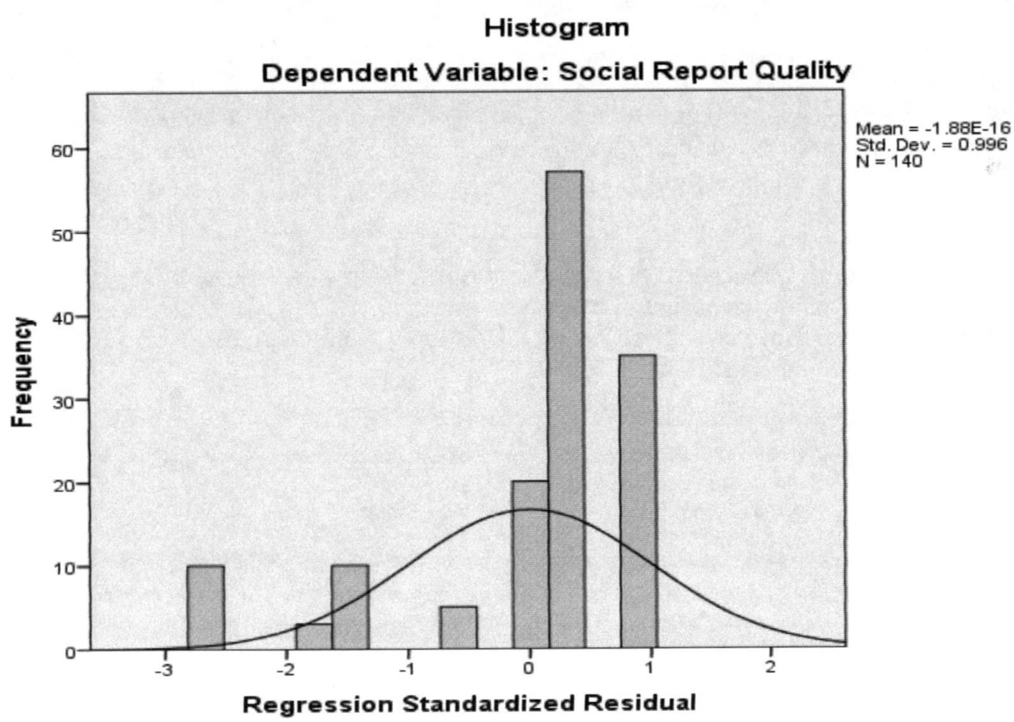

Figure 2: Dependent variable: Social report Quality

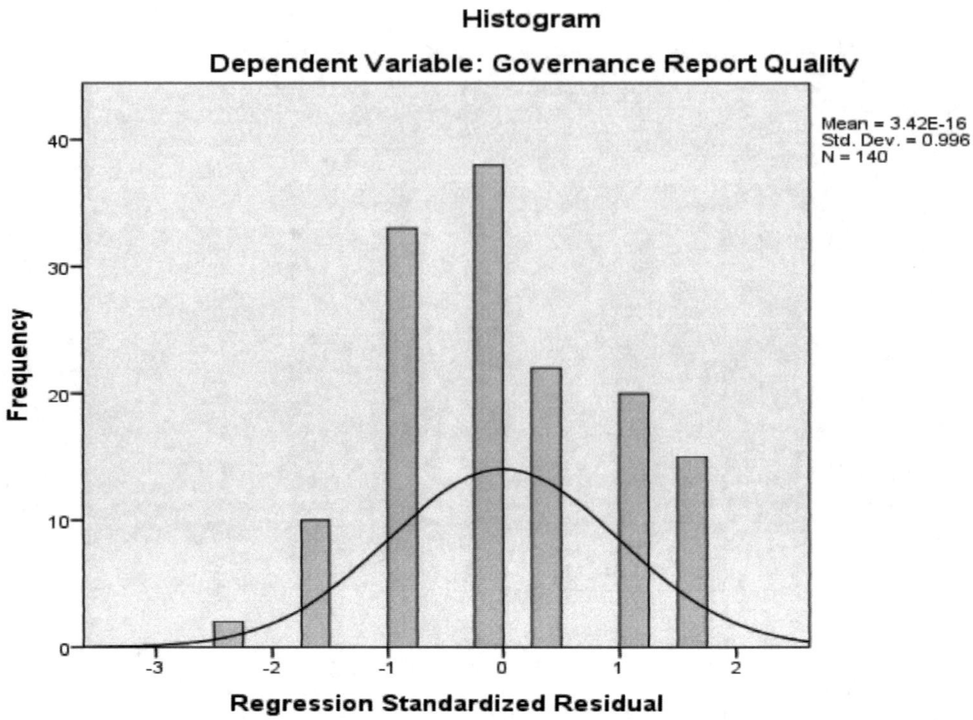

Figure 3: Dependent variable: Governance report quality

4.2 Test of Hypotheses

The first hypothesis was tested with Pearson correlation analysis, while the second hypothesis was tested with multivariate regression analysis.

4.2.1 Hypothesis One

H_0: Corporate ethical standard does not have a significant relationship with the quality of sustainability disclosures published by commercial banks in Nigeria.

H_1: Corporate ethical standard has a significant relationship with the quality of sustainability disclosures published by commercial banks in Nigeria

Table 2. Correlations

Pearson Correlation Sig. (2- tailed)	ERQ	SRQ	GRQ	ETS	ROE	POWS	GI
ERQ	1	.237 .005	.391 .000	.455 .000	.083 .332	-.224 .008	.57 .000
SRQ		1	.304 .000	.287 .001	.015 .861	.043 .615	.138 .103
GRQ			1	.345 .000	.131 .124	.093 .273	.568 .000
ETS				1	-.09 .292	.093 .275	-.326 .000
ROE					1	.434 .000	.316 .000
POWS						1	-.304 .000
GI							1
N	140	140	140	140	140	140	140

The correlation table presents a coefficient of 0.455 between corporate ethical standard and environmental reporting quality (sig. at 0.01 level). The correlation coefficient between corporate ethical standard and social reporting quality is 0.287 (sig. at 0.01). The correlation coefficient between corporate ethical standard and governance reporting quality is 0.345 (significant at 0.01 level). This result confirms that a relationship exists between corporate ethical standard and sustainability reporting quality. Therefore, the decision was taken to reject the null hypothesis while the alternate was accepted that corporate ethical standard has a significant relationship with the quality of sustainability disclosures published by commercial banks in Nigeria.
Table 2 also includes the correlation between ethical standard and power of shareholders, which implies that there is a relationship between the ethical standard of a commercial bank in Nigeria and the investment decision of the bank's shareholders.

4.2.2. Hypothesis Two
H_0: Corporate ethical standard has no impact on the quality of sustainability disclosures published by commercial banks in Nigeria.

H_1: Corporate ethical standard has impact on the quality of sustainability disclosures published by commercial banks in Nigeria.

Table 3. Between-Subjects Factors

		Value Label	N
Ethical Standard	-1	litigation default	72
	1	no litigation default	68

IOP Conf. Series: Earth and Environmental Science **331** (2019) 012067 doi:10.1088/1755-1315/331/1/012067

Of the 140 corporate reports analysed, 72 contained contingent liabilities arising from litigation while 68 had no contingent liabilities.

Table 4. Multivariate Tests

Effect		Value	F	Hypothesis df	Error df	Sig.
Intercept	(a) Pillai's Trace	.860	271.339^b	3	133	.000
	(b) Wilks' Lambda	.140	271.339^b	3	133	.000
	(c) Hotelling's Trace	6.120	271.339^b	3	133	.000
	(d) Roy's Largest Root	6.120	271.339^b	3	133	.000
ROE	(a)	.043	1.972^b	3	133	.121
	(b)	.957	1.972^b	3	133	.121
	(c)	.044	1.972^b	3	133	.121
	(d)	.044	1.972^b	3	133	.121
POWS	(a)	.052	2.409^b	3	133	.07
	(b)	.948	2.409^b	3	133	.07
	(c)	.054	2.409^b	3	133	.07
	(d)	.054	2.409^b	3	133	.07
SIZ	(a)	.383	27.487^b	3	133	.000
	(b)	.617	27.487^b	3	133	.000
	(c)	.620	27.487^b	3	133	.000
	(d)	.620	27.487^b	3	133	.000
ETS	(a)	.192	10.525^b	3	133	.000
	(b)	.808	10.525^b	3	133	.000
	(c)	.237	10.525^b	3	133	.000
	(d)	.237	10.525^b	3	133	.000

a. Design: Intercept + ROE + POWS + SIZ + ETS
b. Exact statistic
Corporate ethical standard (ETS), Return on equity (ROE), Power of shareholders (POWS), and Bank size (GI) jointly and significantly impact the quality of sustainability reporting on commercial banks in Nigeria.

(ERQ- Environmental Report Quality; SRQ- Social Report Quality; GRQ- Governance Report Quality)

Table 5. Tests of Between-Subjects Effects

Source	Dependent Variable	Type III Sum of Squares	df	Mean Square	F	Sig.
Corrected Model	ERQ	92.167[a]	4	23.042	26.889	.000
	SRQ	16.278[b]	4	4.069	3.183	.016
	GRQ	31.338[c]	4	7.834	18.572	.000
Intercept	ERQ	107.292	1	107.292	125.208	.000
	SRQ	482.497	1	482.497	377.370	.000
	GRQ	228.932	1	228.932	542.701	.000
ROE	ERQ	4.965	1	4.965	5.794	.017
	SRQ	.038	1	.038	.030	.864
	GRQ	.098	1	.098	.232	.631
POWS	ERQ	5.773	1	5.773	6.736	.010
	SRQ	.141	1	.141	.111	.740
	GRQ	.093	1	.093	.220	.640
SIZ	ERQ	32.697	1	32.697	38.157	.000
	SRQ	.644	1	.644	.503	.479
	GRQ	19.247	1	19.247	45.626	.000
ETS	ERQ	20.449	1	20.449	23.863	.000
	SRQ	11.164	1	11.164	8.732	.004
	GRQ	2.157	1	2.157	5.114	.025
Error	ERQ	115.683	135	.857		
	SRQ	172.608	135	1.279		
	GRQ	56.948	135	.422		
Total	ERQ	981.000	140			
	SRQ	2812.000	140			
	GRQ	1734.000	140			
Corrected Total	ERQ	207.850	139			
	SRQ	188.886	139			
	GRQ	88.286	139			

	R Squared	Adjusted R Squared
a.	.443	.427
b.	.086	.059
c.	.355	.336

From the adjusted R-squared of 42.7%, 5.9% and 33.6% it interprets that corporate ethical standard causes 42.7% change in environmental reporting quality, 5.9% change in social responsibility reporting quality and 33.6% change in corporate governance reporting quality of commercial banks in Nigeria. Also, the null hypothesis was rejected, and the alternate accepted that corporate ethical standard has a significant impact on the quality of sustainability disclosures of Nigerian banks. The findings of this study agree with the research of Ramakrishan who confirmed the relationship between corporate governance and business ethics (2007).

5. Conclusion

Based on the data analysis and findings of this work, the following conclusions are drawn:

- This research has proved that corporate ethical standard directly impacts the quality of sustainability reporting of commercial banks in Nigeria.
- Disclosure on corporate governance policies and compliance is at a comparably higher quality than social responsibility disclosure and environmental disclosure. However, environmental disclosure reacts to ethical standard better (42.7%) than governance (33.6%) and social responsibility (5.9%) disclosures.
- Of the three disclosure practices in this study, social responsibility reporting practises is the oldest in Nigeria and Nigerian companies have imbibed it over the years. This disclosure by the commercial banks mainly entails a list of donations stating amounts donated and the recipients.
- The environmental reporting practice of the banks is basically policy statements. In most of the analysed reports environmental and social reports were combined under headings like 'Health, safety and environment (HSE)', 'Community Affairs, Safety, Health, Environment and Security (CASHES)'.
- There is a significant long-term effect as anticipated by the ten-year study. Environmental, social responsibility and governance disclosures improved over the years and their relationships with ethical standard became more significant in every additional time period.
- The sustainability reporting practice of commercial banks in Nigeria could be greatly improved. There is a wide gap between the current sustainability disclosure quality and international standard of sustainability disclosure practice.

5.1 Recommendation

The following recommendations are offered:

- Voluntary disclosure of environmental preservation activities and social responsibility activities should be improved. The quality of sustainability reports published by the bank would present the bank as an ethical entity. This would influence the investment decisions of shareholders, as suggested by the findings of this work.
- Business and industry regulators need to educate managers and directors of commercial banks on sustainability accounting and reporting. This would help to improve the reporting quality.
- Regulators and leaders of the commercial banks should be equipped with relevant information on the social and environmental impacts peculiar with the nature of their business.
- Academic institutions in Nigeria should review their curriculum for accounting courses to include subjects and topics that teach sustainability accounting and reporting so the Nigerian accountant would be well equipped for the job.

5.2. Limitation and Suggestion for Further Research

This work focused on the ethical behaviour of the bank, further studies should consider the legal requirements that pertain to aspects of sustainability such as corporate governance policies and impacts on disclosure quality. The Sustainability Disclosure Guidelines (SDGs) proposed by Nigerian Stock Exchange (NSE) and recently approved by Securities and Exchange Commission (SEC) to take effect this year 2019 should be explored for further studies.

Sources of data for corporate ethical standard were secondary. This work did not consider the perceptions of stakeholders on assessing the commercial banks' ethical standard. However, the secondary sources of data are considered valid evaluations of the corporate ethical standard by involvement in litigation, since they were generated from external factors and not within the company. Further studies on corporate ethical standard should gather primary data from stakeholders on their perception of the bank's ethics. These perceptions can be further analysed in experimental research where perceptions of stakeholders are studied in different situations, the company's response is also captured. This experiment is applicable to smaller sample size or case study.

6. References

[1]. Achugamonu, U., Olokoyo, F., Babajide, A., Owalabi, F. and Adetula, D. (2016). Adoption of International Financial Reporting Standards and Its Implication on Bank Performance in Nigeria: A Comparative Approach. *Journal of internet banking and commerce*, [online] Volume
[2]. 21(S5), p. 1- 5. Available at:
[3]. https://www.researchgate.net/publication/324680577_Journal_of_Internet_Banking_and_Commerce_Special_Issue_E-commerce_trends_and_future_of_E-commerce_ [Accessed 9 Apr. 2019].

[4]. Aguilera-Caracuel, J. and Guerrero-Villegas, J. (2017). How Corporate Social Responsibility Helps MNEs to Improve their Reputation. The Moderating Effects of Geographical Diversification and Operating in Developing Regions. *Corporate Social Responsibility and Environmental Management*, [online] Volume 2(1), p. 355- 359. Available at: https://onlinelibrary.wiley.com/doi/10.1002/csr.1465 [Accessed 9 Apr. 2019].

[5]. Asaolu, T., Agboola, A., Ayoola, T., and Salawu, M. (2011). *Sustainability reporting in the Nigerian oil and gas sector*. [online]. Proceedings of the Environmental Management Conference, Federal University of Agriculture, Abeokuta, Nigeria. Available at: http://scholar.oauife.edu.ng/tjayoola/files/238- 475-1-sm.pdf [Accessed 20 Feb. 2018].

[6]. Bassey, B., Effiok, O. and Eton, E. (2013). The impact of environmental accounting and reporting on organizational performance of selected oil and gas companies in Niger Delta region of Nigeria. *Research Journal of Finance and Accounting,* [online] Volume 4(3), p. 57–73. Available at: https://www.google.com.ng/url?sa=tandrct=jandq=andesrc=sandsource=webandcd=3andcad=rjaanduact=8andved=0CC0QFjACahUKEwjm29q4nrXIAhUJp3IKHbnpAI4andurl=http%3A%2F%2Fwww.iiste.org%2FJournals%2Findex.php%2FRJFA%2Farticle%2Fdownload%2F4962%2F5045andusg=AFQjCNGGOOB20cMizADI0kR4WM_NdJ08Tg [Accessed 20 Feb. 2018].

[7]. Chalmers, K. and Godfrey, M. (2004). Reputation costs: The impetus for voluntary Derivative financial Instrument reporting. *Accounting Organizations and Society,* [online] Volume 29(2), p. 95– 125. Available at: http://www.sciencedirect.com/science/article/pii/S036136820200034X [Accessed 20 Feb. 2018].

[8]. Deephouse, D. and Carter, S. (2005). An Examination of differences between organization legitimacy and organizational reputation, *Journal of Management Studies,* [online] Volume
[9]. 42(2), p. 329– 360. Available at: http://onlinelibrary.wiley.com/doi/10.1111/j.14676486.2005.00499.x/epdf [Accessed 20 Feb. 2018].

[10]. Dibia, N. and Onwuchekwa, J. (2015). Determinants of environmental disclosures in Nigeria: A case study of oil and gas companies. *International Journal of Finance and Accounting,* [online]

[11]. Volume 4(3), p. 145– 152. Available at: https://www.google.com.ng/url?sa=tandrct=jandq=andesrc=sandsource=webandcd=6andcad=rjaanduact=8andved=0CEgQFjAFahUKEwjm29q4nrXIAhUJp3IKHbnpAI4andurl=http%3A%2F%2Fwww.sapub.org%2Fglobal%2Fshowpaperpdf.aspx%3Fdoi%3D10.5923%2Fj.ijfa.20150403.01andusg=AFQjCNGFSQrZZzRU7kMaPzHnlO6giVw-sQ [Accessed 9 Apr. 2019].

[12]. Ebiringa, O., Yadirichukwu, E., Chigbu, E. and Ogochukwu, O. (2013). Effect of firm size and profitability on corporate social disclosures: the Nigerian oil and gas sector in focus. *British Journal of Economics, Management and Trade,* [online] Volume 3(4), p. 563– 574. Available at: https://www.google.com.ng/url?sa=tandrct=jandq=andesrc=sandsource=webandcd=9andcad=rjaanduact=8andved=0CGIQFjAIahUKEwjfj7TnurXIAhVCTRoKHYX8CIAandurl=http%3A%2F%2Fsciencedomain.org%2Fdownload%2FMjA0NEBAcGYandusg=AFQjCNGQgWWACnOyTQDxZfGf9EyV0EnNYA [Accessed 9 Apr. 2019]

[13]. Famuyiwa, O. (2013) The Nigerian Financial Crisis: A Reductionist Diagnosis. Journal od Sustainable Development Law and Policy, vol 2 no. 1. Available at: www.ajol.info/index.php/jsdlp/article/view/122583 [Accessed 18 June. 2019]

[14]. Feldman, P. M., Bahamonde, R. A., and Bellido, I. V. (2014). A New Approach for Measuring Corporate Reputation. Revista de Administracao de Empresas. 54(1), 123- 136.

[15]. globalreporting.org (2011). *Sustainability reporting guidelines, Version 3.1.* [online] Available at: https://www.globalreporting.org/resourcelibrary/GRI-Reporting-Trends-2011.pdf [Accessed 20 Feb. 2018].

[16]. globalreporting.org (2015). *Sustainability Reporting Guidelines, Version 4.* [online] Available at: https://www.globalreporting.org/resourcelibrary/GRI-Reporting-Trends-2015.pdf [Accessed 12 Sep. 2018].

[17]. Grahovar, M. (2010). *The Role of Corporate social disclosure-Trust, reputation or fashion tool?* [online] Available at: http://hdl.handle.net/2077/30501 [Accessed 12 Sep. 2018].

[18]. Haniffa, R. and Cooke, T. (2005). The impact of culture and governance on corporate social reporting. *Journal of Accounting and Public Policy,* [online] Volume 24(1), p. 391- 430. Available at: www.elsevier.com/locate/jaccpubpol [Accessed 12 Sep. 2018].

[19]. Irvine, J. (2017) Companies need to be more accountable to stakeholders. *Economia, news 15 august 2017,* ICAEW. Available at: ps://economia.icaew.com [Accessed June 2019]

[20]. Nwobu, O., Iyoha F. and Owolabi, A. (2018). Managerial Perceptions of Corporate Sustainability Reporting Determinants in Nigeria. *Journal of Business and Retail Management*

[21]. *Research,* [online] Volume 12(2), p. 72- 82. Available at: http://www.jbrmr.com/cdn/article_file/content_49144_18-01-19-01-32-47.pdf [Accessed 9 Apr. 2019].

[22]. Oriesek, D. (2004). *Maximizing corporate reputation through effective governance: a study of structures and behaviors.* 1st ed. [ebook] Universal Publishers. Available at:

http://www.universal-publishers.com/book.php?method=ISBNandbook=1581122292 [Accessed 12 Sep. 2018].

[23]. Owolabi, A. (2009). *Environmental disclosures in annual reports: The Nigerian perspective.* [online]

[24]. International Business Review. Available at: https://www.researchgate.net/publication/254833575_Environmental_Disclosures_in_Annual_Reports_the_Nigerian_Perspective [Accessed 12 Sep. 2018].

[25]. Oyewole, OS., Olusanmi, O. and Owolabi, F. (2014), 'Corporate Governance and Credit Risk Management: A Study of the Nigerian Financial Crisis', *Proceedings of the 23rd International Business Information Management Association Conference, IBIMA* (1), 2663-2683.

[26]. Ramakrishan, D. (2007). *Inter-relationship between business ethics and corporate governance among Indian companies.* [online] International Seminar at the Institute of Management, NIRMA

[27]. University. Available at: https://www.academia.edu/1015118/InterRelationship_between_Business_Ethics_and_Corporate_Governance_Among_Indian_Companies [Accessed 12 Sep. 2018].

[28]. Suchman, M. (1995). Managing legitimacy: Strategic and institutional approaches. *Academy of Management Review.* [online] Volume 20(3), p. 571– 610. Available at: http://amr.aom.org/content/20/3/571.full.pdf+html [Accessed 12 Sep. 2018].

[29]. Toms, S., Hasseldine, J. and Salama, A. (2005). Quantity versus quality: The impact of environmental disclosures on the reputations of UK plc's. *British Accounting Review,* [online] Volume 37(2), p. 231– 248. Available at: http://eprints.whiterose.ca.uk/archive/1421/ [Accessed 20 Feb. 2018].

IOP Conf. Series: Earth and Environmental Science **331** (2019) 012068 doi:10.1088/1755-1315/331/1/012068

Air pollution Estimation Over Brikama

Emetere M.E.[1,2], **and Olawole O.C.**[4]
[1]Department of Physics, Covenant University Canaan land, P.M.B 1023, Ota, Nigeria.
[2]Department of Mechanical Engineering and Science, University of Johannesburg, APK, South Africa

E-mail: emetere@yahoo.com

Abstract. Recent reports on the air pollution over Brikama-Gambia by United Nations Environment Programme (UNEP) and the National Environment Agency (NEA) may be shocking. In this paper, fifteen years primary (aerosol optical depth) dataset was obtained from the Multi-angle Imaging Spectro-Radiometer (MISR). The secondary datasets were generated from the primary dataset to understand the short and long term effect of aerosol loading over Brikama. The dataset is important to help researcher understand the dynamics of the aerosol loadings in the area. It was discovered that the air pollution over Brikama is not seasonal and do not come from same source. It was observed that the pollution over Brikama is determined by the massisve influx and migration of human population.

Keywords: air pollution, aerosol, Brikama, aerosol loadings

1. Introduction

In 2008, Departement of State for Forestry and the Environment, the United Nations Environment Programme (UNEP) and the National Environment Agency (NEA) created awarenes on sulphur reduction in vehicule fuels in The Gambia [1]. In most cities of West africa, ground measurement is inadequate. There are no significant contribution from government in the sub-region to invest in ground measurement of air pollution. The Ground measurement that is found in the region are that of DACCIWA, AERONET and some Institutions. This situation is disturbing as satellite imagery had revealed that the West Africa region had one of the highest air pollution in the globe (Figure 1). Cases of respiratory infection or diseaes has become very common. With this projection, it is easy to infer that the health risk is on children and women that resides in the research area.

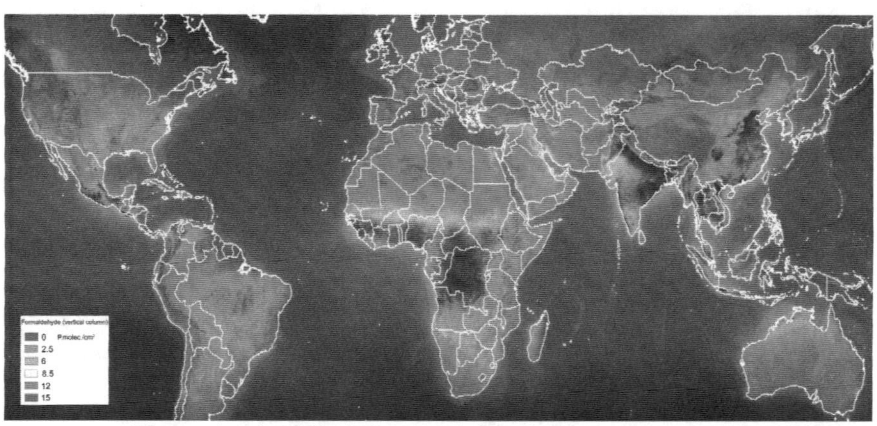

Figure 1: Air pollution around the globe

IOP Conf. Series: Earth and Environmental Science **331** (2019) 012068 doi:10.1088/1755-1315/331/1/012068

At the moment, satellite measurement is the main means of investigating air pollution [2-5] in the research area.

In the past, satellite measurement has shown baise that may be significant – depending on the geographical location. In this paper, the aerosol loading over Brikama was derived from satellite measurements. The quantity of risk on aerosols loading in West Africa cannot be estimated due to the fact that the theoretical validation concerning the aerosol dispersion model between the Stokes' rule has no longer been clarified. There is need to quantify the myriads of challenges dealing with aerosols dataset. The focus of this research is to repoort the current air pollution state for further research work.

2. Experimental Design, Materials and Methods

Brikama is one of the largest cities in the Gambia, lying south of the country's capital, Banjul. Brikama is located on longitude and latitude of 16.6409° W and 13.2748° N respectively (Figure 1). The dataset was obtained from MISR (https://l0dup05.larc.nasa.gov/L3Web/download). The data was processed using excel. The conversion from AOD to aerosol loading was done using West African regional scale dispersion model WASDM. The statistical analysis was carried out to observe the trends and significance of the dataset. Hence, among many other things, the dataset in this work is important to: give a good background for further study on aerosol loading; provide meteorological centers insight towards configuring sun-photometer over Brikama-Gambia; help to quantify the extent of air pollution; and provide modeller necessary insight on aerosol loading and retention challenges over Brikama-Gambia.

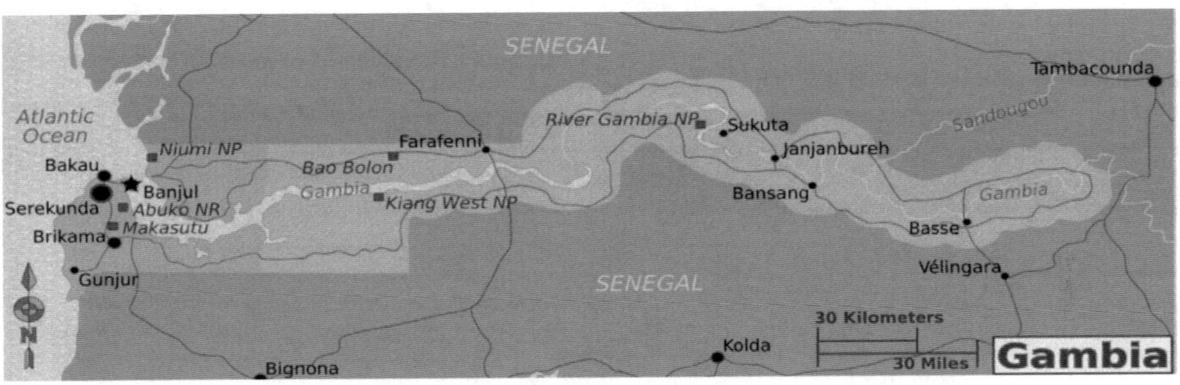

Figure 2: Geographical map of Brikama

WASDM was used aerosol loading over a region [7,8]:

$$\psi(\lambda) = a_1{}^2 \cos\left(\frac{n_1 \pi \tau(\lambda)}{2} x\right) \cos\left(\frac{n_1 \pi \tau(\lambda)}{2} y\right) + \cdots \ldots a_n{}^2 \cos\left(\frac{n_n \pi \tau(\lambda)}{2} x\right) \cos\left(\frac{n_n \pi \tau(\lambda)}{2} y\right) \quad (1)$$

a is atmospheric constant gotten from the fifteen years aerosol optical depth (AOD) dataset from MISR, n is the tunning constant, $\tau(\lambda)$ is the AOD of the area and $\psi(\lambda)$ is the aerosol loading. The validation of the summarized dataset was done using mathematical models and statistical software. The analysis of equations (1) was done using the C++ codes.

3. Results and Discussion

The summarized primary data was obtained from Multi-angle Imaging Spectro-Radiometer (MIS shown in Table 1 for 550 nm wavelength [6]. It was observed the months of highest AOD v significantly across the months over the years (see green boxes in Table 1). This result signifie the air pollution over Brikama is not seasonal and do not come from same source. Hence the poll over Brikama is caused by massisve influx and migration of human population. The missing d was due to biases as discussed above [7].The aerosol loading over the area was obtained usin West African regional scale dispersion model (WASDM) from the primary dataset (Tables 2) highest aerosol loading varies across month i.e. over the years. This means that the aerosol load not highest at months of highest AOD. This shows that the mathematical model was accurate tc an estimate value of aerosol deposition in the atmosphere.

Table 1: Summarized Aerosol Optical Depth Dataset over Brikama

Month	2000	2001	2002	2003	2004	2005	2006	2007	2008	2009	2010	2011	2012	201
Jan	0.493	0.174	0.150	0.201	0.105	0.223	0.336	0.571	0.210	0.203	0.290	0.193	0.276	0.2
Feb	0.569	0.340	0.246	0.382	0.381	0.291	0.175	0.282	0.424	0.396	0.207	0.181	0.408	0.2
Mar	0.322	0.348	0.413	0.411	0.535	0.209	0.423	0.444	0.394	0.373	0.517	0.340	0.729	0.3
Apr	0.456	0.487	0.490	0.419	0.346	0.411	0.489	0.343	0.509	0.363	0.493	0.561	0.428	0.5
May	0.304	0.281	0.486	0.576	0.186	0.351	0.362	0.504	0.562	0.517	0.576	0.366	0.515	0.5
Jun	0.541	0.481	0.425	0.568	0.561	0.577	0.658	0.497	0.331	0.316	0.605	0.398	0.573	0.6
Jul	0.175	0.104	0.384	0.294	-	-	-	0.437	0.145	0.303	1.453	0.664	0.367	0.2
Aug	0.322	-	0.193	0.356	0.235	0.390	0.438	0.262	0.313	0.221	0.280	0.400	0.360	
Sep	0.255	0.249	0.271	0.412	0.177	0.470	0.130	-	0.304	0.546	0.283	0.253	0.185	0.0
Oct	0.219	0.317	-	0.303	0.660	0.399	0.264	0.260	0.097	0.270	0.102	0.507	0.256	
Nov	-	0.244	0.167	0.308	0.204	0.179	0.147	0.297	0.247	0.145	0.224	0.207	0.227	0.1
Dec	-	0.165	0.168	0.185	0.264	0.191	0.131	0.164	0.284	0.137	0.193	0.246	0.124	

Table 2A: Aerosol loading over Brikama

Month	2000	2001	2002	2003	2004	2005	2006	2007	2008	2009	2010	2011	2012	201
Jan	0.845	0.931	0.935	0.927	0.940	0.923	0.897	0.812	0.926	0.927	0.909	0.929	0.912	0.9
Feb	0.813	0.896	0.919	0.884	0.884	0.909	0.931	0.911	0.870	0.879	0.926	0.930	0.875	0.9
Mar	0.901	0.894	0.874	0.874	0.828	0.926	0.870	0.863	0.880	0.887	0.835	0.896	0.736	0.8
Apr	0.859	0.847	0.846	0.872	0.894	0.874	0.847	0.895	0.838	0.889	0.845	0.817	0.868	0.8
May	0.906	0.911	0.848	0.810	0.930	0.893	0.890	0.840	0.816	0.835	0.810	0.889	0.836	0.8
Jun	0.825	0.849	0.870	0.814	0.817	0.810	0.772	0.843	0.899	0.902	0.797	0.878	0.812	0.7
Jul	0.931	0.940	0.883	0.908	-	-	-	0.865	0.935	0.906	0.298	0.769	0.888	0.9
Aug	0.901	-	0.929	0.891	0.921	0.881	0.865	0.915	0.903	0.924	0.911	0.878	0.890	
Sep	0.917	0.918	0.913	0.874	0.931	0.854	0.937	-	0.906	0.823	0.911	0.917	0.930	0.9
Oct	0.924	0.902	-	0.906	0.771	0.878	0.915	0.916	0.940	0.914	0.940	0.839	0.917	
Nov	-	0.919	0.932	0.905	0.927	0.931	0.935	0.907	0.919	0.935	0.923	0.926	0.923	0.9

Dec	-	0.933	0.932	0.930	0.915	0.929	0.937	0.933	0.910	0.936	0.929	0.919	0.938	

The statistical analysis of the summarized primary dataset is shown in Tables 3. The close ranges of values of the variance, average deviation and standard deviation are close indication that the correlation of each year would be natuarally be >0.75. It means the pattern of aersol dispersion in the research area is unique as there are no data spikes over the years.

Table 3: statistics of aerosols content over Brikama

Statistics	**2000**	**2001**	**2002**	**2003**	**2004**	**2005**	**2006**	**2007**	**2008**	**2009**	**2010**	**2011**	**2012**
Number of values	10.000	11.000	11.000	12.000	11.000	11.000	11.000	11.000	12.000	12.000	12.000	12.000	12.000
Minimum	0.175	0.104	0.150	0.185	0.105	0.179	0.130	0.164	0.097	0.137	0.102	0.181	0.124
Maximum	0.569	0.487	0.490	0.576	0.660	0.577	0.658	0.571	0.562	0.546	1.453	0.664	0.729
Mean	0.366	0.290	0.308	0.368	0.332	0.335	0.323	0.369	0.318	0.316	0.435	0.360	0.371
First quartile	0.255	0.192	0.174	0.299	0.190	0.213	0.154	0.267	0.229	0.212	0.215	0.226	0.241
Third quartile	0.493	0.346	0.422	0.415	0.497	0.408	0.434	0.484	0.409	0.384	0.546	0.454	0.472
Standard error	0.044	0.037	0.040	0.035	0.055	0.039	0.052	0.039	0.040	0.038	0.104	0.045	0.050
95% confidence interval	0.100	0.082	0.090	0.078	0.122	0.086	0.115	0.086	0.088	0.084	0.229	0.099	0.111
99% confidence interval	0.143	0.117	0.128	0.110	0.174	0.123	0.164	0.123	0.124	0.118	0.323	0.139	0.156
Variance	0.019	0.015	0.018	0.015	0.033	0.017	0.029	0.017	0.019	0.017	0.130	0.024	0.030
Average deviation	0.119	0.095	0.119	0.093	0.150	0.106	0.139	0.111	0.105	0.103	0.245	0.123	0.133
Standard deviation	0.139	0.122	0.134	0.122	0.182	0.129	0.172	0.128	0.138	0.132	0.361	0.155	0.174
Coefficient of variation	0.381	0.422	0.433	0.333	0.548	0.384	0.531	0.348	0.434	0.417	0.829	0.432	0.470
Skew	0.247	0.322	0.169	0.333	0.683	0.399	0.520	0.075	0.222	0.386	2.282	0.666	0.635
Kurtosis	-1.462	-0.488	-1.837	-0.201	-0.813	-0.656	-0.414	-1.154	-0.366	-0.552	6.200	-0.444	0.119
Kolmogorov-Smirnov stat	0.224	0.135	0.170	0.172	0.191	0.173	0.170	0.168	0.131	0.105	0.239	0.170	0.123
Critical K-S stat, alpha=.10	0.369	0.352	0.352	0.338	0.352	0.352	0.352	0.352	0.338	0.338	0.338	0.338	0.338
Critical K-S stat, alpha=.05	0.409	0.391	0.391	0.375	0.391	0.391	0.391	0.391	0.375	0.375	0.375	0.375	0.375
Critical K-S stat, alpha=.01	0.489	0.468	0.468	0.449	0.468	0.468	0.468	0.468	0.449	0.449	0.449	0.449	0.449

This result is further affirmed by the averages of the AOD over the years i.e. within the range of 0.3±0.05. The year 2010 had the highest AOD average over the years. The standard error is a good indication that the satellite measurement has a very low baise. Hence, the dataset presented is reliable for modelling and optimization process.

4. Conclusion

The research affirm that air pollution in Brikama is not seasonal and do not come from same source. It was observed that the pollution over Brikama is determined by the massive influx and migration of human population. Though the aerosol loading is safe for the inhabitants, however, if the massive vehicular movement is not checked, the continous emission of sulphur dioxide as reported by United Nations Environment Programme (UNEP) and the National Environment Agency (NEA) would be a massive threat to all the life form in the research area.

Acknowledgement

The authors appreciate Covenant University for partial sponsorship. The authors acknowledge NASA for primary dataset.

References

[1]. Abdoulie John, Environment: Efforts to combat air pollution, http://africa.gm/africa/gambia/article/environment-efforts-to-combat-air-pollution (accessed 16th June, 2018)

[2]. Fioletov, V.E., Bodeker, G.E., Miller, A.J., McPeters, R.D. and Stolarski, R., Global and zonal total ozone variations estimated from ground-based and satellite measurements: 1964–2000, Journal of Geophysical Research 107 (D22) (2002) 4647-4651.

[3]. Giglio, L., Randerson, J.T., van der Werf, G.R., Kasibhatla, P.S., Collatz, G.J., Morton, D.C. and DeFries, R.S., Assessing variability and long-term trends in burned area by merging multiple satellite fire products. Biogeosciences 7 (2010) 1171-1186.

[4]. Mohammad Hadi Dehghani Sima Jarahzadeh , Mostafa Hade, Nabiollah Mansouri, Yousef Rashidi, Mahmood Yousefi, The Data on the Dispersion Modeling of Traffic-related PM10 and CO Emissions using CALINE3; A Case Study in Tehran, Iran, Data in Brief 19 (2018) 2284-2290.

[5]. Liu, Y., Wang, Z., Wang, J., Ferrare, R., Newsom, R. and Welton, E., The effect of aerosol vertical profiles on satellite-estimated surface particle sulphate concentrations. Remote Sensing of Environment, 115 (2) (2011) 508-513.

[6]. McGill, M., CATS/ISS L2O D-M7.1 05kmPro HDF-5 File - Version 1.05 [Data set]. NASA Langley Research Center Atmospheric Science Data Center DAAC. (2016). https://doi.org/10.5067/iss/cats/l2o_d-m7.1-v1-05_05kmpro

[7]. M. E. Emetere, Impacts of recirculation event on aerosol dispersion and rainfall patterns in parts of Nigeria, GLOBAL NEST JOURNAL 19 (2) (2017) 344-352

[8]. Moses Eterigho Emetere, (2017). Investigations on aerosols transport over micro- and macro-scale settings of West Africa, Environ. Eng. Res., 22(1), 75-86

Air pollution Estimation Over Ponta

Emetere M.E.[1,2], **and Ahiara M.O.**[1]
[1]Department of Physics, Covenant University Canaan land, P.M.B 1023, Ota, Nigeria.
[2]Department of Mechanical Engineering and Science, University of Johannesburg, APK, South Africa

E-mail: emetere@yahoo.com

Abstract. The principal challenge of air pollution reportage in most communities of West Africa is the unavailability of working data. In this paper, satellite measurements were used to derive salient parameter (aerosol loading) to describe air quality over Ponta-Cape Verde was considered. Fifteen years primary (aerosol optical depth) dataset was obtained from the Multi-angle Imaging Spectro-Radiometer (MISR). The dataset is important to quantify the extent of air pollution in the research area. The statistical analysis of the dataset was also presented. It was observed that air pollution over Ponta is seasonal and comes from same source. Hence the pollution over is sustained due to a same pollution routine over the years.

Keywords: air pollution, aerosol, Brikama, aerosol loadings

1. Introduction

At the moment, not much is known about ground measurement dataset over some areas in Cape Verde [1]. The main source of dataset is satellite measurement only [2-5]. In this paper, we show the aerosol loading measurements that was derived from the aerosol optical depth dataset through the West African regional scale dispersion model (WASDM) [6, 7]. The objective of this paper is to further the sensitization of West African communities to its current air pollution state.

The danger of air pollution is enormous. About 33% of human die from stroke, lung malignant growth cardiovascular, respiratory sickness and coronary illness. This is having a proportional impact to that of smoking tobacco and other societal dangers. Air pollution is firmly connected to environmental change. Environmental change caused by is non-renewable energy source ignition (fossil fuel burning) which contributes largely to anthropogenic emission. The United Nation UN Intergovernmental Panel on Climate Change cautioned that coal-burning as a means of genrating power must finish by 2050. This decision would largely confine worldwide temperature alteration to 1.5C [8].

The mode of air pollution in most developing region can be summarized via Figure 1. From literature, biomass and fossil fuel burning is most prominent. Other sources of pollution are also significant. The accumulation of all these sources of pollution in the atmosphere over time is referred to as aerosol loading. The aerosol loading is a vital parameter that show how long the emitted aerosols interfere with the atmosphere.

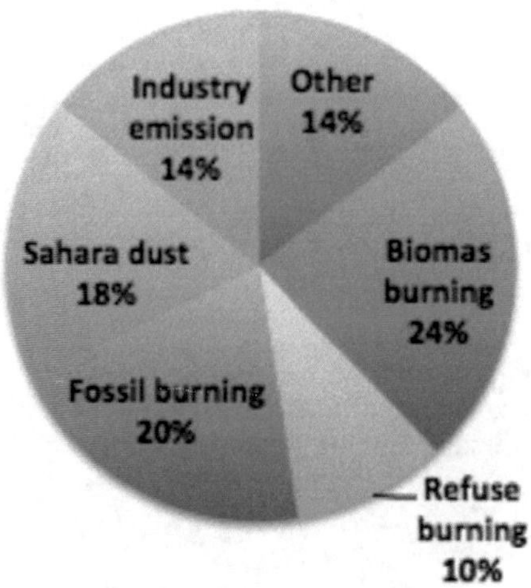

Figure 1: Sources of air pollution

In this research, we documented the dataset of aerosol optical depth (AOD) and aerosol loading. Hence, this research: gives a good background for further study on aerosol loading; provides meteorological centres insight towards configuring sun-photometer over Ponta-Cape Verde; helps to quantify the extent of air pollution; provides modeller necessary insight on aerosol loading and retention challenges over Ponta-Cape Verde.

2. Experimental Design, Materials and Methods

Ponta do Sol is the northernmost town on the island of Santo Antão and Cape Verde. Ponta is located on longitude and latitude of -25° 00' 0.00" W and 17° 06' 60.00" N (Figure 2). The dataset was obtained from MISR (https://l0dup05.larc.nasa.gov/L3Web/download). The data was processed using excel. The conversion from AOD to aerosol loading was done using WASDM.

Figure 2: Geographical map of Ponta

WASDM was used aerosol loading over a region [6-7]:

$$\psi(\lambda) = a_1{}^2\cos\left(\frac{\mathrm{n_1}\pi\tau(\lambda)}{2}x\right)\cos\left(\frac{\mathrm{n_1}\pi\tau(\lambda)}{2}y\right) + \cdots \ldots a_n{}^2\cos\left(\frac{\mathrm{n_n}\pi\tau(\lambda)}{2}x\right)\cos\left(\frac{\mathrm{n_n}\pi\tau(\lambda)}{2}y\right) \quad (1)$$

a is atmospheric constant gotten from the fifteen years aerosol optical depth (AOD) dataset from MISR, n is the tuning constant, $\tau(\lambda)$ is the AOD of the area and $\psi(\lambda)$ is the aerosol loading. The validation of the summarized dataset was done using mathematical models and statistical software. The analysis of equations (1) was done using the C++ codes.

The aerosol loading was calculated from the WASDM. Further analysis were carried-out using the modified Bessel equatio, F distribution and Chi-Square distribution. The modified Bessel functions, $I_n(x)$, (also known as the hyperbolic Bessel Functions) is mathematically respresented as:

$$I_n(x)x = (i)^{-n}J_n(ix) \quad (2)$$

x is the value at which to evaluate the function, n is order of the Bessel function. In this work, n=3 so as to mimic the aerosol loading trend. The chi-squared distribution is commonly used to study variation within a given parameter. It is mathematically written as

$$F(x;n) = \frac{\gamma\left(\frac{n}{2},\frac{x}{2}\right)}{\Gamma\left(\frac{n}{2}\right)} \quad (3)$$

γ is the lower incomplete gamma function.

F probability distribution is used to determine whether two data sets have different degrees of diversity. It is mathematically written as:

$$f(x;d_1,d_2) = \frac{\sqrt{\frac{(d_1x)^{d_1}d_2^{d_2}}{(d_1x+d_2)^{d_1+d_2}}}}{xB\left(\frac{d_1}{2},\frac{d_2}{2}\right)} \quad (4)$$

3. Results and Discussion

The summarized primary data was obtained from Multi-angle Imaging Spectro-Radiometer (MISR) is shown in Table 1 for 550 nm wavelength [8]. It was observed the months of highest AOD varied significantly between May to August over the years (see blue boxes in Table 1). This result signifies that the air pollution over Ponta is seasonal and comes from same source. Hence the pollution over is sustained due to a same pollution routine over the years. The missing dataset was due to biases as discussed above [6].The aerosol loading over the area was obtained using the West African regional scale dispersion model (WASDM) from the primary dataset as presented in Figure 3. The aerosol loading of Ponta is very high at the moment. It further supports the ascertions made with the interpretation of AOD in Table 1. If the sources of anthropogenic emission in the region is not curbed, there would be massive health discomforts in the nearest future. The modified Bessel function shows that the aerosol loading significantly reduce around June of every year. Since high rain rate is one of the factors that reduce aerosol loading, it is infered that the research area has high rain rate in June every year. The Chi distribution shows that the variation of aerosol loading changes every year. Also, it reveals that the effect of aerosol loading is the same all through the year. The F propability distribution shows that there is a significant occurrence in December. Hence, the aerosl loading may not be high in December, however, the significance is highest in December. The statistical analysis of the summarized primary dataset is shown in Tables 2. The close ranges of values of the variance, average deviation and standard deviation are close indication that the correlation of each year would be natuarally be >0.75. It means the pattern of aersol dispersion in the research area is unique as there are no data spikes over the years. This result is further affirmed by the averages of the AOD over the years i.e. within the range of 0.3±0.05. The year 2012 had the highest AOD average over the years. The standard error is a good indication that the satellite measurement has a very low bias. Hence, the dataset presented is reliable for modelling and optimization process.

Table 1: Summarized Aerosol Optical Depth Dataset over Ponta

Month	2000	2001	2002	2003	2004	2005	2006	2007	2008	2009	2010	2011	2012	201
Jan	0.630	0.163	0.244	0.174	0.152	0.233	0.194	0.449	0.454	0.151	0.237	0.128	0.238	0.1
Feb	0.516	0.386	0.360	0.508	0.188	0.120	0.138	0.335	0.184	0.148	0.090	0.198	0.238	0.2
Mar	0.374	0.237	0.184	0.286	0.336	0.118	0.363	0.193	0.358	0.192	0.144	0.132	0.390	0.1
Apr	0.619	0.356	0.281	0.267	0.303	0.266	0.248	0.285	0.471	0.185	0.183	0.212	0.372	0.1
May	0.636	0.212	0.405	0.501	0.282	0.477	0.229	0.293	0.337	0.457	0.640	0.294	0.740	0.3
Jun	0.354	0.731	0.586	0.669	0.563	0.448	0.695	0.428	0.611	0.415	0.653	0.463	0.360	0.5
Jul	0.559	0.444	0.685	0.699	0.425	0.555	0.695	0.534	0.604	0.565	0.414	0.508	0.799	0.4
Aug	0.289	0.533	0.356	0.438	0.431	0.455	0.603	0.535	0.630	0.495	0.608	0.570	0.601	0.5
Sep	0.413	0.359	0.409	0.430	0.531	0.533	0.413	0.443	0.510	0.639	0.503	0.405	0.271	0.3
Oct	0.164	0.288	0.141	0.327	0.388	0.340	0.273	0.346	0.133	0.263	0.175	0.489	0.261	0.2
Nov		0.192	0.205	0.163	0.216	0.117	0.153	0.292	0.241	0.225	0.467	0.212	0.214	0.1
Dec		0.243	0.201	0.222	0.301	0.108	0.383	0.384	0.184	0.096	0.121	0.227	0.173	

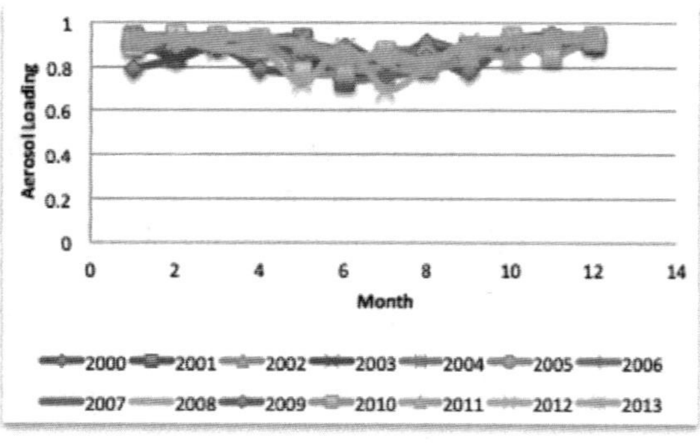

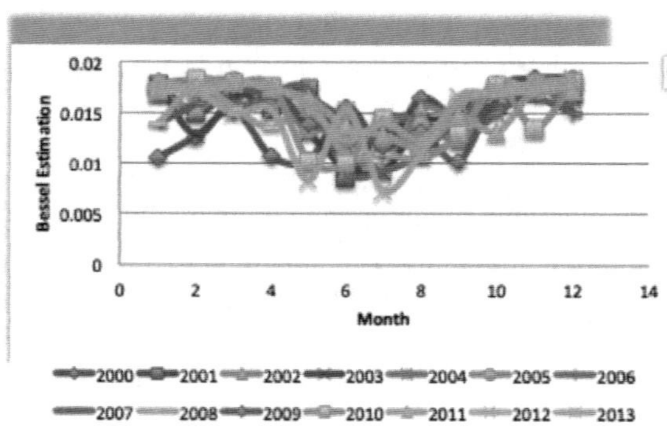

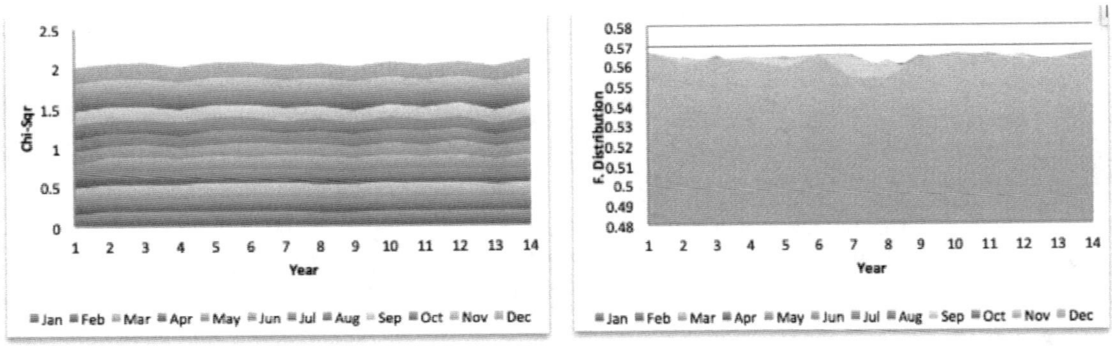

Figure 3: Analysis of aerosol loading (a) aerosol loading (b) Modified Bessel analysis (c) Chi-square distribution (d) F probability distribution

Table 2: statistics of aerosols content over Ponta

Statistics	**2000**	**2001**	**2002**	**2003**	**2004**	**2005**	**2006**	**2007**	**2008**	**2009**	**2010**	**2011**	**2012**	**2**
Number of values	10.000	12.000	12.000	12.000	12.000	12.000	12.000	12.000	12.000	12.000	12.000	12.000	12.000	1
Minimum	0.164	0.163	0.141	0.163	0.152	0.108	0.138	0.193	0.133	0.096	0.090	0.128	0.173	[illegible]
Maximum	0.636	0.731	0.685	0.699	0.563	0.555	0.695	0.535	0.630	0.639	0.653	0.570	0.799	[illegible]
Mean	0.455	0.345	0.338	0.390	0.343	0.314	0.366	0.376	0.393	0.319	0.353	0.320	0.388	[illegible]
First quartile	0.354	0.224	0.203	0.245	0.249	0.119	0.212	0.292	0.213	0.168	0.159	0.205	0.238	[illegible]
Third quartile	0.619	0.415	0.407	0.505	0.428	0.466	0.508	0.446	0.557	0.476	0.555	0.476	0.495	[illegible]
Standard error	0.051	0.047	0.048	0.052	0.037	0.051	0.058	0.030	0.052	0.053	0.063	0.046	0.061	[illegible]
95% confidence interval	0.116	0.104	0.105	0.115	0.082	0.111	0.128	0.067	0.114	0.117	0.138	0.100	0.134	[illegible]
99% confidence interval	0.166	0.147	0.149	0.162	0.116	0.157	0.180	0.094	0.161	0.166	0.194	0.142	0.189	[illegible]
Variance	0.026	0.027	0.028	0.033	0.017	0.031	0.040	0.011	0.032	0.034	0.047	0.025	0.045	[illegible]
Average deviation	0.137	0.123	0.129	0.151	0.104	0.154	0.160	0.086	0.154	0.162	0.194	0.139	0.163	[illegible]
Standard deviation	0.162	0.164	0.166	0.181	0.130	0.175	0.201	0.105	0.179	0.185	0.217	0.158	0.211	[illegible]
Coefficient of variation	0.355	0.475	0.491	0.464	0.378	0.557	0.549	0.279	0.457	0.578	0.614	0.494	0.544	[illegible]
Skew	-0.464	1.249	0.963	0.448	0.279	0.040	0.721	0.053	-0.082	0.521	0.220	0.332	1.123	[illegible]
Kurtosis	-0.841	1.576	0.352	-0.868	-0.714	-1.797	-0.841	-0.646	-1.514	-1.279	-1.794	-1.582	-0.039	-
Kolmogorov-Smirnov stat	0.146	0.152	0.168	0.138	0.120	0.200	0.179	0.120	0.134	0.203	0.203	0.222	0.246	[illegible]
Critical K-S stat, alpha=.10	0.369	0.338	0.338	0.338	0.338	0.338	0.338	0.338	0.338	0.338	0.338	0.338	0.338	[illegible]

Critical K-S stat, alpha=.05	0.409	0.375	0.375	0.375	0.375	0.375	0.375	0.375	0.375	0.375	0.375	0.375	0.375	(
Critical K-S stat, alpha=.01	0.489	0.449	0.449	0.449	0.449	0.449	0.449	0.449	0.449	0.449	0.449	0.449	0.449	(

4. Conclusion

Generally, it observed that the aerosol loading in Ponta is very high at the moment. The research affirms that air pollution over Ponta is seasonal and comes from same source. Hence the pollution over is sustained due to a same pollution routine over the years. Also, the modified Bessel function shows that the aerosol loading significantly reduce around june of every year, hence, infers that that the research area has high rain rate in June every year. It is recommended that government in the region should embark on a sporadic campaign to curb excessive dispersion of aerosols into the atmosphere.

Acknowledgement

The authors appreciate Covenant University for partial sponsorship. The authors acknowledge NASA for primary dataset.

References

[1]. Numbeo, Pollution in Cape Verde, https://www.numbeo.com/pollution/country_result.jsp?country=Cape+Verde (accessed 16 June, 2018)

[2]. Fioletov, V.E., Bodeker, G.E., Miller, A.J., McPeters, R.D. and Stolarski, R., Global and zonal total ozone variations estimated from ground-based and satellite measurements: 1964–2000, Journal of Geophysical Research 107 (D22) (2002) 4647-4651.

[3]. Giglio, L., Randerson, J.T., van der Werf, G.R., Kasibhatla, P.S., Collatz, G.J., Morton, D.C. and DeFries, R.S., Assessing variability and long-term trends in burned area by merging multiple satellite fire products. Biogeosciences 7 (2010) 1171-1186.

[4]. Mohammad Hadi Dehghani Sima Jarahzadeh , Mostafa Hade, Nabiollah Mansouri, Yousef Rashidi, Mahmood Yousefi, The Data on the Dispersion Modeling of Traffic-related PM10 and CO Emissions using CALINE3; A Case Study in Tehran, Iran Data in Brief 19 (2018) 2284-2290.

[5]. Liu, Y., Wang, Z., Wang, J., Ferrare, R., Newsom, R. and Welton, E., The effect of aerosol vertical profiles on satellite-estimated surface particle sulphate concentrations. Remote Sensing of Environment, 115 (2) (2011) 508-513.

[6]. M. E. Emetere, Impacts of recirculation event on aerosol dispersion and rainfall patterns in parts of Nigeria, GLOBAL NEST JOURNAL 19 (2) (2017) 344-352

[7]. M Emetere, SE Sanni, SA Akinwumi, Theoretical Study of Aerosols Loading and Retention Over Bolgatanga, Ghana, Air, Soil and Water Research 10 (2017) 1-8

[8]. WHO, 2019, Air pollution, https://www.who.int/air-pollution/news-and-events/how-air-pollution-is-destroying-our-health (Accessed 23rd March, 2019).

[9]. McGill, M., CATS/ISS L2O D-M7.1 05kmPro HDF-5 File - Version 1.05 [Data set]. NASA Langley Research Center Atmospheric Science Data Center DAAC. (2016). https://doi.org/10.5067/iss/cats/l2o_d-m7.1-v1-05_05kmpro

Air pollution Estimation Over Bouake

Emetere M.E.[1,2], **and Ahiara M.O.**.[4]
[1]Department of Physics, Covenant University Canaan land, P.M.B 1023, Ota, Nigeria.
[2]Department of Mechanical Engineering and Science, University of Johannesburg, APK, South Africa

E-mail: emetere@yahoo.com

Abstract. Air pollution over Bouake-Cote d'Ivoire was considered. Fifteen years primary (aerosol optical depth) dataset was obtained from the Multi-angle Imaging Spectro-Radiometer (MISR). Its aerosol loading dataset were generated from the primary dataset. The dataset depicts the air quality over the area. The statistical analysis of the dataset was also presented. The close ranges of values of the variance, average deviation and standard deviation are close indication that the correlation of each year would be naturally be >0.75. The averages of the AOD over the years i.e. is within the range of 0.4±0.05. It was observed that air pollution over Ponta is seasonal and comes from same source. Hence the pollution over is sustained due to a same pollution routine over the years. If the sources of anthropogenic emission in the region were not curbed, there would be massive health discomforts in the nearest future.

Keywords: air pollution, aerosol, Brikama, aerosol loadings

1. Introduction

Air contamination introduce toxins into the air that are harmful to human wellbeing and the planet overall. Government in developed nations have laws that curbs indiscriminate release of pollution into the atmosphere. Unfortunately, the under developed world have no or weak laws that controls indiscriminate anthropogenic pollutions. For example, in the US, there is a Clean Air Act that is regulated and enforced by the U.S. Ecological Protection Agency (EPA). The importance of clean air acts or laws are to curb the release of pollutions that have grave effect on health and climate system. For example, air contaminants such as carbon dioxide and methane raises the world's temperature and expans warmth. Climate change is reality and are caused by accumulated allergenic air poisons including mold (on account of moist conditions brought about by extraordinary climate and expanded flooding) and dust (because of a more extended dust season and more dust generation).
In this research, the values of air pollution parameters were documented so that the government of the region can start planning for a clean air future for its inhabitants. Hence, the dataset that is documented in this research is meant to: give a good background for further study on aerosol loading; provide meteorological centers insight towards configuring sun-photometer over Bouake-Cote d'Ivoire; to quantify the extent of air pollution; provide modeller necessary insight on aerosol loading and retention challenges over Bouake-Cote d'Ivoire.
The campaign (by the main author) on air pollution documentation over most cities in West Africa has been very succesful. Most importantly is the introduction of the term 'aerosol loading' to describe accumulated air pollutions in the atmosphere for a long time. The life time of these pollutants can be as high as 100 years. This means that aerosol loading may last longer than 300 years before its is technically eradicated. Since air pollution cannot be completely eradicated, it is important to advocate for a safe atmosphere. Aerosol loading calculation using West African regional scale dispersion model (WASDM) has showed great success in relating more with air quality [1-2]. Hence, the importance of

IOP Conf. Series: Earth and Environmental Science **331** (2019) 012070 doi:10.1088/1755-1315/331/1/012070

this dataset can be seen in Ref [2-6]. Dataset on aerosol loading shows the accumulative particulates transport within its lifetime.

2. Experimental Design, Materials and Methods

Bouaké is the second-largest city in Cote d'Ivoire. It is the seat of three levels of subdivision—Vallée du Bandama District, Gbêkê Region, and Bouaké Department. Bouake is located on longitude and latitude of 5.0391° W and 7.6905° N respectively (Figure 1). The dataset was obtained from MISR (https://l0dup05.larc.nasa.gov/L3Web/download). The data was processed using excel. The conversion from AOD to aerosol loading was done using WASDM. The google map is shown in Figure 2.

Figure 1: Geographical map of Bouake

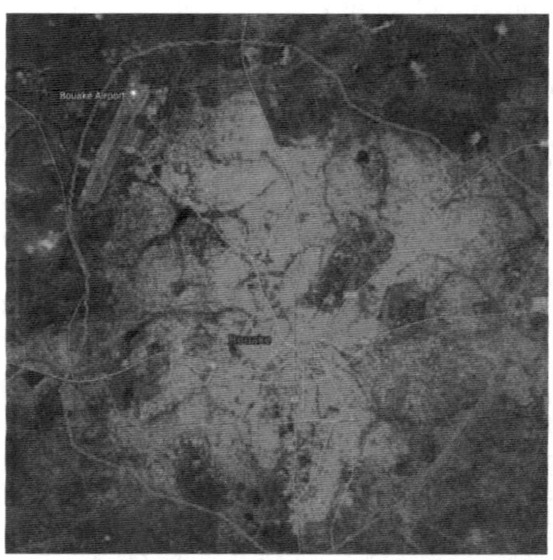

Figure 2: Google map of Bouake

WASDM was used aerosol loading over a region [6-7]:

$$\psi(\lambda) = a_1{}^2\cos\left(\frac{n_1\pi\tau(\lambda)}{2}x\right)\cos\left(\frac{n_1\pi\tau(\lambda)}{2}y\right) + \cdots \ldots a_n{}^2\cos\left(\frac{n_n\pi\tau(\lambda)}{2}x\right)\cos\left(\frac{n_n\pi\tau(\lambda)}{2}y\right) \quad (1)$$

a is atmospheric constant gotten from the fifteen years aerosol optical depth (AOD) dataset from MISR, n is the tuning constant, $\tau(\lambda)$ is the AOD of the area and $\psi(\lambda)$ is the aerosol loading. The validation of the summarized dataset was done using mathematical models and statistical software. The analysis of equations (1) was done using the C++ codes.

The aerosol loading was calculated from the WASDM. Further analysis were carried-out using the modified Bessel equatio, F distribution and Chi-Square distribution. The modified Bessel functions, $I_n(x)$, (also known as the hyperbolic Bessel Functions) is mathematically respresented as:

$$I_n(x)x = (i)^{-n}J_n(ix) \quad (2)$$

x is the value at which to evaluate the function, n is order of the Bessel function. In this work, n=3 so as to mimic the aerosol loading trend. The chi-squared distribution is commonly used to study variation within a given parameter. It is mathematically written as

$$F(x;n) = \frac{\gamma\left(\frac{n}{2},\frac{x}{2}\right)}{\Gamma\left(\frac{n}{2}\right)} \quad (3)$$

γ is the lower incomplete gamma function.

F probability distribution is used to determine whether two data sets have different degrees of diversity. It is mathematically written as:

$$f(x;d_1,d_2) = \frac{\sqrt{\frac{(d_1x)^{d_1}d_2^{d_2}}{(d_1x+d_2)^{d_1+d_2}}}}{xB\left(\frac{d_1}{2},\frac{d_2}{2}\right)} \quad (4)$$

3. Results and Discussion

The summarized primary data was obtained from Multi-angle Imaging Spectro-Radiometer (MISR) is shown in Table 1 for 550 nm wavelength [7]. It was observed the months of highest AOD varied significantly between January to March over the years (see blue boxes in Table 1). This result signifies that the air pollution over Ponta is seasonal and comes from same source. Hence the pollution over is sustained due to a same pollution routine over the years. The missing dataset was due to biases as discussed above [2]. Table 1 shows that the satellite biases over Bouake is high. The aerosol loading over the area was obtained using the West African regional scale dispersion model (WASDM) from the primary dataset as presented in Figure 3. The aerosol loading of Bouake is very high at the moment. It further supports the ascertions made with the interpretation of AOD in Table 1. If the sources of anthropogenic emission in the region is not curbed, there would be massive health discomforts in the nearest future. The modified Bessel function shows that the aerosol loading significantly reduce around February of every year. Since high rain rate is one of the factors that reduce aerosol loading. Due to huge AOD data loss, it cannot be infered that the research area has high rain rate in February every year. The Chi distribution shows that the variation of aerosol loading changes every year. Also, it reveals that the effect of aerosol loading is the same all through the year. The F propability distribution shows that there is a significant occurrence in December. Hence, the aerosl loading may not be high in December, however, the significance is highest in December. The

IOP Conf. Series: Earth and Environmental Science **331** (2019) 012070 doi:10.1088/1755-1315/331/1/012070

statistical analysis of the summarized primary dataset is shown in Tables 2. The close ranges of values of the variance, average deviation and standard deviation are close indication that the correlation of each year would be natuarally be >0.75. The averages of the AOD over the years i.e. is within the range of 0.4±0.05.

Table 1: Summarized Aerosol Optical Depth Dataset over Bouake

Month	2000	2001	2002	2003	2004	2005	2006	2007	2008	2009	2010	2011	2012	20
Jan	0.838	0.454	0.465	0.497	0.476	0.825	0.461	0.645	0.359	0.627	0.492	0.583	0.385	0.4
Feb	0.285	0.658	0.396	0.616	0.714	0.502	0.316	0.417	0.729	0.481	0.254	0.416	0.658	0.4
Mar		0.366	0.865	0.481	0.763	0.521	0.847	0.558	0.518			0.451	0.599	0.3
Apr		0.290	0.446			0.453		0.330	0.336		0.381	0.352	0.285	0.1
May		0.269		0.436		0.139	0.337	0.338	0.257	0.371		0.379		
Jun					0.270								0.260	
Jul														
Aug														
Sep					0.070									
Oct	0.433		0.262								0.136		0.087	0.1
Nov				0.329		0.258	0.413	0.303	0.323	0.329	0.533	0.251	0.375	
Dec			0.322	0.388	0.576	0.355	0.368	0.549	0.431	0.416	0.422	0.419	0.396	

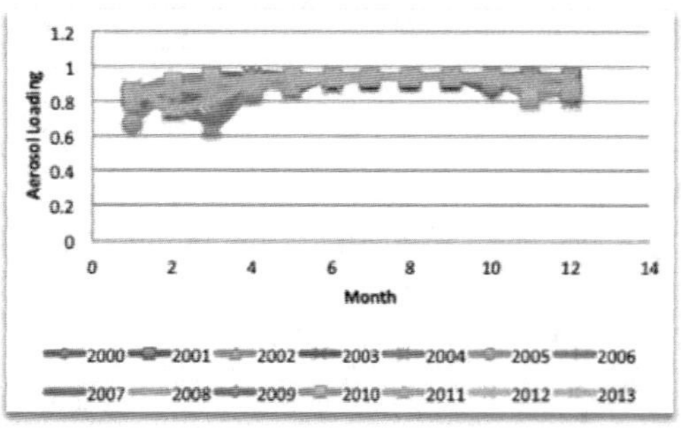

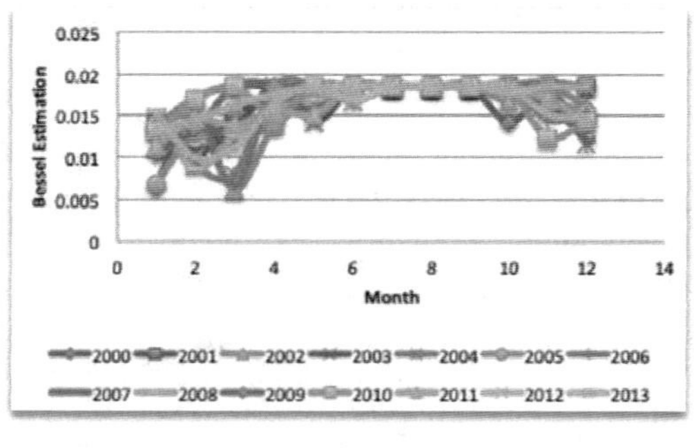

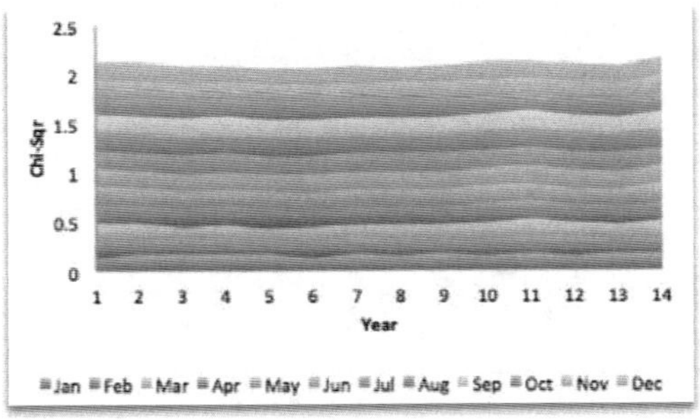

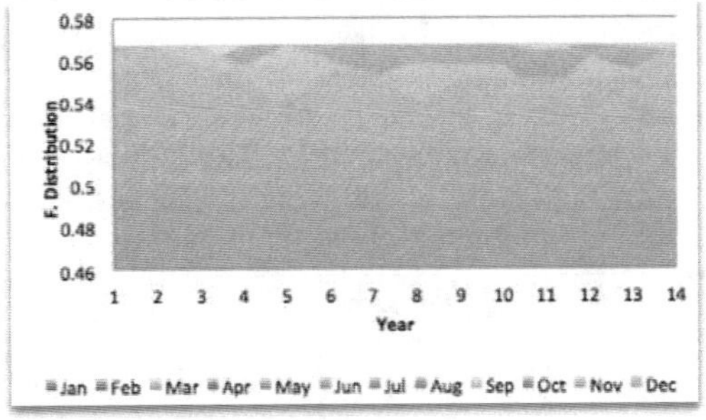

Figure 3: Analysis of aerosol loading (a) aerosol loading (b) Modified Bessel analysis (c) Chi-square distribution (d) F probability distribution

Table 2: statistics of aerosols content over Bouake

Statistics	**2000**	**2001**	**2002**	**2003**	**2004**	**2005**	**2006**	**2007**	**2008**	**2009**	**2010**	**2011**	**2012**	**20**
Number of values	3.000	5.000	6.000	6.000	6.000	7.000	6.000	7.000	7.000	5.000	6.000	7.000	8.000	5.0
Minimum	0.285	0.269	0.262	0.329	0.070	0.139	0.316	0.303	0.257	0.329	0.136	0.251	0.087	0.1
Maximum	0.838	0.658	0.865	0.616	0.763	0.825	0.847	0.645	0.729	0.627	0.533	0.583	0.658	0.4
Mean	0.518	0.407	0.459	0.458	0.478	0.436	0.457	0.448	0.422	0.445	0.370	0.407	0.380	0.3
First quartile	#N/A	0.285	0.322	0.388	0.270	0.282	0.337	0.332	0.326	0.361	0.254	0.359	0.272	0.1
Third quartile	#N/A	0.505	0.465	0.497	0.714	0.516	0.461	0.556	0.496	0.517	0.492	0.443	0.497	0.4
Standard error	0.165	0.071	0.087	0.040	0.109	0.083	0.081	0.051	0.060	0.052	0.061	0.038	0.065	0.0
95% confidence interval	0.711	0.196	0.223	0.104	0.280	0.203	0.208	0.125	0.147	0.144	0.157	0.093	0.153	0.1
99% confidence interval	1.640	0.325	0.350	0.163	0.439	0.308	0.326	0.189	0.223	0.239	0.247	0.142	0.227	0.2
Variance	0.082	0.025	0.045	0.010	0.071	0.048	0.039	0.018	0.025	0.013	0.023	0.010	0.034	0.0
Average deviation	0.213	0.119	0.137	0.073	0.206	0.159	0.131	0.116	0.118	0.087	0.116	0.069	0.129	0.1
Standard deviation	0.286	0.158	0.213	0.099	0.267	0.220	0.198	0.135	0.159	0.116	0.150	0.101	0.183	0.1
Coefficient of variation	0.552	0.387	0.463	0.216	0.558	0.504	0.433	0.301	0.377	0.261	0.406	0.248	0.482	0.4
Skew	1.228	1.219	1.748	0.478	-0.629	0.601	2.084	0.356	1.364	1.088	-0.710	0.340	0.122	0.1
Kurtosis	#N/A	1.085	3.621	0.519	-0.816	0.965	4.569	-1.837	1.802	0.952	-0.576	1.595	-0.053	2.0
Kolmogorov-Smirnov stat	0.285	0.203	0.322	0.181	0.164	0.207	0.326	0.222	0.225	0.197	0.196	0.190	0.217	0.1
Critical K-S stat,	0.636	0.509	0.468	0.468	0.468	0.436	0.468	0.436	0.436	0.509	0.468	0.436	0.410	0.5

alpha=.10														
Critical K-S stat, alpha=.05	0.708	0.563	0.519	0.519	0.519	0.483	0.519	0.483	0.483	0.563	0.519	0.483	0.454	0.5
Critical K-S stat, alpha=.01	0.829	0.669	0.617	0.617	0.617	0.576	0.617	0.576	0.576	0.669	0.617	0.576	0.542	0.6

4. Conclusion

Generally, it observed that the aerosol loading in Bouake is very high at the moment. The research affirms that air pollution over Ponta is seasonal and comes from same source. Hence the pollution over is sustained due to a same pollution routine over the years. The missing dataset shows that the satellite biases over Bouake are high. The close ranges of values of the variance, average deviation and standard deviation are close indication that the correlation of each year would be natuarally be >0.75. The averages of the AOD over the years i.e. is within the range of 0.4±0.05.

It is recommended that government in the region should embark on a sporadic campaign to curb excessive dispersion of aerosols into the atmosphere.

Acknowledgement

The authors appreciate Covenant University for partial sponsorship. The authors acknowledge NASA for primary dataset.

References

[1]. Emetere Moses E., Akinyemi M.L. & Oladimeji T.E., Statistical Examination Of The Aerosols Loading Over Kano-Nigeria: The Satellite Observation Analysis, Scientific Review Engineering and Environmental Sciences, 72 (2016) 167-176

[2]. Emetere Moses E., Akinyemi M.L. & Akinwumi S.A., Aerosols Loading Trends And Its Environmental Threats Over Cotonou-Benin, Nature Environment and Pollution Technology, 15 (3) (2016) 1111

[3]. Emetere M.E., Akinyemi M.L., & Akinojo O., Parametric retrieval model for estimating aerosol size distribution via the AERONET, LAGOS station, Environmental Pollution, 207 (C) (2015) 381-390

[4]. Vijayakumar, K. and Devara, P.C.S., Study of aerosol optical depth, ozone, and precipitable water vapour content over Sinhagad, a high-altitude station in the Western Ghats. International Journal of Remote Sensing, 34 (2) (2013) 613-630.

[5]. Boucher, O. and Quaas, J., Water vapour affects both rain and aerosol optical depth, Nature Geosciences 6 (2013) 4–5

[6]. Mohammad Hadi Dehghani Sima Jarahzadeh , Mostafa Hade, Nabiollah Mansouri, Yousef Rashidi, Mahmood Yousefi, The Data on the Dispersion Modeling of Traffic-related PM10 and CO Emissions using CALINE3; A Case Study in Tehran, Iran, Data in Brief 19 (2018) 2284-2290.

[7]. McGill, M., CATS/ISS L2O D-M7.1 05kmPro HDF-5 File - Version 1.05 [Data set]. NASA Langley Research Center Atmospheric Science Data Center DAAC. (2016). https://doi.org/10.5067/iss/cats/l2o_d-m7.1-v1-05_05kmpro